The 80386, 80486, and Pentium® Processor

Hardware, Software, and Interfacing

WALTER A. TRIEBEL
Intel Corporation

PRENTICE HALL
Upper Saddle River, New Jersey   Columbus, Ohio

To my father, Adolf A. Triebel

Library of Congress Cataloging-in-Publication Data

Triebel, Walter A.
   The 80386, 80486, and Pentium processors: hardware, software, and interfacing/Walter A. Triebel.
      p.   cm.
   Includes bibliographical references and index.
   ISBN 0-13-533225-7
   1. Intel 80386 (Microprocessor) 2. Intel 80486 (Microprocessor)
3. Pentium (Microprocessor) I. Title.
   QA76.8.I2684T75 1998
   621.39'16—dc21   97-16163
                                          CIP

Photo Credit: © 1996 PhotoDisk, Inc.
Editor: Charles E. Stewart, Jr.
Production Editor: Mary M. Irvin
Production Coordination: bookworks
Design Coordinator: Karrie M. Converse
Cover Designer: Ken MacKay
Production Manager: Deidra M. Schwartz
Marketing Manager: Debbie Yarnell

This book was set in Times Roman by Bi-Comp and was printed and bound by Courier/Westford. The cover was printed by Phoenix Color Corp.
Reprinted with corrections May, 1999.

© 1998 by Prentice-Hall, Inc.
Upper Saddle River, New Jersey 07458

Printed in the United States of America

10 9 8 7 6 5 4 3

ISBN: 0-13-533225-7

Prentice-Hall International (UK) Limited, *London*
Prentice-Hall of Australia Pty. Limited, *Sydney*
Prentice-Hall of Canada, Inc., *Toronto*
Prentice-Hall Hispanoamericana, S. A., *Mexico*
Prentice-Hall of India Private Limited, *New Delhi*
Prentice-Hall of Japan, Inc., *Tokyo*
Prentice-Hall Asia Pte. Ltd., *Singapore*
Editora Prentice-Hall do Brasil, Ltda., *Rio de Janeiro*

Preface

Today Intel's 80X86 family of microprocessors is the most widely used architecture in modern microcomputer systems. The family includes both 16-bit microprocessors, such as the 8088, 8086, 80186, 80188, and 80286 processors, and 32-bit microprocessors, such as the 80386, 80486, and Pentium® processors. The 8088, which is the 8-bit bus version of the 8086, is the microprocessor employed in the original IBM personal computer (PC). The 8088 and 8086 microprocessors were also used by many other manufacturers to make personal computers compatible with IBM's original PC. Moreover, IBM's original personal computer advanced technology (PC/AT) was designed with the 80286 microprocessor. Like the PC, PC/AT-compatible personal computers were made by many other manufacturers, and today they are built with Pentium®-processor-based microcomputers. Intel's 80X86 family of microprocessors are also used in a wide variety of other electronic equipment.

The 80386, 80486, and Pentium® Processors: Hardware, Software, and Interfacing is a thorough study of the 32-bit microprocessors of the 80X86 family, their microcomputer system architectures, and the circuitry used in the design of a PC/AT compatible microcomputer. This book is written for use as a textbook in courses on microprocessors at community colleges, colleges, and universities. The intended use is in a one- or two-semester course in microprocessor technology that emphasizes both assembly language software and microcomputer circuit design.

Individuals involved in the design of microprocessor-based electronic equipment need a systems-level understanding of the 80X86 microcomputer—that is, a thorough understanding of both their software and hardware. The first part of this book explores the software architecture of the 80386, 80486, and Pentium® processors and teaches the reader how to write, execute, and debug assembly language programs. In order to be successful at writing assembly language programs for the 80X86 family of microprocessor, one must learn:

1. *Software architecture*: the internal registers, flags, memory organization, and stack, and how they are used from a software point of view.

iii

2. *Software-development tools*: how to use the commands of the program debugger (DEBUG) that is available in DOS to assemble, execute, and debug instructions and programs.

3. *Instruction set*: the function of each of the instructions in the instruction set, the permissible operand variations, and how to write statements using the instructions.

4. *Programming techniques*: basic techniques of programming, such as flowcharting, jumps, loops, strings, subroutines, and parameter passing.

5. *Applications*: a step-by-step study of the process of writing programs for several practical applications.

All this material is developed in detail in Chapters 2 through 8.

The software section includes many practical concepts and practical software applications. For instance, examples are used to demonstrate applications such as 32-bit addition and subtraction, masking of bits, and the use of branch and loop operations to implement IF-THEN-ELSE, REPEAT-UNTIL, and WHILE-DO program structures. In addition, the various steps of the assembly language program development cycle are explored.

The study of software architecture, instruction set, and assembly language programming is closely coupled with use of the DEBUG program on the PC/AT-compatible microcomputer. That is, the line-by-line assembler in DEBUG is used to assemble instructions and programs into the memory of the PC/AT, while other DEBUG commands are used to execute and debug the programs. The use of a practical 80X86 assembler program, the Microsoft MASM Assembler, is also covered. Using MASM and other PC/AT-based software development tools, the student learns how to create a source program; assemble the program; form a run module; and load, run, and debug a program.

Chapter 8 is a thorough study of the 80386DX's protected-address mode architecture. Here we begin by introducing the protected-mode register model. This is followed by a detailed study of the function of the 80386DX's memory-management unit, segmentation and paging, virtual addressing, and the translation of virtual addresses to physical addresses. The various types of descriptor table and page table entries supported by the 80386DX are covered, as is how they relate to memory management and the protection model. The instructions of the protected-mode system control instruction set are described. Finally, the concepts of protection, the task, task switching, the multitasking system environment, and virtual 8086 mode are explored.

The second part of the book examines the hardware architecture of microcomputers built with the 80386, 80486, and Pentium[R] processors. To understand the hardware design of a microcomputer system, the reader must begin by first understanding the function and operation of each of the microprocessor's hardware interfaces: memory, input/output, and interrupt. After this, the role of each of these subsystems can be explored relative to overall microcomputer system operation. It is this material that is presented in Chapters 9 through 14.

We begin in Chapter 9 by examining the architecture of the 80386DX and 80386SX microprocessors from a hardware point of view. Included is information such as pin layout, signal interfaces, signal functions, and clock requirements. The latter part of the chapter covers the memory and input/output interfaces of the 80386DX. This material includes extensive coverage of memory and input/output bus cycles, pipelining, address maps, memory and input/output interface circuits (address latches and buffers, data bus transceivers, and address decoders), use of programmable logic devices in implementing bus-control logic, types of input/output, and input/output instructions and programs.

This hardware introduction is followed by separate studies of the architecture, operation, devices, and typical circuit designs for the memory, input/output, and interrupt interfaces of the 80386DX-based microcomputer in Chapters 10, 11, and 12, respectively. For instance, in Chapter 10, we cover devices and circuits for the program storage memory (ROM, PROM, EPROM, and FLASH), data storage memory (SRAM and DRAM), and cache memory subsystems. Practical bus interface circuit and memory subsystem design techniques are also examined, including parity checker/generator circuitry and wait state generator circuitry.

Chapter 11 covers input/output interface circuits and LSI peripheral devices. The material on core I/O interfaces includes detail studies of discrete parallel input/output circuits, 82C55A, 82C54, and 82C37A peripheral ICs. Practical parallel input/output design techniques are illustrated with examples such as circuits and programs for polling switches, lighting LEDs, scanning displays and keyboards, and printing characters at a parallel printer port. The chapter also explores a number of special purpose peripheral IC devices and interfaces. For instance, serial communication and the 16450 UART controller are studied and keyboard scanning and display driving are demonstrated with the 8279 keyboard/display controller.

The topic of interrupt and exception processing is examined in Chapter 12. This chapter introduces the interrupt context switching mechanism and related topics such as priority, interrupt vectors, the interrupt vector table, interrupt acknowledge bus cycle, and interrupt service routine. External hardware interrupt interface circuits are demonstrated using both discrete circuitry and the 82C59 programmable interrupt controller peripheral IC. Also included is coverage of special interrupt functions such as software interrupts, the nonmaskable interrupt, reset operation, and internal interrupt and exception processing.

The hardware design section continues in Chapter 13 with a thorough study of an 80386DX-based microcomputer design of a main processor board for a PC/AT-compatibile personal computer. The microcomputer design examined employs Intel Corporation's 82340 high-integration PC/AT-compatible chip set. Each of the ICs in the 82340 chip set (82345 data buffer, 82346 system controller, 82344 ISA controller, and 82341 peripheral combo) are described in detail from a hardware point of view. This material includes their block diagram, signal interfaces, and interconnection to implement the PC/AT-compatible microcomputer. Moreover, the circuitry used in the design of the cache memory subsystem, DRAM array and BIOS EPROMs, floppy disk drive interface, IDE hard disk drive interface, serial communication interface, and parallel printer interface is described. This chapter demonstrates a practical implementation of the material presented in the prior chapters on microcomputer interfacing techniques.

The hardware section concludes with Chapter 14, which covers PC bus interfacing and techniques for circuit construction, testing, and troubleshooting.

The last part of the textbook provides detailed coverage of the other 32-bit microprocessors of the 80X86 family, the 80486SX, 80486DX, 80486DX2, 80486DX4, and Pentium® processors. Throughout these chapters, the focus is on how these processors' software and hardware architecture differs from that of the earlier family members. Advanced topics introduced include: RISC, CRISP, and superscaler processor architectures, clock scaling, big and little endian data organization and conversion, dynamic bus sizing, burst, pipelined, and cached bus cycles, address and data parity, and on-chip code and data caches. The Pentium® Pro Processor and Pentium® Processor with MMX™ technology are also introduced.

W.A.T.

Supplements

An extensive package of supplementary materials are available to complement the 80X86 microprocessor program offered by this textbook. They include materials for the student, instructor, and for easy implementation of a practical PC/AT-hosted laboratory program. They include:

1. *80X86 Microprocessor Experiments Laboratory Manual*, ISBN 0-13-367913-6, Prentice Hall, Upper Saddle River, NJ 07458

   Laboratory manual that contains 24 skill-building laboratory exercises that explore the software architecture of the 80X86 microcomputer in the PC/AT, assembly language program development, the internal hardware of the PC/AT, and interface circuit operation, design, testing, and troubleshooting.

2. *Instructor's Manual* to accompany *The 80386, 80486, and Pentium® Processor: Hardware, Software, and Interfacing,* ISBN 0-13-889106-0, Prentice Hall, Upper Saddle River, NJ 07458

   Provides the answers to all the student exercises in the textbook and transparency masters for more than 200 of the illustrations in the textbook.

3. *Instructor's Solution Manual* to accompany the *80X86 Microprocessor Experiments Laboratory Manual,* ISBN 0-13-700154-1, Prentice Hall, Upper Saddle River, NJ 07458

   Contains detailed solutions for each of the 24 exercises in the laboratory manual, including all computer printouts, program listings, and circuit drawings.

4. *Laboratory Program Diskette*, ISBN 0-13-788555-5, Prentice Hall, Upper Saddle River, NJ 07458

   Contains all the programs needed by the student to perform the exercises in the laboratory manual. Included are files that contain the source program, source listing, object code, and run module. These files have been produced by assembling the source program with the *Microsoft Macro Assembler.*

5. *Laboratory Solutions Diskette,* ISBN 0-13-788563-6, Prentice Hall, Upper Saddle River, NJ 07458

Contains all the programs and executable files that are created by the student in the process of performing the 24 exercises in the laboratory manual. Based on the method identified in the exercise, the programs have been created with either the assembler in DEBUG or the *Microsoft Macro Assembler.*

6. *PCμLAB*, Microcomputer Directions, Inc., P.O. Box 15127, Fremont, CA 94539

An easy-to-use and versatile, external hardware expansion environment for any PC or PC/AT-compatible personal computer for experimenting with microcomputer interface circuits. It extends the ISA bus external to the PC/AT, thereby forming a bench top laboratory test unit for building, testing, and troubleshooting interface circuits. It includes a large solderless breadboard area for working with student-constructed circuitry; a single PC/AT-compatible ISA bus slot for installation of commercially available or custom-built add-on cards; and built-in I/O devices, LEDs, switches, and a speaker. The PCμLAB also has a continuity tester and logic probe for testing circuit operation.

7. *Microsoft MACRO Assembler*, Microsoft Corporation, Redmond, WA 98052

A popular assembler for converting 80X86 assembly language programs to machine code. This assembler enables the student to perform professional-caliber assembly language program development for the 80X86 microcomputer architecture. This software development tool runs on PC/AT-compatible microcomputer systems.

Contents

Contents   **ix**

Contents   **xi**

▲ 15 THE 80486 MICROPROCESSOR FAMILY   787

▲ 16 THE PENTIUM^R PROCESSOR FAMILY   833

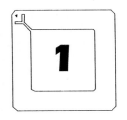

Introduction
to Microprocessors
and Microcomputers

▲ 1.1 INTRODUCTION

In the past decade, most of the important advances in computer system technology have been closely related to the development of high-performance 16-bit and 32-bit microprocessor architectures and the microcomputer systems built with them. During this period, there has been a major change in the direction of businesses away from minicomputers to smaller, lower-cost microcomputers. The *IBM personal computer* (the *PC*, as it has become known), which was introduced in mid-1981, was one of the earliest microcomputers that used a 16-bit microprocessor, the 8088, as its central processing unit. A few years later it was followed by another IBM personal computer, the *PC/AT* (personal computer advanced technology). This new system was implemented using the more powerful 80286 microprocessor.

The PC and PC/AT quickly became cornerstones of the evolutionary process from minicomputer to microcomputer. In 1985 an even more powerful microprocessor, the 80386DX, was introduced. The 80386DX was Intel Corporation's first 32-bit member of the 8086 family of microprocessors. Availability of the 80386DX led to a new generation of higher-performance PC/ATs. In the years that followed, Intel expanded its 32-bit architecture offering with the 80486 and Pentium^R processor families. These processors brought new levels of performance and capabilities to the personal computer marketplace. Today Pentium^R processor–based PC/AT microcomputers represent the industry standard computer platform for the new computer industry.

Since the introduction of the IBM PC, the microprocessor market has matured significantly. Today, several complete families of 16- and 32-bit microprocessors are available. They all include support products, such as *very large-scale integrated* (VLSI) peripherals devices, emulators, and high-level software languages. Over the same period of time, these higher-performance microprocessors have become more widely used in the design of new

1

electronic equipment and computers. This book presents a detailed study of the software and hardware architectures of Intel Corporation's 32-bit 80X86 microprocessors, that is, the 80386, 80486, and Pentium^R processor families.

In this chapter we begin our study with an introduction to microprocessors and microcomputers. The following topics are discussed:

1. The IBM and IBM-compatible personal computers: reprogrammable microcomputers
2. General architecture of a microcomputer system
3. Evolution of the Intel microprocessor architecture

▲ 1.2 THE IBM AND IBM-COMPATIBLE PERSONAL COMPUTERS: REPROGRAMMABLE MICROCOMPUTERS

The IBM personal computer (the PC), which is shown in Fig. 1.1, was IBM's first entry into the microcomputer market. After its introduction in mid-1981, market acceptance of the PC grew by leaps and bounds so that it soon became the leading personal computer architecture. One of the important keys to its success is that an enormous amount of application software became available for the machine. Today, there are more than 50,000 off-the-shelf software packages available for use with the PC. They include business applications, software languages, educational programs, games, and even alternative operating systems.

Another reason for the IBM PC's success was the fact that it was an open system. By *open system*, we mean that the functionality of the PC can be expanded by simply adding boards into the system. Some examples of add-in hardware features are additional memory, a modem, serial communication interfaces, and local area network interfaces. This system expansion is provided by the PC's expansion bus—five card slots in the original PC's chassis. IBM defined an 8-bit expansion bus standard known as the *I/O channel* and provided its specification to other manufacturers so that they could build different types of add-in products for the PC. Just as for software, a wide variety of add-in boards quickly became

Figure 1.1 Original IBM personal computer. (Courtesy of International Business Machines Corporation)

Figure 1.2 PCXT personal computer. (Courtesy of International Business Machines Corporation)

available. The result was a very flexible system that could be easily adapted to a wide variety of applications. For instance, the PC can be enhanced with add-in hardware to permit its use as a graphics terminal, to synthesize music, and even for the control of industrial equipment.

The success of the PC caused IBM to introduce additional family members. IBM's *PCXT* is shown in Fig. 1.2, and an 80286-based PC/AT is illustrated in Fig. 1.3. The PCXT employed the same system architecture as that of the original PC. It was also designed with the 8088 microprocessor, but one of the floppy disk drives was replaced with a 10-M-byte (10,000,000-byte) hard disk drive. The original PC/AT was designed with a 6-MHz 80286 microprocessor and defined a new open-system bus architecture that today is called the *industry standard architecture* (ISA). This architecture provides a 16-bit, higher-performance I/O expansion bus.

Figure 1.3 PC/AT personal computer. (Courtesy of International Business Machines Corporation)

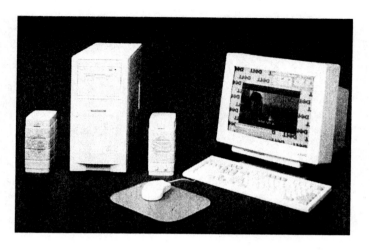

Figure 1.4 Dell Opti Plex GXMT 5133 Pentium[R] processor–based PC/
AT compatible personal computer. (Courtesy of Dell Computer Corp.)

Today, Pentium[R] processor–based ISA PC/ATs are the mainstay of the personal computer marketplace. Figure 1.4 shows a popular Pentium[R] processor–based PC/AT-compatible personal computer. Most of the systems that are implemented with the Pentium[R] processor have an ISA and a second bus known as the *peripheral component interface* (PCI) *bus*. This new bus is a high-speed I/O data bus intended for connection of high-performance I/O interfaces, such as graphics, video, and high-speed local area networks (LAN). The PCI bus supports 32-bit and 64-bit data transfers, and its data-transfer rate is more than 10 times that of the ISA bus. These machines offer a wide variety of computing capabilities, range of performance, and software base for use in business and at home.

The PC/AT-compatible is an example of a *reprogrammable microcomputer*. By reprogrammable microcomputer we mean one that is intended to run programs for a wide variety of applications. For example, one could use the PC with a standard application package for accounting or inventory control. In this type of application, the primary task of the microcomputer is to analyze and process a large amount of data, known as the *database*. Another user could be running a word-processing software package. This is an example of a data input/output–intensive task. The user enters text information, which is reorganized by the microcomputer and then output to a diskette or printer. As a third example, a programmer uses a language, such as C, to write programs for a scientific application. Here the primary function of the microcomputer is to solve complex mathematical problems. The personal computer used for each of these applications is the same; the difference is in the software application that the microcomputer is running. That is, the microcomputer is simply reprogrammed to run the new application.

Let us now look at what a microcomputer is and how it differs from the other classes of computers. Evolution of the computer marketplace over the last 25 years has taken us from very large *mainframe computers* to smaller *minicomputers,* and now to even smaller *microcomputers*. These three classes of computers did not originally replace each other. They all coexisted in the marketplace. Computer users had the opportunity to select the computer that best met their needs. The mainframe computer was used in an environment where it serviced a large number of users. For instance, a large university or institution would select a mainframe computer for its data-processing center. Here it would service many hundreds of users. Mainframes are still widely used today to satisfy large-computer requirements.

The minicomputer had been the primary computer solution for the small, multiuser business environment. In this environment, from several to a hundred users connect to the system with terminals. In this way, they all share the same computer system, and many of these users may be actively working on the computer at the same time. An important characteristic of this computer system configuration is that all computational power resides at the minicomputer. The user terminals are what are known as *dumb terminals*; that is, they are not self-sufficient computers. If the minicomputer is not working, all users are down and cannot do any work at their terminals. Examples of a community that would traditionally use a minicomputer are a department at a university and a business with a multiuser-dedicated need, such as application software development.

Managers in a department may select a microcomputer, such as the PC/AT, for their personal needs, such as word processing and database management. The original IBM PC was called a personal computer because it was initially intended to be a single-user system— that is, the user's personal computer. Several people could use the same computer, but only one at a time. Today, the microcomputer has taken over most of the traditional minicomputer user base. High-feature, high-performance PC/ATs have replaced the minicomputer as a *file server*. Many users have their personal computers attached to the file server through a *local area network*. However, in this more modern computer system architecture, all users also have local computational power in their own PC/ATs. The file server extends the computational power and system resources such as memory available to the user. If the file server is not operating, users can still do work with their individual personal computers.

Along the evolutionary path from mainframes to microcomputers, the basic concepts of computer architecture have not changed. Just like the mainframe and minicomputer, the microcomputer is a general-purpose electronic data-processing system intended for use in a wide variety of applications. The key difference is that microcomputers, such as the PC/AT-compatible, employ the newest *very large-scale integration* (VLSI) circuit technology *microprocessing unit* (MPU) to implement the system. Microcomputers, such as a Pentium[R] processor–based PC/AT, which are designed for the high-performance end of the microcomputer market, are physically smaller computer systems, outperform comparable minicomputer systems, and are available at a much lower cost.

▲ 1.3 GENERAL ARCHITECTURE OF A MICROCOMPUTER SYSTEM

The *hardware* of a microcomputer system can be divided into four functional sections: the *input unit, microprocessing unit, memory unit,* and *output unit.* The block diagram of Fig. 1.5 shows this general microcomputer architecture. Each of these units has a special function in terms of overall system operation. Next we will look at each of these sections in more detail.

The heart of a microcomputer is its microprocessing unit (MPU). The MPU of a microcomputer is implemented with a VLSI device known as a *microprocessor*. A microprocessor is a general-purpose processing unit built into a single integrated circuit (IC). The MPU used most widely in PC/ATs today is Intel Corporation's Pentium[R] processor, which is shown in Fig. 1.6.

The MPU is the part of the microcomputer that executes instructions of the program and processes data. It is responsible for performing all arithmetic operations and making the logical decisions initiated by the computer's program. In addition to arithmetic and logic functions, the MPU controls overall system operation.

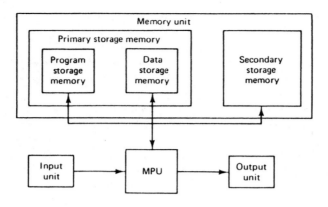

Figure 1.5 General architecture of a microcomputer system.

The input and output units are the means by which the MPU communicates with the outside world. Input units, such as the *keyboard* on the PC/AT, allow the user to input information or commands to the MPU. For instance, a programmer could key in the lines of a BASIC program from the keyboard. Many other input devices are available for the PC/AT, two examples being a *mouse,* for implementing a more user-friendly input interface, and a *joystick,* for use when playing video games.

The most widely used output devices of a PC/AT are the *display* and *printer.* The output unit in a microcomputer is used to give feedback to the user and for producing documented results. For example, key entries from the keyboard are echoed back to the display. By looking at the screen of the display, the user can confirm that the correct entry was made. Moreover, the results produced by the MPU's processing can be either displayed or printed. For our earlier example of a BASIC program, once it is entered and corrected, a listing of the instructions could be printed. Alternative output devices are also available for the microcomputer; for instance, it can be equipped with a color video display instead of the monochrome video display.

The memory unit in a microcomputer is used to *store* information, such as number or character data. By store we mean that memory has the ability to hold this information for processing or for outputting at a later time. Programs that define how the computer is to operate and process data also reside in memory.

Figure 1.6 Pentium[R] processor. (Courtesy of Intel Corporation)

In the microcomputer system, memory can be divided into two different types, called *primary storage memory* and *secondary storage memory*. Secondary storage memory is used for long-term storage of information that is not currently being used. For example, it can hold programs, files of data, and files of information. In the original IBM PC/AT, the *floppy disk drive* is one of the secondary storage memory subsystems. It was a 5¼-inch drive that used double-sided, quad-density *floppy diskette* storage media that could each store up to 1.2 Mbytes (1,200,000 bytes) of data. This floppy diskette is an example of a removable media. That is, to use the diskette it is inserted into the drive and locked in place. If the diskette is either full or one with a different file or program is needed, the diskette is simply unlocked and removed and another diskette installed.

The IBM PC/AT also employs a second type of secondary storage device called a *hard disk drive*. Typical hard disk sizes are 20 Mbyte (20 million bytes), 40 Mbyte, 80 Mbyte, 420 Mbyte, and 1.2 Gbyte (1200 million bytes). Earlier we pointed out that the original IBM PC/AT was equipped with a 20-Mbyte hard disk drive. The hard disk drive differs from the floppy disk drive in that the media is fixed, which means that the media cannot be removed. However, being fixed is not a problem because the storage capacity of the media is so much larger. Today PC/ATs are quipped with a 1.2-Gbyte hard disk drive.

Both the floppy diskette and hard disk are examples of read/write media. That is, a file of data can be read in from or written out to the storage media in the drive. Another secondary storage device that is becoming very popular in personal computers today is a CD drive. Here a removable *compact disk* (CD) is used as the storage media. This media has very large storage capacity, more than 600 Kbytes, but is read-only. This means you cannot write information onto a CD for storage. For this reason, it is normally used for storage of large programs or files of data that are not to be changed.

Primary storage memory is normally smaller in size and is used for temporary storage of active information, such as the operating system of the microcomputer, the program that is currently being run, and the data that it is processing. In Fig. 1.5 we see that primary storage memory is further subdivided into *program-storage memory* and *data-storage memory*. The program section of memory is used to store instructions of the operating system and application programs. The data section normally contains data that are to be processed by the programs as they are executed: for example, text files for a word-processor program or a database for a database-management program.

Typically, primary storage memory is implemented with both *read-only memory* (ROM) and *random-access read/write memory* (RAM) integrated circuits. The original IBM PC/AT had 64 Kbytes of ROM and can be configured with either 256 Kbytes or 512 Kbytes of RAM without adding a memory-expansion board. Modern PC/ATs made with the Pentium[R] processor are typically equipped with 16 Mbyte of RAM.

Data, whether they represent numbers, characters, or instructions of a program, can be stored in either ROM or RAM. In the IBM PC/AT a small part of the operating system and BASIC language are made resident to the computer by supplying them in ROM. By using ROM, this information is made *nonvolatile*—that is, the information is not lost if power is turned off. This type of memory can only be read from; it cannot be written into. On the other hand, data that are to be processed and information that frequently changes must be stored in a type of primary storage memory from which they can be read by the microprocessor, modified through processing, and written back for storage. This requires a type of memory that can be both read from and written into. For this reason, such data are stored in RAM instead of ROM.

Earlier we pointed out that the instructions of a program can also be stored in RAM. In fact, the *DOS 5.0 operating system* for the PC/AT is provided on diskettes, but to be used it must be loaded into the RAM of the microcomputer. Normally the operating system, supplied on floppy diskettes, is first read from the diskettes and written onto the hard disk. This is called *copying* the operating system onto the hard disk. After this, the floppy diskette version of the DOS may not be used again. The PC is set up so that when it is turned on, the DOS program is automatically read from the hard disk, written into the RAM, and then run.

RAM is an example of a *volatile* memory. That is, when power is turned off, the data that it holds are lost. This is why the DOS program must be reloaded from the hard disk each time the PC is turned on.

▲ 1.4 EVOLUTION OF THE INTEL MICROPROCESSOR ARCHITECTURE

The principal way in which microprocessors and microcomputers are categorized is in terms of the maximum number of binary bits in the data they process—that is, their word length. Over time, five standard data widths have evolved for microprocessors and microcomputers: *4-bit*, *8-bit*, *16-bit*, *32-bit*, and *64-bit*.

Figure 1.7 illustrates the evolution of Intel's microprocessors since their introduction in 1972. The first microprocessor, the 4004, was designed to process data arranged as 4-bit words. This organization is also referred to as a *nibble* of data.

The 4004 implemented a very low-performance microcomputer by today's standards. This low performance and limited system capability restricted its use to simpler, special-purpose applications. A common use was in electronic calculators.

Beginning in 1974 a second generation of microprocessors was introduced. These devices, the 8008, 8080, and 8085, were 8-bit microprocessors. That is, they were all designed to process 8-bit (1-byte-wide) data instead of 4-bit data. The 8080, identified in Fig. 1.7, was introduced in 1975.

These newer 8-bit microprocessors were characterized by higher-performance operation, larger system capabilities, and greater ease of programming. They were able to provide the system requirements for many applications that could not be satisfied with the earlier 4-bit microprocessors. These extended capabilities led to widespread acceptance of multichip 8-bit microcomputers for special-purpose system designs. Examples of these dedicated applications are electronic instruments, cash registers, and printers.

Plans for development of third-generation 16-bit microprocessors were announced by many of the leading semiconductor manufacturers in the mid-1970s. Looking at Fig. 1.7, we see that Intel's first 16-bit microprocessor, the 8086, became available in 1979 and was followed the next year by its 8-bit bus version, the 8088. This was the birth of Intel's 8086 family architecture. Other family members, such as the 80286, 80186, and 80188, were introduced in the years that followed.

These 16-bit microprocessors provided higher performance and had the ability to satisfy a broad scope of special-purpose and general-purpose microcomputer applications. They all have the ability to handle 8-bit, 16-bit, and special-purpose data types. Moreover, their powerful instruction sets are more in line with those provided by a minicomputer.

In 1985, Intel Corporation introduced its first 32-bit microprocessor, the 80386DX. The 80386DX microprocessor brought true minicomputer-level performance to the microcomputer system. This device was followed by a 16-bit external bus version, the 80386SX, in 1988. Intel's second generation of 32-bit microprocessors, called the 80486DX and 80486SX, became available in 1989 and 1990, respectively. They were followed by yet a higher-performance family, the Pentium[R] processors, in 1993.

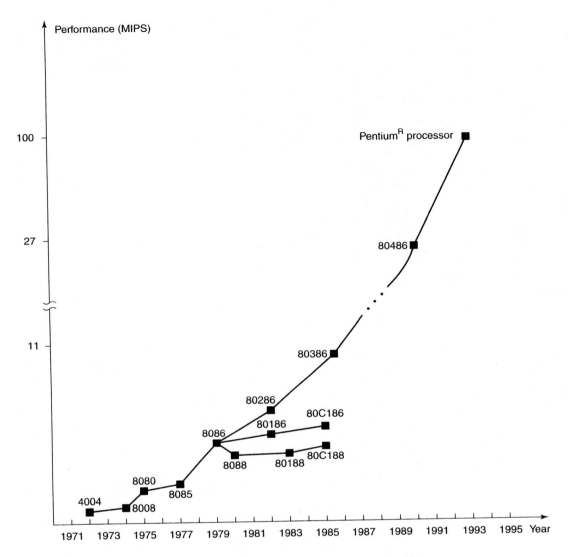

Figure 1.7 Evolution of the Intel microprocessor architecture.

Microprocessor Performance: MIPS and iCOMP

In Fig. 1.7 the 8086 microprocessor families are illustrated relative to their performance. Here performance is measured in what are called *MIPS*—that is, how many million instructions they can execute per second. Today, the number of MIPS provided by a microprocessor is the standard most frequently used to compare performance. Notice that performance has vastly increased with each new generation of microprocessor. For instance, the performance identified for the 80386 corresponds to a 80386DX device operating at 33 MHz and equals approximately 11 MIPS. With the introduction of the 80486, the level of performance capability of the architecture was raised to approximately 27 MIPS. This shows that perfor-

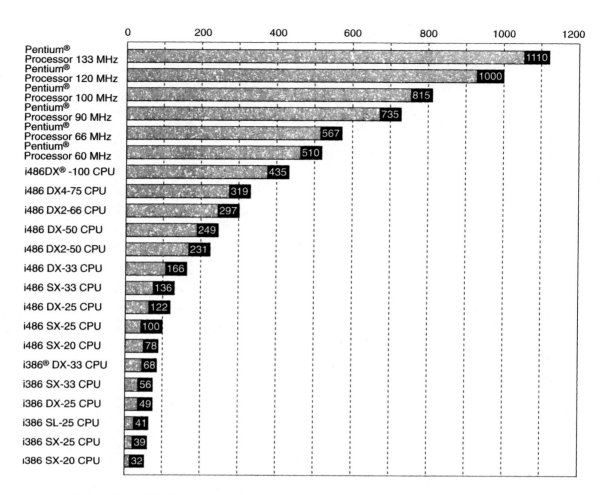

Figure 1.8 iCOMP index rating chart. (Reprinted by permission of Intel Corp. Copyright/ Intel Corp. 1993)

mance of the 8086 architecture was more than doubled with the introduction of the 33-MHz 80486DX microprocessor.

The MIPS used in this chart are known as Drystone V1.1 MIPS. That is, they are measured by running a test program called the *Drystone program*, and the resulting performance measurements are normalized to those of a VAX 1.1 computer (VAX 1.1 was a minicomputer manufactured by Digital Equipment Corporation). Therefore, we say that the 80486DX is capable of delivering up to 27 VAX MIPS of performance.

Another method, called the *iCOMP index*, is provided by Intel Corporation for comparison of the performances of their 32-bit microprocessors in a personal computer application. In the iCOMP index chart of Fig. 1.8, a bar is used to represent a measure of the performance for each of Intel's MPUs. Instead of being related to the performance of a test program, such as the Drystone program, the iCOMP rating of an MPU is based on a variety of 16-bit and 32-bit MPU performance components important to the personal computer. That is, the iCOMP rating encompasses performance components that represent integer mathematics, floating-point mathematics, graphics, and video. The contribution by each of these categories is also weighted based on an estimate of their normal occurrence in widely used software applications. In this way, we see that iCOMP is a more broad-based rating of MPU performance for the personal computer applications.

The higher the iCOMP rating, the higher the performance offered by the MPU. Notice that the members of the 80386 family offer low performance when compared to the newer 80486 and Pentium[R] processor families. In fact, the slowest 80386SX MPU shown in Fig. 1.8, the -20, has a performance rating of 32, whereas the fastest 80386DX, the -33, is rated at 68. In this way, we see that by selecting between the various members of the 80386, 80486, and Pentium[R] processor families, we can achieve a wide range of system-performance levels.

Transistor Density

The evolution of microprocessors is made possible by advances in semiconductor process technology. Semiconductor device geometry decreased from about 5 microns in the early 1970s to submicrons today. Smaller-device geometry permits integration of several orders of magnitude more transistors into the same-size chip and at the same time has led to higher operating speeds. In Fig. 1.9 we see that the 4004 contained about 10,000 transistors.

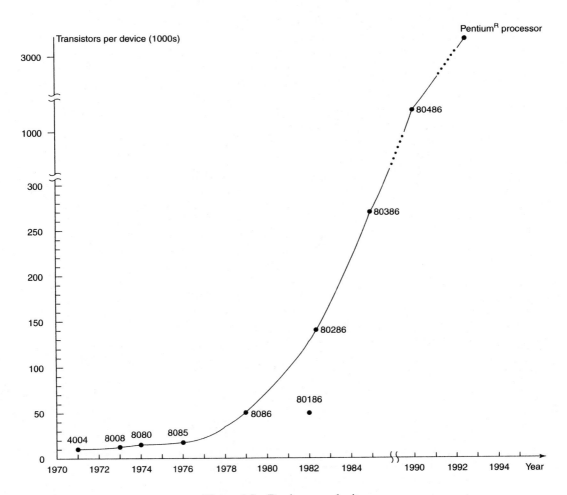

Figure 1.9 Device complexity.

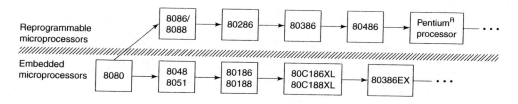

Figure 1.10 Processors for embedded control and reprogrammable applications.

Transistor density was increased to about 30,000 with the development of the 8086 in 1979; with the introduction of the 80286, the transistor count was further increased to approximately 140,000; and the 275,000 transistors of the 80386DX almost doubled transistor density. The 80486DX is the first family member with a density above the 1 million transistor level (1,200,000 transistors); with the Pentium[R] processor's complexity, density has risen to more than 3 million transistors.

Reprogrammable and Embedded Microprocessors

Microprocessors can be classified according to the type of application for which they have been designed. In Fig. 1.10 we have placed Intel microprocessors into two application-oriented categories: *reprogrammable microprocessors* and *embedded microprocessors and microcontrollers*. Initially devices such as the 8080 were most widely used as *special-purpose microcomputers*. By special-purpose microcomputer we mean a system that has been tailored to meet the needs of a specific application. These special-purpose microcomputers were used in *embedded control applications*—that is, applications in which the microcomputer performs a dedicated control function.

Embedded control applications are further divided into those that involve primarily *event control* and those that require *data control*. An example of an embedded control application that is primarily event control is a microcomputer used for industrial process control. Here the program of the microprocessor is used to initiate a timed sequence of events. On the other hand, an application that focuses more on data control than event control is a hard disk controller interface. In this case, a block of data that is to be processed— for example, a file of data—must be quickly transferred from secondary storage memory to primary storage memory.

The spectrum of embedded control applications requires a wide variety of system features and performance levels. Devices developed specifically for the needs of this market-place have stressed low cost and high integration. In Fig. 1.10 we see that the earlier multichip 8080 solutions were initially replaced by highly integrated 8-bit, single-chip microcomputer devices such as the 8048 and 8051. These devices were tailored to work best as event controllers. For instance, the 8051 offers one-order-of-magnitude-higher performance than the 8080, a more powerful instruction set, and special on-chip functions such as ROM, RAM, an interval/event timer, a universal asynchronous receiver/transmitter (UART), and programmable parallel I/O ports. Today these type of embedded control devices are called *microcontrollers*.

Later, devices such as the 80C186XL, 80C188XL, and 80386EX were designed to better meet the needs of data-control applications. They are also highly integrated, but they have additional features, such as string instructions and direct-memory access channels, which better handle the movement of data. They are known as *embedded microprocessors*.

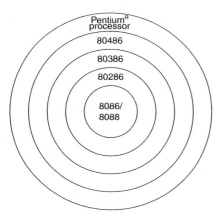

Figure 1.11 Code and system-level compatibility.

The category of reprogrammable microprocessors represents the class of applications in which a microprocessor is used to implement a *general-purpose microcomputer*. Unlike a special-purpose microcomputer, a general-purpose microcomputer is intended to run a wide variety of software applications; that is, while it is in use it can be easily reprogrammed to run a different application program. Two examples of reprogrammable microcomputers are the personal computer and file server. In Fig. 1.10 we see that the 8086, 8088, 80286, 80386, 80486, and Pentium[R] processor are the Intel microprocessors intended for use in this type of application.

Architectural compatibility is a critical need of microprocessors developed for use in reprogrammable applications. As shown in Fig. 1.11, each of the new members of the 8086/8088 family provides a superset of the earlier device's architecture. That is, the features offered by the 80386 microprocessor are a superset of the 80286 architecture, and those of the 80286 are a superset of the original 8086/8088 architecture.

Actually, the 80286, 80386, 80486, and Pentium[R] processors can operate in either of two modes—the *real-address mode* or *protected-address mode*. When in the real mode, they operate like a high-performance 8086/8088. They can execute what is called the *base instruction set*, which is object code compatible with the 8086/8088. For this reason, operating systems and application programs written for the 8086 and 8088 run on the 80286, 80386, 80486, or Pentium[R] processor architectures without modification. Further, a number of new instructions has been added in the instruction sets of the 80286, 80386, 80486, and Pentium[R] processors to enhance their performance and functionality. We say that object code is *upward compatible* within the 8086 architecture. This means that 8086/8088 code will run on the 80286, 80386, 80486, and Pentium[R] processors, but the reverse is not true if any of the new instructions are in use.

Microprocessors designed for implementing general-purpose microcomputers must offer more advanced system features than those of a microcontroller. For example, it needs to support and manage a large memory subsystem. The 80286 is capable of managing a 1 GB (1 gigabyte) address space and the 80386 supports 64 T-byte (64 terabytes) of memory. Moreover, a reprogrammable microcomputer, such as a personal computer, normally runs an operating system. The architectures of the 80286, 80386, 80486, and Pentium[R] processors have been enhanced with on-chip support for operating system functions such as *memory management, protection,* and *multitasking*. These new features become active only when the MPU is operated in the protected mode. The 80386, 80486, and Pentium[R] processors also have a special mode of operation known as *virtual 8086 mode* that permits 8086/8088 code to be run in the protected mode.

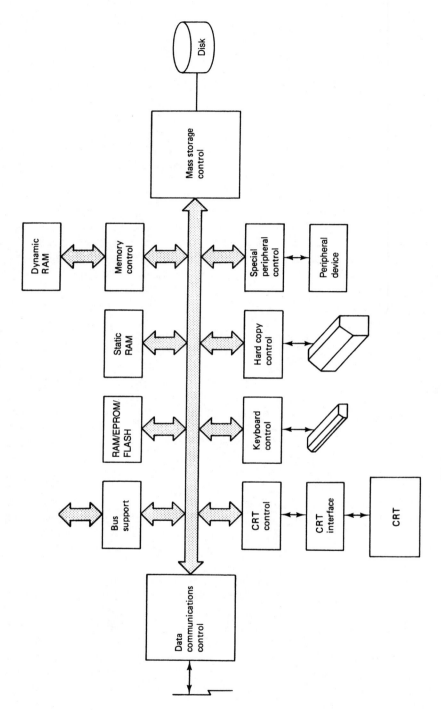

Figure 1.12 Peripheral support for the MPU.

Reprogrammable microcomputers, such as those based on the 8086 family, require a wide variety of input/output resources. Figure 1.12 shows the kinds of interfaces that are frequently implemented in a personal computer or other microcomputer system. A large family of VLSI peripheral ICs is needed to support reprogrammable microprocessors such as the 8086, 80286, 80386, 80486, and Pentium^R processors. Examples are floppy disk controllers, hard disk controllers, local area network controllers, and communication controllers. For this reason, the 8086, 8088, 80286, 80386, 80486, and Pentium^R processors are designed to implement a multichip microcomputer system. In this way a system can easily be configured with the appropriate set of I/O interfaces.

ASSIGNMENTS

Section 1.2

1. Which IBM personal computer employs the 8088 microprocessor?
2. What is meant by the term *open system*?
3. What is the I/O expansion bus of the original IBM PC called?
4. What does PC/AT stand for?
5. What does ISA stand for?
6. What is a reprogrammable microcomputer?
7. Name the three classes of computers.
8. What are the main similarities and differences between the minicomputer and the microcomputer?
9. What does VLSI stand for?

Section 1.3

10. What are the four building blocks of a microcomputer system?
11. What is the heart of the microcomputer system called?
12. Is the 8088 an 8-bit or 16-bit microprocessor?
13. What is the primary input unit of the PC/AT? Give two other examples of input units available for the PC/AT.
14. What are the primary output devices of the PC/AT?
15. Into what two sections is the memory of a PC/AT partitioned?
16. What is the storage capacity of the standard $5\frac{1}{4}$-inch floppy diskette of the original PC/AT? What is the storage capacity of its standard hard disk drive?
17. What do ROM and RAM stand for?
18. How much ROM was provided in the original PC/AT's processor board? What was the maximum amount of RAM that could be implemented on this processor board?
19. Why must DOS be reloaded from the hard disk each time power is turned on?

Section 1.4

20. What are the standard data word lengths for which microprocessors have been developed?
21. What was the first 4-bit microprocessor introduced by Intel Corporation? Eight-bit microprocessor? Sixteen-bit microprocessor? Thirty-two-bit microprocessor?

22. Name five 16-bit members of the 8086 family architecture.
23. What does MIPS stand for?
24. Approximately how many MIPS are delivered by the 33 MHz 80486DX?
25. What is the name of the program that is used to run the MIPS measurement test for the data in Fig. 1.8?
26. What is the iCOMP rating of an 80386SX-25 MPU? An 80386DX-25 MPU?
27. Approximately how many transistors are used to implement the 8088 microprocessor? The 80286 microprocessor? The 80386DX microprocessor? The 80486DX microprocessor? The Pentium^R processor?
28. What is an embedded microcontroller?
29. Name the two groups into which embedded processors are categorized based on applications.
30. What is the difference between a multichip microcomputer and a single-chip microcomputer?
31. Name six 8086 family microprocessors intended for use in reprogrammable microcomputer applications.
32. Give the names for the 80386DX's two modes of operation.
33. What is meant by upward software compatibility relative to 8086 architecture microprocessors?
34. List three advanced architectural features provided by the 80386DX microprocessor.
35. Give three types of VLSI peripheral support devices needed in a reprogrammable microcomputer system.

Real-Addressed Mode Software Architecture of the 80386DX Microprocessor

▲ 2.1 INTRODUCTION

In this chapter we begin our study of the 80386DX microprocessor and its assembly language programming. To program the 80386DX with assembly language, we must understand how the microprocessor and its memory subsystem operate from a software point of view. Remember that the 80386DX can operate in either of two modes, called the *real-addressed mode* (real mode) and the *protected-addressed mode* (protected mode). Moreover, we pointed out in Chapter 1 that when in the real mode, the 80386DX operates like a very high-performance 8086 microprocessor. In fact, a 16-MHz 80386DX provides more than 10 times higher performance than that of the standard 5-MHz 8086. Here we will examine just the real-mode software architecture. The 80386DX's protected mode is presented in Chapter 7. The following topics are covered in this chapter:

1. Internal architecture of the 80386DX microprocessor
2. Real-mode software model of the 80386DX microprocessor
3. Real-mode memory address space and data organization
4. Data types
5. Segment registers and memory segmentation
6. Instruction pointer
7. General-purpose data registers
8. Pointer and index registers
9. Flags register
10. Generating a real-mode memory address
11. The stack
12. Real-mode input/output address space

17

▲ 2.2 INTERNAL ARCHITECTURE
OF THE 80386DX MICROPROCESSOR

The internal architecture of the 8086 family of microprocessors has changed a lot as part of the evolutionary process from the original 8086 to the 80386. All members of the 8086 family employ what is called *parallel processing*. That is, they are implemented with simultaneously operating multiple processing units. Each unit has a dedicated function and they operate at the same time. The more the parallel processing, the higher the performance of the microprocessor.

The 8086 microprocessor contains just two processing units: the bus interface unit and execution unit. In the 80286 microprocessor, the internal architecture was further partitioned into four independent processing elements: the bus unit, the instruction unit, the execution unit, and the address unit. This additional parallel processing provided an important contribution to the higher level of performance achieved with the 80286 architecture.

The 80386DX's internal architecture is illustrated in Fig. 2.1. Here we see that to enhance the performance, more parallel processing elements are provided. Notice that now there are six functional units: the *execution unit*, the *segment unit*, the *page unit*, the *bus unit*, the *prefetch unit*, and the *decode unit*. Let us now look more closely at each of the processing units of the 80386DX.

The bus unit is the 80386DX's interface to the outside world. By interface, we mean the path by which it connects to external devices. The bus interface provides a 32-bit data bus, a 32-bit address bus, and the signals needed to control transfers over the bus. In fact, 8-bit, 16-bit, and 32-bit data transfers are supported. These buses are demultiplexed like those of the 80286. That is, the 80386DX has separate pins for its address and data bus lines. This demultiplexing of address and data results in higher performance and easier hardware design. Expanding the data bus width to 32 bits further improves the performance of the 80386DX's hardware architecture as compared to that of either the 8086 or 80286.

The bus unit is responsible for performing all external bus operations. This processing unit contains the latches and drivers for the address bus, transceivers for the data bus, and

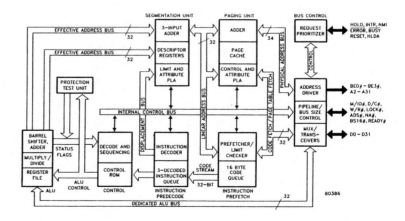

Figure 2.1 Internal architecture of the 80386DX microprocessors. (Reprinted with permission of Intel Corp. Copyright/Intel Corp. 1986)

control logic for signaling whether a memory, input/output, or interrupt-acknowledge bus cycle is being performed. Looking at Fig. 2.1, we find that for data accesses, the address of the storage location that is to be accessed is input from the paging unit, and for code accesses the address is provided by the prefetch unit.

The prefetch unit implements a mechanism known as an *instruction stream queue*. This queue permits the 80386DX to prefetch up to 16 bytes of instruction code. Whenever the queue is not full—that is, it has room for at least 4 more bytes and, at the same time, the execution unit is not asking it to read or write operands from memory—the prefetch unit supplies addresses to the bus interface unit and signals it to look ahead in the program by fetching the next sequential instructions. Prefected instructions are held in the FIFO queue for use by the instruction decoder. Whenever bytes are loaded at the input end of the queue, they are automatically shifted up through the FIFO to the empty locations near the output. With its 32-bit data bus, the 80386DX fetches 4 bytes of instruction code in a single memory cycle. Through this prefetch mechanism, the fetch time for most instructions is hidden.

If the queue in the prefetch unit is full and the execution unit is not requesting access to operands in memory, the bus interface unit does not need to perform any bus cycle. These intervals of no bus activity, which occur between bus cycles, are known as *idle states*.

The prefetch unit prioritizes bus activity. Highest priority is given to operand accesses for the execution unit. However, if the bus unit is already in the process of fetching instruction code when the execution unit requests it to read or write operands from memory or I/O, the current instruction fetch is first completed before the operand read/write cycle is initiated.

In Fig. 2.1, we see that the decode unit accesses the output end of the prefetch unit's instruction queue. It reads machine code instructions from the output side of the prefetch queue and decodes them into the microcode instruction format used by the execution unit. That is, it off-loads the responsibility for instruction decoding from the execution unit. The *instruction queue* within the 80386DX's instruction unit permits three fully decoded instructions to be held waiting for use by the execution unit. Once again the result is improved performance for the MPU.

The execution unit includes the arithmetic/logic unit (ALU), the 80386DX's registers, special multiply, divide, and shift hardware, and a control ROM. By registers, we mean the 80386DX's general-purpose registers, such as EAX, EBX, and ECX. The control ROM contains the microcode sequences that define the operation performed by each of the 80386DX's machine code instructions. The execution unit reads decoded instructions from the instruction queue and performs the operations that they specify. It is the ALU that performs the arithmetic, logic, and shift operations required by an instruction. If necessary, during the execution of an instruction, it requests the segment and page units to generate operand addresses and the bus interface unit to perform read or write bus cycles to access data in memory or I/O devices. The extra hardware that is provided to perform multiply, divide, shift, and rotate operations improves the performance of instructions that employ these functions.

The segment and page units provide the memory-management and protection services for the 80386DX. They off-load the responsibility for address generation, address translation, and segment checking from the bus interface unit, thereby further boosting the performance of the MPU. The segment unit implements the segmentation model of the 80386DX's memory management. That is, it contains dedicated hardware for performing high-speed address calculations, logical-to-linear address translation, and protection checks. For instance, when in the real mode, the execution unit requests the segment unit to obtain the address of the next instruction to be fetched by adding an appended version of the current contents of the code segment (CS) register with the value in the instruction pointer (IP) register to obtain the 20-bit physical address that is to be output on the address bus. This address is passed on to the bus unit.

For protected mode, the segment unit performs the logical-to-linear address translation and various protection checks needed when performing bus cycles. It contains the segment registers and the 6-*word* × 64-*bit cache* that is used to hold the current descriptors within the 80386DX.

The page unit implements the protected mode paging model of the 80386DX's memory management. It contains the *translation lookaside buffer* that stores recently used page directory and page table entries. When paging is enabled, the linear address produced by the segment unit is used as the input of the page unit. Here the linear address is translated into the physical address of the memory or I/O location to be accessed. This physical memory or I/O address is output to the bus interface unit.

▲ 2.3 REAL-MODE SOFTWARE MODEL OF THE 80386DX MICROPROCESSOR

The purpose of a *software model* is to aid the programmer in understanding the operation of the microcomputer system from a software point of view. To be able to program a microprocessor, one does not necessarily need to know all its hardware architecture features. For instance, we do not need to know the function of the signals at its various pins, their electrical connections, or their electrical switching characteristics. The function, interconnection, and operation of the internal circuits of the microprocessor also need not normally be considered. What is important to the programmer is to know the various registers within the device and to understand their purpose, functions, operating capabilities, and limitations. Furthermore, it is essential to know how external memory is organized and how it is accessed to obtain instructions and data.

The software architecture of the 80386DX microprocessor is illustrated with the software model shown in Fig. 2.2. Looking at this diagram, we find that it includes six 16-bit registers and twenty-four 32-bit registers. The registers (or parts of registers) that are important to real-mode application programming are highlighted in the diagram. Nine of them—the data registers (EAX, EBX, ECX, and EDX), the pointer registers (EBP and ESP), the index registers (ESI and EDI), and the flag register (FLAGS)—are identical to the corresponding registers in the 8086 and 80286 software models except that they are now all 32 bits in length. On the other hand, the instruction pointer (IP) and segment registers (CS, DS, SS, and ES) are still 16 bits in length. From a software point of view, all these registers serve functions similar to those they performed in the 8086 and 80286 architectures. For instance, CS:IP points to the next instruction that is to be fetched.

Several new registers are found in the 80386DX's software model. For example, it has two more data segment registers, denoted FS and GS. These registers are not implemented in either the 8086 or 80286 microprocessors. Another new register is called *control register zero* (CR_0). The 5 least significant bits of this register are called the *machine status word* (MSW) and are identical to the MSW of the 80286 microprocessor. However, only one bit, bit 0, in the MSW is active in real mode. This is the *protection enable* (PE) bit. PE is used to switch the 80386DX from real to protected mode. At reset, PE is set to 0 and selects the real-addressed mode of operation.

Looking at the software model in Fig. 2.2, we see that the 80386DX architecture implements independent memory and I/O address spaces. Notice that the memory address space is 1,048,576 bytes (1 Mbyte) in length and the I/O address space is 65,536 bytes (64 Kbytes) in length. We are concerned here with what can be done with this architecture and

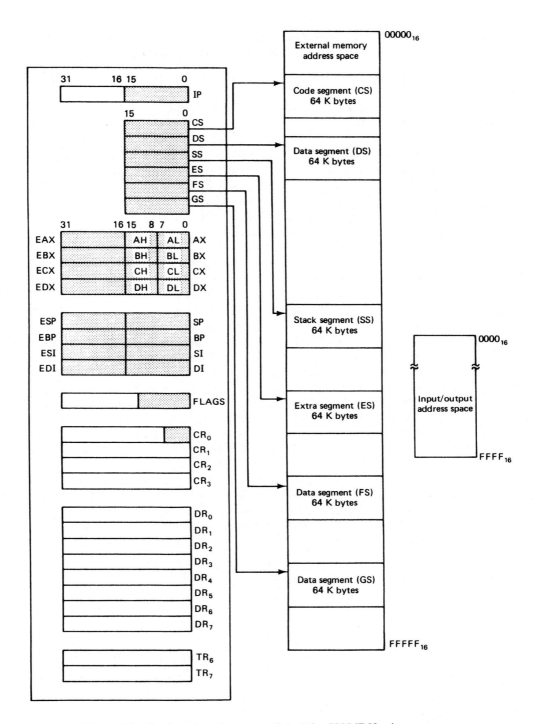

Figure 2.2 Real-mode software model of the 80386DX microprocessor.

how to do it through software. For this purpose, we now begin a detailed study of the elements of the model and their relationship to software.

▲ 2.4 REAL-MODE MEMORY-ADDRESS SPACE AND DATA ORGANIZATION

Now that we have introduced the idea of a software model, let us look at how information such as numbers, character, and instructions are stored in memory. As shown in Fig. 2.3, the 80386DX microcomputer supports 1 Mbyte of external memory. This memory space is organized from a software point of view as individual bytes of data stored at consecutive addresses over the address range 00000_{16} (0H) to $FFFFF_{16}$ (FFFFFH). However, the 80386DX can access any 2 consecutive bytes of data as a word of data or any 4 consecutive bytes as a double word of data.

Any part of the 80386DX microcomputer's 1M-byte address space can be implemented for user's access; however, some address locations have dedicated functions. These dedicated locations should not be used as general memory for storage of data or instructions of a program. The real-mode memory-address space is partitioned into general-use and dedicated-use areas in the same way as for the 8086 microprocessor. For instance, in Fig. 2.3 we find that the first 1024 bytes of memory address space, addresses 0_{16} through $3FF_{16}$, are dedicated. They are again used for storage of the microcomputer's interrupt-vector table. This table contains pointers that define the starting point of interrupt-service routines. Each pointer in this table requires 4 bytes of memory, a double word. Therefore, it can contain up to 256 interrupt pointers. A pointer is a two-word address element. The word of this pointer at the higher address is called the *segment base address,* and the word at the lower address is the *offset. General-use memory* is where data or instructions of the program are stored. In Fig. 2.3, we see that the general-use area of memory is the range from addresses 400_{16} through $FFFFF_{16}$.

To permit efficient use of memory, words and double words of data can be stored at what are called either *aligned double-word* boundaries or *unaligned double-word* boundaries. Aligned double-word boundaries correspond to addresses that are multiples of 4: for instance, 00000_{16}, 00004_{16}, 00008_{16}, and so on. Figure 2.4 shows a number of words and double

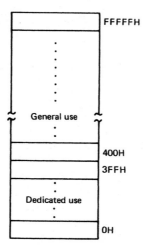

Figure 2.3 Real-mode memory-address space.

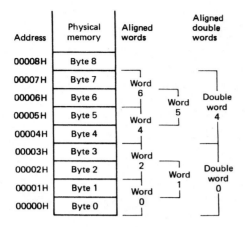

Address	Physical memory	Aligned words	Aligned double words
00008H	Byte 8		
00007H	Byte 7	Word 6	
00006H	Byte 6		Double word 4
00005H	Byte 5	Word 4	Word 5
00004H	Byte 4		
00003H	Byte 3	Word 2	
00002H	Byte 2		Double word 0
00001H	Byte 1	Word 0	Word 1
00000H	Byte 0		

Figure 2.4 Examples of aligned data words and double words.

words that are stored at aligned double-word address boundaries. For each of these pieces of data all the bytes of data exist within the same double word.

For example, the value $5AF0_{16}$ is stored in memory at the double-word aligned address 02000_{16}, as shown in Fig. 2.5(a). Notice that the lower-addressed byte-storage location, 02000_{16}, contains the value $11110000_2 = F0_{16}$. Moreover, the contents of the next-higher-addressed byte, which is in storage location 02001_{16}, are $01011010_2 = 5A_{16}$. These two bytes represent the word $0101101011110000_2 = 5AF0_{16}$.

EXAMPLE 2.1

What is the data word shown in Fig. 2.5(b)? Express the result in hexadecimal form. Is it stored at an aligned double-word boundary?

Solution

The most significant byte of the word is stored at address $0200E_{16}$ and equals

$$00101100_2 = 2C_{16}$$

Its least significant byte is stored at address $0200D_{16}$ and is

$$10010110_2 = 96_{16}$$

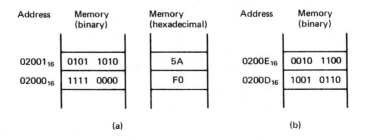

Figure 2.5 (a) Storing an aligned word of data in memory. (b) An example.

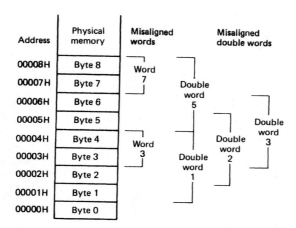

Figure 2.6 Examples of misaligned data words and double words.

Together the two bytes give the word

$$0010110010010110_2 = 2C96_{16}$$

This word is aligned within double-word address boundary $0200C_{16}$.

It is not always possible to have all words or double words of data aligned at double-word boundaries. Figure 2.6 shows some examples of misaligned words and double words of data that can be accessed by the 80386DX. Notice that word 3 consists of byte 3 from aligned double word 0 and byte 4 from aligned double word 4. An example that shows a double word of data that is stored in memory at an unaligned double-word boundary is given in Fig. 2.7(a). Here we see that the higher-addressed word, which equals 0123_{16}, is stored in memory starting at address 02104_{16} in aligned double-word address 02104_{16}, and the lower-addressed word, whose value is $ABCD_{16}$, begins at address 02102_{16} in the aligned double-word address 02100_{16}. The complete double word equals $0123ABCD_{16}$.

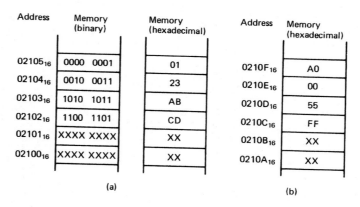

Figure 2.7 (a) Storing a misaligned double word in memory. (b) An example of an aligned double word.

EXAMPLE 2.2

How should the double word $A00055FF_{16}$ be stored in memory starting at address $0210C_{16}$? Is the double word aligned or misaligned?

Solution

Storage of the double word requires 4 consecutive byte locations in memory, starting at address $0210C_{16}$. The least-significant byte is stored at address $0210C_{16}$. This value is FF_{16} in Fig. 2.7(b). The second byte, which equals 55_{16}, is stored at address $0210D_{16}$. These two bytes are followed by the values 00_{16} and $A0_{16}$ at addresses $0210E_{16}$ and $0210F_{16}$, respectively. This double word is aligned at double-word address $0210C_{16}$.

▲ 2.5 DATA TYPES

In the preceding section we identified the fundamental data formats of the 80386DX as the byte (8 bits), word (16 bits), and double word (32 bits). We also showed how each of these elements is stored in memory. We continue by examining the types of data that can be coded into these formats for processing by the 80386DX microprocessor.

The 80386DX microprocessor can directly process data expressed in a number of different data types. Let us begin with the *integer data type*. The 80386DX can process data as either *unsigned* or *signed integer* numbers. Moreover, both types of integer can be byte wide, word wide, or double-word wide. Figure 2.8(a) represents an *unsigned byte integer*. This data type can be used to represent decimal numbers in the range 0 through 255. The *unsigned word integer* is shown in Fig. 2.8(b). It can be used to represent decimal numbers in the range 0 through 65,535. Finally, Fig. 2.8(c) illustrates an *unsigned double-word integer*.

EXAMPLE 2.3

What value does the unsigned double-word integer 00010000_{16} represent?

Solution

First, the hexadecimal integer is converted to binary form:

$$00010000_{16} = 00000000000000010000000000000000_2$$

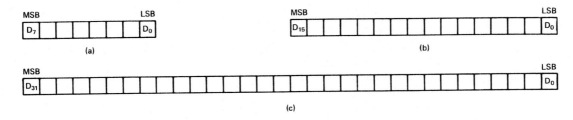

Figure 2.8 (a) Unsigned byte integer. (b) Unsigned word integer. (c) Unsigned double-word integer.

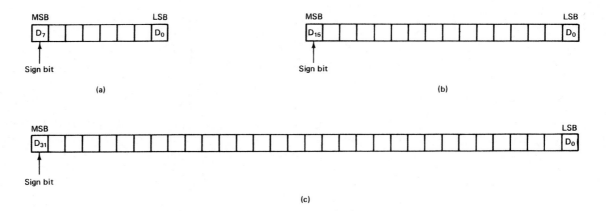

Figure 2.9 (a) Signed byte integer. (b) Signed word integer. (c) Signed double-word integer.

Next, we find the decimal value for the binary number:

$$0000000000000001000000000000000_2 = 2^{16} = 65,536$$

The signed integer byte, word, and double word of Fig. 2.9(a), (b), and (c) are similar to the unsigned integer data types we just introduced; however, here the most significant bit is a sign bit. A zero in this bit position identifies a positive number. For this reason, the *signed integer byte* can represent decimal numbers in the range +127 through −128, and the *signed integer word* permits numbers in the range +32,767 through −32,768. For example, the number +3 expressed as a signed integer byte is 00000011 (03_{16}). On the other hand, the 80386DX always expresses negative numbers in 2's-complement notation. Therefore, −3 is coded as 11111101 (FD_{16}).

EXAMPLE 2.4

A double-word signed integer equals $FFFEFFFF_{16}$. What decimal number does it represent?

Solution

Expressing the hexadecimal number in binary form, we get

$$FFFEFFFF_{16} = 11111111111111101111111111111111_2$$

Since the most significant bit is 1, the number is negative and is in 2's-complement form. Converting to its binary equivalent by subtracting 1 from the least-significant bit and then complementing all bits gives

$$FFFEFFFF_{16} = -00000000000000010000000000000001_2$$
$$= -65537$$

Decimal	BCD
0	0000
1	0001
2	0010
3	0011
4	0100
5	0101
6	0110
7	0111
8	1000
9	1001

(a)

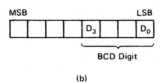

(b)

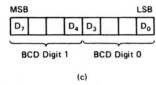

(c)

Figure 2.10 (a) BCD numbers. (b) Unpacked BCD digit. (c) Packed BCD digits.

The 80386DX can also process data coded as *binary-coded decimal* (BCD) *numbers.* Figure 2.10(a) lists the BCD values for decimal numbers 0 through 9. BCD data can be stored in either unpacked or packed form. For instance, the *unpacked BCD byte* in Fig. 2.10(b) shows that a single BCD digit is stored in the 4 least significant bits, and the upper 4 bits are set to 0. Figure 2.10(c) shows a byte with packed BCD digits. Here two BCD numbers are stored in a byte. The upper 4 bits represent the most significant digit of a two-digit BCD number.

EXAMPLE 2.5

The packed BCD data stored at byte address 01000_{16} equals 10010001_2. What is the two-digit decimal number?

Solution

Writing the value 10010001_2 as separate BCD digits gives

$$10010001_2 = 1001_{BCD}\ 0001_{BCD} = 91_{10}$$

Information expressed in *ASCII (American Standard Code for Information Interchange)* can also be directly processed by the 80386DX microprocessor. The chart in Fig.

		b7	0	0	0	0	1	1	1	1
		b6	0	0	0	0	1	0	1	1
		b5	0	1	0	1	0	1	0	1
$b_4\,b_3\,b_2\,b_1$	H1 / H0		0	1	2	3	4	5	6	7
0 0 0 0	0		NUL	DLE	SP	0	@	P	`	p
0 0 0 1	1		SOH	DC1	!	1	A	Q	a	q
0 0 1 0	2		STX	DC2	"	2	B	R	b	r
0 0 1 1	3		ETX	DC3	#	3	C	S	c	s
0 1 0 0	4		EOT	DC4	$	4	D	T	d	t
0 1 0 1	5		ENQ	NAK	%	5	E	U	e	u
0 1 1 0	6		ACK	SYN	&	6	F	V	f	v
0 1 1 1	7		BEL	ETB	'	7	G	W	g	w
1 0 0 0	8		BS	CAN	(	8	H	X	h	x
1 0 0 1	9		HT	EM	)	9	I	Y	i	y
1 0 1 0	A		LF	SUB	*	:	J	Z	j	z
1 0 1 1	B		V	ESC	+	;	K	[	k	}
1 1 0 0	C		FF	FS	,	<	L	\	l	|
1 1 0 1	D		CR	GS	–	=	M	]	m	{
1 1 1 0	E		SO	RS	.	>	N	∧	n	~
1 1 1 1	F		SI	US	/	?	O	-	o	DEL

(a)

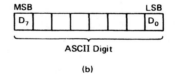

ASCII Digit

(b)

Figure 2.11 (a) ASCII table. (b) ASCII digit.

2.11(a) shows how numbers, letters, and control characters are coded in ASCII. For instance, number 5 is coded as

$$H_1H_0 = 0110101 = 35H$$

where H denotes that the ASCII-coded number is in hexadecimal form. As shown in Fig. 2.11(b), ASCII data are stored one character per byte.

EXAMPLE 2.6

Byte addresses 01100_{16} through 01104_{16} contain the ASCII data 01000001, 01010011, 01000011, 01001001, and 01001001, respectively. What do the data stand for?

Solution

Using the chart in Fig. 2.11(a), the data are converted to ASCII as follows:

$$(01100\text{H}) = 01000001_{\text{ASCII}} = \text{A}$$
$$(01101\text{H}) = 01010011_{\text{ASCII}} = \text{S}$$
$$(01102\text{H}) = 01000011_{\text{ASCII}} = \text{C}$$
$$(01103\text{H}) = 01001001_{\text{ASCII}} = \text{I}$$
$$(01104\text{H}) = 01001001_{\text{ASCII}} = \text{I}$$

▲ 2.6 SEGMENT REGISTERS AND MEMORY SEGMENTATION

Even though the 80386DX has a 1M-byte address space in real mode, not all this memory can be active at one time. Actually, the 1M bytes of memory can be partitioned into 64Kbyte (65,536-byte) *segments*. A segment represents an independently addressable unit of memory consisting of 64K consecutive byte-wide storage locations. Each segment is assigned a *base address* that identifies its starting point—that is, its lowest-addressed byte-storage location.

Only six of these 64K-byte segments can be active at a time: the *code segment, stack segment*, and *data segments D, E, F,* and *G*. The location of the segments of memory that are active, as shown in Fig. 2.12, are identified by the values of addresses held in the 80386DX's six internal segment registers: *CS (code segment), SS (stack segment), DS (data segment), ES (extra segment), FS (data segment F)*, and *GS (data segment G)*. Each of these registers contains a 16-bit base address that points to the lowest-addressed byte of the segment in memory. Six segments give a maximum of 384K bytes of active memory. Of this, 64K bytes are for code (*program storage*); 64K bytes are for a *stack*; and 256K bytes are for *data storage*.

The values held in these registers are usually referred to as the *current-segment register values*. For example, the value in CS points to the first double-word storage location in the current code segment. Code is always fetched as double words, not as words or bytes.

Figure 2.13 illustrates the *segmentation of memory*. In this diagram, we have identified 64K-byte segments with letters such as A, B, and C. The data segment (DS) register contains the value B. Therefore, the second 64K-byte segment of memory from the top, which is labeled B, acts as the current data-storage segment. This is one of the segments in which data that are to be processed by the microcomputer is stored. For this reason this part of the microcomputer's memory address space must contain read/write storage locations that can be accessed by instructions as storage locations for source and destination operands. Segment E is selected by CS as the code segment. It is this segment of memory from which instructions of the program are currently being fetched for execution. The stack segment (SS) register contains H, thereby selecting the 64K-byte segment labeled as H for use as a stack. Finally, the extra segment register, F data segment register, and G data segment register are loaded with the values J, K, and L such that segments J, K, and L of memory can function as additional 64K-byte data storage segments.

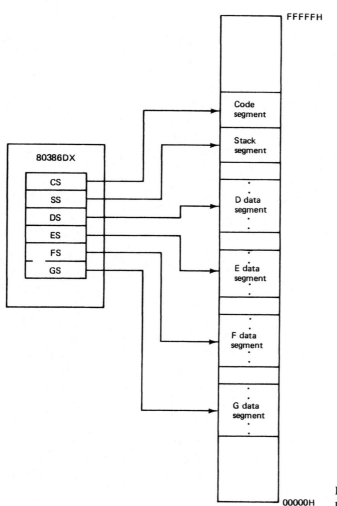

Figure 2.12 Active segments of memory.

The segment registers are said to be *user accessible.* This means that the programmer can change the value they hold through software. Therefore, for a program to gain access to another part of memory, one just has to change the value of the appropriate register or registers. For instance, a new data space, with up to 256K bytes, can be brought in simply by changing the values in DS, ES, FS, and GS.

There is one restriction on the value that can be assigned to a segment as a base address: It must reside on a 16-byte address boundary. This is due to the fact that increasing the 16-bit value in a segment register by 1 actually increases the corresponding memory address by 16. Valid examples are 00000_{16}, 000010_{16}, 00020_{16}, and so on. Other than this restriction, segments can be set up to be contiguous, adjacent, disjointed, or even overlapping. For example, in Fig. 2.13 segments A and B are contiguous, whereas segments B and C are overlapping.

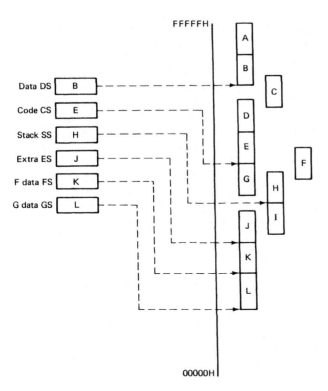

Figure 2.13 Contiguous, adjacent, disjointed, and overlapping segments. (Reprinted by permission of Intel Corp. Copyright/Intel Corp. 1979)

▲ 2.7 INSTRUCTION POINTER

The register that we will consider next in the 80386DX's real-mode software model of Fig. 2.2 is the *instruction pointer* (IP). IP is also 16 bits in length and identifies the location of the next double word of instruction code to be fetched from the current code segment of memory. The IP is similar to a program counter; however, it contains the offset of the next double word of instruction code instead of its actual address. This is because in the real-mode 80386DX, IP and CS are both 16 bits in length, but a 20-bit address is needed to access memory. Internal to the 80386DX, the offset in IP is combined with the current value in CS to generate the address of the instruction code. Therefore, the value of the address for the next code access is often denoted as CS:IP.

During normal operation, the 80386DX fetches instructions from the code segment of memory, stores them in its instruction queue, and executes them one after the other. Every time a double word of instruction code is fetched from memory, the 80386DX updates the value in IP such that it points to the first byte of the next sequential double word of code. That is, IP is incremented by 4. Actually, the 80386DX prefetches up to 16 bytes of instruction code into its internal code queue and holds them there waiting for execution. After an instruction is read from the output of the code queue, it is decoded; if necessary, operands are read from either the data segment of memory or internal registers. Next, the operation specified in the instruction is performed on the operands and the result is written back to either a storage location in memory or an internal register. The 80386DX is now ready to execute the next instruction in the code queue.

The active code segment can be changed simply by executing an instruction that loads a new value into the CS register. For this reason, we can use any 64K-byte segment of memory for storage of instruction code.

▲ 2.8 GENERAL-PURPOSE DATA REGISTERS

As shown in Fig. 2.2, four general-purpose data registers are located within the 80386DX. During program execution, they are used for temporary storage of frequently used intermediate results. Their contents can be read, loaded, or modified through software. Any of the general-purpose registers can be used as the source or destination of an operand during an arithmetic operation, such as ADD, or a logic operation, such as AND. For instance, the values of two pieces of data, called A and B, could be moved from memory into separate data registers and operations such as addition, subtraction, and multiplication performed on them. The advantage of storing these data in internal registers instead of memory during processing is that they can be accessed much faster.

These four registers, known as the *data registers,* are shown in more detail in Fig. 2.14(a). Notice that they are referred to as the *accumulator register (A), base register (B),*

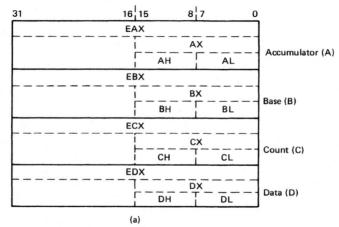

(a)

Register	Operations
EAX, AX, AH, AL	ASCII adjust for addition/subtraction Convert byte to word/word to double word/ double word to quad word Decimal adjust for addition/subtraction Unsigned multiply/divide Signed divide Input/output operations Load/store flags Load/compare/store string operations Table-lookup translations
EBX, BX, BH, BL	Table-lookup translations
ECX, CX, CH, CL	Loop operations Repeat string operations Variable shift/rotate operations
EDX, DX, DH, DL	Indirect input/output operations Input/output string operations Unsigned word/double word multiply Signed word/double word divide Unsigned word/double word divide

(b)

Figure 2.14 (a) General-purpose data registers. (b) Special register functions.

count register (C), and *data register (D)*. These names imply special functions they are meant to perform in the 8086, 80286, and real-mode 80386DX microprocessors. Figure 2.14(b) summarizes these operations. Notice that the C register is used as a count register during string, loop, rotate, and shift operations. For example, the value in the C register is the number of bits by which the contents of an operand must be shifted or rotated during the execution of the multibit shift or rotate instructions. This is the reason it is given the name *count register*. Another example of the dedicated use of data registers is that all I/O operations require the data that are to be input of output to be in the A register, whereas register D holds the address of the I/O port.

Each of these registers can be accessed as a whole (32 bits) for double-word data operations, the lower 16 bits can be used as a word of data, two 8-bit registers can be used for byte-wide data operations, or it can be used as 32 individual bits. References to a register as a word are identified by an X after the register letter. For instance, the 16-bit accumulator register is referenced as AX. Similarly, the other three word registers are referred to as BX, CX, and DX.

On the other hand, when referencing all 32 bits of a register, the word register name is prefixed by the letter E. For example, the extended accumulator is denoted as EAX, and the extended count register as ECX. Finally, the high byte and low byte of a word register are identified by following the register name with the letter H or L, respectively. For the A register, the more significant byte is referred to as AH and the less significant byte as AL.

Actually, some of the data register can also be used to store address information such as a base address or input/output address. Even though they are 32 bits in length, real-mode address operands are always 16 bits in length. For example, BX could hold a 16-bit base address.

▲ 2.9 POINTER AND INDEX REGISTERS

There are four other general-purpose registers shown in Fig. 2.2: two *index registers,* ESI and EDI, and the two *pointer registers,* EBP and ESP. In the 8086, 80286, and real-mode 80386DX architectures, these registers are used to store what are called *offset addresses.* An offset address represents the displacement of a storage location in memory from the segment base address in a segment register. That is, they are used as a pointer or index to select a specific storage location within a 64K-byte segment of memory. Values held in the index registers are used to reference data relative to the data segment or extra segment register, and the pointer registers are used to store offset addresses of memory locations relative to the stack segment register. Just as for the data registers, the values held in these registers can be read, loaded, or modified through software. This is done prior to executing the instruction that references the register for address offset. In this way, to use the offset address in the register, the instruction simply specifies the register that contains the value of the offset address.

Figure 2.15 shows that the two pointer registers are the *extended stack pointer* (ESP) and *extended base pointer* (EBP). When used to hold real-mode address information, the length of the register is always 16 bits. Therefore, the registers are identified as SP and BP. The values in SP and BP are used as offsets from the current value of SS during the execution of instructions that involve the stack segment of memory. In this way, they permit easy access to storage locations in the stack part of memory. In fact, during the execution of an instruction that involves SP or BP as a base address, the value in the base register is automatically combined with the contents of the stack segment register to produce the physical memory address. The value in SP always represents the offset of the next stack

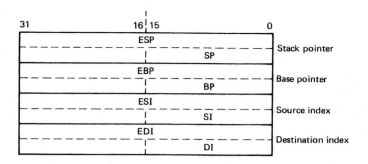

Figure 2.15 Pointer and index registers.

location that is to be accessed. That is, in real mode, combining SP with the value in SS (SS:SP) results in a 20-bit address that points to the *top of the stack* (TOS).

BP also represents an offset relative to the SS register. It is used to access storage locations within the stack segment of memory. To do this it is employed as the offset in an addressing mode called the *based addressing mode*. One common use of BP is to reference parameters that were passed to a subroutine by way of the stack. In this case, instructions are included in the subroutine that use based addressing to read the values of the parameters from the stack.

The *extended source index register* (ESI) and *extended destination index register* (EDI) are used to hold offset addresses for instructions that access data stored in the data segment part of memory. For this reason, they are automatically combined with the value in the DS register during address calculations. Even though these registers are 32 bits in length, in real mode just the lower 16-bit *source index register* (SI) and *destination index register* (DI) are used in indexed addressing. In instructions that involve indexed addressing, SI is used to hold an offset address for a source operand and DI holds an offset that identifies the location of a destination operand.

Earlier we pointed out that any of the data registers can be used as the source or destination of an operand during an arithmetic operation such as an ADD or logic operation such as AND. However, for some operations, an operand that is to be processed may be located in memory instead of internal register. In this case, an index address can be used to identify the location of the operand in memory. For example, string instructions use the index registers to access operands in memory. SI and DI are used as the pointers to the source and destination locations in memory, respectively.

▲ 2.10 FLAGS REGISTER

In Fig. 2.2 we see that the *flags register* (FLAGS) is another 32-bit register within the 80386DX. This register is shown in more detail in Fig. 2.16. Notice that just nine of its bits are active in the real mode. These bits are the same as those implemented in the FLAGS registers of the 8086 and real-mode 80286 microprocessors. Six of these bits represent *status flags*: the *carry flag* (CF), *parity flag* (PF), *auxiliary carry flag*, *zero flag* (ZF), *sign flag* (SF), and *overflow flag* (OF). The logic state of these status flags indicate conditions that are produced as the result of executing an instruction. That is, after executing an instruction such as ADD, specific flag bits are reset (logic 0) or set (logic 1) based on the result that is produced.

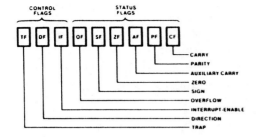

Figure 2.16 Status and control flags. (Reprinted by permission of Intel Corp. Copyright/Intel Corp. 1979)

Let us first summarize the operation of these flags:

1. The carry flag (CF): CF is set if there is a carry-out or a borrow-in for the most significant bit of the result during the execution of an arithmetic instruction. Otherwise, CF is reset.

2. The parity flag (PF): PF is set if the result produced by the instruction has even parity, that is, if it contains an even number of bits at the 1 logic level. If parity is odd, PF is reset.

3. The auxiliary carry flag (AF): AF is set if there is a carry-out from the low nibble into the high nibble or a borrow-in from the high nibble into the low nibble of the lower byte in an 8-, 16-, or 32-bit word. Otherwise, AF is reset.

4. The zero flag (ZF): ZF is set if the result of execution of an arithmetic or logic operation is zero. Otherwise, ZF is reset.

5. The sign bit (SF): The MSB of the result is copied into SF. Thus SF is set if the result is a negative number or reset if it is positive.

6. The overflow flag (OF): When OF is set, it indicates that the signed result is out of range. If the result is not out of range, OF remains reset.

For example, at the completion of execution of a byte-addition instruction, the carry flag (CF) could be set to indicate that the sum of the operands caused a carry-out condition. The auxiliary carry flag (AF) could also set due to the execution of the instruction. This depends on whether or not a carry-out occurred from the least significant nibble to the most significant nibble when the byte operands are added. The sign flag (SF) is also affected, and it will reflect the logic level of the MSB of the result. The overflow flag (OF) is set if there is a carry-out of the sign bit but no carry into the sign bit (an indication of overflow).

The 80386DX provides instructions within its instruction set that are able to use these flags to alter the sequence in which the program is executed. For instance, the condition ZF equal to logic 1 could be tested to initiate a jump to another part of the program. This operation is called *jump on zero*.

The other three implemented flag bits—*direction flag* (DF), *interrupt enable flag* (IF), and *trap flag* (TF)—are control flags. These three flags are provided to control functions of the 80386DX as follows:

1. The trap flag (TF): If TF is set, the 80386DX goes into the *single-step mode*. When in single-step mode, it executes an instruction and then jumps to a special service routine to determine the effect of executing the instruction. This type of operation is very useful for debugging programs.

2. The interrupt flag (IF): For the 80386DX to recognize *maskable interrupt requests* at the INT input, the IF flag must be set. When IF is reset, requests at INT are ignored and the maskable interrupt interface is disabled.

3. The direction flag (DF): The logic level of DF determines the direction in which string operations occur. When set, the string instruction automatically decrements the address. Therefore, the string data transfers proceed from high address to low address. On the other hand, resetting DF causes the string address to be incremented. In this way, data transfers proceed from low address to high address.

The instruction set of the 80386DX includes instructions for saving, loading, and manipulating the flags. For instance, special instructions are provided to permit user software to set or reset CF, DF, and IF at any point in the program. For example, just prior to the beginning of a string operation, DF could be set so that the string address automatically decrements.

▲ 2.11 GENERATING A REAL-MODE MEMORY ADDRESS

A *logical address* in the 80386DX's real-mode software architecture is described by a segment address and an offset address. As shown in Fig. 2.17, both the segment base and offset are 16-bit quantities. This is because all register and memory locations used in address calculations are always 16 bits long. However, the *physical addresses* that are used to access memory are 20 bits in length. The generation of the physical address involves combining a 16-bit offset value that is located in the instruction pointer, a base register, a pointer register, or an index register and a 16-bit segment base value that is located in one of the segment registers.

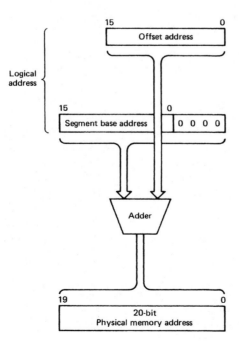

Figure 2.17 Real-mode physical address generation. (Reprinted by permission of Intel Corp. Copyright/Intel 1981)

Type of Reference	Segment Used Register Used	Default Selection Rule
Instructions	Code Segment CS register	Automatic with instruction fetch.
Stack	Stack Segment LSS register	All stack pushes and pops. Any memory reference which uses ESP or EBP as a base register.
Local Data	Data Segment DS register	All data references except when relative to stack or string destination.
Destination Strings	E-Space Segment ES register	Destination of string instructions.

Figure 2.18   Segment register references for memory accesses. (Reprinted by permission of Intel Corp. Copyright/Intel Corp. 1989)

The source of the offset value depends on which type of memory reference is taking place. It can be the base pointer (BP) register, the base (BX) register, the source index (SI) register, the destination index (DI) register, or the instruction pointer (IP). An offset can even be formed from the contents of several of these registers. On the other hand, the segment base value always resides in one of the 80386DX's segment registers: CS, DS, SS, ES, FS, or GS.

During code accesses, data accesses, and stack accesses of memory, the 80386DX selects the appropriate segment register based on the rules shown in Fig. 2.18. For instance, when an instruction acquisition takes place, the source of the segment base value is always CS and the source of the offset value is always IP. This physical address can be denoted as CS:IP. On the other hand, if the value of a variable is being written to memory during execution of an instruction (local data), the segment base address defaults to the DS register, and the offset will be in a register such as DI or BX. An example is the physical address DS:DI. A provision in the processor called the *segment-override prefix* can be used to change the segment from which the variable is accessed. For instance, a prefix could be used to make a data access occur in which the segment base is in the FS register.

Another example is the stack address that is needed when pushing parameters onto the stack. This address is formed from the values of the segment base in the SS register and offset in the SP register and is described as SS:SP.

Remember that the segment base address represents the starting location of the 64K-byte segment in memory, that is, the lowest-addressed byte in the segment. Figure 2.19 shows that the offset identifies the distance in bytes that the storage location of interest resides from this starting address. Therefore, the lowest-addressed byte in a segment has an offset of 0000_{16}, and the highest-addressed byte has an offset of $FFFF_{16}$.

Figure 2.20 illustrates how a segment base value in a segment register and an offset value are combined to form a physical address. What happens is that the value in the segment register is shifted left by 4 bit positions, with its LSBs being filled with zeros. Then the offset value is added to the 16 LSBs of the shifted segment value, called the *segment address*. The result of this addition is the 20-bit physical address. This is identical to how the 8086 and real mode 80286 microprocessors perform the physical address calculation.

The example in Fig. 2.20 represents a segment base value of 1234_{16} and an offset value

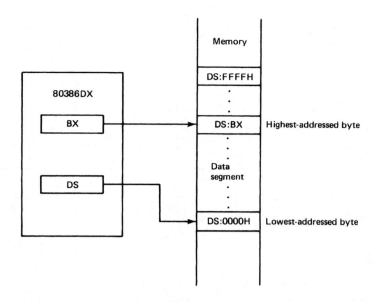

Figure 2.19 Boundaries of a segment.

of 0022_{16}. First let us express the segment base value in binary form. This gives

$$1234_{16} = 0001001000110100_2$$

Shifting left four times and filling with zeros results in the segment address as

$$00010010001101000000_2 = 12340_{16}$$

The offset in binary form is

$$0022_{16} = 0000000000100010_2$$

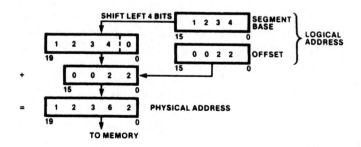

Figure 2.20 Generating a real-mode physical address. (Reprinted by permission of Intel Corp. Copyright/Intel Corp. 1979)

Adding the segment address and the offset, we get

$$00010010001101000000_2 + 0000000000100010_2 = 00010010001101100010_2$$
$$= 12362_{16}$$

This address calculation is done automatically within the 80386DX each time a memory access is initiated.

EXAMPLE 2.7

What would be the offset required to map to physical address location $002C3_{16}$ if the contents of the corresponding segment register are $002A_{16}$?

Solution

The offset value can be obtained by shifting the contents of the segment register left by four bit positions and then subtracting from the physical address. Shifting left gives

$$002A0_{16}$$

Now subtracting, we get the value of the offset:

$$002C3_{16} - 002A0_{16} = 0023_{16}$$

Actually, many different logical addresses can be mapped to the same physical address location in memory. This is done simply by changing the segment base value in the segment register and its corresponding offset. The diagram in Fig. 2.21 demonstrates this idea. Notice

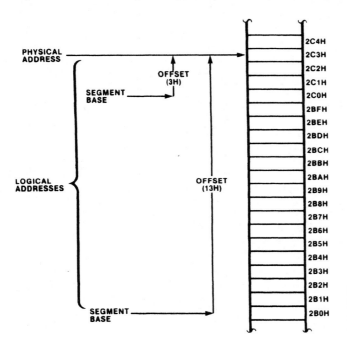

Figure 2.21 Relationship between logical and physical addresses. (Reprinted by permission of Intel Corp. Copyright/Intel Corp. 1979)

that segment base $002B_{16}$ with offset 0013_{16} maps to physical address $002C3_{16}$ in memory. However, if the segment base is changed to $002C_{16}$ with a new offset of 0003_{16}, the physical address is still $002C3_{16}$. In this way, we see that physical address 002BH:0013H is equal to the physical address 002CH:0003H.

▲ 2.12 THE STACK

As indicated earlier, the *stack* is implemented in the memory of the 80386DX microcomputer. A stack is used for temporary storage of information such as data or addresses. For instance, when a *call instruction* is executed, the 80386DX automatically pushes the current values in CS and IP onto the stack. As part of the subroutine, the contents of other registers can also be saved on the stack by executing *push instructions*. An example is the instruction PUSH ESI. When executed it causes the contents of the extended source index register to be pushed onto the stack. At the end of the subroutine, *pop instructions* can be included to pop values from the stack back into their corresponding internal registers. For example, POP ESI causes the value at the top of the stack to be popped back into the extended source index register. At the end of the subroutine, a *return instruction* causes the values of CS and IP to be popped off the stack and put back into the same internal register where they originally resided.

From a real-mode software point of view, stack is 64K bytes long and is organized as 32K words. Figure 2.22 shows that the lowest-addressed word in the current stack is pointed to by the segment base value in the SS register. The contents of the SP and BP registers are used as offsets into the stack segment of memory.

Looking at Fig. 2.22, we see that SP contains an offset value that points to a storage location in the current stack segment. The address obtained from the contents of SS and SP (SS:SP) is the physical address of the last storage location in the stack to which data were pushed. This memory address is known as the *top of the stack*. At the microcomputer's

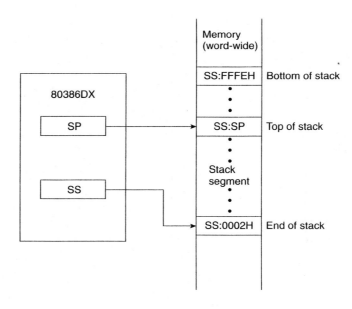

Figure 2.22 Stack segment of memory.

start-up, the value in SP is initialized to $FFFE_{16}$. Combining this value with the current value in SS gives the highest-address word location in the stack (SS:FFFEH)—that is, the *bottom of the stack.*

The 80386DX can push data and address information onto the stack from its internal registers, a storage location in memory, or the immediate operand of an instruction. These data must be either a word or a double word, not a byte. Each time a word or double word is to be pushed onto the top of the stack, the value in SP is first automatically decremented by 2 or 4, respectively, and then the contents of the register are written into the stack part of memory. In this way we see that the stack grows down in memory from the bottom of the stack, which corresponds to the physical address SS:FFFEH, toward the *end of the stack,* which corresponds to the physical address obtained from SS and offset 0002_{16} (SS:0002H).

When a value is popped from the top of the stack, the reverse of this sequence occurs. The physical address defined by SS and SP points to the location of the last value pushed onto the stack. Its contents are first popped off the stack and put into the specified register within the 80386DX; then SP is automatically incremented by 2 or 4, depending on whether a word or double word has been popped. The top of the stack then corresponds to the previous value pushed onto the stack.

An example of how the contents of a word-wide register are pushed onto the stack is shown in Fig. 2.23(a). Here we find the state of the stack prior to execution of the PUSH AX instruction. Notice that the stack segment register contains 0105_{16}. As indicated, the bottom of the stack resides at the physical address derived from SS and offset $FFFE_{16}$. This gives the bottom-of-stack address, A_{BOS}, as

$$A_{BOS} = 1050_{16} + FFFE_{16}$$

$$= 1104E_{16}$$

Furthermore, the stack pointer, which represents the offset from the beginning of the stack specified by the contents of SS to the top of the stack, equals 0008_{16}. Therefore, the current top of the stack is at physical address A_{TOS}, which equals

$$A_{TOS} = 1050_{16} + 0008_{16}$$

$$= 01058_{16}$$

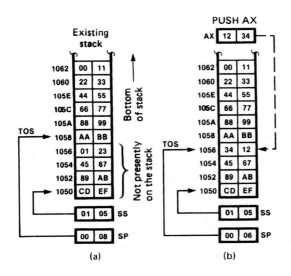

Figure 2.23 (a) Stack just prior to push operation. (Reprinted by permission of Intel Corp. Copyright/Intel Corp. 1979) (b) Stack after execution of the PUSH AX instruction. (Reprinted by permission of Intel Corp. Copyright/Intel Corp. 1979)

Addresses with higher values than that of the top of stack, 01058_{16}, contain valid stack data. Those with lower addresses do not yet contain valid stack data. Notice that the last value pushed to the stack in Fig. 2.23(a) is $BBAA_{16}$.

Figure 2.23(b) demonstrates what happens when the PUSH AX instruction is executed. Here we see that AX contains the value 1234_{16}. Notice that execution of the push instruction causes the stack pointer to be decremented by 2 but does not affect the contents of the stack segment register. Therefore, the next location to be accessed in the stack corresponds to address 1056_{16}. It is to this location that the value in AX is pushed. Notice that the most significant byte of AX, which equals 12_{16}, now resides in memory address 1057_{16}, and the least significant byte of AX, which is 34_{16}, is held in memory address 1056_{16}.

Now let us look at an example in which stack data are popped from the stack back into the register from which they were pushed. Figure 2.24 illustrates this operation. In Fig. 2.24(a) the stack is shown to be in the state that resulted due to our prior PUSH AX example. That is, SP equals 0006_{16}, SS equals 0105_{16}, the address of the top of the stack equals 1056_{16}, and the word at the top of the stack equals 1234_{16}.

Looking at Fig. 2.24(b), we see what happens when the instructions POP AX and POP BX are executed in that order. Execution of the first instruction causes the 80386DX to read the value from the top of the stack and put it into the AX register as 1234_{16}. Next, SP is incremented to give 0008_{16} and another read operation is initiated from the stack.

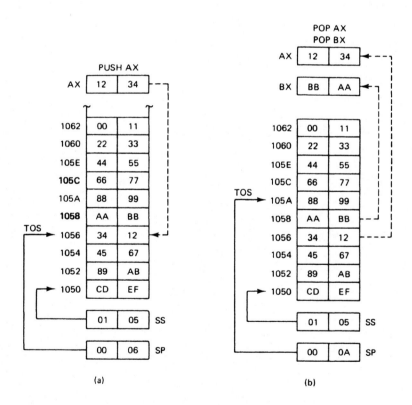

(a)   (b)

Figure 2.24 (a) Stack just prior to pop operation. (Reprinted by permission of Intel Corp. Copyright/Intel Corp. 1979) (b) Stack after execution of the POP AX and POP BX instructions. (Reprinted by permission of Intel Corp. Copyright/Intel Corp. 1979)

This second read corresponds to the POP BX instruction, and it causes the value $BBAA_{16}$ to be loaded into the BX register. SP is incremented once more and now equals $000A_{16}$. Therefore, the new top of stack is at address $105A_{16}$.

In Fig. 2.24(b) we see that the values read out of 1056_{16} and 1058_{16} remain at these addresses. But now they reside at locations that are above the top of the stack. Therefore, they no longer represent valid data. If new information are pushed to the stack, these value are written over.

EXAMPLE 2.8

Assume that the stack is in the state shown in Fig. 2.24(a) and that the instruction POP ECX is executed instead of the instructions POP AX and POP BX. What value would be popped from the stack? Into which register would it be popped? What is the new address of the top of the stack?

Solution

Since a 32-bit register is specified in the POP instruction, the value popped from the stack is $1234BBAA_{16}$. This value is popped into the ECX register. After the value is popped, the contents of SP are incremented by 4. Therefore, it equals $000A_{16}$, and the new top of stack is at address $105A_{16}$.

Any number of stacks may exist in an 80386DX microcomputer. A new stack can be brought in simply by changing the value in the SS register. For instance, executing the instruction MOV SS,DX loads a new value from DX into SS. Even though many stacks can exist, only one can be active at a time.

▲ 2.13 REAL-MODE INPUT/OUTPUT ADDRESS SPACE

The 80386DX has separate memory and *input/output* (I/O) address spaces. The *I/O address space* is the place where I/O interfaces, such as printer and terminal ports, are implemented. Figure 2.25 shows a map of the 80386DX's I/O address space. Notice that this address range is from 0000_{16} through $FFFF_{16}$. This represents just 64K byte addresses; therefore, unlike memory, I/O addresses are only 16 bits long. Each of these addresses corresponds to one byte-wide I/O port.

The part of the map from address 0000_{16} through $00FF_{16}$ is referred to as *page* 0. Page 0 corresponds to the first 256 byte addresses of the I/O address space. Certain of the 80386DX's I/O instructions can perform only input or output data-transfer operations to

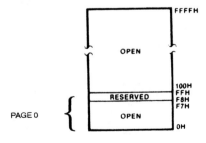

Figure 2.25 I/O address space. (Reprinted by permission of Intel Corp. Copyright/Intel Corp. 1979)

I/O devices located in this part of the I/O address space. Other I/O instructions can perform I/O operations for devices located anywhere in the I/O address space. I/O data transfers can be byte-wide, word-wide, or double-word-wide.

ASSIGNMENTS

Section 2.2

1. Name the six internal processing units of the 80386DX.
2. What are the word lengths of the 80386DX's address bus and data bus?
3. Does the 80386DX have a multiplexed address/data bus or separate address and data buses?
4. How large is the 80386DX's instruction stream queue?
5. In which unit is the instruction stream queue located?
6. How large is the descriptor cache?
7. Where are recently used page directory and page table entries stored?

Section 2.3

8. What is the purpose of a software model of the 80386DX microprocessor?
9. What must an assembly language programmer know about the registers within the 80386DX microprocessor?
10. How many of the 80386DX's internal registers are important to the real-mode application programmer?
11. What are the five least significant bits of CR_0 called?
12. What does PE stand for? What is the purpose of this bit of CR_0?
13. How large is the 80386DX's real-mode memory address space?

Section 2.4

14. What is the highest address in the 80386DX's real-mode address space? The lowest address?
15. Is memory in the 80386DX microcomputer organized as bytes, words, or double words?
16. How much of the 80386DX's real-mode address space is dedicated to storage of the interrupt vector table?
17. The contents of memory location $B0000_{16}$ are FF_{16} and those at $B0001_{16}$ are 00_{16}. What is the data word stored starting at address $B0000_{16}$? Is the word aligned or misaligned?
18. What is the value of the double word stored in memory starting at address $B0003_{16}$ if the contents of memory locations $B0003_{16}$, $B0004_{16}$, $B0005_{16}$, and $B0006_{16}$ are 11_{16}, 22_{16}, 33_{16}, and 44_{16}, respectively? Is this an example of an aligned double word or a misaligned double word?
19. Show how the word $ABCD_{16}$ is stored in memory starting at address $0A002_{16}$. Is the word aligned or misaligned?
20. Show how the double word 12345678_{16} will be stored in memory starting at address $0A001_{16}$. Is the double word aligned or misaligned?

Section 2.5

21. List five data types processed directly by the 80386DX.

22. Express each of the following signed decimal integers as either a byte or word hexadecimal number (use 2's-complement notation for negative numbers).
   (a) +127   **(b)** −10   **(c)** −128   **(d)** +500

23. How would the integer in Problem 22(d) be stored in memory starting at address $0A000_{16}$?

24. How would the decimal number −1000 be expressed for processing by the 80386DX?

25. Express the following decimal numbers as unpacked and packed BCD bytes.
   (a) 29   **(b)** 88

26. How would the number in Problem 22(a) be stored in memory starting at address $0B000_{16}$? Assume that it is coded in packed BCD form and that the least-significant digits are stored at the lower address.

27. What is the statement coded in ASCII by the following binary strings?

$$1001110$$
$$1000101$$
$$1011000$$
$$1010100$$
$$0100000$$
$$1001001$$

28. How would the decimal number 1234 be coded in ASCII and stored in memory starting at address $0C000_{16}$? Assume that the least significant digit is stored at the lower addressed memory location.

Section 2.6

29. How large is a real-mode memory segment?

30. Which of the 80386DX's internal registers are used for memory segmentation?

31. What register defines the beginning of the current code segment in memory?

32. How much memory can be active at a time in the 80386DX microcomputer?

33. How much of the 80386DX's active memory is available as general-purpose data-storage memory?

34. Which range of the 80386DX's memory address space can be used to store program instructions?

Section 2.7

35. What is the function of the instruction pointer register?

36. Provide an overview of the fetch and the execution of an instruction by the 80386DX.

37. What happens to the value in IP each time the 80386DX completes an instruction fetch?

38. How large is the 80386DX's instruction prefetch queue?

Section 2.8

39. Make a list of the data registers of the 80386DX.

40. How is the word value of the base register labeled? The double-word value?

41. How are the upper and lower bytes of the data register denoted?

42. Name two dedicated operations that are assigned to the CX registers.

Section 2.9

43. Name the two pointer registers.

44. For which segment register is the contents of the pointer registers used as an offset?

45. What do *SI* and *DI* stand for?

46. Which sizes of data can be stored in the base and index registers? What size real-mode offset addresses?

Section 2.10

47. Categorize each flag bit of the 80386DX as either a control flag or a flag that monitors the status due to execution of an instruction.

48. Describe the function of each status flag.

49. How are the status flags used by software?

50. What does *TF* stand for?

51. Which flag determines whether the address for a string operation is incremented or decremented?

52. Can the state of the flags be modified through software?

Section 2.11

53. What are the word lengths of the 80386DX's real-mode logical and physical addresses?

54. What two address elements are combined to form a physical address?

55. What is the default segment register for the segment base address in a string instruction destination operand access?

56. Calculate the value of each physical address.
   (a) 1000H:1234H
   (b) 0100H:ABCDH
   (c) A200H:12CFH
   (d) B2C0H:FA12H

57. Find the unknown value for each physical address.
   (a) A000H:? = A0123H
   (b) ?:14DAH = 235DAH
   (c) D765H:? = DABC0H
   (d) ?:CD21H = 32D21H

58. If the current values in the code segment register and the instruction pointer are 0200_{16} and $01AC_{16}$, respectively, what physical address is used in the next instruction fetch?

59. A data segment is to be located from address $A0000_{16}$ to $AFFFF_{16}$. What value must be loaded into DS?

60. If the data segment register contains the value found in Problem 59, what value must be loaded into DI if it is to point to a destination operand stored at address $A1234_{16}$?

Section 2.12

61. What is the function of the stack?

62. If the current values in the stack segment register and stack pointer are $C000_{16}$ and $FF00_{16}$, respectively, what is the address of the top of the stack?

63. For the base and offset addresses in Problem 62, how many words of data are currently held in the stack?

64. Show how the value $EE11_{16}$ from register AX would be pushed onto the top of the stack as it exists in Problem 62.

Section 2.13

65. In the 80386DX software architecture, are the input/output and memory address spaces common or separate?

66. How large is the 80386DX's I/O address space?

67. What is the name given to the part of the I/O address space from 0000_{16} through $00FF_{16}$?

Assembly Language Programming

▲ 3.1 INTRODUCTION

Up to this point we have studied the real-mode software architecture of the 80386DX microprocessor. Here we begin a detailed study of assembly language programming for the 80386DX-based microcomputer. This chapter introduces software and the microcomputer program, the process used to develop an assembly language program, the evolution of the 80386DX's instruction set, and its addressing modes. The operation of the individual instructions of the instruction set are examined in Chapters 4 and 5. The topics covered in this chapter are as follows:

1. Software: the microcomputer program
2. Assembly language program development on the IBM-compatible PC/AT
3. The 80386DX microprocessor instruction set
4. Addressing modes of the 80386DX microprocessor

▲ 3.2 SOFTWARE: THE MICROCOMPUTER PROGRAM

In this section we begin our study of assembly language programming with the topics of software and the microcomputer program. A microcomputer does not know how to process data. It must be told exactly what to do, where to get data, what to do with the data, and where to put the results when it is finished. These jobs are those of *software* in a microcomputer system.

The sequence of commands that are used to tell a microcomputer what to do is called a *program*. Each command in a program is an *instruction*. A program may be simple and include just a few instructions, or it may be very complex and contain more than 100,000 instructions. When the microcomputer is operating, it fetches and executes one instruction of the program after the other. In this way, the instructions of the program guide it step by step through the task that is to be performed.

Software is a general name used to refer to a wide variety of programs that can be run by a microcomputer. Examples are *languages, operating systems, application programs,* and *diagnostics*.

The native language of an 80386DX-based PC/AT is *machine language.* Programs must always be coded in this machine language before they can be executed by the microprocessor. The 80386DX microprocessor understands and performs operations for more than 150 basic instructions. A program written in machine language is often referred to as *machine code*. When expressed in machine code, an instruction is encoded using 0s and 1s. A single machine language instruction can take up 1 or more bytes of code. Even though the 80386DX understands only machine code, it is almost impossible to write programs directly in machine language. For this reason, programs are normally written in other languages, such as *80386DX assembly language* or a high-level language such as *C*.

In 80386DX assembly language each of the operations that can be performed by the 80386DX microprocessor is described with alphanumeric symbols instead of with 0s and 1s. Each instruction in a program is represented by a single *assembly language statement.* This statement must specify which operation is to be performed and what data are to be processed. For this reason, an instruction is divided into two parts: its *operation code* (*opcode*) and its *operands*. The opcode is the part of the instruction that identifies the operation that is to be performed. For example, typical operations are add, subtract, and move. Each opcode is assigned a unique letter combination called a *mnemonic*. The mnemonics for the earlier mentioned operations are ADD, SUB, and MOV. Operands describe the data that are to be processed as the microprocessor carries out the operation specified by the opcode. They identify whether the source and destination of the data are registers within the MPU or storage locations in data memory.

An example of an instruction written in 80386DX assembly language is

```
ADD   EAX,EBX
```

This instruction says, Add the contents of registers EBX and EAX together and put the sum in register EAX. EAX is called the *destination operand*, because it is the place where the result ends up, and EBX is called the *source operand*.

An example of a complete assembly language statement is

```
START:  MOV  EAX,EBX  ;COPY EBX INTO EAX
```

This statement begins with the word START:. START is an address identifier for the instruction MOV EAX, EBX. This type of identifier is known as a *label*. The instruction is followed by ;COPY EBX INTO EAX. This part of the statement is called a *comment*. Thus a general format for an assembly language statement is

```
LABEL:  INSTRUCTION  ;COMMENT
```

Programs written in assembly language are referred to as *source code*. An example of a short 80386DX assembly language program is shown in Fig. 3.1(a). The assembly language instructions are located toward the left. Notice that labels are not used in most statements. On the other hand, a comment describing the statement is usually included on the right. This type of documentation makes it easier for a program to be read and debugged.

```
TITLE   BLOCK-MOVE PROGRAM

        PAGE    ,132

COMMENT *This program moves a block of specified number of bytes
        from one place to another place*


;Define constants used in this program

        N       =       16              ;Bytes to be moved
        BLK1ADDR=       100H            ;Source block offset address
        BLK2ADDR=       120H            ;Destination block offset addr
        DATASEGADDR=    1020H           ;Data segment start address


STACK_SEG       SEGMENT         STACK 'STACK'
                DB              64 DUP(?)
STACK_SEG       ENDS


CODE_SEG        SEGMENT         'CODE'
BLOCK           PROC            FAR
        ASSUME  CS:CODE_SEG,SS:STACK_SEG

;To return to DEBUG program put return address on the stack

        PUSH    DS
        MOV     AX, 0
        PUSH    AX

;Set up the data segment address

        MOV     AX, DATASEGADDR
        MOV     DS, AX

;Set up the source and destination offset addresses

        MOV     SI, BLK1ADDR
        MOV     DI, BLK2ADDR

;Set up the count of bytes to be moved

        MOV     CX, N

;Copy source block to destination block

NXTPT:  MOV     AH, [SI]                ;Move a byte
        MOV     [DI], AH
        INC     SI                      ;Update pointers
        INC     DI
        DEC     CX                      ;Update byte counter
        JNZ     NXTPT                   ;Repeat for next byte
        RET                             ;Return to DEBUG program
BLOCK           ENDP
CODE_SEG        ENDS
        END     BLOCK                   ;End of program

                        (a)
```

Figure 3.1 (a) Example of an 80386DX assembly language program. (b) Assembled version of the program.

Assembly language programs cannot be directly run on the 80386DX. They must still be converted to an equivalent machine language program for execution by the 80386DX. This conversion is done automatically by running the source code through a program known as an *assembler*. The machine language output produced by the assembler is called *object code*.

Figure 3.1(b) is the *listing* produced by assembling the assembly language source code in Fig. 3.1(a) with Microsoft's MASM macroassembler. Reading from left to right, this listing

```
 1
 2
 3                              TITLE   BLOCK-MOVE PROGRAM
 4
 5                                      PAGE    ,132
 6
 7                              COMMENT *This program moves a block of specified number of bytes
 8                                       from one place to another place*
 9
10
11                              ;Define constants used in this program
12
13   = 0010                                 N       =       16              ;Bytes to be moved
14   = 0100                                 BLK1ADDR=       100H            ;Source block offset address
15   = 0120                                 BLK2ADDR=       120H            ;Destination block offset addr
16   = 1020                                 DATASEGADDR=    1020H           ;Data segment start address
17
18
19   0000                       STACK_SEG       SEGMENT         STACK 'STACK'
20   0000    40 [                                DB              64 DUP(?)
21                  ??
22                          ]
23
24   0040                       STACK_SEG       ENDS
25
26
27   0000                       CODE_SEG        SEGMENT         'CODE'
28   0000                       BLOCK           PROC            FAR
29                                              ASSUME  CS:CODE_SEG,SS:STACK_SEG
30
31                              ;To return to DEBUG program put return address on the stack
32
33   0000  1E                                   PUSH    DS
34   0001  B8 0000                              MOV     AX, 0
35   0004  50                                   PUSH    AX
36
37                              ;Set up the data segment address
38
39   0005  B8 1020                              MOV     AX, DATASEGADDR
40   0008  8E D8                                MOV     DS, AX
41
42                              ;Set up the source and destination offset adresses
43
44   000A  BE 0100                              MOV     SI, BLK1ADDR
45   000D  BF 0120                              MOV     DI, BLK2ADDR
46
47                              ;Set up the count of bytes to be moved
48
49   0010  B9 0010                              MOV     CX, N
50
51                              ;Copy source block to destination block
52
53   0013  8A 24             NXTPT:  MOV     AH, [SI]            ;Move a byte
```

(b)

Figure 3.1   (Continued)

```
54      0015  88 25              MOV     [DI], AH
55      0017  46                 INC     SI              ;Update pointers
56      0018  47                 INC     DI
57      0019  49                 DEC     CX              ;Update byte counter
58      001A  75 F7              JNZ     NXTPT           ;Repeat for next byte
59      001C  CB                 RET                     ;Return to DEBUG program
60      001D            BLOCK    ENDP
61      001D            CODE_SEG ENDS
62                               END     BLOCK           ;End of program
```

(b)

Figure 3.1 (Continued)

contains line numbers, addresses of memory locations, the machine language instructions, the original assembly language statements, and comments. For example, line 53, which is

```
0013   8A   24    NXTPT:   MOV   AH,[SI]   ;MOVE A BYTE
```

shows that the assembly language instruction MOV AH,[SI] is encoded as 8A24 in machine language and that this 2-byte instruction is loaded into memory starting at address 0013_{16} and ending at address 0014_{16}. Note that for simplicity the machine language instructions are expressed in hexadecimal notation, not in binary form. Use of assembly language makes it much easier to write a program. But notice that there is still a one-to-one relationship between assembly and machine language instructions.

EXAMPLE 3.1

What instruction is at line 58 of the program in Fig. 3.1(b)? How is this instruction expressed in machine code?

Solution

Looking at the listing in Fig. 3.1(b), we find that the instruction is

```
JNZ   NXTPT
```

and the machine code is

```
75   F7
```

High-level languages make writing programs even easier. In a language such as BASIC, high-level commands, such as FOR, NEXT, and GO, are provided. These commands no longer correspond to a single machine language statement. Instead, they implement functions that may require many assembly language statements. Again, the statements must be converted to machine code before they can be run on the 80386DX. The program that converts high-level-language statements to machine code instructions is called a *compiler*.

Some languages—BASIC, for instance—are not always compiled. Instead, *interpretive* versions of the language are available. When a program written in an interpretive form of BASIC is executed, each line of the program is interpreted just before it is executed and at that moment replaced with a corresponding machine language routine. It is this machine code routine that is executed by the 80386DX.

The question you may be asking yourself right now is: If it is so much easier to write programs with a high-level language, why is it important to know how to program the 80386DX in its assembly language? This question will now be answered.

We just pointed out that if a program is written in a high-level language such as C, it must be compiled into machine code before it can be run on the 80386DX. The general nature with which compilers must be designed usually results in inefficient machine code. That is, the quality of the machine code produced for the program depends on the quality of the compiler program in use. What is found is that a compiled machine code implementation of a program that was written in a high-level language results in many more machine language instructions than an assembled version of an equivalent handwritten assembly language program. This leads us to the two key benefits derived from writing programs in assembly language: first, the machine code program that is produced will take up less memory space than the compiled version of the program; and second, it will execute much faster.

Now we know the benefits attained by writing programs in assembly language, but we still do not know when these benefits are important. To be important, they must outweigh the additional effort that must be put into writing the program in assembly language instead of a high-level language. One of the major uses of assembly language programming is in *real-time applications*. By real-time we mean that the task required by the application must be completed before any other input to the program can occur that will alter its operation.

For example, the *device service routine* that controls the operation of the floppy disk drive of the PC/AT is a good example of the kind of program that is usually written in assembly language. This is because it is a segment of program that must closely control the microcomputer hardware in real time. In this case, a program that is written in a high-level language probably could not respond quickly enough to control the hardware, and even if it could, operations performed with the disk subsystem would be much slower. Some other examples of hardware-related operations typically performed by routines written in assembly language are communication routines such as those that drive the display and printer in a personal computer and the I/O routine that scans the keyboard.

Assembly language is important not only for controlling hardware devices of the microcomputer system, but also when performing pure software operations. For example, applications frequently require the microcomputer to search through a large table of data in memory looking for a special string of characters, such as a person's name. This type of operation can easily be performed by writing a program in a high-level language; however, for very large tables of data the search will take very long. By implementing the search routine through assembly language, the performance of the search operation is greatly improved. Other examples of software operations that may require implementation with high-performance routines derived with assembly language are *code translations*, such as from ASCII to EBCDIC, *table sort routines*, such as a bubble sort, and *mathematical routines*, such as those for floating-point arithmetic.

Not all parts of an application require real-time performance. For this reason, it is a common practice to mix, in the same program, routines developed through a high-level language and routines developed with assembly language. That is, assembly language is used to write those parts of the application that must perform real-time operations, and high-level language is used to write those parts that are not time critical. The machine code obtained by assembling or compiling the two types of program segments is linked together to form the final application program.

In this section we will look at the process by which problems are solved using software. An assembly language program is written to solve a specific problem. This problem is sometimes known as the *application.* To develop a program that implements an application, the programmer goes through a multistep process. A chart that outlines the steps in the *program-development cycle* is shown in Fig. 3.2. Let us next examine each step of the development cycle.

Describing the Problem

Looking at the development cycle sequence in Fig. 3.2, we see that the programmer begins by making a clear description of the problem to be solved and ends with a program that correctly performs this function. The first step in the program-development process is to understand and describe the problem that is to be solved. A clear, concise, and accurate description of the problem is an essential part of the process of obtaining a correct and efficient software solution. This description may be provided in an informal way, such as a verbal description, or in a more formal way with a written document.

The program we use here is an example of a simple software application. Its function is to move a fixed-length block of data, called the *source block,* from one location in memory to another location in memory called the *destination block.* For the block-move program, a verbal or a written list of events may be used to describe this problem to the programmer.

On the other hand, in most practical applications, the problem to be solved is quite complex. The programmer must know what the input data are, what operations must be performed on this information, whether or not these operations need to be performed in a special sequence, whether or not there are time constraints on performing some of the operations, if error conditions can occur during the process, and what results need to be output. For this reason, most applications are described with a written document called an *application specification.* This specification is studied by the programmers before they begin to define a software solution for the problem.

Planning the Solution

From Fig. 3.2, we find that the second step in the program-development process is to plan a solution for the problem. This decision assumes that a complete and clear description of the problem to be solved has been provided. Once we have this description, it needs to be carefully analyzed by the programmer. Typically, the problem is broken down into a series of basic operations, which when performed in a certain sequence will produce a solution to the problem. This solution plan is also known as the algorithm. Usually, the algorithm is described with another document called the *software specification.*

The proposed solution is normally presented in a pictorial form known as a *flowchart.* The flowchart is an outline that both documents the operations that must be performed by software to implement the planned solution and shows the sequence in which they are performed. Figure 3.3(a) is the flowchart for a block-move program.

The operations identified in the flowchart identify functions that can be implemented with assembly language instructions. For example, the first block indicates that operations must be performed to set up a data segment, initialize the starting pointers for the addresses of the source and destination blocks, and specify the count of the number of pieces of data

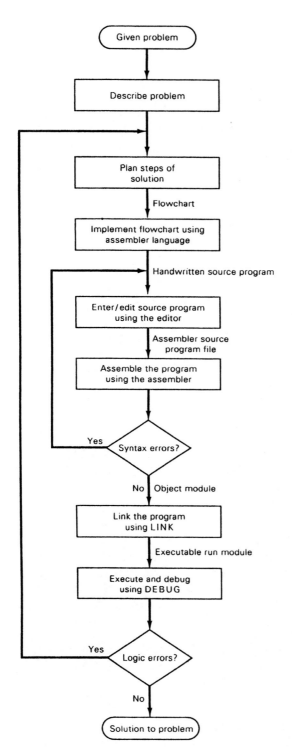

Figure 3.2 General program development cycle.

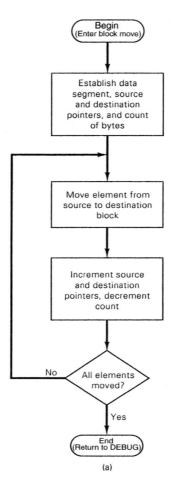

Figure 3.3 (a) Flowchart of a block-move program. (b) Block-move source program.

(a)

that are to be moved. These types of operations can be achieved by moving either immediate data or data from a known memory location into appropriate registers.

A flowchart uses a set of symbols to identify both the operations required in the solution and the sequence in which they are performed. Figure 3.4 lists the most commonly used flowcharting symbols. Notice that symbols are listed for identifying the beginning or end of the flowchart, input or output of data, process functions, making a decision operation, connecting blocks within the flowchart, and connections to other flowcharts. The operation to be performed is written inside the symbol. The flowchart in Fig. 3.3(a) illustrates the use of some of these symbols. Notice that a begin/end symbol, which contains the comment *Enter block-move*, is used to mark the beginning of the program and another, which reads *Return to DEBUG*, marks the end of the sequence. Process function boxes are used to identify each of the tasks (initialize registers, copy source element to destination, and increment source and destination address pointers and decrement data element count) that are performed as part of the block-move routine. Arrows are used to describe the sequence (flow) of these operations as the block-move operation is performed. Flowcharts and their descriptions are included as part of the software specification document.

```
TITLE BLOCK-MOVE PROGRAM

        PAGE        ,132

COMMENT *This program moves a block of specified number of bytes
         from one place to another place*


;Define constants used in this program

        N=                  16          ;Bytes to be moved
        BLK1ADDR=           100H        ;Source block offset address
        BLK2ADDR=           120H        ;Destination block offset addr
        DATASEGADDR=        2000H       ;Data segment start address


STACK_SEG           SEGMENT         STACK 'STACK'
                    DB              64 DUP(?)
STACK_SEG           ENDS
CODE_SEG            SEGMENT         'CODE'
BLOCK               PROC        FAR
    ASSUME          CS:CODE_SEG,SS:STACK_SEG

;To return to DEBUG program put return address on the stack

            PUSH    DS
            MOV     AX, 0
            PUSH    AX

;Setup the data segment address

            MOV     AX, DATASEGADDR
            MOV     DS, AX

;Setup the source and destination offset adresses

            MOV     SI, BLK1ADDR
            MOV     DI, BLK2ADDR

;Setup the count of bytes to be moved

            MOV     CX, N

;Copy source block to destination block

NXTPT:      MOV     AH, [SI]        ;Move a byte
            MOV     [DI], AH
            INC     SI              ;Update pointers
            INC     DI
            DEC     CX              ;Update byte counter
            JNZ     NXTPT           ;Repeat for next byte
            RET                     ;Return to DEBUG program
BLOCK               ENDP
CODE_SEG            ENDS
        END         BLOCK           ;End of program
```

(b)

Figure 3.3   (Continued)

The solution should be tested to verify that it correctly solves the stated problem.
This can be done by specifying test cases with known inputs and outputs. Then, tracing
through the operation sequence defined in the flowchart for these input conditions, the
outputs are found and compared to the known test results. If the results are not the same,
the cause of the error must be found, the algorithm is modified, and the tests are rerun.
When the results match, the algorithm is assumed to be correct, and the programmer is
ready to move on to the next step in the development cycle.

The flowchart representation of the planned solution is a valuable aid to the programmer when coding the solution with the 8088 assembly language instructions. When a problem
is very simple, the flowcharting step may be bypassed. A list of the tasks and the sequence

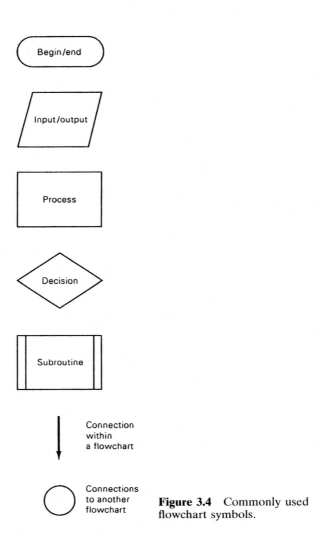

Figure 3.4 Commonly used flowchart symbols.

in which they must be performed may be enough to describe the solution to the problem. However, for complex applications, a flowchart is an essential program-development tool for obtaining an accurate and timely solution.

Coding the Solution with Assembly Language

The third step of the program development cycle, as shown in Fig. 3.2, is the translation of the flowchart solution into its equivalent assembly language program. This requires the programmer to implement the operations described in each symbol of the flowchart with a sequence of 8088 assembly language instructions. These instruction sequences are then combined to form a handwritten assembly language program called the *source program*.

Two types of statements are used in the source program. First, there are the assembly language instructions. They are used to tell the microprocessor what operations are to be performed to implement the application.

The assembly language program in Fig. 3.3(b) implements the block-move operation flowchart of Fig. 3.3(a). Comparing the flowchart to the program, it is easy to see that the initialization block is implemented with the assembly language statements

```
MOV   AX,   DATASEGADDR
MOV   DS,   AX
MOV   SI,   BLK1ADDR
MOV   DI,   BLK2ADDR
MOV   CX,   N
```

The first two move instructions load a segment base address called DATASEGADDR into the data segment register. This defines the data segment in memory where the two blocks of data reside. Next, two more move instructions are used to load SI and DI with the starting offset address of the source block (BLK1ADDR) and destination block (BLK2ADDR), respectively. Finally, the count N of the number of bytes of data to be copied to the destination block is loaded into count register CX.

The source program also contains a second type of statements that are instructions to the assembler program that is used to convert the assembly language program into machine code. Pseudo-op statements are examples of source program statements that are directed at the assembler program. We will discuss these statements later in the chapter. In Fig. 3.3(b), the statements

```
BLOCK   PROC   FAR
```

and

```
BLOCK   ENDP
```

are examples of modular programming pseudo-op statements. They mark the beginning and end, respectively, of the software procedure called BLOCK.

To do this step of the development cycle, the programmer must know the instruction set of the microprocessor, basic assembly language programming techniques, the assembler's instruction statement syntax, and the assembler's pseudo-operations.

Creating the Source Program

After having handwritten the assembly language program, we are ready to enter it into the computer. This step is identified as the enter/edit source program block in the program-development cycle diagram of Fig. 3.2 and is done with a program called an *editor*. We will use either the EDLIN or EDIT editors, which are available as part of the PC/AT's DOS operating system. Using an editor, each of the statements of the program is typed into the computer. If errors are made as the statements are keyed in, the corrections can either be made at the time of entry or edited at a later time. The source program is saved by storing it in a file.

Assembling the Source Program into an Object Module

The fifth step in the flowchart of Fig. 3.2 is the point at which the assembly language source program is converted to its corresponding 80386DX machine language program. To do this, we use a program called an *assembler*. A program available from Microsoft Corporation called *MASM* is an example of an 80386DX assembler that runs in DOS on a PC/AT. The assembler program reads as its input the contents of the *assembler source file*; it converts this program statement by statement to machine code and produces a machine code program as its output. This machine code output is stored in a file called the *object module*.

If during the conversion operation syntax errors are found—that is, violations in the rules of writing 80386DX assembly language instructions for the assembler—they are automatically flagged by the assembler. As shown in the flowchart of Fig. 3.2, before going on, the cause of each error in the source program must be identified and then corrected. The corrections are made using the editor program. After the corrections are made, the source program must be reassembled. This edit-assemble sequence must be repeated until the program assembles with no error.

Producing a Run Module

The object module produced by the assembler cannot be run directly on the 80386DX-based microcomputer. As shown in Fig. 3.2, this module must be processed by the *LINK program* to produce an executable object module, which is known as a *run module*. The linker program converts the object module to a run module by making its address compatible with the microcomputer on which it is to be run. For instance, if our system is implemented with memory at addresses $A000_{16}$ through $FFFF_{16}$, the executable machine code output by the linker will also have addresses in this range.

There is another purpose for the use of a linker: it is used to link together different object modules to generate a single executable object module. This allows program development to be done in modules, which are later combined to form the application program.

Verifying the Solution

Now the executable object module is ready to be run on the microcomputer. Once again, the PC/AT's DOS operating system provides us with a program, which is called DEBUG, to perform this function. DEBUG provides an environment in which we can run the program instruction by instruction or run a group of instructions at a time, look at intermediate results, display the contents of the registers within the microprocessor, and so on.

For instance, we could verify the operation of our earlier block-move program by running it for the data in the cases defined to test the algorithm. DEBUG is used to load the run module for block-move into the memory of the PC/AT. After loading is completed and verified, other DEBUG commands are employed to run the program for the data in the test case. The debug program permits us to trace the operation as instructions are executed and observe each element of data as it is copied from the source to destination block. These results are recorded and compared to those provided with the test case. If the program is found to perform the block-move operation correctly, the program-development process is complete.

On the other hand, Fig. 3.2 shows that if errors are discovered in the logic of the solution, the cause must be determined, corrections must be made to the algorithm, and then the assembly language source program must be corrected using the editor. The edited source file must be reassembled, relinked, and retested by running it with DEBUG. This loop must be repeated until it is verified that the program correctly performs the operation for which it was written.

Programs and Files Involved in the Program Development Cycle

The edit, assemble, link, and debug parts of the general program-development cycle in Fig. 3.2 are performed directly on the PC/AT. Figure 3.5 shows the names of the programs and typical file names with extensions used as inputs and outputs during this process. For example, the *EDIT program* is an editor that is used to create and correct assembly language source files. The program that results is shown to have the name PROG1.ASM. This stands for program 1 assembly source code.

The program that can be used to assemble source files into object modules is called *MASM*, which stands for *macroassembler*. The assembler converts the contents of the source input file PROG1.ASM into three output files called PROG1.OBJ, PROG1.LST, and PROG1.CRF. The file PROG1.OBJ contains the object code module. The other two files provide additional information that is useful for debugging the application program.

Object module PROG1.OBJ can be linked to other object modules with the LINK program. For instance, programs that are available as object modules in a math library could be linked with another program to implement math operations. A library is a collection of prewritten, assembled, and tested programs. Notice that this program produces a run module in file PROG1.EXE and a map file called PROG1.MAP as outputs. The executable object module, PROG1.EXE, is run with the debugger program, which is called DEBUG. PROG1.MAP is supplied as support for the debugging operation by providing additional

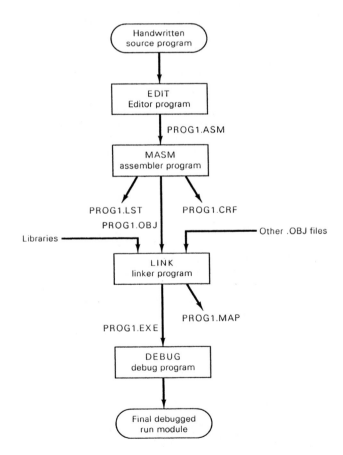

Figure 3.5 The DOS program-development programs and user files.

information such as where the program will be located when loaded into the microcomputer's memory.

▲ 3.4 THE 80386DX MICROPROCESSOR INSTRUCTION SET

The instruction set of a microprocessor defines the basic operations that a programmer can make the device perform. Earlier we pointed out that the 80386DX microprocessor provides a powerful instruction set that contains more than 150 basic instructions. The wide range of operands and addressing modes permitted for use with these instructions further expands the instruction set into many more executable instructions at the machine code level. For instance, the basic MOV instruction expands into more than 30 different machine-level instructions.

Figure 3.6 shows the evolution of the instruction set for the 8086 architectures. The instruction set of the 8086 and 8088 microprocessors, called the *basic instruction set*, was enhanced in the 80286 microprocessor to implement what is known as the *extended instruction set*. This extended instruction set includes several new instructions and implements additional addressing modes for a few of the instructions already available in the basic instruction set. For example, two instructions added as extensions to the basic instruction set are PUSHA *(push all)* and POPA *(pop all)*. Moreover, the PUSH and IMUL instructions have been enhanced to permit the use of immediate operand addressing. In this way we see that the extended instruction set is a super set of the basic instruction set. These instructions are all executable in the real mode.

Finally, Fig. 3.6 shows that in the 80386DX, the real-mode instruction set is further enhanced with a group of instructions called the *80386 specific instruction set*. For example, it includes instructions to load a pointer directly into the FS, GS, and SS registers and bit test and bit scan instructions. Moreover, it contains additional forms of existing instructions that have been added to perform the identical operation in a more general way, on special registers, or on a double word of data. In this way we see that the 80386DX's real-mode instruction set is a super set of the 8086's basic instruction set.

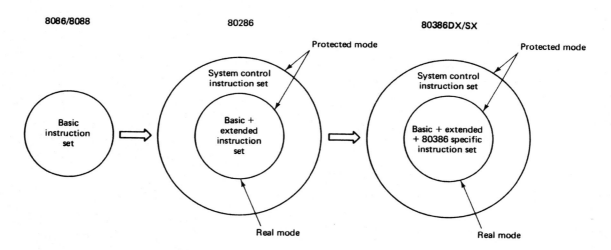

Figure 3.6 Evolution of the 8086 family instruction set.

In Fig. 3.6 we see that the instruction set of the 80286 is further enhanced with system-control instructions to give the *system-control instruction set*. These instructions are provided to control the multitasking, memory management, and protection mechanisms of the protected-mode 80286. For instance, the instruction CLTS *(clear task switched flag)* can be used to clear the task switch flag of the machine status word. The 80386DX's system control instruction set is identical to that of the 80286.

For the purpose of discussion, the 80386DX's real-mode instruction set will be divided into a number of groups of functionally related instructions. In Chapter 5, we consider the data-transfer instructions, arithmetic instructions, logic instructions, shift instructions, rotate instructions, and bit test and bit scan instructions. Advanced instructions such as those for program and processor control are described in Chapter 6. The system-control instructions are covered in Chapter 8.

▲ 3.5 ADDRESSING MODES OF THE 80386DX MICROPROCESSOR

When the 80386DX executes an instruction, it performs the specified function on data. These data, called operands, may be part of the instruction, may reside in one of the internal registers of the microprocessor, may be stored at an address in memory, or may be held at an I/O port. To access these different types of operands, the 80386DX is provided with various *addressing modes*. An addressing mode is a method of specifying an operand. The addressing modes are categorized into three types: *register operand addressing, immediate operand addressing,* and *memory operand addressing*. Let us now consider in detail the addressing modes in each of these categories.

Register Operand Addressing Mode

With the *register addressing mode*, the operand to be accessed is specified as residing in an internal register of the 80386DX. Figure 3.7 lists the internal registers that can be used as a source or destination operand. Notice that only the data registers can be accessed in byte, word, or double-word sizes.

An example of an instruction that uses this addressing mode is

MOV   AX,BX

Register	Operand size		
	Byte (Reg8)	Word (Reg16)	Double word (Reg32)
Accumulator	AL, AH	AX	EAX
Base	BL, BH	BX	EBX
Count	CL, CH	CX	ECX
Data	DL, DH	DX	EDX
Stack pointer	—	SP	ESP
Base pointer	—	BP	EBP
Source index	—	SI	ESI
Destination index	—	DI	EDI
Code segment	—	CS	—
Data segment	—	DS	—
Stack segment	—	SS	—
E data segment	—	ES	—
F data segment	—	FS	—
G data segment	—	GS	—

Figure 3.7 Direct addressing registers and operand sizes.

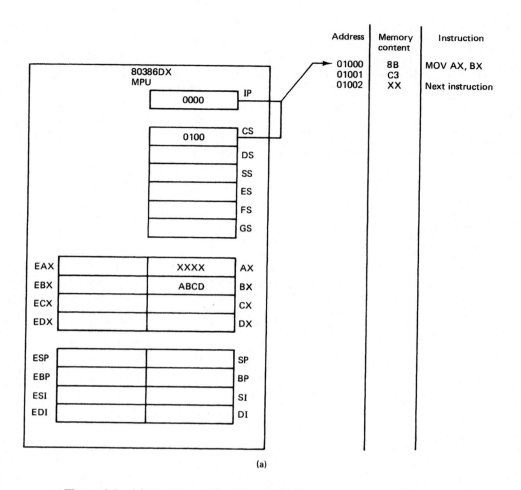

Address	Memory content	Instruction
01000	8B	MOV AX, BX
01001	C3	
01002	XX	Next instruction

(a)

Figure 3.8 (a) Register addressing mode instruction before fetch and execution. (b) After execution.

This stands for move the word-wide contents of EBX, which is the *source operand* BX, to the word location in EAX, which is identified by the *destination operand* AX. Both the source and destination operands have been specified as the contents of internal registers of the 80386DX.

Let us now look at the effect of executing the register addressing mode MOV instruction. In Fig. 3.8(a), we see the state of the 80386DX just prior to fetching the instruction. Notice that the physical address formed from CS and IP (CS:IP) points to the MOV AX,BX instruction at address 01000_{16}. This instruction is fetched into the 80386DX's instruction queue, where it is held waiting to be executed.

Prior to execution of this instruction, the contents of BX are $ABCD_{16}$, and the contents of AX represent a don't-care state. The instruction is read from the output side of the queue, decoded, and executed. As shown in Fig. 3.8(b), the result produced by executing this instruction is that the value $ABCD_{16}$ is copied into AX.

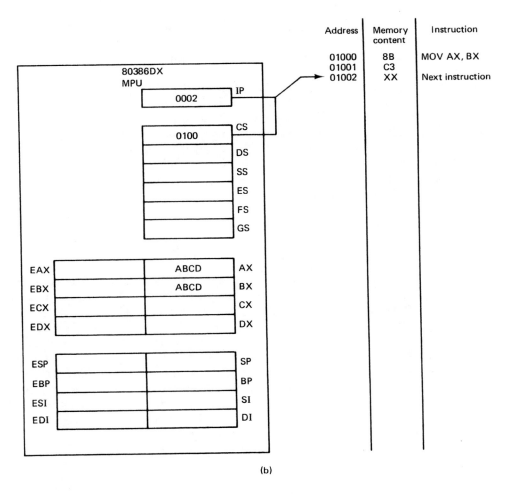

Address	Memory content	Instruction
01000	8B	MOV AX, BX
01001	C3	
01002	XX	Next instruction

(b)

Figure 3.8 (Continued)

EXAMPLE 3.2

What is the destination operand in the instruction? How large is this operand?

    MOV   ECX,EAX

Solution

The destination operand is register ECX, and it holds 32-bit data.

To write software that is compatible with 16-bit members of the 80X86 family, such as the 80C186XL, double-word-size register operands should not be used.

Opcode	Immediate operand

Figure 3.9 Instruction encoded with an immediate operand.

Immediate Operand Addressing Mode

If an operand is part of the instruction instead of the contents of a register or memory location, it represents what is called an *immediate operand* and is accessed using the *immediate addressing mode*. Figure 3.9 shows that the operand, which can be 8 bits (Imm8), 16 bits (Imm16), or 32 bits (Imm32) in length, is encoded as part of the instruction. Since the data are encoded directly into the instruction, immediate operands normally represent constant data. This addressing mode can be used only to specify a source operand.

In the instruction

```
MOV   AL,15H
```

the source operand 15H (15_{16}) is an example of a byte-wide immediate source operand. The destination operand, which is the contents of AL, uses register addressing. Thus this instruction employs both the immediate and register addressing modes.

Figure 3.10(a) and (b) illustrates the fetch and execution of this instruction. Here we find that the immediate operand 15_{16} is stored in the code segment of memory in the byte location immediately following the opcode of the instruction. This value is fetched, along with the opcode for MOV, into the instruction queue with the 80386DX. When it performs the move operation, the source operand is fetched from the queue, not from the memory, and no external memory operations are performed. Notice that the result produced by executing this instruction is that the immediate operand, which equals 15_{16}, is loaded into the lower-byte part of the accumulator (AL).

EXAMPLE 3.3

Write an instruction that will move the immediate value 12345678H into the ECX register.

Solution

The instruction must use immediate operand addressing for the source operand and register operand addressing for the destination operation. This gives

```
MOV   ECX,12345678H
```

To maintain compatibility with 16-bit MPUs of the 80X86 family, the 32-bit immediate operand format cannot be used.

16-Bit Memory Operand Addressing Modes

To reference an operand in memory, the 80386DX must calculate the physical address (PA) of the operand and then initiate a read or write operation for this storage location. The 80386DX MPU is provided with a group of addressing modes known as the *memory*

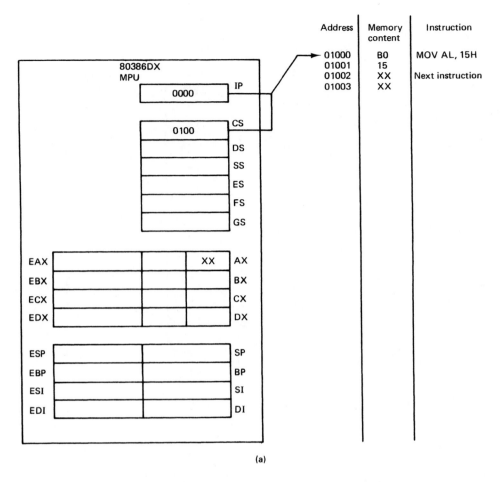

Address	Memory content	Instruction
01000	B0	MOV AL, 15H
01001	15	
01002	XX	Next instruction
01003	XX	

(a)

Figure 3.10 (a) Immediate addressing mode instruction before fetch and execution. (b) After execution.

operand addressing modes for this purpose. The capabilities of these addressing modes have been enhanced significantly in the 80386DX compared to how they operated in the 16-bit members of the 80X86 family. In our examination of register operand addressing and immediate operand addressing, we found that the new 32-bit extensions could not be used if the objective is to write 16-bit compatible software. For this reason, we will first explore a subset of the 80386DX's memory operand addressing modes that when used will result in software that is compatible with the earlier 16-bit 80X86 family MPUs. Then we will describe the differences provided in the 80386DX's 32-bit programming model.

Looking at Fig. 3.11 we see that the physical address is formed from a *segment base address* (SBA) and an *effective address* (EA). SBA identifies the starting location of the segment in memory and EA represents the offset of the operand from the beginning of this segment of memory. Earlier we showed how SBA and EA are combined within the 80386DX to form the real-mode physical address SBA:EA.

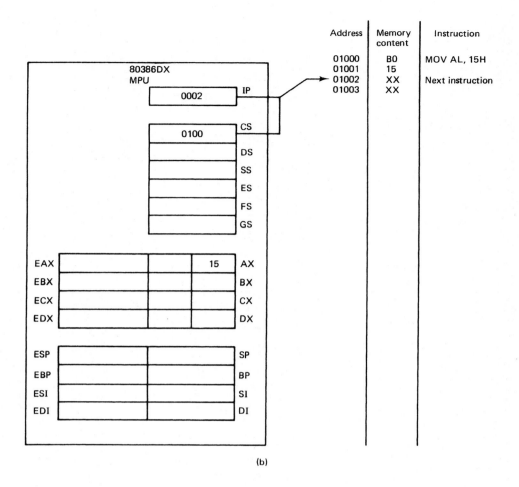

Figure 3.10 (Continued)

The value of the EA can be specified in a variety of ways. One way is to encode the effective address of the operand directly in the instruction. This represents the simplest type of memory addressing, known as the *direct addressing mode*. Figure 3.11 shows that an effective address can be made up from as many as three elements: the *base, index,* and *displacement*. Using these elements, the effective address calculation is made by the general formula

$$EA = base + index + displacement$$

PA = SBA : EA

PA = Segment base : Base + Index + Displacement

$$PA = \begin{Bmatrix} CS \\ SS \\ DS \\ ES \end{Bmatrix} : \begin{Bmatrix} BX \\ BP \end{Bmatrix} + \begin{Bmatrix} SI \\ DI \end{Bmatrix} + \begin{Bmatrix} \text{8-bit displacement} \\ \text{16-bit displacement} \end{Bmatrix}$$

Figure 3.11 Real-mode physical and effective address computation for memory operands.

Assembly Language Programming   Chap. 3

Figure 3.11 also identifies the registers that can be used to hold the values of the segment base, base, and index. For example, it tells us that any of the six segment registers can be the source of the segment base for the physical address calculation and that the value of base for the effective address can be in either the base register (BX) or base pointer register (BP). Also identified in Fig. 3.11 are the sizes permitted for the displacement.

Not all these elements are always used in the effective address calculation. In fact, a number of memory addressing modes are defined by using various combinations of these elements. They are called *register indirect addressing, based addressing, indexed addressing,* and *based-indexed addressing.* For instance, using based addressing mode, the effective address calculation includes just a base. These addressing modes provide the programmer with different ways of computing the effective address of an operand in memory. Next, we examine each of the memory operand addressing modes in detail.

Direct Addressing Mode. *Direct addressing mode* is similar to immediate addressing in that information is encoded directly into the instruction. However, in this case, the instruction opcode is followed by an effective address instead of the data. As shown in Fig. 3.12, this effective address is used directly as the 16-bit offset of the storage location of the operand from the location specified by the current value in the segment register selected. The default segment register is DS. Therefore, the 20-bit physical address of the operand in memory is normally obtained as DS:EA. But, by using a *segment override prefix* (SEG) in the instruction, any of the six segment registers can be referenced.

An example of an instruction that uses direct addressing for its source operand is

```
MOV   CX,[1234H]
```

This stands for move the contents of the memory location with offset 1234_{16} in the current data segment into internal register CX. The offset is encoded as part of the instruction's machine code.

In Fig. 3.13(a), we find that the offset is stored in the two byte locations that follow the instruction's opcode. As the instruction is executed, the 80386DX combines 1234_{16} with 0200_{16} to get the physical address of the source operand as follows:

$$PA = 02000_{16} + 1234_{16}$$
$$= 03234_{16}$$

Then it reads the word of data starting at this address, which is $BEED_{16}$, and loads it into the CX register. This result is illustrated in Fig. 3.13(b).

PA = Segment base:Direct address

$$PA = \begin{Bmatrix} CS \\ DS \\ SS \\ ES \\ FS \\ GS \end{Bmatrix} : \{ Direct\ address \}$$

Figure 3.12 Computation of a direct memory address.

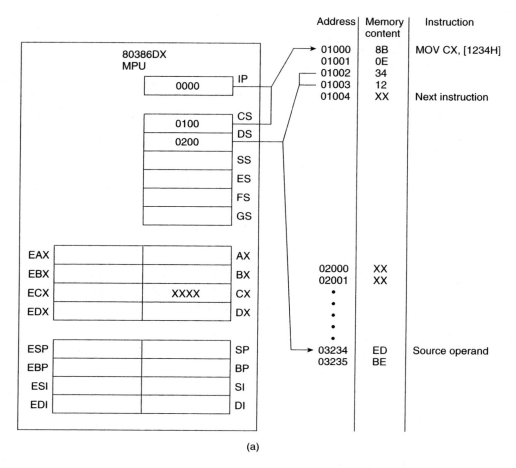

Address	Memory content	Instruction
01000	8B	MOV CX, [1234H]
01001	0E	
01002	34	
01003	12	
01004	XX	Next instruction

80386DX
MPU

0000   IP

0100   CS
0200   DS
       SS
       ES
       FS
       GS

EAX   AX
EBX   BX
ECX XXXX CX
EDX   DX

ESP   SP
EBP   BP
ESI   SI
EDI   DI

| 02000 | XX | |
| 02001 | XX | |

| 03234 | ED | Source operand |
| 03235 | BE | |

(a)

Figure 3.13 (a) Direct addressing mode instruction before fetch and execution. (b) After execution.

Register Indirect Addressing Mode. *Register indirect addressing mode* is similar to the direct addressing we just described, in that an effective address is combined with the contents of DS to obtain a physical address. However, it differs in the way the offset is specified. Figure 3.14 shows that this time the 16-bit EA resides in either a base register or an index register within the 80386DX. The base register can be either base register BX or base pointer register BP, and the index register can be source index register SI or destination index register DI. Another segment register can be referenced by using a segment-override prefix.

An example of an instruction that uses register indirect addressing for its source operand is

```
MOV   AX,[SI]
```

Execution of this instruction moves the contents of the memory location that is offset from the beginning of the current data segment by the value of EA in register SI into the AX register.

For instance, as shown in Fig. 3.15(a) and (b), if SI contains 1234_{16} and DS contains

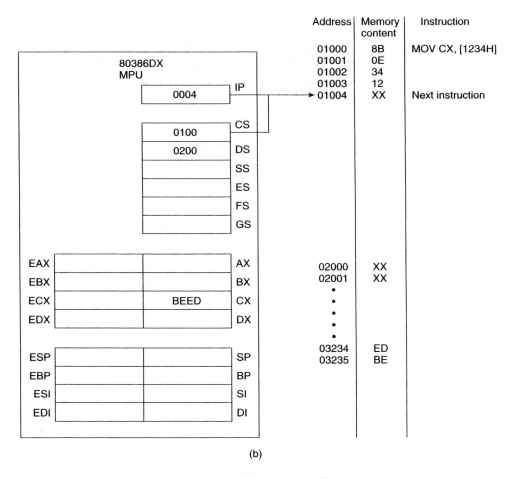

Address	Memory content	Instruction
01000	8B	MOV CX, [1234H]
01001	0E	
01002	34	
01003	12	
01004	XX	Next instruction
02000	XX	
02001	XX	
.		
.		
.		
.		
03234	ED	
03235	BE	

(b)

Figure 3.13 (Continued)

0200_{16}, the result produced by executing the instruction is that the contents of the memory location at address

$$PA = 02000_{16} + 1234_{16}$$

$$= 03234_{16}$$

are moved into the AX register. Notice in Fig. 3.15(b) that this value is $BEED_{16}$. In this example, the value 1234_{16} that was found in the SI register must have been loaded with another instruction prior to executing the move instruction.

PA = Segment base: Indirect address

$$PA = \begin{Bmatrix} CS \\ DS \\ SS \\ ES \\ FS \\ GS \end{Bmatrix} : \begin{Bmatrix} BX \\ BP \\ SI \\ DI \end{Bmatrix}$$

Figure 3.14 Computation of an indirect memory address.

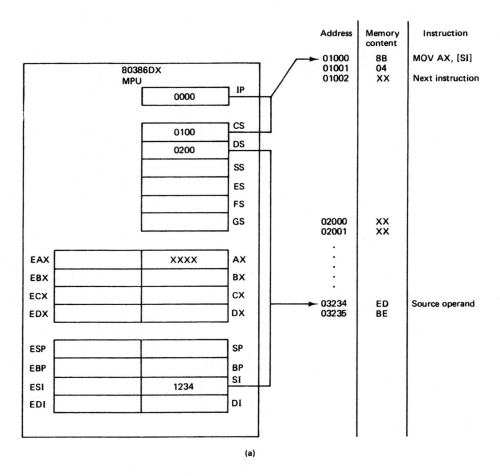

Address	Memory content	Instruction
01000	8B	MOV AX, [SI]
01001	04	
01002	XX	Next instruction
02000	XX	
02001	XX	
03234	ED	Source operand
03235	BE	

(a)

Figure 3.15 (a) Instruction using register indirect addressing mode before fetch and execution. (b) After execution.

The result produced by executing this instruction and the example for the direct addressing mode are the same. However, they differ in the way in which the physical address was generated. The direct addressing method lends itself to applications where the value of EA is a constant. On the other hand, register indirect addressing can be used when the value of EA is calculated and stored, for example, in SI by a previous instruction. That is, EA is a variable. For instance, the instructions executed just before our example instruction could have incremented the value in SI by two.

Based Addressing Mode. In the *based addressing mode*, the effective address of the operand is obtained by adding a direct or indirect displacement to the contents of either base register BX or base pointer register BP. The physical address calculation is shown in Fig. 3.16(a). Looking at Fig. 3.16(b), we see that the value in the base register defines the beginning of a data structure, such as a record, in memory and the displacement selects an element of data within this structure. To access a different element in the record, the programmer simply changes the value of the displacement. On the other hand, to access

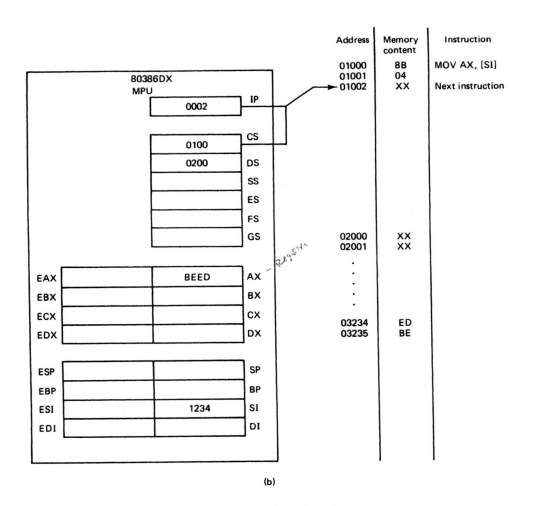

Address	Memory content	Instruction
01000	8B	MOV AX, [SI]
01001	04	
01002	XX	Next instruction
02000	XX	
02001	XX	
03234	ED	
03235	BE	

(b)

Figure 3.15 (Continued)

the same element in another similar record, the programmer can change the value in the base register so that it points to the beginning of the new record.

A move instruction that uses based addressing to specify the location of its destination operand is as follows:

```
MOV  [BX] + 1234H,AL
```

This instruction uses base register BX and direct displacement 1234H to derive the EA of the destination operand. The based addressing mode is implemented by specifying the base register in brackets followed by a + sign and the direct displacement. The source operand in this example is located in byte accumulator AL.

As shown in Fig. 3.17(a) and (b), the fetch and execution of this instruction causes the 80386DX to calculate the physical address of the destination operand from the contents

$$PA = \begin{Bmatrix} CS \\ DS \\ SS \\ ES \\ FS \\ GS \end{Bmatrix} : \begin{Bmatrix} BX \\ BP \end{Bmatrix} + \begin{Bmatrix} \text{8-bit displacement} \\ \text{16-bit displacement} \end{Bmatrix}$$

(a)

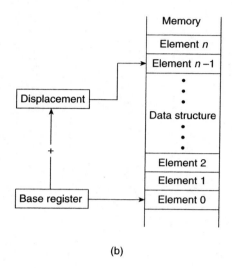

(b)

Figure 3.16 (a) Computation of a based address. (b) Based addressing of a structure of data.

of DS, BX, and the direct displacement. The result is

$$PA = 02000_{16} + 1000_{16} + 1234_{16}$$
$$= 04234_{16}$$

Then it writes the contents of source operand AL into the storage location at 04234_{16}. The result is that ED_{16} is written into the destination memory location. Again, the default segment register for this physical address calculation is DS, but it can be changed to another segment register with the segment-override prefix.

If BP is used instead of BX, the calculation of the physical address is performed using the contents of the stack segment (SS) register instead of DS. This permits access to data in the stack segment of memory.

Indexed Addressing Mode. *Indexed addressing mode* works in a manner similar to that of the based addressing mode we just described. However, as shown in Fig. 3.18(a), indexed addressing mode uses the value of the displacement as a pointer to the starting point of an array of data in memory and the contents of the specified register as an index that selects the specific element in the array that is to be accessed. For instance, for the byte-size element array in Fig. 3.18(a), the index register holds the value *n*. In this way, it selects data element *n* in the array. Figure 3.18(b) shows how the physical address is obtained from the value in a segment register, an index in the SI or DI register, and a displacement.

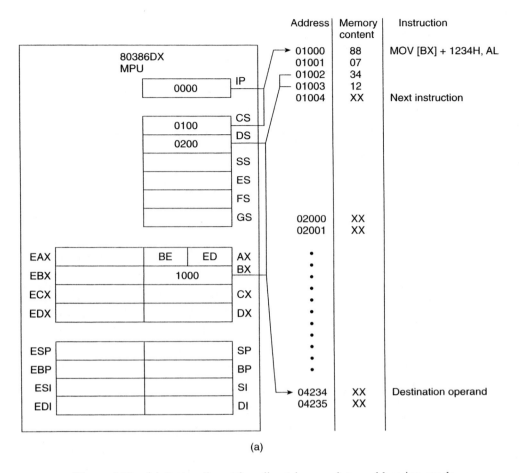

Figure 3.17 (a) Instruction using direct base pointer addressing mode before fetch and execution. (b) After execution.

Here is an example:

```
MOV  AL,[SI] + 2000H
```

The source operand has been specified using direct indexed addressing mode. Notice that the *direct displacement* is 2000H. As with the base register in based addressing, the index register, which is SI, is enclosed in brackets. The effective address is calculated as

$$EA = (SI) + 2000H$$

and the physical address is obtained by combining the contents of DS and EA.

$$PA = DS:(SI) + 2000H$$

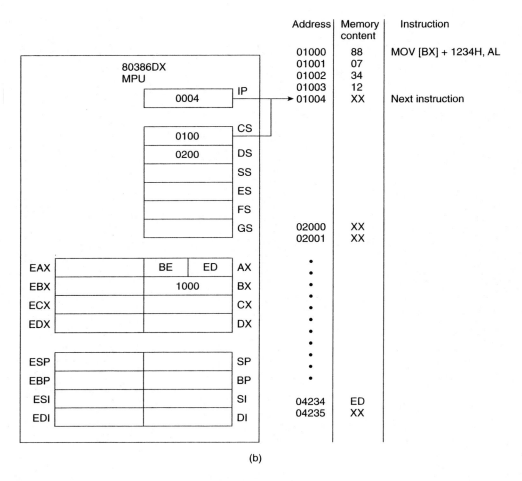

Address	Memory content	Instruction
01000	88	MOV [BX] + 1234H, AL
01001	07	
01002	34	
01003	12	
01004	XX	Next instruction
02000	XX	
02001	XX	
04234	ED	
04235	XX	

80386DX
MPU

0004 IP

0100 CS
0200 DS
SS
ES
FS
GS

EAX BE ED AX
EBX 1000 BX
ECX CX
EDX DX

ESP SP
EBP BP
ESI SI
EDI DI

(b)

Figure 3.17 (Continued)

The example in Fig. 3.19(a) and (b) shows the result of executing the move instruction. First the physical address of the source operand is calculated from the contents of DS, SI, and the direct displacement.

$$PA = 02000_{16} + 2000_{16} + 1234_{16}$$
$$= 05234_{16}$$

Then the byte of data stored at this location, which is BE_{16}, is read into the lower byte (AL) of the accumulator register.

Based-Indexed Addressing Mode. Combining the based addressing mode and the indexed addressing mode results in a new, more powerful mode known as the *based-indexed addressing mode*. This addressing mode can be used to access complex data structures such as two-dimensional arrays. Figure 3.20(a) shows how it can be used to access elements in

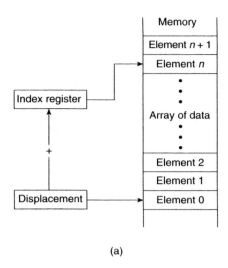

Memory

Element *n* + 1

Element *n*

⋮

Array of data

⋮

Element 2

Element 1

Element 0

Index register

+

Displacement

(a)

PA = Segment base: index + displacement

$$PA = \begin{Bmatrix} CS \\ DS \\ SS \\ ES \\ FS \\ GS \end{Bmatrix} : \begin{Bmatrix} SI \\ DI \end{Bmatrix} + \begin{Bmatrix} \text{8-bit displacement} \\ \text{16-bit displacement} \end{Bmatrix}$$

(b)

Figure 3.18 (a) Indexed addressing of an array of data elements. (b) Computation of an indexed address.

an $m \times n$ array of data. Notice that the displacement, which is a fixed value, locates the array in memory. The base register specifies the m coordinate of the array and the index register identifies the n coordinate. Any element in the array can be accessed simply by changing the values in the base and index registers. The registers permitted in the based-indexed physical address computation are shown in Fig. 3.20(b).

Let us consider an example of a move instruction using this type of addressing.

```
MOV  AH,[BX][SI] + 1234H
```

Notice that the source operand is accessed using based-indexed addressing mode. Therefore, the effective address of the source operand is obtained as

$$EA = (BX) + (SI) + 1234H$$

and the physical address of the operand is obtained from the current contents of DS and the calculated EA.

$$PA = DS:(BX) + (SI) + 1234H$$

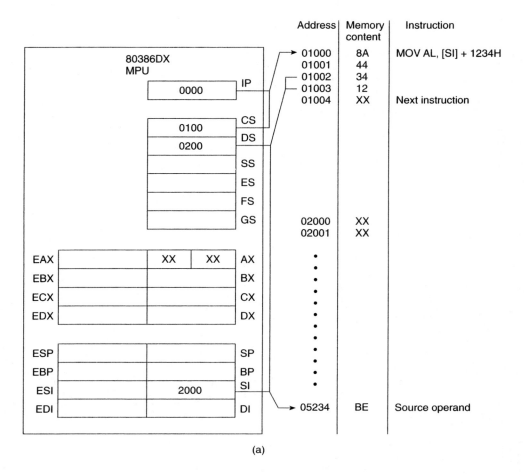

Address	Memory content	Instruction
01000	8A	MOV AL, [SI] + 1234H
01001	44	
01002	34	
01003	12	
01004	XX	Next instruction
02000	XX	
02001	XX	
05234	BE	Source operand

(a)

Figure 3.19 (a) Instruction using direct indexed addressing mode before fetch and execution. (b) After execution.

An example of executing this instruction is illustrated in Fig. 3.21(a) and (b). Using the contents of the various registers in the example, the address of the source operand is calculated as

$$PA = 02000_{16} + 1000_{16} + 2000_{16} + 1234_{16}$$

$$= 6234_{16}$$

Execution of the instruction causes the value stored at this location to be read into AH.

32-Bit Memory Operand Addressing Modes

Earlier we pointed out that extensions have been made to the memory operand addressing modes in the 80386 microprocessor family. They have been enhanced in two

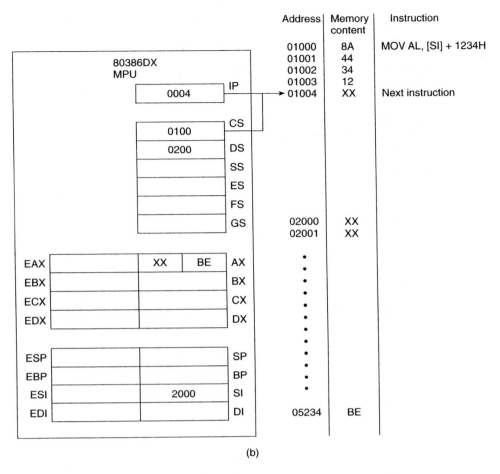

(b)

Figure 3.19 (Continued)

important ways. First, the effective address part of the physical address now has a fourth element called a *scale factor*. Therefore, the new general formula for an effective address calculations is

$$EA = base + (index \times scale\ factor) + displacement$$

Notice that the scale factor is used to multiply the value of the index component of the address.

The 16-bit members of the 80X86 family required that the value of the base or index element of the memory operand addressing modes be in specific registers. For instance, the base element in the based addressing mode could be only in either BX or BP. In the 80386 family, the registers used for the location of a base or index component of an address has been expanded to include any of the data, index, or pointer registers. That is, the 32-bit addressing model of the 80386DX permits physical address computations to be made

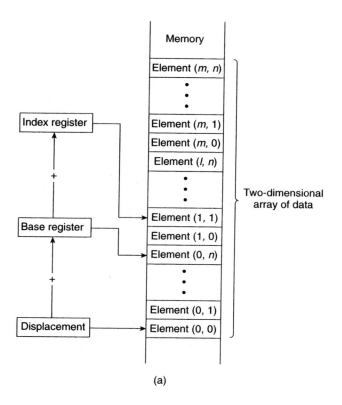

(a)

PA = Segment base: base + index + displacement

$$PA = \begin{Bmatrix} CS \\ DS \\ SS \\ ES \\ FS \\ GS \end{Bmatrix} : \begin{Bmatrix} BX \\ BP \end{Bmatrix} + \begin{Bmatrix} SI \\ DI \end{Bmatrix} + \begin{Bmatrix} \text{8-bit displacement} \\ \text{16-bit displacement} \end{Bmatrix}$$

(b)

Figure 3.20 (a) Based-indexed addressing of a two-dimensional array of data. (b) Computation of a based-indexed address.

using the formula given in Fig. 3.22. Notice that the value of the scale factor is limited to 1, 2, 4, or 8.

Let us now look briefly at how these new addressing capabilities affect the various memory operand addressing modes. The new address computation for register indirect addressing is shown in Fig. 3.23(a). Notice that now any of the data, index, or pointer registers can hold the effective address. An example is the instruction

        MOV   [AX],BX

The destination operand uses the AX register to hold the indirect addressing. Figure 3.23(b) shows the new address computation method for the based addressing mode. For instance, the instruction

        MOV   BL,[AX] + 1234H

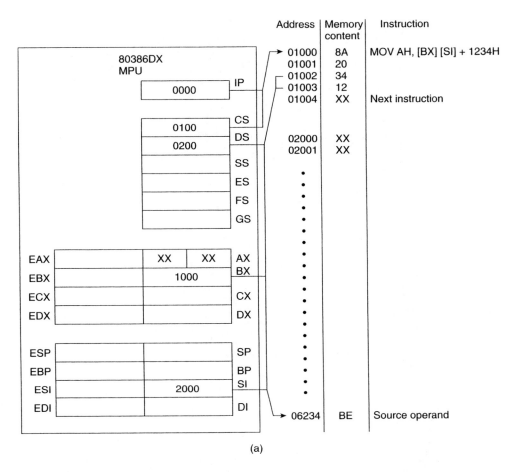

Figure 3.21 (a) Instruction using based-indexed addressing mode before fetch and execution. (b) After execution.

employs based addressing mode for the source operand. In this case, AX has been selected to hold the value of the base address.

A scale factor can be used in the address computation for an operand whose location is specified with either the indexed addressing mode or the based-indexed addressing mode. The physical address computation formulas for these modes are shown in Fig. 3.23(c) and (d), respectively.

EXAMPLE 3.4

What is the value of the physical address of the source operand in the instruction

```
MOV  AL,1234H + [SI × 2]
```

Assume that the values in DS and SI are 0200_{16} and 2000_{16}, respectively.

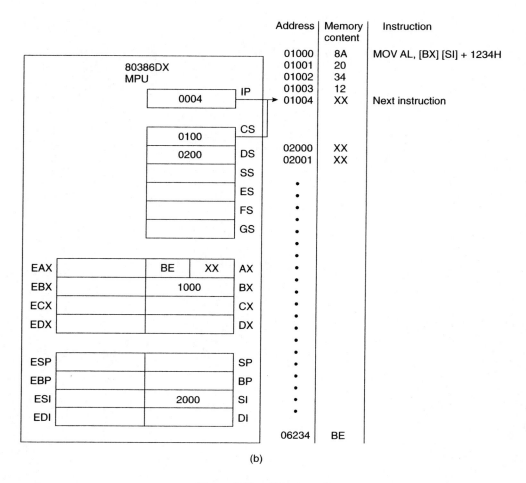

(b)

Figure 3.21 (Continued)

$$PA = \text{Segment base:} \left\{ \text{Base} + (\text{Index} \times \text{Scale factor}) + \text{Displacement} \right\}$$

$$PA = \begin{Bmatrix} CS \\ SS \\ DS \\ ES \\ FS \\ GS \end{Bmatrix} : \begin{Bmatrix} AX \\ BX \\ CX \\ DX \\ SP \\ BP \\ SI \\ DI \end{Bmatrix} + \begin{Bmatrix} AX \\ BX \\ CX \\ DX \\ BP \\ SI \\ DI \end{Bmatrix} \times \begin{Bmatrix} 1 \\ 2 \\ 4 \\ 8 \end{Bmatrix} + \begin{Bmatrix} \text{8-bit displacement} \\ \text{16-bit displacement} \end{Bmatrix}$$

Figure 3.22 32-bit address mode physical and effective address computation for memory operands.

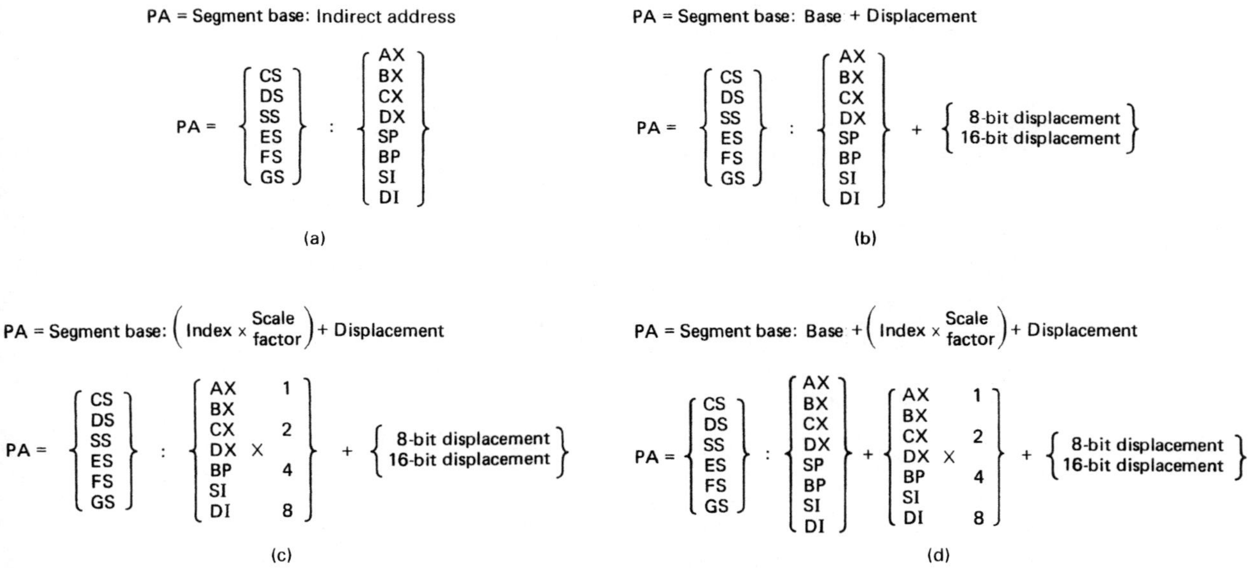

Figure 3.23 (a) 32-bit mode indirect memory-address computation. (b) 32-bit mode-based memory-address computation. (c) 32-bit mode scaled-index memory-address computation. (d) 32-bit mode scaled-based-index memory-address computation.

Solution

The physical address is calculated as follows:

$$PA = DS(0) + 1234_{16} + (SI \times 2)$$
$$= 02000_{16} + 1234_{16} + (2000_{16} \times 2)$$
$$= 07234_{16}$$

ASSIGNMENTS

Section 3.2

1. What tells a microcomputer what to do, where to get data, how to process the data, and where to put the results when done?
2. What is the name given to a sequence of instructions that is used to guide a computer through a task?
3. What is the native language of the 80386DX?
4. How does machine language differ from assembly language?
5. What does opcode stand for? Give two examples.
6. What is an operand? Give two types.

7. In the assembly language statement

```
START:   ADD   EAX,EBX   ;ADD EBX TO EAX
```

what is the label? What is the comment?

8. What is the function of an assembler? A compiler?

9. What is source code? What is object code?

10. Give two benefits derived from writing programs in assembly language instead of a high-level language.

11. What is meant by the phrase real-time application?

12. List two hardware-related applications that require use of assembly language programming. Name two software-related applications.

Section 3.3

13. List the basic steps in the general development cycle for an assembly language program.

14. What document is produced as a result of the problem-description step of the development cycle?

15. Give a name that is used to refer to the software solution planned for a problem. What is the name of the document that is used to describe this solution plan?

16. What is a flowchart?

17. Draw the flowchart symbol used to identify a subroutine.

18. What type of program is EDIT?

19. What type of program is used to produce an object module?

20. What does MASM stand for?

21. What type of program is used to produce a run module?

22. In which part of the development cycle is the EDIT program used? The MASM program? The LINK program? The DEBUG program?

23. Assuming that the file name is PROG_A, give typical names for the files that result from the use of the EDLIN program. The MASM program? The LINK program?

Section 3.4

24. What is the name given to the part of the 80386DX's real-mode instruction set that is common to that of the 8086, 8088, and 80286 microprocessors? Common to the 80286 real-mode instruction set but not the 8086 or 8088?

25. What is the part of the 80386DX's real-mode instruction set that is unique to the 80386DX called?

26. What is the name of the protected-mode instruction set of the 80386DX?

Section 3.5

27. What is meant by an addressing mode?

28. Make a list of the addressing modes available on the 80386DX.

29. What are the three elements that can be used to form the effective address of an operand in memory?
30. Name the five memory operand addressing modes.
31. Identify the addressing modes used for the source and the destination operands in the instructions that follow.
    (a) MOV AL,BL
    (b) MOV AX,0FFH
    (c) MOV [DI],AX
    (d) MOV DI,[SI]
    (e) MOV [BX] + 0400H,CX
    (f) MOV [DI] + 0400H,AH
    (g) MOV [BX][DI] + 0400H,AL
32. Compute the physical address for the specified operand in each of the following instructions. The register contents and variables are as follows: $(CS) = 0A00_{16}$, $(DS) = 0B00_{16}$, $(ESI) = 00000100_{16}$, $(EDI) = 00000200_{16}$, and $(EBX) = 00000300_{16}$.
    (a) Destination operand of the instruction in (c) of Problem 31
    (b) Source operand of the instruction in (d) of Problem 31
    (c) Destination operand of the instruction in (e) of Problem 31
    (d) Destination operand of the instruction in (f) of Problem 31
    (e) Destination operand of the instruction in (g) of Problem 31
33. If $(AX) = 0400_{16}$ and $(DI) = 0200_{16}$, what is the value of the address of the destination operand in the instruction

```
MOV   [AX][DI × 4] + 10H,CX
```

Machine Language Coding and the DEBUG Software Development Program of the PC/AT-Compatible Microcomputer

▲ 4.1 INTRODUCTION

In this chapter, we begin by exploring how the instructions of the base instruction set for the 80386 family of microprocessors are encoded. This is followed by a study of the software development environment provided for this family of microprocessor with the PC/AT-compatible microcomputer. Here we examine the DEBUG program. DEBUG is a program-execution/debug tool that operates in the PC/AT's *disk operating system* (DOS) environment. First DEBUG's command set is thoroughly examined. Then these commands are used to load, assemble, execute, and debug programs. The topics discussed in this chapter are as follows:

1. Converting assembly language instructions to machine code
2. Encoding a complete program in machine code
3. The PC/AT and its DEBUG program
4. Examining and modifying the contents of memory
5. Input and output of data
6. Hexadecimal addition and subtraction
7. Loading, verifying, and saving machine language programs
8. Assembling instructions with the ASSEMBLE command
9. Executing instructions and programs with the TRACE and GO commands
10. Debugging a program

▲ 4.2 CONVERTING ASSEMBLY LANGUAGE INSTRUCTIONS TO MACHINE CODE

To convert an assembly language program to machine code, we must convert each assembly language instruction to its equivalent machine code instruction. In general, the machine code for an instruction specifies things like what operation is to be performed, what operand or operands are to be used, whether the operation is performed on byte, word, or double-word data, whether the operation involves operands that are located in registers or a register and a storage location in memory, and if one of the operands is in memory, how its address is to be generated. All this information is encoded into the bits of the machine code for the instruction.

In this section we will learn how to code the instructions of the base instruction set of the 80386 microprocessor family in machine language. The machine code instructions of the 80386DX vary in the number of bytes used to encode them. Some instructions can be encoded with just 1 byte, others can be done in 2 bytes, and many require even more. Single-byte instructions generally specify a simpler operation with a register or a flag bit. For instance, *complement carry* (CMC) is an example of a single-byte instruction. It is specified by the machine code byte 11110101_2, which equals $F5_{16}$. That is,

$$CMC = 11110101_2 = F5_{16}$$

The machine code for instructions can be obtained by following the formats that are used in encoding the base instructions of the 80386DX. Most multibyte instructions use the *general instruction format* shown in Fig. 4.1. Exceptions to this format exist and are considered later. For now, let us describe the functions of the various bits and fields (groups of bits) in each byte of this format.

Looking at Fig. 4.1, we see that byte 1 contains three kinds of information: the *operation code* (opcode), the *register direction bit* (D), and the *data size bit* (W). Let us summarize the function of each of these pieces of information.

1. **Opcode field (6 bits):** It specifies the operation, such as add, subtract, or move, that is to be performed.

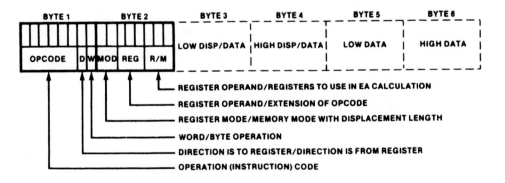

Figure 4.1 General instruction format. (Reprinted by permission of Intel Corp. Copyright/Intel Corp. 1979)

REG	W = 0	W = 1
000	AL	AX
001	CL	CX
010	DL	DX
011	BL	BX
100	AH	SP
101	CH	BP
110	DH	SI
111	BH	DI

Figure 4.2 Register (REG) field encoding. (Reprinted by permission of Intel Corp. Copyright/Intel Corp. 1979)

2. Register direction bit (D bit): It tells whether the register operand that is specified in byte 2 is the source or destination operand. A logic 1 in this bit position indicates that the register operand is a destination operand, and a logic 0 indicates that it is a source operand.

3. Data size bit (W bit): It specifies whether the operation will be performed on 8-bit or 16-bit data. Logic 0 selects 8 bits and 1 selects 16 bits as the data size.

For instance, if a 16-bit value is to be added to register AX, the 6 most significant bits specify the add register operation. This opcode is 000000. The next bit, D, will be at logic 1 to specify that a register, AX in this case, holds the destination operand. Finally, the least significant bit, W, will be logic 1 to specify a 16-bit data operation.

The second byte in Fig. 4.1 has three fields. They are the *mode* (MOD) *field*, the *register* (REG) *field*, and the *register/memory* (R/M) *field*. These fields are used to specify which register is used for the first operand and where the second operand is stored. The second operand can be in either a register or a memory location.

The 3-bit REG field is used to identify the register for the first operand. This operand is the one that was defined as the source or destination by the D bit in byte 1. The encoding for each of the 80386DX's registers is shown in Fig. 4.2. Here we find that the 16-bit register AX and the 8-bit register AL are specified by the same binary code. Notice that the decision whether to use AX or AL is made based on the setting of the operation size (W) bit in byte 1.

For instance, in our earlier example, we said that the first operand, which is the destination operand, is register AX. For this case, the REG field is set to 000.

The 2-bit MOD field and 3-bit R/M field together specify the second operand. Encoding for these two fields is shown in Fig. 4.3(a) and (b), respectively. MOD indicates whether the operand is in a register or memory. Notice that in the case of a second operand that is in a register, the MOD field is always 11. The R/M field along with the W bit from byte 1 selects the register.

For example, if the second operand, the source operand, in our earlier addition example is to be in BX, the MOD and R/M fields will be made 11 and 011, respectively.

EXAMPLE 4.1

The instruction

```
MOV   BL,AL
```

stands for move the byte contents from source register AL to destination register BL. Using the general format in Fig. 4.1, show how to encode the instruction in machine code. Assume that the 6-bit opcode for the move operation is 100010.

MODE	EXPLANATION
00	Memory Mode, no displacement follows*
01	Memory Mode, 8-bit displacement follows
10	Memory Mode, 16-bit displacement follows
11	Register Mode (no displacement)

*Except when R/M = 110, then 16-bit displacement follows

(a)

MOD = 11			EFFECTIVE ADDRESS CALCULATION			
R/M	W = 0	W = 1	R/M	MOD = 00	MOD = 01	MOD = 10
000	AL	AX	000	(BX) + (SI)	(BX) + (SI) + D8	(BX) + (SI) + D16
001	CL	CX	001	(BX) + (DI)	(BX) + (DI) + D8	(BX) + (DI) + D16
010	DL	DX	010	(BP) + (SI)	(BP) + (SI) + D8	(BP) + (SI) + D16
011	BL	BX	011	(BP) + (DI)	(BP) + (DI) + D8	(BP) + (DI) + D16
100	AH	SP	100	(SI)	(SI) + D8	(SI) + D16
101	CH	BP	101	(DI)	(DI) + D8	(DI) + D16
110	DH	SI	110	DIRECT ADDRESS	(BP) + D8	(BP) + D16
111	BH	DI	111	(BX)	(BX) + D8	(BX) + D16

(b)

Figure 4.3 (a) Mode (MOD) field encoding. (Reprinted by permission of Intel Corp. Copyright/Intel Corp. 1979) (b) Register/memory (R/M) field encoding. (Reprinted by permission of Intel Corp. Copyright/Intel Corp. 1979)

Solution

In byte 1 the first 6 bits specify the move operation and thus must be 100010.

$$OPCODE = 100010$$

The next bit, which is D, indicates whether the register that is specified by the REG part of byte 2 is a source or destination operand. Let us say that we will encode AL in the REG field of byte 2; therefore, D is set equal to 0 for source operand.

$$D = 0$$

The last bit (W) in byte 1 must specify a byte operation. For this reason, it is also set to 0.

$$W = 0$$

This leads to

$$\text{BYTE } 1 = 10001000_2 = 88_{16}$$

In byte 2 the source operand, which is specified by the REG field, is AL. The corresponding code from Fig. 4.2 is

$$\text{REG} = 000$$

Since the second operand is also a register, the MOD field is made 11. The R/M field specifies that the destination register is BL, for which the code (from Fig. 4.3(b)) is 011. This gives

$$\text{MOD} = 11$$
$$\text{R/M} = 011$$

Therefore, byte 2 is

$$\text{BYTE } 2 = 11000011_2 = C3_{16}$$

Thus, the hexadecimal machine code for the instruction is given by

```
MOV   BL,AL  =  88C3H
```

For a second operand that is located in memory, there are a number of different ways its location can be specified. That is, any of the addressing modes supported in the base instruction set of the 80386DX microprocessor can be used to generate its address. The addressing mode is selected with the MOD and R/M fields.

Notice in Fig. 4.3(b) that the addressing mode for an operand in memory is indicated by one of the three values 00, 01, or 10 in the MOD field and an appropriate R/M code. The different ways in which the operand's address can be generated are shown in the effective address calculation part of the table in Fig. 4.3(b). These different address-calculation expressions correspond to the addressing modes we introduced in Chapter 3. For instance, if the base register (BX) contains the memory address, this fact is encoded into the instruction by making MOD = 00 and R/M = 111.

EXAMPLE 4.2

The instruction

```
ADD   AX,[SI]
```

stands for add the 16-bit contents of the memory location indirectly specified by SI to the contents of AX. Encode the instruction in machine code. The opcode for add is 000000.

Solution

To specify a 16-bit add operation with a register as the destination, the first byte of machine code will be

$$\text{BYTE 1} = 00000011_2 = 03_{16}$$

The REG field bits in byte 2 are 000 to select AX as the destination register. The other operand is in memory, and its address is specified by the contents of SI with no displacement. In Fig. 4.3(a) and (b), we find that for indirect addressing using SI with no displacement, MOD equals 00 and R/M equals 100. That is,

$$\text{MOD} = 00$$

$$\text{R/M} = 100$$

This gives

$$\text{BYTE 2} = 00000100_2 = 04_{16}$$

Thus the machine code for the instruction is

```
ADD  AX,[SI]  = 0304H
```

Some of the addressing modes of the 80386DX need either data or an address displacement to be coded into the instruction. These types of information are encoded using additional bytes. For instance, looking at Fig. 4.1, we see that byte 3 is needed in the encoding of an instruction if it uses a byte-size address displacement and both byte 3 and byte 4 are needed if the instruction uses a word-size displacement.

The size of the displacement is encoded into the MOD field. For example, if the effective address is to be generated by the expression

$$\text{(BX)} + \text{D8}$$

where D8 stands for 8-bit displacement, MOD is set to 01 to specify memory mode with an 8-*bit displacement* and R/M is set to 111 to select BX.

Bytes 3 and 4 are also used to encode byte-wide immediate operands, word-wide immediate operands, and direct addresses. For example, in an instruction where direct addressing is used to identify the location of an operand in memory, the MOD field must be 00 and the R/M field, 110. The actual value of the operand's address is coded into the bytes that follow.

If both a 16-bit displacement and a 16-bit immediate operand are used in the same instruction, the displacement is encoded into bytes 3 and 4 and the immediate operand is encoded into bytes 5 and 6.

EXAMPLE 4.3

What is the machine code for the instruction

```
XOR  CL,[1234H]
```

This instruction says to exclusive-OR the byte of data at memory address 1234_{16} with the byte contents of CL. The opcode for exclusive-OR is 001100.

Solution

Using 001100 as the opcode bits, 1 to denote the register as the destination operand, and 0 to denote byte data, we get

$$BYTE\ 1 = 00110010_2 = 32_{16}$$

The REG field has to specify CL, which makes it equal to 001. In this case a direct address has been specified for operand 2. This requires MOD = 00 and R/M = 110. Thus,

$$BYTE\ 2 = 00001110_2 = 0E_{16}$$

To specify the address 1234_{16}, we must use byte 3 and byte 4. The least significant byte of the address is encoded first followed by the most significant byte. This gives

$$BYTE\ 3 = 34_{16}$$

and

$$BYTE\ 4 = 12_{16}$$

Thus the machine code form of the instruction is given by

```
XOR   CL,1234H  =  320E3412H
```

EXAMPLE 4.4

The instruction

```
ADD   [BX][DI]+1234H,AX
```

means add the word contents of AX to the contents of the memory location specified by based-indexed addressing mode. The opcode for the add operation is 000000.

Solution

The add opcode, which is 000000, a 0 for source operand, and a 1 for word data gives

$$BYTE\ 1 = 00000001_2 = 01_{16}$$

The REG field in byte 2 is 000 to specify AX as the source register. Since there is a displacement and it needs 16 bits for encoding, the MOD field obtained from Fig. 4.3(a) is 10. The R/M field, which is also obtained from Fig. 4.3(b), is set to 001 for an effective address generated from DI and BX. This gives the second byte as

$$BYTE\ 2 = 10000001_2 = 81_{16}$$

Field	Value	Function
S	0 1	No sign extension Sign extend 8-bit immediate data to 16 bits if W=1
V	0 1	Shift/rotate count is one Shift/rotate count is specified in CL register
Z	0 1	Repeat/loop while zero flag is clear Repeat/loop while zero flag is set

Figure 4.4 Additional 1-bit fields and their functions. (Reprinted by permission of Intel Corp. Copyright/Intel Corp. 1979)

The displacement 1234_{16} is encoded in the next 2 bytes, with the least significant byte first. Therefore, the machine code that results is

$$\text{ADD} \quad [BX][DI]+1234H, AX \ = \ 01813412H$$

As we indicated earlier, the general format in Fig. 4.1 cannot be used to encode all the instructions that can be executed by the 80386DX. There are minor modifications that must be made to this general format to encode a few instructions. In some instructions one or more additional single bit fields need to be added. These 1-bit fields and their functions are shown in Fig. 4.4. For instance, the general format of the *repeat* (REP) instruction is

$$\text{REP} \ = \ 1111001Z$$

Here bit Z is 1 or 0, depending upon whether the repeat operation is to be done when the zero flag is set or when it is reset. Similarly the other two bits, S and V, in Fig. 4.4 are used to encode sign extension for arithmetic instructions and to specify the source of the count for shift or rotate instructions, respectively.

The formats for all the instructions in the 80386DX's base instruction set are shown in Fig. 4.5. This is the information that can be used to encode any of the 80386DX's base instruction.

Instructions that involve a segment register need a 2-bit field to encode which register is to be affected. This field is called the *SR field*. As shown in Fig. 4.5, PUSH and POP are examples of instructions that have an SR field. The four segment registers ES, CS, SS, and DS are encoded according to the table in Fig. 4.6.

EXAMPLE 4.5

The instruction

$$\text{MOV} \quad \text{WORD PTR} \quad [BP][DI]+1234H, 0ABCDH$$

says to move the immediate data word $ABCD_{16}$ into the memory location specified by based-indexed addressing mode. Express the instruction in machine code.

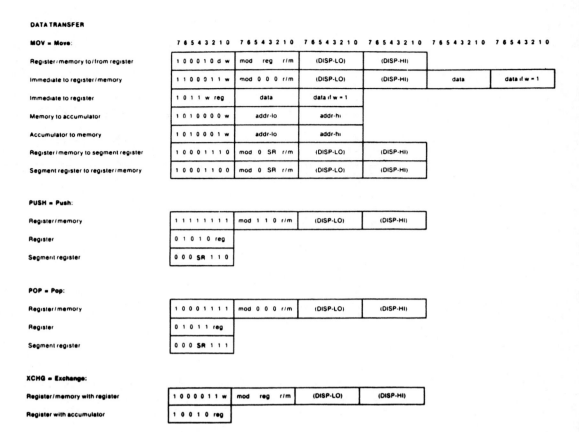

DATA TRANSFER

MOV = Move:

Register/memory to/from register

7 6 5 4 3 2 1 0	7 6 5 4 3 2 1 0	7 6 5 4 3 2 1 0	7 6 5 4 3 2 1 0	7 6 5 4 3 2 1 0	7 6 5 4 3 2 1 0
1 0 0 0 1 0 d w	mod reg r/m	(DISP-LO)	(DISP-HI)		

Immediate to register/memory

| 1 1 0 0 0 1 1 w | mod 0 0 0 r/m | (DISP-LO) | (DISP-HI) | data | data if w = 1 |

Immediate to register

| 1 0 1 1 w reg | data | data if w = 1 |

Memory to accumulator

| 1 0 1 0 0 0 0 w | addr-lo | addr-hi |

Accumulator to memory

| 1 0 1 0 0 0 1 w | addr-lo | addr-hi |

Register/memory to segment register

| 1 0 0 0 1 1 1 0 | mod 0 SR r/m | (DISP-LO) | (DISP-HI) |

Segment register to register/memory

| 1 0 0 0 1 1 0 0 | mod 0 SR r/m | (DISP-LO) | (DISP-HI) |

PUSH = Push:

Register/memory

| 1 1 1 1 1 1 1 1 | mod 1 1 0 r/m | (DISP-LO) | (DISP-HI) |

Register

| 0 1 0 1 0 reg |

Segment register

| 0 0 0 SR 1 1 0 |

POP = Pop:

Register/memory

| 1 0 0 0 1 1 1 1 | mod 0 0 0 r/m | (DISP-LO) | (DISP-HI) |

Register

| 0 1 0 1 1 reg |

Segment register

| 0 0 0 SR 1 1 1 |

XCHG = Exchange:

Register/memory with register

| 1 0 0 0 0 1 1 w | mod reg r/m | (DISP-LO) | (DISP-HI) |

Register with accumulator

| 1 0 0 1 0 reg |

Figure 4.5 80386DX base instruction set encoding tables. (Reprinted by permission of Intel Corp. Copyright/Intel Corp. 1979)

Solution

Since this instruction does not involve one of the registers as an operand, it does not follow the general format that we have been using. From Fig. 4.5 we find that byte 1 in an immediate data to memory move is

$$1100011W$$

In our case, we are moving word-size data; therefore, W equals 1. This gives

$$\text{BYTE 1} = 11000111_2 = \text{C7}_{16}$$

Again from Fig. 4.5, we find that byte 2 has the form

$$\text{BYTE 2} = (\text{MOD})000(\text{R/M})$$

For a memory operand using a 16-bit displacement, Fig. 4.3(a) shows that MOD equals 10, and for based-indexed addressing using BP and DI with a 16-bit displacement, Fig. 4.3(b)

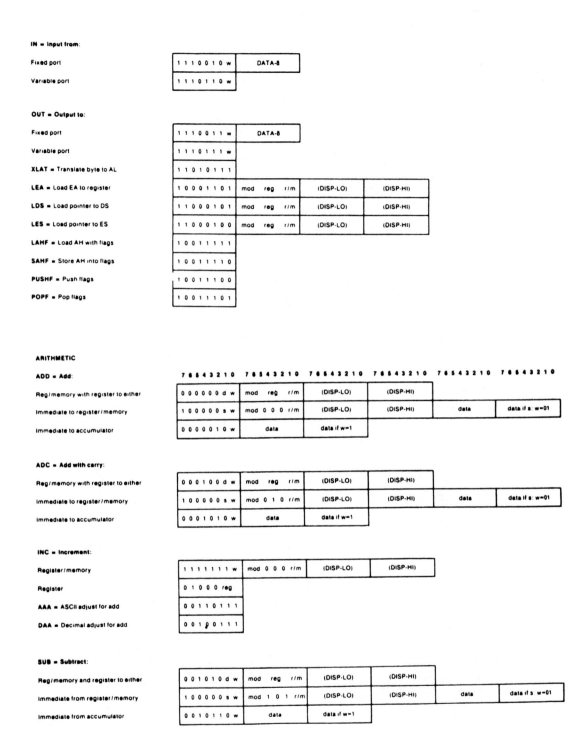

Figure 4.5 (Continued)

SBB = Subtract with borrow:

Reg/memory and register to either	0 0 0 1 1 0 d w	mod reg r/m	(DISP-LO)	(DISP-HI)	
Immediate from register/memory	1 0 0 0 0 0 s w	mod 0 1 1 r/m	(DISP-LO)	(DISP-HI)	data	data if s: w=01
Immediate from accumulator	0 0 0 1 1 1 0 w	data	data if w=1		

DEC Decrement:

Register/memory	1 1 1 1 1 1 1 w	mod 0 0 1 r/m	(DISP-LO)	(DISP-HI)
Register	0 1 0 0 1 reg			
NEG Change sign	1 1 1 1 0 1 1 w	mod 0 1 1 r/m	(DISP-LO)	(DISP-HI)

CMP = Compare:

Register/memory and register	0 0 1 1 1 0 d w	mod reg r/m	(DISP-LO)	(DISP-HI)	
Immediate with register/memory	1 0 0 0 0 0 s w	mod 1 1 1 r/m	(DISP-LO)	(DISP-HI)	data	data if s: w=1
Immediate with accumulator	0 0 1 1 1 1 0 w	data		
AAS ASCII adjust for subtract	0 0 1 1 1 1 1 1			
DAS Decimal adjust for subtract	0 0 1 0 1 1 1 1			
MUL Multiply (unsigned)	1 1 1 1 0 1 1 w	mod 1 0 0 r/m	(DISP-LO)	(DISP-HI)

ARITHMETIC

	7 6 5 4 3 2 1 0	7 6 5 4 3 2 1 0	7 6 5 4 3 2 1 0	7 6 5 4 3 2 1 0
IMUL Integer multiply (signed)	1 1 1 1 0 1 1 w	mod 1 0 1 r/m	(DISP-LO)	(DISP-HI)
AAM ASCII adjust for multiply	1 1 0 1 0 1 0 0	0 0 0 0 1 0 1 0	(DISP-LO)	(DISP-HI)
DIV Divide (unsigned)	1 1 1 1 0 1 1 w	mod 1 1 0 r/m	(DISP-LO)	(DISP-HI)
IDIV Integer divide (signed)	1 1 1 1 0 1 1 w	mod 1 1 1 r/m	(DISP-LO)	(DISP-HI)
AAD ASCII adjust for divide	1 1 0 1 0 1 0 1	0 0 0 0 1 0 1 0	(DISP-LO)	(DISP-HI)
CBW Convert byte to word	1 0 0 1 1 0 0 0			
CWD Convert word to double word	1 0 0 1 1 0 0 1			

LOGIC

NOT Invert	1 1 1 1 0 1 1 w	mod 0 1 0 r/m	(DISP-LO)	(DISP-HI)
SHL/SAL Shift logical/arithmetic left	1 1 0 1 0 0 v w	mod 1 0 0 r/m	(DISP-LO)	(DISP-HI)
SHR Shift logical right	1 1 0 1 0 0 v w	mod 1 0 1 r/m	(DISP-LO)	(DISP-HI)
SAR Shift arithmetic right	1 1 0 1 0 0 v w	mod 1 1 1 r/m	(DISP-LO)	(DISP-HI)
ROL Rotate left	1 1 0 1 0 0 v w	mod 0 0 0 r/m	(DISP-LO)	(DISP-HI)
ROR Rotate right	1 1 0 1 0 0 v w	mod 0 0 1 r/m	(DISP-LO)	(DISP-HI)
RCL Rotate through carry flag left	1 1 0 1 0 0 v w	mod 0 1 0 r/m	(DISP-LO)	(DISP-HI)
RCR Rotate through carry right	1 1 0 1 0 0 v w	mod 0 1 1 r/m	(DISP-LO)	(DISP-HI)

Figure 4.5 (Continued)

AND = And:

Reg/memory with register to either

0 0 1 0 0 0 d w	mod reg r/m	(DISP-LO)	(DISP-HI)		

Immediate to register/memory

1 0 0 0 0 0 0 w	mod 1 0 0 r/m	(DISP-LO)	(DISP-HI)	data	data if w=1

Immediate to accumulator

0 0 1 0 0 1 0 w	data	data if w=1

TEST = And function to flags no result:

Register/memory and register

0 0 0 1 0 0 d w	mod reg r/m	(DISP-LO)	(DISP-HI)		

Immediate data and register/memory

1 1 1 1 0 1 1 w	mod 0 0 0 r/m	(DISP-LO)	(DISP-HI)	data	data if w=1

Immediate data and accumulator

1 0 1 0 1 0 0 w	data

OR = Or:

Reg/memory and register to either

0 0 0 0 1 0 d w	mod reg r/m	(DISP-LO)	(DISP-HI)		

Immediate to register/memory

1 0 0 0 0 0 0 w	mod 0 0 1 r/m	(DISP-LO)	(DISP-HI)	data	data if w=1

Immediate to accumulator

0 0 0 0 1 1 0 w	data	data if w=1

XOR = Exclusive or:

Reg/memory and register to either

0 0 1 1 0 0 d w	mod reg r/m	(DISP-LO)	(DISP-HI)		

Immediate to register/memory

0 0 1 1 0 1 0 w	data	(DISP-LO)	(DISP-HI)	data	data if w=1

Immediate to accumulator

0 0 1 1 0 1 0 w	data	data if w=1

STRING MANIPULATION

7 6 5 4 3 2 1 0   7 6 5 4 3 2 1 0   7 6 5 4 3 2 1 0   7 6 5 4 3 2 1 0   7 6 5 4 3 2 1 0   7 6 5 4 3 2 1 0

REP = Repeat

1 1 1 1 0 0 1 z

MOVS = Move byte/word

1 0 1 0 0 1 0 w

CMPS = Compare byte/word

1 0 1 0 0 1 1 w

SCAS = Scan byte/word

1 0 1 0 1 1 1 w

LODS = Load byte/wd to AL/AX

1 0 1 0 1 1 0 w

STDS = Stor byte/wd from AL/A

1 0 1 0 1 0 1 w

CONTROL TRANL. ...

CALL = Call:

Direct within segment

1 1 1 0 1 0 0 0	IP-INC-LO	IP-INC-HI	

Indirect within segment

1 1 1 1 1 1 1 1	mod 0 1 0 r/m	(DISP-LO)	(DISP-HI)

Direct intersegment

1 0 0 1 1 0 1 0	IP-lo	IP-hi	
	CS-lo	CS-hi	

Indirect intersegment

1 1 1 1 1 1 1 1	mod 0 1 1 r/m	(DISP-LO)	(DISP-HI)

Figure 4.5 (Continued)

JMP = Unconditional Jump:

Direct within segment	1 1 1 0 1 0 0 1	IP-INC-LO	IP-INC-HI
Direct within segment-short	1 1 1 0 1 0 1 1	IP-INC8	
Indirect within segment	1 1 1 1 1 1 1 1	mod 1 0 0 r/m	(DISP-LO)	(DISP-HI)
Direct intersegment	1 1 1 0 1 0 1 0	IP-lo	IP-hi
		CS-lo	CS-hi
Indirect intersegment	1 1 1 1 1 1 1 1	mod 1 0 1 r/m	(DISP-LO)	(DISP-HI)

RET = Return from CALL:

Within segment	1 1 0 0 0 0 1 1		
Within seg adding immed to SP	1 1 0 0 0 0 1 0	data-lo	data-hi
Intersegment	1 1 0 0 1 0 1 1		
Intersegment adding immediate to SP	1 1 0 0 1 0 1 0	data-lo	data-hi
JE/JZ = Jump on equal/zero	0 1 1 1 0 1 0 0	IP-INC8	
JL/JNGE = Jump on less/not greater or equal	0 1 1 1 1 1 0 0	IP-INC8	
JLE/JNG = Jump on less or equal/not greater	0 1 1 1 1 1 1 0	IP-INC8	
JB/JNAE = Jump on below/not above or equal	0 1 1 1 0 0 1 0	IP-INC8	
JBE/JNA = Jump on below or equal/not above	0 1 1 1 0 1 1 0	IP-INC8	
JP/JPE = Jump on parity/parity even	0 1 1 1 1 0 1 0	IP-INC8	
JO = Jump on overflow	0 1 1 1 0 0 0 0	IP-INC8	
JS = Jump on sign	0 1 1 1 1 0 0 0	IP-INC8	
JNE/JNZ = Jump on not equal/not zer0	0 1 1 1 0 1 0 1	IP-INC8	

CONTROL TRANSFER (Cont'd.)   7 6 5 4 3 2 1 0   7 6 5 4 3 2 1 0   7 6 5 4 3 2 1 0   7 6 5 4 3 2 1 0   7 6 5 4 3 2 1 0   7 6 5 4 3 2 1 0

JNL/JGE = Jump on not less/greater or equal	0 1 1 1 1 1 0 1	IP-INC8
JNLE/JG = Jump on not less or equal/greater	0 1 1 1 1 1 1 1	IP-INC8
JNB/JAE = Jump on not below/above or equal	0 1 1 1 0 0 1 1	IP-INC8
JNBE/JA = Jump on not below or equal/above	0 1 1 1 0 1 1 1	IP-INC8
JNP/JPO = Jump on not par/par odd	0 1 1 1 1 0 1 1	IP-INC8
JNO = Jump on not overflow	0 1 1 1 0 0 0 1	IP-INC8
JNS = Jump on not sign	0 1 1 1 1 0 0 1	IP-INC8
LOOP = Loop CX times	1 1 1 0 0 0 1 0	IP-INC8
LOOPZ/LOOPE = Loop while zero/equal	1 1 1 0 0 0 0 1	IP-INC8
LOOPNZ/LOOPNE = Loop while not zero/equal	1 1 1 0 0 0 0 0	IP-INC8
JCXZ = Jump on CX zero	1 1 1 0 0 0 1 1	IP-INC8

Figure 4.5   (Continued)

INT = Interrupt:

Type specified

1 1 0 0 1 1 0 1	DATA-8

Type 3

1 1 0 0 1 1 0 0

INTO = Interrupt on overflow

1 1 0 0 1 1 1 0

IRET = Interrupt return

1 1 0 0 1 1 1 1

PROCESSOR CONTROL

CLC = Clear carry

1 1 1 1 1 0 0 0

CMC = Complement carry

1 1 1 1 0 1 0 1

STC = Set carry

1 1 1 1 1 0 0 1

CLD = Clear direction

1 1 1 1 1 1 0 0

STD = Set direction

1 1 1 1 1 1 0 1

CLI = Clear interrupt

1 1 1 1 1 0 1 0

STI = Set interrupt

1 1 1 1 1 0 1 1

HLT = Halt

1 1 1 1 0 1 0 0

WAIT = Wait

1 0 0 1 1 0 1 1

ESC = Escape (to external device)

1 1 0 1 1 x x x	mod y y y r/m	(DISP-LO)	(DISP-HI)

LOCK = Bus lock prefix

1 1 1 1 0 0 0 0

SEGMENT = Override prefix

0 0 1 reg 1 1 0

Figure 4.5 (Continued)

shows that R/M equals 011. This gives

$$BYTE\ 2 = 10000011_2 = 83_{16}$$

Bytes 3 and 4 encode the displacement with its low byte first. Thus for a displacement of 1234_{16}, we get

$$BYTE\ 3 = 34_{16}$$

and

$$BYTE\ 4 = 12_{16}$$

Register	SR
ES	00
CS	01
SS	10
DS	11

Figure 4.6 Segment register codes.

Lastly, bytes 5 and 6 encode the immediate data also with the least significant byte first. For data word $ABCD_{16}$, we get

$$\text{BYTE } 5 = CD_{16}$$

and

$$\text{BYTE } 6 = AB_{16}$$

Thus the entire instruction in machine code is given by

```
MOV   WORD PTR   [BP][DI]+1234H,ABCDH = C7833412CDABH
```

EXAMPLE 4.6

The instruction

```
MOV   [BP][DI]+1234H,DS
```

says to move the contents of the data segment register to the memory location specified by based-indexed addressing mode. Express the instruction in machine code.

Solution

From Fig. 4.5 we see that this instruction is encoded as

$$10001100(MOD)0(SR)(R/M)(DISP)$$

The MOD and R/M fields are the same as in Example 4.5. That is,

$$MOD = 10$$

and

$$R/M = 011$$

Moreover, the value of DISP is given as 1234_{16}. Finally from Fig. 4.6 we find that to specify DS, the SR field is

$$SR = 11$$

Therefore, the instruction is coded as

$$10001100100110110011010000010010_2 = 8C9B3412_{16}$$

The instructions of the extended instruction set and the 80386 specific instructions are coded in a similar way. However, a few extensions were needed in general format to support them. For instance, many of the new instructions have a 14-bit, instead of 6-bit, opcode field. An example is the instruction move with zero extension (MOVZX). The opcode field for this instruction is 00001111101101. Another extension is that a 3-bit segment register (SEG) field is required for some of the instructions that affect the FS and GS registers. Codes SEG equal to 100 and 101 stand for FS and GS, respectively.

▲ 4.3 ENCODING A COMPLETE PROGRAM IN MACHINE CODE

To encode a complete assembly language program in machine code, we must individually encode each of its instructions. This can be done by using the instruction formats shown in Fig. 4.5 and the information in the tables of Figs. 4.2, 4.3, 4.4, and 4.6. We first identify the general machine code format for the instruction in Fig. 4.5. After determining the format, the bit fields can be evaluated using the tables of Figs. 4.2, 4.3, 4.4, and 4.6. Finally, the binary-coded instruction can be expressed in hexadecimal form.

To execute a program on the PC/AT, we must first store the machine code of the program in the code segment of memory. The bytes of machine code are stored in sequentially addressed locations in memory. The first byte of the program is stored at the lowest address, and it is followed by the other bytes in the order in which they are encoded. That is, the address is incremented by 1 after storing each byte of machine code in memory.

EXAMPLE 4.7

Encode the block-move program in Fig. 4.7(a) and show how it would be stored in memory starting at address 200_{16}.

Solution

To encode this program into its equivalent machine code, we use the base instruction set table in Fig. 4.5. The first instruction,

$$\text{MOV} \quad \text{AX,1020H}$$

is a "move immediate data to register" instruction. In Fig. 4.5, we find it has the form

$$1011(W)(REG)(DATA)(DATA \text{ IF } W = 1)$$

Since the move is to register AX, Fig. 4.2 shows that the W bit is 1 and REG is 000. The immediate data 1020_{16} follows this byte, with the least significant byte coded first. This gives the machine code for the instruction as

$$10111000001000000010000_2 = B82010_{16}$$

The second instruction,

$$\text{MOV} \quad \text{DS,AX}$$

represents a "move register to segment register" operation. This instruction has the general format

$$10001110(MOD)0(SR)(R/M)$$

```
            MOV AX,2000H   ;LOAD AX REGISTER
            MOV DS,AX      ;LOAD DATA SEGMENT ADDRESS
            MOV SI,100H    ;LOAD SOURCE BLOCK POINTER
            MOV DI,120H    ;LOAD DESTINATION BLOCK POINTER
            MOV CX,10H     ;LOAD REPEAT COUNTER
    NXTPT:  MOV AH,[SI]    ;MOVE SOURCE BLOCK ELEMENT TO AH
            MOV [DI],AH    ;MOVE ELEMENT FROM AH TO DESTINATION BLOCK
            INC SI         ;INCREMENT SOURCE BLOCK POINTER
            INC DI         ;INCREMENT DESTINATION BLOCK POINTER
            DEC CX         ;DECREMENT REPEAT COUNTER
            JNZ NXTPT      ;JUMP TO NXTPT IF CX NOT EQUAL TO ZERO
            NOP            ;NO OPERATION
```

(a)

Instruction	Type of instruction	Machine code
MOV AX,2000H	Move immediate data to register	$101110000000000000100000_2 = B80020_{16}$
MOV DS,AX	Move register to segment register	$1000111011011000_2 = 8ED8_{16}$
MOV SI,100H	Move immediate data to register	$1011111000000000000000001_2 = BE0001_{16}$
MOV DI,120H	Move immediate data to register	$1011111100100000000000001_2 = BF2001_{16}$
MOV CX,10H	Move immediate data to register	$10111001000100000000000000_2 = B91000_{16}$
MOV AH,[SI]	Move memory data to register	$1000101000100100_2 = 8A24_{16}$
MOV [DI],AH	Move register data to memory	$1000100000100101_2 = 8825_{16}$
INC SI	Increment register	$01000110_2 = 46_{16}$
INC DI	Increment register	$01000111_2 = 47_{16}$
DEC CX	Decrement register	$01001001_2 = 49_{16}$
JNZ NXTPT	Jump on not equal to zero	$0111010111110111_2 = 75F7_{16}$
NOP	No operation	$10010000_2 = 90_{16}$

(b)

Figure 4.7 (a) Block-move program. (b) Machine coding of the block-move program. (c) Storing the machine code in memory.

From Fig. 4.3(a) and (b), we find that for this instruction MOD = 11 and R/M is 000 for AX. Furthermore, from Fig. 4.6, we find that SR = 11 for data segment. This results in the code

$$1000111011011000_2 = 8ED8_{16}$$

for the second instruction.

The next three instructions have the same format as the first instruction. In the third instruction, REG is 110 for SI and the data is 0100_{16}. This gives the instruction code as

$$1011111000000000000000001_2 = BE0001_{16}$$

Memory address	Contents	Instruction
200H	B8H	MOV AX,2000H
201H	00H	
202H	20H	
203H	8EH	MOV DS,AX
204H	D8H	
205H	BEH	MOV SI,100H
206H	00H	
207H	01H	
208H	BFH	MOV DI,120H
209H	20H	
20AH	01H	
20BH	B9H	MOV CX,10H
20CH	10H	
20DH	00H	
20EH	8AH	MOV AH,[SI]
20FH	24H	
210H	88H	MOV [DI],AH
211H	25H	
212H	46H	INC SI
213H	47H	INC DI
214H	49H	DEC CX
215H	75H	JNZ $-9
216H	F7H	
217H	90H	NOP

(c)

Figure 4.7 (Continued)

The fourth instruction has REG coded as 111 (DI) and the data as 0120_{16}. This results in the code

$$1011111100010000000000001_2 = BF2001_{16}$$

And in the fifth instruction REG is 001 for CX, with 0010_{16} as the data. This gives its code as

$$101110010001000000000000_2 = B91000_{16}$$

Instruction six is a move of byte data from memory to a register. From Fig. 4.5, we find that its general format is

$$100010(D)(W)(MOD)(REG)(R/M)$$

Since AH is the destination and the instruction operates on bytes of data, the D and W bits are 1 and 0, respectively, and the REG field is 100. The contents of SI are used as a pointer to the source operand; therefore, MOD is 00 and R/M is 100. This gives the instruction code as

$$1000101000100100_2 = 8A24_{16}$$

The last MOV instruction has the same form as the last one. However, in this case, AH is the destination and DI is the address pointer. This makes D equal to 0 and R/M equal to 101. Therefore, we get

$$1000100000100101_2 = 8825_{16}$$

The next two instructions increment registers and have the general form

$$01000(\text{REG})$$

For the first one, register SI is incremented. Therefore, REG equals 110. This results in the instruction code as

$$01000110_2 = 46_{16}$$

In the second, REG equals 111 to encode DI. This gives its code as

$$01000111_2 = 47_{16}$$

The two INC instructions are followed by a DEC instruction. Its general form is

$$01001(\text{REG})$$

To encode CX, REG equals 001, which results in the instruction code

$$01001001_2 = 49_{16}$$

The next instruction is a jump to the location NXTPT. Its form is

$$01110101(\text{IP-INC8})$$

We will not yet complete this instruction because it will be easier to determine the number of bytes to be jumped after the data have been coded for storage in memory. The final instruction is NOP, and it is coded as

$$10010000_2 = 90_{16}$$

The entire machine code program is shown in Fig. 4.7(b).

As shown in Fig. 4.7(c), our encoded program will be stored in memory starting from memory address 200H. The choice of program-beginning address establishes the address for the NXTPT label. Notice that the MOV AH,[SI] instruction, which has this label, starts at address $20E_{16}$. This is 9 bytes back from the value in IP after fetching the JNZ instruction. Therefore, the displacement (IP $-$ INC8) in the JNZ instruction is -9, which is $F7_{16}$ as an 8-bit hexadecimal number. Thus the instruction is encoded as

$$0111010111110111_2 = 75F7_{16}$$

▲ 4.4 THE PC/AT AND ITS DEBUG PROGRAM

Now that we know how to convert an assembly language program to machine code and how this machine code is stored in memory, we are ready to enter it into the PC/AT, execute it, examine the results that it produces, and, if necessary, debug any errors in its operation. It is the *DEBUG program*, which is part of the PC/AT's disk operating system (DOS), that permits us to initiate these types of operations from the keyboard of the microcomputer. In this section we will show how to load the DEBUG program from DOS, how to use it to examine or modify the contents of the MPU's internal registers, and how to return back to DOS from DEBUG.

Using DEBUG, the programmer can issue commands to the microcomputer in the PC/AT. Assume that the DOS has already been loaded and that a disk that contains the DEBUG program is in drive A, DEBUG is loaded by simply issuing the command

```
C:\DOS>A:DEBUG    (⏎)
```

Actually, debug can be typed in using either uppercase or lowercase characters. However, for simplicity, we will use all uppercase characters in this book.

EXAMPLE 4.8 ───

Assuming that the DOS has already been loaded and that the DEBUG program is in the DOS directory on drive C, initiate the DEBUG program from the keyboard of the PC/AT. What prompt for command entry is displayed when in the debugger?

Solution

When the operating system has been loaded, DEBUG is brought up by entering

```
C:\DOS>DEBUG    (⏎)
```

Drive C is accessed to load the DEBUG program; DEBUG is then executed and its prompt, which is -, is displayed. DEBUG is now waiting to accept a command. Figure 4.8 shows what is displayed on the screen.

───

The keyboard is the input unit of the debugger and permits the user to enter commands to load data, such as the machine code of a program; examine or modify the state of the MPU's internal registers; or execute a program. All we need to do is type in the command and then depress the enter (⏎) key. These debug commands are the tools a programmer needs to use to enter, execute, and debug programs.

When the command entry sequence is completed, the DEBUG program decodes the entry to determine which operation is to be performed, verifies that it is a valid command, and—if it is valid—passes control to a routine that performs the operation. At the completion of the operation, results are displayed on the screen and the DEBUG prompt (-) is redisplayed. The PC/AT remains in this state until a new entry is made from the keyboard.

```
C:\DOS>DEBUG
-
```
Figure 4.8 Loading the DEBUG program.

Command	Syntax	Function
Register	R [REGISTER NAME]	Examine or modify the contents of an internal register
Quit	Q	End use of the DEBUG program
Dump	D [ADDRESS]	Dump the contents of memory to the display
Enter	E ADDRESS [LIST]	Examine or modify the contents of memory
Fill	F STARTING ADDRESS ENDING ADDRESS LIST	Fill a block in memory with the data in list
Move	M STARTING ADDRESS ENDING ADDRESS DESTINATION ADDRESS	Move a block of data from a source location in memory to a destination location
Compare	C STARTING ADDRESS ENDING ADDRESS DESTINATION ADDRESS	Compare two blocks of data in memory and display the locations that contain different data
Search	S STARTING ADDRESS ENDING ADDRESS LIST	Search through a block of data in memory and display all locations that match the data in list
Input	I ADDRESS	Read the input port
Output	O ADDRESS, BYTE	Write the byte to the output port
Hex Add/Subtract	H NUM1,NUM2	Generate hexadecimal sum and difference of the two numbers
Unassemble	U [STARTING ADDRESS ENDING ADDRESS]	Unassemble the machine code into its equivalent assembler instructions
Name	N FILE NAME	Assign the filename to the data to be written to the disk
Write	W [STARTING ADDRESS [DRIVE STARTING SECTOR NUMBER OF SECTORS]]	Save the contents of memory in a file on a diskette
Load	L [STARTING ADDRESS [DRIVE STARTING SECTOR NUMBER OF SECTORS]]	Load memory with the contents of a file on a diskette
Assemble	A [STARTING ADDRESS]	Assemble the instruction into machine code and store in memory
Trace	T [=ADDRESS] [NUMBER]	Trace the execution of the specified number of instructions
Go	G [=STARTING ADDRESS [BREAKPOINT ADDRESS....]]	Execute the instructions down through the breakpoint address

Figure 4.9 DEBUG program command set.

Six kinds of information are typically entered as part of a command: *a command letter, an address, a register name, a file name, a drive name,* and *data.* The entire command set of DEBUG is shown in Fig. 4.9. This table gives the name for each command, its function, and its general syntax. By *syntax,* we mean the order in which key entries must be made to initiate the command.

With the loading of DEBUG, the state of the microprocessor is initialized. The *initial state* depends upon the DOS version and system configuration at the time the DEBUG command is issued. An example of the initial state is illustrated with the software model in Fig. 4.10. Notice that registers AX, BX, CX, DX, BP, SI, and DI are reset to zero; IP is initialized to 0100_{16}; CS, DS, SS, and ES are all loaded with 1342_{16}; and SP is loaded with $FFEE_{16}$. Finally, all the flags except IF are reset to zero. We can use the register command to verify this initial state.

Let us now look at the syntax for the *REGISTER* (R) *command.* This is the debugger command that allows us to examine or modify the contents of internal registers of the MPU. Notice that the syntax for this command is given in Fig. 4.9 as

```
R   [REGISTER NAME]
```

Here the command letter is R. It is followed by the register name. Figure 4.11 shows what must be entered as the register name for each of the 80386DX's registers. Note that access to the FS and GS segment registers is not supported in DEBUG.

An example of the command entry needed to examine or modify the value in register AX is

```
R   AX   (↵)
```

Notice that backets are not included around the register name. In Fig. 4.9, brackets are simply used to separate the various elements of the DEBUG commands. They are never entered as part of the command. Execution of this register command causes the current value in AX to be displayed as

```
AX   0000
 :-
```

Here we see that AX contains 0000_{16}. The examine register command is not yet complete. Note that a colon (:) followed by the cursor is displayed. We can now either depress (↵) to complete the command, leaving the register contents unchanged, or enter a new value for AX following the colon and then depress (↵). Let us load AX with a new value of $00FF_{16}$. This is done by the entry

```
:00FF   (↵)
 -
```

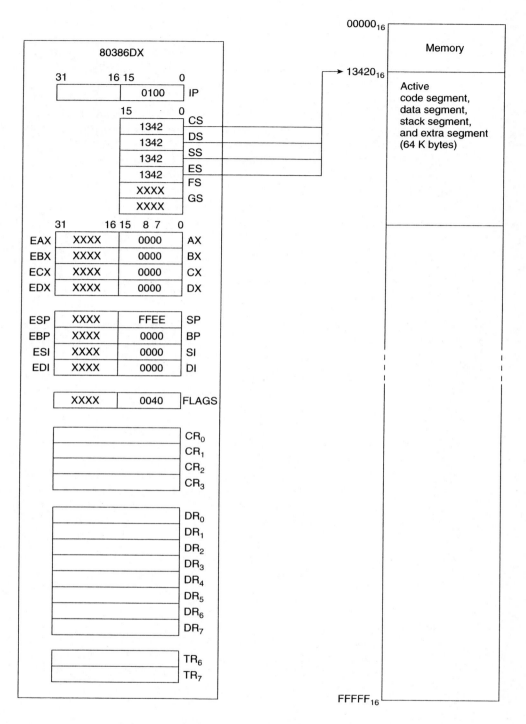

Figure 4.10 Software model of the 80386DX microprocessor.

Symbol	Register
AX	Accumulator register
BX	Base register
CX	Count register
DX	Data register
SI	Source index register
DI	Destination index register
SP	Stack pointer register
BP	Base pointer register
CS	Code segment register
DS	Data segment register
SS	Stack segment register
ES	Extra segment register
F	Flag register
IP	Instruction pointer

Figure 4.11 Register mnemonics for the R command.

EXAMPLE 4.9

Verify the initialized state of the 80386DX by examining the contents of its registers with the register command.

SOLUTION

If we enter the register command without a specific register name, the debugger causes the state of all registers and flags to be displayed. That is, if we enter

$$R \quad (↵)$$

the information displayed is that shown in Fig. 4.12. Looking at Fig. 4.12, we see that all registers were initialized as expected. To verify that all flags other than IF were reset, we can compare the flag settings that are listed to the right of the value for IP with the values in the table of Fig. 4.13. Note that all but IF correspond to the reset state. The last line displays the machine code and assembly language statement of the instruction pointed to by the current values in CS and IP (CS:IP).

EXAMPLE 4.10

Issue commands to the debugger on the PC/AT that will cause the value in BX to be modified to $FF00_{16}$ and then verify that this new value is loaded into BX.

```
-R
AX=0000  BX=0000  CX=0000  DX=0000  SP=FFEE  BP=0000  SI=0000  DI=0000
DS=1342  ES=1342  SS=1342  CS=1342  IP=0100   NV UP EI PL NZ NA PO NC
1342:0100 CD21            INT    21
-
```

Figure 4.12 Displaying the initialized state of the MPU.

Flag	Meaning	Set	Reset
OF	Overflow	OV	NV
DF	Direction	DN	UP
IF	Interrupt	EI	DI
SF	Sign	NG	PL
ZF	Zero	ZR	NZ
AF	Auxiliary carry	AC	NA
PF	Parity	PE	PO
CF	Carry	CY	NC

Figure 4.13 Notations used for displaying the status flags.

Solution

To modify the value in BX, all we need to do is issue the register command with BX and then respond to the :- by entering the value FF00$_{16}$. This is done with the command sequence

```
-R BX      (↵)
BX 0000
:FF00      (↵)
-
```

We can verify that FF00$_{16}$ has been loaded into BX by issuing another register command, as follows:

```
-R   BX    (↵)
BX   FF00
:-         (↵)
-
```

The displayed information for this command sequence is shown in Fig. 4.14.

The way in which the register command is used to modify flags is different than how it is used to modify the contents of a register. If we enter the command

```
R   F   (↵)
```

the flag settings are displayed as

```
NV  UP  EI  PL  NZ  NA  PO  NC-
```

```
-R BX
BX 0000
:FF00
-R BX
BX FF00
:-
-
```

Figure 4.14 Displayed information for Example 4.10.

```
-R F
NV UP EI PL NZ NA PO NC  -PE
-R F
NV UP EI PL NZ NA PE NC  -
-
```

Figure 4.15 Displayed information for Example 4.11.

To modify flags, just type in their new states (using the notations shown in Fig. 4.13) and depress the return key. For instance, to set the carry and zero flags, enter

```
NV   UP   EI   PL   NZ   NA   PO   NC   -CY   ZR   (↵)
```

Note that the new flag states can be entered in any order.

EXAMPLE 4.11

Use the register command to set the parity flag to even parity. Verify that the flag has been changed.

Solution

To set PF for even parity, issue the register command for the flag register and then enter PE as the new flag data. This is done with the command sequence

```
-R   F   (↵)
NV   UP   EI   PL   NZ   NA   PO   NC   -PE   (↵)
```

To verify that PF has been changed to its PE state, just enter another register command for the flag register as follows:

```
-R   F   (↵)
NV   UP   EI   PL   NZ   NA   PE   NC   -   (↵)
```

Notice that the state of the parity flag has changed from PO to PE. Figure 4.15 shows these commands and the displayed flag status that results.

The REGISTER command is very important for debugging programs. For instance, it can be used to check the contents of a register or flag prior to and again just after execution of an instruction. In this way, we can tell whether or not the instruction correctly performed the required operation.

If the command that was entered is identified as being invalid, an *error message* is displayed. Let us look at an example of an invalid command entry. To do this, we repeat our earlier example in which AX was loaded with $00FF_{16}$, but in keying in $00FF_{16}$, we enter the uppercase letter O instead of zeros. The result produced by issuing this command is shown in Fig. 4.16. Here we see that a warning, Error, is displayed, and the symbol $^\wedge$ is

```
-R AX
AX 0000
:OOFF
 ^ Error
-
```

Figure 4.16 Invalid entry.

used to mark the starting location of the error in the command. To correct this error, we simply reenter the command.

We will examine one more command before going on. We now know how to invoke the DEBUG program from the DOS prompt, but we must also be able to return to the DOS once in DEBUG. The debugger contains a command called *QUIT* (Q) to do this. Therefore, to return to the DOS, we simply respond to the debug prompt with

Q   (↵)

▲ 4.5 EXAMINING AND MODIFYING THE CONTENTS OF MEMORY

In Section 4.4 we studied the command that permitted us to examine or modify the contents of the MPU's internal registers. Here we will continue our study of DEBUG's command set with those commands that are used to examine and modify the contents of memory. The ability to do this is essential for debugging programs. For instance, the contents at a memory address can be examined just before and just after execution of an instruction. In this way, we can verify that the instruction performs the operation correctly. Another use of this type of command is to load a program into the code segment of the microcomputer's memory. The complete command set of DEBUG was shown in Fig. 4.9. Six of these commands, DUMP, ENTER, FILL, MOVE, COMPARE, and SEARCH, are provided for use in examining or modifying the contents of storage locations in memory. Let us now look at the operations performed with each of these commands

DUMP Command

The *DUMP* (D) *command* allows us to examine the contents of a memory location or a block of consecutive memory locations. Looking at Fig. 4.9, we see that the general syntax for DUMP is

D   [ADDRESS]

If a segment register is not specified, the value of ADDRESS entered is automatically referenced to the current value in the data segment (DS) register.

DUMP can also be issued without ADDRESS. This gives the command

D   (↵)

Execution of this form of the command causes the 128 consecutive bytes starting at offset 0100_{16} from the current value in DS to be displayed. If DS is initialized with 1342_{16} when DEBUG is started, issuing this command gives the memory dump shown in Fig. 4.17.

```
-D
1342:0100  75 07 05 00 00 00 B2 14-4C 00 2B 04 27 0C 06 74   u.......L.+.'..t
1342:0110  46 74 05 00 00 00 31 13-B2 14 99 00 31 13 31 13   Ft....1.....1.1.
1342:0120  8A 40 8B 1E F2 39 9A 88-97 63 17 8B 1E F2 39 BA   .@...9...c....9.
1342:0130  61 46 E8 37 0B A0 5A 46-0A C0 74 2D 8B 1E AB 42   aF.7..ZF..t-...B
1342:0140  FE C8 75 0B 89 1E 5C 46-C6 06 5A 46 02 EB 1A 3B   ..u...\F..ZF...;
1342:0150  1E 5C 46 74 14 C6 06 5A-46 00 E8 17 01 75 03 E9   .\Ft...ZF....u..
1342:0160  FC 00 3C 59 75 03 E9 27-01 E8 97 01 8A C5 74 22   ..<Yu..'......t"
1342:0170  80 3E CE 09 00 75 10 80-F9 0D 75 0B E8 55 00 74   .>...u....u..U.t
-
```

Figure 4.17   Examining the contents of 128 consecutive bytes in memory.

Notice that sixteen bytes of data are displayed per line, and only the address of the first byte is shown at the left. From Fig. 4.17 we see that the address of the first location in the first line is denoted as 1342:0100. This corresponds to the physical address

$$13420_{16} + 0100_{16} = 13520_{16}$$

The second byte of data displayed in the first line corresponds to the memory address 1342:0101, or 13521_{16}, and the last byte on this line corresponds to the memory address 1342:010F, or $1352F_{16}$. Note that the values of the eighth and ninth bytes are separated by a hyphen.

For all memory dumps, an ASCII version of the memory data is also displayed. It is displayed to the right of the hexadecimal data in Fig. 4.17. All bytes that result in an unprintable ASCII character are displayed as the . symbol.

The results shown in Fig. 4.17 could be obtained with several other forms of the DUMP command. One way is to enter the current value of DS, which is 1342_{16}, and an offset of 0100_{16} in the address field. This results in the command

```
D   1342:100   (↵)
```

Another way is to enter DS instead of its value with the offset. This gives

```
D   DS:100    (↵)
```

In fact, the same results can be obtained by just issuing the command

```
D   100        (↵)
```

EXAMPLE 4.12

What is the physical address range of the bytes of data in the last line of data shown in Fig. 4.17?

Solution

In Fig. 4.17 we see that the first byte is at address 1342:0170. This is the physical address

$$13420_{16} + 0170_{16} = 13590_{16}$$

The last byte is at address 1342:017F, and its physical address is

$$13420_{16} + 017F_{16} = 1359F_{16}$$

EXAMPLE 4.13

What happens if we repeat the entry D (↵) after obtaining the memory dump shown in Fig. 4.17?

```
-D
1342:0180  0B 72 0C 3C 0A 75 08 E8-4A 00 75 03 E9 D9 00 E9   .r.<.u..J.u....
1342:0190  BF FE E8 23 40 75 02 EB-82 80 3E CE 09 00 75 15   ...#@u...>...u.
1342:01A0  80 F9 0D 75 10 E8 2C 00-74 E2 72 18 3C 0A 75 14   ...u..,.t.r.<.u.
1342:01B0  8A C5 E8 03 40 9A 49 C5-63 17 9A B3 B8 63 17 E8   ....@.I.c....c..
1342:01C0  9F 0A EB C3 9A 49 C5 63-17 9A B3 B8 63 17 E8 90   .....I.c....c...
1342:01D0  0A E9 7D FE E8 D1 01 74-66 89 2E D7 09 8A E8 24   ..}....tf......$
1342:01E0  7F 8A C8 3C 1B 74 07 3C-1D 75 3B EB 59 90 E8 B7   ...<.t.<.u;.Y...
1342:01F0  01 74 31 52 8A F0 E8 AF-01 74 38 3C 1C 75 0E C6   .t1R.....t8<.u..
-
```

Figure 4.18 Displayed information for repeat of 128-byte memory-dump command.

Solution

The contents of the next 128 consecutive bytes of memory are dumped to the display. The displayed information is shown in Fig. 4.18.

Frequently, we do not want to examine such a large block of memory. Instead, we may want to look at just a few bytes or a specific-sized block. The dump command can also do this. This time we enter two addresses. The first address defines the starting point of the block and the second address identifies the end of the block. For instance, if we want to examine the two bytes of data that are at offsets equal to 200_{16} and 201_{16} in the current data segment, the command is

$$D \quad DS:200 \quad 201 \quad (\lrcorner)$$

The result obtained by executing this command is given in Fig. 4.19.

EXAMPLE 4.14

Issue a dump command that will display the contents of the 32 bytes of memory that are located at offsets 0300_{16} through $031F_{16}$ in the current data segment.

Solution

The command needed to display the contents of this part of memory is

$$D \quad 300 \quad 31F \quad (\lrcorner)$$

and the information that is displayed is shown in Fig. 4.20.

Up to now, all the data displayed with the DUMP command were contained in the data segment of memory. It is also possible to examine data that are stored in the code segment, stack segment, or extra segment. To do this, we simply use the appropriate segment

```
-D DS:200 201
 1342:0200  06 CE
```

Figure 4.19 Displaying just 2 bytes of data.

```
-D 300 31F
1342:0300   59 5F C3 E8 D0 3E 74 0A-E8 BF 3E E8 5C 3E 75 F3   Y_...>t...>.\>u.
1342:0310   0C 01 C3 80 3E 25 39 00-74 4A 53 52 FF 36 AB 42   ....>%9.tJSR.6.B
-
```

Figure 4.20 Displayed information for Example 4.14.

register name in the command. For instance, the command needed to dump the values in the first 16 bytes of the current code segment is

$$D \quad CS:0 \quad F \quad (\lrcorner)$$

EXAMPLE 4.15 _____

Use the DUMP command to examine the 16 bytes of memory just below the top of the stack.

Solution

The top of the stack is defined by the contents of the SS and SP registers (SS:SP). Earlier we found that SP was initialized to $FFEE_{16}$ when debug was loaded. Therefore, the 16 bytes we are interested in reside at offsets $FFEE_{16}$ through $FFFD_{16}$ from the current value in SS. This part of the stack is viewed with the command

$$D \quad SS:FFEE \quad FFFD \quad (\lrcorner)$$

The result displayed by executing this command is shown in Fig. 4.21.

ENTER Command

The DUMP command allowed us to examine the contents of memory, but we also need to be able to modify or enter the information in memory—for instance, to load a machine code program. It is for this purpose that the *ENTER* (E) *command* is provided in the DEBUG program.

In Fig. 4.9 we find that the syntax of the ENTER command is

$$E \quad [ADDRESS] \quad [LIST]$$

The address part of the E command is entered the same way we just described for the DUMP command. If no segment name is included with the offset, the DS register is assumed. The list that follows the address is the data that gets loaded into memory.

```
-D SS:FFEE FFFD
1342:FFE0                                                    00 00              ..
1342:FFF0   00 00 00 00 00 00 00 00-00 00 00 00 00 00   ..............
-
```

Figure 4.21 Displayed information for Example 4.15.

Sec. 4.5   Examining and Modifying the Contents of Memory   **115**

```
-E DS:100 FF FF FF FF FF
-D DS:100 104
1342:0100  FF FF FF FF FF          . . . . .
-
```

Figure 4.22 Modifying 5 consecutive bytes of memory and verifying the change of data.

As an example, let us write a command that will load five consecutive byte-wide memory locations that start at address DS:100 with the value FF_{16}. This is done with the command

```
E  DS:100  FF  FF  FF  FF  FF  (↵)
```

To verify that the new values of data have been stored in memory, let us dump the contents of these locations to the display. To do this, we issue the command

```
D  DS:100  104  (↵)
```

These commands and the displayed results are shown in Fig. 4.22. Notice that the byte storage locations from addresses DS:100 through DS:104 now all contain the value FF_{16}.

The ENTER command can also be used in a way in which it either examines or modifies the contents of memory. If we issue the command with an address but no data, what happens is that the contents of the addressed storage location are displayed. For instance, the command

```
E  DS:100  (↵)
```

causes the value at this address to be displayed as follows:

```
1342:0100  FF._
```

Notice that the value at address 1342:0100 is FF_{16}.

At this point we have several options; for one, the return key can be depressed. This terminates the ENTER command without changing the contents of the displayed memory location and causes the debug prompt to be displayed. Rather than depressing return, we can depress the space bar. Again, the contents of the displayed memory location remain unchanged, but this time the command is not terminated. Instead, it causes the contents of the next consecutive memory address to be displayed. Let us assume that this was done. Then the display would read

```
1342:0100  FF.  FF._
```

Here we see that the data stored at address 1342:0101 are also FF_{16}. A third type of entry that could be made is to enter a new value of data and then depress the space bar or return key. For example, we can enter 11_{16} and then depress the space bar. This gives the display

```
1342:0100  FF.  FF.11  FF._
```

```
-E DS:100
1342:0100   FF._
1342:00FF   FF._
-
```

Figure 4.23 Using the - key to examine the contents of the previous memory location.

The value pointed to by address 1342:101 has been changed to 11_{16}, and the contents of address 1342:0102, which are FF_{16}, are displayed. Now, depress the return key to finalize the data entry sequence.

EXAMPLE 4.16

Start a data entry sequence by examining the contents of address DS:100 and then, without entering new data, depress the - key. What happens?

Solution

The data-entry sequence is initiated as

```
E  DS:100    (↵)
1342:0100   FF._
```

Entering - causes the following address and data to be displayed:

```
1342:00FF   FF._
```

Notice that these are the address and contents of the storage location at the address equal to 1 less than DS:100—that is, the previous byte storage location. This result is shown in Fig. 4.23.

The ENTER command can also be used to enter ASCII data. This is done by simply enclosing the data entered in quotation marks. An example is the command

```
E  DS:200   ''ASCII''   (↵)
```

This command causes the ASCII data for letters A, S, C, I, and I to be stored in memory at addresses DS:200, DS:201, DS:202, DS:203, and DS:204, respectively. This character data entry can be verified with the command

```
D  DS:200   204   (↵)
```

Looking at the ASCII field of the data dump shown in Fig. 4.24, we see that the correct ASCII data were stored into memory. Actually, either single or double quote marks can be used. Therefore, the entry could also have been made as

```
E  DS:200   'ASCII'   (↵)
```

```
-E DS:200 "ASCII"
-D DS:200 204
1342:0200   41 53 43 49 49                                      ASCII
-
```

Figure 4.24 Loading ASCII data into memory with the ENTER command.

```
-F 100 11F 22
-D 100 11F
1342:0100  22 22 22 22 22 22 22 22-22 22 22 22 22 22 22 22   """"""""""""""""""
1342:0110  22 22 22 22 22 22 22 22-22 22 22 22 22 22 22 22   """"""""""""""""""
-
```

Figure 4.25 Initializing a block of memory with the FILL command.

FILL Command

Frequently, we want to fill a block of consecutive memory locations all with the same data. For example, we may need to initialize storage locations in an area of memory with zeros. To do this by entering the data address by address with the ENTER command would be very time consuming. It is for this type of operation that the *FILL* (F) *command* is provided.

From Fig. 4.9, we see that the general form of the FILL command is

```
F  [STARTING ADDRESS]  [ENDING ADDRESS]  [LIST]
```

Here STARTING ADDRESS and ENDING ADDRESS specify the block of storage locations in memory. They are followed by a LIST of data. An example is the command

```
F  100  11F  22  (↵)
```

Execution of this command causes the 32 byte locations in the range 1342:100 through 1342:11F to be loaded with 22_{16}. The fact that this change in memory contents has happened can be verified with the command

```
D  100  11F  (↵)
```

Figure 4.25 shows the result of executing these two commands.

EXAMPLE 4.17 ───────────────────────────────

Initialize all storage locations in the block of memory from DS:120 through DS:13F with the value 33_{16} and the block of storage locations from DS:140 through DS:15F with the value 44_{16}. Verify that the contents of these ranges of memory are correctly modified.

Solution

The initialization operations can be done with the FILL commands

```
F  120  13F  33  (↵)
F  140  15F  44  (↵)
```

They are then verified with the DUMP command

```
D  120  15F  (↵)
```

The information displayed by the command sequence is shown in Fig. 4.26.

```
-F 120 13F 33
-F 140 15F 44
-D 120 15F
1342:0120  33 33 33 33 33 33 33 33-33 33 33 33 33 33 33 33   3333333333333333
1342:0130  33 33 33 33 33 33 33 33-33 33 33 33 33 33 33 33   3333333333333333
1342:0140  44 44 44 44 44 44 44 44-44 44 44 44 44 44 44 44   DDDDDDDDDDDDDDDD
1342:0150  44 44 44 44 44 44 44 44-44 44 44 44 44 44 44 44   DDDDDDDDDDDDDDDD
-
```

Figure 4.26 Displayed information for Example 4.17.

MOVE Command

The *MOVE* (M) *command* allows us to copy a block of data from one part of memory to another part. For instance, using this command, a 32-byte block of data that resides in memory from addresses DS:100 to DS:11F can be copied to the address range DS:200 through DS:21F with a single operation.

The general form of the MOVE command is given in Fig. 4.9 as

```
M  [STARTING ADDRESS]  [ENDING ADDRESS]  [DESTINATION ADDRESS]
```

Notice that it is initiated by depressing the M key. After that, we must enter three addresses. The first two addresses are the *starting address* and *ending address* of the source block of data, that is, the block of data that is to be copied. The third address is the *destination starting address*, that is, the starting address of the section of memory to which the block of data is to be copied.

The command for our example, which copies a 32-byte block of data located at addresses DS:100 through DS:11F to the block of memory starting at address DS:200, is

```
M  100  11F  200  (↵)
```

EXAMPLE 4.18

Fill each storage location in the block of memory from address DS:100 through DS:11F with the value 11_{16}. Then copy this block of data to a destination block starting at DS:160. Verify that the block move is correctly done.

Solution

First, we fill the source block with 11_{16} using the command

```
F  100  11F  11  (↵)
```

Next it is copied to the destination with the command

```
M  100  11F  160  (↵)
```

Finally, we dump the complete range from DS:100 to DS:17F by issuing the command

```
D  100  17F  (↵)
```

The result of this memory dump is given in Fig. 4.27. It verifies that the block move was successfully performed.

```
-F 100 11F 11
-M 100 11F 160
-D 100 17F
1342:0100  11 11 11 11 11 11 11 11-11 11 11 11 11 11 11 11   ................
1342:0110  11 11 11 11 11 11 11 11-11 11 11 11 11 11 11 11   ................
1342:0120  33 33 33 33 33 33 33 33-33 33 33 33 33 33 33 33   3333333333333333
1342:0130  33 33 33 33 33 33 33 33-33 33 33 33 33 33 33 33   3333333333333333
1342:0140  44 44 44 44 44 44 44 44-44 44 44 44 44 44 44 44   DDDDDDDDDDDDDDDD
1342:0150  44 44 44 44 44 44 44 44-44 44 44 44 44 44 44 44   DDDDDDDDDDDDDDDD
1342:0160  11 11 11 11 11 11 11 11-11 11 11 11 11 11 11 11   ................
1342:0170  11 11 11 11 11 11 11 11-11 11 11 11 11 11 11 11   ................
-
```

Figure 4.27 Displayed information for Example 4.18.

COMPARE Command

Another type of memory operation we sometimes need to perform is to compare the
contents of two blocks of data to determine if they are or are not the same. This operation
can be done with the *COMPARE* (C) *command* of the DEBUG program. Figure 4.9 shows
that the general form of this command is

    C [STARTING ADDRESS] [ENDING ADDRESS] [DESTINATION ADDRESS]

For example, to compare a block of data located from addresses DS:100 through DS:11F
to an equal-size block of data starting at address DS:160, we issue the command

    C 100 10F 160 (↵)

This command causes the contents of corresponding address locations in each block to be
compared to each other. That is, the contents of address DS:100 are compared to those at
address DS:160, the contents at address DS:101 are compared to those at address DS:161,
and so on. Each time unequal elements are found, the address and contents of that byte
in both blocks are displayed.

Since both of these blocks contain the same information, no data are displayed.
However, if this source block is next compared to the destination block starting at address
DS:120 by entering the command

    C 100 10F 120 (↵)

all elements in both blocks are unequal; therefore, the information shown in Fig. 4.28
is displayed.

```
-C 100 10F 120
1342:0100  11  33  1342:0120
1342:0101  11  33  1342:0121
1342:0102  11  33  1342:0122
1342:0103  11  33  1342:0123
1342:0104  11  33  1342:0124
1342:0105  11  33  1342:0125
1342:0106  11  33  1342:0126
1342:0107  11  33  1342:0127
1342:0108  11  33  1342:0128
1342:0109  11  33  1342:0129
1342:010A  11  33  1342:012A
1342:010B  11  33  1342:012B
1342:010C  11  33  1342:012C
1342:010D  11  33  1342:012D
1342:010E  11  33  1342:012E
1342:010F  11  33  1342:012F
-
```

Figure 4.28 Results produced
when unequal data are found with
a COMPARE command.

SEARCH Command

The *SEARCH* (S) *command* can be used to scan through a block of data in memory to determine whether or not it contains specific data. The general form of this command as given in Fig. 4.9 is

```
S   [STARTING ADDRESS]   [ENDING ADDRESS]   [LIST]
```

When the command is issued, the contents of each storage location in the block of memory between the starting address and the ending address are compared to the data in LIST. The address is displayed for each memory location where a match is found.

EXAMPLE 4.19 ───────────────────────────────

Perform a search of the block of data from addresses DS:100 through DS:17F to determine which memory locations contain 33_{16}.

Solution

The search command that must be issued is

```
S   100   17F   33   (↵)
```

Figure 4.29 shows that all addresses in the range 1342:120 through 1342:13F contain this value of data.

───

```
-S 100 17F 33
1342:0120
1342:0121
1342:0122
1342:0123
1342:0124
1342:0125
1342:0126
1342:0127
1342:0128
1342:0129
1342:012A
1342:012B
1342:012C
1342:012D
1342:012E
1342:012F
1342:0130
1342:0131
1342:0132
1342:0133
1342:0134
1342:0135
1342:0136
1342:0137
1342:0138
1342:0139
1342:013A
1342:013B
1342:013C
1342:013D
1342:013E
1342:013F
-
```

Figure 4.29 Displayed information for Example 4.19.

▲ 4.6 INPUT AND OUTPUT OF DATA

The commands studied in the last section allowed examination or modification of information in the memory of the microcomputer, but not in its input/output address space. To access data at I/O ports, we use the *input* (I) and *output* (O) *commands*. These commands can be used to input or output data for any of the 64K byte-wide ports in the 80386DX's I/O address space. Let us now look at how these two commands are used to read data at an input port or write data to an output port.

The general format of the input command as shown in Fig. 4.9 is

I   [ADDRESS]

Here ADDRESS identifies the byte wide I/O port that is to be accessed. When the command is executed, the data are read from the port and displayed. For instance, if the command

I   61   (↵)

is issued and if the result displayed on the screen is

4D

the contents of the port at I/O address 0061_{16} are $4D_{16}$.

EXAMPLE 4.20

Write a command that will display the byte contents of the input port at I/O address $00FE_{16}$.

Solution

To input the contents of the byte-wide port at address FE_{16}, the command is

I   FE   (↵)

Figure 4.9 gives the general format of the output command as

O   [ADDRESS]   [BYTE]

Here we see that both the address of the output port and the byte of data that is to be written to the port must be specified. An example of the command is

O   61   4F   (↵)

This command causes the value $4F_{16}$ to be written into the byte-wide output port at address 0061_{16}.

The DEBUG program also provides the ability to add and subtract hexadecimal numbers. Both operations are performed with a single command known as the *hexadecimal* (H) *command*. In Fig. 4.9, we see that the general format of the H command is

H   [NUM1]   [NUM2]

When executed, both the sum and difference of NUM1 and NUM2 are formed. These results are displayed as follows

[NUM1+NUM2]   [NUM1-NUM2]

Both numbers and the result are limited to four digits.

This hexadecimal arithmetic capability is useful when debugging programs. One example of a use of the H command is for the calculation of the physical address of an instruction or data in memory. For instance, if the current value in the code segment register is $0ABC_{16}$ and that in the instruction pointer is $0FFF_{16}$, the physical address is found with the command

H   ABC0   0FFF   (⏎)
BBBF   9BC1

Notice that the sum of these two hexadecimal numbers is $BBBF_{16}$, and their difference is $9BC1_{16}$. The sum $BBBF_{16}$ is the value of the physical address CS:IP.

The subtraction operation performed with the H command is also valuable in address calculations. For instance, a frequently used software operation is to jump a number of bytes of instruction code backward in the code segment of memory. In this case, the physical address of the new location can be found by subtraction. Let us start with the physical address just found, $BBBF_{16}$, and assume that we want to jump to a new location 10_{10} bytes back in memory. First, 10_{10} is expressed in hexadecimal form as A_{16}. Then the new address is calculated as

H   BBBF   A   (⏎)
BBC9   BBB5

Therefore, the new physical address is $BBB5_{16}$. Because the hexadecimal numbers are limited to four digits, physical address calculations with the H command are limited to the address range 00000_{16} through $0FFFF_{16}$.

EXAMPLE 4.21

Use the H command to find the negative of the number 0009_{16}.

Solution

The negative of a hexadecimal number can be found by subtracting it from 0. Therefore, the difference produced by the command

H   0   9   (⏎)
0009   FFF7

is $FFF7_{16}$, and is the negative of 9_{16} expressed in 2's-complement form.

EXAMPLE 4.22

If a byte of data is located at physical address $02A34_{16}$ and the data segment register contains 0150_{16}, what value must be loaded into the source index register such that DS:SI points to the byte storage location?

Solution

The offset required in SI can be found by subtracting the data segment base address from the physical address. Using the H command, we get

```
H  2A34  1500  (↵)
3F34  1534
```

This shows that SI must be loaded with the value 1534_{16}.

▲ 4.8 LOADING, VERIFYING, AND SAVING MACHINE LANGUAGE PROGRAMS

Up to this point we have learned how to use the register, memory, and I/O commands of DEBUG to examine or modify the contents of the MPU's internal registers, data stored in memory, or information at an input or output port. Let us now look at how we can load machine code instructions and programs into the memory of the PC/AT.

In Section 4.5 we found that the ENTER command can be used to load either a single or a group of memory locations with data, such as the machine code for instructions. As an example, let us load the machine code $88C3_{16}$ for the instruction MOV BL,AL. This instruction is loaded into memory starting at address CS:100 with the ENTER command

```
E  CS:100  88  C3  (↵)
```

We can verify that it has been loaded correctly with the DUMP command

```
D  CS:100  101  (↵)
```

This command displays the data

```
1342:0100  88  C3
```

Let us now introduce another command that is important for debugging programs on the PC/AT. It is the *UNASSEMBLE* (U) *command*. By *unassemble* we mean the process of converting machine code instructions to their equivalent assembly language source statements. The U command lets us specify a range in memory, and execution of the command causes the source statements for the memory data in this range to be displayed on the screen. Looking at Fig. 4.9, we find that the syntax of the UNASSEMBLE command is

```
U  [STARTING ADDRESS]  [ENDING ADDRESS]
```

```
-E CS:100 88 C3
-D CS:100 101
1342:0100  88 C3
-U CS:100 101
1342:0100 88C3      MOV     BL,AL
-
```

Figure 4.30 Loading, verifying, and disassembly of an instruction.

We can use this command to verify that the machine code entered for an instruction is correct. To do this for our earlier example, we use the command

$$U \quad CS:100 \quad 101 \quad (\lrcorner)$$

This command results in display of the starting address for the instruction followed by both the machine code and assembly forms of the instruction. This gives

$$1342:0100 \quad 88C3 \quad MOV \quad BL,AL$$

The entry sequence and displayed information for loading, verification, and unassembly of the instruction are shown in Fig. 4.30.

EXAMPLE 4.23

Use a sequence of commands to load, verify loading, and unassemble the machine code instruction 0304H. Load the instruction at address CS:200.

Solution

The machine code instruction is loaded into the code segment of the microcomputer's memory with the command

$$E \quad CS:200 \quad 03 \quad 04 \quad (\lrcorner)$$

Next, we can verify that it was loaded correctly with the command

$$D \quad CS:200 \quad 201 \quad (\lrcorner)$$

and, finally, unassemble the instruction with

$$U \quad CS:200 \quad 201 \quad (\lrcorner)$$

The results produced by this sequence of commands are shown in Fig. 4.31. Here we see that the instruction entered is

$$ADD \quad AX,[SI]$$

```
-E CS:200 03 04
-D CS:200 201
1342:0200  03 04
-U CS:200 201
1342:0200 0304      ADD     AX,[SI]
-
```

Figure 4.31 Displayed information for Example 4.23.

Before going further we will cover two more commands that are useful for loading and saving programs. They are the *WRITE* (W) *command* and *LOAD* (L) *command*. These commands give the ability to save data stored in memory on a diskette and to reload memory from a diskette, respectively. We can load the machine code of a program into memory with the E command the first time we use it and then save it on a diskette. In this way, the next time the program is needed it can be simply reloaded from the diskette.

Figure 4.9 shows that the general forms of the W and L commands are

```
W   [STARTING ADDRESS]   [DRIVE]   [STARTING SECTOR]   [NUMBER OF SECTORS]
L   [STARTING ADDRESS]   [DRIVE]   [STARTING SECTOR]   [NUMBER OF SECTORS]
```

For instance, to save the ADD instruction we just loaded at address CS:200 in Example 4.23, we can issue the write command

```
W  CS:200  1  10  1  (↵)
```

Notice that we have selected 1 (drive B) for the disk drive specification, 10 as an arbitrary starting sector on the diskette, and an arbitrary length of 1 sector. Before the command is issued, a formatted data diskette must be inserted into drive B. Then issuing the command causes one sector of data starting at address CS:200 to be read from memory and written into sector 10 on the diskette in drive B. Unlike the earlier commands we have studied, the W command automatically references the CS register instead of the DS register. For this reason, the command

```
W  200  1  10  1  (↵)
```

will perform the same operation.

Let us digress for a moment to examine the file specification of the W command in more detail. The diskettes for PC/AT that have double-sided, double-density drives are organized into 10,001 sectors that are assigned sector numbers over the range 0_{16} through $27F_{16}$. Each sector is capable of storing 512 bytes of data. With the file specification in a W command, we can select any one of these sectors as the starting sector. The value of the number of sectors should be specified based on the number of bytes of data that are to be saved. The specification that we made earlier for our example of a write command selected one sector (sector number 10_{16}) and for this reason could only save up to 512 bytes of data. The maximum value of sectors that can be specified with a write command is 80_{16}.

The LOAD command can be used to reload a file of data stored on a diskette anywhere in memory. As an example, let us load the instruction that we just saved on a diskette with a W command at a new address (CS:300). This is done with the L command

```
L  300  1  10  1  (↵)
```

The reloading of the instruction can be verified by issuing the U command

```
U  CS:300  301  (↵)
```

This command causes the display

```
1342:300  301  ADD AX,[SI]
```

EXAMPLE 4.24

Show the sequence of keyboard entries needed to enter the machine code program of Fig. 4.32 into memory of the PC/AT. The program is to be loaded into memory starting at address CS:100. Verify that the hexadecimal machine code was entered correctly and then unassemble the machine code to assure that it represents the source program. Save the program in sector 100 of a formatted data diskette.

Solution

We will use the ENTER command to load the program.

```
E CS:100 B8 00 20 8E D8 BE 0 01 BF 20 01 B9 10 0 8A 24 88 25 46
47 49 75 F7 90  (↵)
```

First, we verify that the machine code has been loaded correctly with the command

```
D  CS:100  117  (↵)
```

Comparing the displayed source data in Fig. 4.33 to the machine code in Fig. 4.32, we see that it has been loaded correctly. Now the machine code can be unassembled by the command

```
U  CS:100  117  (↵)
```

Machine code	Instruction
B8H	MOV AX,2000H
00H	
20H	
8EH	MOV DS,AX
D8H	
BEH	MOV SI,100H
00H	
01H	
BFH	MOV DI,120H
20H	
01H	
B9H	MOV CX,10H
10H	
00H	
8AH	MOV AH,[SI]
24H	
88H	MOV [DI],AH
25H	
46H	INC SI
47H	INC DI
49H	DEC CX
75H	JNZ $-9
F7H	
90H	NOP

Figure 4.32 Machine code and assembly language instructions of a block-move program.

```
-E CS:100 B8 00 20 8E D8 BE 0 01 BF 20 01 B9 10 0 8A 24 88 25 46 47 49 75 F7 90
-D CS:100 117
1342:0100  B8 00 20 8E D8 BE 00 01-BF 20 01 B9 10 00 8A 24    . ........ .....$
1342:0110  88 25 46 47 49 75 F7 90                            .%FGIu..
-U CS:100 117
1342:0100 B80020       MOV     AX,2000
1342:0103 8ED8         MOV     DS,AX
1342:0105 BE0001       MOV     SI,0100
1342:0108 BF2001       MOV     DI,0120
1342:010B B91000       MOV     CX,0010
1342:010E 8A24         MOV     AH,[SI]
1342:0110 8825         MOV     [DI],AH
1342:0112 46           INC     SI
1342:0113 47           INC     DI
1342:0114 49           DEC     CX
1342:0115 75F7         JNZ     010E
1342:0117 90           NOP
-W CS:100 1 100 1
```

Figure 4.33 Displayed information for Example 4.24.

Comparing the displayed source program of Fig. 4.33 to that in Fig. 4.32, it again verifies correct entry. Finally, the program is saved on the data diskette with the command

$$W \quad CS:100 \quad 1 \quad 100 \quad 1 \quad (\lrcorner)$$

At this time, it is important to mention that using the W command to save a program can be quite risky. For instance, if the command is written with the wrong disk specifications, by mistake, some other program or data on the diskette may be written over. Moreover, the diskette should not contain files that were created in any other way. This is because the locations of these files will not be known and may accidentally be written over by the selected file specification. Overwriting a file like this will ruin its contents. Even more important is to never issue the command to the hard disk (c:). This action could destroy the installation of the disk operating system.

Another method of saving and loading programs is available in DEBUG, and this alternative approach eliminates the overwrite problem. We will now look at how a program can be saved using a file name instead of with a file specification.

By using the *name* (N) *command* along with the write command, a program can be saved on the diskette under a file name. In Fig. 4.9 we see that the N command is specified in general as

$$N \quad [FILE \ NAME]$$

FILE NAME has the form

$$NAME.EXT$$

Here the name of the file can be up to five characters but must start with a letter. On the other hand, the extension (EXT) is from 0 to 3 characters. Neither EXE nor COM is a valid extension. Some examples of valid file names are BLOCK, TEMP.1, BLOCK1.ASM, and BLK_1.R1.

As part of the process of using the file name command, the BX and CX registers must be updated to identify the size of the program that is to be saved in the file. The size of the program in bytes is given as

$$BX \ CX = number \ of \ bytes$$

Together CX and BX specify an 8-digit hexadecimal number that identifies the number of bytes in the file. In general, the programs we will work with are small. For this reason, the upper four digits will always be zero. That is, the contents of BX will be 0000_{16}. Just using CX permits a file to be up to 64K bytes long.

After the name command has been issued and the CX and BX registers have been initialized, the write command form

W   [STARTING ADDRESS]

is used to save the program on the diskette. To reload the program into memory, we begin by naming the file and then simply issuing a load command with the address at which it is to start. This gives the sequence

N   [FILE NAME]
L   [STARTING ADDRESS]

As an example let us look at how the name command is set up to save the machine program used in Example 4.24 in a file called BLK.1 on a diskette in drive A. First, the name command

N   A:BLK.1   (↵)

is entered. Looking at Fig. 4.33, we see that the program is stored in memory from addresses CS:100 to CS:117. This gives a size of 18_{16} bytes. Therefore, CX and BX are initialized as follows:

R   CX   (↵)
CX   XXXX
:18   (↵)
R   BX   (↵)
BX   XXXX
:0   (↵)

Now the program is saved on the diskette with the command

W   CS:100   (↵)

To reload the program into memory, we simply perform the command sequence

N   A:BLK.1   (↵)
L   CS:100   (↵)

In fact, the program can be loaded starting at another address by just specifying that address in the load command.

Once saved on a diskette, the file can be changed to an executable file (that is, a file with the extension .EXE) by using the DOS RENAME operation. To do this we must first return to the DOS with the command

Q   (↵)

and then issue the command

$$\text{C:\DOS> REN A:BLK.1 BLK.EXE (⏎)}$$

Programs that are in an executable file can be directly loaded when the DEBUG program is brought up. For our example, the program command is

$$\text{C:\DOS> DEBUG A:BLK.EXE (⏎)}$$

Execution of this command loads the program at address CS:100 and then displays the DEBUG prompt.

▲ 4.9 ASSEMBLING INSTRUCTIONS WITH THE ASSEMBLE COMMAND

All the instructions we have worked with up to this point have been hand assembled into machine code. The DEBUG program has a command that lets us automatically assemble the instructions of a program, one after the other, and store them in memory. It is called the *ASSEMBLE* (A) *command.*

The general syntax of ASSEMBLE is given in Fig. 4.9 as

$$\text{A [STARTING ADDRESS]}$$

Here STARTING ADDRESS is the address at which the machine code of the first instruction of the program is to be stored. For example, to assemble the instruction ADD [BX+SI+1234H],AX and store its machine code in memory starting at address CS:100, we start with the command entry

$$\text{A CS:100 (⏎)}$$

The response to this command input is the display of the starting address in the form

$$\text{1342:0100 \_}$$

The instruction to be assembled is typed in following this address, and when the (⏎) key is depressed, the instruction is assembled into machine code; it is stored in memory; and the starting address of the next instruction is displayed. As shown in Fig. 4.34, for our

```
-A CS:100
1342:0100 ADD [BX+SI+1234],AX
1342:0104
-D CS:100 103
1342:0100  01 80 34 12                    . . 4 .
-N A:INST.1
-R CX
:4
-R BX
:0
-W CS:100
-
```

Figure 4.34 Assembling the instruction ADD [BX+SI+1234H],AX.

example we have

```
1342:0100  ADD  [BX+SI+1234],AX  (↵)
1342:0104  _
```

Now either the next instruction is entered or the (↵) key is depressed to terminate the ASSEMBLE command.

Assuming that the assemble operation just performed was terminated by entering (↵), we can view the machine code that was produced for the instruction by issuing a DUMP command. Notice that the address displayed as the starting point of the next instruction is 1342:0104. Therefore, the machine code for the ADD instruction took up 4 bytes of memory, CS:100, CS:101, CS:102, and CS:103. The command needed to display this machine code is

```
D  CS:100  103  (↵)
```

In Fig. 4.34, we find that the machine code stored for the instruction is 01803412H.

At this point, the instruction can be executed or saved on a diskette. For instance, to save the machine code on a diskette in file INST.1, we issue the commands

```
N  A:INST.1  (↵)
R  CX        (↵)
:4           (↵)
R  BX        (↵)
:0           (↵)
W  CS:100    (↵)
```

Now that we have shown how to assemble an instruction, view its machine code, and save the machine code on a data diskette, let us look into how a complete program can be assembled with the A command. For this purpose, we will use the program shown in Fig. 4.35(a). The same program was entered as hand-assembled machine code in Example 4.24.

We begin by assuming that the program is to be stored in memory starting at address CS:200. For this reason, the *line-by-line assembler* is invoked with the command

```
A  CS:200  (↵)
```

This gives the response

```
1342:0200  _
```

Now we type in the instructions of the program as follows:

```
1342:0200  MOV  AX,2000  (↵)
1342:0203  MOV  DS,AX    (↵)
1342:0205  MOV  SI,100   (↵)
    .        .     .      .
    .        .     .      .
1342:0217  NOP           (↵)
1342:0218                (↵)
```

The details of the instruction entry sequence are shown in Fig. 4.35(b).

```
MOV    AX,2000H           -A CS:200
                          1342:0200 MOV AX,2000
MOV    DS,AX              1342:0203 MOV DS,AX
                          1342:0205 MOV SI,100
MOV    SI,0100H           1342:0208 MOV DI,120
                          1342:020B MOV CX,10
MOV    DI,0120H           1342:020E MOV AH,[SI]
                          1342:0210 MOV [DI],AH
MOV    CX,010H            1342:0212 INC SI
                          1342:0213 INC DI
MOV    AH,[SI]            1342:0214 DEC CX
                          1342:0215 JNZ 20E
MOV    [DI],AH            1342:0217 NOP
                          1342:0218
INC    SI

INC    DI

DEC    CX

JNZ    20EH                        (b)

NOP

         (a)
```

```
-U CS:200 217
1342:0200 B80020          MOV    AX,2000
1342:0203 8ED8            MOV    DS,AX
1342:0205 BE0001          MOV    SI,0100
1342:0208 BF2001          MOV    DI,0120
1342:020B B91000          MOV    CX,0010
1342:020E 8A24            MOV    AH,[SI]
1342:0210 8825            MOV    [DI],AH
1342:0212 46              INC    SI
1342:0213 47              INC    DI
1342:0214 49              DEC    CX
1342:0215 75F7            JNZ    020E
1342:0217 90              NOP
```

                              (c)

Figure 4.35 (a) Block-move program. (b) Assembling the program.
(c) Verifying the assembled program with the U command.

Now that the complete program has been entered, let us verify that it has been assembled correctly. This can be done with an UNASSEMBLE command. Notice in Fig. 4.35(b) that the program resides in memory over the address range CS:200 through CS:217. To unassemble the machine code in this part of memory, we issue the command

                    U   CS:200   217   (↵)

The results produced with this command are shown in Fig. 4.35(c). Comparing the instructions to those in Fig. 4.35(a) confirms that the program has been assembled correctly.

The ASSEMBLE command allows us to assemble instructions involving any of the various addressing modes. For instance, the instruction we used in our earlier example,

                    MOV   AX,2000H

employs immediate addressing mode for the source operand. Instructions such as

                    MOV   AX,[2000H]

which uses direct addressing mode for the source operand, can also be assembled into memory.

ASSEMBLE also supports two psuedo-instructions that can be used to assemble data directly into memory. They are *data byte* (DB) and *data word* (DW). An example is

```
DB  1,2,3,'JASSI'
```

With this command, the byte-size representation of numbers 1, 2, and 3 and the ASCII code for letters J, A, S, S, and I are assembled into memory.

▲ 4.10 EXECUTING INSTRUCTIONS AND PROGRAMS WITH THE TRACE AND GO COMMANDS

Once the program has been entered into the memory of the PC/AT, it is ready to be executed. The DEBUG program allows us to execute the entire program with one *GO* (G) *command* or to execute the program in several segments of instructions by using *breakpoints* in the GO command. Moreover, by using the *TRACE* (T) *command*, the program can be stepped through by executing one or more instructions at a time.

Let us begin by examining the operation of the TRACE command in more detail. Trace provides the programmer with the ability to execute one instruction at a time. This mode of operation is also known as *single-stepping* the program; it is very useful during early phases of program debugging. This is because the contents of registers or memory can be viewed both before and after the execution of each instruction to determine whether or not the correct operation was performed.

The general form of the command as shown in Fig. 4.9 is

```
T  =[STARTING ADDRESS]   [NUMBER]
```

Notice that a *starting address* is specified as part of the command. This is the address of the instruction at which execution is to begin. It is followed by a *number* that tells how many instructions are to be executed. The use of the equal sign before the starting address is very important. If it is left out, the microcomputer usually hangs up and will have to be restarted with a power on reset.

If an instruction count is not specified in the command, just one instruction is executed. For instance, the command

```
T  =CS:100  (↵)
```

causes the instruction starting at address CS:100 to be executed. At completion of the instruction's execution, the complete state of the MPU's internal registers is automatically displayed. At this point, other DEBUG commands can be issued—for instance, to display the contents of memory—or the next instruction can be executed.

This TRACE command can also be issued as

```
T  (↵)
```

In this case the instruction pointed to by the current values of CS and IP (CS:IP) is executed. This is the form of the TRACE command that is used to execute the next instruction.

If we want to step through several instructions, the TRACE command must include the number of instructions to be executed. This number is included after the address. For example, to trace through three instructions, the command is issued as

$$T \quad =CS:100 \quad 3 \quad (\dashv)$$

Again, the internal state of the MPU is displayed after each instruction is executed.

EXAMPLE 4.25

Load the instruction stored at file specification 1 10 1 at offset 100 of the current code segment. Unassemble the instruction. Then initialize AX with 1111H, SI with 1234H, and the word contents of memory address 1234_{16} to the value 2222_{16}. Next, display the internal state of the MPU and the contents of address 1234_{16} to verify their initialization. Finally execute the instruction with the TRACE command. What operation is performed by the instruction?

Solution

First, the instruction is loaded at CS:100 from the diskette with the command

$$L \quad CS:100 \quad 1 \quad 10 \quad 1 \quad (\dashv)$$

Now the machine code is unassembled to verify that the instruction has loaded correctly.

$$U \quad 100 \quad 101 \quad (\dashv)$$

Looking at the displayed information in Fig. 4.36, we see that it is an ADD instruction. Next we initialize the internal registers and memory with the command sequence

```
R   AX              (↵)
AX  0000
:1111               (↵)
R   SI              (↵)
SI  0000
:1234               (↵)
E   DS:1234   22   22   (↵)
```

```
-L CS:100 1 10 1
-U 100 101
1342:0100 0304          ADD    AX,[SI]
-R AX
AX 0000
:1111
-R SI
SI 0000
:1234
-E DS:1234 22 22
-R
AX=1111  BX=0000  CX=0000  DX=0000  SP=FFEE  BP=0000  SI=1234  DI=0000
DS=1342  ES=1342  SS=1342  CS=1342  IP=0100   NV UP EI PL NZ NA PO NC
1342:0100 0304          ADD    AX,[SI]
-D DS:1234 1235
1342:1230               22 22                           " "
-T =CS:100

AX=3333  BX=0000  CX=0000  DX=0000  SP=FFEE  BP=0000  SI=1234  DI=0000
DS=1342  ES=1342  SS=1342  CS=1342  IP=0102   NV UP EI PL NZ NA PE NC
1342:0102 0000          ADD    [BX+SI],AL                DS:1234=22

-
```

Figure 4.36 Displayed information for Example 4.25.

Now the initialization is verified with the commands

```
R               (↵)
D  DS:1234  1235  (↵)
```

In Fig. 4.36, we see that AX, SI, and the word contents of address 1234_{16} were correctly initialized. Therefore, we are ready to execute the instruction. This is done with the command

```
T  =CS:100  (↵)
```

From the displayed trace information in Fig. 4.36, we find that the value 2222_{16} at address 1234_{16} was added to the value 1111_{16} held in AX. Therefore, the new contents of AX are 3333_{16}.

The GO command is typically used to run programs that are already working or to execute programs in the latter stages of debugging. For example, if the beginning part of a program is already operating correctly, a GO command can be used to execute this group of instructions and then stop execution at a point in the program where additional debugging is to begin.

The table in Fig. 4.9 shows that the general form of the GO command is

```
G  =[STARTING ADDRESS]  [BREAKPOINT ADDRESS LIST]
```

The first address is the *starting address* of the segment of program that is to be executed, that is, the address of the instruction at which execution is to begin. The second address, the *breakpoint address*, is the address of the end of the program segment, that is, the address of the instruction at which execution is to stop. The breakpoint address that is specified must correspond to the first byte of an instruction. A list of up to ten breakpoint addresses can be supplied with the command.

An example of the GO command is

```
G  =CS:200  217  (↵)
```

This command loads the IP register with 0200_{16}, sets a breakpoint at address CS:217, and then begins program execution at address CS:200. Instruction execution proceeds until address CS:217 is accessed. When the breakpoint address is reached, program execution is terminated, the complete internal status of the MPU is displayed, and control is returned to DEBUG.

Sometimes we just want to execute a program without using a breakpoint. This can also be done with the GO command. For instance, to execute a program that starts at offset 100_{16} in the current CS, we can issue the GO command without a breakpoint address as follows:

```
G  =CS:100  (↵)
```

This command will cause the program to run to completion provided there are appropriate instructions in the program to initiate a normal termination, such as those needed to return to DEBUG. In the case of a program where CS and IP are already initialized with the correct values, we can just enter

$$G \quad (\lrcorner)$$

However, it is recommended that the GO command always include a breakpoint address. If a GO is issued without a breakpoint address and the value of CS and IP are not already set up or the program is not correctly prepared for normal termination, the microcomputer can lock up. This is because the program execution may go beyond the end of the program into an area with data that represents invalid instructions.

EXAMPLE 4.26

In Section 4.8, we saved the block move program in file BLK.EXE on a data diskette in drive A. Load this program into memory starting at address CS:200. Then initialize the microcomputer by loading the DS register with 2000_{16}; fill the block of memory from DS:100 through DS:10F with FF_{16} and the block of memory from DS:120 through DS:12F with 00_{16}. Verify that the blocks of memory were initialized correctly. Load DS with 1342_{16}, and display the state of the MPU's registers. Display the assembly language version of the program from CS:200 through CS:217. Use a GO command to execute the program through address CS:20E. What changes have occurred in the contents of the registers? Now execute down through address CS:215. What changes are found in the blocks of data? Next execute the program down to address CS:217. What new changes are found in the blocks of data?

Solution

The commands needed to load the program are

```
N   A:BLK.EXE   (⌐)
L   CS:200      (⌐)
```

Next we initialize the DS register and memory with the commands

```
R   DS                  (⌐)
DS  1342
:2000                   (⌐)
F   DS:100   10F   FF   (⌐)
F   DS:120   12F   00   (⌐)
```

The blocks of data in memory are displayed using the commands

```
D   DS:100   10F   (⌐)
D   DS:120   12F   (⌐)
```

The displayed information is shown in Fig. 4.37. DS is restored with 1342_{16} using the command

```
R   DS           (⌐)
DS  2000
:1342            (⌐)
```

```
-N A:BLK.EXE
-L CS:200
-R DS
DS 1342
:2000
-F DS:100 10F FF
-F DS:120 12F 00
-D DS:100 10F
2000:0100  FF FF FF FF FF FF FF FF-FF FF FF FF FF FF FF FF   ................
-D DS:120 12F
2000:0120  00 00 00 00 00 00 00 00-00 00 00 00 00 00 00 00   ................
-R DS
DS 2000
:1342
-R
AX=0000  BX=0000  CX=0020  DX=0000  SP=FFEE  BP=0000  SI=0000  DI=0000
DS=1342  ES=1342  SS=1342  CS=1342  IP=0100   NV UP EI PL NZ NA PO NC
1342:0100 0000           ADD     [BX+SI],AL                  DS:0000=CD
-U CS:200 217
1342:0200 B80020         MOV     AX,2000
1342:0203 8ED8           MOV     DS,AX
1342:0205 BE0001         MOV     SI,0100
1342:0208 BF2001         MOV     DI,0120
1342:020B B91000         MOV     CX,0010
1342:020E 8A24           MOV     AH,[SI]
1342:0210 8825           MOV     [DI],AH
1342:0212 46             INC     SI
1342:0213 47             INC     DI
1342:0214 49             DEC     CX
1342:0215 75F7           JNZ     020E
1342:0217 90             NOP
-G =CS:200 20E

AX=2000  BX=0000  CX=0010  DX=0000  SP=FFEE  BP=0000  SI=0100  DI=0120
DS=2000  ES=1342  SS=1342  CS=1342  IP=020E   NV UP EI PL NZ NA PO NC
1342:020E 8A24           MOV     AH,[SI]                     DS:0100=FF
-G =CS:20E 215

AX=FF00  BX=0000  CX=000F  DX=0000  SP=FFEE  BP=0000  SI=0101  DI=0121
DS=2000  ES=1342  SS=1342  CS=1342  IP=0215   NV UP EI PL NZ AC PE NC
1342:0215 75F7           JNZ     020E
-D DS:100 10F
2000:0100  FF FF FF FF FF FF FF FF-FF FF FF FF FF FF FF FF   ................
-D DS:120 12F
2000:0120  FF 00 00 00 00 00 00 00-00 00 00 00 00 00 00 00   ................
-G =CS:215 217

AX=FF00  BX=0000  CX=0000  DX=0000  SP=FFEE  BP=0000  SI=0110  DI=0130
DS=2000  ES=1342  SS=1342  CS=1342  IP=0217   NV UP EI PL ZR NA PE NC
1342:0217 90             NOP
-D DS:100 10F
2000:0100  FF FF FF FF FF FF FF FF-FF FF FF FF FF FF FF FF   ................
-D DS:120 12F
2000:0120  FF FF FF FF FF FF FF FF-FF FF FF FF FF FF FF FF   ................
-
```

Figure 4.37 Displayed information for Example 4.26.

and the state of the MPU's registers is displayed with the command

$$R \quad (\lrcorner)$$

Before beginning to execute the program, we will display the source code with the command

$$U \quad CS:200 \quad 217 \quad (\lrcorner)$$

The program that is displayed is shown in Fig. 4.37.

Now the first segment of program is executed with the command

$$G \quad =CS:200 \quad 20E \quad (\lrcorner)$$

Looking at the displayed state of the MPU in Fig. 4.37, we see that DS was loaded with 2000_{16}, AX was loaded with 2000_{16}, SI was loaded with 0100_{16}, and CX was loaded with 0010_{16}.

Next, another GO command is used to execute the program down through address CS:215.

```
G  =CS:20E  215  (↵)
```

We can check the state of the blocks of memory with the commands

```
D  DS:100  10F  (↵)
D  DS:120  12F  (↵)
```

From the displayed information in Fig. 4.37, we see that FF_{16} was copied from the first element of the source block to the first element of the destination block.

Now we execute through CS:217 with the command

```
G  =CS:215  217  (↵)
```

and examine the blocks of data with the commands

```
D  DS:100  10F  (↵)
D  DS:120  12F  (↵)
```

We find that the complete source block has been copied to the destination block.

▲ 4.11 DEBUGGING A PROGRAM

In Sections 4.8, 4.9, and 4.10 we learned how to use DEBUG to load a machine code program into the memory of the PC/AT, assemble a program, and execute the program. However, we did not determine if the program when run performed the operation for which it was written. It is common to have errors in programs, and even a single error can render the program useless. For instance, if the address to which a "jump" instruction passes control is wrong, the program may get hung up. Errors in a program are also referred to as *bugs*; the process of removing them is called *debugging*.

The two types of errors that can be made by a programmer are the *syntax error* and the *execution error*. A syntax error is an error caused by not following the rules for coding or entering an instruction. These types of errors are typically identified by the microcomputer and signaled to the user with an error message. For this reason, they are usually easy to find and correct. For example, if a DUMP command is keyed in as

```
D  DS:100120  (↵)
```

an error condition exists. This is because the space between the starting and ending address is left out. This incorrect entry is signaled by the warning Error in the display; the spot where the error begins, in this case, the 1 in 120, is marked with the symbol $^\wedge$ to indentify the position of the error.

An execution error is an error in the logic behind the development of the program. That is, the program is correctly coded and entered, but it still does not perform the operation for which it was written. This type of error can be identified by entering the program into the microcomputer and observing its operation. Even when an execution error has been identified, it is usually not easy to find the exact cause of the problem.

Our ability to debug execution errors in a program is aided by the commands of the DEBUG program. For instance, the TRACE command allows us to step through the program by executing just one instruction at a time. We can use the display of the internal register state produced by TRACE and the memory dump command to determine the state of the MPU and memory prior to execution of an instruction and again after its execution. This information will tell us whether the instruction has performed the operation planned for it. If an error is found, its cause can be identified and corrected.

To demonstrate the process of debugging a program, let us once again use the program that we stored in file A:BLK.EXE. We load it into the code segment at address CS:200 with the command

```
N   A:BLK.EXE   (↵)
L   200         (↵)
```

Now the program resides in memory at addresses CS:200 through CS:217. The program is displayed with the command

```
U   200   217   (↵)
```

The program that is displayed is shown in Fig. 4.38. This program implements a block data-transfer operation. The block of data to be moved starts at memory address DS:100 and is 16 bytes in length. It is to be moved to another block of storage locations starting at address DS:120. DS equals 2000_{16}; therefore, it points to a data segment starting at physical address 20000_{16}.

Before executing the program, let us issue commands to initialize the source block of memory locations from addresses 100_{16} through $10F_{16}$ with FF_{16} and the bytes in the destination block starting at 120_{16} with 00_{16}. To do this, we issue the command sequence

```
F   2000:100   10F   FF   (↵)
F   2000:120   12F   00   (↵)
```

The first two instructions of the program in Fig. 4.38 are

```
MOV   AX,2000H
```

and

```
MOV   DS,AX
```

These two instructions, when executed, load the data segment register with the value 2000_{16}. In this way they define a data segment starting at address 20000_{16}. The next three instructions are used to load the SI, DI, and CX registers with 100_{16}, 120_{16}, and 10_{16}, respectively. Let us now show how to execute these instructions and then determine if they perform the correct function. They are executed by issuing the command

```
T   =CS:200   5   (↵)
```

```
C:\DOS>DEBUG
-N A:BLK.EXE
-L 200
-U 200 217
1342:0200 B82010        MOV     AX,2000
1342:0203 8ED8          MOV     DS,AX
1342:0205 BE0001        MOV     SI,0100
1342:0208 BF2001        MOV     DI,0120
1342:020B B91000        MOV     CX,0010
1342:020E 8A24          MOV     AH,[SI]
1342:0210 8825          MOV     [DI],AH
1342:0212 46            INC     SI
1342:0213 47            INC     DI
1342:0214 49            DEC     CX
1342:0215 75F7          JNZ     020E
1342:0217 90            NOP
-F 2000:100 10F FF
-F 2000:120 12F 00
-T =CS:200 5

AX=2000  BX=0000  CX=0010  DX=0000  SP=FFEE  BP=0000  SI=0100  DI=0120
DS=1020  ES=1342  SS=1342  CS=1342  IP=0203   NV UP EI PL NZ NA PO NC
1342:0203 8ED8          MOV     DS,AX

AX=2000  BX=0000  CX=0010  DX=0000  SP=FFEE  BP=0000  SI=0100  DI=0120
DS=2000  ES=1342  SS=1342  CS=1342  IP=0205   NV UP EI PL NZ NA PO NC
1342:0205 BE0001        MOV     SI,0100

AX=2000  BX=0000  CX=0010  DX=0000  SP=FFEE  BP=0000  SI=0100  DI=0120
DS=2000  ES=1342  SS=1342  CS=1342  IP=0208   NV UP EI PL NZ NA PO NC
1342:0208 BF2001        MOV     DI,0120

AX=2000  BX=0000  CX=0010  DX=0000  SP=FFEE  BP=0000  SI=0100  DI=0120
DS=2000  ES=1342  SS=1342  CS=1342  IP=020B   NV UP EI PL NZ NA PO NC
1342:020B B91000        MOV     CX,0010

AX=2000  BX=0000  CX=0010  DX=0000  SP=FFEE  BP=0000  SI=0100  DI=0120
DS=2000  ES=1342  SS=1342  CS=1342  IP=020E   NV UP EI PL NZ NA PO NC
1342:020E 8A24          MOV     AH,[SI]                          DS:0100=FF
-D DS:120 12F
2000:0120  00 00 00 00 00 00 00 00-00 00 00 00 00 00 00 00   ................
-T 2

AX=FF00  BX=0000  CX=0010  DX=0000  SP=FFEE  BP=0000  SI=0100  DI=0120
DS=2000  ES=1342  SS=1342  CS=1342  IP=0210   NV UP EI PL NZ NA PO NC
1342:0210 8825          MOV     [DI],AH                          DS:0120=00

AX=FF00  BX=0000  CX=0010  DX=0000  SP=FFEE  BP=0000  SI=0100  DI=0120
DS=2000  ES=1342  SS=1342  CS=1342  IP=0212   NV UP EI PL NZ NA PO NC
1342:0212 46            INC     SI
-D DS:120 12F
2000:0120  FF 00 00 00 00 00 00 00-00 00 00 00 00 00 00 00   ................
-T 3

AX=FF00  BX=0000  CX=0010  DX=0000  SP=FFEE  BP=0000  SI=0101  DI=0120
DS=2000  ES=1342  SS=1342  CS=1342  IP=0213   NV UP EI PL NZ NA PO NC
1342:0213 47            INC     DI

AX=FF00  BX=0000  CX=0010  DX=0000  SP=FFEE  BP=0000  SI=0101  DI=0121
DS=2000  ES=1342  SS=1342  CS=1342  IP=0214   NV UP EI PL NZ NA PE NC
```

Figure 4.38 Program debugging demonstration.

To determine if the instructions that were executed performed the correct operation, we just need to look at the trace display that they produce. This display trace is shown in Fig. 4.38. Here we see that the first instruction loads AX with 2000_{16} and the second moves this value into the DS register. Also notice in the last trace that is displayed for this command that SI contains 0100_{16}, DI contains 0120_{16}, and CX contains 0010_{16}.

The next two instructions copy the contents of memory location 100_{16} into the storage location at address 120_{16}. Let us first check the contents of the destination block with the D command

$$D \quad DS:120 \quad 12F \quad (\lrcorner)$$

```
1342:0214 49              DEC    CX

AX=FF00  BX=0000  CX=000F  DX=0000  SP=FFEE  BP=0000  SI=0101  DI=0121
DS=2000  ES=1342  SS=1342  CS=1342  IP=0215      NV UP EI PL NZ AC PE NC
1342:0215 75F7            JNZ    020E
-T

AX=FF00  BX=0000  CX=000F  DX=0000  SP=FFEE  BP=0000  SI=0101  DI=0121
DS=2000  ES=1342  SS=1342  CS=1342  IP=020E      NV UP EI PL NZ AC PE NC
1342:020E 8A24            MOV    AH,[SI]                            DS:0101=FF
-G =CS:20E 215

AX=FF00  BX=0000  CX=000E  DX=0000  SP=FFEE  BP=0000  SI=0102  DI=0122
DS=2000  ES=1342  SS=1342  CS=1342  IP=0215      NV UP EI PL NZ NA PO NC
1342:0215 75F7            JNZ    020E
-D DS:120 12F
2000:0120  FF FF 00 00 00 00 00 00-00 00 00 00 00 00 00 00   ................
-T

AX=FF00  BX=0000  CX=000E  DX=0000  SP=FFEE  BP=0000  SI=0102  DI=0122
DS=2000  ES=1342  SS=1342  CS=1342  IP=020E      NV UP EI PL NZ NA PO NC
1342:020E 8A24            MOV    AH,[SI]                            DS:0102=FF
-G =CS:20E 217

AX=FF00  BX=0000  CX=0000  DX=0000  SP=FFEE  BP=0000  SI=0110  DI=0130
DS=2000  ES=1342  SS=1342  CS=1342  IP=0217      NV UP EI PL ZR NA PE NC
1342:0217 90              NOP
-D DS:120 12F
2000:0120  FF FF FF FF FF FF FF FF-FF FF FF FF FF FF FF FF   ................
-
```

Figure 4.38 (Continued)

Looking at the dump display in Fig. 4.38, we see that the original contents of these locations are 00_{16}. Now the two instructions are executed with the command

$$T \quad 2 \quad (\lrcorner)$$

and the contents of address DS:120 are checked once again with the command

$$D \quad DS:120 \quad 12F \quad (\lrcorner)$$

The display dump in Fig. 4.38 shows that the first element of the source block was copied to the location of the first element of the destination block. Therefore, both address 100_{16} and address 120_{16} now contain the value FF_{16}.

The next three instructions are used to increment pointers SI and DI and decrement block counter CX. To execute them, we issue the command

$$T \quad 3 \quad (\lrcorner)$$

Referring to the trace display in Fig. 4.38 to verify their operation, we find that the new values in SI and DI are 0101_{16} and 0121_{16}, respectively, and CX is now $000F_{16}$.

The jump instruction is next, and it transfers control to the instruction 8 bytes back if CX did not become zero. It is executed with the command

$$T \quad (\lrcorner)$$

Notice that the result of executing this instruction is that the value in IP is changed to $020E_{16}$. This corresponds to the location of the instruction

```
MOV    AH,[SI]
```

In this way we see that control has been returned to the part of the program that performs the data-move operation.

The move operation performed by this part of the program was already checked; however, we must still determine if it runs to completion when the count in CX decrements to zero. Therefore, we will execute another complete loop with the GO command

```
G   =CS:20E   215   (↵)
```

Correct operation is verified because the trace shows that CX has been decremented by 1 more and equals E. The fact that the second element has been moved can be verified by dumping the destination block with the command

```
D  DS:120   12F   (↵)
```

Now we are again at address CS:215. To execute the jump instruction at this location, we can again use the T command

```
T   (↵)
```

This returns control to the instruction at CS:20E. The previous two commands can be repeated until the complete block is moved and CX equals 0_{16}. Or we can use the GO command to execute to the address CS:217, which is the end of the program.

```
G   =CS:20E   217   (↵)
```

At completion, the overall operation of the program can be verified by examining the contents of the destination block with the command sequence

```
D  DS:120   12F   (↵)
```

FF_{16} should be displayed as the data held in each storage location.

ASSIGNMENTS

Section 4.2

1. Encode the following instruction using the infomation in Figs. 4.1 through 4.4.

```
ADD   AX,DX
```

Assume that the opcode for the add operation is 000000.

2. Encode the following instructions using the information in the tables of Figs. 4.2 through 4.6.
   (a) MOV [DI], DX
   (b) MOV [BX][SI], BX
   (c) MOV DL, [BX]+10H

3. Encode the instructions that follow using the tables in Figs. 4.2 through 4.6.
   (a) PUSH DS
   (b) ROL BL,CL
   (c) ADD AX, [1234H]

Section 4.3

4. How many bytes are required to encode the instruction MOV SI,0100H?

5. How many bytes of memory are required to store the machine code for the program in Fig. 4.7(a)?

Section 4.4

6. What purpose is served by the DEBUG program?

7. Can DEBUG be brought up by typing the command using lowercase letters?

8. If the DEBUG command R AXBX is entered to a PC/AT, what happens?

9. Write the REGISTER command needed to change the value in CX to 10_{16}.

10. Write the command needed to change the state of the parity flag to PE.

11. Write a command that will dump the state of the MPU's internal registers.

Section 4.5

12. Write a DUMP command that will display the contents of the first 16 bytes of the current code segment.

13. Show an ENTER command that can be used to examine the contents of the same 16 bytes of memory that were displayed in Problem 12.

14. Show the ENTER command needed to load 5 consecutive bytes of memory starting at address CS:100 of the current code segment with FF_{16}.

15. Show how an ENTER command can be used to initialize the first 32 bytes at the top of the stack to 00_{16}.

16. Write a sequence of commands that will fill the first six storage locations starting at address CS:100 with 11_{16}, the second six with 22_{16}, the third six with 33_{16}, the fourth six with 44_{16}, and the fifth six with 55_{16}, change the contents of storage locations CS:105 and CS:113 to FF_{16}, display the first 30 bytes of memory starting at CS:100, and then use a search command on this 30-byte block of memory to find those storage locations that contain FF_{16}.

Section 4.6

17. What DEBUG commands do I and O stand for?

18. What operation is performed by the command

$$\text{I} \quad 123 \quad (\lrcorner)$$

19. Write an output command that will load the byte-wide output port at I/O address 0124_{16} with the value $5A_{16}$.

Section 4.7

20. What two results are produced by the hexadecimal command?

21. How large can the numbers in an H command be?

22. The difference $FA_{16} - 5A_{16}$ is to be found. Write the H command.

Section 4.8

23. Show the sequence of commands needed to load the machine code instruction 320E3412H starting at address CS:100, unassemble it to verify that the correct instruction was loaded, and save it on a data diskette at file specification 1 50 1.

24. Write commands that will reload the instruction saved on the data diskette in Problem 23 into memory at offset 400 in the current code segment and unassemble it to verify correct loading.

Section 4.9

25. Show how the instruction MOV [DI],DX can be assembled into memory at address CS:100.

26. Write a sequence of commands that will first assemble the instruction ROL BL,CL into memory starting at address CS:200 and then verify its entry by disassembling the instruction.

Section 4.10

27. Show a sequence of commands that will load the instruction saved on the data diskette in Problem 23 at address CS:300, unassemble it to verify correct loading, initialize the contents of register CX to $000F_{16}$ and the contents of the word memory location starting at DS:1234 to $00FF_{16}$, execute the instruction with the TRACE command, and verify its operation by examining the contents of CX and the word of data stored starting at DS:1234 in memory.

28. Write a sequence to repeat Example 4.26; however, this time execute the complete program with one GO command.

Section 4.11

29. What is the difference between a syntax error and an execution error?

30. Give another name for an error in a program.

31. What is the name given to the process of removing errors in a program?

32. Write a sequence of commands to repeat the debug demonstration presented in Section 4.11, but this time use only GO commands to execute the program.

Real-Mode 80386DX Microprocessor Programming I

▲ 5.1 INTRODUCTION

Up to this point we have studied the real-mode software architecture of the 80386DX microprocessor, the assembly language program-develpment cycle, the evolution of the 80X86 microprocesssor family instruction set, the addressing modes available in the real mode, and the software development tools provided by the DEBUG program of DOS. In this chapter we begin a detailed study of the 80386DX's real-mode instruction set. A large part of the instruction set is covered in this chapter. These instructions provide the ability to write straight-line programs. The rest of the real-mode instruction set and some more sophisticated programming concepts are covered in Chapter 6. The following topics are presented in this chapter:

1. Data-transfer instructions
2. Arithmetic instructions
3. Logic instructions
4. Shift instructions
5. Rotate instructions
6. Bit test and bit scan instructions

▲ 5.2 DATA-TRANSFER INSTRUCTIONS

The 80386DX microprocessor has a group of *data-transfer instructions* that are provided to move data either between its internal registers or between an internal register and a storage location in memory. This group includes the *move* (MOV) instruction, *sign-extend and move* (MOVSX) instruction, *zero-extend and move* (MOVZX) instruction, *exchange* (XCHG) instruction, *translate* (XLAT) instruction, *load effective address* (LEA) instruction, *load data segment* (LDS) instruction, *load extra segment* (LES) instruction, *load register and SS* (LSS) instruction, *load register and FS* (LFS) instruction, and *load register and GS* (LGS) instruction. These instructions are all discussed in this section.

The MOV Instruction

The move (MOV) instruction shown in Fig. 5.1(a) is used to transfer a byte, a word, or a double word of data from a source operand to a destination operand. These operands can be internal registers of the 80386DX and storage locations in memory. Figure 5.1(b) shows the valid source and destination operand variations. This large choice of operands results in many different MOV instructions. Looking at this list of operands, we see that data can be moved between general-purpose registers, between a general-purpose register and a segment register, between a general-purpose register or segment register and memory, between a memory location and the accumulator, or between a general-purpose register and a special register.

Notice that the MOV instruction cannot transfer data directly between a source and a destination that both reside in external memory. Instead, the data must first be moved from memory into an internal register, such as to the accumulator (EAX), with one move instruction and then moved to the new location in memory with a second move instruction.

All transfers between general-purpose registers and memory can involve either a byte, word, or double word of data. The fact that the instruction corresponds to byte, word, or double-word data is designated by the way in which its operands are specified. For instance, AL or AH would be used to specify a byte operand, AX, a word operand, and EAX, a double-word operand. On the other hand, data moved between one of the general-purpose registers and a segment register or between a segment register and a memory location must always be word-wide.

In Fig. 5.1(a), we also find additional important information. For instance, flag bits within the 80386DX are not modified by execution of a MOV instruction.

An example of a segment register to general-purpose register MOV instruction shown in Fig. 5.1(c) is

```
MOV    DX,CS
```

In this instruction the code segment register is the source operand, and the data register is the destination. It stands for move the contents of CS into DX. That is.

$$(CS) \rightarrow (DX)$$

For example, if the contents of CS are 0100_{16}, execution of the instruction MOV DX,CS as shown in Fig. 5.1(d), makes

$$(DX) = (CS) = 0100_{16}$$

In all memory reference MOV instructions, the machine code for the instruction includes an offset address relative to the current contents of the data segment register. An example of this type of instruction is

```
MOV    [SUM],AX
```

Mnemonic	Meaning	Format	Operation	Flags affected
MOV	Move	MOV D, S	(S) → (D)	None

(a)

Destination	Source
Memory	Accumulator
Accumulator	Memory
Register	Register
Register	Memory
Memory	Register
Register	Immediate
Memory	Immediate
Seg-reg	Reg16
Seg-reg	Mem16
Reg16	Seg-reg
Mem16	Seg-reg
Register	Spec-reg
Spec-reg	Register

(b)

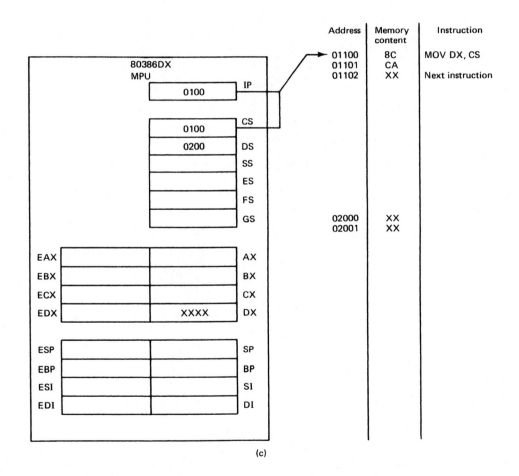

(c)

Figure 5.1 (a) Move-data transfer instruction. (b) Allowed operands. (c) MOV DX,CS instruction before fetch and execution. (d) After execution. (e) Special registers.

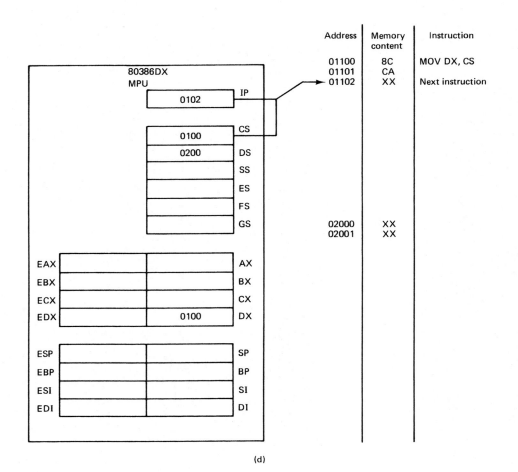

Address	Memory content	Instruction
01100	8C	MOV DX, CS
01101	CA	
01102	XX	Next instruction
02000	XX	
02001	XX	

(d)

Mnemonic	Name
CR_0	Control register 0
CR_2	Control register 2
CR_3	Control register 3
DR_0	Debug register 0
DR_1	Debug register 1
DR_2	Debug register 2
DR_3	Debug register 3
DR_6	Debug register 6
DR_7	Debug register 7
TR_6	Test register 6
TR_7	Test register 7

(e)

Figure 5.1 (Continued)

In this instruction, the memory location identified by the variable SUM is specified using direct addressing. That is, the value of the offset SUM is encoded in the two byte locations that follow its opcode.

Let us assume that the contents of DS equal 0200_{16} and that SUM equals 1212_{16}. Then this instruction means move the contents of accumulator AX to the memory location offset

148   Real-Mode 80386DX Microprocessor Programming I   Chap. 5

by 1212_{16} from the starting location of the current data segment. The physical address of this location is obtained as

$$PA = 02000_{16} + 1212_{16} = 03212_{16}$$

Thus the effect of the instruction is

$$(AL) \rightarrow (\text{memory location } 03212_{16})$$

and

$$(AH) \rightarrow (\text{memory location } 03213_{16})$$

EXAMPLE 5.1

What is the effect of executing the instruction

```
MOV   CX, [SOURCE_MEM]
```

where SOURCE_MEM equal to 20_{16} is a memory location offset relative to the current data segment starting at address $1A000_{16}$?

Solution

Execution of this instruction results in the following:

$$((DS)0 + 20_{16}) \rightarrow (CL)$$
$$((DS)0 + 20_{16} + 1_{16}) \rightarrow (CH)$$

In other words, CL is loaded with the contents held at memory address

$$1A000_{16} + 20_{16} = 1A020_{16}$$

and CH is loaded with the contents of memory address

$$1A000_{16} + 20_{16} + 1_{16} = 1A021_{16}$$

EXAMPLE 5.2

Use the DEBUG program on the PC/AT to verify the operation of the instruction in Example 5.1. Initialize the word storage location pointed to by SOURCE_MEM to the value $AA55_{16}$ before executing the instruction.

Solution

First, the DEBUG program is invoked by entering the command

```
C:\DOS>DEBUG    (↵)
```

```
C:\DOS>DEBUG
-R
AX=0000  BX=0000  CX=0000  DX=0000  SP=FFEE  BP=0000  SI=0000  DI=0000
DS=1342  ES=1342  SS=1342  CS=1342  IP=0100   NV UP EI PL NZ NA PO NC
1342:0100 0F            DB     0F
-A
1342:0100 MOV    CX,[20]
1342:0104
-R DS
DS 1342
:1A00
-E 20 55 AA
-T

AX=0000  BX=0000  CX=AA55  DX=0000  SP=FFEE  BP=0000  SI=0000  DI=0000
DS=1A00  ES=1342  SS=1342  CS=1342  IP=0104   NV UP EI PL NZ NA PO NC
1342:0104 FFF3          PUSH   BX
-Q

C:\DOS>
```

Figure 5.2 Display sequence for Example 5.2.

As shown in Fig. 5.2, this results in the display of the debugger's prompt

To determine the memory locations the debugger assigns for use in entering instructions
and data, we can examine the state of the internal registers with the command

$$-R \quad (\llcorner)$$

Looking at the displayed information for this command in Fig. 5.2, we find that the contents
of CS and IP indicate that the starting address in the current code segment is 1342:0100
and the current data segment starts at address 1342:0000. Also note that the initial value
in CX is 0000H.

To enter the instruction from Example 5.1 at location 1342:0100, we use the ASSEM-
BLE command

```
-A                      (⌐)
1342:0100   MOV  CX,[20]  (⌐)
1342:0104                 (⌐)
```

Note that we must enter the offset address value instead of SOURCE_MEM and that it
must be enclosed in brackets to indicate that it is a direct address.

Let us now redefine the data segment so that it starts at $1A000_{16}$. This is done by
loading the DS register with $1A00_{16}$ using the REGISTER command. As shown in Fig. 5.2,
we do this with the entries

```
-R   DS    (⌐)
DS   1342
:1A00      (⌐)
```

Now we initialize the memory locations at addresses 1A00:20 and 1A00:21 to 55_{16} and AA_{16}, respectively, with the ENTER command

```
-E   20   55   AA   (↵)
```

Notice that the bytes of the word of data must be entered in the reverse order.

Finally, to execute the instruction, we issue the trace command

```
-T   (↵)
```

The result of executing the instruction is shown in Fig. 5.2. Note that CX has been loaded with $AA55_{16}$.

The move instruction can also be used to load or save the contents of one of the 80386DX's special registers. The names and mnemonics of the special registers are listed in Fig. 5.1(e). Notice that they include four control registers (CR_0 through CR_3) and eight debug registers (DR_0 through DR_7). These registers provide special functions for the 80386DX microcomputer. For instance, control register CR_0 contains a number of system-control flags. Two examples are the emulation (EM) control bit and the protection enable (PE) bit. An example is the instruction

```
MOV   CR0,EAX
```

When executed, this instruction causes the 32-bit value in EAX to be loaded into control register 0. On the other hand, the instruction

```
MOV   EAX,CR0
```

saves the contents of CR_0 in the accumulator.

A use of the move instruction is to load initial address and data values into the registers of the MPU. This can be done by loading immediate data using instructions. For instance, the instruction sequence in Fig. 5.3 uses immediate data to initialize the values in the segment, index, and data registers. In Fig. 5.1(b) we see that the immediate data cannot be directly loaded into a segment register. For this reason, the initial values of the segment base addresses for DS, ES, and SS are first loaded into AX and then copied into the

```
MOV AX,2000H
MOV DS, AX
MOV ES, AX
MOV AX,3000H
MOV SS,AX
MOV AX,0H
MOV BX,AX
MOV CX,0AH
MOV DX,100H
MOV SI,200H
MOV DI,300H
```

Figure 5.3 Initializing the internal registers of the 80386DX.

appropriate segment registers. The result produced by executing the first five instructions in this sequence is to initialize these segment registers.

The next six instructions are used to initialize the data and index registers. First, the AX register is cleared to 0000_{16} and then BX is also cleared by copying this value from AX to BX. Finally, CX, DX, SI, and DI are loaded with the immediate values $000A_{16}$, 0100_{16}, 0200_{16}, and 0300_{16}, respectively.

Sign-Extend and Zero-Extend Move Instructions: MOVSX and MOVZX

In Fig. 5.4(a) we find that a number of special-purpose move instructions have been included in the instruction set of the 80386DX. The instructions *move with sign-extend* (MOVSX) and *move with zero-extend* (MOVZX) are used to sign extend or zero extend, respectively, a source operand as it is moved to the destination operand location. Figure 5.4(b) shows that the source operand is either a byte or a word of data in a register or a storage location in memory, whereas the destination operand is either a 16-bit or 32-bit register. For example, the instruction

```
MOVSX   EBX,AX
```

is used to copy the 16-bit value in AX into EBX. As the copy is performed, the value in the sign bit, which is bit 15 of AX, is extended into the 16 higher-order bits of EBX. If AX contains $FFFF_{16}$, the sign bit is logic 1. Therefore, after execution of the MOVSX instruction, the value that results in EBX is $FFFFFFFF_{16}$. The MOVZX instruction performs a similar function to the MOVSX instruction except that it extends the value moved to the destination operand location with zeros.

Mnemonic	Meaning	Format	Operation	Flags affected
MOVSX	Move with sign-extend	MOVSX D, S	(S) → (D) MSBs of D are filled with sign bit of S	None
MOVZX	Move with zero-extend	MOVZX D, S	(S) → (D) MSBs of D are filled with 0	None

(a)

Destination	Source
Reg16	Reg8
Reg32	Reg8
Reg32	Reg16
Reg16	Mem8
Reg32	Mem8
Reg32	Mem16

(b)

Figure 5.4 (a) Sign-extend and zero-extend move instructions. (b) Allowed operands.

EXAMPLE 5.3

Explain the operation performed by the instruction

```
MOVZX   CX, BYTE POINTER [DATA_BYTE]
```

if the value of data at memory address DATA_BYTE is FF_{16}.

Solution

When the MOVZX instruction is executed, the value FF_{16} is copied into the lower byte of CX and the upper 8 bits are filled with 0s. This gives

$$CX = 00FF_{16}$$

The XCHG Instruction

In our study of the move instruction, we found that it could be used to copy the contents of a register or memory location into another register or contents of a register into a storage location in memory. In all these cases the original contents of the source location are preserved and the original contents of the destination are destroyed. In some applications we need to exchange the contents of two registers. For instance, we might want to exchange the data in the AX and BX registers.

This could be done using multiple move instructions and storage of the data in a temporary register, such as DX. However, to perform the exchange function more efficiently, a special instruction has been provided in the instruction set of the 80386DX. This is the exchange (XCHG) instruction. The form of the XCHG instruction and its allowed operands are shown in Fig. 5.5(a) and (b). Here we see that it can be used to swap data between two general-purpose registers or between a general-purpose register and a storage location in memory. In particular, it allows for the exchange of a word or double word of data between one of the general-purpose registers and the accumulator (AX or EAX), exchange of a byte, word, or double word of data between one of the general-purpose registers and a storage location in memory, or between two of the general-purpose registers.

Let us consider an example of an exchange between two internal registers. Here is a typical instruction.

```
XCHG   AX,DX
```

Its execution by the 80386DX swaps the contents of AX with that of DX. That is,

$$(AX\ original) \rightarrow (DX)$$
$$(DX\ original) \rightarrow (AX)$$

or

$$(AX) \leftrightarrow (DX)$$

Mnemonic	Meaning	Format	Operation	Flags affected
XCHG	Exchange	XCHG D, S	(D) ↔ (S)	None

(a)

Destination	Source
Accumulator	Reg16
Accumulator	Reg32
Memory	Register
Register	Register

(b)

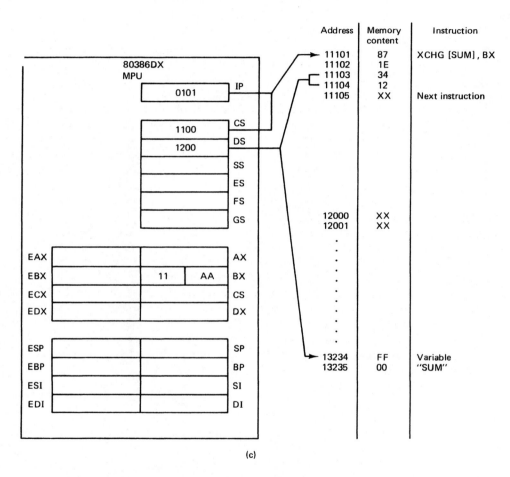

(c)

Figure 5.5 (a) Exchange data-transfer instruction. (b) Allowed operands.
(c) XCHG [SUM],BX before fetch and execution. (d) After execution.

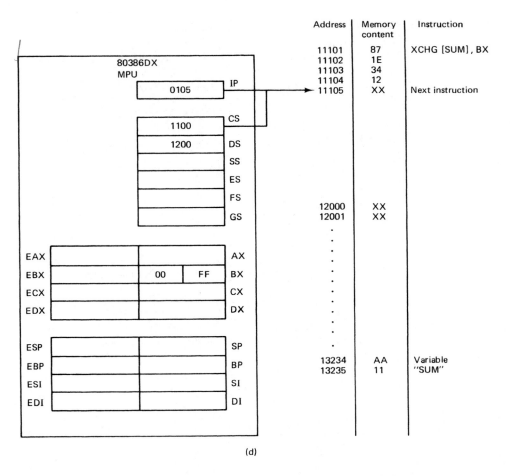

Address	Memory content	Instruction
11101	87	XCHG [SUM] , BX
11102	1E	
11103	34	
11104	12	
11105	XX	Next instruction
12000	XX	
12001	XX	
13234	AA	Variable
13235	11	"SUM"

(d)

Figure 5.5 (Continued)

EXAMPLE 5.4

For the data shown in Fig. 5.5(c), what is the result of executing the following instruction:

$$\text{XCHG} \quad [\,\text{SUM}\,]\,, \text{BX}$$

Solution

Execution of this instruction performs the function

$$((DS)0 + SUM) \leftrightarrow (BX)$$

In Fig. 5.5(c), we see that $(DS) = 1200_{16}$ and the direct address $SUM = 1234_{16}$. Therefore, the corresponding physical address is

$$PA = 12000_{16} + 1234_{16} = 13234_{16}$$

Notice that this location contains FF_{16} and the address that follows contains 00_{16}. Moreover, note that BL contains AA_{16} and BH contains 11_{16}.

Execution of the instruction performs the following 16-bit swap.

$$(13234_{16}) \leftrightarrow (BL)$$

$$(13235_{16}) \leftrightarrow (BH)$$

As shown in Fig. 5.5(d), we get

$$(BX) = 00FF_{16}$$

$$(SUM) = 11AA_{16}$$

EXAMPLE 5.5

Use the PC/AT's DEBUG program to verify the operation of the instruction in Example 5.4.

Solution

The DEBUG commands needed to enter the instruction, enter the data, execute the instruction, and verify the result of its operation are shown in Fig. 5.6. Here we see that, after

```
C:\DOS>DEBUG
-R
AX=0000  BX=0000  CX=0000  DX=0000  SP=FFEE  BP=0000  SI=0000  DI=0000
DS=1342  ES=1342  SS=1342  CS=1342  IP=0100   NV UP EI PL NZ NA PO NC
1342:0100 0F            DB      0F
-A 1100:101
1100:0101 XCHG [1234],BX
1100:0105
-R BX
BX 0000
:11AA
-R DS
DS 1342
:1200
-R CS
CS 1342
:1100
-R IP
IP 0100
:101
-R
AX=0000  BX=11AA  CX=0000  DX=0000  SP=FFEE  BP=0000  SI=0000  DI=0000
DS=1200  ES=1342  SS=1342  CS=1100  IP=0101   NV UP EI PL NZ NA PO NC
1100:0101 871E3412      XCHG    BX,[1234]                 DS:1234=0000
-E 1234 FF 00
-U 101 104
1100:0101 871E3412      XCHG    BX,[1234]
-T

AX=0000  BX=00FF  CX=0000  DX=0000  SP=FFEE  BP=0000  SI=0000  DI=0000
DS=1200  ES=1342  SS=1342  CS=1100  IP=0105   NV UP EI PL NZ NA PO NC
1100:0105 8946FE        MOV     [BP-02],AX                SS:FFFE=0000
-D 1234 1235
1200:1230             AA 11                               . .
-Q

C:\DOS>
```

Figure 5.6 Display sequence for Example 5.5.

invoking DEBUG and displaying the initial state of the 80386DX's registers, the instruction is loaded into memory with the command

```
-A  1100:101                    (↵)
1100:0101   XCHG   [1234],BX    (↵)
1100:0105                       (↵)
-
```

Next, as shown in Fig. 5.6, R commands are used to initialize the contents of registers BX, DS, CS, and IP to $11AA_{16}$, 1200_{16}, 1100_{16} and 0101_{16}, respectively, and then the updated register states are verified with another R command. Now memory locations DS:1234H and DS:1235H are loaded with the values FF_{16} and 00_{16}, respectively, with the E command

```
-E  1234  FF  00  (↵)
-
```

Before executing the instruction, its loading is verified with an unassemble command. Looking at Fig. 5.6, we see that it has been correctly loaded. Therefore, the instruction is executed by issuing the TRACE command

```
-T  (↵)
```

The displayed trace information in Fig. 5.6 shows that BX now contains $00FF_{16}$. To verify that the memory location was loaded with data from BX, we must display the data held at addresses DS:1234H and DS:1235H. This is done with the DUMP command

```
-D  1234  1235  (↵)
1200:1230              AA  11
```

In this way, we see that the word contents of memory location DS:1234H have been exchanged with the contents of the BX register.

The XLAT and XLATB Instruction

The translate instructions have been provided in the instruction set of the 80386DX to simplify implementation of the lookup-table operation. These instructions are described in Fig. 5.7. When using XLAT, the contents of register BX represent the offset of the starting address of the lookup table from the beginning of the current data segment. Also, the contents of AL represent the offset of the element to be accessed from the beginning of the lookup table. This 8-bit element address permits a table with up to 256 elements. The values in both of these registers must be initialized prior to execution of the XLAT instruction.

Mnemonic	Meaning	Format	Operation	Flags affected
XLAT	Translate	XLAT Source-table	$((AL) + (BX) + (DS)0) \rightarrow (AL)$	None
XLATB	Translate	XLATB	$((AL) + (BX) + (DS)0) \rightarrow (AL)$	None

Figure 5.7 Translate data-transfer instructions.

Execution of XLAT replaces the contents of AL by the contents of the accessed lookup table location. The physical address of this element in the table is derived as

$$PA = (DS)0 + (BX) + (AL)$$

An example of the use of this instruction is the software code conversions. For instance, an ASCII-to-EBCDIC conversion can be performed with the translate instruction. This requires an EBCDIC table in memory. The individual EBCDIC codes are located in the table at element displacements (AL) equal to their equivalent ASCII character values. That is, the EBCDIC code $C1_{16}$, which represents letter A, would be positioned at displacement 41_{16}, which equals ASCII A, from the start of the table. The start of this ASCII-to-EBCDIC table in the current data segment is specified by the contents of BX.

As an illustration of XLAT, let us assume that $(DS) = 0300_{16}$, $(BX) = 0100_{16}$, and $(AL) = 0D_{16}$. Here $0D_{16}$ represents the ASCII character CR (carriage return). Execution of XLAT replaces the contents of AL by the contents of the memory location given by

$$PA = (DS)0 + (BX) + (AL)$$
$$= 03000_{16} + 0100_{16} + 0D_{16} = 03100D_{16}$$

Thus the execution can be described by

$$(03100D_{16}) \rightarrow (AL)$$

Assuming that this memory location contains 52_{16} (EBCDIC carriage return), this value is placed in AL. That is,

$$(AL) = 52_{16}$$

When using the XLAT instruction, the use of the value in DS in the address calculation can be overridden with a segment prefix. This leads to the difference between XLAT and XLATB. The only difference between the two forms of the instruction is that XLATB does not permit the use of the SEG prefix. It must always use the contents of DS in the address calculation.

The Load Effective Address and Load Full Pointer Instructions

Another type of data-transfer operation that is important is to load a segment and a general-purpose register with an address directly from memory. Special instructions are provided in the instruction set of the 80388DX to give a programmer this capability. These instructions are described in Fig. 5.8. They are load register with effective address (LEA), load register and data segment register (LDS), and load register and extra segment register (LES), load register and the F segment register (LFS), and load register and the G segment register (LGS).

Looking at Fig. 5.8(a), we see that these instructions provide the ability to manipulate memory addresses by loading either a 16-bit or 32-bit offset address into a general-purpose register or a 16- or 32-bit offset address into a general-purpose register together with a 16-bit segment address into a specific segment register.

Mnemonic	Meaning	Format	Operation	Flags affected
LEA	Load effective address	LEA Reg16, EA LEA Reg32, EA	(EA) → (Reg16) (EA) → (Reg32)	None None
LDS	Load register and DS	LDS Reg16, EA	(EA) → (Reg16) (EA + 2) → (DS)	None
		LDS Reg32, EA	(EA) → (Reg32) (EA + 4) → (DS)	None
LSS	Load register and SS	LSS Reg16, EA	(EA) → (Reg16) (EA + 2) → (SS)	None
		LSS Reg32, EA	(EA) → (Reg32) (EA + 4) → (SS)	None
LES	Load register and ES	LES Reg16, EA	(EA) → (Reg16) (EA + 2) → (ES)	None
		LES Reg32, EA	(EA) → (Reg32) (EA + 4) → (DS)	None
LFS	Load register and FS	LFS Reg16, EA	(EA) → (Reg16) (EA + 2) → (FS)	None
		LFS Reg32, EA	(EA) → (Reg32) (EA + 4) → (FS)	None
LGS	Load register and GS	LGS Reg16, EA	(EA) → (Reg16) (EA + 2) → (GS)	None
		LGS Reg32, EA	(EA) → (Reg32) (EA + 4) → (GS)	None

(a)

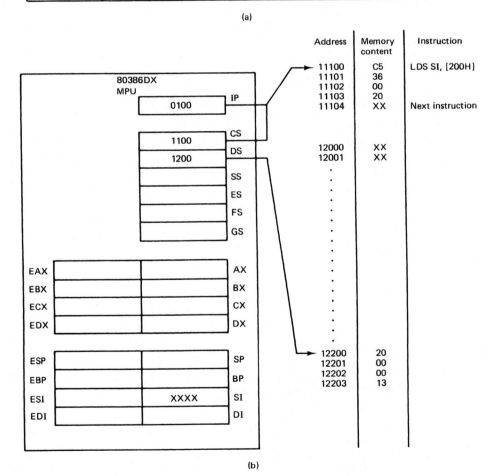

(b)

Figure 5.8 (a) Load effective address and full pointer data-transfer instructions. (b) LDS SI,[200H] before fetch and execution. (c) After execution.

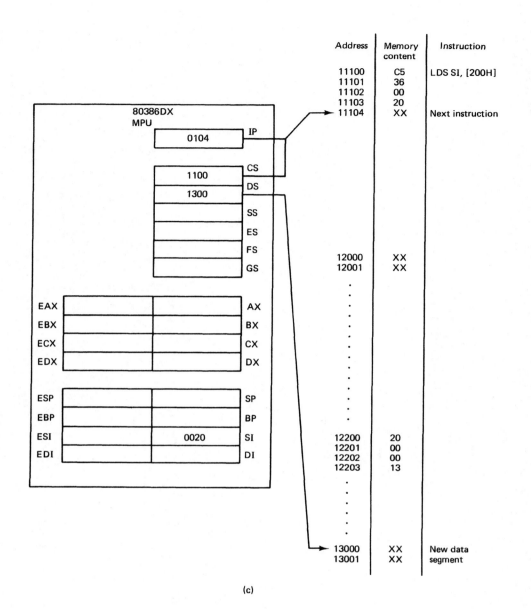

Address	Memory content	Instruction
11100	C5	LDS SI, [200H]
11101	36	
11102	00	
11103	20	
11104	XX	Next instruction
12000	XX	
12001	XX	
12200	20	
12201	00	
12202	00	
12203	13	
13000	XX	New data
13001	XX	segment

(c)

Figure 5.8   (Continued)

The LEA instruction is used to load a specified register with a 16- or 32-bit offset address. An example of this instruction is

$$\text{LEA} \quad \text{SI,} \quad [\text{EA}]$$

When executed, it loads the SI register with an offset address value. The value of this offset is represented by the effective address EA. The value of EA can be specified by any valid addressing mode. For instance, if the value in DI equals 1000H and that in BX is 20H, then executing the instruction

$$\text{LEA} \quad \text{SI,} \quad [\text{DI + BX + 5H}]$$

will load SI with the value

$$\text{EA} = 1000\text{H} + 20\text{H} + 5\text{H} = 1025\text{H}$$

That is,

$$(\text{SI}) = 1025\text{H}$$

The other five instructions, LDS, LSS, LES, LFS and LGS, are similar to LEA except that they load the specified register as well as the DS, SS, ES,FS, or GS segment register, respectively. That is, they are able to load a complete address pointer that is stored in memory. In this way, a new data segment can be activated by executing a single instruction. Notice that executing the *load register and SS* (LSS) instruction causes both the register specified in the instruction and the stack segment register to be loaded from the source operand. For example, the instruction

$$\text{LSS} \quad \text{ESP,} \quad [\text{STACK\_POINTER}]$$

causes the first 32 bits starting at memory address STACK_POINTER to be loaded into the 32-bit register ESP and the next 16 bits into the SS register.

EXAMPLE 5.6

Assuming that the 80386DX is set up as shown in Fig. 5.8(b), what is the result of executing the following instruction:

$$\text{LDS} \quad \text{SI,[200H]}$$

Solution

Execution of the instruction loads the SI register from the word location in memory whose offset address with respect to the current data segment is 200_{16}. Figure 5.8(b) shows that the contents of DS are 1200_{16}. This gives a physical address of

$$\text{PA} = 12000_{16} + 0200_{16} = 12200_{16}$$

It is the contents of this location and the one that follows that are loaded into SI. Therefore, in Fig. 5.8(c) we find that SI contains 0020_{16}. The next two bytes, that is, the contents of addresses 12202_{16} and 12203_{16}, are loaded into the DS register. As shown, this defines a new data segment address of 13000_{16}.

EXAMPLE 5.7

Verify the execution of the instruction in Example 5.6 using the DOS's debug program. The memory and register contents are to be those shown in Fig. 5.8(b).

```
C:\DOS>DEBUG
-R IP
IP 0100
:
-R CS
CS 1342
:1100
-R DS
DS 1342
:1200
-R SI
SI 0000
:
-A CS:100
1100:0100 LDS   SI,[200]
1100:0104
-E 200 20 00 00 13
-T

AX=0000  BX=0000  CX=0000  DX=0000  SP=FFEE  BP=0000  SI=0020  DI=0000
DS=1300  ES=1342  SS=1342  CS=1100  IP=0104    NV UP EI PL NZ NA PO NC
1100:0104 C0          DB      C0
-Q

C:\DOS>
```

Figure 5.9 Display sequence for Example 5.7.

Solution

As shown in Fig. 5.9, DEBUG is first brought up and then REGISTER commands are used to initialize registers IP, CS, DS, and SI with values 0100_{16}, 1100_{16}, 1200_{16}, and 0000_{16}, respectively. Next the instruction is assembled at address CS:100 with the command

```
-A  CS:100                  (↵)
1100:0100  LDS  SI,[200]    (↵)
1100:0104                   (↵)
```

Before executing the instruction, we need to initialize two words of data starting at location DS:200 in memory. As shown in Fig. 5.9, this is done with an E command.

```
-E  200  20  00  00  13  (↵)
```

Then the instruction is executed with the TRACE command

```
-T  (↵)
```

Looking at the displayed register status in Fig. 5.9, we see that SI has been loaded with the value 0020_{16} and DS with the value 1300_{16}.

Earlier we showed how the segment registers, index registers, and data registers of the MPU can be initialized with immediate data. Another way of initializing them is from a table of data in memory. Using the LDS and LES instructions along with the MOV instruction provides an efficient method for performing register initialization from a data

```
MOV  AX,[INIT_TABLE]
MOV  SS, AX
LDS  SI,[INIT_TABLE+02H]
LES  DI,[INIT_TABLE+06H]
MOV  AX,[INIT_TABLE+0AH]
MOV  BX,[INIT_TABLE+0CH]
MOV  CX,[INIT_TABLE+0EH]
MOV  DX,[INIT_TABLE+10H]
```

Figure 5.10 Initializing the internal registers of the 80386DX from a table in memory.

table. The sequence of instructions in Fig. 5.10 will load the SS, DS, and ES segment registers, index registers SI and DI, and data registers AX, BX, CX, and DX of the MPU with initial addresses and data from a table located at starting address INIT_TABLE in the memory. The table is 18 bytes long and spans the address range INIT_TABLE through INIT_ TABLE+11H.

Looking at the source addresses in the instructions, we can determine the location of each of the addresses or data elements in the table. For example, the 16-bit base address for register SS is held in the table at addresses INIT_TABLE and INIT_TABLE+1 and the word of data for DX is held at addresses INIT_TABLE+10H and INIT_TABLE+11H. This table of information must be loaded into memory before the instruction sequence is executed.

In general, the base addresses or data values are fetched from the table in memory and loaded into the appropriate registers. The table location that is to be accessed is identified by the direct address specified for the source operand. For instance, the first instruction reads a base address, which is the word content of the memory location at address INIT_TABLE, into the AX register and then this value is copied into SS with a second MOV instruction. Notice that an LDS instruction is used instead of MOV instructions to load the SI and DS registers. When this instruction is executed, the word of data for SI is loaded from table locations INIT_TABLE+2H and INIT_TABLE+3H, whereas the base address for DS is loaded from INIT_TABLE+4H and INIT_TABLE+5H.

▲ 5.3 ARITHMETIC INSTRUCTIONS

The instruction set of the 80386DX microprocessor contains a variety of *arithmetic instructions*. They include instructions for the *addition, subtraction, multiplication,* and *division* operations. These operations can be performed on numbers expressed in a variety of numeric data formats. These formats include *unsigned* or *signed binary bytes, words, or double words, unpacked* or *packed decimal bytes,* or *ASCII numbers*. Remember that by *packed decimal* we mean that two BCD digits are packed into a byte-size register or a memory location. *Unpacked decimal numbers* are stored one BCD digit per byte. The BCD numbers are unsigned decimal numbers. ASCII numbers are expressed in ASCII code and stored one number per byte.

The status that results from the execution of an arithmetic instruction is recorded in the flags of the microprocessor. The flags that are affected by the arithmetic instructions are carry flag (CF), auxiliary flag (AF), sign flag (SF), zero flag (ZF), parity flag (PF), and overflow flag (OF). Each of these flags was discussed in Chapter 2.

For the purpose of discussion, we will divide the arithmetic instructions into the subgroups shown in Fig. 5.11.

	Addition	
ADD	Add byte or word	
ADC	Add byte or word with carry	
INC	Increment byte or word by 1	
AAA	ASCII adjust for addition	
DAA	Decimal adjust for addition	
	Subtraction	
SUB	Subtract byte or word	
SBB	Subtract byte or word with borrow	
DEC	Decrement byte or word by 1	
NEG	Negate byte or word	
AAS	ASCII adjust for subtraction	
DAS	Decimal adjust for subtraction	
	Multiplication/Division	
MUL	Multiply byte or word unsigned	
IMUL	Integer multiply byte or word	
AAM	ASCII adjust for multiply	
DIV	Divide byte or word unsigned	
IDIV	Integer divide byte or word	
AAD	ASCII adjust for division	
CBW	Convert byte to word	
CWDE	Convert word to double word in EAX	
CWD	Convert word to double word in DX and AX	
CDQ	Convert double word to quad word	

Figure 5.11 Arithmetic instructions.

Mnemonic	Meaning	Format	Operation	Flags affected
ADD	Addition	ADD D, S	$(S) + (D) \rightarrow (D)$ carry $\rightarrow$ (CF)	OF, SF, ZF, AF, PF, CF
ADC	Add with carry	ADC D, S	$(S) + (D) + (CF) \rightarrow (D)$ carry $\rightarrow$ (CF)	OF, SF, ZF, AF, PF, CF
INC	Increment by 1	INC D	$(D) + 1 \rightarrow (D)$	OF, SF, ZF, AF, PF
DAA	Decimal adjust for addition	DAA		SF, ZF, AF, PF, CF OF undefined
AAA	ASCII adjust for addition	AAA		AF, CF OF, SF, ZF, PF undefined

(a)

Destination	Source
Register	Register
Register	Memory
Memory	Register
Register	Immediate
Memory	Immediate
Accumulator	Immediate

(b)

Destination
Reg16
Reg8
Memory

(c)

Figure 5.12 (a) Addition arithmetic instructions. (b) Allowed operands for ADD and ADC instructions. (c) Allowed operands for INC instruction.

Addition Instructions: ADD, ADC, INC, AAA, and DAA

The form of each of the instructions in the *addition subgroup* is shown in Fig. 5.12(a); the allowed operand variations, for all but the INC instruction, are shown in Fig. 5.12(b), and the allowed operands for the INC instruction are shown in Fig. 5.12(c). Let us begin by looking more closely at the operation of the *add* (ADD) instruction. Notice in Fig. 5.12(b) that it can be used to add an immediate operand to the contents of the accumulator, the contents of another register, or the contents of a storage location in memory. It also allows us to add the contents of two registers together or the contents of a register and a memory location.

In general, the result of executing the instruction is expressed as

$$(S) + (D) \rightarrow (D)$$

That is, the contents of the source operand are added to those of the destination operand and the sum that results is put into the location of the destination operand.

EXAMPLE 5.8

Assume that the AX and BX registers contain 1100_{16} and $0ABC_{16}$, respectively. What is the result of executing the instruction ADD AX,BX?

Solution

Execution of the ADD instruction causes the contents of source operand BX to be added to the contents of destination register AX. This gives

$$(BX) + (AX) = 0ABC_{16} + 1100_{16} = 1BBC_{16}$$

This sum ends up in destination register AX. That is,

$$(AX) = 1BBC_{16}$$

Execution of this instruction is illustrated in Fig. 5.13(a) and (b).

EXAMPLE 5.9

Use the debug program on the PC/AT to verify the execution of the instruction in Example 5.8. Assume that the registers are to be initialized with the values shown in Fig. 5.13(a).

Solution

The debug sequence for this is shown in Fig. 5.14. After the debug program is brought up, the instruction is assembled into memory with the command

```
-A  1100:0100         (↵)
1100:0100   ADD   AX,BX   (↵)
1100:0102             (↵)
-
```

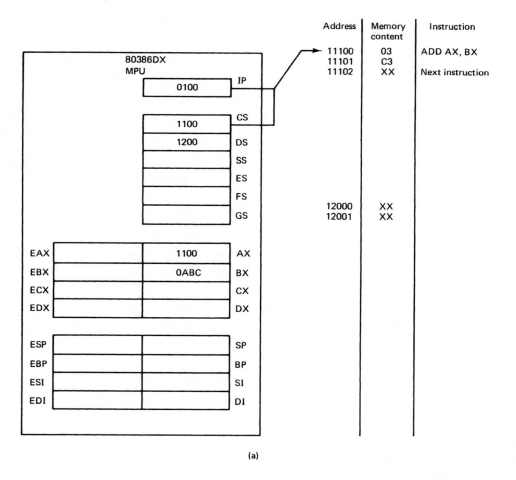

Address	Memory content	Instruction
11100	03	ADD AX, BX
11101	C3	
11102	XX	Next instruction
12000	XX	
12001	XX	

Figure 5.13 (a) ADD instruction before fetch and execution. (b) After execution.

Next, as shown in Fig. 5.14, the AX and BX registers are loaded with the values 1100_{16} and $0ABC_{16}$, respectively, using R commands.

```
-R    AX      (↵)
AX    0000
:1100         (↵)
-R    BX      (↵)
BX    0000
:0ABC         (↵)
```

Next, the loading of the instruction is verified with the unassemble command

```
-U   1100:0100   0100   (↵)
```

and is shown in Fig. 5.14 to be correct.

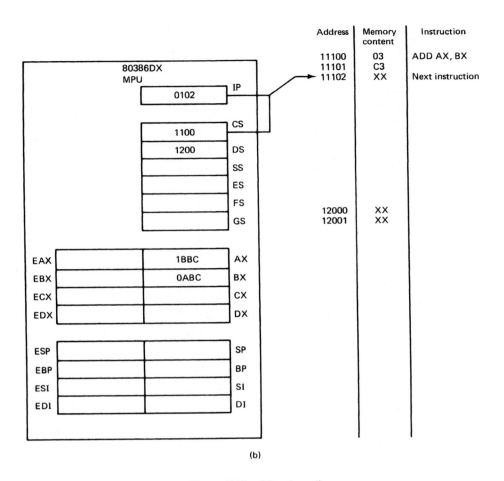

Address	Memory content	Instruction
11100	03	ADD AX, BX
11101	C3	
11102	XX	Next instruction
12000	XX	
12001	XX	

(b)

Figure 5.13 (Continued)

```
C:\DOS>DEBUG
-A 1100:0100
1100:0100 ADD AX,BX
1100:0102
-R AX
AX 0000
:1100
-R BX
BX 0000
:0ABC
-U 1100:100 100
1100:0100 01D8          ADD       AX,BX
-T =1100:100

AX=1BBC  BX=0ABC  CX=0000  DX=0000  SP=FFEE  BP=0000  SI=0000  DI=0000
DS=1342  ES=1342  SS=1342  CS=1100  IP=0102  NV UP EI PL NZ NA PO NC
1100:0102 0002          ADD       [BP+SI],AL                  SS:0000=CD
-Q

C:\DOS>
```

Figure 5.14 Display sequence for Example 5.9.

Sec. 5.3   Arithmetic Instructions

167

We are now ready to execute the instruction with the TRACE command

$$-T \quad =1100:0100 \quad (\lrcorner)$$

From the trace dump in Fig. 5.14, we see that the sum of AX and BX, which equals $1BBC_{16}$, is now held in destination register AX. Also note that no carry (NC) has occurred.

The instruction *add with carry* (ADC) works similarly to ADD. But in this case, the content of the carry flag is also added; that is.

$$(S) + (D) + (CF) \rightarrow (D)$$

The valid operand combinations are the same as those for the ADD instruction. ADC is primarily used for multiword add operations.

Another instruction that can be considered as part of the addition subgroup of arithmetic instructions is the *increment* (INC) instruction. As shown in Fig. 5.12(c), its operands can be the contents of an 8-, 16-, or 32-bit internal register or a byte, word, or double word storage location in memory. Execution of the INC instruction adds 1 to the specified operand. An example of an instruction that increments the high byte of AX is

```
INC   AH
```

Looking at Fig. 5.12(a), we see how execution of these three instructions affects the earlier mentioned flags.

EXAMPLE 5.10

The original contents of AX, BL, word-size memory location SUM, and carry flag (CF) are 1234_{16}, AB_{16}, $00CD_{16}$, and 0_{16}, respectively. Describe the results of executing the following sequence of instructions:

```
ADD   AX,[SUM]
ADC   BL,05H
INC   WORD   PTR   [SUM]
```

Solution

By executing the first instruction, we add the word in the accumulator and the word in the memory location identified as SUM. The result is placed in the accumulator. That is,

$$(AX) \leftarrow (AX) + (SUM) = 1234_{16} + 00CD_{16} = 1301_{16}$$

The carry flag remains reset.

The second instruction adds to the lower byte of the base register (BL), the immediate operand 5_{16}, and the carry flag, which is 0_{16}. This gives

$$(BL) \leftarrow (BL) + IOP + (CF) = AB_{16} + 5_{16} + 0_{16} = B0_{16}$$

Since no carry is generated, CF stays reset.

Instruction	(AX)	(BL)	(SUM)	(CF)
Initial state	1234	AB	00CD	0
ADD AX, [SUM]	1301	AB	00CD	0
ADC BL, 05H	1301	B0	00CD	0
INC WORD PTR [SUM]	1301	B0	00CE	0

Figure 5.15 Results due to execution of arithmetic instructions in Example 5.10.

The last instruction increments the contents of memory location SUM by 1. That is,

$$(SUM) \leftarrow (SUM) + 1_{16} = 00CD_{16} + 1_{16} = 00CE_{16}$$

These results are summarized in Fig. 5.15.

EXAMPLE 5.11

Verify the operation of the instruction sequence in Example 5.10 by executing it with the debug program on the PC/AT. A source program that includes this sequence of instructions is shown in Fig. 5.16(a), and the source listing produced when the program is assembled is shown in Fig. 5.16(b). A run module that was produced by linking this program is stored in file EX511.EXE.

Solution

The DEBUG program is brought up and at the same time the run module from file EX511.EXE is loaded with the command

        C:\DOS>DEBUG A:EX511.EXE   (↵)

Next, we will verify the loading of the program by unassembling it with the command

            -U 0 12   (↵)

Looking at the displayed instruction sequence in Fig. 5.16(c) and comparing it to the source listing in Fig. 5.16(b), we find that the program has loaded correctly.

Notice in Fig. 5.16(c) that the instructions for which we are interested in verifying operation start at address 0D03:000A. For this reason, a GO command will be used to execute down to this point in the program. This command is

            -G   A   (↵)

Notice in the trace information displayed for this command that now CS contains $0D03_{16}$ and IP contains $000A_{16}$; therefore, the next instruction to be executed is at address 0D03:000A. This is the ADD instruction.

Now we need to initialize registers AX and BX and the memory location pointed to by SUM (WORD PTR [0000]). We must also ensure that the CF status flag is set to NC

```
        TITLE    EXAMPLE 5.11
             PAGE      ,132
        STACK_SEG        SEGMENT          STACK 'STACK'
                         DB               64 DUP(?)
        STACK_SEG        ENDS

        DATA_SEG         SEGMENT
        SUM              DW               0CDH
        DATA_SEG         ENDS

        CODE_SEG         SEGMENT          'CODE'
        EX511    PROC    FAR
             ASSUME  CS:CODE_SEG, SS:STACK_SEG, DS:DATA_SEG

        ;To return to DEBUG program put return address on the stack

             PUSH    DS
             MOV     AX, 0
             PUSH    AX

        ;Following code implements the Example 5.11

             MOV     AX, DATA_SEG    ;Establish data segment
             MOV     DS, AX

             ADD     AX, SUM
             ADC     BL, 05H
             INC     WORD PTR SUM

             RET                     ;Return to DEBUG program
        EX511 ENDP

        CODE_SEG         ENDS

             END     EX511
```

(a)

Figure 5.16 (a) Source program for Example 5.11. (b) Source listing produced by assembler. (c) Debug session for execution of program EX511.EXE.

(no carry). In Fig. 5.16(c), we find that these operations are done with the following sequence of commands:

```
        -R   AX             (↵)
        AX   0D05
        :1234               (↵)
        -R   BX             (↵)
        BX   0000
        :AB                 (↵)
        -R   F              (↵)
        NV   UP   EI   PL   NZ   NA   PO   NC   -   (↵)
        -E   0   CD   00    (↵)
        -D   0   1          (↵)
        0D03:0000   CD   00
```

```
                     TITLE    EXAMPLE 5.11

                          PAGE       ,132

0000                              STACK_SEG      SEGMENT        STACK 'STACK'
0000        40 [                                 DB             64 DUP(?)
              ??
                ]

0040                              STACK_SEG      ENDS


0000                              DATA_SEG       SEGMENT
0000    00CD                      SUM            DW             0CDH
0002                              DATA_SEG       ENDS

0000                              CODE_SEG       SEGMENT        'CODE'
0000                              EX511  PROC    FAR
                          ASSUME  CS:CODE_SEG, SS:STACK_SEG, DS:DATA_SEG

                    ;To return to DEBUG program put return address on the stack

0000    1E                            PUSH       DS
0001    B8 0000                       MOV        AX, 0
0004    50                            PUSH       AX

                    ;Following code implements the Example 5.11

0005    B8  ---- R                    MOV        AX, DATA_SEG    ;Establish data segment
0008    8E D8                         MOV        DS, AX

000A    03 06 0000 R                  ADD        AX, SUM
000E    80 D3 05                      ADC        BL, 5H
0011    FF 06 0000 R                  INC        SUM

0015    CB                            RET                        ;Return to DEBUG program
0016                              EX511    ENDP

0016                              CODE_SEG       ENDS

                          END     EX511

Segments and groups:

            N a m e                    Size    align    combine class

CODE_SEG . . . . . . . . . . . .       0016    PARA     NONE     'CODE'
DATA_SEG . . . . . . . . . . . .       0002    PARA     NONE     NONE
STACK_SEG. . . . . . . . . . . .       0040    PARA     STACK    'STACK'

Symbols:

            N a m e                    Type    Value    Attr

EX511. . . . . . . . . . . . . .       F PROC  0000     CODE_SEG        Length =0016
SUM. . . . . . . . . . . . . . .       L WORD  0000     DATA_SEG

Warning Severe
Errors  Errors
0       0
```

(b)

Figure 5.16 (Continued)

Now we are ready to execute the ADD instruction. This is done by issuing the TRACE command

$$-T \quad (\lrcorner)$$

From the state information displayed for this command in Fig. 5.16(c), notice that the value CD_{16} has been added to the original value in AX, which was 1234_{16}, and the sum that results in AX is 1301_{16}.

Next the ADC instruction is executed with another T command.

$$-T \quad (\lrcorner)$$

```
C:DOS>DEBUG A:EX511.EXE
-U 0 12
0D03:0000 1E             PUSH    DS
0D03:0001 B80000         MOV     AX,0000
0D03:0004 50             PUSH    AX
0D03:0005 B8050D         MOV     AX,0D05
0D03:0008 8ED8           MOV     DS,AX
0D03:000A 03060000       ADD     AX,[0000]
0D03:000E 80D305         ADC     BL,05
0D03:0011 FF060000       INC     WORD PTR [0000]
-G A

AX=0D03  BX=0000  CX=0000  DX=0000  SP=003C  BP=0000  SI=0000  DI=0000
DS=0D05  ES=0CF3  SS=0D06  CS=0D03  IP=000A   NV UP EI PL NZ NA PO NC
0D03:000A 03060000       ADD     AX,[0000]                    DS:0000=00CD
-R AX
AX 0D03
:1234
-R BX
BX 0000
:AB
-R F
NV UP EI PL NZ NA PO NC  -
-E 0 CD 00
-D 0 1
0D05:0000  CD 00
-T

AX=1301  BX=00AB  CX=0000  DX=0000  SP=003C  BP=0000  SI=0000  DI=0000
DS=0D05  ES=0CF3  SS=0D06  CS=0D03  IP=000E   NV UP EI PL NZ AC PO NC
0D03:000E 80D305         ADC     BL,05
-T

AX=1301  BX=00B0  CX=0000  DX=0000  SP=003C  BP=0000  SI=0000  DI=0000
DS=0D05  ES=0CF3  SS=0D06  CS=0D03  IP=0011   NV UP EI NG NZ AC PO NC
0D03:0011 FF060000       INC     WORD PTR [0000]              DS:0000=00CD
-T

AX=1301  BX=00B0  CX=0000  DX=0000  SP=003C  BP=0000  SI=0000  DI=0000
DS=0D05  ES=0CF3  SS=0D06  CS=0D03  IP=0015   NV UP EI PL NZ NA PO NC
0D03:0015 CB             RETF
-D 0 1
0D05:0000  CE 00
-G

Program terminated normally
-Q

C:\DOS>
```

(c)

Figure 5.16   (Continued)

It causes the immediate operand value 05_{16} to be added to the original contents of BL, which is AB_{16}, and the sum that is produced in BL is $B0_{16}$.

The last instruction is executed with one more T command, and it causes the SUM (WORD PTR [0000]) to be incremented by 1. This can be verified by issuing the DUMP command

$$-D \quad 0 \quad 1 \quad (\lrcorner)$$

Notice that value of SUM is identified as a WORD PTR. This assembler directive means that the memory location for SUM is to be treated as a word-wide storage location. Similarly if a byte-wide storage location, say BYTE_LOC, is to be accessed, it would be identified as BYTE PTR [BYTE_LOC].

The addition instructions we just covered can also be used to add numbers expressed in ASCII code provided the binary result that is produced is converted back to its equivalent

ASCII representation. This eliminates the need for doing a code conversion on ASCII data prior to processing it with addition operations. Whenever the 80386DX does an addition on ASCII format data, an adjustment must be performed on the binary result to convert it to the equivalent decimal number. It is specifically for this purpose that the *ASCII adjust for addition* (AAA) instruction is provided in the instruction set. The AAA instruction should be executed immediately after the ADD instruction that adds ASCII data.

Assuming that AL contains the result produced by adding two ASCII coded numbers, execution of the AAA instruction causes the contents of AL to be replaced by its equivalent decimal value. If the sum is greater than 9, AL contains the LSD and AH is incremented by 1. Otherwise, AL contains the sum and AH is unchanged. Figure 5.12(a) shows that the AF and CF flags can be affected. Since AAA can adjust only data that are in AL, the destination register for ADD instructions that process ASCII numbers should be AL.

EXAMPLE 5.12

What is the result of executing the following instruction sequence?

```
ADD    AL,BL
AAA
```

Assume that AL contains 32_{16}, which is the ASCII code for number 2, BL contains 34_{16}, which is the ASCII code for number 4, and AH has been cleared.

Solution

Executing the ADD instruction gives

$$(AL) \leftarrow (AL) + (BL) = 32_{16} + 34_{16} = 66_{16}$$

Next, the result is adjusted to give its equivalent decimal number. This is done by execution of the AAA instruction. The equivalent of adding 2 and 4 is decimal 6 with no carry. Therefore, the result after the AAA instruction is

$$(AL) = 06_{16}$$

$$(AH) = 00_{16}$$

and both AF and CF remain cleared.

The instruction set of the 80386DX includes another instruction, called *decimal adjust for addition* (DAA). This instruction is used to perform an adjust operation similar to that performed by AAA but for the addition of packed BCD numbers instead of ASCII numbers. Information about this instruction is also provided in Fig. 5.12(a). Similar to AAA, DAA performs an adjustment on the value in AL. A typical instruction sequence is

```
ADD    AL,BL
DAA
```

Remember that the contents of AL and BL must be packed BCD numbers, that is, two BCD digits packed into a byte. The adjusted result in AL is again a packed BCD byte.

As an example of the use of the instructions covered in this section, let us perform a 32-bit binary add operation on the contents of the processor's registers. We will implement the addition

$$(DX,CX) \leftarrow (DX,CX) + (BX,AX)$$

for the following data in the registers

$$(DX,CX) = FEDCBA98_{16}$$
$$(BX,AX) = 01234567_{16}$$

We first initialize the registers with the data using move instructions as follows:

```
MOV   DX,0FEDCH
MOV   CX,0BA98H
MOV   BX,0123H
MOV   AX,4567H
```

Next, the 16 least significant bits of the 32-bit number are added with the instruction

```
ADD   CX,AX
```

Note that the result from this part of the addition is in CX and the carry flag.

To add the most significant 16 bits, we must account for the possibility of a carry out from the addition of the lower 16 bits. For this reason, the ADC instruction must be used. Thus the last instruction is

```
ADC   DX,BX
```

Execution of this instruction produces the upper 16 bits of the 32-bit sum in register DX.

Subtraction Instructions: SUB, SBB, DEC, AAS, DAS, and NEG

The instruction set of the 80386DX includes an extensive set of instructions provided for implementing subtraction. As shown in Fig. 5.17(a), the subtraction subgroup is similar to the addition subgroup. It includes instructions for subtracting a source and a destination operand, decrementing an operand, and adjusting the result of subtractions of ASCII and BCD data. An additional instruction in this subgroup is negate.

The *subtract* (SUB) instruction is used to subtract the value of a source operand from a destination operand. The result of this operation in general is given as

$$(D) \leftarrow (D) - (S)$$

As shown in Fig. 5.17(b), it can employ operand combinations similar to the ADD instruction.

The *subtract with borrow* (SBB) instruction is similar to SUB; however, it also subtracts the contents of the carry flag from the destination. That is,

$$(D) \leftarrow (D) - (S) - (CF)$$

Mnemonic	Meaning	Format	Operation	Flags affected
SUB	Subtract	SUB D,S	$(D) - (S) \to (D)$ Borrow $\to$ (CF)	OF, SF, ZF, AF, PF, CF
SBB	Subtract with borrow	SBB D,S	$(D) - (S) - (CF) \to (D)$	OF, SF, ZF, AF, PF, CF
DEC	Decrement by 1	DEC D	$(D) - 1 \to (D)$	OF, SF, ZF, AF, PF
NEG	Negate	NEG D	$0 - (D) \to (D)$ $1 \to (CF)$	OF, SF, ZF, AF, PF, CF
DAS	Decimal adjust for subtraction	DAS		SF, ZF, AF, PF, CF OF undefined
AAS	ASCII adjust for subtraction	AAS		AF, CF OF, SF, ZF, PF undefined

(a)

Destination	Source
Register	Register
Register	Memory
Memory	Register
Accumulator	Immediate
Register	Immediate
Memory	Immediate

(b)

Destination
Register
Memory

(c)

Figure 5.17 (a) Subtraction arithmetic instructions. (b) Allowed operands for SUB and SBB instructions. (c) Allowed operands for DEC and NEG instructions.

EXAMPLE 5.13

Assuming that the contents of registers BX and CX are 1234_{16} and 0123_{16}, respectively, and the carry flag is 0, what will be the result of executing the following instruction?

$$\text{SBB} \quad \text{BX,CX}$$

Solution

Since the instruction implements the operation

$$(BX) - (CX) - (CF) \to (BX)$$

we get

$$(BX) = 1234_{16} - 0123_{16} - 0_{16}$$
$$= 1111_{16}$$

Since no borrow was needed, the carry flag will stay cleared.

```
C:\DOS>DEBUG
-R
AX=0000  BX=0000  CX=0000  DX=0000  SP=FFEE  BP=0000  SI=0000  DI=0000
DS=1342  ES=1342  SS=1342  CS=1342  IP=0100   NV UP EI PL NZ NA PO NC
1342:0100 0F             DB      0F
-R BX
BX 0000
:1234
-R CX
CX 0000
:0123
-R F
NV UP EI PL NZ NA PO NC  -
-A
1342:0100 SBB BX,CX
1342:0102
-R
AX=0000  BX=1234  CX=0123  DX=0000  SP=FFEE  BP=0000  SI=0000  DI=0000
DS=1342  ES=1342  SS=1342  CS=1342  IP=0100   NV UP EI PL NZ NA PO NC
1342:0100 19CB           SBB     BX,CX
-U 100 101
1342:0100 19CB           SBB     BX,CX
-T

AX=0000  BX=1111  CX=0123  DX=0000  SP=FFEE  BP=0000  SI=0000  DI=0000
DS=1342  ES=1342  SS=1342  CS=1342  IP=0102   NV UP EI PL NZ NA PE NC
1342:0102 B98AFF         MOV     CX,FF8A
-Q

C:\DOS>
```

Figure 5.18 Display sequence for Example 5.14.

EXAMPLE 5.14

Verify the operation of the subtract instruction in Example 5.13 by repeating the example using the debug program on the PC/AT.

Solution

As shown in Fig. 5.18, we first bring up the debug program and then dump the initial state of the 80386DX with a REGISTER command. Next, we load registers BX, CX, and flag CF with the values 1234_{16}, 0123_{16}, and NC, respectively. Notice in Fig. 5.18 that this is done with three more R commands.

Now the instruction is assembled at address CS:100 with the command

```
-A                           (↵)
1342:0100   SBB   BX,CX       (↵)
1342:0102                     (↵)
-
```

Before executing the instruction, we can verify the initialization of the registers and the entry of the instruction by issuing the commands

```
-R   (↵)
```

and

```
-U  100  101  (↵)
```

Looking at Fig. 5.18, we find that the registers have been correctly initialized and that the instruction SBB BX,CX has been correctly entered.

Finally, the instruction is executed with a TRACE command. As shown in Fig. 5.18, the result of executing the instruction is that the contents of CX are subtracted from the contents of BX. The difference, which is 1111_{16}, resides in destination register BX.

Just as the INC instruction can be used to add 1 to an operand, the *decrement* (DEC) instruction can be used to subtract 1 from its operand. The allowed operands for DEC are shown in Fig. 5.17(c).

In Fig. 5.17(d), we see that the *negate* (NEG) instruction can operate on operands in a general-purpose register or a storage location in memory. Execution of this instruction causes the value of its operand to be replaced by its negative. The way this is actually done is through subtraction. That is, the contents of the specified operand are subtracted from zero and the result is returned to the operand location. The subtraction is actually performed by the processor hardware using 2's-complement arithmetic. To obtain the correct value of the carry flag that results from a NEG operation, the carry generated by the add operation used in the 2's-complement subtraction calculation must be complemented.

EXAMPLE 5.15

Assuming that register BX contains $003A_{16}$, what is the result of executing the following instruction?

$$\text{NEG} \quad \text{BX}$$

Solution

Executing the NEG instruction causes the 2's-complement subtraction that follows:

$$(BX) = 0000_{16} - (BX) = 0000_{16} + 2\text{'s-complement of } 003A_{16}$$

$$= 0000_{16} + FFC6_{16}$$

$$= FFC6_{16}$$

Since no carry is generated in this add operation, the carry flag is complemented to give

$$(CF) = 1$$

EXAMPLE 5.16

Verify the operation of the NEG instruction in Example 5.15 by executing it with the debugger on the PC/AT.

Solution

After loading the DEBUG program, we must initialize the contents of the BX register.

This is done with the command

```
-R   BX      (↵)
BX   0000
:3A          (↵)
-
```

Next the instruction is assembled with the command

```
-A                      (↵)
1342:0100   NEG   BX    (↵)
1342:0102               (↵)
-
```

At this point, we can verify the initialization of BX by issuing the command

```
-R   BX      (↵)
BX   003A
:            (↵)
-
```

To check the assembly of the instruction, we can unassemble it with the command

```
-U   100   101   (↵)
1342:0100   F7DB   NEG   BX
-
```

Now the instruction is executed with the command

```
-T   (↵)
```

The information that is dumped by issuing this command is shown in Fig. 5.19. Here the new contents in register BX are verified as FFC6$_{16}$, which is the negative of 003A$_{16}$. Also note that the carry flag is set (CY).

```
C:\DOS>DEBUG
-R BX
BX 0000
:3A
-A
1342:0100 NEG BX
1342:0102
-R BX
BX 003A
:
-U 100 101
1342:0100 F7DB          NEG       BX
-T

AX=0000  BX=FFC6  CX=0000  DX=0000  SP=FFEE  BP=0000  SI=0000  DI=0000
DS=1342  ES=1342  SS=1342  CS=1342  IP=0102    NV UP EI NG NZ AC PE CY
1342:0102 B98AFF          MOV       CX,FF8A
-Q

C:\DOS>
```

Figure 5.19   Display sequence for Example 5.15.

In our study of the addition instruction subgroup, we found that the 80386DX is capable of directly adding ASCII and BCD numbers. The SUB and SBB instructions can subtract numbers represented in these formats as well. Just as for addition, the results that are obtained must be adjusted to produce the corresponding decimal numbers. In the case of ASCII subtraction, we use the *ASCII adjust for subtraction* (AAS) instruction, and for packed BCD subtraction we use the *decimal adjust for subtract* (DAS) instruction.

An example of an instruction sequence for direct ASCII subtraction is

```
SUB   AL,BL
AAS
```

ASCII numbers must be loaded into AL and BL before execution of the subtract instruction. Notice that the destination of the subtraction should be AL. After execution of AAS, AL contains the difference of the two numbers, and AH is unchanged if no borrow takes place or is decremented by 1 if a borrow occurs.

As an example, let us implement a 32-bit subtraction of two numbers X and Y that are stored in memory as

$$X = (DS:203H)(DS:202H)(DS:201H)(DS:200H)$$

| MS byte of MS word | LS byte of MS word | MS byte of LS word | LS byte of LS word |

$$Y = (DS:103H)(DS:102H)(DS:101H)(DS:100H)$$

The result of X − Y is to be saved in the place where X is stored in memory.

First, we subtract the least significant 16 bits of the 32-bit words using the instructions

```
MOV   AX,[200H]
SUB   AX,[100H]
MOV   [200H],AX
```

Next, the most significant words of X and Y are subtracted. In this part of the 32-bit subtraction, we must use the borrow that might have been generated by the subtraction of the least significant words. Therefore, SBB must be used to perform the subtraction operation. The instructions to do this are

```
MOV   AX,[200H]
SBB   AX,[100H]
MOV   [200H],AX
```

These instructions use direct addressing to access the data in memory. The 32-bit subtract operation can also be done with indirect addressing as follows:

```
MOV  SI,200H  ;Initialize pointer for X
MOV  DI,100H  ;Initialize pointer for Y
MOV  AX,[SI]  ;Subtract LS words
SUB  AX,[DI]
MOV  [SI],AX  ;Save the LS word of result
ADD  SI,2     ;Update pointer for X
ADD  DI,2     ;Update pointer for Y
MOV  AX,[SI]  ;Subtract MS words
SBB  AX,[DI]
MOV  [SI],AX  ;Save the MS word of result
```

Multiplication and Division Instructions: MUL, DIV, IMUL, IDIV, AAM, AAD, CBW, CWDE, CWD, and CDQ

The 80386DX has instructions to support multiplication and division of binary and BCD numbers. Two basic types of multiplication and division instructions, for the processing of unsigned numbers and signed numbers, are available. To do these operations on unsigned numbers, the instructions are MUL and DIV. On the other hand, to multiply or divide 2's-complement signed numbers, the instructions are IMUL and IDIV.

Figure 5.20(a) describes these instructions. Notice in Fig. 5.20(b) that a single byte-wide word-wide, or double-word-wide operand is specified in a multiplication instruction. It is the source operand. As shown in Fig. 5.20(a), the other operand, which is the destination, is assumed already to be in AL for 8-bit multiplication, in AX for 16-bit multiplication, or in EAX for 32-bit multiplication. The allowed operands for IMUL are shown in Fig. 5.20(c).

The result of executing a MUL or IMUL instruction on byte data can be represented as

$$(AX) \leftarrow (AL) \times (8\text{-bit operand})$$

That is, the resulting 16-bit product is produced in the AX register. On the other hand, for multiplication of data words, the 32-bit result is given by

$$(DX,AX) \leftarrow (AX) \times (16\text{-bit operand})$$

where AX contains the 16 LSBs and DX the 16 MSBs. Finally, multiplying double-word data produces the result

$$(EDX,EAX) \leftarrow (EAX) \times (32\text{-bit operand})$$

Here EAX contains the 32 LSBs of the product, and EDX contains the 32 MSBs.

For the division operation, again just the source operand is specified. The other operand is either the contents of AX for 16-bit dividends, the contents of both DX and AX for 32-bit dividends, or EDX and EAX for 64-bit dividends. The result of a DIV or IDIV instruction for an 8-bit divisor is represented by

$$(AH),(AL) \leftarrow (AX)/(8\text{-bit operand})$$

Mnemonic	Meaning	Format	Operation	Flags affected
MUL	Multiply (unsigned)	MUL S	$(AL) \cdot (S8) \rightarrow (AX)$ $(AX) \cdot (S16) \rightarrow (DX), (AX)$ $(EAX) \cdot (S32) \rightarrow (EDX), (EAX)$	OF, CF SF, ZF, AF, PF undefined
DIV	Division (unsigned)	DIV S	(1) $Q ((AX)/(S8)) \rightarrow (AL)$ $R ((AX)/(S8)) \rightarrow (AH)$ (2) $Q((DX, AX)/(S16)) \rightarrow (AX)$ $R ((DX, AX)/(S16)) \rightarrow (DX)$ (3) $Q ((EDX, EAX)/(S32)) \rightarrow (EAX)$ $R ((EDX, EAX)/(S32)) \rightarrow (EDX)$ If Q is FF_{16} in case (1), $FFFF_{16}$ in case (2), or $FFFFFFFF_{16}$ in case (3), then type 0 interrupt occurs	All flags undefined
IMUL	Integer multiply (signed)	IMUL S	$(AL) \cdot (S8) \rightarrow (AX)$ $(AX) \cdot (S16) \rightarrow (DX), (AX)$ $(EAX) \cdot (S32) \rightarrow (EDX), (EAX)$	OF, CF SF, ZF, AF, PF undefined
		IMUL R, I	$(R16) \cdot (Imm8) \rightarrow (R16)$ $(R32) \cdot (Imm8) \rightarrow (R32)$ $(R16) \cdot (Imm16) \rightarrow (R16)$ $(R32) \cdot (Imm32) \rightarrow (R32)$	OF, CF SF, ZF, AF, PF undefined
		IMUL R, S, I	$(S16) \cdot (Imm8) \rightarrow (R16)$ $(S32) \cdot (Imm8) \rightarrow (R32)$ $(S16) \cdot (Imm16) \rightarrow (R16)$ $(S32) \cdot (Imm32) \rightarrow (R32)$	OF, CF SF, ZF, AF, PF undefined
		IMUL R, S	$(R16) \cdot (S16) \rightarrow (R16)$ $(R32) \cdot (S32) \rightarrow (R32)$	OF, CF SF, ZF, AF, PF undefined
IDIV	Integer divide (signed)	IDIV S	(1) $Q ((AX)/(S8)) \rightarrow (AL)$ $R ((AX)/(S8)) \rightarrow (AH)$ (2) $Q ((DX, AX)/(S16)) \rightarrow (AX)$ $R ((DX, AX)/(S16)) \rightarrow (DX)$ (3) $Q ((EDX, EAX)/(S32)) \rightarrow (EAX)$ $R ((EDX, EAX)/(S32)) \rightarrow (EDX)$ If Q is positive and exceeds $7FFF_{16}$ or if Q is negative and becomes less than 8001_{16}, then type 0 interrupt occurs	All flags undefined
AAM	Adjust AL after multiplication	AAM	$Q ((AL)/10) \rightarrow (AH)$ $R ((AL)/10) \rightarrow (AL)$	SF, ZF, PF OF, AF, CF undefined
AAD	Adjust AX before division	AAD	$(AH) \cdot 10 + (AL) \rightarrow (AL)$ $00 \rightarrow (AH)$	SF, ZF, PF OF, AF, CF undefined
CBW	Convert byte to word	CBW	(MSB of AL) $\rightarrow$ (All bits of AH)	None
CWDE	Convert word to double word	CWDE	(MSB of AX) $\rightarrow$ (16 MSBs of EAX)	None
CWD	Convert word to double word	CWD	(MSB of AX) $\rightarrow$ (All bits of DX)	None
CDQ	Convert double word to quad word	CDQ	(MSB of EAX) $\rightarrow$ (All bits of EDX)	None

(a)

Figure 5.20   (a) Multiplication and division arithmetic instructions. (b) Allowed operands for MUL, DIV, and IDIV instructions. (c) Allowed operands for IMUL instruction.

Destination	Source
	Reg8
	Reg16
	Reg32
Reg16	Imm8
Reg32	Imm8
Reg16	Imm16
Reg32	Imm32
Reg16	Reg16, Imm8
Reg16	Mem16, Imm8
Reg32	Reg32, Imm8
Reg32	Mem32, Imm8
Reg16	Reg16, Imm16
Reg16	Mem16, Imm16
Reg32	Reg16, Imm32
Reg32	Mem16, Imm32
Reg16	Reg16
Reg16	Mem16
Reg32	Reg32
Reg32	Mem32

Source
Reg8
Reg16
Reg32
Mem8
Mem16
Mem32

(b) (c)

Figure 5.20 (Continued)

where AH contains the remainder and AL contains the quotient. For 16-bit division, we get

$$(DX),(AX) \leftarrow (DX,AX)/(\text{16-bit operand})$$

Here AX contains the quotient and DX contains the remainder. Finally, for a 32-bit division, the result is given by

$$(EDX),(EAX) \leftarrow (EDX,EAX)/(\text{32-bit operand})$$

The quotient is in EAX and the remainder is in EDX.

EXAMPLE 5.17

The 2's-complement signed data contents of AL equal -1 and the contents of CL are -2. What result will be produced in AX by executing the following instructions?

```
MUL   CL
```

and

```
IMUL   CL
```

Solution

As binary data, the contents of AL and CL are

$$(AL) = -1 \ (\text{as 2's complement}) = 11111111_2 = FF_{16}$$

$$(CL) = -2 \ (\text{as 2's complement}) = 11111110_2 = FE_{16}$$

Executing the MUL instruction gives

$$(AX) = 11111111_2 \times 11111110_2 = 1111110100000010_2$$
$$= FD02_{16}$$

The second instruction multiplies the same two numbers as signed numbers to generate the signed result. That is,

$$(AX) = -1_{16} \times -2_{16}$$
$$= 2_{16} = 0002H$$

EXAMPLE 5.18

Verify the operation of the MUL instruction in Example 5.17 by performing the same operation on the PC/AT with the debug program.

Solution

First the DEBUG program is loaded and then registers AX and CX are initialized with the values FF_{16} and FE_{16}, respectively. These registers are loaded as follows:

```
-R   AX      (↵)
AX   0000
:FF          (↵)
-R   CX      (↵)
CX   0000
:FE          (↵)
-
```

Next, the instruction is loaded with the command

```
-A              (↵)
1342:0100   MUL  CL
1342:0102       (↵)
-
```

Before executing the instruction, let us verify the loading of AX, CX, and the instruction. To do this, we use the commands as follows:

```
-R   AX         (↵)
AX   00FF
:               (↵)
-R   CX         (↵)
CX   00FE
:               (↵)
-U  100   101  (↵)
1342:0100   F6E1   MUL   CL
-
```

```
C:\DOS>DEBUG
-R AX
AX 0000
:FF
-R CX
CX 0000
:FE
-A
1342:0100 MUL CL
1342:0102
-R AX
AX 00FF
:
-R CX
CX 00FE
:
-U 100 101
1342:0100 F6E1          MUL    CL
-T

AX=FD02  BX=0000  CX=00FE  DX=0000  SP=FFEE  BP=0000  SI=0000  DI=0000
DS=1342  ES=1342  SS=1342  CS=1342  IP=0102   OV UP EI NG NZ AC PE CY
1342:0102 B98AFF          MOV    CX,FF8A
-Q

C:\DOS>
```

Figure 5.21 Display sequence for Example 5.18.

To execute the instruction, we issue the T command.

$$-T \quad (\hookleftarrow)$$

The displayed result in Fig. 5.21 shows that AX now contains $FD02_{16}$, which is the unsigned product of FF_{16} and FE_{16}.

As shown in Fig. 5.20(a), adjust instructions for BCD multiplication and division are also provided. They are *adjust AX for multiply* (AAM) and *adjust AX for divide* (AAD). The multiplication performed just before execution of the AAM instruction is assumed to have been performed on two unpacked BCD numbers with the product produced in AL. The AAD instruction converts the result in AH and AL to unpacked BCD numbers.

The division instructions can also be used to divide an 8-bit dividend in AL by an 8-bit divisor. However, to do this, the sign of the dividend must first be extended to fill the AX register. That is, AH is filled with 0s if the number in AL is positive or with 1s if it is negative. This conversion is automatically done by executing the *convert byte to word* (CBW) instruction. Notice that the sign extension does not change the signed value for the data. It simply allows data to be represented using more bits.

In a similar way, the 32-bit by 16-bit division instructions can be used to divide a 16-bit dividend in AX by a 16-bit divisor. In this case, the sign bit of AX must be extended by 16 bits into the DX register. This can be done by another instruction, which is known as *convert word to double word* (CWD). These two sign-extension instructions are also shown in Fig. 5.20(a).

Notice that the CBW, CWDE, CWD, and CDQ instructions are provided to handle operations where the result or intermediate results of an operation cannot be held in the correct word length for use in other arithmetic operations. Using these instructions, we can extend a byte, word, or double word of data to its equivalent word, double word, or quad word.

EXAMPLE 5.19 _____

What is the result of executing the following sequence of instructions?

```
MOV  AL,0A1H
CBW
CWD
```

Solution

The first instruction loads AL with $A1_{16}$. This gives

$$(AL) = A1_{16} = 10100001_2$$

Executing the second instruction extends the most significant bit of AL, which is 1, into all bits of AH. The result is

$$(AH) = 11111111_2 = FF_{16}$$

or

$$(AX) = 1111111110100001_2 = FFA1_{16}$$

This completes conversion of the byte in AL to a word in AX.

The last instruction loads each bit of DX with the most significant bit of AX. This bit is also 1. Therefore, we get

$$(DX) = 1111111111111111_2 = FFFF_{16}$$

Now the word in AX has been extended to the double word. That is,

$$(AX) = FFA1_{16}$$
$$(DX) = FFFF_{16}$$

EXAMPLE 5.20 _____

Use an assembled version of the program in Example 5.19 to verify the results obtained when it is executed.

Solution

The source program is shown in Fig. 5.22(a). Notice that this program differs from that described in Example 5.19 in that it includes the pseudo-op statements that are needed to assemble it and some additional instructions so that it can be executed using the debug program.

The source file is assembled with the macroassembler and then linked with the LINK program to give a run module stored in the file EX520.EXE. The source listing (EX520.LST) produced by assembling the source file, which is called EX520.ASM, is shown in Fig. 5.22(b).

```
                TITLE    EXAMPLE 5.20

                    PAGE        ,132

    STACK_SEG              SEGMENT       STACK 'STACK'
                          DB            64 DUP(?)
    STACK_SEG             ENDS


    CODE_SEG              SEGMENT       'CODE'
    EX520   PROC          FAR
            ASSUME  CS:CODE_SEG, SS:STACK_SEG

    ;To return to DEBUG program put return address on the stack

            PUSH    DS
            MOV     AX, 0
            PUSH    AX

    ;Following code implements Example 5.20

            MOV     AL, 0A1H
            CBW
            CWD

            RET                         ;Return to DEBUG program
    EX520 ENDP

    CODE_SEG        ENDS

            END     EX520
```

(a)

```
                TITLE    EXAMPLE 5.20

                    PAGE        ,132

0000                          STACK_SEG              SEGMENT       STACK 'STACK'
0000    40 [                                        DB            64 DUP(?)
        ??
          ]

0040                          STACK_SEG             ENDS


0000                          CODE_SEG              SEGMENT       'CODE'
0000                          EX520   PROC          FAR
                              ASSUME  CS:CODE_SEG, SS:STACK_SEG

                  ;To return to DEBUG program put return address on the stack

0000  1E                              PUSH    DS
0001  B8 0000                         MOV     AX, 0
0004  50                              PUSH    AX

                  ;Following code implements Example 5.20

0005  B0 A1                           MOV     AL, 0A1H
0007  98                              CBW
0008  99                              CWD

0009  CB                              RET                         ;Return to DEBUG program
000A                          EX520   ENDP

000A                          CODE_SEG        ENDS

                              END     EX520
```

```
Segments and groups:

            N a m e                    Size    align   combine class

CODE_SEG . . . . . . . . . . . .       000A    PARA    NONE    'CODE'
STACK_SEG. . . . . . . . . . . .       0040    PARA    STACK   'STACK'

Symbols:

            N a m e                    Type    Value   Attr

EX520. . . . . . . . . . . . .         F PROC  0000    CODE_SEG        Length =000A

Warning Severe
Errors  Errors
0       0
```

(b)

Figure 5.22 (a) Source program for Example 5.20. (b) Source listing produced by assembler. (c) Debug session for execution of program EX520.EXE.

```
C:\DOS>DEBUG A:EX520.EXE
-U 0 9
0D03:0000 1E          PUSH    DS
0D03:0001 B80000      MOV     AX,0000
0D03:0004 50          PUSH    AX
0D03:0005 B0A1        MOV     AL,A1
0D03:0007 98          CBW
0D03:0008 99          CWD
0D03:0009 CB          RETF
-G 5

AX=0000  BX=0000  CX=0000  DX=0000  SP=003C  BP=0000  SI=0000 DI=0000
DS=0CF3  ES=0CF3  SS=0D04  CS=0D03  IP=0005   NV UP EI PL NZ NA PO NC
0D03:0005 B0A1          MOV     AL,A1
-T

AX=00A1  BX=0000  CX=0000  DX=0000  SP=003C  BP=0000  SI=0000 DI=0000
DS=0CF3  ES=0CF3  SS=0D04  CS=0D03  IP=0007   NV UP EI PL NZ NA PO NC
0D03:0007 98           CBW
-T

AX=FFA1  BX=0000  CX=0000  DX=0000  SP=003C  BP=0000  SI=0000 DI=0000
DS=0CF3  ES=0CF3  SS=0D04  CS=0D03  IP=0008   NV UP EI PL NZ NA PO NC
0D03:0008 99           CWD
-T

AX=FFA1  BX=0000  CX=0000  DX=FFFF  SP=003C  BP=0000  SI=0000 DI=0000
DS=0CF3  ES=0CF3  SS=0D04  CS=0D03  IP=0009   NV UP EI PL NZ NA PO NC
0D03:0009 CB           RETF
-G

Program terminated normally
-Q

C:\DOS>
```

(c)

Figure 5.22   (Continued)

As shown in Fig. 5.22(c), the run module is loaded for execution as part of calling up the DEBUG program. This is done with the command

$$C:\DOS>DEBUG \quad A:EX520.EXE \quad (\lrcorner)$$

The loading of the program can now be verified with the UNASSEMBLE command

$$-U \quad 0 \quad 9 \quad (\lrcorner)$$

Looking at the instructions displayed in Fig. 5.22(c), we see that the program is correct.

From the unassembled version of the program in Fig. 5.22(c), we find that the instructions in which we are interested start at address 0D03:0005. Thus we execute the instructions prior to the MOV AL,A1 instruction by issuing the command

$$-G \quad 5 \quad (\lrcorner)$$

The state information that is displayed in Fig. 5.22(c) shows that $(AX) = 0000_{16}$ and $(DX) = 0000_{16}$. Moreover, $(IP) = 0005_{16}$ and points to the first instruction in which we are interested. This instruction is executed with the command

$$-T \quad (\lrcorner)$$

In the trace dump information of Fig. 5.22(c), we see that AL has been loaded with $A1_{16}$ and DX contains 0000_{16}.

Now the second instruction is executed with the command

$$-T \quad (\lrcorner)$$

Again looking at the trace information, we see that AX now contains the value $FFA1_{16}$ and DX still contains 0000_{16}. This shows that the byte in AL has been extended to a word in AX.

To execute the third instruction, the command is

$$-T \quad (\lrcorner)$$

Then, looking at the trace information produced, we find that AX still contains $FFA1_{16}$ and the value in DX has changed to $FFFF_{16}$. This shows that the word in AX has been extended to a double word in DX and AX.

To run the program to completion, enter the command

$$-G \quad (\lrcorner)$$

This executes the remaining instructions, which cause control to be returned to the DE-BUG program.

▲ 5.4 LOGIC INSTRUCTIONS

The 80386DX has instructions for performing the logic operations *AND, OR, exclusive-OR*, and *NOT*. As shown in Fig. 5.23(a), the AND, OR, and XOR instructions perform their respective logic operations bit by bit on the specified source and destination operands,

Mnemonic	Meaning	Format	Operation	Flags Affected
AND	Logical AND	AND D,S	$(S) \cdot (D) \rightarrow (D)$	OF, SF, ZF, PF, CF AF undefined
OR	Logical Inclusive-OR	OR D,S	$(S) + (D) \rightarrow (D)$	OF, SF, ZF, PF, CF AF undefined
XOR	Logical Exclusive-OR	XOR D,S	$(S) \oplus (D) \rightarrow (D)$	OF, SF, ZF, PF, CF AF undefined
NOT	Logical NOT	NOT D	$(\overline{D}) \rightarrow (D)$	None

(a)

Destination	Source
Register	Register
Register	Memory
Memory	Register
Register	Immediate
Memory	Immediate
Accumulator	Immediate

(b)

Destination
Register
Memory

(c)

Figure 5.23 (a) Logic instructions. (b) Allowed operands for the AND, OR, and XOR instructions. (c) Allowed operands for NOT instruction.

the result being represented by the final contents of the destination operand. Figure 5.23(b) shows the allowed operand combinations for the AND, OR, and XOR instructions. For example, the instruction

```
AND   AX,BX
```

causes the contents of BX to be ANDed with the contents of AX. The result is reflected by the new contents of AX. For instance, if AX contains 1234_{16} and BX contains $000F_{16}$, the result produced by the instruction is

$$1234_{16} \bullet 000F_{16} = 0001001000110100_2 \bullet 0000000000001111_2$$

$$= 0000000000000100_2$$

$$= 0004_{16}$$

This result is stored in the destination operand. This result is

$$(AX) = 0004_{16}$$

Notice that the 12 most significant bits are all 0. In this way we see how the AND instruction is used to mask the 12 most significant bits of the destination operand.

The NOT logic instruction differs from those for AND, OR, and exclusive-OR in that it operates on a single operand. Looking at Fig. 5.23(c), which shows the allowed operands for the NOT instruction, we see that this operand can be the contents of an internal register or a location in memory.

EXAMPLE 5.21

Describe the result of executing the following sequence of instructions.

```
MOV   AL,01010101B
AND   AL,00011111B
OR    AL,11000000B
XOR   AL,00001111B
NOT   AL
```

Here B is used to specify a binary number.

Solution

The first instruction moves the immediate operand 01010101_2 into the AL register. This loads the data that are to be manipulated with the logic instructions. The next instruction performs a bit-by-bit AND operation of the contents of AL with immediate operand 00011111_2. This gives

$$01010101_2 \bullet 00011111_2 = 00010101_2$$

This result is placed in destination register AL:

$$(AL) = 00010101_2 = 15_{16}$$

Note that this operation has masked off the 3 most significant bits of AL.

Instruction	(AL)
MOV AL,01010101B	01010101
AND AL,00011111B	00010101
OR AL,11000000B	11010101
XOR AL,00001111B	11011010
NOT AL	00100101

Figure 5.24 Results of example program using logic instructions.

The third instruction performs a bit-by-bit logical OR of the present contents of AL with immediate operand $C0_{16}$. This gives

$$00010101_2 + 11000000_2 = 11010101_2$$

$$(AL) = 11010101_2 = D5_{16}$$

This operation is equivalent to setting the two most significant bits of AL.

The fourth instruction is an exclusive-OR operation of the contents of AL with immediate operand 00001111_2. We get

$$11010101_2 \oplus 00001111_2 = 11011010_2$$

$$(AL) = 11011010_2 = DA_{16}$$

Note that this operation complements the logic state of those bits in AL that are 1s in the immediate operand.

The last instruction, NOT AL, inverts each bit of AL. Therefore, the final contents of AL become

$$(AL) = \overline{11011010}_2 = 00100101_2 = 25_{16}$$

These results are summarized in Fig. 5.24.

EXAMPLE 5.22

Use the PC/AT's debug program to verify the operation of the program in Example 5.21.

Solution

After the debug program is brought up, the line-by-line assembler is used to enter the program, as shown in Fig. 5.25. The first instruction is executed by issuing the T command

$$-T \quad (\llcorner\lrcorner)$$

The trace dump given in Fig. 5.25 shows that the value 55_{16} has been loaded into the AL register.

The second instruction is executed by issuing another T command.

$$-T \quad (\llcorner\lrcorner)$$

```
C:\DOS>DEBUG
-A
1342:0100 MOV AL,55
1342:0102 AND AL,1F
1342:0104 OR AL,C0
1342:0106 XOR AL,0F
1342:0108 NOT AL
1342:010A
-T

AX=0055  BX=0000  CX=0000  DX=0000  SP=FFEE  BP=0000  SI=0000  DI=0000
DS=1342  ES=1342  SS=1342  CS=1342  IP=0102    NV UP EI PL NZ NA PO NC
1342:0102 241F          AND     AL,1F
-T

AX=0015  BX=0000  CX=0000  DX=0000  SP=FFEE  BP=0000  SI=0000  DI=0000
DS=1342  ES=1342  SS=1342  CS=1342  IP=0104    NV UP EI PL NZ NA PO NC
1342:0104 0CC0          OR      AL,C0
-T

AX=00D5  BX=0000  CX=0000  DX=0000  SP=FFEE  BP=0000  SI=0000  DI=0000
DS=1342  ES=1342  SS=1342  CS=1342  IP=0106    NV UP EI NG NZ NA PO NC
1342:0106 340F          XOR     AL,0F
-T

AX=00DA  BX=0000  CX=0000  DX=0000  SP=FFEE  BP=0000  SI=0000  DI=0000
DS=1342  ES=1342  SS=1342  CS=1342  IP=0108    NV UP EI NG NZ NA PO NC
1342:0108 F6D0          NOT     AL
-T

AX=0025  BX=0000  CX=0000  DX=0000  SP=FFEE  BP=0000  SI=0000  DI=0000
DS=1342  ES=1342  SS=1342  CS=1342  IP=010A    NV UP EI NG NZ NA PO NC
1342:010A 2B04          SUB     AX,[SI]                   DS:0000=20CD
-Q

C:\DOS>
```

Figure 5.25 Display sequence for Example 5.22.

Execution of this instruction causes $1F_{16}$ to be ANDed with the value 55_{16} in AL. Looking at the trace information displayed in Fig. 5.25, we see that the 3 most significant bits of AL have been masked off to produce the result 15_{16}.

The third instruction is executed in the same way.

$$-T \quad (\lrcorner)$$

It causes the value $C0_{16}$ to be ORed with the value $1F_{16}$ in AL. This gives the result $D5_{16}$ in AL.

A fourth T command is used to execute the XOR instruction.

$$-T \quad (\lrcorner)$$

and the trace dump that results shows that the new value in AL is DA_{16}.

The last instruction is a NOT instruction, and its execution with the command

$$-T \quad (\lrcorner)$$

causes the bits of DA_{16} to be inverted. This produces 25_{16} as the final result in AL.

A common use of logic instructions is to mask a group of bits of a byte, word, or double word of data. By *mask*, we mean to clear the bit or bits to zero. Remember that when a bit is ANDed with another bit that is at logic 0, the result will always be 0. On the other hand, if a bit is ANDed with a bit that is at logic 1, its value will remain unchanged. Thus we see that the bits that are to be masked must be set to 0 in the *mask*, which is the source operand, and those that are to remain unchanged are set to 1. For instance, in the instruction

```
AND   AX,  000FH
```

the *mask* equals $000F_{16}$; therefore, it would mask off the upper 12 bits of the word of data in destination AX. Let us assume that the original value in AX is $FFFF_{16}$. Then executing the instruction performs the operation

$$0000000000001111_2 \cdot 1111111111111111_2 = 0000000000001111_2$$

$$(AX) = 000F_{16}$$

This shows that just the lower 4 bits in AX remain intact.

The OR instruction can be used to set a bit or bits in a register or a storage location in memory to logic 1. If a bit is ORed with another bit that is 0, the value of the bit remains unchanged; however, if it is ORed with another bit that is 1, the bit will become 1. For instance, let us assume that we want to set bit B_4 of the byte at the offset address CONTROL_FLAGS in the current data segment of memory to logic 1. This can be done with the following instruction sequence.

```
MOV   AL,[CONTROL_FLAGS]
OR    AL,10H
MOV   [CONTROL_FLAGS],AL
```

First the value of the flags are copied into AL and the logic operation

$$(AL) = XXXXXXXX_2 + 00010000_2 = XXX1XXXX_2$$

is performed. Finally, the new byte in AL, which has bit B_4 set to 1, is written back to the memory location called CONTROL_FLAGS.

▲ 5.5 SHIFT INSTRUCTIONS

The six *shift instructions* of the 80386DX can perform two basic types of shift operations. They are the *logical shift* and the *arithmetic shift*. Moreover, each of these operations can be performed to the right or to the left. The shift instructions are *shift logical left* (SHL), *shift arithmetic left* (SAL), *shift logical right* (SHR), *shift arithmetic right* (SAR), *double-precision shift left* (SHLD), and *double-precision shift right* (SHRD).

The logical shift instructions, SHL and SHR, are described in Fig. 5.26(a). Notice in Fig. 5.26(b) that the destination operand, the data whose bits are to be shifted, can be either the contents of an internal register or a storage location in memory. Moreover, the source operand can be specified in three ways. If it is assigned the value of 1, a 1-bit shift will take place. For instance, as illustrated in Fig. 5.27(a), executing

```
SHL   AX,1
```

Mnemonic	Meaning	Format	Operation	Flags affected
SAL/SHL	Shift arithmetic left/shift logical left	SAL/SHL D, Count	Shift the (D) left by the number of bit positions equal to Count and fill the vacated bit positions on the right with zeros. The last bit shifted out at MSB end is in CF.	SF, ZF, PF, CF AF undefined OF undefined if Count ≠1
SHR	Shift logical right	SHR D, Count	Shift the (D) right by the number of bit positions equal to Count and fill the vacated bit positions on the left with zeros. The last bit shifted out at the LSB end is in CF.	SF, ZF, PF, CF AF undefined OF undefined if Count ≠1
SAR	Shift arithmetic right	SAR D, Count	Shift the (D) right by the number of bit positions equal to Count and fill the vacated bit positions on the left with the original most significant bit. The last bit shifted out of the LSB end is in CF.	SF, ZF, PF, CF AF undefined OF undefined if Count ≠1
SHLD	Double precision shift left	SHLD D1, D2, Count	Shift (D1) left by the number of bit positions equal to Count. (D2) supplies the bits to be loaded into D1. Bits are shifted from the MSB of D2 to the LSB of D1. The value in D2 is unchanged. The last bit shifted out of D1 is in CF.	SF, ZF, PF, CF OF, AF undefined
SHRD	Double precision shift right	SHLD D1, D2, Count	Shift (D1) right by the number of bit positions equal to Count. (D2) supplies the bits to be loaded into D1. Bits are shifted from the LSB of D2 into the MSB of D1. The value in D2 is unchanged. The last bit shifted out of D1 is in CF.	SF, ZF, PF, CF OF, AF undefined

(a)

Destination	Count
Register	1
Register	CL
Register	Imm8
Memory	1
Memory	CL
Memory	Imm8

(b)

Destination 1	Destination 2	Count
Reg16	Reg16	Imm8
Reg16	Reg16	CL
Reg32	Reg32	Imm8
Reg32	Reg32	CL
Mem16	Reg16	Imm8
Mem16	Reg16	CL
Mem32	Reg32	Imm8
Mem32	Reg32	CL

(c)

Figure 5.26 (a) Shift instructions. (b) Allowed operands for basic shift instructions. (c) Allowed operands for double precision shift instructions.

causes the 16-bit contents of the AX register to be shifted 1 bit position to the left. Here we see that the vacated LSB location is filled with zero and the bit shifted out of the MSB is saved in CF.

On the other hand, if the source operand is specified as CL instead of 1, the count in this register represents the number of bit positions the contents of the operand are to be

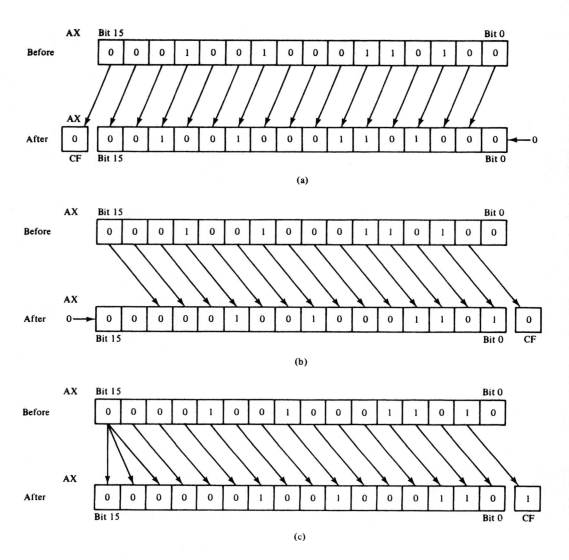

Figure 5.27 Results of executing SHL AX,1. (b) Results of executing SHR AX,CL. (c) Results of executing SAR AX,CL.

shifted. This permits the count to be defined under software control and allows a range of shifts from 1 to 255 bits.

An example of an instruction specified in this way is

```
SHR    AX,CL
```

Assuming that CL contains the value 02_{16}, the logical shift right that occurs is as shown in Fig. 5.27(b). Notice that the two MSBs have been filled with 0s, and the last bit shifted out at the LSB, which is 0, is placed in the carry flag.

The third way of specifying the count is with an 8-bit immediate operand. Again this permits a shift range of 1 to 255 bits to be specified.

In an arithmetic shift to the left, SAL operation, the vacated bits at the right of the operand are filled with 0s, whereas in an arithmetic shift to the right, SAR operation, the vacated bits at the left are filled with the value of the original MSB of the operand. Thus, in an arithmetic shift to the right, the original sign of the number is maintained. This operation is equivalent to division by powers of 2 as long as the bits shifted out of the LSB are 0s.

EXAMPLE 5.23

Assume that CL contains 02_{16} and AX contains $091A_{16}$. Determine the new contents of AX and the carry flag after the instruction

$$SAR \quad AX,CL$$

is executed.

Solution

Figure 5.27(c) shows the effect of executing the instruction. Here we see that since CL contains 02_{16}, a shift right by two bit locations takes place, and the original sign bit, which is logic 0, is extended to the two vacated bit positions. Moreover, the last bit shifted out from the LSB location is placed in CF. This makes CF equal to 1. Therefore, the results produced by execution of the instruction are

$$(AX) = 0246_{16}$$

and

$$(CF) = 1_2$$

EXAMPLE 5.24

Verify the operation of the SAR instruction in Example 5.23 by executing with the debug program on the PC/AT.

Solution

After invoking the debug program, we enter the instruction by assembling it with the command

```
-A                        (↵)
1342:0100   SAR   AX,CL   (↵)
1342:0102                 (↵)
-
```

Next, registers AX and CL are loaded with data, and the carry flag is reset. This is done with the command sequence

```
-R   AX            (↵)
AX  0000
:091A              (↵)
-R   CX            (↵)
CX  0000
:2                 (↵)
-R   F             (↵)
NV   UP  EI  PL  NZ  NA  PO  NC  -   (↵)
-
```

Notice that the carry flag was already clear, so no status entry was made.
Now the instruction is executed with the T command

```
-T   (↵)
```

Note in Fig. 5.28 that the value in AX has become 0246_{16} and a carry (CY) has occurred. These results are identical to those obtained in Example 5.23.

Now that we have described the basic shift instructions of the 80386DX, let us continue with the double-precision shift instructions. Looking at Fig. 5.26(a) we find two double-precision instructions: *double-precision shift left* (SHLD) and *double-precision shift right* (SHRD). These instructions employ two destination operands. In Fig. 5.26(c), we see that D_1 is either a register or storage location in memory. This is the operand that has its contents shifted. D_2 identifies a register that contains the bits that are to be shifted into D_1 during the shift operation. Finally, the count can be specified by either an 8-bit immediate operand or a count in CL. Even though the immediate operand or CL register is 8 bits in length, the shift performed with these instructions is limited to from 0 to 31 bits.

```
C:\DOS>DEBUG
-A
1342:0100 SAR AX,CL
1342:0102
-R AX
AX 0000
:091A
-R CX
CX 0000
:2
-R F
NV UP EI PL NZ NA PO NC  -
-T

AX=0246  BX=0000  CX=0002  DX=0000  SP=FFEE  BP=0000  SI=0000  DI=0000
DS=1342  ES=1342  SS=1342  CS=1342  IP=0102   NV UP EI PL NZ AC PO CY
1342:0102 B98AFF        MOV     CX,FF8A
-Q

C:\DOS>
```

Figure 5.28 Display sequence for Example 5.23.

For instance, when the instruction

```
SHLD    EAX,EBX,3
```

is executed, the contents of EAX are shifted left three bit positions, the three vacated bits in EAX (bits 0, 1, and 2) are filled with values from the three MSBs of EBX (bits 29, 30, and 31), and the last value shifted out of the MSB end of EAX (bit 29) is saved in CF. The value held in EBX remains unchanged during the shift operation.

A frequent need in programming is to isolate the value in one bit of a byte, word, or double word of data by shifting it into the carry flag. The shift instructions can be used to perform this operation on data either in a register or a storage location in memory. The instructions that follow perform this type of operation on a byte of data stored in memory at address CONTROL_FLAGS.

```
MOV    AL,[CONTROL_FLAGS]
MOV    CL,04H
SHR    AL, CL
```

The first instruction reads the value of the byte of data at address CONTROL_FLAGS into AL. Next a shift count of 4 is loaded into CL and then the value in AL is shifted to the right four bit positions. Since the MSBs of AL are reloaded with 0s as part of the shift operation, the results are

$$(AL) = 0000B_7B_6B_5B_4$$

and

$$(CF) = B_3$$

In this way, we see that bit B_3 of CONTROL_FLAGS has been isolated by placing it in CF.

▲ 5.6 ROTATE INSTRUCTIONS

Another group of instructions, known as the *rotate instructions*, are similar to the shift instructions we just introduced. This group, as shown in Fig. 5.29(a), includes the *rotate left* (ROL), *rotate right* (ROR), *rotate left through carry* (RCL), and *rotate right through carry* (RCR) instructions.

As shown in Fig. 5.29(b), the rotate instructions are similar to the shift instructions in several ways. They have the ability to rotate the contents of either an internal register or storage location in memory. Also, the rotation that takes place can be from 1 to 255 bit positions to the left or to the right. Moreover, in the case of a multibit rotate, the number of bit positions to be rotated is specified by the contents of CL or an 8-bit immediate operand. Their difference from the shift instructions lies in the fact that the bits moved out at either the MSB or LSB end are not lost; instead, they are reloaded at the other end.

As an example, let us look at the operation of the ROL instruction. Execution of ROL causes the contents of the selected operand to be rotated left the specified number of bit positions. Each bit shifted out at the MSB end is reloaded at the LSB end. Moreover,

Mnemonic	Meaning	Format	Operation	Flags affected
ROL	Rotate left	ROL D, Count	Rotate the (D) left by the number of bit positions equal to Count. Each bit shifted out from the leftmost bit goes back into the rightmost bit position	CF OF undefined if Count ≠1
ROR	Rotate right	ROR D, Count	Rotate the (D) right by the number of bit positions equal to Count. Each bit shifted out from the rightmost bit goes into the leftmost bit position	CF OF undefined if Count ≠1
RCL	Rotate left through carry	RCL D, Count	Same as ROL except carry is attached to (D) for rotation	CF OF undefined if Count ≠1
RCR	Rotate right through carry	RCR D, Count	Same as ROR except carry is attached to (D) for rotation	CF OF undefined if Count ≠1

(a)

Destination	Count
Register	1
Register	CL
Register	Imm8
Memory	1
Memory	CL
Memory	Imm8

(b)

Figure 5.29 (a) Rotate instructions. (b) Allowed operands.

the content of CF reflects the state of the last bit that was shifted out. For instance, the instruction

```
ROL   AX,1
```

causes a 1-bit rotate to the left. Figure 5.30(a) shows the result produced by executing this instruction. Notice that the original value of bit 15 is zero. This value has been rotated into CF and bit 0 of AX. All other bits have been rotated one bit position to the left.

The ROR instruction operates the same way as ROL except that it causes data to be rotated to the right instead of to the left. For example, execution of

```
ROR   AX,CL
```

causes the contents of AX to be rotated right by the number of bit positions specified in CL. The result for CL equal to four is illustrated in Fig. 5.30(b).

The other two rotate instructions, RCL and RCR, differ from ROL and ROR in that the bits are rotated through the carry flag. Figure 5.31 illustrates the rotation that takes place due to execution of the RCL instruction. Notice that the value returned to bit 0 is the prior content of CF and not bit 31. The value shifted out of bit 31 goes into the carry flag. Thus the bits rotate through carry.

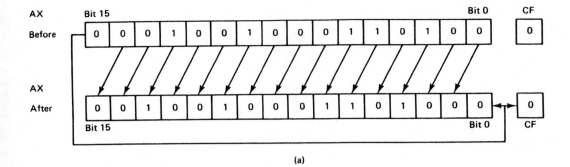

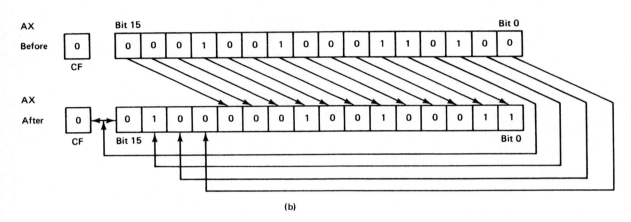

(b)

Figure 5.30 Results of executing ROL AX,1. (b) Results of executing ROR AX,CL.

EXAMPLE 5.25

What is the result in BX and CF after execution of the following instruction?

RCR   BX,CL

Assume that prior to execution of the instruction, $(CL) = 04_{16}$, $(BX) = 1234_{16}$, and $(CF) = 0$.

Solution

The original contents of BX are

$$(BX) = 0001001000110100_2 = 1234_{16}$$

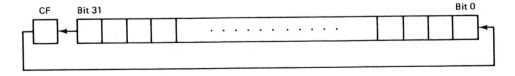

Figure 5.31 Rotation caused by execution of the RCL instruction.

Execution of the RCR instruction causes a 4-bit rotate right through carry to take place on the data in BX. The resulting contents of BX and CF are

$$(BX) = 1000000100100011_2 = 8123_{16}$$

$$(CF) = 0_2$$

In this way, we see that the original content of bit 3, which was zero, resides in carry flag and 1000_2 has been reloaded from the bit-15 end of BX.

EXAMPLE 5.26

Use the PC/AT's debug program to verify the operation of the RCR instruction in Example 5.25.

Solution

After loading DEBUG, the instruction is assembled into memory with the command

```
-A              (↵)
1342:0100   RCR   BX,CL
1342:0102       (↵)
```

Next, BX and CX are loaded with data and CF is cleared by issuing the commands

```
-R   BX         (↵)
BX   0000
:1234           (↵)
-R   CX         (↵)
CX   0000
:4              (↵)
-R   F          (↵)
NV   UP   EI   PL   NZ   NA   PO   NC   -   (↵)
-
```

Notice that CF is already cleared (NC); therefore, no entry is made for the flag register command.

Now we can execute the instruction with the command

```
-T   (↵)
```

Looking at the trace information displayed in Fig. 5.32, we see that the new contents of BX are 8123_{16} and CF equals NC. These are the same results as obtained in Example 5.25.

The rotate instructions can be used to perform many of the same programming functions as the shift instructions. An example of a software operation that can be performed with the rotate instructions is the disassembly of the two hexadecimal digits in a byte of data in memory so that they can be added. The instructions in Fig. 5.33 will perform this operation. First the byte containing the two hexadecimal digits is read into AL. Then a

```
C:\DOS>DEBUG
-A
1342:0100 RCR BX,CL
1342:0102
-R BX
BX 0000
:1234
-R CX
CX 0000
:4
-R F
NV UP EI PL NZ NA PO NC   -
-T

AX=0000  BX=8123  CX=0004  DX=0000  SP=FFEE  BP=0000  SI=0000  DI=0000
DS=1342  ES=1342  SS=1342  CS=1342  IP=0102   OV UP EI PL NZ NA PO NC
1342:0102 B98AFF        MOV     CX,FF8A
-Q

C:\DOS>
```

Figure 5.32   Display sequence for Example 5.25.

copy is made in BL. Next the four most significant bits in BL are moved to the four least significant bit locations with a rotate operation. This repositions the most significant hexadecimal digit of the original byte into the least significant digit position in BL. Now the most significant hexadecimal digits in both AL and BL are masked off. This isolates one hexadecimal digit in the lower 4 bits of AL and the other in the lower 4 bits of BL. Finally, the two digits are added together in AL.

▲ 5.7 BIT TEST AND BIT SCAN INSTRUCTIONS

The *bit test and bit scan instructions* of the 80386DX enable a programmer to test the logic value of a bit in either a register or a storage location in memory. Let us begin by examining the bit test instructions. They are used to test the state of a single bit in a register or memory location. When the instruction is executed, the value of the tested bit is saved in the carry flag. Instructions are provided that can also reset, set, or complement the contents of the tested bit during the execution of the instruction.

In Fig. 5.34(a), we see that the *bit test* (BT) instruction has two operands. The first operand identifies the register or memory location that contains the bit that is to be tested. The second operand contains an index that selects the bit that is to be tested. Notice in Fig. 5.34(b) that the index may be either an 8-bit immediate operand or the value in a 16- or 32-bit register. When this instruction is executed, the state of the tested bit is simply copied into the carry flag.

Once the state of the bit is saved in CF, it can be tested through software. For instance, a conditional jump instruction could be used to test the value in CF, and if CF equals 1,

```
MOV AL,[HEX_DIGITS]
MOV BL,AL
MOV CL,04H
ROR BL,CL
AND   AL,0FH
AND   BL,0FH
ADD   AL,BL
```

Figure 5.33   Program for disassembly and addition of 2 hexadecimal digits stored as a byte in memory.

Mnemonic	Meaning	Format	Operation	Flags affected
BT	Bit test	BT D, S	Saves the value of the bit in D specified by the value in S in CF.	CF OF, SF, ZF, AF, PF undefined
BTR	Bit test and reset	BTR D, S	Saves the value of the bit in D specified by the value in S in CF and then resets the bit in D.	CF OF, SF, ZF, AF, PF undefined
BTS	Bit test and set	BTS D, S	Saves the value of the bit in D specified by the value in S in CF and then sets the bit in D.	CF OF, SF, ZF, AF, PF undefined
BTC	Bit test and complement	BTC D, S	Saves the value of the bit in D specified by the value in S in CF and then complements the bit in D.	CF OF, SF, ZF, AF, PF undefined
BSF	Bit scan forward	BSF D, S	Scan the source operand starting from bit 0. ZF = 0 if all bits are 0, else ZF = 1 and the destination operand is loaded with the bit index of the first set bit.	ZF OF, SF, AF, PF, CF undefined
BSR	Bit scan reverse	BSR D, S	Scan the source operand starting from the MSB. ZF = 0 if all bits are 0, else ZF = 1 and the destination operand is loaded with the bit index of the first set bit.	ZF OF, SF, AF, PF, CF undefined

(a)

Destination	Source
Reg16	Reg16
Reg16	Imm8
Reg32	Reg32
Reg32	Imm8
Mem16	Reg16
Mem16	Imm8
Mem32	Reg32
Mem32	Imm8

(b)

Destination	Source
Reg16	Reg16
Reg16	Mem16
Reg32	Reg32
Reg32	Mem32

(c)

Figure 5.34 (a) Bit-test and bit-scan instructions. (b) Allowed operands for bit-test instructions. (c) Allowed operands for bit-scan instructions.

program control could be passed to a service routine. On the other hand, if CF equals 0, the value of the index could be incremented, a jump performed back to the BT instruction, and the next bit in the operand tested.

Another example is the instruction

```
BTR    EAX, EDI
```

Execution of this instruction causes the bit in 32-bit register EAX that is selected by the index in EDI to be tested. The value of the tested bit is first saved in the carry flag and then it is reset in the register EAX.

EXAMPLE 5.27

Describe the operation that is performed by the instruction

```
BTC   BX,7
```

Assume that register BX contains the value $03F0_{16}$.

Solution

Let us first express the value in BX in binary form. This gives

$$(BX) = 0000001111110000_2$$

Execution of the *bit test and complement* instruction causes the value of the B_7 to be first tested and then complemented. Since this bit is logic 1, CF is set to 1. This gives

$$(CF) = 1$$
$$(BX) = 0000001101110000_2 = 0370_{16}$$

The *bit scan forward* (BSF) and *bit scan reverse* (BSR) instructions are used to scan through the bits of a register or storage location in memory to determine whether or not they are all 0. For example, by executing the instruction

```
BSF   ESI,   EDX
```

the bits of 32-bit register EDX are tested one after the other, starting from bit 0. If all bits are found to be 0, the ZF is cleared. On the other hand, if the contents of EDX are not zero, ZF is set to 1 and the index value of the first bit tested as 1 is copied into ESI.

ASSIGNMENTS

Section 5.2

1. Explain what operation is performed by each of the instructions that follows.
   (a) MOV AX,0110H
   (b) MOV DI,AX
   (c) MOV BL,AL
   (d) MOV [0100H],AX
   (e) MOV [BX+DI],AX
   (f) MOV [DI]+4,AX
   (g) MOV [BX][DI] +4,AX
2. Assume that registers EAX, EBX, and EDI are all initialized to 00000000_{16} and that all the data storage memory has been cleared. Determine the location and value of the destination operand as instructions (a) through (g) in Problem 1 are executed as a sequence.

3. Write an instruction sequence that will initialize the ES register with the immediate value 1010_{16}.

4. Write an instruction that will save the contents of the ES register in memory at address DS:1000H.

5. Why will the instruction MOV CL,AX result in an error when it is assembled?

6. Describe the operation performed by the instruction

```
MOVSX    EAX,BL
```

7. Write an instruction that will zero-extend the word of data at address DATA_WORD and copy it into register EAX.

8. Describe the operation performed by each of the instructions that follows.
   (a) XCHG AX,BX
   (b) XCHG BX,DI
   (c) XCHG [DATA],AX
   (d) XCHG [BX+DI],AX

9. If register EBX contains the value 00000100_{16}, register EDI contains 00000010_{16}, and register DS contains 1075_{16}, what physical memory location is swapped with AX when the instruction in Problem 8(d) is executed?

10. Assuming that $(EAX) = 00000010_{16}$, $(EBX) = 00000100_{16}$, and $(DS) = 1000_{16}$, what happens if the XLAT instruction is executed?

11. Write a single instruction that will load AX from address 0200_{16} and DS from address 0202_{16}.

12. What operation is performed using the instruction

```
LFS   EDI,[DATA_F_ADDRESS]
```

13. Two code-conversion tables starting with offsets TABL1 and TABL2 in the current data segment are to be accessed. Write an instruction sequence that initializes the needed registers and then replaces the contents of memory locations MEM1 and MEM2 (offsets in the current data segment) by the equivalent converted codes from the respective code-conversion tables.

Section 5.3

14. What operation is performed by each of the following instructions?
   (a) ADD   AX,00FFH
   (b) ADC   SI,AX
   (c) INC   BYTE PTR [0100H]
   (d) SUB   DL,BL
   (e) SBB   DL,[0200H]
   (f) DEC   BYTE PTR [DI+BX]
   (g) NEG   BYTE PTR [DI]+0010H
   (h) MUL   DX
   (i) IMUL BYTE PTR [BX+SI]
   (j) DIV   BYTE PTR [SI]+0030H
   (k) IDIV BYTE PTR [BX] [SI]+0030H

15. Assume that the state of 80386DX's registers and memory just prior to the execution of each instruction in Problem 14 is as follows:

$$(EAX) = 00000010H$$
$$(EBX) = 0000020H$$
$$(ECX) = 00000030H$$
$$(EDX) = 00000040H$$
$$(ESI) = 00000100H$$
$$(EDI) = 00000200H$$
$$(CF) = 1$$
$$(DS:100H) = 10H$$
$$(DS:101H) = 00H$$
$$(DS:120H) = FFH$$
$$(DS:121H) = FFH$$
$$(DS:130H) = 08H$$
$$(DS:131H) = 00H$$
$$(DS:150H) = 02H$$
$$(DS:151H) = 00H$$
$$(DS:200H) = 30H$$
$$(DS:201H) = 00H$$
$$(DS:210H) = 40H$$
$$(DS:211H) = 00H$$
$$(DS:220H) = 30H$$
$$(DS:221H) = 00H$$

What is the result produced in the destination operand by executing instructions (a) through (k)?

16. Write an instruction that will add the immediate value $111F_{16}$ and the carry flag to the contents of the extended data register EDX.

17. Write an instruction that will subtract the word contents of the storage location pointed to by the base register BX and the carry flag from the accumulator.

18. Two word-wide unsigned integers are stored at the physical memory addresses $00A00_{16}$ and $00A02_{16}$, respectively. Write an instruction sequence that computes and stores their sum, difference, product, and quotient. Store these results at consecutive memory locations starting at physical address $00A10_{16}$ in memory. To obtain the difference, subtract the integer at $00A02_{16}$ from the integer at $00A00_{16}$. For the division, divide the integer at $00A00_{16}$ by the integer at $00A02_{16}$. Use register indirect relative addressing mode to store the various results.

Assignments

19. Assuming that $(AX) = 0123_{16}$ and $(BL) = 10_{16}$, what will be the new contents of AX after executing the instruction DIV BL?

20. What instruction is used to adjust the result of an addition that processed packed BCD numbers?

21. Which instruction is provided in the instruction set of the 80386DX to adjust the result of a subtraction that involved ASCII coded numbers?

22. If AL contains $A0_{16}$, what happens when the instruction CBW is executed?

23. If the value in AX is $7FFF_{16}$, what happens when the instruction CWD is executed?

24. Two byte-sized BCD integers are stored at the symbolic offset addresses NUM1 and NUM2, respectively. Write an instruction sequence to generate their difference and store it at NUM3. The difference is to be formed by subtracting the value at NUM1 from that at NUM2. Assume that all storage locations are in the current data segment.

Section 5.4

25. Describe the operation performed by each of the following instructions.
  (a) AND BYTE PTR [0300H] , 0FH
  (b) AND DX,[SI]
  (c) OR   [BX+DI],AX
  (d) OR   BYTE PTR [BX][DI]+10H,0F0H
  (e) XOR AX,[SI + BX]
  (f) NOT BYTE PTR [0300H]
  (g) NOT WORD PTR [BX+DI]

26. Assume that the state of 80386DX's registers and memory just prior to execution of each instruction in Problem 25 is as follows:

$$(EAX) = 00005555H$$

$$(EBX) = 00000010H$$

$$(ECX) = 00000010H$$

$$(EDX) = 0000AAAAH$$

$$(ESI) = 00000100H$$

$$(EDI) = 00000200H$$

$$(DS:100H) = 0FH$$

$$(DS:101H) = F0H$$

$$(DS:110H) = 00H$$

$$(DS:111H) = FFH$$

$$(DS:200H) = 30H$$

$$(DS:201H) = 00H$$

$$(DS:210H) = AAH$$

$$(DS:211H) = AAH$$

$$(DS:220H) = 55H$$

$$(DS:221H) = 55H$$

$$(DS:300H) = AAH$$

$$(DS:301H) = 55H$$

What are the results produced in the destination operands by executing instructions (a) through (g)?

27. Write an instruction that when executed will mask off all but bit 7 of the contents of the extended data register.

28. Write an instruction that will mask off all but bit 7 of the word of data stored at address DS:0100H.

29. Specify the relation between the old and new contents of AX after executing the following sequence of instructions.

```
NOT   AX
ADD   AX,1
```

30. Write an instruction sequence that generates a byte-size integer in the memory location defined as RESULT. The value of the integer is to be calculated from the logic equation:

$$(RESULT) = (AL \cdot NUM1) + ((\overline{NUM2} \cdot AL) + BL)$$

Assume that all parameters are byte-sized. NUM1, NUM2, and RESULT are the offset addresses of memory locations in the current data segment.

31. Write an instruction sequence that will read the byte of control flags from the storage location at offset address CONTROL_FLAGS in the current data segment into register AL, mask off all but the most significant and least significant flag bits, and then save the result back in the original storage location.

32. Describe the operation that is performed by the following instruction sequence.

```
MOV   BL,[CONTROL_FLAGS]
AND   BL,08H
XOR   BL,08H
MOV   [CONTROL_FLAGS],BL
```

Section 5.5

33. Explain the operation performed by each of the following instructions.
    (a) SHL DX, CL
    (b) SHL EDX,7
    (c) SHL BYTE PTR [0400H],CL
    (d) SHR BYTE PTR [DI],1
    (e) SHR DWORD PTR [DI],3
    (f) SHR BYTE PTR [DI+BX],CL
    (g) SAR WORD PTR [BX+DI],1
    (h) SAR WORD PTR [BX] [DI]+10H,CL

34. Assume that the state of 80386DX's registers and memory just prior to execution of each instruction in Problem 33 is as follows:

$$(EAX) = 00000000H$$
$$(EBX) = 00000010H$$
$$(ECX) = 00000105H$$
$$(EDX) = 00001111H$$
$$(ESI) = 00000100H$$
$$(EDI) = 00000200H$$
$$(CF) = 0$$
$$(DS:100H) = 0FH$$
$$(DS:200H) = 22H$$
$$(DS:201H) = 44H$$
$$(DS:202H) = 00H$$
$$(DS:203H) = 00H$$
$$(DS:210H) = 55H$$
$$(DS:211H) = AAH$$
$$(DS:220H) = AAH$$
$$(DS:221H) = 55H$$
$$p(DS:400H) = AAH$$
$$DS:401H) = 55H$$

What are the results produced in the destination operands by executing instructions (a) through (h)?

35. Write an instruction that shifts the contents of the extended count register left by one bit position.

36. Write an instruction sequence that, when executed, shifts left by eight bit positions the contents of the word-wide memory location pointed to by the address in the destination index register.

37. Identify the condition under which the contents of AX would remain unchanged after execution of the instructions that follow.

```
MOV   CL,4
SHL   AX,CL
SHR   AX,CL
```

38. Implement the following operation using shift and arithmetic instructions.

$$7(AX) - 5(BX) - (BX)/8 \rightarrow (AX)$$

Assume that all parameters are word-sized.

39. What instruction does the mnemonic SHRD stand for?

40. What happens when the instruction

```
SHRD   DOUBLE   WORD   PTR   [DI],EAX,CL
```

is executed?

41. Describe the operation performed by the instruction sequence that follows

```
MOV   AL,[CONTROL_FLAGS]
AND   AL,80H
SHL   AL,   1
```

What is the result in AL after the shift is complete?

42. Write a routine that will read the word of data from the offset address ASCII_DATA in the current data segment of memory. Assume that this word storage location contains two ASCII-coded characters, one character in the upper byte and the other in the lower byte. Disassemble the two bytes and save them as separate characters in the lower byte location of the word storage locations with offsets ASCII_CHAR_L and ASCII_CHAR_H in the current data segment. The upper 8 bits in each of these character storage locations should be made zero. Use a SHR instruction to relocate the most significant bits.

Section 5.6

43. Describe what happens as each of the instructions that follows is executed by the 80386DX.
- **(a)** ROL DX, CL
- **(b)** RCL BYTE PTR [0400H],CL
- **(c)** ROR BYTE PTR [DI],1
- **(d)** ROR BYTE PTR [DI+BX],CL
- **(e)** RCR WORD PTR [BX+DI],1
- **(f)** RCR WORD PTR [BX] [DI]+10H,CL

44. Assume that the state of 80386DX's registers and memory just prior to execution of each of the instructions in Problem 43 is as follows:

$$(EAX) = 00000000H$$
$$(EBX) = 00000010H$$
$$(ECX) = 00000105H$$
$$(EDX) = 00001111H$$
$$(ESI) = 00000100H$$
$$(EDI) = 00000200H$$
$$(CF) = 1$$
$$(DS:100H) = 0FH$$
$$(DS:200H) = 22H$$
$$(DS:201H) = 44H$$

$$(DS:210H) = 55H$$

$$(DS:211H) = AAH$$

$$(DS:220H) = AAH$$

$$(DS:221H) = 55H$$

$$(DS:400H) = AAH$$

$$(DS:401H) = 55H$$

What are the results produced in the destination operands by executing instructions (a) through (f)?

45. Write an instruction sequence that, when executed, rotates left through carry by 1 bit position the contents of the word-wide memory location pointed to by the address in the base register.

46. Write a program that saves the content of bit 5 in AL in BX as a word.

47. Repeat Problem 42, but this time use an ROR instruction to perform the bit shifting operation.

Section 5.7

48. What does BTS stand for?

49. If the values in EAX and ECX are $0000F0F0_{16}$ and 00000004_{16}, what is the result in AX and CF after execution of each of the following instructions?
(a) BT AX,CX
(b) BTR AX,CX
(c) BTC AX,CX

50. Describe the operation performed by executing the instruction BTR WORD PTR [0100H],3.

51. If the word contents of DS:100H are $00FF_{16}$ and CF = 0 just before the instruction in Problem 50 is executed, what is the new value of the word in memory and the carry flag?

52. Write an instruction that will test bit 7 of the word storage location DS:DI in memory and save this value in the carry flag.

53. Write an instruction that will scan the double-word contents of the memory location pointed to by SI and save the index of the MSB that is logic 1 in register EAX.

Real-Mode 80386DX Microprocessor Programming II

▲ 6.1 INTRODUCTION

In Chapter 5 we discussed many of the instructions that can be executed by the real-mode 80386DX microprocessors. Furthermore, we used these instructions in simple programs. In this chapter, we introduce the rest of the instruction set and at the same time cover some more complicated programming techniques. The following topics are discussed in this chapter:

1. Flag-control instructions
2. Compare and set instructions
3. Jump instructions
4. Subroutines and subroutine-handling instructions
5. The loop and loop-handling instructions
6. Strings and string-handling instructions

▲ 6.2 FLAG-CONTROL INSTRUCTIONS

The 80386DX microprocessor has a set of flags that either monitors the status of executing instructions or control options available in its operation. These flags were described in detail in Chapter 2. The instruction set includes a group of instructions that, when executed, directly affect the state of the flags. These instructions, shown in Fig. 6.1(a), are *load AH from flags* (LAHF), *store AH into flags* (SAHF), *clear carry* (CLC), *set carry* (STC), *complement carry* (CMC), *clear interrupt* (CLI), and *set interrupt* (STI). A few more instructions exist that can directly affect the flags; however, we will not cover them until later in the chapter when we introduce the subroutine and string instructions.

Mnemonic	Meaning	Operation	Flags affected
LAHF	Load AH from flags	$(AH) \leftarrow (Flags)$	None
SAHF	Store AH into flags	$(Flags) \leftarrow (AH)$	SF,ZF,AF,PF,CF
CLC	Clear carry flag	$(CF) \leftarrow 0$	CF
STC	Set carry flag	$(CF) \leftarrow 1$	CF
CMC	Complement carry flag	$(CF) \leftarrow (\overline{CF})$	CF
CLI	Clear interrupt flag	$(IF) \leftarrow 0$	IF
STI	Set interrupt flag	$(IF) \leftarrow 1$	IF

(a)

```
  7                    0
AH │SF│ZF│ — │AF│ — │PF│ — │CF│
```

SF = Sign flag
ZF = Zero flag
AF = Auxiliary
PF = Parity flag
CF = Carry flag
 — = Undefined (do not use)

(b)

Figure 6.1 (a) Flag-control instructions. (b) Format of the AH register for the LAHF and SAHF instructions.

Looking at Fig. 6.1(a), we see that the first two instructions, LAHF and SAHF, can be used either to read the flags or to change them, respectively. Notice that the data transfer that takes place is always between the AH register and the flag register. Figure 6.1(b) shows the format of the flag information in AH. Notice that bits 1, 3, and 5 are undefined. For instance, we may want to start an operation with certain flags set or reset. Assume that we want to preset all flags to logic 1. To do this, we can first load AH with FF_{16} and then execute the SAHF instruction.

EXAMPLE 6.1

Write an instruction sequence to save the current contents of the 80386DX's flags in the memory location at offset MEM1 of the current data segment and then reload the flags with the contents of the storage location at offset MEM2.

Solution

To save the current flags, we must first load them into the AH register and then move them to the location MEM1. The instructions that do this are

```
        LAHF
        MOV   [MEM1], AH
```

Similarly, to load the flags with the contents of MEM2, we must first copy the contents of MEM2 into AH and then store the contents of AH into the flags. The instructions for this are

```
        MOV   AH, [MEM2]
        SAHF
```

The entire instruction sequence is shown in Fig. 6.2.

```
         LAHF
         MOV     [MEM1],AH
         MOV     AH,[MEM2]
         SAHF
```

Figure 6.2 Instruction sequence for saving the contents of the flag register and reloading it from memory.

EXAMPLE 6.2

Use the DEBUG program on the PC/AT to enter the instruction sequence in Example 6.1 starting at memory address 00110_{16}. Assign memory addresses 00150_{16} and 00151_{16} to symbols MEM1 and MEM2, respectively. Then initialize the contents of MEM1 and MEM2 to FF_{16} and 01_{16}, respectively. Verify the operation of the instructions by executing them one after the other with the TRACE command.

Solution

As shown in Fig. 6.3, the DEBUG program is called up with the DOS command

C:\DOS>DEBUG   (↵)

```
C:\DOS>DEBUG
-A 0:0110
0000:0110 LAHF
0000:0111 MOV     [0150],AH
0000:0115 MOV     AH,[0151]
0000:0119 SAHF
0000:011A
-E 0:150 FF 01
-R CS
CS 1342
:0
-R IP
IP 0100
:0110
-R DS
DS 1342
:0
-R
AX=0000  BX=0000  CX=0000  DX=0000  SP=FFEE  BP=0000  SI=0000  DI=0000
DS=0000  ES=1342  SS=1342  CS=0000  IP=0110   NV UP EI PL NZ NA PO NC
0000:0110 9F          LAHF
-T

AX=0200  BX=0000  CX=0000  DX=0000  SP=FFEE  BP=0000  SI=0000  DI=0000
DS=0000  ES=1342  SS=1342  CS=0000  IP=0111   NV UP EI PL NZ NA PO NC
0000:0111 88265001    MOV     [0150],AH                       DS:0150=FF
-T

AX=0200  BX=0000  CX=0000  DX=0000  SP=FFEE  BP=0000  SI=0000  DI=0000
DS=0000  ES=1342  SS=1342  CS=0000  IP=0115   NV UP EI PL NZ NA PO NC
0000:0115 8A265101    MOV     AH,[0151]                       DS:0151=01
-D 150 151
0000:0150  02 01
-T

AX=0100  BX=0000  CX=0000  DX=0000  SP=FFEE  BP=0000  SI=0000  DI=0000
DS=0000  ES=1342  SS=1342  CS=0000  IP=0119   NV UP EI PL NZ NA PO NC
0000:0119 9E          SAHF
-T

AX=0100  BX=0000  CX=0000  DX=0000  SP=FFEE  BP=0000  SI=0000  DI=0000
DS=0000  ES=1342  SS=1342  CS=0000  IP=011A   NV UP EI PL NZ NA PO CY
0000:011A 00F0        ADD     AL,DH
-Q

C:\DOS>
```

Figure 6.3 Display sequence for Example 6.2.

Now we are ready to assemble the program into memory. This is done by using the ASSEMBLE command, as follows

```
-A  0:0110                    (↵)
0000:0110  LAHF               (↵)
0000:0111  MOV  [0150], AH    (↵)
0000:0115  MOV  AH, [0151]    (↵)
0000:0119  SAHF               (↵)
```

Now the contents of MEM1 and MEM2 are initialized with the ENTER command

```
-E  0:0150  FF  01  (↵)
```

Next, the registers CS and IP must be initialized with the values 0000_{16} and 0110_{16} to provide access to the program. Also, The DS register must be initialized to permit access to the data memory locations. This is done with the commands

```
-R  CS    (↵)
CS  1342
:0
-R  IP    (↵)
IP  0100
:0110     (↵)
-R  DS    (↵)
DS  1342
:0        (↵)
```

Before going further, let us verify the initialization of the internal registers. This is done by displaying their state with the R command

```
-R  (↵)
```

Looking at the information displayed in Fig. 6.3, we see that all three registers have been correctly initialized.

Now we are ready to step through the execution of the program. The first instruction is executed with the command

```
-T  (↵)
```

Notice from the displayed trace information in Fig. 6.3 that the contents of the status register, which are 02_{16}, have been copied into the AH register.

The second instruction is executed by issuing another T command

```
-T  (↵)
```

This instruction causes the status, which is now in AH, to be saved in memory at address 0000:0150. The fact that this operation has occurred is verified with the D command

```
-D  150  151  (↵)
```

In Fig. 6.3, we see that the data held at address 0000:0150 is displayed by this command as 02_{16}. This verifies that status was saved at MEM1.

The third instruction is now executed with the command

$$-T \quad (\ \lrcorner)$$

Its function is to copy the new status from MEM2 (0000:0151) into the AH register. From the data displayed in the earlier D command, we see that this value is 01_{16}. Looking at the displayed information for the third instruction, we find that 01_{16} has been copied into AH.

The last instruction is executed with another T command and, as shown by its trace information in Fig. 6.3, it has caused the carry flag to set. That is, CF is displayed with the value CY.

The next three instructions, CLC, STC, and CMC, as shown in Fig. 6.1, are used to manipulate the carry flag. They permit CF to be cleared, set, or complemented, respectively. For example, if CF is 1 and the CMC instruction is executed, it becomes 0.

The last two instructions are used to manipulate the interrupt flag. Executing the clear interrupt (CLI) instruction sets IF to logic 0 and disables the interrupt interface. On the other hand, executing the STI instruction sets IF to 1, and the microprocessor is enabled to accept interrupts from that point on.

EXAMPLE 6.3

Of the three carry flag instructions CLC, STC, and CMC, only one is really an independent instruction. That is, the operation that it provides cannot be performed by a series of the other two instructions. Determine which one of the carry instructions is the independent instruction.

Solution

Let us begin with the CLC instruction. The clear-carry operation can be performed by an STC instruction followed by a CMC instruction. Therefore, CLC is not an independent instruction. The operation of the set-carry (STC) instruction is equivalent to the operation performed by a CLC instruction, followed by a CMC instruction. Thus, STC is also not an independent instruction. On the other hand, the operation performed by the last instruction, complement carry (CMC), cannot be expressed in terms of the CLC and STC instructions. Therefore, it is the independent instruction.

EXAMPLE 6.4

Verify the operation of the following instructions that affect the carry flag,

```
CLC
STC
CMC
```

by executing them with the debug program of the PC/AT. Start with CF set to one (CY).

Solution

After bringing up the debug program, we enter the instructions with the ASSEMBLE command as

```
-A                  (↵)
1342:0100   CLC     (↵)
1342:0101   STC     (↵)
1342:0102   CMC     (↵)
1342:0103           (↵)
-
```

These inputs are shown in Fig. 6.4.

Next, the carry flag is initialized to CY with the R command

```
-R   F   (↵)
NV   UP   EI   PL   NZ   NA   PO   NC   -CY   (↵)
```

and in Fig. 6.4 the updated status is displayed with another R command to verify that CF is set to the CY state.

Now the first instruction is executed with the TRACE command

```
-T   (↵)
```

Looking at the displayed state information in Fig. 6.4, we see that CF has been cleared, and its new state is NC.

The other two instructions are also executed with T commands and, as shown in Fig. 6.4, the STC instruction sets CF (CY in the state dump) and CMC inverts CF (NC in the state dump).

```
C:\DOS>DEBUG
-A
1342:0100  CLC
1342:0101  STC
1342:0102  CMC
1342:0103
-R F
NV UP EI PL NZ NA PO NC  -CY
-R F
NV UP EI PL NZ NA PO CY  -
-T

AX=0000  BX=0000  CX=0000  DX=0000  SP=FFEE  BP=0000  SI=0000  DI=0000
DS=1342  ES=1342  SS=1342  CS=1342  IP=0101   NV UP EI PL NZ NA PO NC
1342:0101 F9              STC
-T

AX=0000  BX=0000  CX=0000  DX=0000  SP=FFEE  BP=0000  SI=0000  DI=0000
DS=1342  ES=1342  SS=1342  CS=1342  IP=0102   NV UP EI PL NZ NA PO CY
1342:0102 F5              CMC
-T

AX=0000  BX=0000  CX=0000  DX=0000  SP=FFEE  BP=0000  SI=0000  DI=0000
DS=1342  ES=1342  SS=1342  CS=1342  IP=0103   NV UP EI PL NZ NA PO NC
1342:0103 8AFF            MOV      BH,BH
-Q

C:\DOS>
```

Figure 6.4 Display sequence for Example 6.4.

An instruction is included in the instruction set of the 80386DX that can be used to compare two 8-bit, 16-bit, or 32-bit numbers. It is the *compare* (CMP) instruction of Fig. 6.5(a). The compare operation enables us to determine the relationship between two numbers—that is whether they are equal or unequal, and when they are unequal, which one is larger. Figure 6.5(b) shows that the operands for this instruction can reside in a storage location in memory, a register within the MPU, or be part of the instruction. For instance, a byte-wide number in a register such as BL can be compared to a second byte-wide number that is supplied as immediate data.

The result of the comparison is reflected by changes in six of the status flags of the 80386DX. Notice in Fig. 6.5(a) that it affects the overflow flag, sign flag, zero flag, auxiliary carry flag, parity flag, and carry flag. The new logic state of these flags can be used by the instructions that follow to make a decision whether or not to alter the sequence in which the program executes.

The process of comparison performed by the CMP instruction is basically a subtraction operation. The source operand is subtracted from the destination operand. However, the result of this subtraction is not saved. Instead, based on the result of the subtraction operation, the appropriate flags are set or reset. The importance of the flags lies in the fact that they lead us to an understanding of the relationship between the two numbers. For instance, if 5 is compared to 7 by subtracting 5 from 7, the ZF and CF both become logic 0. These conditions indicate that a smaller number was compared to a larger one. On the other hand, if 7 is compared to 5, we are comparing a large number to a smaller number. This comparison results in ZF and CF equal to 0 and 1, respectively. Finally, if two equal numbers—for instance, 5 and 5—are compared, ZF is set to 1 and CF cleared to 0 to indicate the equal condition.

For example, let us assume that the destination operand equals $10011001_2 = -103_{10}$ and that the source operand equals $00011011_2 = +27_{10}$. Subtracting the source operand

Mnemonic	Meaning	Format	Operation	Flags affected
CMP	Compare	CMP D,S	(D) − (S) is used in setting or resetting the flags	CF,AF,OF,PF,SF,ZF

(a)

Destination	Source
Register	Register
Register	Memory
Memory	Register
Register	Immediate
Memory	Immediate
Accumulator	Immediate

(b)

Figure 6.5 (a) Compare instructions. (b) Allowed operands.

from the destination operand, we get

$$
\begin{array}{rcl}
10011001_2 = & -103_{10} \\
-\ 00011011_2 = & -\ (+27_{10}) \\
\hline
101111110_2 = & +126_{10}
\end{array}
$$

In the process of subtraction, we get the status that follows:

1. Borrow is needed from bit 4 to bit 3; therefore, the auxiliary carry flag, AF, is set.
2. There is no borrow to bit 7. Thus, the carry flag, CF, is reset.
3. Even though there is no borrow to bit 7, there is a borrow from bit 7 to bit 6. This is an indication of the overflow condition. Therefore, the OF flag is set.
4. There is an even number of 1s; therefore, this sets the parity flag, PF.
5. Bit 7 is 0, so the sign flag, SF, is reset.
6. The result that is produced is nonzero, which resets the zero flag, ZF.

Notice that the result produced is not correct as an 8-bit signed number. This condition was indicated by setting the overflow flag.

EXAMPLE 6.5

Describe what happens to the status flags as the sequence of instructions that follows is executed.

```
MOV   AX,1234H
MOV   BX,0ABCDH
CMP   AX,BX
```

Assume that flags ZF, SF, CF, AF, OF, and PF are all initially reset.

Solution

The first instruction loads AX with 1234_{16}. No status flags are affected by the execution of a MOV instruction.

The second instruction puts $ABCD_{16}$ into the BX register. Again, status is not affected. Thus, after execution of these two move instructions, the contents of AX and BX are

$$(AX) = 1234_{16} = 0001001000110100_2$$

and

$$(BX) = ABCD_{16} = 1010101111001101_2$$

The third instruction is a 16-bit comparison, with AX representing the destination and BX the source. Therefore, the contents of BX are subtracted from that of AX.

$$(AX) - (BX) = 0001001000110100_2 - 1010101111001101_2 = 0110011001100111_2$$

Instruction	ZF	SF	CF	AF	OF	PF
Initial state	0	0	0	0	0	0
MOV AX,1234H	0	0	0	0	0	0
MOV BX,0ABCDH	0	0	0	0	0	0
CMP AX,BX	0	0	1	1	0	0

Figure 6.6 Effect on flags of executing instructions.

The flags are either set or reset based on the result of this subtraction. Notice that the result is nonzero and positive. This makes ZF and SF equal to zero. Moreover, the overflow condition has not occurred. Therefore, OF is also at logic 0. The carry and auxiliary carry conditions have occurred; therefore, CF and AF are 1. Finally, the result has odd parity; therefore, PF is 0. These results are summarized in Fig. 6.6.

EXAMPLE 6.6

Verify the execution of the instruction sequence in Example 6.5 on the PC/AT. Use DEBUG to load and run the instruction sequence, which is provided in run module EX66.EXE.

Solution

A source program written to implement a procedure that contains the instruction sequence executed in Example 6.5 is shown in Fig. 6.7(a). This program was assembled with MASM and linked with LINK to form a run module in file EX66.EXE. The source listing produced by the assembler is shown in Fig. 6.7(b).

To execute this program with DEBUG, we bring up the debug program and load the file from a data diskette in drive A with the DOS command

```
C:\DOS>DEBUG   A:EX66:EXE   (↵)
```

```
TITLE    EXAMPLE 6.6

        PAGE    ,132

STACK_SEG       SEGMENT         STACK 'STACK'
                DB              64 DUP(?)
STACK_SEG       ENDS


CODE_SEG        SEGMENT         'CODE'
EX66    PROC    FAR
        ASSUME  CS:CODE_SEG, SS:STACK_SEG
;To return to DEBUG program put return address on the stack

        PUSH    DS
        MOV     AX, 0
        PUSH    AX

;Following code implements Example 6.6

        MOV     AX, 1234H
        MOV     BX, 0ABCDH
        CMP     AX, BX

        RET                     ;Return to DEBUG program
EX66    ENDP

CODE_SEG        ENDS

        END     EX66
                        (a)
```

Figure 6.7(a) Source program for Example 6.6.

```
                                    PAGE      ,132
0000                                          STACK_SEG        SEGMENT        STACK 'STACK'
0000      40 [                                                 DB             64 DUP(?)
          ??
            ]

0040                                          STACK_SEG        ENDS


0000                                          CODE_SEG       SEGMENT        'CODE'
0000                                          EX66   PROC    FAR
                                         ASSUME  CS:CODE_SEG, SS:STACK_SEG

                                ;To return to DEBUG program put return address on the stack

0000      1E                                                  PUSH     DS
0001      B8 0000                                             MOV      AX, 0
0004      50                                                  PUSH     AX

                                ;Following code implements Example 6.6

0005      B8 1234                                             MOV      AX, 1234H
0008      BB ABCD                                             MOV      BX, 0ABCDH
000B      3B C3                                               CMP      AX, BX

000D      CB                                                  RET                      ;Return to DEBUG program
000E                                          EX66   ENDP

000E                                          CODE_SEG       ENDS

                                              END     EX66
```

Segments and groups:

N a m e	Size	align	combine	class
CODE_SEG	000E	PARA	NONE	'CODE'
STACK_SEG.	0040	PARA	STACK	'STACK'

Symbols:

N a m e	Type	Value	Attr	
EX66	F PROC	0000	CODE_SEG	Length =000E

```
Warning Severe
Errors  Errors
0       0
```

(b)

Figure 6.7(b)   Source listing produced by assembler.

To verify its loading, the following UNASSEMBLE command can be used:

$$-U \quad 0 \quad D \quad (↵)$$

As shown in Fig. 6.7(c), the instructions of the source program are correctly displayed.
    First, we execute the instructions up to the CMP instruction. This is done with the GO command

$$-G \quad B \quad (↵)$$

Note in Fig. 6.7(c) that AX has been loaded with 1234_{16}, and BX with the value $ABCD_{16}$.
    Next, the compare instruction is executed with the command

$$-T \quad (↵)$$

```
C:\DOS>DEBUG A:EX66.EXE
-U 0 D
0F50:0000 1E              PUSH    DS
0F50:0001 B80000          MOV     AX,0000
0F50:0004 50              PUSH    AX
0F50:0005 B83412          MOV     AX,1234
0F50:0008 BBCDAB          MOV     BX,ABCD
0F50:000B 3BC3            CMP     AX,BX
0F50:000D CB              RETF
-G B

AX=1234  BX=ABCD  CX=000E  DX=0000  SP=003C  BP=0000  SI=0000  DI=0000
DS=0F40  ES=0F40  SS=0F51  CS=0F50  IP=000B  NV UP EI PL NZ NA PO NC
0F50:000B 3BC3            CMP     AX,BX
-T

AX=1234  BX=ABCD  CX=000E  DX=0000  SP=003C  BP=0000  SI=0000  DI=0000
DS=0F40  ES=0F40  SS=0F51  CS=0F50  IP=000D  NV UP EI PL NZ AC PO CY
0F50:000D CB              RETF
-G

Program terminated normally
-Q

C:\DOS>
```

(c)

Figure 6.7(c)   Execution of the program with DEBUG.

By comparing the state information before and after execution of the CMP instruction, we find that auxiliary carry flag and carry flag are the only flags that have changed states, and they have both been set. Their new states are identified as AC and CY, respectively. These results are identical to those found in Example 6.5.

Byte Set on Condition: SETcc

Earlier we pointed out that the flag bits set or reset by the compare instruction are examined through software to decide whether or not branching should take place in the program. One way of using these bits is to test them directly with the jump instruction. Another approach is to test them for a specific condition and then save a flag value representing whether the tested condition is true or false. An instruction that performs this operation is *byte set on condition* (SETcc). The flag value can then be used later for program-branching decisions.

The byte set on condition instruction can be used to test for various states of the flags. In Fig. 6.8(a) we see that the general form of the instruction is denoted as

$$SETcc \quad D$$

Here the cc part of the mnemonic stands for a general flag relationship and must be replaced with a specific relationship when writing the instruction. Figure 6.8(b) is a list of the mnemonics that can be used to replace cc and their corresponding flag relationship. For instance, replacing cc by A gives the mnemonic SETA. This stands for *set byte if above* and tests the flags to determine if

$$CF = 0 \cdot ZF = 0$$

Sec. 6.3   Compare and Set Instructions

221

Mnemonic	Meaning	Format	Operation	Affected flags
SETcc	Byte set on condition	SETcc D	11111111 → D if cc is true 00000000 → D if cc is false	None

(a)

Instruction	Meaning	Conditions code relationship
SETA r/m8	Set byte if above	CF = 0 · ZF = 0
SETAE r/m8	Set byte if above or equal	CF = 0
SETB r/m8	Set byte if below	CF = 1
SETBE r/m8	Set byte if below or equal	CF = 1 + ZF = 1
SETC r/m8	Set if carry	CF = 1
SETE r/m8	Set byte if equal	ZF = 1
SETG r/m8	Set byte if greater	ZF = 0 + SF = OF
SETGE r/m8	Set byte if greater	SF = OF
SETL r/m8	Set byte if less	SF <> OF
SETLE r/m8	Set byte if less or equal	ZF = 1 · SF <> OF
SETNA r/m8	Set byte if not above	CF = 1
SETNAE r/m8	Set byte if not above	CF = 1
SETNB r/m8	Set byte if not below	CF = 0
SETNBE r/m8	Set byte if not below	CF = 0 · ZF = 0
SETNC r/m8	Set byte if not carry	CF = 0
SETNE r/m8	Set byte if not equal	ZF = 0
SETNG r/m8	Set byte if not greater	ZF = 1 + SF <> OF
SETNGE r/m8	Set if not greater or equal	SF <> OF
SETNL r/m8	Set byte if not less	SF = OF
SETNLE r/m8	Set byte if not less or equal	ZF = 1 · SF <> OF
SETNO r/m8	Set byte if not overflow	OF = 0
SETNP r/m8	Set byte if not parity	PF = 0
SETNS r/m8	Set byte if not sign	SF = 0
SETNZ r/m8	Set byte if not zero	ZF = 0
SETO r/m8	Set byte if overflow	OF = 1
SETP r/m8	Set byte if parity	PF = 1
SETPE r/m8	Set byte if parity even	PF = 1
SETPO r/m8	Set byte if parity odd	PF = 0
SETS r/m8	Set byte if sign	SF = 1
SETZ r/m8	Set byte if zero	ZF = 1

(b)

Source
Reg8
Mem8

(c)

Figure 6.8 (a) Byte set on condition instruction. (b) Flag relationships. (c) Allowed operands.

If these conditions are satisfied, a byte of 1s is written to the register or memory location specified as the destination operand. Notice in Fig. 6.8(c) that the allowed operands are a byte-wide internal register (Reg8) or a byte storage location in memory (Mem8). On the other hand, if the conditions are not valid, a byte of 0s is written to the destination operand.

An example is the instruction

```
SETE   AL
```

Looking at Fig. 6.8(b), we find that execution of this instruction causes the ZF to be tested. If ZF equals 1, 11111111_2 is written into AL; otherwise, it is loaded with 00000000_2.

EXAMPLE 6.7

Write an instruction that will load memory location EVEN_PARITY with the value FF_{16} if the result produced by the last instruction had even parity.

Solution

In Fig. 6.8(b), the instruction that tests for PF equal to 1 is SETPE, and making its destination operand the memory location EVEN_PARITY gives

```
SETPE   [EVEN_PARITY]
```

▲ 6.4 JUMP INSTRUCTIONS

The purpose of a *jump* instruction is to alter the execution path of instructions in the program. In the 80386DX microprocessor, the code segment register and instruction pointer keep track of the next instruction to be fetched for execution. Thus a jump instruction involves altering the contents of these registers. In this way, execution continues at an address other than that of the next sequential instruction. That is, a jump occurs to another part of the program. Typically, program execution is not intended to return to the next sequential instruction after the jump instruction. Therefore, no return linkage is saved when the jump takes place.

The Unconditional and Conditional Jump

The 80386DX microprocessor allows two different types of jump instructions. They are the *unconditional jump* and the *conditional jump* instructions. In an unconditional jump, no status requirements are imposed for the jump to occur. That is, as the instruction is executed, the jump always takes place to change the execution sequence.

The unconditional jump concept is illustrated in Fig. 6.9(a). Notice that when the instruction JMP AA in part I is executed, program control is passed to a point in part III identified by the label AA. Execution resumes with the instruction corresponding to AA. In this way, the instructions in part II of the program are bypassed; that is, they are jumped over.

On the other hand, for a conditional jump instruction, status conditions that exist at the time the jump instruction is executed decide whether or not the jump will occur. If the condition or conditions are met, the jump takes place; otherwise, execution continues with the next sequential instruction of the program. The conditions that can be referenced by a conditional jump instruction are status flags such as carry (CF), parity (PF), and overflow (OF) flags.

Looking at Fig. 6.9(b), we see that execution of the conditional jump instruction Jcc AA in part I causes a test to be initiated. If the conditions of the test are not met, the NO path is taken and execution continues with the next sequential instruction. This corresponds to the first instruction in part II. However, if the result of the conditional test is YES, a jump is initiated to the segment of program identified as part III and the instructions in part II are bypassed.

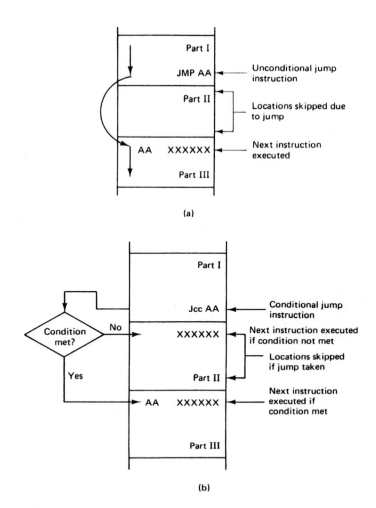

Figure 6.9 (a) Unconditional jump program sequence. (b) Conditional jump program sequence.

This type of software operation is referred to as making a *branch*, or a point in the program where a choice is made between two paths of execution. If the conditions specified by the jump instruction are met, program control is passed to the part of the program identified by the label. On the other hand, if they are not met, the next sequential instruction is executed.

The branch is a frequently used program structure and is sometime referred to as an *IF-THEN-ELSE* structure. By this, we means that IF the conditions specified in the jump instruction are met, THEN program control is passed to the point identified by the label; ELSE the program continues with the instruction following the jump instruction.

Unconditional Jump Instruction

The unconditional jump instruction of the 80386DX is shown in Fig. 6.10(a), together with its valid operand combinations in Fig. 6.10(b). There are two basic kinds of uncondi-

Mnemonic	Meaning	Format	Operation	Affected flags
JMP	Unconditional jump	JMP Operand	Jump is initiated to the address specified by the operand	None

(a)

Operands
Short-label
Near-label
Far-label
Memptr16
Regptr16
Memptr32
Regptr32

(b)

Figure 6.10 (a) Unconditional jump instruction. (b) Allowed operands.

tional jumps. The first, called an *intrasegment jump*, is limited to addresses within the current code segment. This type of jump is achieved by just modifying the value in IP. The second kind of jump, the *intersegment jump*, permits jumps from one code segment to another. Implementation of this type of jump requires modification of the contents of both CS and IP.

Jump instructions specified with a *short-label, near-label, memptr16,* or *regptr16 operands* represent intrasegment jumps. The short-label and near-label operands specify the jump relative to the address of the jump instruction itself. For example, in a short-label jump instruction, an 8-bit number is coded as an immediate operand to specify the *signed displacement* of the next instruction to be executed from the location of the jump instruction. When the jump instruction is executed, IP is reloaded with a new value equal to the updated value in IP, which is (IP) + 2, plus the signed displacement. The new value of IP and current value in CS give the address of the next instruction to be fetched and executed. With an 8-bit displacement, the short-label operand can only be used to initiate a jump in the range from −126 to +129 bytes from the location of the jump instruction.

On the other hand, the near-label operand specifies a new value for IP with a 16-bit immediate operand. This size of offset corresponds to the complete range of the current code segment. The value of the offset is automatically added to IP upon execution of the instruction. In this way, program control is passed to the location identified by the new IP. Consider the following example of an unconditional jump instruction

```
JMP    1234H
```

It means jump to address 1234H. However, the value of the address encoded in the instruction is not 1234H; instead, it is the difference between the incremented value in IP and 1234_{16}. This offset is encoded as either an 8-bit constant (short label) or a 16-bit constant (near label), depending on the size of the difference.

The jump-to address can also be specified indirectly by the contents of a memory location or the contents of a register. These two types correspond to the memptr16 and regptr16 operands, respectively. Just as for the near-label operand, they both permit a jump to any address in the current code segment.

For example,

```
JMP   BX
```

uses the contents of register BX for the offset in the current code segment. That is, the value in BX is copied into IP.

EXAMPLE 6.8

Verify the operation of the instruction JMP BX using the DEBUG program on the PC/AT. Let the contents of BX be 0010_{16}.

Solution

As shown in Fig. 6.11, DEBUG is invoked, and then the line-by-line assembler is used to load the instruction with the command

```
-A                        (↵)
1342:0100   JMP   BX      (↵)
1342:0102                 (↵)
```

Next, BX is initialized with the command

```
-R   BX      (↵)
BX   0000    (↵)
:10          (↵)
```

Let us check the value in IP before executing the JMP instruction. This is done with another R command as

```
-R   (↵)
```

Looking at the state information displayed in Fig. 6.11, we see that IP contains 0100_{16} and BX contains 0010_{16}.

```
C:\DOS>DEBUG
-A
1342:0100 JMP BX
1342:0102
-R BX
BX 0000
:10
-R
AX=0000  BX=0010  CX=0000  DX=0000  SP=FFEE  BP=0000  SI=0000  DI=0000
DS=1342  ES=1342  SS=1342  CS=1342  IP=0100    NV UP EI PL NZ NA PO NC
1342:0100 FFE3           JMP     BX
-T

AX=0000  BX=0010  CX=0000  DX=0000  SP=FFEE  BP=0000  SI=0000  DI=0000
DS=1342  ES=1342  SS=1342  CS=1342  IP=0010    NV UP EI PL NZ NA PO NC
1342:0010 8B09           MOV     CX,[BX+DI]                DS:0010=098B
-Q

C:\DOS>
```

Figure 6.11 Display sequence for Example 6.8.

Executing the instruction with the command

$$-\text{T} \quad (\lrcorner)$$

and then looking at Fig. 6.11, we see that the value in IP has become 10_{16}. Therefore, the address at which execution picks up is 1342:0010.

To specify an operand to be used as a pointer, the various addressing modes available with the 80386DX can be used. For instance

$$\text{JMP} \quad [\text{BX}]$$

uses the contents of BX as the address of the memory location that contains the IP offset address (memptr16 operand). This offset is loaded into IP, where it is used together with the current contents of CS to compute the jump-to address.

EXAMPLE 6.9

Use the DEBUG program to observe the operation of the instruction

$$\text{JMP} \quad [\text{BX}]$$

Assume that the pointer held in BX is 1000_{16} and the value held at memory location DS:1000 is 200_{16}. What is the address of the next instruction to be executed?

Solution

Figure 6.12 shows that first the debugger is brought up and then an ASSEMBLE command is issued to load the instruction. This assemble command is

```
-A                        (⌐)
1342:0100   JMP   [BX]    (⌐)
1342:0102                 (⌐)
```

```
C:\DOS>DEBUG
-A
1342:0100 JMP [BX]
1342:0102
-R BX
BX 0000
:1000
-E 1000 00 02
-D 1000 1001
1342:1000  00 02                                    . .
-R
AX=0000  BX=1000  CX=0000  DX=0000  SP=FFEE  BP=0000  SI=0000  DI=0000
DS=1342  ES=1342  SS=1342  CS=1342  IP=0100   NV UP EI PL NZ NA PO NC
1342:0100 FF27        JMP    [BX]                         DS:1000=0200
-T

AX=0000  BX=1000  CX=0000  DX=0000  SP=FFEE  BP=0000  SI=0000  DI=0000
DS=1342  ES=1342  SS=1342  CS=1342  IP=0200   NV UP EI PL NZ NA PO NC
1342:0200 4D          DEC    BP
-Q
```

Figure 6.12 Display sequence for Example 6.9.

Next BX is loaded with the pointer address using the R command

```
-R   BX     (⏎)
BX   0000
:1000       (⏎)
```

and the memory location is initialized with the command

```
-E   1000   00   02   (⏎)
```

As shown in Fig. 6.12, the loading of memory location DS:1000 and the BX register are next verified with D and R commands, respectively.

Now the instruction is executed with the command

```
-T   (⏎)
```

Notice from the state information displayed in Fig. 6.12 that the new value in IP is 0200_{16}. This value was loaded from memory location 1342:1000. Therefore, program execution continues with the instruction at address 1342:0200.

The intersegment unconditional jump instructions correspond to the *far-label, regptr32,* and *memptr32 operands* that are shown in Fig. 6.10(b). Far-label uses a 32-bit immediate operand to specify the jump-to address. The first 16 bits of this 32-bit pointer are loaded into IP and are an offset address relative to the contents of the code-segment register. The next 16 bits are loaded into the CS register and define the new code segment.

An indirect way to specify the offset and code segment address for an intersegment jump is by using the memptr32 operand. This time four consecutive memory bytes starting at the specified address contain the offset address and the new code segment address, respectively. Just like the memptr16 operand, the memptr32 operand may be specified using any one of the various addressing modes of the 80386DX.

An example is the instruction

```
JMP   DWORD PTR   [DI]
```

It uses the contents of DS and DI to calculate the address of the memory location that contains the first word of the pointer that identifies the location to which the jump will take place. The two-word pointer starting at this address is read into IP and CS to pass control to the new point in the program.

Conditional Jump Instruction

The second type of jump instruction performs conditional jump operations. Figure 6.13(a) shows a general form of this instruction; Fig. 6.13(b) is a list of each of the conditional jump instructions in the 80386DX's instruction set. Notice that each of these instructions tests for the presence or absence of certain status conditions.

For instance, the *jump on carry* (JC) instruction makes a test to determine if carry flag (CF) is set. Depending on the result of the test, the jump to the location specified by its operand either takes place or does not. If CF equals 0, the test fails and execution continues with the instruction at the address following the JC instruction. On the other hand, if CF equals 1, the test condition is satisfied and the jump is performed.

Mnemonic	Meaning	Format	Operation	Flags Affected
Jcc	Conditional jump	Jcc Operand	If the specific condition cc is true, the jump to the address specified by the Operand is initiated; otherwise, the next instruction is executed	None

(a)

Mnemonic	Meaning	Condition
JA	above	CF = 0 and ZF = 0
JAE	above or equal	CF = 0
JB	below	CF = 1
JBE	below or equal	CF = 1 or ZF = 1
JC	carry	CF = 1
JCXZ	CX register is zero	CX = 0000H
JECXZ	ECX register is zero	ECX = 00000000H
JE	equal	ZF = 1
JG	greater	ZF = 0 and SF = OF
JGE	greater or equal	SF = OF
JL	less	(SF xor OF) = 1
JLE	less or equal	((SF xor OF) or ZF) = 1
JNA	not above	CF = 1 or ZF = 1
JNAE	not above nor equal	CF = 1
JNB	not below	CF = 0
JNBE	not below nor equal	CF = 0 and ZF = 0
JNC	not carry	CF = 0
JNE	not equal	ZF = 0
JNG	not greater	((SF xor OF) or ZF) = 1
JNGE	not greater nor equal	(SF xor OF) = 1
JNL	not less	SF = OF
JNLE	not less nor equal	ZF = 0 and SF = OF
JNO	not overflow	OF = 0
JNP	not parity	PF = 0
JNS	not sign	SF = 0
JNZ	not zero	ZF = 0
JO	overflow	OF = 1
JP	parity	PF = 1
JPE	parity even	PF = 1
JPO	parity odd	PF = 0
JS	sign	SF = 1
JZ	zero	ZF = 1

(b)

Figure 6.13 (a) Conditional jump instruction. (b) Types of conditional jump instructions.

Notice that for some of the instructions in Fig. 6.13(b) two different mnemonics can be used. This feature can be used to improve program readability. That is, for each occurrence of the instruction in the program, it can be identified with the mnemonic that best describes its function.

For instance, the instruction *jump on parity* (JP) or *jump on parity even* (JPE) can be

Sec. 6.4   Jump Instructions

229

used to test parity flag PF for logic 1. Since PF is set to 1 if the result from a computation has even parity, this instruction can initiate a jump based on the occurrence of even parity. The reverse instruction JNP/JPO is also provided. It can be used to initiate a jump based on the occurrence of a result with odd instead of even parity.

In a similar manner, the instructions *jump if equal* (JE) and *jump if zero* (JZ) have the same function. Either notation can be used in a program to determine if the result of a computation was zero.

All other conditional jump instructions work in a similar way except that they test different conditions to decide whether or not the jump is to take place. Examples of these conditions are: the contents of CX are 0, an overflow has occurred, or the result is negative.

To distinguish between comparisons of signed and unsigned numbers by jump instructions, two different names, which seem to imply the same, have been devised. They are *above* and *below* for comparison of unsigned numbers and *less* and *greater* for comparison of signed numbers. For instance, the number $ABCD_{16}$ is above the number 1234_{16} if they are considered as unsigned numbers. On the other hand, if they are considered as signed numbers, $ABCD_{16}$ is negative and 1234_{16} is positive. Therefore, $ABCD_{16}$ is less than 1234_{16}.

The Branch Program Structure: IF-THEN

Let us now look at some simple examples of how the conditional jump instruction can be used to implement the software branch program structure called *IF-THEN*. One example is a branch that is made based on the flag settings that result after the contents of two registers are compared to each other. Figure 6.14 shows a program structure that implements this software operation.

First, the CMP instruction subtracts the value in BX from that in AX and adjusts the flags based on the result. Next the jump on equal instruction tests the zero flag to see if it is 1. If ZF is 1, it means that the contents of AX and BX are equal and a jump is made to the location in the program identified by the label EQUAL. Otherwise, if ZF is 0, which means that the contents of AX and BX are not equal, the instruction after the CMP instruction gets executed.

Similar instruction sequences can be used to initiate branch operations for other conditions. For instance, by using the instruction JG GREATER, the branch is taken if the value in BX is greater than that in AX.

Another common use of a conditional jump is to branch based on the setting of a specific bit in a register. When this is done, a logic operation is normally used to mask off the values of all of the other bits in the register. For example, we may want to mask off all bits of the value in AL other than bit 2 and then make a conditional jump if the unmasked bit is logic 1.

```
        CMP  AX, BX
        JE   EQUAL
        ---  ---        ; Next instruction if (AX) ≠ (BX)
         .
         .
EQUAL:  ---  ---        ; Next instruction if (AX) = (BX)
         .
         .
        ---  ---
```

Figure 6.14 IF-THEN-ELSE branch program structure using a flag-condition test.

```
            AND   AL, 04H
            JNZ   BIT2_ONE
            ---   ---              ; Next instruction if B2 of AL = 0
                   .
                   .
            ---   ---
BIT2_ONE:   ---   ---              ; Next instruction if B2 of AL = 1
                   .
                   .
            ---   ---
```

Figure 6.15   IF-THEN-ELSE branch program structure using a register-bit test.

This operation can be done with the instruction sequence in Fig. 6.15. First the contents of AL are ANDed with 04_{16} to give

$$(AL) = XXXXXXXX_2 \cdot 00000100_2 = 00000X00_2$$

Now the content of AL is 0 if bit 2 is 0 and the resulting value in the ZF is 1. On the other hand, if bit 2 is 1, the content of AL is nonzero and ZF is 0. Remember we want to make the jump when bit 2 is 1. Therefore, a jump on not zero instruction is used to make the conditional test. When ZF is 1, the next instruction is executed, but if ZF is 0 the JNZ instruction passes control to the instruction identified by the label BIT2_ONE.

Let us look at how to perform this exact same branch operation in another way. Instead of masking off all of the bits in AL, we could shift bit 2 into the carry flag and then make a conditional jump if CF equals 1. The program structure in Fig. 6.16 uses this method to test for logic 1 in bit 2 of AL. Notice that this implementation takes one extra instruction.

The Loop Program Structure: REPEAT-UNTIL and WHILE-DO

In many practical applications, we frequently need to repeat a part of the program many times. That is, a group of instructions may need to be executed over and over again until a condition is met. Before initiating the program sequence, a parameter must be assigned to keep track of how many times the sequence of instructions has been repeated. This parameter is tested each time the sequence is performed to verify whether or not it

```
            MOV   CL,03H
            SHR   AL, CL
            JC    BIT2_ONE
            ---   ---              ; Next instruction if B2 of AL = 0
                   .
                   .
            ---   ---
BIT2_ONE:   ---   ---              ; Next instruction if B2 of AL = 1
                   .
                   .
            ---   ---
```

Figure 6.16   IF-THEN-ELSE branch program structure using an alternative register-bit test.

is to be repeated again. For instance, to repeat a part of a program ten times, we can begin by loading the count register CL with the value $0A_{16}$; execute the series of instructions; decrement the count in CL by 1; and after decrementing CL, test the zero flag to see if it has reached 0. If ZF is 0, which means that (CL) $\neq$ 0, program control is passed back to the first instruction of the sequence. This is repeated until CL becomes 0 so that ZF tests as 1. This type of software structure is known as a *loop*.

Figure 6.17(a) shows a loop structure that is called *REPEAT-UNTIL*. Here we see that the sequence of instructions from label AGAIN to the conditional jump instruction

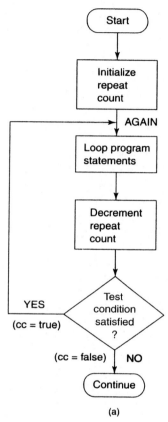

(a)

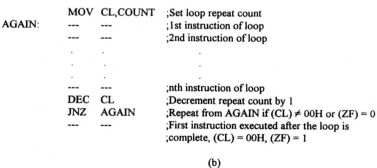

(b)

Figure 6.17 (a) REPEAT-UNTIL program sequence. (b) Typical REPEAT-UNTIL instruction sequence.

Jcc represents the loop. Notice that the label for the instruction that is to be jumped to is located before the jump instruction that makes the conditional test. In this way, if the test result is true, program control returns to AGAIN and the segment of program repeats. This continues until the condition specified by cc is false. Before entering the loop, the register that is to be used for the conditional test must be loaded with the appropriate count.

The instruction sequence in Fig. 6.17(b) implements a simple REPEAT-UNTIL loop. Notice that first CL is initialized with the count value; therefore, the loop will repeat COUNT times. Then the operation performed by instructions 1 through n of the loop is performed. Next, the value in CL is decremented by 1 to indicate that the instructions in the loop are done, and then ZF is tested with the JNZ instruction to determine if the value in CL has reached 0. That is, the question Should I repeat again? is asked with software. If CL has not reached 0, the answer is Yes, repeat again, and program control is returned to the instruction labeled AGAIN and the loop instruction sequence repeats. This continues until the value in CL reaches 0 to identify that the loop is done. When this happens, the answer to the conditional test is No, do not repeat again, and the jump is not taken. Instead, the instruction following JNZ AGAIN is executed.

Another loop structure is shown in Fig. 6.18(a). It differs in that the conditional test that is used to decide whether or not the loop will repeat is made before entering the instruction sequence that is to be repeated. This sequence is known as a *WHILE-DO loop*. To implement this loop, we will need to use both a conditional and unconditional jump instruction. A typical WHILE-DO instruction sequence is shown in Fig. 6.18(b).

The *no operation* (NOP) instruction is sometimes used in conjunction with loop routines. As its name implies, it performs no operation. That is, no register contents are changed and the flags are not affected when it is executed. However, a period of time is needed to perform the NOP function. In some practical applications, for instance, a time-delay loop, the duration it takes to execute NOP is used to extend the time-delay interval.

Applications Using the Loop and Branch Software Structures

As a practical application of the use of a conditional jump operation, let us write a program known as a *block-move program*. The purpose of this program is to move a block of N consecutive bytes of data starting at offset address BLK1ADDR in memory to another block of memory locations starting at offset address BLK2ADDR. We will assume that both blocks are in the same data segment, whose starting point is defined by the data segment value DATASEGADDR.

The steps to be implemented to solve this problem are outlined in the flowchart of Fig. 6.19(a). It has four basic operations. The first operation is initialization. Initialization involves establishing the initial address of the data segment. This is done by loading the DS register with the value DATASEGADDR. Furthermore, source index register SI and destination index register DI are initialized with offset addresses BLK1ADDR and BLK2ADDR, respectively. In this way, they point to the beginning of the source block and the beginning of the destination block, respectively. To keep track of the count, register CX is initialized with N, the number of bytes to be moved. This leads us to the following assembly language statements:

```
MOV   AX,  DATASEGADDR
MOV   DS,  AX
MOV   SI,  BLK1ADDR
MOV   DI,  BLK2ADDR
MOV   CX,  N
```

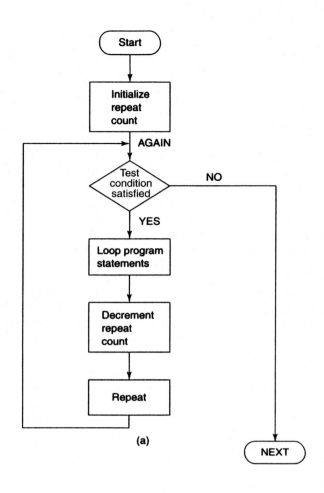

(a)

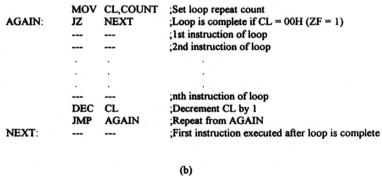

```
         MOV   CL,COUNT    ;Set loop repeat count
AGAIN:   JZ    NEXT        ;Loop is complete if CL = 00H (ZF = 1)
         ---   ---         ;1st instruction of loop
         ---   ---         ;2nd instruction of loop
          .     .               .
          .     .               .
          .     .               .
         ---   ---         ;nth instruction of loop
         DEC   CL          ;Decrement CL by 1
         JMP   AGAIN       ;Repeat from AGAIN
NEXT:    ---   ---         ;First instruction executed after loop is complete
```

(b)

Figure 6.18 (a) WHILE-DO program sequence. (b) Typical WHILE-DO instruction sequence.

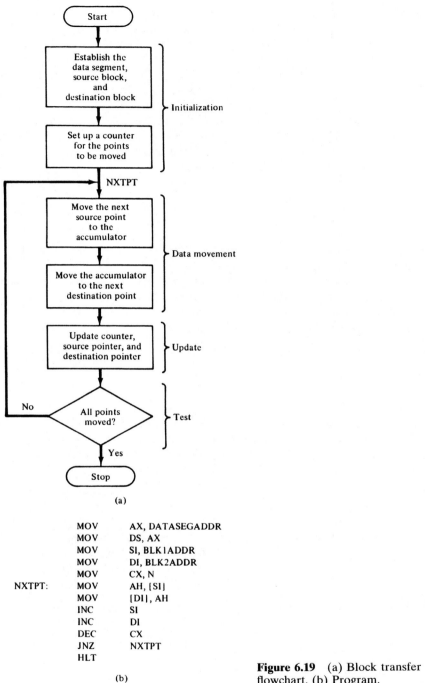

```
                MOV     AX, DATASEGADDR
                MOV     DS, AX
                MOV     SI, BLK1ADDR
                MOV     DI, BLK2ADDR
                MOV     CX, N
        NXTPT:  MOV     AH, [SI]
                MOV     [DI], AH
                INC     SI
                INC     DI
                DEC     CX
                JNZ     NXTPT
                HLT
```

(b)

Figure 6.19 (a) Block transfer flowchart. (b) Program.

Notice that DS cannot be directly loaded by immediate data with a MOV instruction. Therefore, the segment address is first loaded into AX and then moved to DS. SI, DI, and CX can be loaded directly with immediate data.

The next operation that must be performed is the actual movement of data from the source block of memory to the destination block. The offset addresses are already loaded into SI and DI; therefore, move instructions that employ indirect addressing can be used to accomplish the data-transfer operation. Remember that the 80386DX does not allow direct memory-to-memory moves. For this reason, AX will be used as a temporary storage location for data. The source byte is moved into AX with one instruction, and then another instruction is needed to move it from AX to the destination location. Thus, the data move is accomplished by the following instructions.

```
NXTPT:  MOV  AH, [SI]
        MOV  [DI], AH
```

Notice that for a byte move only the higher 8 bits of AX are used. Therefore, the operand is specified as AH instead of AX.

Now the pointers in SI and DI must be updated so that they are ready for the next byte-move operation. Also, the counter must be decremented so that it corresponds to the number of bytes that remains to be moved. These updates can be done by the following sequence of instructions.

```
INC  SI
INC  DI
DEC  CX
```

The test operation involves determining whether or not all the data points have been moved. The contents of CX represent this condition. When its value is not 0, there still are points to be moved, whereas a value of 0 indicates that the block move is complete. This 0 condition is reflected by 1 in ZF. The instruction needed to perform this test is

```
JNZ  NXTPT
```

Here NXTPT is a label that corresponds to the first instruction in the data move operation. The last instruction in the program can be a *halt* (HLT) instruction to indicate the end of the block move operation. The entire program is shown in Fig. 6.19(b).

EXAMPLE 6.10

The program

```
        CMP  AX, BX
        JC   DIFF2
DIFF1:  MOV  DX, AX
        SUB  DX, BX  ;(DX) = (AX) - (BX)
        JMP  DONE
DIFF2:  MOV  DX, BX
        SUB  DX, AX  ;(DX) = (BX) - (AX)
DONE:   ----
```

implements an instruction sequence that calculates the absolute difference between the contents of AX and BX and places it in DX. Use the run module produced by assembling and linking the source program in Fig. 6.20(a) to verify the operation of the program for the two cases that follow:

$$\text{(a) } (AX) = 6, (BX) = 2$$

$$\text{(b) } (AX) = 2, (BX) = 6$$

Solution

The source program in Fig. 6. 20(a) can be assembled with MASM and linked with LINK to produce a run module called EX610.EXE. The MASM and LINK programs will be discussed in Chapter 7. At this point, the only thing to note is that they are used to generate a run module that can be used in conjunction with the debug program. The source listing produced as part of the assembly process is shown in Fig. 6.20(b).

As shown in Fig. 6.20(c), the run module can be loaded as part of calling up the debug program by issuing the DOS command

```
C:\DOS>DEBUG  A:EX610.EXE  (↵)
```

Next, the loading of the program is verified with the UNASSEMBLE command

```
-U  0  15  (↵)
```

Notice in Fig. 6.20(c) that the CMP instruction, which is the first instruction of the sequence that generates the absolute difference, is located at address 0D03:0005. Let us execute down

```
TITLE     EXAMPLE 6.10

        PAGE    ,132

STACK_SEG       SEGMENT         STACK 'STACK'
                DB              64 DUP(?)
STACK_SEG       ENDS


CODE_SEG        SEGMENT         'CODE'
EX610     PROC    FAR
        ASSUME   CS:CODE_SEG, SS:STACK_SEG

;To return to DEBUG program put return address on the stack

        PUSH    DS
        MOV     AX, 0
        PUSH    AX

;Following code implements Example 6.10

        CMP     AX, BX
        JC      DIFF2
DIFF1:  MOV     DX, AX
        SUB     DX, BX          ; DX = AX - BX
        JMP     DONE
DIFF2:  MOV     DX, BX
        SUB     DX, AX          ; DX = BX - AX
DONE:   NOP

        RET                     ;Return to DEBUG program
EX610     ENDP

CODE_SEG        ENDS

        END     EX610
```

Figure 6.20(a) Source program for Example 6.10.

```
                            TITLE    EXAMPLE 6.10
                                    PAGE        ,132

        0000                              STACK_SEG        SEGMENT         STACK 'STACK'
        0000      40 [                                     DB              64 DUP(?)
                  ??
                     ]

        0040                              STACK_SEG        ENDS


        0000                              CODE_SEG         SEGMENT         'CODE'
        0000                              EX610    PROC    FAR
                                            ASSUME   CS:CODE_SEG, SS:STACK_SEG

                            ;To return to DEBUG program put return address on the stack

        0000   1E                              PUSH     DS
        0001   B8 0000                         MOV      AX, 0
        0004   50                              PUSH     AX

                            ;Following code implements Example 6.10

        0005   3B C3                           CMP      AX, BX
        0007   72 07                           JC       DIFF2
        0009   8B D0              DIFF1:       MOV      DX, AX
        000B   2B D3                           SUB      DX, BX          ; DX = AX - BX
        000D   EB 05 90                        JMP      DONE
        0010   8B D3              DIFF2:       MOV      DX, BX
        0012   2B D0                           SUB      DX, AX          ; DX = BX - AX
        0014   90                 DONE:        NOP

        0015   CB                              RET                     ;Return to DEBUG program
        0016                              EX610    ENDP

        0016                              CODE_SEG         ENDS

                                            END      EX610

Segments and groups:

             N a m e                    Size    align    combine class

CODE_SEG . . . . . . . . . . . .        0016    PARA     NONE     'CODE'
STACK_SEG. . . . . . . . . . . .        0040    PARA     STACK    'STACK'

Symbols:

             N a m e                    Type    Value    Attr

DIFF1. . . . . . . . . . . . . .        L NEAR  0009     CODE_SEG
DIFF2. . . . . . . . . . . . . .        L NEAR  0010     CODE_SEG
DONE . . . . . . . . . . . . . .        L NEAR  0014     CODE_SEG
EX610 . . . . . . . . . . . . .         F PROC  0000     CODE_SEG     Length =0016

Warning Severe
Errors  Errors
0       0
```

Figure 6.20(b)   Source listing produced by assembler.

to this statement with the GO command

$$-G \quad 5 \quad (\leftarrow\!\!\!\lrcorner)$$

Now we will load AX and BX with the case (a) data. This is done with the R commands

```
              -R   AX        (↵)
              AX   0000
              :6             (↵)
              -R   BX        (↵)
              BX   0000
              :2             (↵)
```

```
C:\DOS>DEBUG A:EX610.EXE
-U 0 15
0D03:0000 1E          PUSH    DS
0D03:0001 B80000      MOV     AX,0000
0D03:0004 50          PUSH    AX
0D03:0005 3BC3        CMP     AX,BX
0D03:0007 7207        JB      0010
0D03:0009 8BD0        MOV     DX,AX
0D03:000B 2BD3        SUB     DX,BX
0D03:000D EB05        JMP     0014
0D03:000F 90          NOP
0D03:0010 8BD3        MOV     DX,BX
0D03:0012 2BD0        SUB     DX,AX
0D03:0014 90          NOP
0D03:0015 CB          RETF
-G 5


AX=0000  BX=0000  CX=0016  DX=0000  SP=003C  BP=0000  SI=0000  DI=0000
DS=0DD6  ES=0DD6  SS=0DE8  CS=0D03  IP=0005   NV UP EI PL NZ NA PO NC
0D03:0005 3BC3        CMP     AX,BX
-R AX
AX 0000
:6
-R BX
BX 0000
:2
-T


AX=0006  BX=0002  CX=0016  DX=0000  SP=003C  BP=0000  SI=0000  DI=0000
DS=0DD6  ES=0DD6  SS=0DE8  CS=0D03  IP=0007   NV UP EI PL NZ NA PO NC
0D03:0007 7207        JB      0010
-G 14

AX=0006  BX=0002  CX=0016  DX=0004  SP=003C  BP=0000  SI=0000  DI=0000
DS=0DD6  ES=0DD6  SS=0DE8  CS=0D03  IP=0014   NV UP EI PL NZ NA PO NC
0D03:0014 90          NOP
-G

Program terminated normally
-R
AX=0006  BX=0002  CX=0016  DX=0004  SP=003C  BP=0000  SI=0000  DI=0000
DS=0DD6  ES=0DD6  SS=0DE8  CS=0D03  IP=0014   NV UP EI PL NZ NA PO NC
0D03:0014 90          NOP
-R IP
IP 0014
:0
-G 5

AX=0000  BX=0002  CX=0016  DX=0004  SP=0038  BP=0000  SI=0000  DI=0000
DS=0DD6  ES=0DD6  SS=0DE8  CS=0D03  IP=0005   NV UP EI PL NZ NA PO NC
0D03:0005 3BC3        CMP     AX,BX
-R AX
AX 0000
:2
-R BX
BX 0002
:6
-T


AX=0002  BX=0006  CX=0016  DX=0004  SP=0038  BP=0000  SI=0000  DI=0000
DS=0DD6  ES=0DD6  SS=0DE8  CS=0D03  IP=0007   NV UP EI NG NZ AC PE CY


0D03:0007 7207        JB      0010
-G 14

AX=0002  BX=0006  CX=0016  DX=0004  SP=0038  BP=0000  SI=0000  DI=0000
DS=0DD6  ES=0DD6  SS=0DE8  CS=0D03  IP=0014   NV UP EI PL NZ NA PO NC
0D03:0014 90          NOP
-G

Program terminated NORMALLY
-Q

C:\DOS>
```

Figure 6.20(c)   Execution of the program with DEBUG.

Next, we execute the compare instruction with the command

-T   (↵)

Note in the trace information display in Fig. 6.20(c) that the carry flag is reset (NC). Therefore, no jump will take place when the JB instruction is executed.

The rest of the program can be executed by inputting the command

-G   14   (↵)

From Fig. 6.20(c), we find that DX contains 4. This result was produced by executing the SUB instruction at 0D03:000B. Before executing the program for the (b) set of data, the command

-G   (↵)

is issued. This command causes the program to terminate normally.

The R command shows that the value in IP must be reset, and then we can execute down to the CMP instruction. This is done with the commands

```
-R   IP      (↵)
IP   0014
:0           (↵)
```

and

-G   5   (↵)

Notice in Fig. 6.20(c) that IP again contains 0005_{16} and points to the CMP instruction. Next, the data for case (b) are loaded with R commands. This gives

```
-R   AX      (↵)
AX   0000
:2           (↵)
-R   BX      (↵)
BX   0002
:6           (↵)
```

Now a T command is used to execute the CMP instruction. Notice that CY is set this time. Therefore, control is passed to the instruction at 0D03:0010.

A GO command is now used to execute down to the instruction at 0D03:0014. This command is

-G   14   (↵)

Notice that DX again contains 4; however, this time it was calculated with the SUB instruction at 0D03:0012.

▲ 6.5 SUBROUTINES AND SUBROUTINE-HANDLING INSTRUCTIONS

A *subroutine* is a special segment of program that can be called for execution from any point in a program. Figure 6.21(a) illustrates the concept of a subroutine. Here we see a program structure where one part of the program is called the *main program*. In addition to this, we find a segment attached to the main program, known as a *subroutine*. The subroutine is written to provide a function that must be performed at various points in the main program. Instead of including this piece of code in the main program each time the function is needed, it is put into the program just once as a subroutine. A subroutine is also known as a *procedure*.

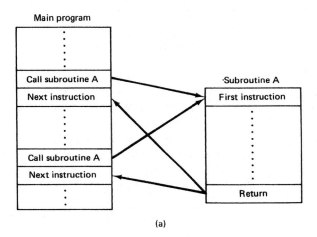

(a)

Mnemonic	Meaning	Format	Operation	Affected flags
CALL	Subroutine call	CALL Operand	Execution continues from the address of the subroutine specified by the operand. Information required to return back to the main program such as IP and CS are saved on the stack.	None

(b)

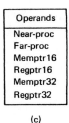

Operands
Near-proc
Far-proc
Memptr16
Regptr16
Memptr32
Regptr32

(c)

Figure 6.21 (a) Subroutine concept. (b) Subroutine call instruction. (c) Allowed operands.

Wherever the function must be performed, a single instruction is inserted into the main body of the program to "call" the subroutine. Remember that the physical address CS:IP identifies the next instruction to be fetched for execution. Thus, to branch to a subroutine that starts elsewhere in memory, the value in either IP or CS and IP must be modified. After executing the subroutine, we want to return control to the instruction that follows the one that called the subroutine. In this way, program execution resumes in the main program at the point where it left off due to the subroutine call. A return instruction must be included at the end of the subroutine to initiate the *return sequence* to the main program environment.

The instructions provided to transfer control from the main program to a subroutine and return control back to the main program are called *subroutine-handling instructions*. Let us now examine the instructions provided for this purpose.

CALL and RET Instructions

There are two basic instructions in the instruction set of the 80386DX for subroutine handling. They are the *call* (CALL) and *return* (RET) instructions. Together they provide the mechanism for calling a subroutine into operation and returning control back to the main program at its completion. We will first discuss these two instructions and later introduce other instructions that can be used in conjunction with subroutines.

Just like the JMP instruction, CALL allows implementation of two types of operations, the *intrasegment call* and the *intersegment call*. The CALL instruction is shown in Fig. 6.21(b), and its allowed operand variations are shown in Fig. 6.21(c).

It is the operand that initiates either an intersegment or an intrasegment call. The operands near-proc, memptr16, and regptr16 all specify intrasegment calls to a subroutine. In all three cases, execution of the instruction causes the contents of IP to be saved on the stack. Then the stack pointer (SP) is decremented by two. This is the push operation that was introduced in the section covering the stack in Chapter 2. The saved values of IP is the offset address of the instruction that follows the CALL instruction. After saving this return address, a new 16-bit value, which is specified by the instruction's operand and corresponds to the storage location of the first instruction in the subroutine, is loaded into IP.

The types of operands represent different ways of specifying a new value of IP. Using a near-proc operand, a subroutine located in the same code segment can be called. An example is

```
CALL   1234H
```

Here 1234H identifies the starting address of the subroutine. It is encoded as the difference between 1234H and the updated value of IP—that is, the IP for the instruction following the CALL instruction.

The memptr16 and regptr16 operands provide indirect subroutine addressing by specifying a memory location or an internal register, respectively, as the source of a new value for IP. The value specified is the actual offset that is loaded into IP. An example of a regptr16 operand is

```
CALL   BX
```

When this instruction is executed, the contents of BX are loaded into IP and execution continues with the subroutine starting at a physical address derived from the current CS and the new value of IP.

By using various addressing modes of the 80386DX, an operand that resides in memory can be used as the call to offset address. This represents a memptr16 type of operand. For instance, the instruction

```
CALL    [BX]
```

has its subroutine offset address at the memory location whose physical address is derived from the contents of DS and BX. The value stored at this memory location is loaded into IP. Again the current contents of CS and the new value in IP point to the first instruction of the subroutine.

Notice that in both intrasegment call examples the subroutine was located within the same code segment as the call instruction. The other type of CALL instruction, the intersegment call, permits the subroutine to reside in another code segment. It corresponds to the far-proc, regptr32, and memptr32 operands. These operands specify both a new offset address for IP and a new segment address for CS. In each case, execution of the call instruction causes the contents of the CS and IP registers to be saved on the stack, and then new values are loaded into IP and CS. The saved values of CS and IP permit return to the main program from a different code segment.

Far-proc represents a 32-bit immediate operand that is stored in the four bytes that follow the opcode of the call instruction in program memory. These two words are loaded directly from code segment memory into IP and CS with execution of the CALL instruction.

On the other hand, when the operand is memptr32, the pointer for the subroutine is stored as four consecutive bytes in data memory. The location of the first byte of the pointer can be specified indirectly by one of the 80386DX's memory addressing modes. An example is

```
CALL   DWORD   PTR [DI]
```

Here the physical address of the first byte of the 4-byte pointer in memory is derived from the contents of DS and DI.

Every subroutine must end by executing an instruction that returns control to the main program. This is the return (RET) instruction. It is described in Fig. 6.22(a) and (b). Notice that its execution causes the value of IP or both the values of IP and CS that were saved on the stack to be returned back to their corresponding registers and the stack pointer to be adjusted appropriately. This is the pop operation that was discussed in Chapter 2. In general, an intrasegment return results from an intrasegment call and an intersegment return results from an intersegment call. In this way, program control is returned to the instruction that follows the call instruction in program memory.

There is an additional option with the return instruction. It is that a 2-byte constant can be included with the return instruction. This constant gets added to the stack pointer after restoring the return address. The purpose of this stack pointer displacement is to provide a simple means by which the *parameters* that were saved on the stack before the call to the subroutine was initiated can be discarded. For instance, the instruction

```
RET   2
```

when executed adds 2 to SP. This discards one word parameter as part of the return sequence.

Mnemonic	Meaning	Format	Operation	Affected flags
RET	Return	RET or RET Operand	Return to the main program by restoring IP (and CS for far-procedure). If operand is present it is added to the content of SP.	None

(a)

Operands
None
Disp16

(b)

Figure 6.22 (a) Return instruction. (b) Allowed operands.

EXAMPLE 6.11

The source program in Fig. 6.23(a) can be used to demonstrate the use of the call and return instructions to implement a subroutine. This program was assembled and linked on the PC/AT to produce a run module in file EX611.EXE. Its source listing is provided for reference in Fig. 6.23(b). Trace the operation of the program by executing it with DEBUG for data (AX) = 2 and (BX) = 4.

```
TITLE    EXAMPLE 6.11

        PAGE     ,132

STACK_SEG        SEGMENT      STACK 'STACK'
                 DB           64 DUP(?)
STACK_SEG        ENDS


CODE_SEG         SEGMENT      'CODE'
EX611   PROC     FAR
        ASSUME   CS:CODE_SEG, SS:STACK_SEG

;To return to DEBUG program put return address on the stack

        PUSH     DS
        MOV      AX, 0
        PUSH     AX

;Following code implements Example 6.11

        CALL     SUM
        RET

SUM     PROC     NEAR
        MOV      DX, AX
        ADD      DX, BX          ; (DX) = (AX) + (BX)
        RET
SUM     ENDP

EX611 ENDP
CODE_SEG         ENDS

        END      EX611
```

Figure 6.23(a) Source program for Example 6.11.

```
                          TITLE    EXAMPLE 6.11

                          PAGE      ,132

0000                               STACK_SEG     SEGMENT          STACK 'STACK'
0000        40 [                   DB                             64 DUP(?)
            ??
                    ]

0040                               STACK_SEG     ENDS


0000                               CODE_SEG      SEGMENT        'CODE'
0000                               EX611   PROC  FAR
                                     ASSUME  CS:CODE_SEG, SS:STACK_SEG

                     ;To return to DEBUG program put return address on the stack

0000   1E                          PUSH    DS
0001   B8 0000                     MOV     AX, 0
0004   50                          PUSH    AX

                     ;Following code implements Example 6.11

0005   E8 0009 R                   CALL    SUM
0008   CB                          RET

0009                               SUM    PROC   NEAR
0009   8B D0                              MOV    DX, AX
000B   03 D3                              ADD    DX, BX        ; (DX) = (AX) + (BX)
000D   C3                                 RET
000E                               SUM    ENDP

000E                               EX611   ENDP
000E                               CODE_SEG      ENDS

                                     END     EX611

Segments and groups:

                N a m e                 Size   align    combine class

CODE_SEG . . . . . . . . . . .          000E   PARA     NONE     'CODE'
STACK_SEG. . . . . . . . . . .          0040   PARA     STACK    STACK'

Symbols:

                N a m e                 Type   Value    Attr

EX611. . . . . . . . . . . . .          F PROC  0000    CODE_SEG    Length =000E
SUM. . . . . . . . . . . . . .          N PROC  0009    CODE_SEG    Length =0005

Warning Severe
Errors  Errors
0       0
```

Figure 6.23(b) Source listing produced by assembler.

Solution

We begin by calling up DEBUG and loading the program with the DOS command

$$C:\backslash DOS>DEBUG \ A:EX611.EXE \quad (\dashv)$$

The loading of the program is now verified with the UNASSEMBLE command

$$-U \quad 0 \quad D \quad (\dashv)$$

Looking at Fig. 6.23(c), we see that the program has correctly loaded. Moreover, we find that the CALL instruction is located at offset 0005_{16} of the current code segment. The command

$$-G \quad 5 \quad (\dashv)$$

```
C:\DOS>DEBUG A:EX611.EXE
-U 0 D
0D03:0000 1E              PUSH    DS
0D03:0001 B80000          MOV     AX,0000
0D03:0004 50              PUSH    AX
0D03:0005 E80100          CALL    0009
0D03:0008 CB              RETF
0D03:0009 8BD0            MOV     DX,AX
0D03:000B 03D3            ADD     DX,BX
0D03:000D C3              RET
-G 5

AX=0000  BX=0000  CX=000E  DX=0000  SP=003C  BP=0000  SI=0000  DI=0000
DS=0F41  ES=0F41  SS=0F52  CS=0D03  IP=0005    NV UP EI PL NZ NA PO NC
0D03:0005 E80100          CALL    0009
-R AX
AX 0000
:2
-R BX
BX 0000
:4
-T

AX=0002  BX=0004  CX=000E  DX=0000  SP=003A  BP=0000  SI=0000  DI=0000
DS=0F41  ES=0F41  SS=0F52  CS=0D03  IP=0009    NV UP EI PL NZ NA PO NC
0D03:0009 8BD0            MOV     DX,AX
-D SS:3A 3B
0F52:0030                                      08 00                    ..
-T

AX=0002  BX=0004  CX=000E  DX=0002  SP=003A  BP=0000  SI=0000  DI=0000
DS=0F41  ES=0F41  SS=0F52  CS=0D03  IP=000B    NV UP EI PL NZ NA PO NC
0D03:000B 03D3            ADD     DX,BX
-T

AX=0002  BX=0004  CX=000E  DX=0006  SP=003A  BP=0000  SI=0000  DI=0000
DS=0F41  ES=0F41  SS=0F52  CS=0D03  IP=000D    NV UP EI PL NZ NA PE NC
0D03:000D C3              RET
-T

AX=0002  BX=0004  CX=000E  DX=0006  SP=003C  BP=0000  SI=0000  DI=0000
DS=0F41  ES=0F41  SS=0F52  CS=0D03  IP=0008    NV UP EI PL NZ NA PE NC
0D03:0008 CB              RETF
-G

Program terminated normally
-Q

C:\DOS>
```

Figure 6.23(c) Execution of the program with DEBUG.

executes the program down to the CALL instruction. The state information displayed in Fig. 6.23(c) shows that $(CS) = 0D03_{16}$, $(IP) = 0005_{16}$, and $(SP) = 003C_{16}$. Now let us load the AX and BX registers with R commands:

```
-R  AX    (↵)
AX  0000
:2        (↵)
-R  BX    (↵)
BX  0000
:4        (↵)
```

Next, the CALL instruction is executed with the T command

$$-T (↵)$$

and, looking at the displayed state information in Fig. 6.23(c), we find that CS still contains $0D03_{16}$, IP has been loaded with 0009_{16}, and SP has been decremented to $003A_{16}$. This information tells us that the next instruction to be executed is the move instruction at address 0D03:0009, and a word of data has been pushed to the stack.

Before executing another instruction, let us look at what got pushed onto the stack. This is done by issuing the memory dump command

$$-D \quad SS:3A \quad 3B \quad (\lrcorner)$$

Note from Fig. 6.23(c) that the value 0008_{16} has been pushed onto the stack. This is the address offset of the RETF instruction that follows the CALL instruction and is the address of the instruction to which control is to be returned at the completion of the subroutine.

Two more T commands are used to execute the move and add instructions of the subroutine. From the state information displayed in Fig. 6.23(c), we see that their execution causes the value 2_{16} in AX to be copied into DX and then the value 4_{16} in BX to be added to the value in DX. This results in the value 6_{16} in DX.

Now the RET instruction is executed by issuing another T command. In Fig. 6.23(c), we see that execution of this instruction causes the value 0008_{16} to be popped off the stack and put back into the IP register. Therefore, the next instruction to be executed is the one located at address 0D03:0008; this is the RETF instruction. Moreover, notice that, as the word is popped from the stack back into IP, the value in SP is incremented by 2. After this, the program is run to completion by issuing a GO command.

PUSH and POP Instructions

After the context switch to a subroutine, we find that it is usually necessary to save the contents of certain registers or some other main program parameters. These values are saved by pushing them onto the stack. Typically, these data correspond to registers and memory locations that are used by the subroutine. In this way, their original contents are kept intact in the stack segment of memory during the execution of the subroutine. Before a return to the main program takes place, the saved registers and main program parameters are restored. This is done by popping the saved values from the stack back into their original locations. Thus, a typical structure of a subroutine is that shown in Fig. 6.24.

The instruction that is used to save parameters on the stack is the *push* (PUSH) instruction, and that used to retrieve them is the *pop* (POP) instruction. Notice in Fig. 6.25(a) and (b) that the standard PUSH and POP instructions can be written with a general-

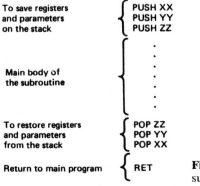

Figure 6.24 Structure of a subroutine.

Mnemonic	Meaning	Format	Operation	Flags Affected
PUSH	Push word onto stack	PUSH S	((SP)) ← (S) (SP) ← (SP) − 2	None
POP	Pop word off stack	POP D	(D) ← ((SP)) (SP) ← (SP) + 2	None

(a)

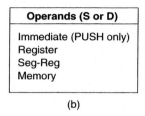

Operands (S or D)
Immediate (PUSH only) Register Seg-Reg Memory

(b)

Figure 6.25   (a) PUSH and POP instructions. (b) Allowed operands.

purpose register, a segment register (excluding CS), or a storage location in memory as their operand. The PUSH instruction also allows for use of a double-word, word-wide, or sign-extended byte immediate source operand.

Execution of a PUSH instruction causes the data corresponding to the operand to be pushed onto the top of the stack. For instance, if the instruction is

$$\text{PUSH} \quad \text{AX}$$

the result is as follows:

$$((SP) - 1) \leftarrow (AH)$$
$$((SP) - 2) \leftarrow (AL)$$
$$(SP) \leftarrow (SP) - 2$$

This shows that the two bytes of AX are saved in the stack part of memory and the stack pointer is decremented by two so that it points to the new top of the stack.

On the other hand, if the instruction is

$$\text{POP} \quad \text{AX}$$

its execution results in

$$(AL) \leftarrow ((SP))$$
$$(AH) \leftarrow ((SP) + 1)$$
$$(SP) \leftarrow (SP) + 2$$

In this manner, the saved contents of AX are restored in the register. When a 32-bit operand, such as EAX, is pushed to or popped from the stack, the value in the stack pointer register is decremented or incremented, respectively, by 4.

EXAMPLE 6.12 _____

Write a procedure named SQUARE that squares the contents of BL and places the result in BX. Assume that this procedure is called from another procedure in the same code segment.

Solution

The beginning of the procedure is defined with the pseudo-op statement

```
SQUARE   PROC   NEAR
```

To square the number in BL, we can use the 8-bit signed multiply instruction, IMUL. This instruction requires the use of register AX for its operation. Therefore, at entry of the procedure, we must save the value currently held in AX. This is done by pushing its contents to the stack with the instruction

```
PUSH   AX
```

Now we load AX with the contents of BL using the instruction

```
MOV   AL,BL
```

To square the contents of AL, we use the instruction

```
IMUL   BL
```

which multiplies the contents of AL with the contents of BL and places the result in AX. The result is the square of the original contents of BL. To place the result in BX, we use the instruction

```
MOV   BX,AX
```

This completes the square operation; but before we return to the main part of the program, the original contents of AX that were saved on the stack are restored with the pop instruction

```
POP   AX
```

Then a return instruction is used to pass control back to the main program.

```
RET
```

The procedure must be terminated with the end procedure pseudo-op statement that follows:

```
SQUARE   ENDP
```

The complete instruction sequence is shown in Fig. 6.26.

```
;Subroutine:   SQUARE
;Description:  (BX) = square of (BL)

SQUARE PROC   NEAR
        PUSH AX          ;Save the register to be used
        MOV  AL,BL       ;Place the number in AL
        IMUL BL          ;Multiply with itself
        MOV  BX,AX       ;Save the result
        POP  AX          ;Restore the register used
        RET
SQUARE ENDP
```

Figure 6.26 Program for Example 6.12.

EXAMPLE 6.13

A source program that can be used to demonstrate execution of the procedure written in Example 6.12 is shown in Fig. 6.27(a). The source listing produced when this program was assembled is given in Fig. 6.27(b), and the run module produced when it was linked is stored in file EX613.EXE. Use the DEBUG program to load this run module and verify its operation by executing it on the PC/AT.

Solution

The DEBUG program and run module can be loaded with the command

$$C:\backslash DOS>DEBUGA:EX613.EXE \quad (\downarrow)$$

After loading is completed, the instructions of the program are unassembled with the command

$$-U \quad 0 \quad 18 \quad (\downarrow)$$

The displayed information in Fig. 6.27(c) shows that the program did load correctly.

From the instruction sequence in Fig. 6.27(c), we find that the part of the program whose operation we are interested in observing starts with the CALL instruction at address 0DEC:000C. Now we execute down to this point in the program with the GO command

$$-G \quad C \quad (\downarrow)$$

Notice that BL contains the number 12H that will be squared by the subroutine.

Now the call instruction is executed with the command

$$-T \quad (\downarrow)$$

Notice from the displayed information for this command in Fig. 6.27(c) that the value held in IP has been changed to 0010_{16}. Therefore, control has been passed to address 0DEC:0010, which is the first instruction of procedure SQUARE. Moreover, note that stack pointer (SP) has been decremented to the value $003A_{16}$. The new top of the stack is at address 0DE7:003A. The word held at the top of the stack can be examined with the command

$$-D \quad SS:3A \quad 3B \quad (\downarrow)$$

```
TITLE    EXAMPLE 6.13

      PAGE      ,132

STACK_SEG          SEGMENT              STACK 'STACK'
                   DB                   64 DUP(?)
STACK_SEG          ENDS


DATA_SEG           SEGMENT
TOTAL              DW                   1234H
DATA_SEG           ENDS


CODE_SEG           SEGMENT              'CODE'
EX613    PROC      FAR
                   ASSUME  CS:CODE_SEG, SS:STACK_SEG, DS:DATA_SEG

;To return to DEBUG program put return address on the stack

         PUSH      DS
         MOV       AX, 0
         PUSH      AX

;Setup the data segment

         MOV       AX, DATA_SEG
         MOV       DS, AX

;Following code implements Example 6.13

         MOV       BL,12H              ;BL contents = the number
                                       ;to be squared
         CALL      SQUARE              ;Call the procedure to
                                       ;square BL contents
         RET                           ;Return to DEBUG program
EX613    ENDP

;Subroutine:   SQUARE
;Description:  (BX) = square of (BL)

SQUARE  PROC      NEAR
         PUSH      AX                  ;Save the register to be
                                       ;used
         MOV       AL,BL               ;Place the number in AL
         IMUL      BL                  ;Multiply with itself
         MOV       BX,AX               ;Save the result
         POP       AX                  ;Restore the register used
         RET
SQUARE  ENDP

CODE_SEG           ENDS

         END       EX613
```

Figure 6.27(a)   Source program for Example 6.13.

Notice in Fig. 6.27(c) that its value is $000F_{16}$. From the instruction sequence in Fig. 6.27(c), we see that this is the address of the RETF instruction. This is the instruction to which control is to be returned at the completion of the procedure.

Next, the PUSH AX instruction is executed with the command

$$-T \quad (\dashv)$$

and again, looking at the displayed state information, we find that SP has been decremented to the value 0038_{16}. Displaying the word at the top of the stack with the command

$$-D \quad SS:38 \quad 39 \quad (\dashv)$$

we find that it is the same as the contents of AX. This confirms that the original contents of AX are saved on the stack.

```
                           PAGE      ,132

0000                               STACK_SEG       SEGMENT         STACK 'STACK'
0000    0040[                                      DB              64 DUP(?)
        ??
                  ]

0040                               STACK_SEG       ENDS


0000                               DATA_SEG        SEGMENT
0000    1234                       TOTAL           DW              1234H
0002                               DATA_SEG        ENDS


0000                               CODE_SEG        SEGMENT         'CODE'
0000                               EX613   PROC    FAR
                                        ASSUME  CS:CODE_SEG, SS:STACK_SEG, DS:DATA_SEG

                            ;To return to DEBUG program put return address on the stack

0000    1E                                         PUSH    DS
0001    B8 0000                                    MOV     AX, 0
0004    50                                         PUSH    AX

                            ;Setup the data segment

0005    B8 ---- R                                  MOV     AX, DATA_SEG
0008    8E D8                                      MOV     DS, AX

                            ;Following code implements Example 6.13

000A    B3 12                                      MOV     BL,12H          ;BL contents = the number
                                                                           ;to be squared
000C    E8 0010 R                                  CALL    SQUARE          ;Call the procedure to
                                                                           ;square BL contents
000F    CB                                         RET                     ;Return to DEBUG program
0010                               EX613   ENDP
```

Figure 6.27(b)   Source listing produced by assembler.

Now we execute down to the POP AX instruction with the command

$$-G \quad 17 \quad (\dashv)$$

Looking at the displayed state information in Fig. 6.27(c), we find that the square of the contents of BL has been formed in BX.

Next, the pop instruction is executed with the command

$$-T \quad (\dashv)$$

and the displayed information shows that the original contents of AX have been popped off the stack and put back into AX. Moreover, the value in SP has been incremented to $003A_{16}$ so that once again the return address is at the top of the stack.

Finally, the RET instruction is executed with the command

$$-T \quad (\dashv)$$

As shown in Fig. 6.27(c), this causes the value 0010_{16} to be popped from the top of the stack back into IP. Therefore, IP now equals $000F_{16}$. In this way, we see that control has been returned to the instruction at address 0DEC:000F of the main program.

```
                        ;Subroutine:     SQUARE
                        ;Description:    (BX) = square of (BL)

0010                            SQUARE   PROC    NEAR
0010  50                                 PUSH    AX        ;Save the register to
                                                           ;used
0011  8A C3                              MOV     AL,BL     ;Place the number in A
0013  F6 EB                              IMUL    BL        ;Multiply with itself
0015  8B D8                              MOV     BX,AX     ;Save the result
0017  58                                 POP     AX        ;Restore the register
0018  C3                                 RET
0019                            SQUARE   ENDP

0019                            CODE_SEG         ENDS


                        END      EX613
```

Segments and Groups:

N a m e	Length	Align	Combine	Class
CODE_SEG	0019	PARA	NONE	'CODE'
DATA_SEG	0002	PARA	NONE	
STACK_SEG	0040	PARA	STACK	'STACK'

Symbols:

N a m e	Type	Value	Attr	
EX613	F PROC	0000	CODE_SEG	Length = 0010
SQUARE	N PROC	0010	CODE_SEG	Length = 0009
TOTAL	L WORD	0000	DATA_SEG	
@CPU	TEXT	0101h		
@FILENAME	TEXT	EX613		
@VERSION	TEXT	613		

```
        53 Source  Lines
        53 Total   Lines
        13 Symbols

     48016 + 440523 Bytes symbol space free

         0 Warning Errors
         0 Severe  Errors
```

Figure 6.27(b) (Continued)

At times, we also want to save the contents of the flag register, and if we save them, we will later have to restore them. These operations can be accomplished with *push flags* (PUSHF) and *pop flags* (POPF) instructions, respectively. They are shown in Fig. 6.28. Notice that PUSHF saves the contents of the flag register on the top of the stack. On the other hand, POPF returns the flags from the top of the stack to the flag register.

When writing the compiler for a high-level language such as C, it is very common to push the contents of all the general registers of the 80386DX to the stack before calling a subroutine. If we use the PUSH instruction to perform this operation, many instructions are needed. To simplify this operation, special instructions are provided in the instruction set of the 80386DX. They are called *push all* (PUSHA) and *pop all* (POPA).

Looking at Fig. 6.29 we see that execution of PUSHA causes the values in AX, CX, DX, BX, OLD SP, BP, SI, and DI to be pushed, in that order, onto the top of the stack. Figure 6.30 shows the state of the stack before and after execution of the instruction. As shown in Fig. 6.31, executing a POPA instruction at the end of the subroutine restores the old state of the 80386DX.

Sec. 6.5   Subroutines and Subroutine-Handling Instructions   **253**

```
C:\DOS>DEBUG A:EX613.EXE
-U 0 18
0DEC:0000 1E              PUSH    DS
0DEC:0001 B80000          MOV     AX,0000
0DEC:0004 50              PUSH    AX
0DEC:0005 B8EB0D          MOV     AX,0DEB
0DEC:0008 8ED8            MOV     DS,AX
0DEC:000A B312            MOV     BL,12
0DEC:000C E80100          CALL    0010
0DEC:000F CB              RETF
0DEC:0010 50              PUSH    AX
0DEC:0011 8AC3            MOV     AL,BL
0DEC:0013 F6EB            IMUL    BL
0DEC:0015 8BD8            MOV     BX,AX
0DEC:0017 58              POP     AX
0DEC:0018 C3              RET
-G C

AX=0DEB  BX=0012  CX=0069  DX=0000  SP=003C  BP=0000  SI=0000  DI=0000
DS=0DEB  ES=0DD7  SS=0DE7  CS=0DEC  IP=000C   NV UP EI PL NZ NA PO NC
0DEC:000C E80100          CALL    0010
-T

AX=0DEB  BX=0012  CX=0069  DX=0000  SP=003A  BP=0000  SI=0000  DI=0000
DS=0DEB  ES=0DD7  SS=0DE7  CS=0DEC  IP=0010   NV UP EI PL NZ NA PO NC
0DEC:0010 50              PUSH    AX
-D SS:3A 3B
0DE7:0030                                     0F 00                         . .
-T

AX=0DEB  BX=0012  CX=0069  DX=0000  SP=0038  BP=0000  SI=0000  DI=0000
DS=0DEB  ES=0DD7  SS=0DE7  CS=0DEC  IP=0011   NV UP EI PL NZ NA PO NC
0DEC:0011 8AC3            MOV     AL,BL
-D SS:38 39
0DE7:0030                                 EB 0D                             . .
-G 17

AX=0144  BX=0144  CX=0069  DX=0000  SP=0038  BP=0000  SI=0000  DI=0000
DS=0DEB  ES=0DD7  SS=0DE7  CS=0DEC  IP=0017   OV UP EI PL NZ NA PE CY
0DEC:0017 58              POP     AX
-T

AX=0DEB  BX=0144  CX=0069  DX=0000  SP=003A  BP=0000  SI=0000  DI=0000
DS=0DEB  ES=0DD7  SS=0DE7  CS=0DEC  IP=0018   OV UP EI PL NZ NA PE CY
0DEC:0018 C3              RET
-T

AX=0DEB  BX=0144  CX=0069  DX=0000  SP=003C  BP=0000  SI=0000  DI=0000
DS=0DEB  ES=0DD7  SS=0DE7  CS=0DEC  IP=000F   OV UP EI PL NZ NA PE CY
0DEC:000F CB              RETF
-G

Program terminated normally
-Q

C:\DOS>
```

Figure 6.27(c) Execution of the program with DEBUG.

It is also possible to push the contents of all of the 80386DX's 32-bit registers to the stack with a single instruction. Looking at Fig. 6.29, we see that this is done with the PUSHAD instruction. The contents of these registers can be restored from the stack by executing a POPAD instruction.

Mnemonic	Meaning	Operation	Flags Affected
PUSHF	Push flags onto stack	$((SP)) \leftarrow (Flags)$ $(SP) \leftarrow (SP) - 2$	None
POPF	Pop flags from stack	$(Flags) \leftarrow ((SP))$ $(SP) \leftarrow (SP) + 2$	OF, DF, IF, TF, SF, ZF, AF, PF, CF

Figure 6.28 Push flags and pop flags instructions.

Mnemonic	Meaning	Operation	Flags Affected
PUSHA	Push all 16-bit general registers onto stack	((SP)) ← (AX) ((SP–2)) ← (CX) ((SP–4)) ← (DX) ((SP–6)) ← (BX) ((SP–8)) ← (OLD SP) ((SP–A)) ← (BP) ((SP–C)) ← (SI) ((SP–E)) ← (DI)	None
PUSHAD	Push all 32-bit general registers onto stack	((SP)) ← (EAX) ((SP–4)) ← (ECX) ((SP–8)) ← (EDX) ((SP–C)) ← (EBX) ((SP–10)) ← (OLD ESP) ((SP–14)) ← (EBP) ((SP–18)) ← (ESI) ((SP–1C)) ← (EDI)	None
POPA	Pop all 16-bit general registers from stack	(DI) ← ((SP + 2)) (SI) ← ((SP + 4)) (BP) ← ((SP + 6)) (OLD SP) ← ((SP + 8)) (BX) ← ((SP + A)) (DX) ← ((SP + C)) (CX) ← ((SP + E)) (AX) ← ((SP + 10))	None
POPAD	Pop all 32-bit general registers from stack	(EDI) ← ((SP + 4)) (ESI) ← ((SP + 8)) (EBP) ← ((SP + C)) (OLD ESP) ← ((SP + 10)) (EBX) ← ((SP + 14)) (EDX) ← ((SP + 18)) (ECX) ← ((SP + 1C)) (EAX) ← ((SP + 20))	

Figure 6.29   Push all and pop all instructions.

Stack Frame Instructions: ENTER and LEAVE

Before the main program calls a subroutine, quite often it is necessary for the calling program to pass the values of some *variables* (parameters) to the subroutine. It is a common practice to push these variables onto the stack before calling the routine. Then during the execution of the subroutine, they are accessed by reading from the stack and used in computations. Two instructions are provided in the extended instruction set of the 80386DX to allocate and deallocate a data area called a *stack frame*. This data area, which is located in the stack part of memory, provides local storage of parameters and other data used by the subroutine.

Normally, high-level languages allocate a stack frame for each procedure in a program. The stack frame provides a dynamically allocated local storage space for the procedure and contains data such as variables, pointers to the stack frames of the previous procedures from which the current procedure was called, and a return address for linkage to the stack frame of the calling procedure. This mechanism also permits access to data in stack frames of the calling procedures.

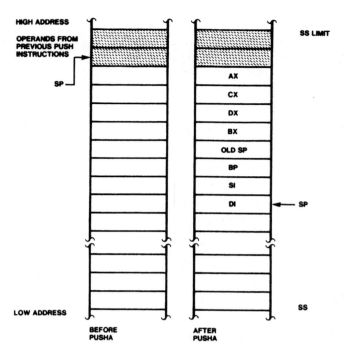

Figure 6.30 State of the stack before and after executing PUSHA. (Reprinted by permission of Intel Corp. Copyright/Intel Corp. 1987)

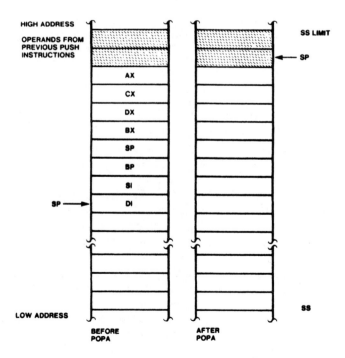

Figure 6.31 State of the stack before and after executing POPA. (Reprinted by permission of Intel/ Corp. Copyright/Intel Corp. 1987)

Mnemonic	Meaning	Format	Operation

ENTER	Make stack frame	ENTER Imm16,Imm8	Push BP
			TEMP ← (SP)
			If Imm8 > 0 then
			   Repeat (Imm8 − 1) times
			(BP) ← (BP) − 2
			  Push ((BP))
			  End repeat
			  PUSH TEMP
			End if
			(BP) ← TEMP
			(SP) ← (SP) − Imm16

LEAVE	Release stack frame	LEAVE	(SP) ← (BP) Pop to BP

Figure 6.32 Enter and leave instructions.

The instructions used for allocation and deallocation of stack frames are given in Fig. 6.32 as *enter* (ENTER) and *leave* (LEAVE). Execution of an ENTER instruction allocates a stack frame; it is deallocated by executing the LEAVE instruction. For this reason, as shown in Fig. 6.33, the ENTER instruction is used at the beginning of a subroutine, and LEAVE is used at the end, just before the return instruction.

Looking at Fig. 6.32, we find that ENTER has two operands. The first operand, identified as *dw*, is a word-size immediate operand. This operand specifies the number of bytes to be allocated on the stack for local data storage of the procedure. The second operand, which is a byte-size immediate operand, specifies what is called the *lexical nesting level* of the routine. This lexical level defines how many pointers to previous stack frames can be stored in the current stack frame. This list of previous stack frame pointers is called a *display*. The value of the lexical level byte must be limited to a maximum of 32 in 80386DX programs.

An example of an ENTER instruction is

```
ENTER   12, 2
```

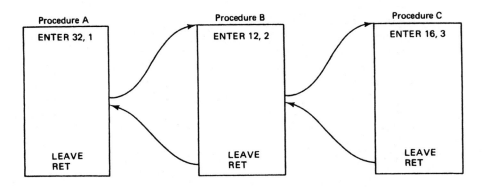

Figure 6.33 Enter/leave example.

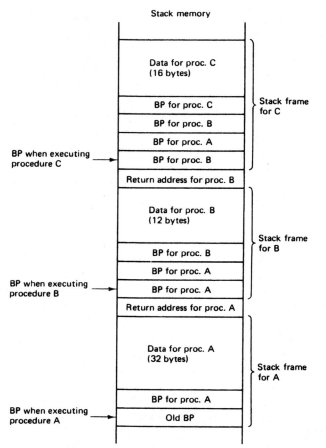

Stack memory

Data for proc. C
(16 bytes)

BP for proc. C

BP for proc. B

BP for proc. A

BP when executing
procedure C → BP for proc. B

} Stack frame
for C

Return address for proc. B

Data for proc. B
(12 bytes)

BP for proc. B

BP for proc. A

BP when executing
procedure B → BP for proc. A

} Stack frame
for B

Return address for proc. A

Data for proc. A
(32 bytes)

BP for proc. A

BP when executing
procedure A → Old BP

} Stack frame
for A

Figure 6.34 Stack after execution of ENTER instructions for procedures A, B, and C.

Execution of this instruction allocates 12 bytes of local storage on the stack for use as a stack frame. It does this by decrementing SP by 12. This defines a new top of stack at the address formed from the contents of SS and SP − 12. Also, the base pointer (BP) that identifies the beginning of the previous stack frame is copied into the stack frame created by the ENTER instruction. This value is called the *dynamic link* and is held in the first storage location of the stack frame. The number of stack frame pointers that can be saved in a stack frame is equal to the value of the byte that specifies the lexical level of the procedure. Therefore, in our example, just two levels of nesting are specified.

The BP register is used as a pointer into the stack segment of memory. When a procedure is called, the value in BP points to the stack frame location that contains the previous stack frame pointer (dynamic link). Therefore, based-indexed addressing can be used to access variables in the stack frame by referencing the BP register.

The LEAVE instruction reverses the process of an ENTER instruction. That is, its execution deallocates the stack frame. This is done by first automatically loading SP from the BP register. This returns the storage locations of the current stack frame to the stack. Now SP points to the location where the dynamic link (pointer to the previous stack frame) is stored. Next, popping the contents of the stack into BP returns the pointer to the stack frame of the previous procedure.

To illustrate the operation of the stack frame instructions, let us consider the example of Fig. 6.33. Here we find three procedures. Procedure A is used to call procedure B, which in turn calls procedure C. It is assumed that the lexical levels for these procedures are 1, 2, and 3, respectively. The ENTER, LEAVE, and RET instructions for each procedure are shown in the diagram. Notice that the ENTER instructions specify the lexical levels for the procedures.

The stack frames created by executing the ENTER instructions in the three procedures are shown in Fig. 6.34. As the ENTER instruction in procedure A is executed, the old BP from the procedure that called procedure A is pushed on the stack. BP is loaded from SP to point to the location of the old BP. Since the second operand is 1, only the current BP, that is, the BP for procedure A, is pushed onto the stack. Finally, to allocate 32 bytes for the stack frame, 32 is subtracted from the current value in SP.

After entering procedure B, a second ENTER instruction is encountered. This time the lexical level is 2. Therefore, the instruction first pushes the old BP, that is, the BP for procedure A, onto the stack, then pushes the BP previously stored on the stack frame for A to the stack, and lastly it pushes the current BP for procedure B to the stack. This mechanism provides access to the stack frame for procedure A from procedure B. Next, 12 bytes, as specified by the instruction, are allocated for local storage.

The ENTER instruction in procedure C has the lexical level 3; therefore, it pushes the BPs for the two previous procedures, B and A, to the stack in addition to the BP for C and the BP for the calling procedure B.

▲ 6.6 THE LOOP AND THE LOOP-HANDLING INSTRUCTIONS

The 80386DX microprocessor has three instructions specifically designed for implementing *loop operations*. These instructions can be used in place of certain conditional jump instructions and give the programmer a simpler way of writing loop sequences. The loop instructions are listed in Fig. 6.35.

The first instruction, *loop* (LOOP), works with respect to the contents of the CX register. CX must be preloaded with a count that represents the number of times the loop

Mnemonic	Meaning	Format	Operation
LOOP	Loop	LOOP Short-label	$(CX) \leftarrow (CX) - 1$ Jump is initiated to location defined by Short-label if $(CX) \neq 0$; otherwise, execute next sequential instruction.
LOOPE/LOOPZ	Loop while equal/ loop while zero	LOOPE/LOOPZ Short-label	$(CX) \leftarrow (CX) - 1$ Jump to location defined by Short-label if $(CX) \neq 0$ and $(ZF) = 1$; otherwise, execute next sequential instruction.
LOOPNE/LOOPNZ	Loop while not equal/ loop while not zero	LOOPNE/LOOPNZ Short-label	$(CX) \leftarrow (CX) - 1$ Jump to location defined by Short-label if $(CX) \neq 0$ and $(ZF) = 0$; otherwise, execute next sequential instruction.

Figure 6.35   Loop instructions.

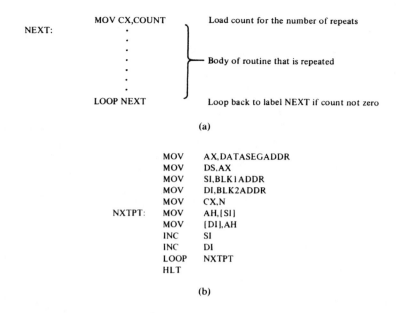

```
NEXT:     MOV CX,COUNT          Load count for the number of repeats
            .
            .
            .                  ── Body of routine that is repeated
            .
            .
            .
LOOP NEXT                        Loop back to label NEXT if count not zero
```

(a)

```
               MOV     AX,DATASEGADDR
               MOV     DS,AX
               MOV     SI,BLK1ADDR
               MOV     DI,BLK2ADDR
               MOV     CX,N
     NXTPT:    MOV     AH,[SI]
               MOV     [DI],AH
               INC     SI
               INC     DI
               LOOP    NXTPT
               HLT
```

(b)

Figure 6.36 (a) Typical loop routine structure. (b) Block-move program employing the LOOP instructions.

is to be repeated. Whenever LOOP is executed, the contents of CX are first decremented by 1 and then checked to determine if they are equal to 0. If equal to 0, the loop is complete and the instruction following LOOP is executed; otherwise, control is returned to the instruction at the label specified in the loop instruction. In this way, we see that LOOP is a single instruction that functions the same as a decrement CX instruction followed by a JNZ instruction.

For example, the LOOP instruction sequence shown in Fig. 6.36(a) will cause the part of the program from the label NEXT through the instruction LOOP to be repeated a number of times equal to the value stored in CX. For example, if CX contains $000A_{16}$, the sequence of instructions included in the loop is executed 10 times.

Figure 6.36(b) shows a practical implementation of a loop. Here we find the block move program that was developed in Section 6.4 rewritten using the LOOP instruction. Comparing this program with the one in Fig. 6.19(b), we see that the instruction LOOP NXTPT has replaced both the DEC and JNZ instructions.

EXAMPLE 6.14

The source program in Fig. 6.37(a) demonstrates the use of the LOOP instruction to implement a software loop operation. This program was assembled by the macroassembler on the PC/AT and linked to produce a run module called EX614.EXE. The source listing is shown in Fig. 6.37(b). Observe the operation of the loop by executing the program using DEBUG.

```
TITLE    EXAMPLE 6.14

     PAGE      ,132

STACK_SEG              SEGMENT          STACK 'STACK'
                       DB               64 DUP(?)
STACK_SEG              ENDS


CODE_SEG               SEGMENT          'CODE'
EX614     PROC         FAR
          ASSUME CS:CODE_SEG, SS:STACK_SEG

;To return to DEBUG program put return address on the stack

          PUSH         DS
          MOV          AX, 0
          PUSH         AX

;Following code implements Example 6.14

          MOV          CX, 5H
          MOV          DX, 0H
AGAIN:    NOP
          INC          DX
          LOOP         AGAIN

          RET                           ;Return to DEBUG program
EX614     ENDP
CODE_SEG               ENDS

          END          EX614
```

Figure 6.37(a)   Source program for
Example 6.14.

Solution

Looking at Fig. 6.37(c), we see that the run module is loaded with the DOS command

$$C:\backslash DOS>DEBUG\ A:EX614.EXE\quad (\hookleftarrow)$$

Now the loading of the program is verified by unassembling it with the command

$$-U\quad 0\quad F\quad (\hookleftarrow)$$

By comparing the instruction sequence displayed in Fig. 6.37(c) with the listing in Fig. 6.37(b), we find that the program has loaded correctly.

Also, we find from the instruction sequence in Fig. 6.37(c) that the loop whose operation we want to observe is located from address 0D03:000B through 0D03:000D. Therefore, we will begin by executing the instructions down to address 0D03:000B with the GO command

$$-G\quad B\quad (\hookleftarrow)$$

Again, looking at Fig. 6.37(c), we see that these instructions initialize the count in CX to 0005_{16} and the contents of DX to 0000_{16}.

Now we execute the loop down to address 0D03:000D with another GO command,

$$-G\quad D\quad (\hookleftarrow)$$

and, looking at the displayed information, we find that the pass count in DX has been incremented by 1 to indicate that the first pass through the loop is about to be completed.

```
                              PAGE    ,132

        0000                           STACK_SEG      SEGMENT        STACK 'STACK'
        0000      40 [                 DB             64 DUP(?)
                  ??
                      ]

        0040                           STACK_SEG      ENDS


        0000                           CODE_SEG       SEGMENT        'CODE'
        0000                 EX614     PROC    FAR
                                       ASSUME  CS:CODE_SEG, SS:STACK_SEG

                     ;To return to DEBUG program put return address on the stack

        0000  1E                       PUSH    DS
        0001  B8 0000                  MOV     AX, 0
        0004  50                       PUSH    AX

                     ;Following code implements Example 6.14

        0005  B9 0005                  MOV     CX, 5H
        0008  BA 0000                  MOV     DX, 0H
        000B  90            AGAIN:     NOP
        000C  42                       INC     DX
        000D  E2 FC                    LOOP    AGAIN

        000F  CB                       RET                    ;Return to DEBUG program
        0010                 EX614     ENDP
        0010                 CODE_SEG  ENDS

                                       END     EX614
```

Segments and groups:

N a m e	Size	align	combine	class
CODE_SEG	0010	PARA	NONE	'CODE'
STACK_SEG.	0040	PARA	STACK	'STACK'

Symbols:

N a m e	Type	Value	Attr	
AGAIN.	L NEAR	000B	CODE_SEG	
EX614.	F PROC	0000	CODE_SEG	Length =0010

```
Warning Severe
Errors  Errors
0       0
```

Figure 6.37(b)   Source listing produced by assembler.

```
C:\DOS>DEBUG A:EX614.EXE
-U 0 F
0D03:0000 1E            PUSH    DS
0D03:0001 B80000        MOV     AX,0000
0D03:0004 50            PUSH    AX
0D03:0005 B90500        MOV     CX,0005
0D03:0008 BA0000        MOV     DX,0000
0D03:000B 90            NOP
0D03:000C 42            INC     DX
0D03:000D E2FC          LOOP    000B
0D03:000F CB            RETF
-G B

AX=0000  BX=0000  CX=0005  DX=0000  SP=003C  BP=0000  SI=0000  DI=0000
DS=0DD7  ES=0DD7  SS=0DE8  CS=0D03  IP=000B  NV UP EI PL NZ NA PO NC
0D03:000B 90            NOP
-G D

AX=0000  BX=0000  CX=0005  DX=0001  SP=003C  BP=0000  SI=0000  DI=0000
DS=0DD7  ES=0DD7  SS=0DE8  CS=0D03  IP=000D  NV UP EI PL NZ NA PO NC
0D03:000D E2FC          LOOP    000B
-T

AX=0000  BX=0000  CX=0004  DX=0001  SP=003C  BP=0000  SI=0000  DI=0000
DS=0DD7  ES=0DD7  SS=0DE8  CS=0D03  IP=000B  NV UP EI PL NZ NA PO NC
0D03:000B 90            NOP
-G F

AX=0000  BX=0000  CX=0000  DX=0005  SP=003C  BP=0000  SI=0000  DI=0000
DS=0DD7  ES=0DD7  SS=0DE8  CS=0D03  IP=000F  NV UP EI PL NZ NA PE NC
0D03:000F CB            RETF
-G

Program terminated normally
-Q

C:\DOS>
```

Figure 6.37(c)   Execution of the
program with DEBUG.

Next the LOOP instruction is executed with the TRACE command

$$-T \quad (\dashv)$$

From Fig. 6.37(c), we find that the loop count CX has decremented by 1, meaning that the first pass through the loop is complete, and the value in IP has been changed to $000B_{16}$. Therefore, control has been returned to the NOP instruction that represents the beginning of the loop.

Now that we have observed the basic loop operation, let us execute the loop to completion with the command

$$-G \quad F \quad (\dashv)$$

The displayed information for this command in Fig. 6.37(c) shows us that at completion of the program the loop count in CX has been decremented to 0000_{16} and the pass count in DX has been incremented to 0005_{16}.

The other two loop instructions in Fig. 6.35 operate in a similar way except that they check for two conditions. For instance, the instruction *loop while equal* (LOOPE)/*loop while zero* (LOOPZ) checks the contents of both CX and the ZF flag. Each time the loop instruction is executed, CX decrements by 1 without affecting the flags, its contents are checked for 0, and the state of ZF that results from execution of the previous instruction is tested for 1. If CX is not equal to 0 and ZF equals 1, a jump is initiated to the location specified with the short-label operand and the loop continues. If either CX or ZF is 0, the loop is complete, and the instruction following the loop instruction is executed.

Instruction *loop while not equal* (LOOPNE)/*loop while not zero* (LOOPNZ) works in a similar way to the LOOPE/LOOPZ instruction. The difference is that it checks ZF and CX looking for ZF equal to 0 together with CX not equal to 0. If these conditions are met, the jump back to the location specified with the short-label operand is performed and the loop continues.

EXAMPLE 6.15

Given the following sequence of instructions, explain what happens as they are executed.

```
          MOV    DL,05
          MOV    AX,0A00H
          MOV    DS, AX
          MOV    SI,0
          MOV    CX, 0FH
AGAIN:    INC    SI
          CMP    [SI],DL
          LOOPNE  AGAIN
```

Solution

The first five instructions are for initializing internal registers. Data register DL is loaded with 05_{16}; data segment register DS is loaded via AX with the value $0A00_{16}$; source index register SI is loaded with 0000_{16}; and count register CX is loaded with $0F_{16}$ (15_{10}). After initialization, a data segment is set up at address $0A000_{16}$ and SI points to the memory location at address 0000_{16} in this data segment. DL contains the data 5_{10} and the CX register contains the loop count 15_{10}.

The part of the program that starts at the label AGAIN and ends with the LOOPNE instruction is a software loop. The first instruction in the loop increments SI by 1. Therefore, the first time through the loop SI points to the memory address $A001_{16}$. The next instruction compares the contents of this memory location with the contents of DL, which are 5_{10}. If the data held at $A001_{16}$ are 5_{10}, the zero flag is set; otherwise, it is reset. The LOOPNE instruction then decrements CX (making it E_{16}) and then checks for CX = 0 or ZF = 1. If neither of these two conditions is satisfied, program control is returned to the instruction with the label AGAIN. This causes the comparison to be repeated for the examination of the contents of the next byte in memory. On the other hand, if either condition is satisfied, the loop is complete. In this way, we see that the loop is repeated until either a number 5_{10} is found or all locations in the address range $A001_{16}$ through $A00F_{16}$ have been tested and found not to contain 5_{10}.

EXAMPLE 6.16

Figure 6.38(a) shows the source version of the program that is written in Example 6.15. This program was assembled by the macroassembler on the PC/AT and linked to produce run module EX616.EXE. The source listing that resulted from the assembly process is shown in Fig. 6.38(b). Verify the operation of the program by executing it with GO commands.

Solution

As shown in Fig. 6.38(a), the run module is loaded with the command

```
C:\DOS>DEBUG   A:EX616.EXE   (↵)
```

```
TITLE     EXAMPLE 6.16

          PAGE      ,132

STACK_SEG          SEGMENT        STACK 'STACK'
                   DB             64 DUP(?)
STACK_SEG          ENDS


CODE_SEG           SEGMENT        'CODE'
EX616    PROC      FAR
         ASSUME    CS:CODE_SEG, SS:STACK_SEG

;To return to DEBUG program put return address on the stack

         PUSH      DS
         MOV       AX, 0
         PUSH      AX

;Following code implements Example 6.16

         MOV       DL, 5H
         MOV       AX, 0A00H
         MOV       DS, AX
         MOV       SI, 0H
         MOV       CX, 0FH
AGAIN:   INC       SI
         CMP       [SI], DL
         LOOPNE    AGAIN

         RET                        ;Return to DEBUG program
EX616    ENDP
CODE_SEG           ENDS

         END       EX616
```

Figure 6.38(a) Source program for Example 6.16.

```
                          PAGE    ,132

0000                      STACK_SEG        SEGMENT        STACK 'STACK'
0000      40 [                            DB             64 DUP(?)
          ??
              ]

0040                      STACK_SEG        ENDS


0000                      CODE_SEG         SEGMENT        'CODE'
0000                      EX616    PROC    FAR
                          ASSUME   CS:CODE_SEG, SS:STACK_SEG

         ;To return to DEBUG program put return address on the stack

0000  1E                                   PUSH    DS
0001  B8 0000                              MOV     AX, 0
0004  50                                   PUSH    AX

         ;Following code implements Example 6.16

0005  B2 05                                MOV     DL, 5H
0007  B8 0A00                              MOV     AX, 0A00H
000A  8E D8                                MOV     DS, AX
000C  BE 0000                              MOV     SI, 0H
000F  B9 000F                              MOV     CX, 0FH
0012  46            AGAIN:                 INC     SI
0013  38 14                                CMP     [SI], DL
0015  E0 FB                                LOOPNE  AGAIN

0017  CB                                   RET                    ;Return to DEBUG program
0018                      EX616    ENDP
0018                      CODE_SEG         ENDS

                          END      EX616
```

Segments and groups:

N a m e	Size	align	combine	class
CODE_SEG	0018	PARA	NONE	'CODE'
STACK_SEG.	0040	PARA	STACK	'STACK'

Symbols:

N a m e	Type	Value	Attr	
AGAIN.	L NEAR	0012	CODE_SEG	
EX616.	F PROC	0000	CODE_SEG	Length =0018

```
Warning  Severe
Errors   Errors
0        0
```

Figure 6.38(b)   Source listing produced by assembler.

The program loaded is verified with the command

$$-U \quad 0 \quad 17 \quad (↵)$$

Comparing the sequence of instructions displayed in Fig. 6.38(c) with those in the source listing of Fig. 6.38(b), we find that the program has loaded correctly.

Notice that the loop that performs the memory-compare operation starts at address 0D03:0012. Let us begin by executing down to this point with the GO command

$$-G \quad 12 \quad (↵)$$

From the state information displayed in Fig. 6.38(c), we see that DL has been loaded with 05_{16}, AX with $0A00_{16}$, DS with $0A00_{16}$, SI with 0000_{16}, and CX with $000F_{16}$.

```
C:\DOS>DEBUG A:EX616.EXE
-U 0 17
0D03:0000 1E               PUSH    DS
0D03:0001 B80000           MOV     AX,0000
0D03:0004 50               PUSH    AX
0D03:0005 B205             MOV     DL,05
0D03:0007 B8000A           MOV     AX,0A00
0D03:000A 8ED8             MOV     DS,AX
0D03:000C BE0000           MOV     SI,0000
0D03:000F B90F00           MOV     CX,000F
0D03:0012 46               INC     SI
0D03:0013 3814             CMP     [SI],DL
0D03:0015 E0FB             LOOPNZ  0012
0D03:0017 CB               RETF
-G 12

AX=0A00  BX=0000  CX=000F  DX=0005  SP=003C  BP=0000  SI=0000  DI=0000
DS=0A00  ES=0DD7  SS=0DE9  CS=0D03  IP=0012   NV UP EI PL NZ NA PO NC
0D03:0012 46               INC     SI
-E A00:0 4,6,3,9,5,6,D,F,9
-D A00:0 F
0A00:0000  04 06 03 09 05 06 0D 0F-09 75 09 80 7C 02 54 75   .........u..|.Tu
-G 17

AX=0A00  BX=0000  CX=000B  DX=0005  SP=003C  BP=0000  SI=0004  DI=0000
DS=0A00  ES=0DD7  SS=0DE9  CS=0D03  IP=0017   NV UP EI PL ZR NA PE NC
0D03:0017 CB               RETF
-G

Program terminated normally
-Q

C:\DOS>
```

Figure 6.38(c)   Execution of the program with DEBUG.

Next, the table of data is loaded with the E command

$$-E \quad A00:0 \quad 4, \; 6, \; 3, \; 9, \; 5, \; 6, \; D, \; F, \; 9 \quad (\dashv)$$

The nine values in this list are loaded into consecutive bytes of memory over the range 0A00:0000 through 0A00:0008. The compare routine actually also checks the storage locations from 0A00:0009 through 0A00:000F. Let us dump the data held in this part of memory to verify that it has been initialized correctly. In Fig. 6.38(c), we see that this is done with the command

$$-D \quad A00:0 \quad F \quad (\dashv)$$

and, looking at the displayed data, we find that it has loaded correctly.

Now the loop is executed with the command

$$-G \quad 17 \quad (\dashv)$$

In the display dump for this command in Fig. 6.38(c), we find that SI has incremented to the value 0004_{16}; therefore, the loop was run only 4 times. The fourth time through the loop SI equals 4 and the memory location pointed to by SI, which is the address 0A00:0005, contains the value 5. This value is equal to the value in DL; therefore, the instruction CMP [SI],DL results in a difference of 0, and the zero flag is set. Notice in Fig. 6.38(c) that this flag is identified as ZR in the display dump. For this reason, execution of the LOOPNZ instruction causes the loop to be terminated, and control is passed to the RETF instruction.

The 80386DX microprocessor is equipped with special instructions to handle *string opera-tions*. By *string* we mean a series of data bytes, words, or double words that reside in consecutive memory locations. The string instructions of the 80386DX permit a programmer to implement operations such as to move data from one block of memory to a block elsewhere in memory. A second type of operation that is easily performed is to scan a string of data elements stored in memory looking for a specific value. Other examples are to compare the elements of two strings in order to determine whether they are the same or different, and to initialize a group of consecutive memory locations. Complex operations such as these require several nonstring instructions to be implemented.

There are five basic string instructions in the instruction set of the 8038DX. These instructions, as listed in Fig. 6.39, are *move byte, word,* or *double-word string* (MOVSB/MOVSW/MOVSD), *compare string* (CMPSB/CMPSW/CMPSD), *scan string* (SCASB/SCASW/SCASD), *load string* (LODSB/LODSW/LODSD), and *store string* (STOSB/STOSW/STOSD). They are called the *basic string instructions* because each defines an operation for one element of a string. Thus, these operations must be repeated to handle a string of more than one element. Let us first look at the operations performed by these in-structions.

Move String: MOVSB, MOVSW

The instructions MOVSB, MOVSW, and MOVSD all perform the same basic opera-tion. An element of the string specified by the source index (SI) register with respect to the current data segment (DS) register is moved to the location specified by the destination index (DI) register with respect to the current extra segment (ES) register. The move can be performed on a byte, word, or double word of data. After the move is complete, the contents of both SI and DI are automatically incremented or decremented by 1 for a byte move, by 2 for a word move, and by 4 for a double-word move. The address pointers in SI and DI increment or decrement, depending on how the direction flag DF is set.

Mnemonic	Meaning	Format	Operation	Flags affected
MOVSB/ MOVSW/ MOVSD	Move string	MOVSB MOVSW MOVSD	((ES)0 + (DI) ← ((DS)0 + (SI)) (SI) ← (SI) ± 1, 2, or 4 (DI) ← (DI) ± 1, 2, or 4	None
CMPSB/ CMPSW/ CMPSD	Compare strings	CMPSB CMPSW CMPSD	Set flags as per ((DS)0 + (SI)) − ((ES)0 + (DI)) (SI) ← (SI) ± 1, 2, or 4 (DI) ← (DI) ± 1, 2, or 4	CF, PF, AF, ZF, SF, OF
SCASB/ SCASW/ SCASD	Scan string	SCASB SCASW SCASD	Set flags as per (AL, AX, or EAX) − ((ES)0 + (DI)) (DI) ← (DI) ± 1, 2, or 4	CF, PF, AF, ZF, SF, OF
LODSB/ LODSW/ LODSD	Load string	LODSB LODSW LODSD	(AL, AX, or EAX) ← ((DS)0 + (SI) (SI) ← (SI) ± 1, 2, or 4	None
STOSB/ STOSW/ STOSD	Store string	STOSB STOSW STOSD	((ES)0 + (DI) ← (AL, AX, or EAX) ± 1, 2, or 4 (DI) ← (DI) ± 1, 2, or 4	None

Figure 6.39 Basic string instructions.

```
                MOV     AX,DATASEGADDR
                MOV     DS,AX
                MOV     ES,AX
                MOV     SI,BLK1ADDR
                MOV     DI,BLK2ADDR
                MOV     CX,N
                CLD
    NXTPT:      MOVSB
                LOOP    NXTPT
                HLT
```

Figure 6.40 Block-move program using the move-string instruction.

For example, the instruction

MOVSB

can be used to move a byte.

An example of a program that uses MOVSB is shown in Fig. 6.40. This program is a modified version of the block-move program of Fig. 6.36(b). Notice that the two MOV instructions that perform the data transfer and two INC instructions that update the pointer have been replaced with one move-string byte instruction. We have also made DS equal to ES.

Compare String and Scan String: CMPSB/CMPSW/CMPSD and SCASB/SCASW/SCASD

The compare strings instruction can be used to compare two elements in the same or different strings. It subtracts the destination operand from the source operand and adjusts the flags accordingly. The result of subtraction is not saved; therefore, the operation does not affect the operands in any way.

An example of a compare strings instruction for bytes of data is

CMPSB

Again, the source element is pointed to by the address in SI with respect to the current value in DS, and the destination element is specified by the contents of DI relative to the contents of ES. When executed, the operands are compared, the flags are adjusted, and both SI and DI are updated so that they point to the next elements in their respective strings.

The scan-string instruction is similar to compare strings; however, it compares the byte, word, or double-word element of the destination string at the physical address derived from DI and ES to the contents of AL, AX, or EAX, respectively. The flags are adjusted based on this result and DI incremented or decremented.

A program using the SCASB instruction that implements a string scan operation similar to that described in Example 6.15 is shown in Fig. 6.41. Notice that we have made DS equal to ES.

Load and Store String: LODSB/LODSW/LODSD and STOSB/STOSW/STOSD

The last two instructions in Fig. 6.39, load string and store string, are specifically provided to move string elements between the accumulator and memory. LODSB loads a

```
                MOV       AX,0
                MOV       DS,AX                  MOV       AX,0
                MOV       ES,AX                  MOV       DS,AX
                MOV       AL,05                  MOV       ES,AX
                MOV       DI,0A000H              MOV       AL,05
                MOV       CX,0FH                 MOV       DI,0A000H
                CLD                              MOV       CX,0FH
    AGAIN:      SCASB                            CLD
                LOOPNE    AGAIN      AGAIN:      STOSB
    NEXT:                                        LOOP      AGAIN
```

Figure 6.41 Block scan
operation using the SCAS
instruction.

Figure 6.42 Initializing a
block of memory with a store
string operation.

byte from a string in memory into AL. The address in SI is used relative to DS to determine the address of the memory location of the string element; SI is incremented by 1 after loading. Similarly, the instruction LODSW indicates that the word string element at the physical address derived from DS and SI is to be loaded into AX. Then the index in SI is automatically incremented by 2.

On the other hand, STOSB stores a byte from AL into a string location in memory. This time the contents of ES and DI are used to form the address of the storage location in memory. For example, the program in Fig. 6.42 will load the block of memory locations from $0A000_{16}$ through $0A00F_{16}$ with number 5.

Repeat String: REP

In most applications, the basic string operations must be repeated in order to process arrays of data. This is done by inserting a repeat prefix before the instruction that is to be repeated. The *repeat prefixes* of the 80386DX are shown in Fig. 6.43.

The first prefix, REP, causes the basic string operation to be repeated until the contents of register CX become equal to zero. Each time the instruction is executed, it causes CX to be tested for 0. If CX is found not to be 0, it is decremented by 1 and the basic string operation is repeated. On the other hand, if it is 0, the repeat string operation is done and the next instruction in the program is executed. The repeat count must be loaded into CX prior to executing the repeat string instruction. As indicated in Fig. 6.43, the REP prefix is used with the MOVS and STOS instructions. Figure 6.44 is the memory load routine of Fig. 6.42 modified by using the REP prefix.

The prefixes REPE and REPZ stand for the same function. They are meant for use with the CMPS and SCAS instructions. With REPE/REPZ, the basic compare or scan

Prefix	Used with	Meaning
REP	MOVS STOS	Repeat while not end of string $CX \neq 0$
REPE/REPZ	CMPS SCAS	Repeat while not end of string and strings are equal $CX \neq 0$ and $ZF = 1$
REPNE/REPNZ	CMPS SCAS	Repeat while not end of string and strings are not equal $CX \neq 0$ and $ZF = 0$

Figure 6.43 Prefixes for use with the basic string operations.

```
MOV        AX,0
MOV        DS,AX
MOV        ES,AX
MOV        AL,05
MOV        DI,0A000H
MOV        CX,0FH
CLD
REPSTOSB
```

Figure 6.44 Initializing a block of memory by repeating the STOSB instruction.

operation can be repeated as long as both the contents of CX are not equal to 0 and the zero flag is 1. The first condition, CX not equal to 0, indicates that the end of the string has not yet been reached, and the second condition, ZF = 1, indicates that the elements that were compared are equal.

The last prefix, REPNE/REPNZ, works similarly as the REPE/REPZ except that now the operation is repeated as long as CX is not equal to 0 and ZF is 0. That is, the comparison or scanning is to be performed as long as the string elements are unequal and the end of the string is not yet found.

Autoindexing for String Instructions

Earlier we pointed out that during the execution of a string instruction the address indices in SI and DI are either automatically incremented or decremented. Moreover, we indicated that the decision to increment or decrement is made based on the setting of the direction flag DF. The 80386DX provides two instructions, clear direction flag (CLD) and set direction flag (STD), to permit selection between *autoincrement* and *autodecrement* *mode* of operation. These instructions are shown in Fig. 6.45. When CLD is executed, DF is set to 0. This selects the autoincrement mode, and each time a string operation is performed SI and/or DI are incremented by 1 if byte data are processed, by 2 if word data are processed, and by 4 if double-word data are processed.

EXAMPLE 6.17

Describe what happens as the following sequence of instructions is executed.

```
CLD
MOV    AX,DATA_SEGMENT
MOV    DS,   AX
MOV    AX,   EXTRA_SEGMENT
MOV    ES,AX
MOV    CX,20H
MOV    SI,OFFSET   MASTER
MOV    DI,OFFSET   COPY
REPMOVSB
```

Mnemonic	Meaning	Format	Operation	Affected flags
CLD	Clear DF	CLD	(DF) ← 0	DF
STD	Set DF	STD	(DF) ← 1	DF

Figure 6.45 Instructions for selecting autoincrementing and autodecrementing in string instructions.

Solution

The first instruction clears the direction flag and selects autoincrement mode of operation for string addressing. The next two instructions initialize DS with the value DATA_SEGMENT. It is followed by two instructions that load ES with the value EXTRA_SEGMENT. Then the number of repeats, 20_{16}, is loaded into CX. The next two instructions load SI and DI with offset addresses MASTER and COPY that point to the beginning of the source and destination strings, respectively. Now we are ready to perform the string operation. Execution of REPMOVSB moves a block of 32 consecutive bytes from the block of memory locations starting at address MASTER in the current data segment (DS) to a block of locations starting at address COPY in the current extra segment (ES).

EXAMPLE 6.18

The source program in Fig. 6.46(a) implements the block move operation of Example 6.17. This program was assembled with the macroassembler on the PC/AT and linked to produce a run module called EX618.EXE. The source listing that was produced during the assembly process is shown in Fig. 6.46(b). Execute the program using DEBUG and verify its operation.

```
TITLE     EXAMPLE 6.18

          PAGE      ,132

STACK_SEG          SEGMENT          STACK 'STACK'
                   DB               64 DUP(?)
STACK_SEG          ENDS


DATA_SEG           SEGMENT          'DATA'
MASTER             DB               32 DUP(?)
COPY               DB               32 DUP(?)
DATA_SEG           ENDS


CODE_SEG           SEGMENT          'CODE'
EX618    PROC      FAR
         ASSUME    CS:CODE_SEG, SS:STACK_SEG, DS:DATA_SEG, ES:DATA_SEG

;To return to DEBUG program put return address on the stack

         PUSH      DS
         MOV       AX, 0
         PUSH      AX

;Following code implements Example 6.18

         MOV       AX, DATA_SEG     ;Set up data segment
         MOV       DS, AX
         MOV       ES, AX           ;Set up extra segment

         CLD
         MOV       CX, 20H
         MOV       SI, OFFSET MASTER
         MOV       DI, OFFSET COPY
REP      MOVS      COPY, MASTER

         RET                        ;Return to DEBUG program
EX618 ENDP
CODE_SEG           ENDS

         END       EX618
```

Figure 6.46(a)   Source program for Example 6.18.

```
                             TITLE    EXAMPLE 6.18

                                 PAGE    ,132

  0000                                      STACK_SEG      SEGMENT      STACK 'STACK'
  0000    40 [                                             DB           64 DUP(?)
          ??
               ]

  0040                                      STACK_SEG      ENDS


  0000                                      DATA_SEG       SEGMENT      'DATA'
  0000    20 [                              MASTER         DB           32 DUP(?)
          ??
               ]

  0020    20 [                              COPY           DB           32 DUP(?)
          ??
               ]

  0040                                      DATA_SEG       ENDS


  0000                                      CODE_SEG       SEGMENT      'CODE'
  0000                                      EX618   PROC   FAR
                                    ASSUME  CS:CODE_SEG, SS:STACK_SEG, DS:DATA_SEG, ES:DATA_SEG

                            ;To return to DEBUG program put return address on the stack

  0000    1E                                               PUSH    DS
  0001    B8 0000                                          MOV     AX, 0
  0004    50                                               PUSH    AX

                            ;Following code implements Example 6.18

  0005    B8  ---- R                                       MOV     AX, DATA_SEG   ;Set up data segment
  0008    8E D8                                            MOV     DS, AX
  000A    8E C0                                            MOV     ES, AX         ;Set up extra segment

  000C    FC                                               CLD
  000D    B9 0020                                          MOV     CX, 20H
  0010    BE 0000 R                                        MOV     SI, OFFSET MASTER
  0013    BF 0020 R                                        MOV     DI, OFFSET COPY
  0016    F3/ A4                              REP          MOVS    COPY, MASTER

  0018    CB                                               RET                    ;Return to DEBUG program
  0019                                       EX618   ENDP
  0019                                       CODE_SEG      ENDS

                                                 END     EX618


Segments and groups:

                'N a m e                     Size     align    combine  class

CODE_SEG . . . . . . . . . . . .             0019     PARA     NONE     'CODE'
DATA_SEG . . . . . . . . . . . .             0040     PARA     NONE     'DATA'
STACK_SEG. . . . . . . . . . . .             0040     PARA     STACK    'STACK'


Symbols:

               N a m e                       Type     Value    Attr

COPY . . . . . . . . . . . . .               L BYTE   0020     DATA_SEG      Length =0020
EX618. . . . . . . . . . . . .               F PROC   0000     CODE_SEG      Length =0019
MASTER . . . . . . . . . . . .               L BYTE   0000     DATA_SEG      Length =0020

Warning Severe
Errors  Errors
0       0
```

Figure 6.46(b)   Source listing produced by assembler.

Solution

The program is loaded with the DEBUG command

```
        C:\DOS>DEBUG A:EX618.EXE   (↵)
```

and verified by the UNASSEMBLE command

```
        -U  0  18  (↵)
```

Comparing the displayed instruction sequence in Fig. 6.46(c) with the source listing in Fig. 6.46(b), we find that the program has loaded correctly.

```
C:\DOS>DEBUG A:EX618.EXE
-U 0 18
0DE7:0000 1E              PUSH    DS
0DE7:0001 B80000          MOV     AX,0000
0DE7:0004 50              PUSH    AX
0DE7:0005 B8E90D          MOV     AX,0DE9
0DE7:0008 8ED8            MOV     DS,AX
0DE7:000A 8EC0            MOV     ES,AX
0DE7:000C FC              CLD
0DE7:000D B92000          MOV     CX,0020
0DE7:0010 BE0000          MOV     SI,0000
0DE7:0013 BF2000          MOV     DI,0020
0DE7:0016 F3              REPZ
0DE7:0017 A4              MOVSB
0DE7:0018 CB              RETF
-G 16

AX=0DE9  BX=0000  CX=0020  DX=0000  SP=003C  BP=0000  SI=0000  DI=0020
DS=0DE9  ES=0DE9  SS=0DED  CS=0DE7  IP=0016   NV UP EI PL NZ NA PO NC
0DE7:0016 F3              REPZ
0DE7:0017 A4              MOVSB
-F DS:0 1F FF
-F DS:20 3F 00
-D DS:0 3F
0DE9:0000  FF FF FF FF FF FF FF FF-FF FF FF FF FF FF FF FF   ................
0DE9:0010  FF FF FF FF FF FF FF FF-FF FF FF FF FF FF FF FF   ................
0DE9:0020  00 00 00 00 00 00 00 00-00 00 00 00 00 00 00 00   ................
0DE9:0030  00 00 00 00 00 00 00 00-00 00 00 00 00 00 00 00   ................
-G 18

AX=0DE9  BX=0000  CX=0000  DX=0000  SP=003C  BP=0000  SI=0020  DI=0040
DS=0DE9  ES=0DE9  SS=0DED  CS=0DE7  IP=0018   NV UP EI PL NZ NA PO NC
0DE7:0018 CB              RETF
-D DS:0 3F
0DE9:0000  FF FF FF FF FF FF FF FF-FF FF FF FF FF FF FF FF   ................
0DE9:0010  FF FF FF FF FF FF FF FF-FF FF FF FF FF FF FF FF   ................
0DE9:0020  FF FF FF FF FF FF FF FF-FF FF FF FF FF FF FF FF   ................
0DE9:0030  FF FF FF FF FF FF FF FF-FF FF FF FF FF FF FF FF   ................
-G

Program terminated normally
-Q

C:\DOS>
```

Figure 6.46(c) Execution of the program with DEBUG.

First, we execute down to the REPZ instruction with the GO command

$$-G \quad 16 \quad (\downarrow)$$

and, looking at the state information displayed in Fig. 6.46(c), we see that CX has been loaded with 0020_{16}, SI with 0000_{16}, and DI with 0020_{16}.

Now the storage locations in the 32-byte source block that starts at address DS:0000 are loaded with the value FF_{16} using the FILL command

$$-F \quad DS:0 \quad 1F \quad FF \quad (\downarrow)$$

and each of the 32 bytes of the destination block, which start at DS:0020, are loaded with the value 00_{16} by the FILL command

$$-F \quad DS:20 \quad 3F \quad 00 \quad (\downarrow)$$

Now a memory dump command is used to verify the initialization of memory

$$-D \quad DS:0 \quad 3F \quad (\downarrow)$$

Looking at the displayed information in Fig. 6.46(c), we see that memory has been initialized correctly.

Now we execute the string move operation with the command

$$-G \quad 18 \quad (\lrcorner)$$

Again looking at the display in Fig. 6.46(c), we see that the repeat count in CX has been decremented to 0 and that the source and destination pointers have been incremented to $(SI) = 0020_{16}$ and $(DI) = 0040_{16}$. In this way, we see that the string move instruction was executed 32 times and that the source and destination addresses were correctly incremented to complete the block transfer. The block transfer operation is verified by repeating the DUMP command

$$-D \quad DS:0 \quad 3F \quad (\lrcorner)$$

Notice that both the source and destination blocks now contain FF_{16}.

ASSIGNMENTS

Section 6.2

1. Explain what happens when the instruction sequence

```
LAHF
MOV   [BX+DI],  AH
```

is executed.

2. What operation is performed by the instruction sequence that follows?

```
MOV   AH,   [BX+SI]
SAHF
```

3. What instruction should be executed to ensure that the carry flag is in the set state? The reset state?

4. Which instruction when executed disables the interrupt interface?

5. Write an instruction sequence to configure the 80386DX as follows: interrupts not accepted; save the original contents of flags SF, ZF, AF, PF, and CF at the address $A000_{16}$; and then clear CF.

Section 6.3

6. Describe the difference in operation and the effect on status flags due to the execution of the subtract words and compare words instructions.

7. Describe the operation performed by each of the instructions that follow.
(a) CMP [0100H], AL
(b) CMP AX, [SI]
(c) CMP DWORD PTR [DI], 1234H

8. What is the state of the 80386DX's flags after executing the instructions in parts (a) through (c) of Problem 7? Assume that the following initial state exists before executing the instructions.

$$(EAX) = 00008001H$$

$$(ESI) = 00000200H$$

$$(EDI) = 00000300H$$

$$(DS:100H) = F0H$$

$$(DS:200H) = F0H$$

$$(DS:201H) = 01H$$

$$(DS:300H) = 34H$$

$$(DS:301H) = 12H$$

$$(DS:302H) = 00H$$

$$(DS:303H) = 00H$$

9. What happens to the ZF and CF status flags as the following sequence of instructions is executed? Assume that they are both initially cleared.

```
MOV    BX,1111H
MOV    AX,0BBBBH
CMP    BX,AX
```

10. What does the mnemonic SETNC stand for?
11. What flag condition does the instruction SETNZ test for?
12. Which byte set on condition instruction tests for the conditional relationship CF = 1 + ZF = 1?
13. Describe the operation performed by the instruction

```
SETO    OVERFLOW
```

14. Write an instruction that will load FF_{16} into memory location NEGATIVE if the result of the preceding arithmetic instruction was a negative number.

Section 6.4

15. What is the key difference between the unconditional jump instruction and conditional jump instruction?
16. Which registers have their contents changed during an intrasegment jump? Intersegment jump?
17. How large is a short-label displacement? Near-label displacement? Memptr16 operand? Regptr32 operand?
18. Is a far-label used to initiate an intrasegment jump or an intersegment jump?

Assignments

19. Identify the type of jump, the type of operand, and operation performed by each of the following instructions.
   (a) JMP 10H
   (b) JMP 1000H
   (c) JMP WORD PTR [SI]

20. If the state of the 80386DX is as follows before executing each instruction in Problem 19, to what address is program control passed?

$$(CS) = 1075H$$
$$(IP) = 0300H$$
$$(SI) = 0100H$$
$$(DS:100H) = 00H$$
$$(DS:101H) = 10H$$

21. Which flags are tested by the various conditional jump instructions?

22. What flag condition is tested for by the instruction JNS?

23. What flag conditions are tested for by the instruction JA?

24. Identify the type of jump, the type of operand, and operation performed by each of the following instructions.
   (a) JNC 10H
   (b) JNP 1000H
   (c) JO DWORD PTR [BX]

25. What value must be loaded into BX such that execution of the instruction JMP BX transfers control to the memory location offset from the beginning of the current code segment by 256_{10}?

26. The program that follows implements what is known as a *delay loop*.

```
            MOV   CX,1000H
    DLY:    DEC   CX
            NOP
            JNZ   DLY
    NXT:    ---   ---
```

   (a) How many times does the JNZ DLY instruction get executed?
   (b) Change the program so that JNZ DLY is executed just 17 times.
   (c) Change the program so that JNZ DLY is executed 2^{32} times.

27. Given a number N in the range $0 < N \leq 5$, write a program that computes its factorial and saves the result in memory location FACT ($N! = 1 \cdot 2 \cdot 3 \cdots N$).

28. Write a program that compares the elements of two arrays, A(I) and B(I). Each array contains 100 16-bit signed numbers. The comparison is to be done by comparing the corresponding elements of the two arrays until either two elements are found to be unequal or all elements of the arrays have been compared and found to be equal. Assume that the arrays start in the current data segment at offset addresses $A000_{16}$ and $B000_{16}$, respectively. If the two arrays are found to be unequal, save the address of the first unequal element of A(I) in memory location with offset address FOUND in the current data segment; otherwise, write all 0s into this location.

29. Given an array A(I) of 100 16-bit signed numbers that are stored in memory starting at address $A000_{16}$, write a program to generate two arrays from the given array such that one P(J) consists of all the positive numbers and the other N(K) contains all the negative numbers. Store the array of positive numbers in memory starting at address $B000_{16}$ and the array of negative numbers starting at address $C000_{16}$.

30. Given a 16-bit binary number in DX, write a program that converts it to its equivalent BCD number in DX. If the result is bigger than 16 bits, place all 1s in DX.

31. Given an array A(I) with 100 16-bit signed integer numbers, write a program to generate a new array B(I) as follows:

$$B(I) = A(I), \quad \text{for } I = 1, 2, 99, \text{ and } 100$$

and

$$B(I) = \text{median value of } A(I - 2), A(I - 1), A(I), A(I + 1),$$
$$\text{and } A(I + 2), \quad \text{for all other Is}$$

Section 6.5

32. Describe the difference between a jump and call instruction.

33. Why are intersegment and intrasegment call instructions provided in the 80386DX?

34. What is saved on the stack when a call instruction with a memptr16 operand is executed? A memptr32 operand?

35. Identify the type of call, the type of operand, and operation performed by each of the instructions that follows.
    (a) CALL 1000H
    (b) CALL WORD PTR [100H]
    (c) CALL DWORD PTR [BX+SI]
    (d) CALL EDX

36. The state of the 80386DX is as follows. To what address is program control passed after executing each instruction in parts (a), (b), and (c) of Problem 35?

$$(CS) = 1075H$$
$$(IP) = 0300H$$
$$(EBX) = 00000100H$$
$$(ESI) = 00000100H$$
$$(DS:100H) = 00H$$
$$(DS:101H) = 10H$$
$$(DS:200H) = 00H$$
$$(DS:201H) = 01H$$
$$(DS:202H) = 00H$$
$$(DS:203H) = 10H$$

37. What function is performed by the RET instruction?

38. Describe the operation performed by each of the instructions that follows.
   (a) PUSH 1000H
   (b) PUSH DS
   (c) PUSH [SI]
   (d) POP EDI
   (e) POP [BX + DI]
   (f) POPF

39. At what addresses will the bytes of the immediate operand in Problem 38(a) be stored after the instruction is executed?

40. What operation is performed by the following sequence of instructions?

```
PUSH   AX
PUSH   BX
POP    AX
POP    BX
```

41. Write a subroutine that converts a given 16-bit BCD number to its equivalent binary number. The BCD number is in register DX. Replace it with the equivalent binary number.

42. When must PUSHF and POPF instructions be included in a subroutine?

43. Given an array A(I) of 100 16-bit signed integer numbers, write a program to generate a new array B(I) so that

$$B(I) = A(I), \qquad \text{for I} = 1 \text{ and } 100$$

and

$$B(I) = \tfrac{1}{4}[A(I - 1) - 5A(I) + 9A(I + 1)], \qquad \text{for all other Is}$$

For the calculation of B(I), the values of A(I − 1), A(I), and A(I + 1) are to be passed to a subroutine in registers AX, BX, and CX and the subroutine returns the result B(I) in register AX.

44. Write a segment of main program and show its subroutine structure to perform the following operations. The program is to check continuously the three most significant bits in register DX and, depending on their setting, execute one of three subroutines: SUBA, SUBB, or SUBC. The subroutines are selected as follows:
   (a) If bit 15 of DX is set, initiate SUBA.
   (b) If bit 14 of DX is set and bit 15 is not set, initiate SUBB.
   (c) If bit 13 of DX is set and bits 14 and 15 are not set, initiate SUBC.
   If a subroutine is executed, the corresponding bit of DX is to be cleared and then control returned to the main program. After returning from the subroutine, the main program is repeated.

Section 6.6

45. Which flags are tested by the various conditional loop instructions?

46. What two conditions can terminate the operation performed by the instruction LOOPNE?

47. How large a jump can be employed in a loop instruction?

48. What is the maximum number of repeats that can be implemented with a loop instruction?

49. Using loop instructions, implement the program in Problem 27.

50. Using loop instructions, implement the program in Problem 28.

Section 6.7

51. What determines whether the SI and DI registers increment or decrement during a string operation?

52. Which segment register is used to form the destination address for a string instruction?

53. Write equivalent instruction sequences using string instructions for each of the following:

   (a) MOV AL, [SI]
       MOV [DI], AL
       INC SI
       INC DI

   (b) MOV AX, [SI]
       INC SI
       INC SI

   (c) MOV AL, [DI]
       CMP AL, [SI]
       DEC SI
       DEC DI

54. Use string instructions to implement the program in Problem 28.

55. Write a program to convert a table of 100 ASCII characters that are stored starting at offset address ASCII_CHAR into their equivalent table of EBCDIC characters and store them at offset address EBCDIC_CHAR. The translation is to be done using an ASCII_TO_EBCDIC conversion table starting at offset address ASCII_TO_EBCDIC. Assume that all three tables are located in different segments of memory.

Assembly Langauge Program Development with the Microsoft MASM Assembler

▲ 7.1 INTRODUCTION

In Chapter 4 we learned how to use the DEBUG program development tool that is available in the PC/AT's disk operating system. We found that there is a line-by-line assembler included in the DEBUG program; however, this assembler is not practical to use when writing longer programs. Other assembly language–development tools, such as Microsoft's macroassembler (MASM) and the linker (LINK) programs, are available for DOS. These are the kind of software-development tools that are used to develop programs for larger applications. In this chapter, we will learn how to use MASM, LINK, and DEBUG to develop and debug assembly language programs for practical applications. The topics covered in the chapter are as follows:

1. Statement syntax for the source program
2. Pseudo operations
3. Generating a source file with an editor
4. Assembling a source program with the MASM program
5. Creating a run module with the LINK program
6. Loading and executing a run module

▲ 7.2 STATEMENT SYNTAX FOR THE SOURCE PROGRAM

In Chapter 3, we found that a source program is a series of assembly language and pseudo-operation statements that solve a specific problem. The *assembly language statements* tell the microprocessor the operations to be performed. On the other hand, the *pseudo-operation statements* direct the assembler program on how the program is to be assembled. The structure of the assembly language statement was briefly introduced in Chapter 3. Here we continue with a detail study of the rules that must be used when writing assembly language and pseudo-operation statements for the Microsoft MASM assembler.

Syntax of an Assembly Language Statement

A source program must be written using the syntax understood by the assembler program. By *syntax*, we mean the rules according to which statements must be written. The general format of an assembly language statement for the MASM assembler is

```
LABEL   OPCODE   OPERAND(S)   COMMENT(S)
```

Note that it contains four separate fields: the *label field, opcode field, operand field*, and *comment field*. An example is the instruction

```
START:  MOV  CX,10  ;Load a count of 10 into register CX
```

Here START is in the label field; MOV (for *move operation*) is the opcode field; the operands are CX and decimal number 10; and the comment field tells us that execution of the instruction loads a count of 10 into register CX. Note that the label field ends with a colon (:) and the comment field begins with a semicolon (;).

Not all of the fields may be present in an instruction. In fact, the only part of the format that is always required is the opcode. For instance, the instruction

```
MOV  CX,10  ;Initialize the count in CX
```

has nothing in the label field. Other instructions may not need anything in the operand field. The instruction

```
CLC  ;Clear the carry flag
```

is an example.

One rule that must be followed when writing assembly language statements for MASM is that the fields must be separated by at least one blank space, and if no label is used, the opcode field must be preceded by at least one space.

Let us now look at each field of an assembly language source statement in more detail. We will begin with the label field. It is used to give a *symbolic name* to an assembly language statement. When a symbolic name is given to an assembly language statement, other instructions in the program can reference the instruction by simply referring to this symbol instead of the actual memory address where the instruction is stored.

For example, it is common to use a jump instruction to pass control to an instruction elsewhere in the program. An example is the instruction sequence

```
        .
        .
        .
        JMP  START
        .
        .
        .
START:  MOV  CX,10 ;Initialize the count in CX
        .
```

Execution of the jump instruction causes program execution to pass to the point in the program corresponding to label START—that is, the MOV instruction that is located further down in the program.

The label is an arbitrarily selected string of alphanumeric characters followed by a colon (:). Some examples of valid labels are START, LOOPA, SUBROUTINE_A, and COUNT_ROUTINE. As in our earlier example, the names used for labels are typically selected to help document what is happening at that point of the program.

There are some limitations on the selection of labels. One limitation is that only the first 31 characters of the label are recognized by the assembler. Moreover, the first character of the label must be either a letter or the symbol period (.). A period cannot be used at any other point in the label except as the first character.

Another restriction is that *reserved symbols*, such as those used to refer to the internal registers of the 80386DX (AH, AL, AX, etc.), cannot be used. Still another restriction is that a label cannot include embedded blanks. This is the reason that the earlier example COUNT_ROUTINE has an underscore character (_) separating the two words. Use of the underscore makes the assembler view the character string as a single label.

Each of the basic operations that can be performed by the 80386DX microprocessor is identified with a three- to six-letter *mnemonic*, which is called its *operation code* (opcode). For example, the mnemonics for the add, subtract, and move operations are ADD, SUB, and MOV, respectively. It is these mnemonics that are entered into the opcode field during the writing of the assembly language statements.

The entries in the operand field tell where the data that is to be processed is located and how it is to be accessed. An instruction may have zero, one, or two operands. For example, the move instruction

```
MOV   AX,BX
```

has two operands. Here the two operands are the accumulator (AX) register and the base (BX) register of the 80386DX. Note that the operands are separated by a comma. Furthermore, 80386DX assembly language instructions are written with the destination operand first. Therefore, BX is the source operand (the *move from* location) and AX is the destination operand (the *move to* location). In this way, we see that the operation performed by the instruction is to move the value held in BX into AX. The notations used to identify the 80386DX's internal registers are shown in Fig. 7.1.

A large number of addressing modes are provided for the 80386DX to help us in specifying the location of operands. Examples using each of the addressing modes are provided in Fig. 7.2. For example, the instruction that specifies an immediate data operand simply includes the value of the piece of data in the operand location. On the other hand, if the operand is a direct address it is specified using a label for the memory address.

The comment field can be used to describe the operation performed by the instruction. It is always preceded by a semicolon (;). For instance, in the instruction

```
MOV   AX,BX   ;Copy BX into AX
```

the *comment* tells us that execution of the instruction causes the value in source register BX to be copied into destination register AX.

The assembler program ignores comments when it assembles a program into an object

Symbol	Register
AX	Accumulator register
AH	Accumulator register high byte
AL	Accumulator register lower byte
BX	Base register
BH	Base register high byte
BL	Base register low byte
CX	Count register
CH	Count register high byte
CL	Count register low byte
DX	Data register
DH	Data register high byte
DL	Data register low byte
SI	Source index register
DI	Destination index register
SP	Stack pointer register
BP	Base pointer register
CS	Code segment register
DS	Data segment register
SS	Stack segment register
ES	Extra segment register
FS	F segment register
GS	G segment register

Figure 7.1 Symbols for specifying register operands.

Addressing mode	Operand	Example	Segment
Register	Destination	MOV AX, LABEL	—
Immediate	Source	MOV AL,15H	—
Direct	Destination	MOV LABEL,AX	Data
Register indirect	Source	MOV AX,[SI]	Data
		MOV AX,[BP]	Stack
		MOV AX,[DI]	Data
		MOV AX,[BX]	Data
Based	Destination	MOV [BX]+DISP,AL	Data
		MOV [BP]+DISP,AL	Stack
Indexed	Source	MOV AL,[SI]	Data
		MOV AL,[DI]	Data
Based indexed	Destination	MOV [BX][SI]+DISP,AH	Data
		MOV [BX][DI]+DISP,AH	Data
		MOV [BP][SI]+DISP,AH	Stack
		MOV [BP][DI]+DISP,AH	Stack

Figure 7.2 Examples using the various 80386DX addressing modes.

module. Comments are produced only in the source listing. This does not mean that comments are not important. In fact, they are very important, because they document the operation of the source program. If a program were picked up a long time after it was written—for instance, for a software update—the comments would permit the programmer to quickly understand its operation.

Syntax of a Pseudo-Operation Statement

The syntax used to write pseudo-operation statements for MASM is essentially the same as that for an assembly language statement. The general format is

```
LABEL   PSEUDO-OPCODE   OPERAND(S)   COMMENT(S)
```

Notice that the only difference is that the instruction opcode is replaced with a *pseudo-opcode*. It tells the assembler which type of operation is to be performed. For example, the pseudo-opcode DB stands for *define byte*, and if a statement is written as

```
DB   0FFH   ;Allocate a byte location initialized to FFH
```

it causes the next byte location in memory to be loaded with the value FF_{16}. This type of command can be used to initialize memory locations with data.

Another difference between the pseudo-operation statement and an assembly language statement is that pseudo-operations frequently have more than two operands. For instance, the statement

```
DB   0FFH,0FFH,0FFH,0FFH,0FFH
```

causes the assembler to load the next five consecutive bytes in memory with the value FF_{16}.

Constants in a Statement

Constants in an instruction or pseudo-operation, such as an immediate value of data or an address, can be expressed in any of five data types: *binary, decimal, hexadecimal, octal,* or *character*. The first four types of data are defined by including the letter B, D, H, or Q, respectively, after the number. For example, decimal number 9 is expressed in each of these four data forms as follows:

```
1001B
  9D
  9H
 11Q
```

One exception is that decimal numbers do not have to be followed by a D. Therefore, 9D can also be written simply as 9.

Another variation is that the first digit of a hexadecimal number must always be one of the numbers in the range 0 through 9. For this reason, hexadecimal A must be written as 0AH instead of AH.

Typically, data and addresses are expressed in hexadecimal form. On the other hand, it is more common to express the count for shift, rotate, and string instructions in decimal form.

EXAMPLE 7.1

The repeat count in CX for a string instruction is to be equal to decimal 255. Assume that the instruction that is to load the count has the form

```
MOV   CX,XX
```

where XX stands for the count, which is an immediate operand, that is to be loaded into CX. Show how the instruction would be written, first using decimal notation for the immediate operand and then using hexadecimal notation.

Solution

Using decimal notation, we get

```
MOV   CX,255D
```

or just

```
MOV   CX,255
```

In hexadecimal form, 255 is represented by FF_{16}. Therefore, the instruction becomes

```
MOV   CX,0FFH
```

The numbers used as operands in assembly language and pseudo-op statements can also be *signed* (positive or negative) *numbers*. For decimal numbers, this is simply done by preceding them with a + or − sign. For example, an immediate count of −10 that is to be loaded into the CX register with a MOV instruction can be written as

```
MOV   CX,-10
```

However, for negative numbers expressed in binary, hexadecimal, or octal form, the 2's complement of the number must be entered.

EXAMPLE 7.2

The count in a MOV instruction that is to load CX is to be −10. Write the instruction and express the immediate operand in binary form.

Solution

The binary form of 10_{10} is 01010_2. Forming the 2's complement, we get

$$
\begin{array}{r}
10101 \\
+1 \\
\hline
10110
\end{array}
$$

Therefore, the instruction is written as

```
MOV   CX,10110B
```

285

Character data can also be used as an operand. For instance, a string search operation may be used to search through a block of ASCII data in memory looking for a specific ASCII character, such as the letter A. When ASCII data are used as an operand, the character or string of characters must be enclosed by double quotes. For example, if the number 1 is to be expressed as character data, instead of numeric data, it is written as "1". In a string-compare operation, the data in memory are always compared to the contents of the AL register. Therefore, the character being searched for must be loaded into this register. For instance, to load the ASCII value of 1 into AL, we use the instruction

```
MOV  AL, "1"
```

A second kind of operand specifies a storage location in memory. Such operands are written using the memory-addressing modes of the 80386DX, which are shown in Fig. 7.2. For instance, to specify that an operand is held in a storage location that is the tenth byte from the beginning of a source block of data located in the current data segment, we can use indirect addressing through source-index register SI. In this way, the location of the operand is specified as

$$10[SI] \text{ or } [SI] + 10 \text{ or } [SI + 10]$$

SI must be loaded with an offset that points to the beginning of the source-data block in memory.

Certain instructions require operands that are a memory address instead of data. Two examples are the *jump* (JMP) instruction and the *call* (CALL) instruction. Labels can be used to identify these addresses. For instance, in the instruction

```
JMP  LOOP
```

LOOP is a label that specifies the *jump to* address. *Attributes* may also be assigned to the label. An attribute specifies whether or not a given label is a *near, far, external,* or *internal label.*

Operand Expressions Using the Arithmetic, Relational, and Logical Operators

The operands we have used up to this point have all been either constants, variables, or labels. However, it is also possible to have an expression as an operand. For example, the instruction

```
MOV  AH,A + 2
```

has an expression for its source operand. That is, the source operand is written as the sum of variable *A* and the number 2.

Figure 7.3 lists the *arithmetic, relational,* and *logical operators* that can be used to form operand expressions for use with the assembler. Expressions that are used for operands are evaluated as part of the assembly process. As the source program is assembled into an object module, the numeric values for the terms in the operand expressions are combined together based on a *precedence* of the operators in the expression, and then the expression is replaced with the resulting operand value in the final object code.

Type	Operator	Example	Function
Arithmetic	*	A * B	Multiplies A with B and makes the operand equal to the product
	/	A / B	Divides A by B and makes the operand equal to the quotient
	MOD	A MOD B	Divides A by B and assigns the remainder to the operand
	SHL	A SHL n	Shifts the value in A left by n bit positions and assigns this shifted value to the operand
	SHR	A SHR n	Shifts the value in A right by n bit positions and assigns this shifted value to the operand
	+	A + B	Adds A to B and makes the operand equal to the sum
	−	A − B	Subtracts B from A and makes the operand equal to the difference
Relational	EQ	A EQ B	Compares value of A to that of B. If A equals B, the operand is set to FFFFH and if they are not equal it is set to 0H
	NE	A NE B	Compares value of A to that of B. If A is not equal to B, the operand is set to FFFFH and if they are equal it is set to 0H
	LT	A LT B	Compares value of A to that of B. If A is less than B, the operand is set to FFFFH and if it is equal or greater than it is set to 0H
	GT	A GT B	Compares value of A to that of B. If A is greater than B, the operand is set to FFFFH and if it is equal or less than it is set to 0H
	LE	A LE B	Compares value of A to that of B. If A is less than or equal to B, the operand is set to FFFFH and if it is greater than it is set to 0H
	GE	A GE B	Compares value of A to that of B. If A is greater than or equal to B, the operand is set to FFFFH and if it is less than it is set to 0H
Logical	NOT	NOT A	Takes the logical NOT of A and makes the value that results equal to the operand
	AND	A AND B	A is ANDed with B and makes the value that results equal to the operand
	OR	A OR B	A is ORed with B and makes the value that results equal to the operand
	XOR	A XOR B	A is XORed with B and makes the value that results equal to the operand

Figure 7.3   Arithmetic, relational, and logical operators.

In Fig. 7.3 the operators are listed in the order of their precedence. By precedence we mean the order in which the assembler performs operations as it evaluates an expression. For instance, if the expression for an operand is

$$A + B * 2/D$$

when the assembler evaluates the expression, the multiplication is performed first, the division second, and the addition third.

The order of precedence can be overcome by using parentheses. When parentheses are in use, whatever is enclosed within them is evaluated first. For example, if we modify the example we just used as follows:

$$(A + B * 2)/D$$

the multiplication still takes place first, but now it is followed by the addition and then the division. Use of the set of parentheses has changed the order of precedence.

In Fig. 7.3 we have shown a simple expression using each of the operators and described the function performed by the assembler for these expressions. For example, the operand expression

```
A    SHL    n
```

causes the assembler to shift the value of A to the left by n bits.

EXAMPLE 7.3

Find the value the assembler assigns to the source operand for the instruction

```
MOV   BH,(A * 4 - 2)/(B - 3)
```

for A = 8 and B = 5.

Solution

The expression is calculated as

$$(8 * 4 - 2)/(5 - 3) = (32 - 2)/(2)$$
$$= (30)/(2)$$
$$= 15$$

and using hexadecimal notation, we get the instruction

```
MOV   BH,0FH
```

All the examples we have considered so far have used arithmetic operators. Let us now take an example of a relational operator. In Fig. 7.3, we find that there are six relational operators: *equal (EQ), not equal (NE), less than (LT), greater than (GT), less than or equal (LE),* and *greater than or equal (GE)*. The example expression given for equal is

A EQ B

When the assembler evaluates this relational expression, it determines whether or not the value of A equals that of B. If they are equal to each other, the operand is made equal to $FFFF_{16}$; if they are unequal, the operand is made equal to 0_{16}.

This result is true for all relational operators. If by evaluating a relational expression we find that the conditions it specifies are satisfied, the operand expression is replaced with $FFFF_{16}$; if the conditions are not met, it is replaced by 0_{16}.

EXAMPLE 7.4

What value is used for the source operand in the expression

MOV   AX,A   LE   (B - C)

if A = 234, B = 345, and C = 111?

Solution

Substituting into the expression, we get

$$234 \, LE \, (345 - 111)$$

$$234 \, LE \, 234$$

Since the relational operator is satisfied, the instruction is equivalent to

MOV   AX,0FFFFH

The logical operators are similar to the arithmetic and relational operator; however, when they are used, the assembler performs the appropriate sequence of logic operations and then assigns the result that is produced to the operand.

The Value-Returning and Attribute Operators

Two other types of operators are available for use with operands; the *value-returning operators* and the *attribute operators*. The operators in each group, along with an example expression and description of their function, are given in Fig. 7.4.

The value-returning operators return the attribute (segment, offset, or type) value of a variable or label operand. For instance, assuming that the variable A is in a data segment, the instructions

```
MOV   AX,SEG A
MOV   SI,OFFSET A
MOV   CL,TYPE A
```

when assembled cause the 16-bit value in the DS register to replace SEG A, the 16-bit offset of the location of variable A in the current data segment to replace OFFSET A, and the type number of the variable to replace TYPE A, respectively. Assuming that A is a data byte, the value 1 will be assigned to TYPE A.

The attribute operators give the programmer the ability to change the attributes of an operand or label. For example, operands that use the BX, SI, or DI registers to hold the offset to their storage locations in memory are automatically referenced with respect to the contents of the DS register. An example is the instruction

```
MOV   AX, [SI]
```

Type	Operator	Example	Function
Value-returning	SEG	SEG A	Assigns the contents held in the segment register corresponding to the segment in which A resides to the operand
	OFFSET	OFFSET A	Assigns the offset of the location A in its corresponding segment to the operand
	TYPE	TYPE A	Returns to the operand a number representing the type of A; 1 for a byte variable and 2 for a word variable; NEAR or FAR for the label
	SIZE	SIZE A	Returns the byte count of variable A to the operand
	LENGTH	LENGTH A	Returns the number of units (as specified by TYPE) allocated for the variable A to the operand
Attribute	PTR	NEAR PTR A	Overrides the current type of label operand A and assigns a new pointer type: BYTE, WORD, NEAR, or FAR to A
	DS:,ES:,SS:	ES:A	Overrides the normal segment for operand A and assigns a new segment to A
	SHORT	JMP SHORT A	Assigns to operand A an attribute that indicates that it is within +127 or −128 bytes of the next instruction. This lets the instruction be encoded with the minimum number of bytes
	THIS	THIS BYTE A	Assigns to operand A a distance or type attribute: BYTE, WORD, NEAR, or FAR, and the corresponding segment attribute
	HIGH	HIGH A	Returns to the operand A the high byte of the word of A
	LOW	LOW A	Returns to the operand A the low byte of the word of A

Figure 7.4 Value-returning and attribute operators.

We can use the *segment-override* attribute operator to select another segment register. For instance, to select the extra segment register the instruction is written as

```
MOV  AX,ES: [SI]
```

▲ 7.3 PSEUDO-OPERATIONS

The primary function of an assembler program is to convert the assembly language instructions of the source program to their corresponding machine instructions. However, practical assembly language source programs do not consist of assembly language statements only; they also contain what are called *pseudo-operation statements* (*pseudo-op statements* for short). In this section, we will look more closely at what a pseudo-op is and what pseudo-ops are available in MASM, as well as how they are used as part of an assembly language source program.

The Pseudo-OP

In Section 7.2 we introduced the syntax of the pseudo-op statements and found that they differ from assembly language instruction statements in that they are directions to tell

Type	Pseudo-ops		
Data	ASSUME	ENDS	NAME
	COMMENT	EQU	ORG
	DB	= (Equal	PROC
	DD	Sign)	PUBLIC
	DQ	EVEN	.RADIX
	DT	EXTRN	RECORD
	DW	GROUP	SEGMENT
	END	INCLUDE	STRUC
	ENDP	LABEL	
Conditional	ELSE	IFDEF	IFNB
	ENDIF	IFDIF	IFNDEF
	IF	IFE	IF1
	IFB	IFIDN	IF2
Macro	ENDM	IRPC	PURGE
	EXITM	LOCAL	REPT
	IRP	MACRO	
Listing	.CREF	PAGE	TITLE
	.LALL	.SALL	.XALL
	.LFCOND	.SFCOND	.XCREF
	.LIST	SUBTTL	.XLIST
	%OUT	.TFCOND	

Figure 7.5 Pseudo-ops of the macroassembler.

the assembler how to assemble the source program instead of instructions to be processed by the microprocessor. That is, pseudo-ops are statements written in the source program but are meant only for use by the assembler program. The assembler program follows these *directives* (directions) during the assembling of the program, but does not produce any machine code for them.

A list of the pseudo-ops provided in MASM is shown in Fig. 7.5. Notice that the pseudo-ops are grouped into categories based on the type of operation they specify to the assembler. These categories are the *data pseudo-ops, conditional pseudo-ops, macro pseudo-ops,* and *listing pseudo-ops.* Notice that each category contains a number of different pseudo-ops. Here we will consider only a most frequently used subset of the pseudo-ops in these categories. For information on those pseudo-ops not covered here, the reader should consult the manual provided with Microsoft's MASM assembler.

Data Pseudo-Ops

The primary function of the pseudo-ops in the data group is to define values for constants, variables, and labels. Other functions that can be performed by them are to assign a size to variables and to reserve storage locations for them in memory. The most commonly used pseudo-ops to handle these types of data operations are those listed in Fig. 7.6.

The first two data pseudo-ops in Fig. 7.6 are *equate* and *equal to.* Their pseudo-opcodes are EQU and =, respectively. Both of these pseudo-ops can be used to assign a constant value to a symbol. For example, the symbol AA can be set equal to 0100_{16} with the equate statement

AA EQU 0100H

Pseudo-op	Meaning	Function
EQU	Equate	Assign a permanent value to a symbol
=	Equal to	Set or redefine the value of a symbol
DB	Define byte	Define or initialize byte size variables or locations
DW	Define word	Define or initialize word size (2 byte) variables or locations
DD	Define double word	Define or initialize double word size (4 byte) variables or locations

Figure 7.6 Data pseudo-ops.

The value of the operand can also be assigned using the arithmetic, relational, or logic expressions discussed in Section 7.2. An example of the EQU pseudo-op, which uses an arithmetic expression to define an operand, is

$$\text{BB EQU AA+5H}$$

In this statement, the symbol BB is assigned the value of symbol AA plus 5. These two operations can also be done using the = pseudo-op. This gives the statements

$$\text{AA} = 0100\text{H}$$

$$\text{BB} = \text{AA+5H}$$

Once these values are assigned to AA and BB, they can be referenced elsewhere in the program by just using the symbol.

The difference between the EQU and = pseudo-ops lies in the fact that the value assigned to the symbol using EQU cannot be changed, whereas when = is used to define the symbol its value can be changed later in the program.

$$\text{AA EQU 0100H}$$

$$\text{BB EQU AA+5H}$$

.

.

.

$$\text{BB EQU AA+10H ;This is illegal}$$

Here AA is set equal to 0100_{16} and BB to 0105_{16}; the value of BB cannot be changed with the third EQU statement. On the other hand, if we use the = pseudo-ops as follows,

$$\text{AA} = 0100\text{H}$$

$$\text{BB } = \text{AA+5H}$$

.

.

.

$$\text{BB } = \text{AA+10H ;This is legal}$$

BB is assigned the new value of AA+10H as the third = pseudo-op is processed by the assembler.

The other three pseudo-ops given in Fig. 7.6 are *define byte* (DB), *define word* (DW), and *define double word* (DD). The function of these pseudo-ops is to define the size of variables as being byte, word, or double word in length, allocate space for them, and assign them initial values. If the initial value of a variable is not known, the DB, DW, or DD statement simply allocates a byte, word, or double word of memory to the variable name.

An example of the DB pseudo-op is

```
CC   DB   7
```

Here variable CC is defined as byte size and assigned the value 7. It is important to note that the value assigned with a DB, DW, or DD statement must not be larger than the maximum number that can be stored in the specified-size storage location. For instance, for a byte-size variable, the maximum decimal value is 255 for an unsigned number and +127 or −128 for a signed number.

Here is another example:

```
EE   DB   ?
```

In this case, a byte of memory is allocated to the variable EE, but no value is assigned to it. Notice that use of a ? as the operand means that an initial value is not to be assigned.

Look at another example:

```
MESSAGE   DB   "JASBIR"
```

Here each character in the string JASBIR is allocated a byte in memory, and these bytes are initialized with the ASCII values for the characters. This is the way ASCII data are assigned to a name.

If we need to initialize a large block of storage locations with the same value, the assembler provides a way of using the define byte, word, or double-word pseudo-op to repeat a value. An example is the statement

```
TABLE_A   DB   10 DUP(?),5 DUP(7)
```

This statement causes the assembler to allocate 15 bytes of memory for TABLE_A. The first 10 bytes of this block of memory are left uninitialized, and the next 5 bytes are all initialized with the value 7. Notice that use of *duplicate* (DUP) tells the assembler to duplicate the value enclosed in parentheses a number of times equal to the number that precedes DUP.

If each element of the table was to be initialized to a different value, the DB command would be written in a different way. For example, the command

```
TABLE_B   DB   0,1,2,3,4,5,6,7,8,9
```

sets up a table called TABLE_B and assigns to its 10 storage locations the decimal values 0 through 9.

Pseudo-op	Function
SEGMENT	Defines the beginning of a segment and specifies its kind, at what type of address boundary it is to be stored in memory, and how it is to be positioned with respect to other similar segments in memory
ENDS	Specifies the end of a segment
ASSUME	Specifies the segment address register for a given segment

Figure 7.7 Segment pseudo-ops.

Segment Control Pseudo-Ops

Memory of the 80386DX-based microcomputer is partitioned into three kinds of segments: the code segment, data segments, and stack segment. The code segment is where machine-code instructions are stored, the data segments are for storage of data, and the stack segment is for a temporary storage location called the *stack*. Using the *segment-control pseudo-ops* in Fig. 7.7, the statements of a source program written for the 80386DX can be partitioned and assigned to a specific memory segment. These pseudo-ops can be used to specify the beginning and end of a segment in a source program and assign to them attributes such as a starting address boundary, the kind of segment, and how the segment is to be combined with other segments of the same name.

The beginning of a segment is identified by the *segment* (SEGMENT) pseudo-op, and its end is marked by the *end-of-segment* (ENDS) pseudo-op. Here is an example of a segment definition:

```
SEGA    SEGMENT   PARA PUBLIC 'CODE'
        MOV   AX,BX
          .
          .
          .
SEGA    ENDS
```

As shown, the information between the two pseudo-ops statements would be the instructions of the assembly language program.

In this example, SEGA is the name given to the segment. The pseudo-op SEGMENT is followed by the operand PARA PUBLIC 'CODE'. Here PARA (*paragraph*) defines that this segment is to be aligned in memory on a 16-byte address boundary. This part of the operand is called the *align-type* attribute. There are other align-type attributes that can be used instead of PARA. They are given in Fig. 7.8 with brief descriptions of their functions.

Attribute	Function
PARA	Segment begins on a 16 byte address boundary in memory (4 LSBs of the address are equal to 0)
BYTE	Segment begins anywhere in memory
WORD	Segment begins on a word (2 byte) address boundary in memory (LSB of the address is 0)
PAGE	Segment begins on a 256 byte address boundary in memory (8 LSBs of the address are equal to 0)

Figure 7.8 Align-type attributes.

Attribute	Function
PUBLIC	Concatenates segments with the same name
COMMON	Overlaps from the beginning segments with the same name
AT [expression]	Locates the segment at the 16-bit paragraph number evaluated from the expression
STACK	The segment is part of the run-time stack segment
MEMORY	Locates the segment at an address above all other segments

Figure 7.9 Combine-type attributes.

PUBLIC, which follows PARA in the operand of the example SEGMENT pseudo-op, defines what is called a *combine-type* attribute. It specifies that this segment is to be concatenated with all other segments that are assigned the name SEGA to generate one physical segment called SEGA. Other combine-type attributes are given in Fig. 7.9.

The last part of the operand in the SEGMENT statement is 'CODE'; it specifies that the segment is a code segment. This entry is called a *class* attribute. All of the allowed segment classes are shown in Fig. 7.10.

At the end of the group of statements that are to be assigned to the code segment, there must be an ENDS pseudo-op. In Fig. 7.7, we find that this statement is used to mark the end of the segment. ENDS must also be preceded by the segment name, which is SEGA in our example.

The third pseudo-op in Fig. 7.7 is *assume* (ASSUME). It is used to assign the segment registers that hold the base addresses to the program segments. For instance, with the statement

```
ASSUME   CS:SEGA,DS:SEGB,SS:SEGC
```

we specify that register CS holds the base address for segment SEGA, register DS holds the base address for segment SEGB, and register SS holds the base address for segment SEGC. The ASSUME pseudo-op is written at the beginning of a code segment just after the SEGMENT pseudo-op.

Figure 7.11 shows the general structure of a code segment definition using the segment control pseudo-ops.

Modular Programming Pseudo-Ops

For the purpose of development, larger programs are broken down into smaller segments called *modules*. Typically, each module implements a specific function and has

Attribute	Function
CODE	Specifies the code segment
DATA	Specifies the data segment
STACK	Specifies the stack segment
EXTRA	Specifies the extra segment

Figure 7.10 Class attributes.

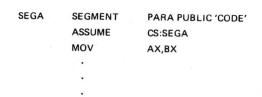

```
SEGA    SEGMENT    PARA PUBLIC 'CODE'
        ASSUME     CS:SEGA
        MOV        AX,BX
          .
          .
          .
SEGA    ENDS
```

Figure 7.11 Example using segment control pseudo-ops.

its own code segment and data segment. However, it is common that during the execution of a module a part of some other module may need to be called for execution or that data that resides in another module may need to be accessed for processing. To support capabilities such that a section of code in one module can be executed from another module or for data to be passed between modules, MASM provides *modular programming pseudo-ops*. The most frequently used modular programming pseudo-ops are listed in Fig. 7.12.

A section of a program that can be called for execution from other modules is called a *procedure*. Similar to the definition of a segment of a program, the beginning and end of a procedure must be marked with pseudo-statements. The beginning of the procedure is marked by the *procedure* (PROC) pseudo-op, and its end is marked by the *end of procedure* (ENDP) pseudo-op.

There are two kinds of procedures: a *near procedure* and a *far procedure*. Whenever a procedure is called, the return address has to be saved on the stack. After completing the code in the called procedure, this address is used to return execution back to the point of its initiation. When a near procedure is called into operation, only the code offset address (IP) is saved on the stack. Therefore, a near procedure can only be called from the same code segment. On the other hand, when a far procedure is called, both the contents of the code segment (CS) register and the code offset (IP) are saved on the stack. For this reason, a far procedure can be called from any code segment. Depending upon the kind of procedure, the *return* (RET) instruction at the end of the procedure restores either IP or both IP and CS from the stack.

If a procedure can be called from other modules, its name must be made *public* by using the PUBLIC pseudo-op. Typically this is done before entering the procedure. Thus,

Pseudo-op			Function
proc-name	PROC	[NEAR]	Defines the beginning of a near-proc procedure
proc-name	PROC	FAR	Defines the beginning of a far-proc procedure
proc-name	ENDP		Defines the end of a procedure
	PUBLIC	Symbol[. . . . . . .]	The defined symbols can be referenced from other modules
	EXTRN	name:type[. . . .]	The specified symbols are defined in other modules and are to be used in this module

Figure 7.12 Modular programming pseudo-ops.

the structure for defining a procedure with name SUB_NEAR will be

```
                         PUBLIC    SUB_NEAR
              SUB_NEAR   PROC
                         .
                         .
                         .
                         RET
              SUB_NEAR   ENDP
```

In this example the PROC does not have either the NEAR or FAR attribute in its operand field. When no attribute is specified as an operand, NEAR is assumed as the default attribute by the assembler. In this way, we see that this procedure can only be called from other modules in the same code segment.

The structure of a far procedure that can be called from modules in other segments would be

```
                        PUBLIC    SUB_FAR
             SUB_FAR    PROC      FAR
                        .
                        .
                        .
                        RET
             SUB_FAR    ENDP
```

If a procedure in another module is to be called from the current module, its name must be declared external in the current procedure by using the *external reference* (EXTRN) pseudo-op. It is also important to know whether this call is to a module in the same code segment or in a different code segment. Depending upon the code segments, the name of the procedure must be assigned either a NEAR or FAR attribute as part of the EXTRN statement.

The example in Fig. 7.13 illustrates the use of the EXTRN pseudo-op. Here we see

```
              PUBLIC    SUB                     EXTRN     SUB:FAR
   CSEG1      SEGMENT                  CSEG2    SEGMENT
              .                                 .
              .                                 .
   SUB        PROC      FAR                      CALL      SUB
              MOV       AX,BX                    .
              .                                  .
              .                                  .
              RET                                .
   SUB        ENDP                               .
              .                        CSEG2    ENDS
              .
              .
   CSEG1      ENDS

           Module 1                             Module 2
```

Figure 7.13 An example showing use of the EXTRN pseudo-op.

Pseudo-op	Function
ORG [expression]	Specifies the memory address starting from which the machine code must be placed
END [expression]	Specifies the end of the source program

Figure 7.14 ORG and END pseudo-op.

that an external call is made from module 2 to the procedure SUB that resides in module 1. Therefore, the procedure SUB is defined as PUBLIC in module 1. Notice that the two modules have different code segments, which are called CSEG1 and CSEG2. Thus the external call in module 2 is to a far procedure. Therefore, the PROC pseudo-op for SUB in module 1 and the external label definition of SUB in module 2 have the FAR attribute attached to them.

Pseudo-Op for Memory Usage Control

If the machine code generated by the assembler must reside in a specific part of the memory address space, an *origin* (ORG) pseudo-op can be used to specify the starting point of that memory area. In Fig. 7.14 we see that the operand in the ORG statement can be an expression. The value that results from this expression is the address at which the machine code is to begin loading. For example, the statement

```
ORG   100H
```

simply tells the assembler that the machine code for subsequent instructions is to be placed in memory starting at address 100_{16}. This pseudo-op statement is normally located at the beginning of the program.

If specific memory locations must be skipped—for example, because they are in a read-only area of memory—one can use ORG pseudo-ops as follows:

```
ORG   100H
ORG   $+200H
```

in which case memory locations 100_{16} to 300_{16} are skipped and the machine code of the program can start at address 301_{16}.

The End of Program Pseudo-Op

The *end* (END) pseudo-op, which is also described in Fig. 7.14, tells the assembler when to stop assembling. It must always be included at the end of the source program. Optionally, we can specify the starting point of the program with an expression in the operand field of the END statement. For instance, an END statement can be written as

```
END   PROG_BLOCK
```

where PROG_BLOCK identifies the beginning address of the program.

Pseudo-Ops for Program Listing Control

The last group of pseudo-ops we will consider is called the *listing control pseudo-ops*. The most widely used pseudo-ops in this group are shown in Fig. 7.15. The purpose of the

Pseudo-op	Function
PAGE operand_1 operand_2	Selects the number of lines printed per page and the maximum number of characters printed per line in the listing
TITLE text	Prints 'text' on the second line of each page of the listing
SUBTTL text	Prints 'text' on the third line of each page of the listing

Figure 7.15   Listing control pseudo-ops.

listing control pseudo-ops is to give the programmer some options related to the way in which source program listings are produced by the assembler. For instance, we may want to set up the print output such that a certain number of lines are printed per page or we may want to title the pages of the listing with the name of the program.

The *page* (PAGE) pseudo-op lets us set the page width and length of the source listing produced as part of the assembly process. For example, if the PAGE pseudo-op

```
PAGE   50    100
```

is encountered at the beginning of a source program, then each printed page will have 50 lines and up to 100 characters in a line. The first operand, which specifies the number of lines per page, can be any number from 10 through 255. The second operand, which specifies the maximum number of characters per line, can range from 60 to 132. The default values for these parameters are 66 lines per page and 80 characters per line. The default parameters are selected if no operand is included with the pseudo-op.

Chapter and page numbers are always printed at the top of each page in a source listing. They are in the form

```
[chapter number] − [page number]
```

As the source listing is produced by the assembler, the page number automatically increments each time a full page of listing information is generated. On the other hand, the chapter number does not change as the listing is generated. The only way to change the chapter number is by using the pseudo-op

```
PAGE    +
```

When this form of the PAGE pseudo-op is processed by the assembler, it increments the chapter count and at the same time resets the page number to 1.

The second pseudo-op in Fig. 7.15 is *title* (TITLE). When this pseudo-op is included in a program, it causes the text in the operand field to be printed on the second line of each page of the source listing. Similarly, the third pseudo-op, *subtitle* (SUBTTL), prints the text included in the pseudo-op statement on the third line of each page.

An Example of a Source Program Using Pseudo-Ops

In order to have our first experience in using pseudo-ops in a source program, let us look at the program in Fig. 7.16. This program is written to copy a given block of data from

```
TITLE BLOCK-MOVE PROGRAM

        PAGE          ,132

COMMENT *This program moves a block of specified number of bytes
         from one place to another place*


;Define constants used in this program

        N=                16            ;Bytes to be moved
        BLK1ADDR=         100H          ;Source block offset address
        BLK2ADDR=         120H          ;Destination block offset addr
        DATASEGADDR=      2000H         ;Data segment start address


STACK_SEG         SEGMENT              STACK 'STACK'
                  DB                   64 DUP(?)
STACK_SEG         ENDS
CODE_SEG          SEGMENT              'CODE'
BLOCK             PROC         FAR
     ASSUME       CS:CODE_SEG,SS:STACK_SEG

;To return to DEBUG program put return address on the stack

            PUSH  DS
            MOV   AX, 0
            PUSH  AX

;Setup the data segment address

            MOV   AX, DATASEGADDR
            MOV   DS, AX

;Setup the source and destination offset adresses

            MOV   SI, BLK1ADDR
            MOV   DI, BLK2ADDR

;Setup the count of bytes to be moved

            MOV   CX, N

;Copy source block to destination block

NXTPT:      MOV   AH, [SI]           ;Move a byte
            MOV   [DI], AH
            INC   SI                 ;Update pointers
            INC   DI
            DEC   CX                 ;Update byte counter
            JNZ   NXTPT              ;Repeat for next byte
            RET                      ;Return to DEBUG program
BLOCK             ENDP
CODE_SEG          ENDS
     END          BLOCK              ;End of program
```

Figure 7.16 Block-move source program.

a location in memory known as the source block or another block location called the destination block. In the sections that follow, we will use this program to learn various aspects of assembly language program development, such as creating a source file, assembling, linking, and debugging. For now we will just look at the pseudo-ops used in the program.

The program starts with a TITLE pseudo-op statement. The text BLOCK_MOVE PROGRAM that is included in this statement will be printed on the second line of each page of the source listing. This text should be limited to 60 characters. The second and third statements in the program are also pseudo-ops. The third statement is a *comment* pseudo-op and is used to place descriptive comments in the program. Note that it begins with the pseudo-opcode COMMENT and is followed by the comment enclosed within the delimiter asterisk (*). This comment gives a brief description of the function of the program.

There is another way of including comments in a program. This is by using a semicolon (;) followed by the text of the comment. The next line in the program is an example of this type of comment. It indicates that the next part of the program is used to define variables that are used in the program. Four equal-to (=) pseudo-op statements follow the comment. Notice that they equate N to the value 16_{10}, BLK1ADDR to the value 100_{16}, BLK2ADDR to the value 120_{16}, and DATASEGADDR to the value 1020_{16}.

There are two segments in the program: the stack segment and the code segment. The next three pseudo-op statements define the stack segment. They are

```
STACK_SEG    SEGMENT   STACK   'STACK
             DB    64      DUP(?)
STACK_SEG    ENDS
```

In the first statement, the stack segment is assigned the name STACK_SEG; the second statement allocates a block of 64 bytes of memory for use as stack and leaves this area uninitialized. The third statement defines the end of the stack.

The code segment is defined between the statements

```
CODE_SEG    SEGMENT    'CODE'
```

and

```
CODE_SEG    ENDS
```

Here CODE_SEG is the name we have used for the code segment. At the beginning of the code segment an ASSUME pseudo-op is used to specify the base registers for the code and stack segments. Notice that CS is the base register for the code segment and SS is the base register for the stack segment.

We also find an END pseudo-op at the end of the program. It identifies the end of the program, and BLOCK in this statement defines the starting address of the source program. Processing of this statement tells the assembler that the assembly is complete.

A PROC pseudo-op is included at the beginning of the source program, and the ENDP pseudo-op is included at the end of the program. This makes the program segment a procedure that can be referenced as a module in a larger program.

▲ 7.4 CREATING A SOURCE FILE WITH AN EDITOR

Now that we have introduced assembly language syntax, the pseudo-ops, and an example of an assembly language program, let us continue by looking at how the source-program file is created on the PC/AT. Source program files are generated using a program called an *editor*. Basically, two types of editors are available: the *screen editor* and the *line editor*. They differ in that a screen editor allows us to work on a full screen of text at a time, whereas a line editor enables us to work on one text line at a time. Both types of editor program are provided in the PC/AT's DOS. The line editor is called *EDLIN* and the screen editor is called *EDIT*. In this book we have assumed that the reader is already familiar with the commands of the DOS and the use of either the EDLIN or EDIT program. For this reason, we will just briefly describe the use of EDLIN and EDIT in creating a source-program file. If additional details are required, the DOS manual that is provided with the operating system software should be consulted.

Using the EDLIN Line-Editor

The diagram in Fig. 7.17(a) outlines the sequence of steps that take place during a typical editing session with EDLIN. We are interested in creating a source program in a file called BLOCK.SRC. This file will be stored on a diskette in drive A. We start by entering the following command to bring up the line-editor program:

```
C:\DOS>EDLIN  A:BLOCK.SRC  (↵)
```

In response to this input, the EDLIN program is first loaded from the operating system and then executed. Once EDLIN is running, it checks to determine whether or not the file BLOCK.SRC already exists on the diskette in drive A. If it does exist, the file is loaded into memory and the response "End of input file" is displayed; on the next line the prompt *_ is displayed. On the other hand, if the file does not exist, "New File" is displayed, followed by the prompt *_.

Let us assume that BLOCK.SRC is a new file. Then the system will respond with the prompt

```
New file
*_
```

As shown in Fig. 7.17(a), the next input should be I followed by (↵). This is the *insert line* command; its entry causes line number 1:* to be displayed. We are now in the *line input mode* of editor operation. This is the mode used when creating new source-program files. Next the text for the first line of the program is keyed in following the line number; then the (↵) key is depressed. For instance, in Fig. 7.17(a) we entered the statement

```
1:*  MOV  AX,DATASEGADDR  (↵)
```

When (↵) is depressed, the entry of line 1 is completed and line number 2:* is displayed. Its text is now entered and (↵) is depressed once more. For example, in Fig. 7.17(a) we find

```
2:*  MOV  DS,AX  (↵)
```

Repeating this sequence, the complete program is entered line by line. Here are some additional line entries for the program:

```
 3:*  MOV  SI,BLK1ADDR  (↵)
 4:*  MOV  DI,BLK2ADDR  (↵)
          .
          .
          .
11:*  JNZ  NXTPT        (↵)
```

After the last line of the program is entered, we must come out of the line input mode. This is done by depressing the Ctrl (control) and Break (break) keys together. On

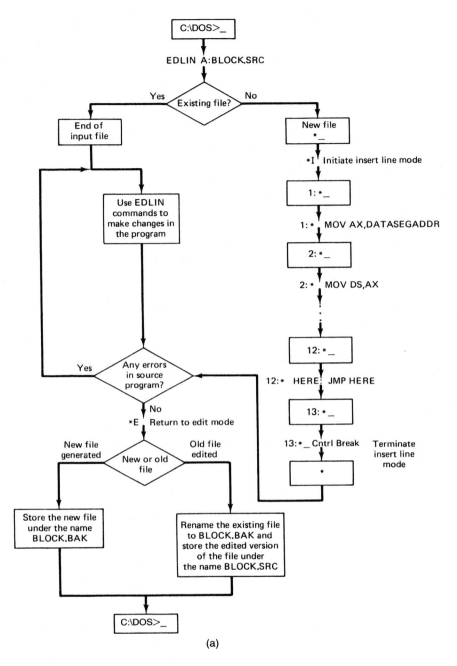

Figure 7.17(a) Flowchart for creating and editing of source files with EDLIN.

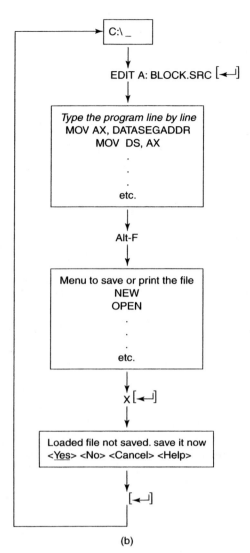

Figure 7.17(b) Flowchart for creating and editing of source files with EDIT.

(b)

the keyboard of the PC/AT the break key may be marked *Scroll Lock* on the top and *Break* on the front. Depression of these two keys together takes EDLIN out of the line-input mode and puts it into the *edit mode*. Figure 7.18 shows the information displayed during the entry of this segment of program

Once we are in the edit mode, the program should be looked over closely for errors. If errors are found, they can be corrected using other editor commands. The edit commands provided in EDLIN and the syntax in which they must be entered are shown in Fig. 7.19. Notice that commands are provided that let us delete a line or lines, insert a new line or lines, list lines, delete or insert characters in a line, and replace or search for a string of characters. For instance, after loading a program from a diskette, the List Lines command can be used to display the lines of the program. The file created in our earlier example can

```
C:\DOS>EDLIN A:BLOCK.SRC
New file
*I
        1:*        MOV     AX,DATASEGADDR
        2:*        MOV     DS,AX
        3:*        MOV     SI,BLK1ADDR
        4:*        MOV     DI,BLK2ADDR
        5:*        MOV     CX,N
        6:*NXTPT:  MOV     AH,[SI]
        7:*        MOV     [DI],AH
        8:*        INC     SI
        9:*        INC     DI
       10:*        DEC     CX
       11:*        JNZ     NXTPT
       12:*^C
*E
C:\DOS
```

Figure 7.18 Entry of example
program with the EDLIN editor.

be reloaded and displayed with the command sequence

$$
\text{C:\DOS>EDLIN   A:BLOCK.SRC   (↵)}
$$
$$
\text{* 1,11 L   (↵)}
$$

The resulting display is shown in Fig. 7.20.

After all corrections have been made, the editing session is complete, and it must be ended. The End Edit command is used for this purpose. Looking at Fig. 7.17(a), we see that it is entered as

$$
\text{*E   (↵)}
$$

The response of EDLIN to this entry depends on whether the current editing was done on an existing file or a new file that was just being created. In Fig. 7.17(a) we find that if it is a new file, such as in our earlier example, the edit mode is terminated and the lines of text that were entered are stored in a file called BLOCK.SRC and then the DOS prompt C:\DOS is displayed.

Command	Format
Append Lines	[n] A
Copy Lines	[line] ,[line] ,line,[count] C
Delete Lines	[line] [,line] D
Edit Line	[line]
End Edit	E
Insert Lines	[line] I
List Lines	[line] [,line] L
Quit Edit	Q
Move Lines	[line] ,[line] ,lineM
Page	[line] [,line] P
Replace Text	[line] [,line] [?] R [string] [<F6>string]
Search Text	[line] [,line] [?] S [string]
Transfer Lines	[line] T filename [.ext]
Write Lines	[n] W

Figure 7.19 EDLIN commands.

```
C:\DOS>EDLIN A:BLOCK.SRC
End of input file
*1,11L
        1:*         MOV     AX,DATASEGADDR
        2:          MOV     DS,AX
        3:          MOV     SI,BLK1ADDR
        4:          MOV     DI,BLK2ADDR
        5:          MOV     CX,N
        6: NXTPT:    MOV     AH,[SI]
        7:          MOV     [DI],AH
        8:          INC     SI
        9:          INC     DI
       10:          DEC     CX
       11:          JNZ     NXTPT
*Q
Abort edit (Y/N)? Y
C:\DOS>
```

Figure 7.20 Listing of the example source program produced with the List Lines command.

On the other hand, if we were editing an existing file called BLOCK.SRC, the original file (before editing) would be renamed using the original file name with the extension BAK. This file, BLOCK.BAK, is called the backup file. Next, the edited version of the program is saved in the file BLOCK.SRC. Then the DOS prompt C:\DOS is displayed. The creation of this *backup file* provides us with a way to get back to the original version of the program if necessary. It should be noted that a file with the extension BAK cannot be edited with EDLIN. However, this backup file can either be copied to another file or have its extension changed with a RENAME command, and then it can be edited.

Using the EDIT Screen Editor

A screen editor, such as EDIT, is easier to use than a line editor. When working on a document such as an assembly language source program, the complete contents of the file can be viewed on the screen during the edit session. This makes it possible to quickly read any part of the program and randomly make changes to its lines of code.

EDIT is a *menu-driven* text editor. That is, a simple key sequence is performed to display a *menu* of operations that can be performed. The desired operation is picked from the menu and then the return key is depressed. At this point, a *dialog box* is displayed to assist in describing the operation that is to be performed. The dialog box is filled out with the appropriate information by typing it in at the keyboard. Then the return key is again depressed to initiate the defined operation.

Figure 7.17(b) shows the sequence of events that take place when a source program file is created with EDIT. For an example, we will simply repeat the creation of the block-move program of Fig. 7.18, but this time we will save it as the file BLOCK.ASM. Again, the sequence begins with a command to load and run the editor program. In this case, the command is

```
C:\DOS>EDIT  A:BLOCK.ASM  (↵)
```

As with EDLIN, the EDIT program first checks to see if the file BLOCK.ASM already exists. If it does, the file is read from drive A, loaded into memory, and displayed on the screen. This would be the case if an existing source program is to be corrected or changed. At this point, we have already demonstrated one of the benefit of EDIT over EDLIN. That is, after loading, the complete source program is in view on the screen ready to be edited.

Looking at the flowchart, we see that the next step is to make the changes in the program. Then, the program should be carefully examined to verify that no additional errors were made during the edit process. The last step is to save the modified program in either the old file or as a file with a new name. Note that a backup file is not automatically made as part of the file-save process of EDIT.

Let us assume that the file BLOCK.ASM does not already exist and go through the sequence of events that must take place to create a source program. In this case, the same command can be used to bring up the editor. However, when it comes up, no program can be loaded, so the screen remains blank. The lines of the assembly language program are simply typed in one after the other. The TAB key is used to make the appropriate indents in the instructions. For instance, to enter the first instruction of the program in Fig. 7.18, the TAB key is depressed once for a single indent and then MOV is keyed in. Another TAB is needed to indent again and then the operand AX,DATASEGADDR is entered. The first instruction is now complete, so the RETURN key is depressed to position the cursor for entry of the next instruction. This sequence is repeated until the whole block-move program has been entered. Notice that when entering the sixth instruction the label, NXTPT: is entered at the left margin and then TAB is depressed to indent to the position for MOV.

The ease with which corrections can be made in the source file is another benefit of EDIT. If errors are made as the instructions of the block-move program is entered, the arrow keys can be used to simply reposition the cursor to the spot that needs to be changed. Next the DELETE key (for characters to the right of the cursor) or Backspace key (for characters to the left of the cursor) is used to remove the incorrect characters. Then, the correct characters are typed. Actually, the editing capability of EDIT is more versatile than just described. Commands are provided to move, copy, delete, find, or find and replace a character, string of characters, or block of text. For instance, we may need to change the name of an operand each time it occurs in a large program. Instead of having to go through the program instruction by instruction to find each occurrence of the operand, this operation can be done with a single find-and-replace command.

Assuming that all corrections have been made, we are now ready to save the program. At the top of the EDIT screen is a *menu bar* with four menus: *File, Edit, Search,* and *Options.* The save operations are located under the File menu. To select this menu, hold down the ALT key and depress the F key. This causes a *pop-down menu* that lists the file commands to appear at the top of the screen. When the menu appears, the *New* command is highlighted. Use the arrow-down (↓) key to move the highlighted area down to the *Save As* operation. Now depress the return (↵) key to initiate the file-save operation. This displays the Save As dialog box. In the dialog box, A:BLOCK.ASM will automatically be filled in as the file name. If the file name is to remain the same, simply depress the return key to save the file. On the other hand, if the name is to be changed, type in the new drive designator and file name information before depressing (↵).

At this moment, the source program has been created, verified, and saved, but we are still in the EDIT program. We are ready to exit the EDIT program. *EXIT* is another operation that is in the File menu. To exit EDIT, depress ALT and then F to display the File pop-down menu. Use the (↓) key to select Exit (or type X) and then depress (↵). The EDIT program terminates and the DOS prompt reappears on the screen.

▲ 7.5 ASSEMBLING SOURCE PROGRAMS WITH MASM

Up to this point, we have studied the assembly language syntax and pseudo-op statements provided in the MASM, the structure of an assembly language program, and how to create a source program using the EDLIN and EDIT editors. Now is the time to learn how to

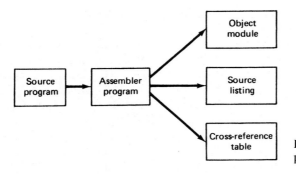

Figure 7.21 Assembling a source program.

bring up MASM, use it to assemble a source-program file into an object-code module, and examine the outputs that are produced by the assembler.

Earlier we said that an assembler is the program used to convert a file that contains an assembly language source program to its equivalent object of 80386DX machine code. Figure 7.21 shows that the input to the assembler program is the assembly language source program. This is the program that is to be assembled. The source file is read by the assembler program from its file on the diskette and translated into three outputs. As shown in Fig. 7.21, these outputs are the *object module, the source listing,* and a *cross-reference table.*

Initiating the Assembly Process

To assemble a program, we must first change to the directory on drive C that contains the MASM program and then insert the diskette that contains the source file into drive A. The source file diskette should not be write-protected. Now the assembler program is loaded by keying in

```
C:\MASM>MASM   (↵)
```

In response to this entry, the following prompt is displayed:

```
Source filename[.ASM]:
```

Next we must enter the name of the source file with its extension and then depress (↵). Let us assume that we are assembling the source program that was shown in Fig. 7.16 and that its file was assigned the name BLOCK.SRC. Therefore, the input is

```
Source filename[.ASM]:A:BLOCK.SRC   (↵)
```

Next the assembler program displays a second prompt, which asks for entry of the file name (with extension) that is to be used to save the object-code module that it creates during the assembly process. This prompt is

```
Object filename[BLOCK.OBJ]:
```

Notice that the assembler automatically fills in the file name entered for the source file and appends this file name with the default extension OBJ. In our example, the file is to be created on the diskette in drive A. For this reason, we will reenter the name for the object file with drive A specified.

```
Object filename[BLOCK.OBJ]:A:BLOCK.OBJ   (↵)
```

Actually, the response can simply be

```
Object filename[BLOCK.OBJ]:A:   (↵)
```

After this entry is made, a third prompt is displayed. This prompt is

```
Source listing[NUL.LST]:
```

Note that a file name of NUL.LST is automatically assigned for the source listing if the (↵) key is depressed without entry of an alternative file name. This default file name causes suppression of a source listing, and none will be produced. If we want the assembler to create a source listing during the assembly process, we can enter a name for a listing file. Let us use a file BLOCK.LST to save our source listing. This is done by entering

```
Source listing[NUL.LST]:A:BLOCK.LST   (↵)
```

Another way of issuing the source listing response lets us just print the listing instead of saving it in a file. This is done by simply entering LPT1 in place of a file name.

The last prompt displayed is for the name of a cross-reference table file. This prompt is

```
Cross reference[NUL.CRF]:
```

Again, depressing (↵) without entering a file name causes creation of a cross reference file to be suppressed. However, for our example, let us create one by assigning the name BLOCK.CRF to the file. This is done by making the entry

```
Cross reference[NUL.CRF]:A:BLOCK.CRF   (↵)
```

This completes the inputs needed to assemble a source file.

As soon as the cross-reference file name entry is made, the assembly process begins and runs through to completion without any additional entries. First the contents of the source file are read from the file BLOCK.SRC on the diskette in drive A; the data are translated and information for the listing and cross-reference table is produced; finally, the object code, source listing, and cross-reference table are saved on the diskette in drive A in files BLOCK.OBJ, BLOCK.LST, and BLOCK.CRF, respectively. The contents of these files can be printed with the TYPE command of the DOS operating system. Displayed information for the start-up and execution of the assembler for our example program is shown in Fig. 7.22(a), the contents of the source-code file are shown in Fig. 7.22(b), and the source listing file is shown in Fig. 7.22(c).

The cross-reference-table file produced by the MASM assembler is not in the correct form to be printed. This is because it is not yet an ASCII file. To convert it to ASCII form,

```
C:\MASM>MASM
Microsoft (R) Macro Assembler Version 5.10
Copyright (C) Microsoft Corp 1981, 1988.  All rights reserved.

Source filename [.ASM]: A:BLOCK.SRC
Object filename [BLOCK.OBJ]: A:BLOCK.OBJ
Source listing  [NUL.LST]: A:BLOCK.LST
Cross-reference [NUL.CRF]: A:BLOCK.CRF

  47218 + 427466 Bytes symbol space free

      0 Warning Errors
      0 Severe  Errors
```

Figure 7.22(a) Display sequence for assembly of a program.

```
TITLE BLOCK-MOVE PROGRAM

        PAGE          ,132

COMMENT *This program moves a block of specified number of bytes
         from one place to another place*


;Define constants used in this program

        N=                   16          ;Bytes to be moved
        BLK1ADDR=            100H         ;Source block offset address
        BLK2ADDR=            120H         ;Destination block offset addr
        DATASEGADDR=         2000H        ;Data segment start address

STACK_SEG        SEGMENT              STACK 'STACK'
                 DB                   64 DUP(?)
STACK_SEG        ENDS
CODE_SEG         SEGMENT              'CODE'
BLOCK            PROC        FAR
     ASSUME      CS:CODE_SEG,SS:STACK_SEG

;To return to DEBUG program put return address on the stack

           PUSH  DS
           MOV   AX, 0
           PUSH  AX

;Setup the data segment address

           MOV   AX, DATASEGADDR
           MOV   DS, AX

;Setup the source and destination offset adresses

           MOV   SI, BLK1ADDR
           MOV   DI, BLK2ADDR

;Setup the count of bytes to be moved

           MOV   CX, N

;Copy source block to destination block

NXTPT:     MOV   AH, [SI]         ;Move a byte
           MOV   [DI], AH
           INC   SI               ;Update pointers
           INC   DI
           DEC   CX               ;Update byte counter
           JNZ   NXTPT            ;Repeat for next byte
           RET                    ;Return to DEBUG program
BLOCK            ENDP
CODE_SEG         ENDS
     END         BLOCK            ;End of program
```

Figure 7.22(b) Source program file.

```
  1
  2
  3                               TITLE BLOCK-MOVE PROGRAM
  4
  5                               PAGE         ,132
  6
  7                     COMMENT *This program moves a block of specified number of bytes
  8                               from one place to another place*
  9
 10
 11                     ;Define constants used in this program
 12
 13 = 0010                        N=              16          ;Bytes to be moved
 14 = 0100                        BLK1ADDR=       100H        ;Source block offset address
 15 = 0120                        BLK2ADDR=       120H        ;Destination block offset addr
 16 = 2000                        DATASEGADDR=        2000H       ;Data segment start address
 17
 18
 19 0000                          STACK_SEG   SEGMENT         STACK 'STACK'
 20 0000  0040[                               DB          64 DUP(?)
 21       ??
 22                     ]
 23
 24 0040                          STACK_SEG   ENDS
 25
 26
 27 0000                          CODE_SEG    SEGMENT         'CODE'
 28 0000                          BLOCK       PROC        FAR
 29                               ASSUME      CS:CODE_SEG,SS:STACK_SEG
 30
 31                     ;To return to DEBUG program put return address on the stack
 32
 33 0000  1E                      PUSH    DS
 34 0001  B8 0000                 MOV     AX, 0
 35 0004  50                      PUSH    AX
 36
 37                     ;Setup the data segment address
 38
 39 0005  B8 2000                 MOV     AX, DATASEGADDR
 40 0008  8E D8                   MOV     DS, AX
 41
 42                     ;Setup the source and destination offset adresses
 43
 44 000A  BE 0100                 MOV     SI, BLK1ADDR
 45 000D  BF 0120                 MOV     DI, BLK2ADDR
 46
 47                     ;Setup the count of bytes to be moved
 48
 49 0010  B9 0010                 MOV     CX, N
 50
 51                     ;Copy source block to destination block
 52
 53 0013  8A 24        NXTPT: MOV  AH, [SI]                  ;Move a byte
 54 0015  88 25               MOV  [DI], AH
 55 0017  46                  INC  SI                        ;Update pointers
 56 0018  47                  INC  DI
 57 0019  49                  DEC  CX                        ;Update byte counter
```

Figure 7.22(c)   Source listing file.

we can use the CREF program that is provided with the MASM. For instance, to convert the cross-reference file for our example to an ASCII file, we issue the command

        C:\DOS>CREF A.BLOCK.CRF A:BLOCK.REF   (↵)

This command reads the file BLOCK.CRF on the diskette in drive A as its input, converts it to an ASCII file, and outputs the ASCII file to a new file called BLOCK.REF, also on

```
58 001A  75 F7                    JNZ    NXTPT              ;Repeat for next byte
59 001C  CB                       RET                       ;Return to DEBUG program
60 001D                           BLOCK      ENDP
61 001D                           CODE_SEG   ENDS
62                                END        BLOCK           ;End of program
```

Segments and Groups:

```
                N a m e            Length     Align      Combine Class

CODE_SEG . . . . . . . . . . . .   001D  PARA  NONE  'CODE'
STACK_SEG  . . . . . . . . . . .   0040  PARA  STACK 'STACK'
```

Symbols:

```
                N a m e            Type   Value      Attr

BLK1ADDR . . . . . . . . . . . .   NUMBER    0100
BLK2ADDR . . . . . . . . . . . .   NUMBER    0120
BLOCK  . . . . . . . . . . . . .   F PROC    0000   CODE_SEG    Length = 001D

DATASEGADDR  . . . . . . . . . .   NUMBER    2000

N  . . . . . . . . . . . . . . .   NUMBER    0010
NXTPT  . . . . . . . . . . . . .   L NEAR    0013   CODE_SEG

@CPU . . . . . . . . . . . . . .   TEXT   0101h
@FILENAME  . . . . . . . . . . .   TEXT   block
@VERSION . . . . . . . . . . . .   TEXT   510
```

```
   59 Source  Lines
   59 Total   Lines
   15 Symbols
```

47222 + 347542 Bytes symbol space free

```
    0 Warning Errors
    0 Severe  Errors
```

Figure 7.22(c)   (Continued)


```
Microsoft Cross-Reference  Version 5.10         Sun May 17 18:17:20 1992
BLOCK-MOVE PROGRAM

    Symbol Cross-Reference          (# definition, + modification)  Cref-1

    @CPU . . . . . . . . . . . .     1#
    @VERSION . . . . . . . . . .     1#

    BLK1ADDR . . . . . . . . . .     14#    44
    BLK2ADDR . . . . . . . . . .     15#    45
    BLOCK. . . . . . . . . . . .     28#    60      62

    CODE . . . . . . . . . . . .     27
    CODE_SEG . . . . . . . . . .     27#    29      61

    DATASEGADDR. . . . . . . . .     16#    39

    N. . . . . . . . . . . . . .     13#    49
    NXTPT. . . . . . . . . . . .     53#    58

    STACK. . . . . . . . . . . .     19
    STACK_SEG. . . . . . . . . .     19#    24      29


    12 Symbols
```

Figure 7.22(d)   Cross-reference table file.

the diskette in drive A. Now the file BLOCK.REF can be displayed or printed with the TYPE command. Figure 7.22(d) shows the contents of the file.

In the assembly process, the default file extensions specified in the prompts can also be used. Assuming that the source file has the extension. ASM, the responses to the prompts would be

```
Source filename[.ASM]:A:BLOCK    (↵)
Object filename[BLOCK.OBJ]:A:     (↵)
Source listing[NUL.LST]:          (↵)
Cross reference[NUL.CRF]:          (↵)
```

Initiating the assembler with this sequence results in no source listing or cross reference file. If the files reside in the MASM directory on drive C instead of a diskette in drive A, the responses to the prompts will be

```
Source filename[.ASM]:BLOCK       (↵)
Object filename[BLOCK.OBJ]:        (↵)
Source listing[NUL.LST]:LPT1       (↵)
Cross reference[NUL.CRF]:          (↵)
```

In this example, we have elected to print the source listing to the printer connected at the LPT1 port of the PC/AT instead of saving it in a file.

Once familiar with the use of the assembler, a command line entry sequence can be used to call it up. This is done by specifying the file names in the entry sequence that calls MASM from the DOS prompt. For our original example, the command would be issued as

```
C:\MASM>MASM A:BLOCK.SRC,A:BLOCK.OBJ,A:BLOCK.LST,A:BLOCK.CRF    (↵)
```

If any of the default names are to be used, simply enter a comma in place of a name.

Syntax Errors in an Assembled File

If the assembler program identifies *syntax errors* in the source file while it is being assembled, the locations of the errors are marked in the source listing file with an *error number* and *error message*. Moreover, the total number of errors is displayed on the screen at the end of the assembly process. Looking at the displayed information for our assembly example in Fig. 7.22(a), we find that no errors occurred.

Figure 7.23(a) shows the response on the display when syntax errors are found during the assembly process. The listing for the program, which is shown in Fig. 7.23(b), contains four syntax errors. Notice how the errors are marked in the source listing. For instance, in Fig. 7.23(b) we find error number A2105 following the instruction DCR CX. Looking at Fig. 7.23(a), we find that this error message stands for a syntax error. The error that was made is that the mnemonic of the instruction is spelled wrong. It should read DEC CX. The source program must first be edited to correct this error and the other three errors, and then it must be reassembled.

```
C:\MASM>MASM
Microsoft (R) Macro Assembler Version 5.10
Copyright (C) Microsoft Corp 1981, 1988.  All rights reserved.

Source filename [.ASM]: A:EBLOCK.SRC
Object filename [EBLOCK.OBJ]: A:
Source listing  [NUL.LST]: A:EBLOCK
Cross-reference [NUL.CRF]: A:EBLOCK
A:EBLOCK.SRC(13): error A2105: Expected: instruction, directive, or label
A:EBLOCK.SRC(34): error A2105: Expected: instruction, directive, or label
A:EBLOCK.SRC(46): error A2009: Symbol not defined: N
A:EBLOCK.SRC(54): error A2105: Expected: instruction, directive, or label

   47200 + 427484 Bytes symbol space free

       0 Warning Errors
       4 Severe  Errors
```

Figure 7.23(a) Displayed information for a source file with assembly syntax errors.

EXAMPLE 7.5

What is the meaning of the error code at line 49 of the program source listing in Fig. 7.23(b)?

Solution

Looking at Fig. 7.23(b) we get the error number as A2009, and in Fig. 7.23(a) we see that it means that a symbol was not defined. The undefined symbol is *N*.

Object Module

The most important output produced by the assembler is the object code file. The contents of this file are called the object module. The object module is a machine language version of the program. Even through the object module is the machine code for the source program, it cannot be directly run on the microcomputer. It must first be processed by the linker to create an executable run module.

Source Listing

Let us now look more closely at the source listing in Fig. 7.22(c). Notice that the first column assigns a number to every statement line used when the source file for the program was created with the editor. The second column is the starting offset address of the machine language instruction from the beginning of the current code segment. In the next column we find the bytes of machine code for the instructions. They are expressed in hexadecimal form. If an *R* is listed after a number, it means that an *external reference* exists and that the link operation may modify this value. The machine code instructions are followed by the original source-code instructions in the next column and the comments in the last column.

For instance, at line number 39, the machine code for the assembly language instruction MOV AX,DATASEGADDR is found. This instruction is encoded with 3 bytes and is

```
MOV  AX,DATASEGADDR = B81020H
```

```
    1
    2
    3                              TITLE BLOCK-MOVE PROGRAM
    4
    5                              PAGE      ,132
    6
    7                              COMMENT *This program moves a block of specified number of bytes
    8                                       from one place to another place*
    9
   10
   11                              ;Define constants used in this program
   12
   13                                      N              16          ;Bytes to be moved
eblock.src(13): error A2105: Expected: instruction, directive, or label
   14 = 0100                              BLK1ADDR=   100H            ;Source block offset address
   15 = 0120                              BLK2ADDR=   120H            ;Destination block offset addr
   16 = 2000                              DATASEGADDR=2000H           ;Data segment start address
   17
   18
   19 0000                        STACK_SEG    SEGMENT          STACK 'STACK'
   20 0000   0040[                             DB               64 DUP(?)
   21        ??
   22                    ]
   23
   24 0040                        STACK_SEG    ENDS
   25
   26
   27 0000                        CODE_SEG     SEGMENT          'CODE'
   28 0000                        BLOCK        PROC             FAR
   29                             ASSUME       CS:CODE_SEG,SS:STACK_SEG
   30
   31                              ;To return to DEBUG program put return address on the stack
   32
   33 0000   1E                            PUSH   DS
   34 0001   B8 0000                       MOV    AX, 0
   35 0004   50                            PUSH   AX
   36
   37                              Setup the data segment address
eblock.src(34): error A2105: Expected: instruction, directive, or label
   38
   39 0005   B8 2000                       MOV    AX, DATASEGADDR
   40 0008   8E D8                         MOV    DS, AX
   41
   42                              ;Setup the source and destination offset adresses
   43
   44 000A   BE 0100                       MOV    SI, BLK1ADDR
   45 000D   BF 0120                       MOV    DI, BLK2ADDR
   46
   47                              ;Setup the count of bytes to be moved
   48
   49 0010   8B 0E 0000 U                  MOV    CX, N
eblock.src(46): error A2009: Symbol not defined: N
   50
   51                              ;Copy source block to destination block
   52
   53 0014   8A 24            NXTPT:        MOV    AH, [SI]          ;Move a byte
   54 0016   88 25                         MOV    [DI], AH
```

Figure 7.23(b) Source listing for a file with syntax errors.

A symbol table is also produced as part of the source listing. It is a list of all the symbols defined in the program. The symbol table for our example program is shown at the bottom of the source listing in Fig. 7.22(c). Notice that the name of each symbol is listed along with its type (and length for data), value, and attribute. The types of symbols identified are *label, variable, number,* and *procedure.* For example, in Fig. 7.22(c) we find that the symbol BLK1ADDR is a number and its value is 0100_{16}. For this symbol, no

```
      55 0018  46                   INC   SI                 ;Update pointers
      56 0019  47                   INC   DI
      57                            DCR   CX                 ;Update byte counter
eblock.src(54): error A2105: Expected: instruction, directive, or label
      58 001A  75 F8                JNZ   NXTPT              ;Repeat for next byte
      59 001C  CB                   RET                      ;Return to DEBUG program
      60 001D                 BLOCK       ENDP
      61 001D                 CODE_SEG    ENDS
      62                      END         BLOCK              ;End of program
```
Microsoft (R) Macro Assembler Version 5.10   2/7/93 00:16:25

BLOCK-MOVE PROGRAM   Symbols-1

Segments and Groups:

                N a m e   Length   Align   Combine Class

CODE_SEG . . . .   001D PARA NONE 'CODE'
STACK_SEG . . . .   0040 PARA STACK 'STACK'

Symbols:

                N a m e   Type   Value   Attr

BLK1ADDR . . . .   NUMBER   0100
BLK2ADDR . . . .   NUMBER   0120
BLOCK . . . .   F PROC   0000   CODE_SEG   Length = 001D

DATASEGADDR . . . .   NUMBER   2000

NXTPT . . . .   L NEAR   0014   CODE_SEG

@CPU . . . .   TEXT   0101h
@FILENAME . . . .   TEXT   eblock
@VERSION . . . .   TEXT   510

     59 Source Lines
     59 Total Lines
     14 Symbols

  47200 + 427052 Bytes symbol space free

      0 Warning Errors
      4 Severe Errors

Figure 7.23(b)   (Continued)

attribute is indicated. On the other hand, for the symbol NXTPT, which is a near-label with value 00013_{16}, the attribute is CODE_SEG.

The source listing is a valuable aid in correcting errors in the program. For instance, earlier we found that syntax errors are marked in the source listing. Therefore, they can easily be found and corrected. Since both the source and corresponding machine code are provided in the source listing, it also serves as a valuable tool for identifying and correcting logical errors in the writing of the program.

EXAMPLE 7.6 _____

What is the cause of the error A2105 that is located between lines 37 and 38 of the source listing in Fig. 7.23(b)?

Solution

In Fig. 7.23(b), we find that this error is in the comment and that the cause is a missing semicolon (;) at the start of the statement.

Cross-Reference Table

The cross-reference table is also useful when debugging programs that contain logical errors. The cross-reference table is a table that tells the number of the line in the source program at which each symbol is defined and the number of each line in which it is referenced. The line number followed by the symbol # is the location at which the symbol is defined. For instance, in the table of Fig. 7.22(d), we find that the label NXTPT is defined in line 53 and referenced in line 58.

Another use of the cross-reference table is when a program is to be modified in such a way that the name of a symbol must be changed. The cross-reference table can be used to find all the locations where the symbol is used. In this way, the locations can easily be found and changed with the editor.

EXAMPLE 7.7

Use the cross-reference information in Fig. 7.22(d) to determine in which lines of the source listing in Fig. 7.22(c) the symbol BLK1ADDR is referenced.

Solution

The table in Fig. 7.22(d) shows that BLK1ADDR is referenced twice in the source program. These references are identified as lines 14 and 44 of the listing.

▲ 7.6 CREATING A RUN MODULE WITH THE LINK PROGRAM

In Section 7.5 we indicated that the object module produced by the assembler is not yet an executable file. That is, in its current state it cannot be loaded with the DEBUG program and run on the 80386DX microprocessor in the PC/AT. To convert an object module to an executable machine code file (run module), we must process it with the linker. The link operation for 80386DX object code is performed by the LINK program, which is part of the DOS.

Modular Programming

At this point we may ask the question, Why doesn't the assembler directly produce an executable run module? To answer this question, let us look into an important idea behind the use of the LINK program and that is *modular programming*. The program we have been using as an example in this chapter is quite simple. For this reason, it can be easily contained in a single source file. However, most practical application programs are very large. For example, a source program may contain 2000 assembly language statements.

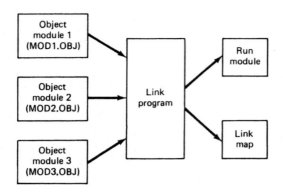

Figure 7.24 Linking object modules.

When assembled, this can result in as much as 4000 to 8000 bytes of machine code. For development purpose, programs of this size are frequently broken down into a number of parts called *modules,* and the individual modules are worked on by different programmers. Each module is written independently and then all modules are combined to form a single executable run module. This idea is illustrated in Fig. 7.24. Notice that the LINK program is the software tool that is used to combine the modules. Its inputs are the object code for modules 1, 2, and 3. Execution of the linker combines these programs into a single run module.

The technique of writing larger programs as a series of modules has several benefits. First, since several programmers are working on the project in parallel, the program can be completed in a much shorter period of time. Another benefit is that the smaller sizes of the modules require less time to edit and assemble. For instance, if we were not using modular programming, to make a change in just one statement the complete program would have to be edited, reassembled, and relinked. On the other hand, when using modular programming, just the module containing the statement that needs to be changed can be edited and reassembled. Then the new object module is relinked with the rest of the old object modules to give a new run module.

A third benefit derived from modular programming is that it makes it easier to reuse old software. For instance, a new software design may need some functions for which modules have already been written as part of an old design. If these modules were integrated into a single large program, we would need to edit them out carefully and transfer them to the new source file. However, if these segments of program exist as separate modules, we may need to do no additional editing to integrate them into the new application.

Initiating the Link Program

Even though our example program has just one object module, we must still perform the link operation on it to obtain a run module. This is because the actual physical addresses to be used by the program when loaded on a PC/AT need to be resolved. Let us now look at how this is done using the LINK program. First the MASM directory on the C drive must be entered and then the diskette that contains object file BLOCK.OBJ is put into drive A. Now we invoke the linker with the command

```
C:\MASM>LINK    (↵)
```

The linker program is loaded and starts to run. It begins by prompting for input of the object file name. This prompt is

```
Object Modules[.OBJ]:
```

In response to this prompt, we enter the names of all object modules that are to be linked together. The file names are listed, separated by the + sign. For example, to link the three object modules named MOD1.OBJ, MOD2.OBJ, and MOD3.OBJ the input would be

```
Object Modules[.OBJ]:A:MOD1.OBJ+A:MOD2.OBJ+A:MOD3.OBJ   (↵)
```

For our earlier example, we just have one object file, which is called BLOCK.OBJ. Therefore, the input is

```
Object Modules[.OBJ]:A:BLOCK   (↵)
```

Here, entry of the extension OBJ is optional. That is, we could have entered just A:BLOCK followed by (↵).

Next, the linker program asks for input of the name of the file into which it is to save the run module. It does this with the prompt

```
Run file [BLOCK.EXE]:
```

Notice that it automatically creates a default file name by appending the .EXE extension to the first file name entered as part of the object module prompt. For this prompt, the file name can be changed, but the .EXE extension should be kept for compatibility with the DOS and DEBUG program. Let us use the default file name but assign it to drive A instead of C. To do this, the input entry is

```
Run File[BLOCK.EXE]:A:   (↵)
```

At the completion of this entry, a third prompt is displayed by the program. It is

```
List File[NUL.MAP]:
```

Selection of the default name NUL.MAP causes the linker to suppress formation of a link map file. For our example, we will request one to be generated and assign it the file name BLOCK.MAP. This is done by issuing the command

```
List File[NUL.MAP]:A:BLOCK.MAP   (↵)
```

The last prompt is

```
Libraries[.LIB]:
```

It asks us whether or not we want to include subroutines from a *library* in the link process. If so, just enter the name of the file that contains the libraries and depress (↵). Libraries are more widely used with high-level languages. For our example, we will not require use of a library; therefore, (↵) is simply depressed.

Libraries[.LIB]: (↵)

This completes the start-up of the linker.

After the response to the library prompt, the linker program reads the object files, combines them, and resolves all external references. Moreover, it produces the information for the link map. Finally, it saves the run module on the diskette in drive A in file BLOCK.EXE and the link map information in BLOCK.MAP. The sequence of entries made to invoke the linker for our example is shown in Fig. 7.25(a), and the map file BLOCK.MAP is shown in Fig. 7.25(b).

The link map shows the start address, stop address, and length for each memory segment employed by the program that was linked. Looking at the link map for our program, which is shown in Fig. 7.25(b), we find that the code segment starts at address 00040_{16} and ends at address $0005C_{16}$. The program is $1D_{16}$, or 30 bytes, long.

EXAMPLE 7.8

From the link map in Fig. 7.22(b), find the range of addresses used for the stack segment. How many bytes are in use?

Solution

The link map indicates that addresses from 00000_{16} to $0003F_{16}$ are used for storage of stack data. From the length we find that this represents 40H bytes.

```
C:\MASM>LINK

Microsoft (R) Segmented-Executable Linker  Version 5.13
Copyright (C) Microsoft Corp 1984-1991.  All rights reserved.

Object Modules [.OBJ]: A:BLOCK
Run File [C:BLOCK.exe]: A:
List File [NUL.MAP]: A:BLOCK.MAP
Libraries [.LIB]:
Definitions File [NUL.DEF]:
```

(a)

```
Start  Stop   Length Name             Class
00000H 0003FH 00040H STACK_SEG        STACK
00040H 0005CH 0001DH CODE_SEG         CODE

Program entry point at 0004:0000
```

(b)

Figure 7.25 (a) Display sequence for initiating linking of object files. (b Link map file.

▲ 7.7 LOADING AND EXECUTING A RUN MODULE

In Chapter 4 we learned how to bring up the DEBUG program; how to use its commands; and how to load, execute, and debug the operation of a program. At that time, we loaded the machine code for the program and data with memory-modify commands. Up to this point in this chapter, we have learned how to form a source program using assembly language and pseudo-op statements, how to assemble the program into an object module, and how to use the linker to produce a run module. Here we will load and execute the run module BLOCK.EXE that was produced in Section 7.6 for the source program BLOCK.SRC.

When the DEBUG program was loaded in Chapter 4, we did not have a run module available. For this reason, we just brought up the debugger by typing in DEBUG and depressing (↵). Now that we do have a run module, the debugger will be brought up in a different way and the run module will be loaded at the same time. This is done by issuing the command

```
C:\DOS>DEBUG A:BLOCK.EXE   (↵)
```

In response to this command, both the DEBUG program and the run module BLOCK.EXE are loaded into the PC/AT's memory. As shown in Fig. 7.26, after loading, the debug prompt - is displayed. Next the register status is dumped with an R command. Notice that DS is initialized with the value $11C6_{16}$.

Let us now verify that the program has loaded correctly by using the command

```
U  CS:000  01C   (↵)
```

Comparing the program displayed in Fig. 7.26 as a result of the command to the source program in Fig. 7.16, we see that they are essentially the same. Therefore, the program has been loaded correctly.

Now we will execute the first eight instructions of the program and verify the operation they perform. To do this, we issue the command

```
G  =CS:000  013   (↵)
```

The information displayed at the completion of this command is also shown in Fig. 7.26. Here we find that the registers have been initialized as follows: DS contains 1020_{16}, AX contains 1020_{16}, SI contains 0100_{16}, DI contains 0120_{16}, and CX contains 0010_{16}.

The FILL command is used to initialize the bytes of data in the source and destination blocks. The storage locations in the source block are loaded with FF_{16} with the command

```
F  DS:100  10F  FF   (↵)
```

and the storage locations in the destination block are loaded with 00_{16} with the command

```
F  DS:120  12F  00   (↵)
```

Next, we execute down through the program to the JNZ instruction, address $01A_{16}$:

```
G  =CS:013  01A   (↵)
```

```
C:\DOS>DEBUG A:BLOCK.EXE
-R
AX=0000  BX=0000  CX=005D  DX=0000  SP=0040  BP=0000  SI=0000  DI=0000
DS=11C5  ES=11C5  SS=11D5  CS=11D9  IP=0000   NV UP EI PL NZ NA PO NC
11D9:0000 1E          PUSH    DS
-U CS:0 1C
11D9:0000 1E          PUSH    DS
11D9:0001 B80000      MOV     AX,0000
11D9:0004 50          PUSH    AX
11D9:0005 B80020      MOV     AX,2000
11D9:0008 8ED8        MOV     DS,AX
11D9:000A BE0001      MOV     SI,0100
11D9:000D BF2001      MOV     DI,0120
11D9:0010 B91000      MOV     CX,0010
11D9:0013 8A24        MOV     AH,[SI]
11D9:0015 8825        MOV     [DI],AH
11D9:0017 46          INC     SI
11D9:0018 47          INC     DI
11D9:0019 49          DEC     CX
11D9:001A 75F7        JNZ     0013
11D9:001C CB          RETF
-G =CS:0 13

AX=2000  BX=0000  CX=0010  DX=0000  SP=003C  BP=0000  SI=0100  DI=0120
DS=2000  ES=11C5  SS=11D5  CS=11D9  IP=0013   NV UP EI PL NZ NA PO NC
11D9:0013 8A24        MOV     AH,[SI]                            DS:0100=50
-F DS:100 10F FF
-F DS:120 12F 00
= -G =CS:13 1A


AX=FF20  BX=0000  CX=000F  DX=0000  SP=003C  BP=0000  SI=0101  DI=0121
DS=2000  ES=11C5  SS=11D5  CS=11D9  IP=001A   NV UP EI PL NZ AC PE NC
11D9:001A 75F7        JNZ     0013
-D DS:100 10F
2000:0100  FF FF FF FF FF FF FF FF-FF FF FF FF FF FF FF FF    ................
-D DS:120 12F
2000:0120  FF 00 00 00 00 00 00 00-00 00 00 00 00 00 00 00    ................
-G

Program terminated normally
-D DS:100 10F
2000:0100  FF FF FF FF FF FF FF FF-FF FF FF FF FF FF FF FF    ................
-D DS:120 12F
2000:0120  FF FF FF FF FF FF FF FF-FF FF FF FF FF FF FF FF    ................
-Q

C:\DOS>
```

Figure 7.26 Loading and executing the run module BLOCK.EXE.

To check the state of the data blocks, we use the commands

$$D \quad DS:100 \quad 10F \quad (\lrcorner)$$
$$D \quad DS:120 \quad 12F \quad (\lrcorner)$$

Looking at the displayed blocks of data in Fig. 7.26, we see that the source block is unchanged and that FF_{16} has been copied into the first element of the destination block.

Finally, the program is run to completion with the command

$$G \quad (\lrcorner)$$

By once more looking at the two blocks of data with the commands

$$D \quad DS:100 \quad 10F \quad (\lrcorner)$$
$$D \quad DS:120 \quad 12F \quad (\lrcorner)$$

we find that the contents of the source block have been copied into the destination block.

322 Assembly Language Program Development Chap. 7

DEBUG is used to run a program when we must either debug the operation of the program or want to understand its execution step by step. If we just want to run a program instead of observing its operation, another method can be used. The run module (.EXE file) can be executed at the DOS prompt by simply entering its name followed by (↵). This will cause the program to be loaded and then executed to completion. For instance, to execute the run module BLOCK.EXE that resides on a diskette in drive A, we enter

```
C:\DOS>A:BLOCK    (↵)
```

ASSIGNMENTS

Section 7.2

1. What are the two types of statements in a source program?
2. What is the function of an assembly language instruction?
3. What is the function of a pseudo-operation?
4. What are the four elements of an assembly language statement?
5. What part of the instruction format is always required?
6. What are the two limitations on format when writing source statements for the MASM?
7. What is the function of a label?
8. What is the maximum number of characters of a label that will be recognized by the MASM?
9. What is the function of the opcode?
10. What is the function of operands?
11. In the instruction statement

```
SUB_A:   MOV   CL,0FFH
```

what are the source operand and destination operand?
12. What is the purpose of the comment field? How are comments processed by an assembler?
13. Give two differences between an assembly language statement and a pseudo-operation statement.
14. Write the instruction MOV AX,[32728D] with the source operand expressed both in binary and hexadecimal forms.
15. Rewrite the jump instruction JMP +25D with the operand expressed both in binary and hexadecimal forms.
16. Repeat Example 7.4 with the values A = 345, B = 234, and C = 111.

Section 7.3

17. Give another name for a pseudo-op.
18. List the names of the pseudo-op categories.

19. What is the function of the data pseudo-ops?

20. What happens when the statements

```
SRC_BLOCK = 0100H
DEST_BLOCK = SRC_BLOCK + 20H
```

are processed by the MASM assembler?

21. Describe the difference between the EQU and = pseudo-ops.

22. What does the statement

```
SEG_ADDR   DW   1234H
```

do when processed by the assembler?

23. What happens when the statement

```
BLOCK_1   DB   128 DUP(?)
```

is processed by the assembler?

24. Write a data pseudo-op statement to define INIT_COUNT as word size and assign it the value $F000_{16}$.

25. Write a data pseudo-op statement to allocate a block of 16 words in memory called SOURCE_BLOCK, but do not initialize them with data.

26. Write a pseudo-op statement that will initialize the block of memory-storage locations allocated in Problem 25 with the data values 0000H, 1000H, 2000H, 3000H, 4000H, 5000H, 6000H, 7000H, 8000H, 9000H, A000H, B000H, C000H, D000H, E000H, and F000H.

27. What does the statement

```
DATA_SEG   SEGMENT   BYTE   MEMORY   'DATA'
```

mean?

28. Show how the segment-control pseudo-ops are used to define a segment called DATA_SEG that is aligned on a word address boundary, overlaps other segments with the same name, and is a data segment.

29. What is the name of the smaller segments in which modular programming techniques specify that programs should be developed?

30. What is a procedure?

31. Show the general structure of a far procedure called BLOCK that is to be accessible from other modules.

32. What is the function of an ORG pseudo-op?

33. Write an origin statement that causes machine code to be loaded at offset 1000H of the current code segment.

34. Write a page statement that will set up the printout for 55 lines per page and 80 characters per line and a title statement that will title pages of the source listing with BLOCK-MOVE PROGRAM.

Section 7.4

35. What type of editor is EDLIN? EDIT?

36. Write an EDLIN command that will initiate editing of line 20.

37. What operation would the EDLIN command 15,17D perform?

38. Describe the operation performed by the EDLIN command 10,12,15C.

39. List the basic editing operations that can be performed using EDIT.

40. What are the names of the four menus in the EDIT menu bar?

41. How could a backup copy of the program in file BLOCK.ASM be created from the file menu?

Section 7.5

42. What is the input to the assembler program?

43. What are the outputs of the assembler? Give a brief description of each.

44. What name does the command entry

```
C:\MASM>MASM A:BLOCK, ,BLOCK.LST,BLOCK.CRF
```

assign to each of the input and output files of the assembler?

45. What is the cause of the first error statement in the source listing of Fig. 7.23(b)?

46. Use the cross-reference table for the program in Fig. 7.22(d) to find in which line a value is assigned to N and in which lines it is referenced.

Section 7.6

47. Can the output of the assembler be directly executed by the 80386DX microprocessor in the PC/AT?

48. Give three benefits of modular programming.

49. What is the input to the LINK program?

50. What are the outputs of the LINK program? Give a brief description of each.

51. If three object modules called MAIN.OBJ, SUB1.OBJ, and SUB2.OBJ are to be combined with LINK, write the response that must be made to the linker's Object Modules[.OBJ]: prompt. Assume that all three files are on a diskette in drive A.

Section 7.7

52. Write a DOS command that will load run module LAB.EXE while bringing up the DEBUG program. Assume that the run module file is on a diskette in drive A.

Application Problems

53. Upgrade the programs written for the assignment problems 13, 18, 24, 30, 38, 42, 46, and 47 in Chapter 5 so that they can be assembled using the MASM and debugged using DEBUG.

54. Upgrade the programs written for the assignment problems 5, 27, 28, 29, 30, 31, 41, 43, 44, 49, 50, 54, and 55 in Chapter 6 so that they can be assembled using the MASM and debugged using DEBUG.

Protected-Mode Software Architecture of the 80386DX

▲ 8.1 INTRODUCTION

Having completed our study of the real-mode operation, instruction set, and assembly language programming of the 80386DX microprocessor, we are now ready to turn our attention to its protected-address mode (protected mode) of operation. Earlier we indicated that whenever the 80386DX microprocessor is reset, it comes up in real mode. Moreover, we identified that the PE bit of control register zero (CR_0) can be used to switch the 80386DX into the protected mode under software control. When configured for protected-mode operation, the 80386DX provides an advanced software architecture that supports memory management, virtual addressing, paging, protection, and multitasking. This is the mode of operation used by operating systems such as Windows95™, OS/2™, and UNIX™. In this chapter we will examine the 80386DX's protected-mode software architecture and advanced system concepts. The topics covered are as follows:

1. Protected-mode register model
2. Protected-mode memory management and address translation
3. Descriptor and page table entries
4. Protected-mode system-control instruction set
5. Multitasking and protection
6. Virtual 8086 mode

▲ 8.2 PROTECTED-MODE REGISTER MODEL

We will begin our study of the 80386DX's protected-mode software architecture with its register model. The protected-mode register set of the 80386DX microprocessor is shown

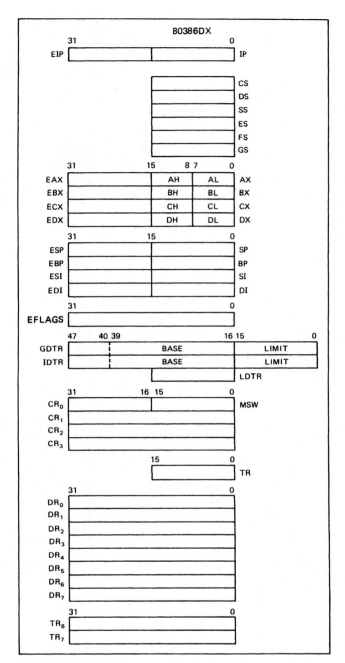

Figure 8.1 Protected-mode register model.

in Fig. 8.1. Looking at this diagram, we see that its application register model is a superset of the real-mode register set shown in Fig. 2.2. Comparing these two diagrams, we find four new registers in the protected-mode model: the *global descriptor table register* (GDTR), *interrupt descriptor table register* (IDTR), *local descriptor table register* (LDTR), and *task register* (TR). Furthermore, the functions of a few registers have been extended. For example, the instruction pointer, which is now called EIP, is 32 bits in length; more bits of the flag

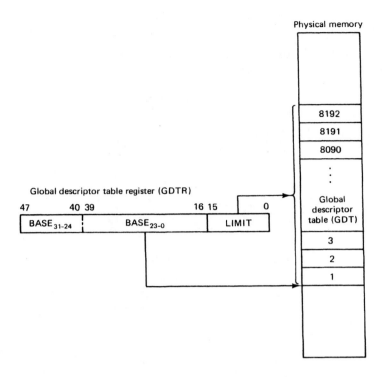

Figure 8.2 Global-descriptor table mechanism.

register (EFLAGS) are active; and all four control registers, CR_0 through CR_3, are functional. Let us next discuss the purpose of each new and extended register and how they are used in the segmented memory protected-mode operation of the microprocessor.

Global Descriptor Table Register

As shown in Fig. 8.2, the contents of the global descriptor table register defines a table in the 80386DX's physical memory address space called the *global descriptor table* (GDT). This global descriptor table is one important element of the 80386DX's memory management system.

GDTR is a 48-bit register that is located inside the 80386DX. The lower 2 bytes of this register, which are identified as LIMIT in Fig. 8.2, specify the size in bytes of the GDT. The value of LIMIT is 1 less than the actual size of the table. For instance, if LIMIT equals $00FF_{16}$, the table is 256 bytes in length. Since LIMIT has 16 bits, the GDT can be up to 65,536 bytes long. The upper 4 bytes of the GDTR, which are labeled BASE in Fig. 8.2, locate the beginning of the GDT in physical memory. This 32-bit base address allows the table to be positioned anywhere in the 80386DX's linear address space.

EXAMPLE 8.1 _____

If the limit and base in the global descriptor table register are $0FFF_{16}$ and 00100000_{16}, respectively, what is the beginning address of the descriptor table, size of the table in bytes, and the ending address of the table?

Solution

The starting address of the global descriptor table in physical memory is given by the base. Therefore,

$$GDT_{START} = 00100000_{16}$$

The limit is the offset to the end of the table. This gives

$$GDT_{END} = 00100000_{16} + 0FFF_{16} = 00100FFF_{16}$$

Finally, the size of the table is equal to the decimal value of the LIMIT plus 1.

$$GDT_{SIZE} = FFF_{16} + 1_2 = 4096 \text{ bytes}$$

The GDT provides a mechanism for defining the characteristics of the 80386DX's *global memory* address space. Global memory is a general system resource that is shared by many or all software tasks. That is, storage locations in global memory are accessible by any task that runs on the microprocessor.

This table contains what are called *system segment descriptors*. These descriptors identify the characteristics of the segments of global memory. For instance, a segment descriptor provides information about the size, starting point, and access rights of a global memory segment. Each descriptor is 8 bytes long; thus our earlier example of a 256-byte table provides storage space for just 32 descriptors. Remember that the size of the global descriptor table can be expanded by simply changing the value of LIMIT in the GDTR under software control. If the table is increased to its maximum size of 65,536 bytes, it can hold up to 8192 descriptors.

EXAMPLE 8.2

How many descriptors can be stored in the global descriptor table defined in Example 8.1?

Solution

Each descriptor takes up 8 bytes; therefore, a 4096-byte table can hold

$$4096/8 = 512 \text{ descriptors}$$

The value of the BASE and LIMIT must be loaded into the GDTR before the 80386DX is switched from the real mode of operation to the protected mode. Special instructions are provided for this purpose in the system control instruction set of the 80386DX. These instructions will be introduced later in this chapter. Once the 80386DX is in protected mode, the location of the table is typically not changed.

Interrupt Descriptor Table Register

Just like the global descriptor table register, the interrupt descriptor table register (IDTR) defines a table in physical memory. However, this table contains what are called

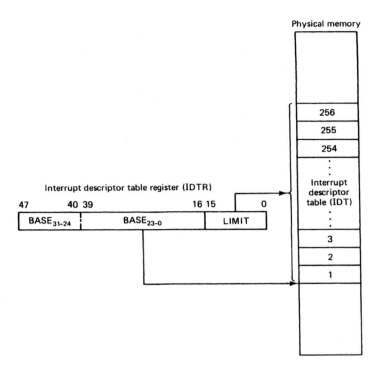

Figure 8.3 Interrupt-descriptor table mechanism.

interrupt descriptors, not segment descriptors. For this reason, it is known as the *interrupt descriptor table* (IDT). This register and table of descriptors provide the mechanism by which the microprocessor passes program control to interrupt and exception service routines.

As shown in Fig. 8.3, just like the GDTR, the IDTR is 48 bits in length. Again, the lower two bytes of the register (LIMIT) define the table size. That is, the size of the table equals LIMIT+1 bytes. Since 2 bytes define the size, the IDT can also be up to 65,536 bytes long. But the 80386DX only supports up to 256 interrupts and exceptions; therefore, the size of the IDT should not be set to support more than 256 interrupts. The upper 4 bytes of IDTR (BASE) identify the starting address of the IDT in physical memory.

The type of descriptor used in the IDT are called *interrupt gates*. These gates provide a means for passing program control to the beginning of an interrupt service routine. Each gate is 8 bytes long and contains both attributes and a starting address for the service routine.

EXAMPLE 8.3

What is the maximum value that should be assigned to LIMIT in the IDTR?

Solution

The maximum number of interrupt descriptors that can be used in an 80386DX microcomputer system is 256. Therefore, the maximum table size in bytes is

$$IDT_{SIZE} = 8 \times 256 = 4096 \text{ bytes}$$

Thus,

$$\text{LIMIT} = \text{0FFF}_{16}$$

This table can also be located anywhere in the linear address space addressable with the 80386DX's 32-bit address. Just like the GDTR, the IDTR needs to be loaded before the 80386DX enters protected mode. Special instructions are provided for loading and saving the contents of the IDTR. Once the location of the table is set, it is typically not changed after entering protected mode.

EXAMPLE 8.4

What is the address range of the last descriptor in the interrupt descriptor table defined by base address 00011000_{16} and limit $01FF_{16}$?

Solution

From the values of the base and limit, we find that the table is located in the address range defined by

$$\text{IDT}_{\text{START}} = 00011000_{16}$$

and

$$\text{IDT}_{\text{END}} = 000111FF_{16}$$

The last descriptor in this table takes up the 8 bytes of memory from address $000111F8_{16}$ through $000111FF_{16}$.

Local Descriptor Table Register

The *local descriptor table register* (LDTR) is also part of the 80386DX's memory-management support mechanism. As shown in Fig. 8.4(a), each task can have access to its own private descriptor table in addition to the global descriptor table. This private table is called the *local descriptor table* (LDT) and defines a *local memory* address space for use by the task. The LDT holds segment descriptors that provide access to code and data in segments of memory that are reserved for the current task. Since each task can have its own segment of local memory, the protected-mode software system may contain many local descriptor tables. For this reason we have identified LDT_0 through LDT_N in Fig. 8.4(a).

Figure 8.4(b) shows us that the contents of the 16-bit LDTR does not directly define the local descriptor table. Instead, it holds a selector that points to an *LDT descriptor* in the GDT. Whenever a selector is loaded into the LDTR, the corresponding descriptor is transparently read from global memory and loaded into the *local descriptor table cache* within the 80386DX. It is this descriptor that defines the local descriptor table. As shown in Fig. 8.4(b), the 32-bit BASE value identifies the starting point of the table in physical memory, and the value of the 16-bit LIMIT determines the size of the table. Loading of this descriptor into the cache creates the LDT for the current task. That is, every time a

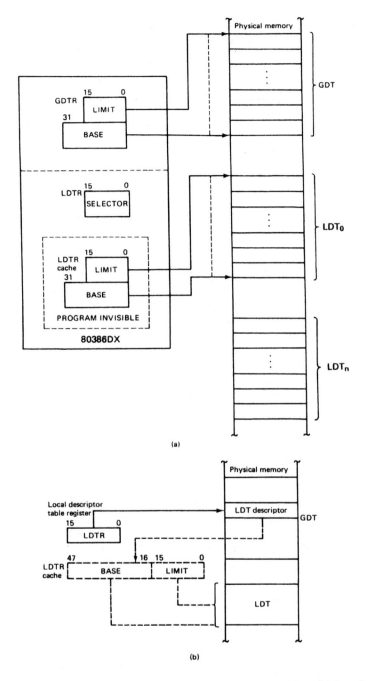

(a)

(b)

Figure 8.4 (a) Task with global and local descriptor tables. (b) Loading the local descriptor table register to define a local descriptor table.

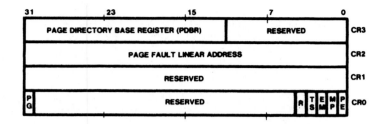

Figure 8.5 Control registers. (Reprinted by permission of Intel Corp. Copyright/Intel Corp. 1986)

selector is loaded into the LDTR, a local descriptor-table descriptor is cached and a new LDT is activated.

Control Registers

The protected-mode model includes the four system-control registers, identified as CR_0 through CR_3 in Fig. 8.1. Figure 8.5 shows these registers in more detail. Notice that the lower 5 bits of CR_0 are system-control flags. These bits make up what is known as the *machine status word* (MSW). The most significant bit of CR_0 and registers CR_2 and CR_3 are used by the 80386DX's paging mechanism.

Let us continue by examining the machine status word bits of CR_0. They contain information about the 80386DX's protected-mode configuration and status. The 4 bits labeled PE, MP, EM, and R are control bits that define the protected-mode system configuration. The fifth bit, TS, is a status bit. These bits can be examined or modified through software.

The *protected-mode enable* (PE) bit determines if the 80386DX is in the real or protected mode. At reset, PE is cleared. This enables real-mode operation. To enter protected mode, we simply switch PE to 1 through software. Once in protected mode, the 80386DX cannot be switched back to real mode under software control by clearing the PE bit. The only way to return to real mode is by initiating a hardware reset.

The *math present* (MP) bit is set to 1 to indicate that a numeric coprocessor is present in the microcomputer system. On the other hand, if the system is to be configured so that a software emulator is used to perform numeric operations, the *emulate* (EM) bit is set to 1. Only one of these two bits can be set at a time. Finally, the *extension type* (R) bit, is used to indicate whether an 80287 or 80387 numeric coprocessor is in use. Logic 1 in R indicates that an 80387 is installed. The last bit in the MSW, *task switched* (TS), automatically gets set whenever the 80386DX switches from one task to another. It can be cleared under software control.

The protected-mode software architecture of the 80386DX also supports paged memory operation. Paging is turned on by switching the PG bit in CR_0 to logic 1. Now addressing of physical memory is implemented with an address translation mechanism that consists of a page directory and page table that are both held in physical memory. Looking at Fig. 8.5, we see that CR_3 contains the *page directory base register* (PDBR). This register holds a 20-bit *page directory base address* that points to the beginning of the page directory. A page-fault error occurs during the page-translation process if the page is not present in memory. In this case, the 80386DX saves the address at which the page fault occurred in register CR_2. This address is denoted as *page-fault linear address* in Fig. 8.5.

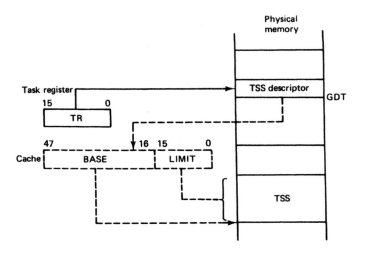

Figure 8.6 Task register and the task-switching mechanism.

Task Register

The *task register* (TR) is a key element in the protected-mode task switching mechanism of the 80386DX microprocessor. This register holds a 16-bit index value called a *selector*. The initial selector must be loaded into TR under software control. This starts the initial task. After this is done, the selector is changed automatically whenever the 80386DX executes an instruction that performs a task switch.

As shown in Fig. 8.6, the selector in TR is used to locate a descriptor in the global descriptor table. Notice that when a selector is loaded into the TR, the corresponding *task state segment (TSS) descriptor* automatically gets read from memory and loaded into the on-chip *task descriptor cache*. This descriptor defines a block of memory called the *task state segment* (TSS). It does this by providing the starting address (BASE) and the size (LIMIT) of the segment. Every task has its own TSS. The TSS holds the information needed to initiate the task, such as initial values for the user-accessible registers.

EXAMPLE 8.5

What is the maximum size of a TSS? Where can it be located in the linear address space?

Solution

Since the value of LIMIT is 16 bits in length, the TSS can be as long as 64K bytes. Moreover, the base is 32 bits in length. Therefore, the TSS can be located anywhere in the 80386DX's 4G-byte address space.

EXAMPLE 8.6

Assume that the base address of the global descriptor table is 00011000_{16} and the selector in the task register is 2108_{16}. What is the address range of the TSS descriptor?

Solution

The beginning address of the TSS descriptor is

$$\text{TSS\_DESCRIPTOR}_{\text{START}} = 00011000_{16} + 2108_{16}$$
$$= 00013108_{16}$$

Since the descriptor is 8 bytes long, it ends at

$$\text{TSS\_DESCRIPTOR}_{\text{END}} = 0001310F_{16}$$

Registers with Changed Functionality

Earlier we pointed out that the function of a few of the registers that are common to both the real- and protected-mode register models changes as the 80386DX is switched into the protected mode of operation. For instance, the segment registers are now called the *segment selector registers,* and instead of holding a base address they are loaded with what is known as a *selector.* The selector does not directly specify a storage location in memory. Instead, it selects a descriptor that defines the size and characteristics of a segment of memory.

The format of a selector is shown in Fig. 8.7. Here we see that the two least significant bits are labeled RPL, which stands for *requested privilege level.* These bits contain either $00 = 0$, $01 = 1$, $10 = 2$, or $11 = 3$ and assign a request protection level to the selector. The next bit, which is identified as *task indicator* (TI) in Fig. 8.7, selects the table to be used when accessing a segment descriptor. Remember that in protected mode two descriptor tables are active at a time, the global descriptor table and a local descriptor table. Looking at Fig. 8.7, we find that if TI is 0, the selector corresponds to a descriptor in the global descriptor table. Finally, the 13 most significant bits contain an *index* that is used as a pointer to a specific descriptor entry in the table selected by the TI bit.

EXAMPLE 8.7

Assume that the base address of the LDT is 00120000_{16} and the GDT base address is 00100000_{16}. If the value of the selector loaded into the CS register is 1007_{16}, what is the

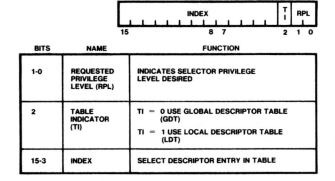

Figure 8.7 Selector format. (Reprinted by permission of Intel Corp. Copyright/Intel Corp. 1986)

request privilege level? Is the segment descriptor in the GDT or LDT? What is the address of the segment descriptor?

Solution

Expressing the selector in binary form, we get

$$(CS) = 0001000000000111_2$$

Since the two least significant bits are both 1,

$$RPL = 3$$

The next bit, bit 2, is also 1. This means that the segment descriptor is in the LDT. Finally, the value in the 13 most significant bits must be scaled by 8 to give the offset of the descriptor from the base address of the table. Therefore,

$$OFFSET = 0001000000000_2 \times 8 = 512 \times 8 = 4096$$
$$= 1000_{16}$$

and the address of the segment descriptor is

$$DESCRIPTOR_{ADDRESS} = 00120000_{16} + 1000_{16}$$
$$= 00121000_{16}$$

Another register whose function changes when the 80386DX is switched to protected mode is the flag register. As shown in Fig. 8.1, the flag register is now identified as EFLAGS and expands to 32 bits in length. The functions of the bits in EFLAGS are given in Fig. 8.8. Comparing this illustration to the real-mode 80386DX flag register in Fig. 2.16, we see

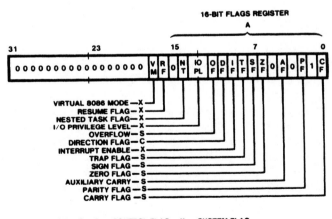

Figure 8.8 Protected-mode flag register. (Reprinted by permission of Intel Corp. Copyright/Intel Corp. 1986)

that five additional bits are implemented. These bits are active only when the 80386DX is in protected mode. They are the two-bit *input/output privilege level* (IOPL) code, the *nested task* (NT) flag, the *resume* (RF) flag, and the *virtual 8086 mode* (VM) flag.

Notice in Fig. 8.8 that each of these flags is identified as a system flag. That is, they represent protected-mode system operations. For example, the IOPL bits are used to assign a maximum privilege level to input/output. For instance, if 00 is loaded into IOPL, I/O can be performed only when the 80386DX is in the highest privilege level, which is called *level 0*. On the other hand, if IOPL is 11, I/O is assigned to the least privileged level, *level 3*.

The NT flag identifies whether or not the current task is a nested task, that is, if it was called from another task. This bit is automatically set whenever a nested task is initiated and can only be reset through software.

▲ 8.3 PROTECTED-MODE MEMORY MANAGEMENT AND ADDRESS TRANSLATION

Up to this point in the chapter, we have introduced the register set of the protected-mode software model for the 80386DX microprocessor. However, the software model of a microprocessor also includes its memory structure. Because of the memory-management capability of the 80386DX, the organization of protected-mode memory appears quite complex. Here we will examine how the *memory-management unit* (MMU) of the 80386DX implements the address space and how it translates virtual (logical) addresses to physical addresses. We begin here with what is called the *segmented* and *paged models* of memory.

The Virtual Address and Virtual Address Space

The protected-mode memory-management unit employs memory pointers that are 48 bits in length and consists of two parts, the *selector* and the *offset*. This 48-bit memory pointer is called a *virtual address* and is used by the program to specify the memory locations of instructions or data. As shown in Fig. 8.9, the selector is 16 bits in length and the offset is 32 bits long. Earlier we pointed out that one source of selectors is the segment selector registers within the 80386DX. For instance, if code is being accessed in memory, the active segment selector will be the one held in CS. This part of the pointer selects a unique segment of the 80386DX's *virtual address space*.

The offset is held in one of the 80386DX's other user-accessible registers. For our example of a code access, the offset would be in the EIP register. This part of the pointer is the displacement of the memory location that is to be accessed within the selected segment of memory. In our example it points to the first byte of the double word of instruction code that is to be fetched for execution. Since the offset is 32 bits in length, segment size can be as large as 4G bytes. We say as large as 4G bytes because segment size is actually variable and can be defined to be as small as 1 byte to as large as 4G bytes.

Figure 8.10 shows that the 16-bit selector breaks down into a 13-*bit index*, *table-select bit*, and two bits used for a *request privilege level*. The two RPL bits are not used in the selection of the memory segment. That is, just 14 of its 16 bits are employed in addressing

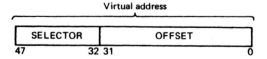

Figure 8.9 Protected-mode memory pointer.

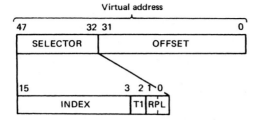

Figure 8.10 Segment-selector format.

memory. Therefore, the virtual address space can consist of 2^{14} (16,384 = 16K) unique segments of memory, each of which has a maximum size of 4G bytes. These segments are the basic elements into which the memory management unit of the 80386DX organizes the virtual address space.

Another way of looking at the size of the virtual address space is that by combining the 14-bit segment selector with the 32-bit offset, we get a 46-bit virtual address. Therefore, the 80386DX's virtual address space can contain 2^{46} equals 64T bytes (64 terabytes).

Segmented Partitioning of the Virtual Address Space

The memory-management unit of the 80386DX implements both a segmented model and paged model of virtual memory. In the segmented model the 80386DX's 64T-byte virtual address space is partitioned into a 32T-byte *global memory address space* and a 32T-byte *local memory address space*. This partitioning is illustrated in general in Fig. 8.11. The T1 bit of the selector shown in Fig. 8.10 is used to select between the global or local descriptor tables that define the virtual address space. Within each of these address spaces, as many as 8192 segments of memory may exist. This assumes that every descriptor in both

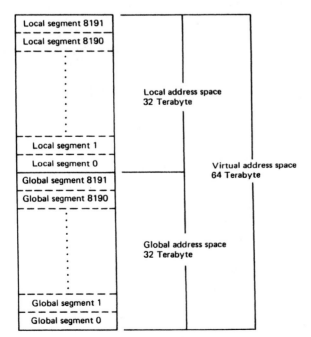

Figure 8.11 Partitioning the virtual address space.

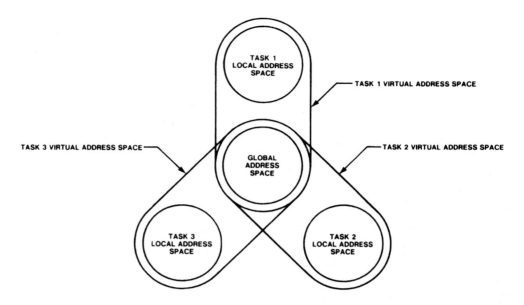

Figure 8.12   Global and local memory for a task. (Reprinted by permission of Intel Corp. Copyright/Intel Corp. 1987)

the global descriptor table and local descriptor table is in use and set for maximum size. These descriptors define the attributes of the corresponding segment. However, in practical system applications not all of the descriptors are normally in use. Let us now look briefly at how global and local segments of memory are used by software.

In the multiprocessing software environment of the 80386DX, an application is expressed as a collection of tasks. By *task* we mean a group of program routines that together perform a specific function. When the 80386DX initiates a task, it can activate both global and local segments of memory. This idea is illustrated in Fig. 8.12. Notice that tasks 1, 2, and 3 each have a reserved segment of the local address space. This part of memory stores data or code that can only be accessed by the corresponding task. That is, task 2 cannot access any of the information in the local address space of task 1. On the other hand, all the tasks are shown to share the same segment of the global address space. This segment typically contains operating system resources and data that are to be shared by all or many tasks.

Physical Address Space and Virtual-to-Physical Address Translation

We have just found that the virtual address space available to the programmer is 64T bytes in length. However, the 32-bit protected-mode address bus of the 80386DX supports just a 4G-byte *physical address space*. Just a small amount of the information in virtual memory can reside in physical memory at a time. For this reason systems that employ a virtual address space that is larger than the implemented physical memory are equipped with a secondary storage device such as a hard disk. The segments that are not currently in use are stored on disk.

If a segment of memory that is not present in physical memory is accessed by a program and space is available in physical memory, the segment is simply read from the

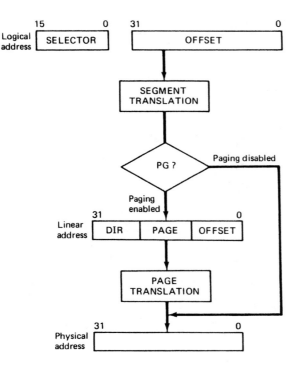

Figure 8.13 Virtual to physical address translation. (Reprinted by permission of Intel Corp. Copyright/Intel Corp. 1986)

hard disk and copied in physical memory. On the other hand, if the physical memory address space is full, another segment must first be sent out to the hard disk to make room for the new information. The memory-manager part of the operating system controls the allocation and deallocation of physical memory and the swapping of data between the hard disk and physical memory of the computer. In this way, the memory address space of the computer appears much larger than the physical memory in the computer.

The segmentation and paging memory management units of the 80386DX provide the mechanism by which 48-bit virtual addresses are mapped into the 32-bit physical addresses needed by hardwar They employ a memory-based lookup table address-translation process. This address translation is illustrated in general by the diagram in Fig. 8.13. Notice that first a *segment translation* is performed on the virtual (logical) address. Then, if paging is disabled, the *linear address* produced is equal to the physical address. However, if paging is enabled, the linear address goes through a second translation process, known as *page translation*, to produce the physical address.

As part of the translation process, the MMU determines whether or not the corresponding segment or page of the virtual address space currently exists in physical memory. If the segment or page already resides in memory, the operation is performed on the information. However, if the segment or page is not present, it signals this condition as an error. Once this condition is identified, the memory-manager software initiates loading of the segment or page from the external storage device to physical memory. This operation is called a *swap*. That is, an old segment or page gets swapped out to disk to make room in physical memory, and then the new segment is swapped into this space. Even though a swap has taken place, it appears to the program that all segments or pages are available in physical memory.

Segmentation Virtual to Physical Address Translation

Let us now look more closely at the address-translation process. We begin by assuming that paging is turned off. In this case, the address-translation sequence that takes place is the one highlighted in Fig. 8.14(a). Figure 8.14(b) details the operations that take place

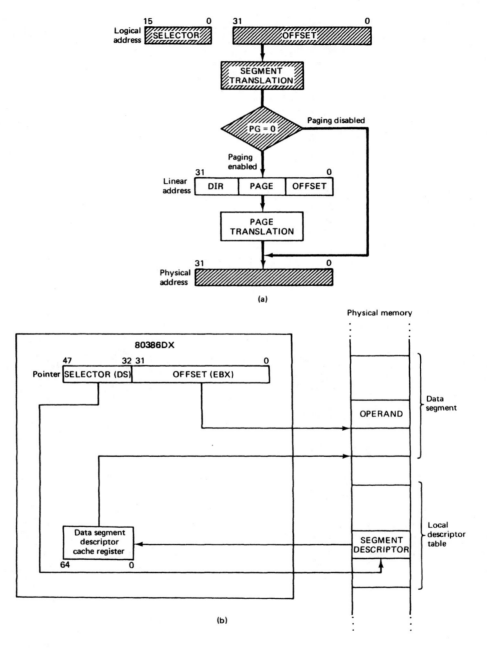

Figure 8.14 (a) Virtual to linear address translation. (Reprinted by permission of Intel Corp. Copyright/Intel Corp. 1986) (b) Translating a virtual address into a physical (linear) address.

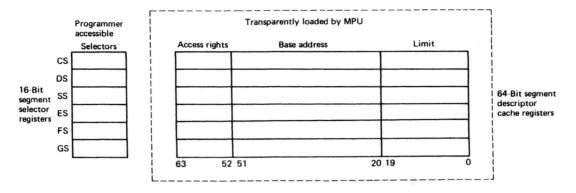

Figure 8.15 Segment-selector registers and the segment-descriptor cache registers.

during the segment translation process. Earlier we found that the 80386DX's segment selector registers, CS, DS, ES, FS, GS, and SS, provide the segment selectors that are used to index into either the global descriptor table or the local descriptor table. Whenever a selector value is loaded into a segment register, the descriptor pointed to by the index in the table selected by the T1 bit is automatically fetched from memory and loaded into the corresponding *segment descriptor cache register*. It is the contents of this descriptor, not the selector, that defines the location, size, and characteristics of the segment of memory.

Notice in Fig. 8.15 that the 80386DX has one 64-bit internal segment descriptor cache register for each of the segment selector registers. These cache registers are not accessible by the programmer. Instead, they are transparently loaded with a complete descriptor whenever an instruction is executed that loads a new selector into a segment register. For instance, if an operand is to be accessed from a new data segment, a local memory data segment selector would be first loaded into DS with the instruction

```
MOV   DS,AX
```

As this instruction is executed, the selector in AX is loaded into DS and then the corresponding descriptor in the local descriptor table is read from memory and loaded into the data segment descriptor cache register. The MMU looks at the information in the descriptor and performs checks to determine whether or not it is valid.

In this way, we see that the segment descriptors held in the caches dynamically change as a task is performed. At any one time, the memory-management unit permits just six segments of memory to be active. These segments correspond to the six segment selector registers, CS, DS, ES, FS, GS, and SS, and can reside in either local or global memory. Once the descriptors are cached, subsequent references to them are performed without any overhead for loading of the descriptor.

In Fig. 8.15 we find that this data segment descriptor has three parts: 12 bits of *access rights* information, a 32-bit *segment base address*, and a 20-bit *segment limit*. The value of the 32-bit base address identifies the beginning of the data segment that is to be accessed. The loading of the data segment descriptor cache completes the table lookup that maps the 16-bit selector to its equivalent 32-bit data segment base address.

The location of the operand in this data segment is determined by the offset part of the virtual address. For example, let us assume that the next instruction to be executed needs to access an operand in this data segment and that the instruction uses based addressing mode to specify the operand. Then the EBX register holds the offset of the operand from the base address of the data segment. Figure 8.14(a) shows that the base address is added directly to the offset to produce the 32-bit physical address of the operand. This addition completes the translation of the 48-bit virtual address into the 32-bit linear address. As shown in Fig. 8.14(a), when paging is disabled, PG = 0, the linear address is the physical address of the storage location to be accessed in memory.

EXAMPLE 8.8

Assume that in Fig. 8.14(b), the virtual address is made up of a segment selector equal to 0100_{16} and offset to equal 00002000_{16} and that paging is disabled. If the segment base address read in from the descriptor is 00030000_{16}, what is the physical address of the operand?

Solution

The virtual address is

$$\text{virtual address} = 0100{:}00002000_{16}$$

This virtual address translates to the physical address

$$\text{linear address} = \text{base address} + \text{offset}$$
$$= 00030000_{16} + 00002000_{16}$$
$$= 00032000_{16}$$

Paged Partitioning of the Virtual Address Space and Virtual-to-Physical Address Translation

Earlier we pointed out that the protected-mode architecture of the 80386DX also supports paged organization of the memory address space. The paging memory-management unit works beneath the segmentation memory management unit, and when enabled it organizes the 80386DX's address space in a different way. We just found that when paging is not in use, the 4G-byte physical address space is organized into segments that can be any size from 1 byte to 4G bytes. However, when paging is turned on, the paging unit arranges the physical address space into 1,048,496 pages that are each 4096 bytes long. Figure 8.16

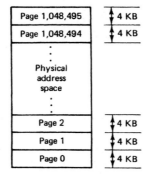

Figure 8.16 Paged organization of the physical address space.

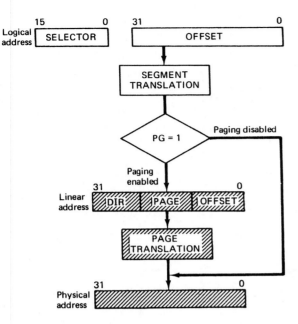

Figure 8.17 Paged translation of a linear address to a physical address. (Reprinted by permission of Intel Corp. Copyright/Intel Corp. 1986)

shows how the physical address space may be organized in this way. The fixed-size blocks of paged memory is a disadvantage in that 4K addresses are allocated by the memory manager even though not all may be used. This creation of unused section of memory is called *fragmentation*. Fragmentation results in less efficient use of memory. However, paging greatly simplifies the implementation of the memory-manager software. Let us continue by looking at what happens to the address translation process when paging is enabled.

In Fig. 8.17 we see that the linear address produced by the segment-translation process is no longer used as the physical address. Instead, it undergoes a second translation called the *page translation*. Figure 8.18 shows the format of a linear address. Notice that it is composed of three elements: a 20-bit offset field, a 10-bit page field, and a 10-bit directory field.

The diagram in Fig. 8.19 illustrates how a linear address is translated into its equivalent physical address. The location of the *page directory table* in memory is identified by the address in the page directory base register (PDBR) in CR_3. These 20 bits are actually the MSBs of the base address. The 12 lower bits are assumed to start at 000_{16} at the beginning of the directory and range to FFF_{16} at its end. Therefore, the page directory contains 4K byte memory locations and is organized as 1K, 32-bit addresses. These addresses each point to a separate page table, which is also in physical memory.

Notice that the 10-bit directory field of the linear address is the offset from the value in PDBR that selects one of the 1K, 32-bit *page directory entries* in the page directory table. This pointer is cached inside the 80386DX in what is called the *translation lookaside buffer*. Its value is used as the base address of a *page table* in memory. Just like the page directory,

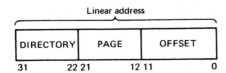

Figure 8.18 Linear address format.

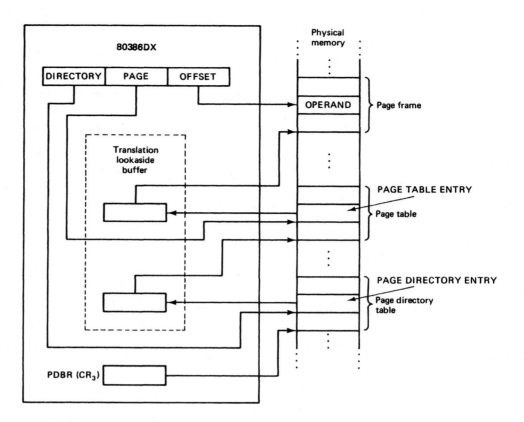

Figure 8.19 Translating a linear address to a physical address.

each page table is also 4K bytes long and contains 1K, 32-bit addresses. However, these addresses are called *page frame addresses*. Each page frame address points to a 4K frame of data storage locations in physical memory.

Next, the 10-bit page field of the linear address selects one of the 1K, 32-bit *page table entries* from the page table. This table entry is also cached in the translation lookaside buffer. In Fig. 8.19, we see that it is another base address and selects a 4K-byte *page frame* in memory. This frame of memory locations is used for storage of data. The 12-bit offset part of the linear address identifies the location of the operand in the active page frame.

The 80386DX's translation lookaside buffer is actually capable of maintaining 32 sets of table entries. In this way we see that 128K bytes of paged memory are always directly accessible. Operands in this part of memory can be accessed without first reading new entries from the page tables. If an operand to be accessed is not in one of these pages, overhead is required to first read the page table entry into the translation lookaside buffer.

▲ 8.4 DESCRIPTOR AND PAGE TABLE ENTRIES

In the previous section we frequently used the terms *descriptor* and *page table entry*. We talked about the descriptor as an element of the global descriptor, local descriptor, and interrupt descriptor tables. Actually, there are several kinds of descriptors supported by the 80386DX and they all serve different functions relative to overall system operation. Some examples are the *segment descriptor, system segment descriptor, local descriptor table descriptor, call gate descriptor, task state segment descriptor,* and *task gate descriptor.* We also discussed page table entries in our description of the 80386DX's page translation of virtual addresses. There are just two types of page table entries: the *page directory entry* and the *page table entry.* Let us now explore the structure of descriptors and page table entries.

Descriptors are the elements by which the on-chip memory manager hardware manages the segmentation of the 80386DX's 64T-byte virtual memory address space. One descriptor exists for each segment of memory in the virtual address space. Descriptors are assigned to the local descriptor table, global descriptor table, task state segment, call gate, task gate, and interrupts. The contents of a descriptor provides mapping from virtual addresses to linear addresses for code, data, stack, and the task state segments and assigns attributes to the segment.

Each descriptor is 8 bytes long and contains three kinds of information. Earlier we identified the 20-bit *LIMIT* field and showed that its value defines the size of the segment or the table. Moreover, we found that the 32-bit *BASE* value provides the beginning address for the segment or the table in the 64G-byte linear address space. The third element of a descriptor, which is called the *access rights byte,* is different for each type of descriptor. Let us now look at the format of just two types of descriptors, the segment descriptor and system segment descriptor.

The segment descriptor is the type of descriptor that is used to describe code, data, and stack segments. Figure 8.20(a) shows the general structure of a segment descriptor.

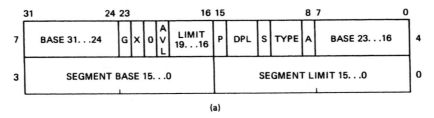

(a)

Bit Position	Name		Function	
7	Present (P)	P = 1	Segment is mapped into physical memory.	
		P = 0	No mapping to physical memory exists, base and limit are not used.	
6–5	Descriptor Privilege Level (DPL)		Segment privilege attribute used in privilege tests.	
4	Segment Descriptor (S)	S = 1	Code or Data (includes stacks) segment descriptor	
		S = 0	System Segment Descriptor or Gate Descriptor	
3	Executable (E)	E = 0	Data segment descriptor type is:	If
2	Expansion Direction (ED)	ED = 0	Expand up segment, offsets must be ≤ limit.	Data
		ED = 1	Expand down segment, offsets must be > limit.	Segment
1	Writeable (W)	W = 0	Data segment may not be written into.	(S = 1,
		W = 1	Data segment may be written into.	E = 0)
3	Executable (E)	E = 1	Code Segment Descriptor type is:	If
2	Conforming (C)	C = 1	Code segment may only be executed when CPL ≥ DPL and CPL remains unchanged.	Code Segment
1	Readable (R)	R = 0	Code segment may not be read.	(S = 1,
		R = 1	Code segment may be read.	E = 1)
0	Accessed (A)	A = 0	Segment has not been accessed.	
		A = 1	Segment selector has been loaded into segment register or used by selector test instructions.	

Type Field Definition (bracket spanning bit positions 3, 2, 1 rows)

(b)

Figure 8.20 (a) Segment-descriptor format. (b) Access byte-bit definitions. (Reprinted by permission of Intel Corp. Copyright/Intel Corp. 1987)

Here we see that the 2 lowest-addressed bytes, byte 0 and 1, hold the 16 least significant bits of the limit, the next 3 bytes contain the 24 least significant bits of the base address, byte 5 is the access rights byte, the lower 4 bits of byte 6 are the 4 most significant bits of the limit, the upper 4 bits include the *granularity* (G) and the *programmer available* (AVL) bits, and byte 7 is the 8 most significant bits of the 32-bit base. Segment descriptors are only found in the local and global descriptor tables.

Figure 8.21 shows how a descriptor is loaded from the local descriptor table in global memory to define a code segment in local memory. Notice that the LDTR descriptor defines

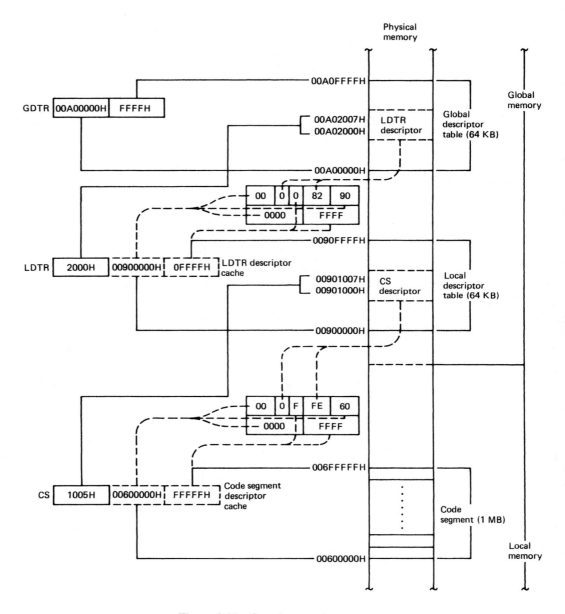

Figure 8.21 Creating a code segment.

a local descriptor table between address 00900000_{16} and $0090FFFF_{16}$. The value 1005_{16}, which is held in the code segment selector register, causes the descriptor at offset 1000_{16} in the local descriptor table to be cached into the code segment descriptor cache. In this way, a 1M-byte code segment is activated starting at address 00600000_{16} in local memory.

The bits of the access rights byte define the operating characteristics of a segment. For example, they contain information about a segment such as whether the descriptor has been accessed, if it is a code or data segment descriptor, its privilege level, if it is readable or writeable, and if it is currently loaded into internal memory. Let us next look at the function of each of these bits in detail.

The function of each bit in the access rights byte is listed in Fig. 8.20(b). Notice that if bit 0 is logic 1, the descriptor has been accessed. A descriptor is marked this way to indicate that its value has been cached on the 80386DX. The memory-manager software checks this information to find out if the segment is already in physical memory. Bit 4 identifies whether the descriptor represents a code/data segment or is a control descriptor. Let us assume that this bit is 1 to identify a segment descriptor. Then the type bits, bits 1 through 3, determine whether the descriptor describes a code segment or a data segment. For instance, 000 means that it is a read/write data segment that grows upward from the base to the limit. The DPL bits, bits 5 and 6, assign a descriptor privilege level to the segment. For example, 00 selects the most privileged level, level 0. Finally, the present bit indicates whether or not the segment is currently loaded into physical memory. This bit can be tested by the operating system software to determine if the segment should be loaded from a secondary storage device such as a hard disk. For example, if the access rights byte has logic 1 in bit 7, the segment is already available in physical memory and does not have to be loaded from an external device. Figure 8.22(a) shows the general form of a code segment descriptor, and Fig. 8.22(b) shows a general data/stack segment descriptor.

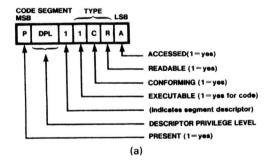

(a)

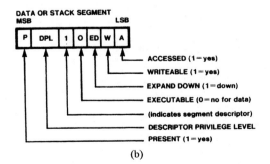

(b)

Figure 8.22 (a) Code segment descriptor access byte configuration. (Reprinted by permission of Intel Corp. Copyright/Intel Corp. 1987) (b) Data or stack segment access byte configuration. (Reprinted by permission of Intel Corp. Copyright/Intel Corp. 1987)

EXAMPLE 8.9

The access rights byte of a segment descriptor contains FE_{16}. What type of segment descriptor does it describe, and what are its characteristics?

Solution

Expressing the access rights byte in binary form, we get

$$FE_{16} = 11111110_2$$

Since bit 4 is 1, the access rights byte is for a code/data segment descriptor. This segment has the characteristics that follow:

$$P = 1 = \text{segment is mapped into physical memory}$$
$$DPL = 11 = \text{privilege level 3}$$
$$E = 1 = \text{executable code segment}$$
$$C = 1 = \text{conforming code segment}$$
$$R = 1 = \text{readable code segment}$$
$$A = 0 = \text{segment has not been accessed}$$

An example of a system segment descriptor is the descriptor used to define the local descriptor table. This descriptor is located in the GDT. Looking at Fig. 8.23, we find that the format of a system segment descriptor is similar to the segment descriptor we just discussed. However, the type field of the access rights byte takes on new functions.

EXAMPLE 8.10

If a system segment descriptor has an access rights byte equal to 82_{16}, what type of descriptor does it represent? What is its privilege level? Is the descriptor present?

Solution

First, we will express the access rights byte in binary form. This gives

$$82_{16} = 10000010_2$$

Now we see that the bits that describe the type of the descriptor are given as

$$TYPE = 0010 = \text{local descriptor table descriptor}$$

The privilege level is given by

$$DPL = 00 = \text{privilege level 0}$$

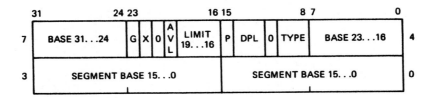

Name	Value	Description
TYPE	0	Reserved by Intel
	1	Available 80286 TSS
	2	LDT
	3	Busy 80286 TSS
	4	Call gate
	5	Task gate
	6	80286 interrupt gate
	7	80286 trap gate
	8	Reserved by Intel
	9	Available 80386 TSS
	A	Reserved
	B	Busy 80386 TSS
	C	80386 call gate
	D	Reserved by Intel
	E	80386 interrupt gate
	F	80386 trap gate
P	0	Descriptor contents are not valid
	1	Descriptor contents are valid
DPL	0–3	Descriptor privilege level 0, 1, 2 or 3
BASE	32-bit number	Base address of special system data segment in memory
LIMIT	20-bit number	Offset of last byte in segment from the base

Figure 8.23 System-segment descriptor format and field definitions.

and since

$$P = 1$$

the descriptor is present in physical memory.

Now that we have explained the format and use of descriptors, let us continue with page table entries. The format of either a page directory or page table entry is shown in Fig 8.24. Here we see that the 20 most significant bits are either the base address of the page table if the entry is in the page directory table or the base address of the page frame if the entry is in the page table. Notice that only bits 12 through 31 of the base address are supplied by the entry. The 12 least significant bits are assumed to be equal to zero. In this way we see that page tables and page frames are always located on a 4K-byte address

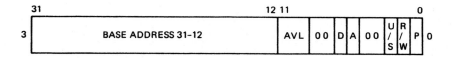

Figure 8.24 Directory or page table entry format.

boundary. In Fig. 8.19, we found that these entries are cached into the translation looka-side buffer.

The 12 lower bits of the entry supply protection characteristics or statistical information about the use of the page table or page frame. For example, the *user/supervisor* (U/S) and *read/write* (R/W) bits implement a two-level page protection mechanism. Setting U/S to 1 selects user-level protection. User is the low privilege level and is the same as protection level 3 of the segmentation model. That is, user is the protection level assigned to pages of memory that are accessible by application software. On the other hand, making U/S equal to 0 assigns supervisor-level protection to the table or frame. Supervisor corresponds to levels 0, 1, and 2 of the segmentation model and is the level assigned to operating system resources. The *read/write* (R/W) bit is used to make a user-level table or frame read-only or read/write. Logic 1 in R/W selects read-only operation. Figure 8.25 summarizes the access characteristics for each setting of U/S and R/W.

Protection characteristics assigned by a page directory entry are applied to all page frames defined by the entries in the page table. On the other hand, the attributes assigned to a page table entry apply only to the page frame that it defines. Since two sets of protection characteristics exist for all page frames, the page-protection mechanism of the 80386DX is designed always to enforce the higher privileged (more restricting) of the two protection rights.

EXAMPLE 8.11

If the page directory entry for the active page frame is F1000007$_{16}$ and its page table entry is 010000005$_{16}$, is the frame assigned to the user or supervisor? What access is permitted to the frame from user mode and from supervisor mode?

Solution

First, the page directory entry is expressed in binary form. This gives

$$F1000003_{16} = 11110001000000000000000000000111_2$$

Therefore, the page-protection bits are

$$U/S \ R/W = 11$$

U/S	R/W	User	Supervisor
0	0	None	Read/write
0	1	None	Read/write
1	0	Read-only	Read/write
1	1	Read/write	Read/write

Figure 8.25 User- and supervisor-level access rights.

This assigns user-mode and read/write accesses to the complete page frame. Next the page table entry for the frame is expressed in binary form as

$$01000005_{16} = 00000001000000000000000000000101_2$$

Here we find that

$$\text{U/S R/W} = 10$$

This defines the page frame as a user-mode, read-only page. Since the page frame attributes are the more restrictive, they apply. Looking at Fig. 8.25, we see that user software (application software) can only read data in this frame. On the other hand, supervisor software (operating system software) can either read data from or write data into the frame.

The other implemented bits in the directory and page table entry of Fig. 8.24 provide statistical information about the table or frame usage. For instance, the *present* (P) bit identifies whether or not the entry can be used for page address translation. P equal to logic 1 indicates that the entry is valid and is available for use in address translation. On the other hand, if P equals 0, the entry is either undefined or not present in physical memory. If an attempt is made to access a page table or page frame that has its P bit marked 0, a page fault results. This page fault needs to be serviced by the memory-manager software of the operating system.

The 80386DX also records the fact that a page table or page frame has been accessed. Just before a read or write is performed to any address in a table or frame, the *accessed* (A) bit of the entry is set to 1. This marks it as having been accessed. For page frame accesses, it also records whether the access was for a read or a write operation. The *dirty* (D) bit is defined only for a page table entry, and it gets set if a write is performed to any address in the corresponding page frame. In a virtual demand paged memory system, the operating system can check the state of these bits to determine if a page in physical memory needs to be updated on the virtual storage device (hard disk) when a new page is swapped into its physical memory address space. The last three bits are labeled AVL and are available for use by the programmer.

▲ 8.5 PROTECTED-MODE SYSTEM-CONTROL INSTRUCTION SET

In Chapters 5 and 6, we studied the real-mode instruction set of the 80386DX microprocessor. The instructions introduced in these chapters represent the base instruction set of the 8088/8086 microprocessor, a number of new instructions called the extended instruction set, and 80386DX specific instruction set. In protected mode, the 80386DX executes all the instructions that are available in the real mode. Moreover, it is enhanced with a number of additional instructions that either apply only to protected-mode operation or are used in the real mode to prepare the 80386DX for entry into the protected mode. As shown in Fig. 8.26, these instructions are known as the *system-control instruction set*.

The instructions of the system control instruction set are listed in Fig. 8.27. Here we find the format of each instruction along with a description of its operation. Moreover, the

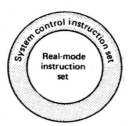

Figure 8.26 Protected mode instruction set.

Instruction	Description	Mode
LGDT S	Load the global descriptor table register. S specifies the memory location that contains the first byte of the 6 bytes to be loaded into the GDTR.	Both
SGDT D	Store the global descriptor table register. D specifies the memory location that gets the first byte of the 6 bytes to be stored from the GDTR.	Both
LIDT S	Load the interrupt descriptor table register. S specifies the memory location that contains the first byte of the 6 bytes to be loaded into the IDTR.	Both
SIDT D	Store the interrupt descriptor table register. D specifies the memory location that gets the first byte of the 6 bytes to be stored from the IDTR.	Both
LMSW S	Load the machine status word. S is an operand to specify the word to be loaded into the MSW.	Both
SMSW D	Store the machine status word. D is an operand to specify the word location or register where the MSW is to be stored.	Both
LLDT S	Load the local descriptor table register. S specifies the operand to specify a word to be loaded into the LDTR.	Protected
SLDT D	Store the local descriptor table register. D is an operand to specify the word location where the LDTR is to be saved.	Protected
LTR S	Load the task register. S is an operand to specify a word to be loaded into the TR.	Protected
STR D	Store the task register. D is an operand to specify the word location where the TR is to be stored.	Protected
LAR D, S	Load access rights byte. S specifies the selector for the descriptor whose access byte is loaded into the upper byte of the D operand. The low byte specified by D is cleared. The zero flag is set if the loading completes successfully; otherwise it is cleared.	Protected
LSL R16, S	Load segment limit. S specifies the selector for the descriptor whose limit word is loaded into the word register operand R16. The zero flag is set if the loading completes successfully; otherwise it is cleared.	Protected
ARPL D, R16	Adjust RPL field of the selector. D specifies the selector whose RPL field is increased to match the PRL field in the register. The zero flag is set if successful; otherwise it is cleared.	Protected
VERR S	Verify read access. S specifies the selector for the segment to be verified for read operation. If successful the zero flag is set; otherwise it is reset.	Protected
VERW S	Verify write access. S specifies the selector for the segment to be verified for write operation. If successful the zero flag is set; otherwise it is reset.	Protected
CLTS	Clear task switched flag.	Protected

Figure 8.27 Protected-mode system control instruction set.

mode or modes in which the instruction is available are identified. Let us now look at the operation of some of these instructions in detail.

Looking at Fig. 8.27, we see that the first six instructions can be executed in either the real or protected mode. They provide the ability to load (L) or store (S) the contents of the global descriptor table (GDT) register, interrupt descriptor table (IDT) register, and machine status word (MSW) part of CR_0. Notice that the instruction *load global descriptor table register* (LGDT) is used to load the GDTR from memory. Operand S specifies the location of the 6 bytes of memory that hold the limit and base that specifies the size and beginning address of the GDT. The first word of memory contains the limit and the next 4 bytes contain the base. For instance, executing the instruction

```
LGDT    [INIT_GDTR]
```

loads the GDTR with the base and limit stored at address INIT_GDTR to create a global descriptor table in memory. This instruction is meant to be used during system initialization and before switching the 80386DX to the protected mode.

Once loaded, the current contents of the GDTR can be saved in memory by executing the *store global descriptor table* (SGDT) instruction. An example is the instruction

```
SGDT    [SAVE_GDTR]
```

The instructions LIDT and SIDT perform similar operations for the interrupt descriptor table register. The IDTR is also set up during initialization.

The instructions *load machine status word* (LMSW) and *store machine status word* (SMSW) are provided to load and store the contents of the machine status word, respectively. These are the instructions that are used to switch the 80386DX from real to protected mode. To do this we must set the least significant bit in the MSW to 1. This can be done by first reading the contents of the machine status word, modifying the LSB (PE), and then writing the modified value back into the MSW part of CR_0. The instruction sequence that follows will switch an 80386DX operating in real mode to the protected mode:

```
SMSW   AX   ;read from the MSW
  OR   AX,1;set the PE bit
LMSW   AX   ;write to the MSW
```

The next four instructions in Fig. 8.27 are also used to initialize or save the contents of protected-mode registers. However, they can be used only when the 80386DX is in the protected mode. To load and to save the contents of the LDTR, we have the instructions LLDT and SLDT, respectively. Moreover, for loading and saving the contents of the TR, the equivalent instructions are LTR and STR.

The rest of the instructions in Fig. 8.27 are for accessing the contents of descriptors. For instance to read a descriptor's access rights byte, the *load access rights byte* (LAR) instruction is executed. An example is the instruction

```
LAR   AX,[LDIS_1]
```

Execution of this instruction causes the access rights byte of the specified local descriptor to be loaded into AH. To read the segment limit of a descriptor, we use the *load segment limit* (LSL) instruction. For instance, to copy the segment limit for the specified local descriptor into register EBX, the instruction

```
LSL    EBX,LDIS_1
```

is executed. In both cases ZF is set to 1 if the operation is performed correctly.

The instruction *adjust RPL field of selector* (ARPL) can be used to increase the RPL field of a selector in memory or a register, destination (D), to match the protection level of the selector in a register, source (S). If an RPL-level increase takes place, ZF is set to 1. Finally, the instructions VERR and VERW are provided to test the accessibility of a segment for a read or write operation, respectively. If the descriptor permits the type of access tested for by executing the instruction, ZF is set to 1.

▲ 8.6 MULTITASKING AND PROTECTION

We say that the 80386DX microprocessor implements a *multitasking* software architecture. By this we mean that it contains on-chip hardware that both permits multiple tasks to exist in a software system and allows them to be scheduled for execution in a time-shared manner. That is, program control is switched from one task to another after a fixed interval of time elapses. For instance, the tasks can be executed in a round-robin fashion. This means that the most recently executed task is returned to the end of the list of tasks being executed. Even though the processes are executed in a time-shared fashion, an 80386DX microcomputer has the performance to make it appear to the user that they are all running simultaneously.

Earlier we defined a task as a collection of program routines that performs a specific function. This function is also called a *process*. Software systems typically need to perform many processes. In the protected-mode 80386DX-based microcomputer, each process is identified as an independent task. The 80386DX provides an efficient mechanism, called the *task-switching mechanism*, for switching between tasks. For instance, an 80386DX running at 16 MHz can perform a task switch operation in just 19 μs.

We also indicated earlier that when a task is called into operation, it can have both global and local memory resources. The local memory address space is divided between tasks. This means that each task normally has its own private segments of local memory. Segments in global memory can be shared by all tasks. Therefore, a task can have access to any of the segments in global memory. As shown in Fig. 8.28, task A has both a private address space and a global address space available for its use.

Protection and the Protection Model

Safeguards can be built into the protected-mode software system to deny unauthorized or incorrect accesses of a task's memory resources. The concept of safeguarding memory is known as *protection*. The 80386DX includes on-chip hardware that implements a *protection mechanism*. This mechanism is designed to put restrictions on the access of local and system resources by a task and to isolate tasks from each other in a multitasking environment.

Segmentation, paging, and descriptors are the key elements of the 80386DX's protection mechanism. We already know that, when using a segmented memory model, a segment is the smallest element of the virtual memory address space that has unique protection attributes. These attributes are defined by the access rights information and limit fields in

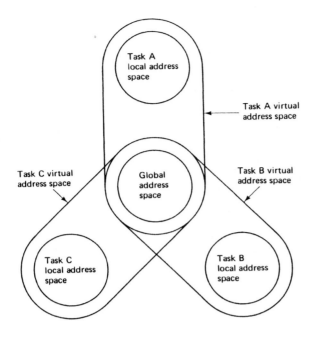

Figure 8.28 Virtual address space of a task. (Reprinted by permission of Intel Corp. Copyright/Intel Corp. 1987)

the segment's descriptor. As shown in Fig. 8.29(a), the on-chip protection hardware performs a number of checks during all memory accesses. Figure 8.29(b) is a list of the protection checks and restrictions imposed on software by the 80386DX. For example, when a data-storage location in memory is written to, the type field in the access rights byte of the segment is tested to assure that its attributes are consistent with the register cache being loaded, and the offset is checked to verify that it is within the limit of the segment.

Let us just review the attributes that can be assigned to a segment with the access rights information in its descriptor. Figure 8.30 shows the format of a data segment descriptor and an executable (code) segment descriptor. The P bit defines whether a segment of memory is present in physical memory. Assuming that a segment is present, bit 4 of the type field makes it either a code segment or data segment. Notice that this bit is 0 if the descriptor is for a data segment and it is 1 for code segments. Segment attributes such as readable, writeable, conforming, expand up or down, and accessed are assigned by other bits in the type field. Finally, a privilege level is assigned with the DPL field.

Earlier we showed that whenever a segment is accessed, the base address and limit are cached inside the 80386DX. In Fig. 8.29(a), we find that the access rights information is also loaded into the cache register. However, before loading the descriptor the MMU verifies that the selected segment is currently present in physical memory, that it is at a privilege level accessible from the privilege level of the current program, that the type is consistent with the target segment selector register (CS = code segment, DS, ES, FS, GS, or SS = data segment), and that the reference into the segment does not exceed the address limit of the segment. If a violation is detected, an error condition is signaled. The memory manager software can determine the cause, correct the problem, and then reinitiate the operation.

Let us now look at some examples of memory accesses that result in protection violations. For example, if the selector loaded into the CS register points to a descriptor that defines a data segment, the type check leads to a protection violation. Another example

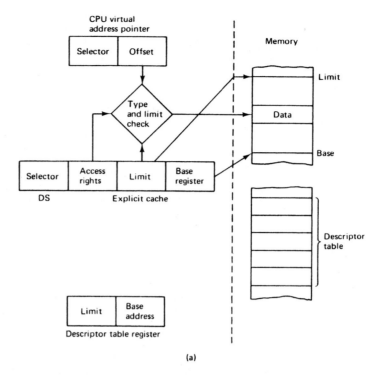

CPU virtual
address pointer

| Selector | Offset |

Type and limit check

Memory

Limit

Data

Base

| Selector | Access rights | Limit | Base register |

DS

Explicit cache

Descriptor table

| Limit | Base address |

Descriptor table register

(a)

Type check
Limit check
Restriction of addressable domain
Restriction of procedure entry point
Restriction of instruction set

(b)

Figure 8.29 (a) Testing the access rights of a descriptor. (Reprinted by permission of Intel Corp. Copyright/Intel Corp. 1982) (b) Protection checks and restrictions.

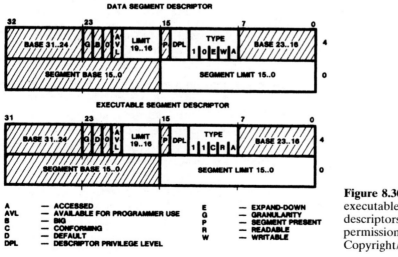

DATA SEGMENT DESCRIPTOR

EXECUTABLE SEGMENT DESCRIPTOR

A	— ACCESSED
AVL	— AVAILABLE FOR PROGRAMMER USE
B	— BIG
C	— CONFORMING
D	— DEFAULT
DPL	— DESCRIPTOR PRIVILEGE LEVEL

E	— EXPAND-DOWN
G	— GRANULARITY
P	— SEGMENT PRESENT
R	— READABLE
W	— WRITABLE

Figure 8.30 Data segment and executable (code) segment descriptors. (Reprinted by permission of Intel Corp. Copyright/Intel Corp. 1986)

of an invalid memory access is an attempt to read an operand from a code segment that is not marked as readable. Finally, any attempt to access a byte of data at an offset greater than LIMIT, a word at an offset equal to or greater than LIMIT, or a double word at an offset equal to or greater than LIMIT−2 extends beyond the end of the data segment and results in a protection violation.

The 80386DX's protection model provides four possible privilege levels for each task. They are called *levels* 0, 1, 2, and 3 and can be illustrated by concentric circles, as in Fig. 8.31. Here level 0 is the most privileged level and level 3 is the least privileged level.

System and application software are typically partitioned in the manner shown in Fig. 8.31. The kernel represents application-independent software that provides microprocessor-oriented functions such as I/O control, task sequencing, and memory management. For this reason it is kept at the most privileged level, level 0. Level 1 contains processes that provide system services such as file accessing. Level 2 is used to implement custom routines to support special-purpose system operations. Finally, the least privileged level, level 3, is the level at which user applications are run. This example also demonstrates how privilege levels are used to isolate system-level software (operating system software in levels 0 through 2) from the user's application software (level 3). Tasks at a level can use programs from the more privileged levels but cannot modify the contents of these routines in any way. In this way, applications are permitted to use system software routines from the three higher privilege levels without affecting their integrity.

Earlier we indicated that protection restrictions are put on the instruction set. One example of this is that the system control instructions can be executed only in a code segment that is at protection level 0. We also pointed out that each task is assigned its own local descriptor table. Therefore, as long as none of the descriptors in a task's local descriptor table reference code or data is available to another task, it is isolated from all other tasks.

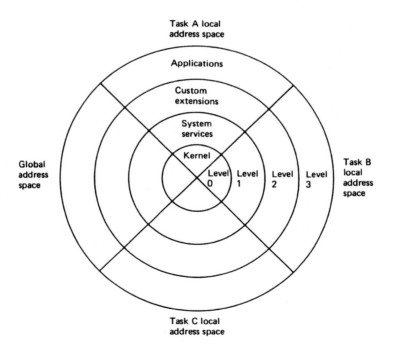

Figure 8.31 Protection model.

That is, it has been assigned a unique part of the virtual address space. For example, in Fig. 8.31 multiple applications running at level 3 are isolated from each other by assigning them different local resources. This shows that segments, privilege levels, and the local descriptor table provide protection for both code and data within a task. These types of protection result in improved software reliability because errors in one application will not affect the operating system or other applications.

Let us now look more closely at how the privilege level is assigned to a code or data segment. Remember that when a task is running, it has access to both local and global code segments, local and global data segments, and stack segments. A privilege level is assigned to each of these segments through the access rights information in its descriptor. A segment may be assigned to any privilege level simply by entering the number for the level into the DPL bits.

To provide more flexibility, input/output has two levels of privilege. First, the I/O drivers, which are normally system resources, are assigned to a privilege level. For the software system of Fig. 8.31, we indicated that the I/O control routines were part of the kernel and are at level 0.

The IN, INS, OUT, OUTS, CLI, and STI instructions are what are called *trusted instructions*. This is because the protection model of the 80386DX puts additional restrictions on their use in protected mode. They can be executed only at a privilege level that is equal to or more privileged than the *input/output privilege level* (IOPL) code. IOPL supplies the second level of I/O privilege. Remember that the IOPL bits exist in the protected-mode flag register. These bits must be loaded with the value of the privilege level that is to be assigned to input/output instructions through software. The value of IOPL may change from task to task. Assigning the I/O instructions to a level higher than 3 restricts applications from directly performing I/O. Therefore, to perform an I/O operation, the application must request service by an I/O driver through the operating system.

Accessing Code and Data Through the Protection Model

During the running of a task, the 80386DX may need either to pass control to program routines at another privilege level or to access data in a segment that is at a different privilege level. Accesses to code or data in segments at a different privilege level are governed by strict rules. These rules are designed to protect the code or data at the more privileged level from contamination by the less privileged routine.

Before looking at how accesses are made for routines or data at the same or different privilege levels, let us first look at some terminology used to identify privilege levels. We have already been using the terms descriptor privilege level (DPL) and I/O privilege level (IOPL). However, when discussing the protection mechanisms by which processes access data or code, two new terms come into play: *current privilege level* (CPL) and *requested privilege level* (RPL). CPL is defined as the privilege level of the code or data segment that is currently being accessed by a task. For example, the CPL of an executing task is the DPL of the access rights byte in the descriptor cache for the CS register. This value normally equals the DPL of the code segment. RPL is the privilege level of the new selector loaded into a segment register. For instance, in the case of code, it is the privilege level of the code segment that contains the routine that is being called. That is, RPL is the DPL of the code segment to which control is to be passed.

As a task in an application runs, it may require access to program routines that reside in segments at any of the four privilege levels. Therefore, the current privilege level of the task changes dynamically with the programs it executes. This is because the CPL of the task is normally switched to the DPL of the code segment currently being accessed.

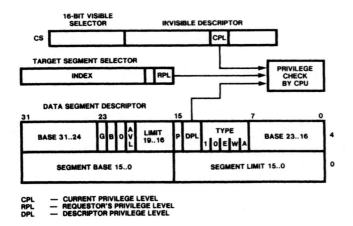

CPL — CURRENT PRIVILEGE LEVEL
RPL — REQUESTOR'S PRIVILEGE LEVEL
DPL — DESCRIPTOR PRIVILEGE LEVEL

Figure 8.32 Privilege-level checks for a data access. (Reprinted by permission of Intel Corp. Copyright/Intel Corp. 1986)

The protection rules of the 80386DX determine what code or data can be accessed by a program. Before looking at how control is passed to code at different protection levels, let us first look at how data segments are accessed by code at the current privilege level. Figure 8.32 illustrates the protection-level checks that are made for a data access. The general rule is that code can access only data that are at the same or a less privileged level. For instance, if the current privilege level of a task is 1, it can access operands that are in data segments with DPL equal to 1, 2, or 3. Whenever a new selector is loaded into the DS, ES, FS, or GS register, the DPL of the target data segment is checked to make sure that it is equal to or less privileged than the most privileged of either CPL or RPL. As long as DPL satisfies this condition, the descriptor is cached inside the 80386DX and the data access takes place.

One exception to this rule is when the SS register is loaded. In this case, the DPL must always equal the CPL. That is, the active stack (one is required for each privilege level) is always at the CPL.

EXAMPLE 8.12

Assuming that, in Fig. 8.32, DPL = 2, CPL = 0, and RPL = 2, will the data access take place?

Solution

DPL of the target segment is 2, and this value is less privileged than CPL = 0, which is the more privileged of CPL and RPL. Therefore, the protection criteria are satisfied and the access will take place.

Different rules apply to how control is passed between code at the same privilege level and between code at different privilege levels. To transfer program control to another instruction in the same code segment, one can simply use a near jump or call instruction. In either case, just a limit check is made to ensure that the destination of the jump or call does not exceed the limit of the current code segment.

To pass control to code in another segment that is at the same or a different privilege level, a far jump or call instruction is used. For this transfer of program control, both type and limit checks are performed and privilege-level rules are applied. Figure 8.33 shows the privilege checks made by the 80386DX. There are two conditions under which the transfer

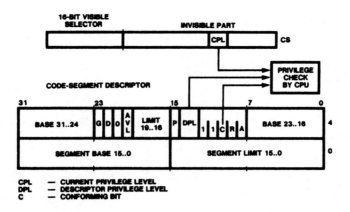

Figure 8.33 Privilege-level checks when directly passing program control to a program in another segment. (Reprinted by permission of Intel Corp. Copyright/Intel Corp. 1986)

in program control will take place. First, if CPL equals the DPL, the two segments are at the same protection level and the transfer occurs. Second, if the CPL represents a more privileged level than DPL but the conforming code (C) bit in the type field of the new segment is set, the routine is executed at the CPL.

The general rule that applies when control is passed to code in a segment that is at a different privilege level is that the new code segment must be at a more privileged level. A special kind of descriptor called a *gate descriptor* comes into play to implement the change in privilege level. An attempt to transfer control to a routine in a code segment at a higher privilege level is still initiated with either a far call or far jump instruction. This time the instruction does not directly specify the location of the destination code; instead, it references a gate descriptor. In this case the 80386DX goes through a much more complex program-control-transfer mechanism.

The structure of a gate descriptor is shown in Fig. 8.34. Notice that there are four types of gate descriptors: the *call gate*, *task gate*, *interrupt gate*, and *trap gate*. The call gate implements an indirect transfer of control within a task from code at the CPL to code at a higher privilege level. It does this by defining a valid entry point into the more privileged segment. The contents of a call gate are the virtual address of the entry point: the *destination selector* and the *destination offset*. In Fig. 8.34 we see that the destination selector identifies the code segment that contains the program to which control is to be redirected. The destination offset points to the instruction in this segment where execution is to resume. Call gates can reside in either the GDT or an LDT.

The operation of the call gate mechanism is illustrated in Fig. 8.35. Here we see that the call instruction includes an offset and a selector. When the instruction is executed, this selector is loaded into CS and points to the call gate. In turn, the call gate causes its destination selector to be loaded into CS. This leads to the caching of the descriptor for the called code segment (executable segment descriptor). The executable segment descriptor provides the base address for the executable segment (code segment) of memory. Notice that the offset in the call gate descriptor locates the entry point of the procedure in the executable segment.

Whenever the task's current privilege level is changed, a new stack is activated. As part of the program context-switching sequence, the old ESP and SS are saved on the new stack along with any parameters and the old EIP and CS. This information is needed to preserve linkage for return to the old program environment.

Now the procedure at the higher privilege level begins to execute. At the end of the routine, the RET instruction must be included to return program control back to the calling

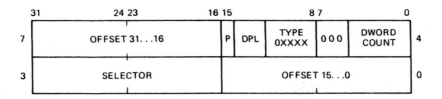

Name	Value	Description
TYPE	4 5 6 7	— Call gate — Task gate — Interrupt gate — Trap gate
P	0 1	— Descriptor contents are not valid — Descriptor contents are valid
DPL	0–3	Descriptor privilege level
WORD COUNT	0–31	Number of double words to copy from callers stack to called procedures stack. Only used with call gate
DESTINATION SELECTOR	16-Bit selector	Selector to the target code segment (call interrupt or trap gate) Selector to the target task state segment (task gate)
DESTINATION OFFSET	32-Bit offset	Entry point within the target code segment

Figure 8.34 Gate descriptor format. (Reprinted by permission of Intel Corp. Copyright/Intel Corp. 1987)

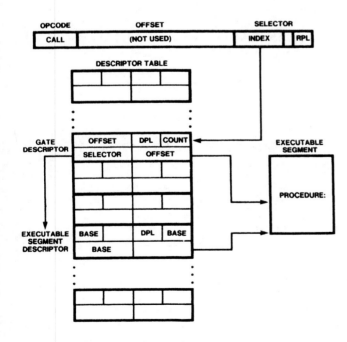

Figure 8.35 Call-gate operation. (Reprinted by permission of Intel Corp. Copyright/Intel Corp. 1986)

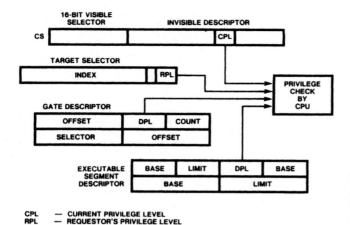

CPL — CURRENT PRIVILEGE LEVEL
RPL — REQUESTOR'S PRIVILEGE LEVEL
DPL — DESCRIPTOR PRIVILEGE LEVEL

Figure 8.36 Privilege-level checks for program-control transfer with a call gate. (Reprinted by permission of Intel Corp. Copyright/Intel Corp. 1986)

program. Execution of RET causes the old values of EIP, CS, the parameters, ESP, and SS to be popped from the stack. This restores the original program environment. Now program execution resumes with the instruction following the call instruction in the lower privileged code segment. Figure 8.36 shows the privilege checks that are performed when program control transfer is initiated through a call gate. For the call to be successful, the DPL of the gate must be the same as the CPL, and the RPL of the called code must be higher than the CPL.

Task Switching and the Task State Segment Table

Earlier we identified the task as the key program element of the 80386DX's multitasking software architecture and indicated that another important feature of this architecture is the high-performance task-switching mechanism. A task can be invoked either directly or indirectly. This is done by executing either the intersegment jump or intersegment call instruction. When a jump instruction is used to initiate a task switch, no return linkage to the prior task is supported. On the other hand, if a call is used to switch to the new task instead of a jump, back linkage information is saved automatically. This information permits a return to be performed to the instruction that follows the calling instruction in the old task at completion of the new task.

Each task that is to be perform by the 80386DX is assigned a unique selector called a *task state selector*. This selector is an index to a corresponding *task state segment descriptor* in the global descriptor table. The format of a task state segment descriptor is given in Fig. 8.37.

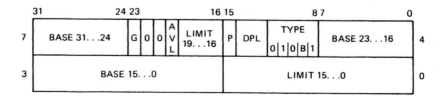

Figure 8.37 TSS descriptor format. (Reprinted by permission of Intel Corp. Copyright/Intel Corp. 1986)

If a jump or call instruction has a task state selector as its operand, a direct entry is performed to the task. As shown in Fig. 8.38, when a call instruction is executed, the selector is loaded into the 80386DX's task register (TR). Then the corresponding task state segment descriptor is read from the GDT and loaded into the task register cache. This happens only if the criteria specified by the access rights information of the descriptor are satisfied. That

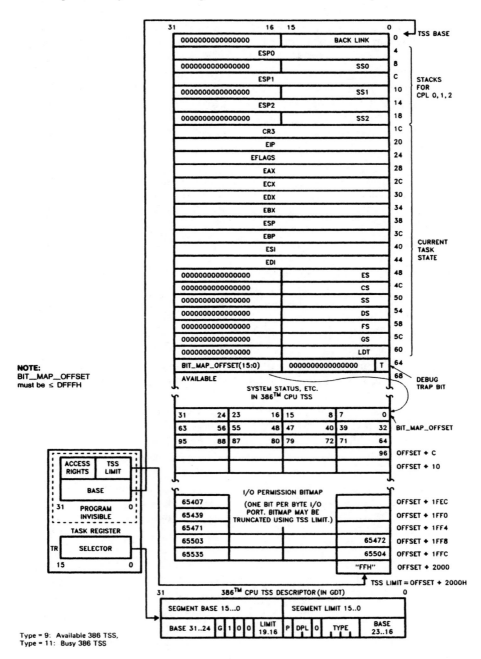

Figure 8.38 Task-state segment table. (Reprinted by permission of Intel Corp. Copyright/Intel Corp. 1987)

is, the descriptor is present (P = 1); the task is not busy (B = 0); and protection is not violated (CPL must be equal to DPL). Looking at Fig. 8.38, we see that once loaded, the base address and limit specified in the descriptor define the starting point and size of the task's *task state segment* (TSS). This TSS contains all of the information that is needed to either start or stop a task.

Before explaining the rest of the task-switch sequence, let us first look more closely at what is contained in the task state segment. A typical TSS is shown in Fig. 8.38. Its minimum size is 103 bytes. For this reason, the minimum limit that can be specified in a TSS descriptor is 00067_{16}. Notice that the segment contains information such as the state of the microprocessor (general registers, segment selectors, instruction pointer, and flags) needed to initiate the task, a back-link selector to the TSS of the task that was active when this task was called, the local descriptor table register selector, a stack selector and pointer for privilege levels 0, 1, and 2, and an I/O permission bit map.

Now we will continue with the procedure by which a task is invoked. Let us assume that a task was already active when a new task was called. Then the new task is what is called a *nested task*, and it causes the NT bit of the flag word to be set to 1. In this case the current task is first suspended and then the state of the 80386DX's user-accessible registers is saved in the old TSS. Next, the B bit in the new task's descriptor is marked busy; the TS bit in the machine status word is set to indicate that a task is active; the state information from the new task's TSS is loaded into the MPU; and the selector for the old TSS is saved as the back-link selector in the new task state segment. The task switch operation is now complete and execution resumes with the instruction identified by the new contents of the code segment selector (CS) and instruction pointer (EIP).

The old program context is preserved by saving the selector for the old TSS as the back-link selector in the new TSS. By executing a return instruction at the end of the new task, the back-link selector for the old TSS is automatically reloaded into TR. This activates the old TSS and restores the prior program environment. Now program execution resumes at the point where it left off in the old task.

The indirect method of invoking a task is by jumping to or calling a *task gate*. This is the method used to transfer control to a task at an RPL that is higher than the CPL. Figure 8.39 shows the format of a task gate. This time the instruction includes a selector that points to the task gate, which is in either the LDT or GDT, instead of a task state selector. The TSS selector held in this gate is loaded into TR to select the TSS and initiate the task. Figure 8.40 illustrates a task initiated through a task gate.

Let us consider an example to illustrate the principle of task switching. In Fig. 8.41 we have a table that contains TSS descriptors SELECT0 through SELECT3. These descriptors contain access rights and selectors for tasks 0 through 3, respectively. To invoke the task corresponding to selector SELECT2 in the data segment where these selectors are stored, we can use the following procedure. First, the data segment register is loaded with the address

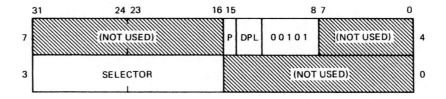

Figure 8.39 Task-gate descriptor format. (Reprinted by permission of Intel Corp. Copyright/Intel Corp. 1986)

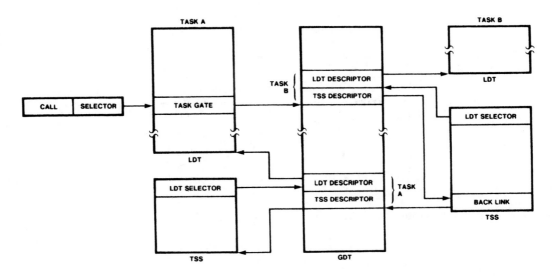

Figure 8.40 Task switch through a task gate. (Reprinted by permission of Intel Corp. Copyright/Intel Corp. 1987)

SELECTOR_DATA_SEG_START to point to the segment that contains the selectors. This is done with the instructions

```
MOV    AX,SELECTOR_DATA_SEG_START
MOV    DS,AX
```

Since each selector is 8 bytes long, SELECT2 is offset from the beginning of the segment by 16 bytes. Let us load the offset into register EBX.

```
MOV    EBX,0F
```

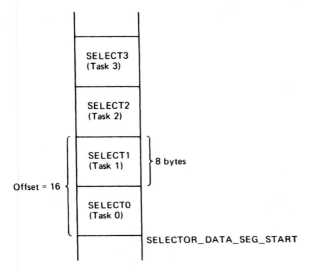

Figure 8.41 Task-gate selectors.

At this point we can use SELECT2 to implement an intersegment jump with the instruction

```
JMP    DWORDPTR[EBX]
```

Execution of this instruction switches program control to the task specified by the selector in descriptor SELECT2. In this case, no program linkage is preserved. On the other hand, by calling the task with the instruction

```
CALL    DWORDPTR[EBX]
```

the linkage is maintained.

▲ 8.7 VIRTUAL 8086 MODE

8086 and 8088 application programs, such as those written for the PC's DOS operating system, can be directly run on the 80386DX in real mode. A protected-mode operating system, such as UNIX, can also run DOS applications without change. This is done through what is called *virtual 8086 mode*. When in this mode, the 80386DX supports an 8086 microprocessor programming model and can directly run programs written for the 8086. That is, it creates a virtual 8086 machine for executing programs.

In this kind of application, the 80386DX is switched back and forth between protected mode and virtual 8086 mode. The UNIX operating system and UNIX applications are run in protected mode, and when the DOS operating system and a DOS application are to be run, the 80386DX is switched to the virtual 8086 mode. This mode switching is controlled by a program known as a *virtual 8086 monitor*.

Virtual 8086 mode of operation is selected by the bit called *virtual mode* (VM) in the extended flag register. VM must be switched to 1 to enable virtual 8086 mode of operation. Actually, the VM bit in EFLAGS is not directly switched to 1 by software. This is because virtual 8086 mode is normally entered as a protected-mode task. Therefore, the copy of EFLAGS in the TSS for the task would include VM equal 1. These EFLAGS are loaded as part of the task switching process. In turn, virtual 8086 mode of operation is initiated. The virtual 8086 program is run at privilege level 3. The virtual 8086 monitor is responsible for setting and resetting the VM bit in the task's copy of EFLAGS and permits both protected-mode tasks and virtual 8086 mode tasks to coexist in a multitasking program environment.

Another way of initiating virtual 8086 mode is through an interrupt return. In this case, the EFLAGS are reloaded from the stack. Again the copy of EFLAGS must have the VM bit set to 1 to enter virtual 8086 mode of operation.

ASSIGNMENTS

Section 8.2

1. List the protected-mode registers that are not part of the real-mode model.
2. What are the two parts of the GDTR called?
3. What function is served by the GDTR?
4. If the contents of the GDTR are $0021000001FF_{16}$, what are the starting and ending addresses of the table? How large is the table? How many descriptors can be stored in the table?

5. What is stored in the GDT?

6. What do IDTR and IDT stand for?

7. What is the maximum limit that should be used in the IDTR?

8. What is stored in the IDT?

9. What descriptor table defines the local memory address space?

10. What gets loaded into the LDTR? What happens when it gets loaded?

11. Which control register contains the MSW?

12. Which bit is used to switch the 80386DX from real-address mode to protected-address mode?

13. What MSW bit settings identify that floating-point operations are to be performed by an 80387 coprocessor?

14. What does TS stand for?

15. What must be done to turn on paging?

16. Where is the page directory base register located?

17. How large is the page directory?

18. What is held in the page table?

19. What gets loaded into TR? What is its function?

20. What is the function of the task descriptor cache?

21. What determines the location and size of a task state segment?

22. What is the name of the CS register in the protected mode? The DS register?

23. What are the names and sizes of the three fields in a selector?

24. What does TI equal 1 mean?

25. If the GDT register contains $0013000000FF_{16}$ and the selector loaded into the LDTR is 0040_{16}, what is the starting address of the LDT descriptor that is to be loaded into the cache?

26. What does NT stand for? RF?

27. If the IOPL bits of the flag register contain 10, what is the privilege level of the I/O instructions?

Section 8.3

28. What is the size of the 80386DX's virtual address?

29. What are the two parts of a virtual address called?

30. How large can a data segment be? How small?

31. How large is the 80386DX's virtual address space? What is the maximum number of segments that can exist in the virtual address space?

32. How large is the global memory address space? How many segments can it contain?

33. In Fig. 8.12, which segments of memory does task 3 have access to? Which segments does it not have access to?

34. What part of the 80386DX is used to translate virtual addresses to physical addresses?

35. What happens when the following instruction sequence is executed?

```
MOV  AX,[SI]
MOV  CS,AX
```

36. If the descriptor accessed in Problem 35 has the value $00200000FFFF_{16}$ and IP contains 0100_{16}, what is the physical address of the next instruction to be fetched?

37. Into how many pages is the 80386DX's address space mapped when paging is enabled? What is the size of a page?

38. What are the three elements of the linear address that is produced by page translation? Give the size of each element.

39. What is the purpose of the translation lookaside buffer?

40. How large is a page frame? What selects the specific storage location in the page frame?

Section 8.4

41. How many bytes are in a descriptor? Name each of its fields and give their size.

42. With which registers are segment descriptors associated? System segment descriptors?

43. The selector 0224_{16} is loaded into the data segment register. This value points to a segment descriptor starting at address 00100220_{16} in the local descriptor table. If the words of the descriptor are

$$00100220H = 0110H$$

$$00100222H = 0000H$$

$$00100224H = 1A20H$$

$$00100226H = 0000H$$

what are the LIMIT and BASE?

44. Is the segment of memory identified by the descriptor in Problem 43 already loaded into physical memory? A code segment or data segment?

45. If the current value of EIP is 00000226_{16}, what is the physical address of the next instruction to be fetched from the code segment of Problem 43?

46. What do the 20 most significant bits of a page directory or page table entry stand for?

47. The page-mode protection of a page frame is to provide no access from the user protection level and read/write operation at the supervisor protection level. What are the settings of R/W and U/S?

48. What happens when an attempt is made to access a page frame that has P = 0 in its page table entry?

49. What does the D bit in a page directory entry stand for?

Section 8.5

50. If the instruction LGDT [INIT_GDTR] is to load the limit $FFFF_{16}$ and base 00300000_{16}, show how the descriptor must be stored in memory.

51. Write an instruction sequence that can be used to clear the task-switched bit of the MSW.

52. Write an instruction sequence that will load the local descriptor table register with the selector $02F0_{16}$ from register BX.

Section 8.6

53. Define the term *multitasking*.

54. What is a task?

55. What two safeguards are implemented by the 80386DX's protection mechanism?

56. What happens if either the segment limit check or segment attributes check fails?

57. What is the highest privilege level of the 80386DX protection model called? What is the lowest level called?

58. At what protection level are applications run?

59. What is the protection mechanism used to isolate local and global resource?

60. What protection mechanism is used to isolate tasks?

61. What is the privilege level of the segment defined by the descriptor in Problem 43?

62. What does CPL stand for? RPL?

63. State the data access protection rule.

64. Which privilege-level data segments can be accessed by an application running at level 3?

65. Summarize the code access protection rules.

66. If an application is running at privilege level 3, what privilege-level operating system software is available to it?

67. What purpose does a call gate serve?

68. Explain what happens when the instruction CALL [NEW_ROUTINE] is executed within a task. Assume that NEW_ROUTINE is at a privilege level that is higher than the CPL.

69. What is the purpose of the task state descriptor?

70. What is the function of a task state segment?

71. Where is the state of the prior task saved? Where is the linkage to the prior task saved?

72. Into which register is the TSS selector loaded to initiate a task?

73. Give an overview of the task switch sequence illustrated in Fig. 8.40.

Section 8.7

74. Which bit position in EFLAGS is VM?

75. Is 80386DX protection active or inactive in virtual 8086 mode? If so, what is the privilege level of a virtual 8086 program?

76. Can both protected-mode and virtual 8086 tasks coexist is an 80386DX multitasking environment?

77. Can multiple virtual 8086 tasks be active in an 80386DX multitasking environment?

9

The 80386DX Microprocessor and Its Memory and Input/Output Interfaces

▲ 9.1 INTRODUCTION

Up to this point in the book, we have studied the 80386DX microprocessor from a software point of view. We have covered its software architecture, its instruction set, and how to write, execute, and debug programs in assembly language. Now we will continue by examining the hardware architecture of the 80386DX-based microcomputer system. In this chapter we begin with the 80386DX microprocessor's signal interfaces, memory interface, types of input/output, input/output interface, and the input/output instructions. In the chapters that follow we cover memory devices, circuits, and subsystems, input/output interface circuits, and large-scale integrated peripheral devices, interrupts and exception processing, and the design of the microcomputer of a practical PC/AT-compatible personal computer. For this purpose, we have included the following topics in the chapter:

1. The 80386DX microprocessor
2. Interfaces of the 80386DX
3. System clock
4. Bus states and pipelined and nonpipelined bus cycles
5. Read and write bus cycle timing
6. Hardware organization of the memory address space
7. Memory interface circuitry
8. Programmable logic arrays: bus-control logic
9. Types of input/output
10. The isolated input/output interface
11. Input and output bus cycle timing
12. I/O instructions

372

The 80386DX, first announced in 1985, was the sixth member of Intel Corporation's 8086 family of microprocessors. We have learned that from the software point of view, the 80386DX offers several modes of operation: real mode, for compatibility with the large existing 8088/8086 software base, protected mode, which offers the user enhanced system-level features such as memory management, multitasking, and protection, and virtual 8086 mode, which provides 8086 real-mode compatibility while in the protected mode.

Hardware compatibility of the 80386 family of microprocessors with either the 8086 or 80286 microprocessors is much less of a concern than software compatibility. In fact, a number of changes have been made to the hardware architecture of the 80386DX to improve both its versatility and performance. For example, additional pipelining has been provided within the 80386DX and the address and data buses have both been made 32 bits in length. These two changes in the hardware result in increased performance for 80386DX-based microcomputers. Another feature, *dynamic bus sizing* for the data bus, provides more versatility in system hardware design.

The original 80386DX was manufactured using Intel's complementary high-performance metal-oxide-semiconductor III (CHMOSIII) process. Its circuitry is equivalent to approximately 275,000 transistors, more than twice those used in the design of the 80286 MPU and almost 10 times that of the 8086.

The 80386DX is housed in a 132-pin ceramic *pin grid array* (PGA) package. An 80386DX in this package is shown in Fig. 9.1. This package can be mounted in a socket that is soldered to the circuit board or have its leads inserted through holes in the board and soldered. The signal pinned out to each lead is shown in Fig. 9.2(a). Notice that all the 80386DX's signals are supplied at separate pins on the package. This is intended to simplify the microcomputer circuit design.

Looking at Fig. 9.2(a), we see that the rows of pins on the package are identified by row numbers 1 through 14 and the columns of pins are labeled A through P. Therefore, the location of the pin for each signal is uniquely defined by a column and row coordinate.

Figure 9.1 80386DX microprocessor. (Courtesy of Intel Corp.)

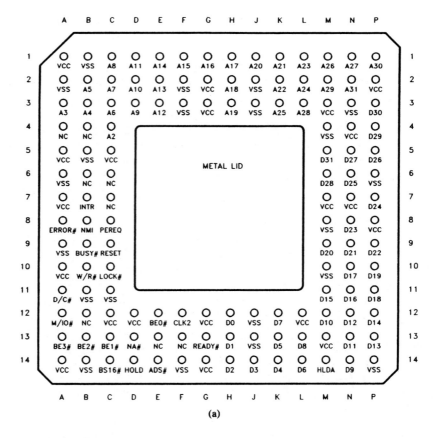

Figure 9.2(a) Pin layout of the 80386DX. (Reprinted by permission of Intel Corp. Copyright/Intel Corp. 1987)

For example, in Fig. 9.2(a) address line A_{31} is at the junction of column N and row 2. That is, it is at pin N2. The pin locations for all the 80386DX's signals are listed in Fig. 9.2(b).

EXAMPLE 9.1

At what pin location is the signal D_0?

Solution

Looking at Fig. 9.2(a), we find that the pin for D_0 is located in column H at row 12. Therefore, its pin is identified as H12.

The 80386SX MPU is not packaged in a PGA. It is available in a 100-lead *plastic quad flat package* (PQFP). A plastic package is used to permit lower cost for the device. This type of package is meant for *surface-mount* installation. That is, its pins do not go through the board; instead, the device is laid on top of the circuit board and then soldered in place.

Pin / Signal		Pin / Signal		Pin / Signal		Pin / Signal	
N2	A31	M5	D31	A1	V_{CC}	A2	V_{SS}
P1	A30	P3	D30	A5	V_{CC}	A6	V_{SS}
M2	A29	P4	D29	A7	V_{CC}	A9	V_{SS}
L3	A28	M6	D28	A10	V_{CC}	B1	V_{SS}
N1	A27	N5	D27	A14	V_{CC}	B5	V_{SS}
M1	A26	P5	D26	C5	V_{CC}	B11	V_{SS}
K3	A25	N6	D25	C12	V_{CC}	B14	V_{SS}
L2	A24	P7	D24	D12	V_{CC}	C11	V_{SS}
L1	A23	N8	D23	G2	V_{CC}	F2	V_{SS}
K2	A22	P9	D22	G3	V_{CC}	F3	V_{SS}
K1	A21	N9	D21	G12	V_{CC}	F14	V_{SS}
J1	A20	M9	D20	G14	V_{CC}	J2	V_{SS}
H3	A19	P10	D19	L12	V_{CC}	J3	V_{SS}
H2	A18	P11	D18	M3	V_{CC}	J12	V_{SS}
H1	A17	N10	D17	M7	V_{CC}	J13	V_{SS}
G1	A16	N11	D16	M13	V_{CC}	M4	V_{SS}
F1	A15	M11	D15	N4	V_{CC}	M8	V_{SS}
E1	A14	P12	D14	N7	V_{CC}	M10	V_{SS}
E2	A13	P13	D13	P2	V_{CC}	N3	V_{SS}
E3	A12	N12	D12	P8	V_{CC}	P6	V_{SS}
D1	A11	N13	D11			P14	V_{SS}
D2	A10	M12	D10				
D3	A9	N14	D9	F12	CLK2	A4	N.C.
C1	A8	L13	D8			B4	N.C.
C2	A7	K12	D7	E14	ADS#	B6	N.C.
C3	A6	L14	D6			B12	N.C.
B2	A5	K13	D5	B10	W/R#	C6	N.C.
B3	A4	K14	D4	A11	D/C#	C7	N.C.
A3	A3	J14	D3	A12	M/IO#	E13	N.C.
C4	A2	H14	D2	C10	LOCK#	F13	N.C.
A13	BE3#	H13	D1				
B13	BE2#	H12	D0	D13	NA#	C8	PEREQ
C13	BE1#			C14	BS16#	B9	BUSY#
E12	BE0#			G13	READY#	A8	ERROR#
		D14	HOLD				
C9	RESET	M14	HLDA	B7	INTR	B8	NMI

(b)

Figure 9.2(b) Signal pin numbering (Reprinted by permission of Intel Corp. Copyright/Intel Corp. 1987)

▲ 9.3 INTERFACES OF THE 80386DX

A block diagram of the 80386DX microprocessor is shown in Fig. 9.3. Here we have grouped its signal lines into four interfaces: the *memory/IO interface, interrupt interface, DMA interface,* and *coprocessor interface.* Figure 9.4 lists each of the signals at the 80386DX's interfaces. Included in this table are a mnemonic, function, type, and active level for each signal. For instance, we find that the signal with the mnemonic M/$\overline{\text{IO}}$ stands for memory/IO indication. This signal is an output produced by the 80386DX that is used to tell external circuitry whether the current address available on the address bus is for memory or an input/output device. Its active level is listed as 1/0, which means that logic 1 on this line

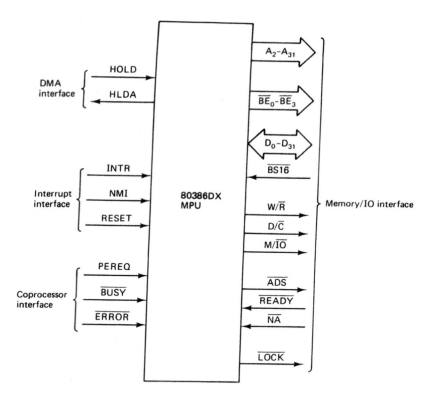

Figure 9.3 Block diagram of the 80386DX.

identifies a memory bus cycle and logic 0 an input/output bus cycle. On the other hand, the signal INTR at the interrupt interface is the maskable interrupt request input of the 80386DX. This input is active when at logic 1. By using this input, external devices can signal the 80386DX that they need to be serviced.

Memory/IO Interface

In a microcomputer system, the address bus and data bus signal lines form a parallel path over which the MPU talks with its memory and input/output subsystems. Like the 80286 microprocessor, but unlike the older 8086 and 8088, the 80386DX has a demultiplexed address/data bus. Notice in Fig. 9.2(b) that the address bus and data bus lines are located at different pins of the IC.

From a hardware point of view, there is only one difference between an 80386DX configured for the real-address mode or protected virtual-address mode. This difference is the size of the address bus. When in real mode, just the lower 18 address lines, A_2 through A_{19}, are active, whereas in the protected mode all 30 lines, A_2 through A_{31}, are functional. Of these, A_{19} and A_{31} are the most significant address bits, respectively. Actually, real-mode addresses are 20 bits long and protected mode addresses are 32 bits long. The other two bits, A_0 and A_1, are decoded inside the 80386DX, along with information about the size of the data to be transferred, to produce *byte-enable* outputs, $\overline{BE_0}$, $\overline{BE_1}$, $\overline{BE_2}$, and $\overline{BE_3}$.

As shown in Fig. 9.4, the address lines are outputs. They are used to carry address information from the 80386DX to memory and input/output ports. In real-address mode,

Name	Function	Type	Level
CLK2	System clock	I	–
$A_{31}-A_2$	Address bus	O	1
BE_3-BE_0	Byte enables	O	0
$D_{31}-D_0$	Data bus	I/O	1
$\overline{BS}_{16}$	Bus size 16	I	0
$W/\overline{R}$	Write/read indication	O	1/0
$D/\overline{C}$	Data/control indication	O	1/0
$M/\overline{IO}$	Memory I/O indication	O	1/0
$\overline{ADS}$	Address status	O	0
$\overline{READY}$	Transfer acknowledge	I	0
$\overline{NA}$	Next address request	I	0
$\overline{LOCK}$	Bus lock indication	O	0
INTR	Interrupt request	I	1
NMI	Nonmaskable interrupt request	I	1
RESET	System reset	I	1
HOLD	Bus hold request	I	1
HLDA	Bus hold acknowledge	O	1
PEREQ	Coprocessor request	I	1
$\overline{BUSY}$	Coprocessor busy	I	0
$\overline{ERROR}$	Coprocessor error	I	0

Figure 9.4   Signals of the 80386DX.

the 20-bit address gives the 80386DX the ability to address a 1M-byte physical memory address space. On the other hand, in protected mode the extended 32-bit address results in a 4G-byte physical memory address space. Moreover, when in protected mode, virtual addressing is provided through software. This results in a 64T-byte virtual memory address space.

In both the real and protected mode, the 80386DX microcomputer has an independent I/O address space. This I/O address space is 64K bytes in length. Therefore, just address lines A_2 through A_{15} and the $\overline{BE}$ outputs are used when addressing I/O devices.

The 80386SX has the same number of active address lines as the 80386DX in the real mode but fewer address lines in the protected mode. Its protected-mode address is just 25 bits long; therefore, the address space is limited to 16M byte. This is the first key difference between the 80386SX and 80386DX MPUs. The 80386SX accesses the I/O address space in exactly the same way as the 80386DX.

Since the 80386DX is a 32-bit microprocessor, its data bus is formed from the 32 data lines D_0 through D_{31}. D_{31} is the most significant bit and D_0 the least significant bit. These lines are identified as bidirectional in Fig. 9.4. This is because they have the ability to carry data either in or out of the MPU. The kinds of data transferred over these lines are read/write data and instructions for memory, input/output data for input/output devices, and interrupt type codes from an interrupt controller.

The 80386SX has a 16-bit data bus instead of a 32-bit data bus. This is the second important difference between the 80386SX and 80386DX. This results in lower performance, but has the advantage of lower cost for the microcomputer system.

Byte Enable	Data Bus Lines
$\overline{BE}_0$	$D_0 - D_7$
$\overline{BE}_1$	$D_8 - D_{15}$
$\overline{BE}_2$	$D_{16} - D_{23}$
$\overline{BE}_3$	$D_{24} - D_{31}$

Figure 9.5 Byte-enable outputs and data bus lines.

Earlier we indicated that the 80386DX supports dynamic bus sizing. Even though the 80386DX has 32 data lines, the size of the bus can be dynamically switched to 16 bits. This is done by simply switching the *bus size* 16 ($\overline{BS\ 16}$) input to logic 0. When in this mode, 32-bit data transfers are performed as two successive 16-bit transfers over bus lines D_0 through D_{15}. Since the 80386SX has a 16-bit data bus, it does not have this input.

Remember that the 80386DX supports byte, word, and double word data transfers over its data bus during a single bus cycle. Therefore, it must signal external circuitry what type of data transfer is taking place and over which part of the data bus the data will be carried. The MPU does this by activating the appropriate byte enable ($\overline{BE}$) output signals.

Figure 9.5 lists each byte enable output signal and the part of the data bus it is intended to enable. For instance, here we see that $\overline{BE}_0$ corresponds to data bus lines D_0 through D_7. If a byte of data is being read from memory, just one of the $\overline{BE}$ outputs is made active. For instance, if the most significant byte of an aligned double word is read from memory, $\overline{BE}_3$ is switched to logic 0 and the data moves on data lines D_{24} through D_{31}. On the other hand, if a word of data is being read, two outputs become active. An example would be to read the most significant word of an aligned double word from memory. In this case, $\overline{BE}_2$ and $\overline{BE}_3$ are both switched to logic 0 and the data is carried by lines D_{16} through D_{31}. Finally, if an aligned double word read is taking place, all four $\overline{BE}$ outputs are made active and all data lines are used to transfer the data.

EXAMPLE 9.2

What code is output on the byte-enable lines whenever the address on the bus is for an instruction-acquisition bus cycle?

Solution

Since code is always fetched as 32-bit words (aligned double words), all the byte-enable outputs are made active. Therefore,

$$\overline{BE}_3\overline{BE}_2\overline{BE}_1\overline{BE}_0 = 0000_2$$

The byte-enable lines work exactly the same when write data transfers are performed over the bus. Figure 9.6(a) identifies what type of data transfer takes place for all the possible variations of the byte-enable outputs. For instance, we find that $\overline{BE}_3\overline{BE}_2\overline{BE}_1\overline{BE}_0 = 1110_2$ means that a byte of data is written over data bus lines D_0 through D_7.

EXAMPLE 9.3

What type of data transfer takes place and over which data bus lines are data transferred if the byte enable code output is

$$\overline{BE}_3\overline{BE}_2\overline{BE}_1\overline{BE}_0 = 1100_2$$

$\overline{BE}_3$	$\overline{BE}_2$	$\overline{BE}_1$	$\overline{BE}_0$	$D_{31}-D_{24}$	$D_{23}-D_{16}$	$D_{15}-D_8$	D_7-D_0
1	1	1	0				XXXXXXXX
1	1	0	1			XXXXXXXX	
1	0	1	1		XXXXXXXX		
0	1	1	1	XXXXXXXX			
1	1	0	0			XXXXXXXX	XXXXXXXX
1	0	0	1		XXXXXXXX	XXXXXXXX	
0	0	1	1	XXXXXXXX	XXXXXXXX		
1	0	0	0		XXXXXXXX	XXXXXXXX	XXXXXXXX
0	0	0	1	XXXXXXXX	XXXXXXXX	XXXXXXXX	
0	0	0	0	XXXXXXXX	XXXXXXXX	XXXXXXXX	XXXXXXXX

(a)

$\overline{BE}_3$	$\overline{BE}_2$	$\overline{BE}_1$	$\overline{BE}_0$	$D_{31}-D_{24}$	$D_{23}-D_{16}$	$D_{15}-D_8$	D_7-D_0
1	1	1	0				XXXXXXXX
1	1	0	1			XXXXXXXX	
1	0	1	1		XXXXXXXX		DDDDDDDD
0	1	1	1	XXXXXXXX		DDDDDDDD	
1	1	0	0			XXXXXXXX	XXXXXXXX
1	0	0	1		XXXXXXXX	XXXXXXXX	
0	0	1	1	XXXXXXXX	XXXXXXXX	DDDDDDDD	DDDDDDDD
1	0	0	0		XXXXXXXX	XXXXXXXX	XXXXXXXX
0	0	0	1	XXXXXXXX	XXXXXXXX	XXXXXXXX	
0	0	0	0	XXXXXXXX	XXXXXXXX	XXXXXXXX	XXXXXXXX

(b)

Figure 9.6 (a) Types of data transfers for the various byte-enable combinations. (b) Data transfers that include duplication.

Solution

Looking at the table in Fig. 9.6(a), we see that a word of data is transferred over data bus lines D_0 through D_{15}.

With its 16-bit data bus, the 80386SX can transfer only a byte or word of data over the bus during a single bus cycle. For this reason, the four byte-enable signals of the 80386DX are replaced by just two signals, *byte high enable* ($\overline{BHE}$) and *byte low enable* ($\overline{BLE}$) on the 80386SX. Logic 0 at $\overline{BLE}$ tells that a byte of data is being transferred over data bus line D_0 through D_7 and logic 0 at $\overline{BHE}$ means a byte transfer is taking place over data bus lines D_8 through D_{15}. When a word of data is transferred, both these signals are at their active 0 logic level.

The 80386DX performs what is called *data duplication* during certain types of write cycles. Data duplication is provided in the 80386DX to optimize the performance of the data bus when it is set for 16-bit mode. Notice that whenever a write cycle is performed in which data are transferred only over the upper part of the 32-bit data bus, the data are duplicated on the corresponding lines of the lower part of the bus. For example, looking at Fig. 9.6(b), we see that when $\overline{BE}_3\overline{BE}_2\overline{BE}_1\overline{BE}_0 = 1011_2$, data (denoted as XXXXXXXX) are actually being written over data bus lines D_{16} through D_{23}. However, at the same time, the data (denoted as DDDDDDDD in Fig. 9.6(b)) are automatically duplicated on data bus lines D_0 through D_7. Despite the fact that the byte is available on the lower eight data bus lines, $\overline{BE}_0$ stays inactive. The same thing happens when a word of data is transferred over D_{16} through D_{31}. In this example, $\overline{BE}_3\overline{BE}_2\overline{BE}_1\overline{BE}_0 = 0011_2$, and Fig. 9.6(b) shows that the word is duplicated on data lines D_0 through D_{15}.

EXAMPLE 9.4

If a word of data that is being written to memory is accompanied by the byte-enable code 1001_2, over which data bus lines are the data carried? Is data duplication performed for this data transfer?

Solution

In the tables of Fig. 9.6, we find that for the byte-enable code 1001_2 the word of data is transferred over data bus lines D_8 through D_{23}. For this transfer, data duplication does not occur.

Control signals are required to support information transfers over the 80386DX's address and data buses. They are needed to signal when a valid address is on the address bus, in which direction data are to be transferred over the data bus, when valid write data are on the data bus, and when an external device can put read data on the data bus. The 80386DX does not directly produce signals for all these functions. Instead, it outputs bus cycle definition and control signals at the beginning of each bus cycle. These bus cycle indication signals must be decoded in external circuitry to produce the needed memory and I/O control signals.

Three signals are used to identify the type of 80386DX bus cycle that is in progress. In Figs. 9.3 and 9.4, they are labeled *write/read indication* ($W/\overline{R}$), *data/control indication* ($D/\overline{C}$), and *memory/input-output indication* ($M/\overline{IO}$). The table in Fig. 9.7 lists all possible combinations of the bus cycle indication signals and the corresponding type of bus cycle. Here we find that the logic level of memory/input-output ($M/\overline{IO}$) tells whether a memory or input/output cycle is to take place over the bus. Logic 1 at this output signals a memory operation, and logic 0 signals an I/O operation. The next signal in Fig. 9.7, data/control indication ($D/\overline{C}$), identifies whether the current bus cycle is a data or control cycle. In the table we see that it signals *control cycle* (logic 0) for instruction fetch, interrupt acknowledge, and halt/shutdown operations and *data cycle* (logic 1) for memory and I/O read and write operations. Looking more closely at the table in Fig. 9.7, we find that if the code on these

$M/\overline{IO}$	$D/\overline{C}$	$W/\overline{R}$	Type of Bus Cycle
0	0	0	Interrupt acknowledge
0	0	1	Idle
0	1	0	I/O data read
0	1	1	I/O data write
1	0	0	Memory code read
1	0	1	Halt/shutdown
1	1	0	Memory data read
1	1	1	Memory data write

Figure 9.7 Bus-cycle indication signals and the types of bus cycles.

two lines, M/$\overline{\text{IO}}$ D/$\overline{\text{C}}$, is 00, an interrupt is to be acknowledged; if it is 01, an input/output operation is in progress; if it is 10, instruction code is being fetched; and finally, if it is 11, a data memory read or write is taking place.

The last signal identified in Fig. 9.7, write/read indication (W/$\overline{\text{R}}$), identifies the specific type of memory or input/output operation that will occur during a bus cycle. For example, when W/$\overline{\text{R}}$ is logic 0 during a bus cycle, data are to be read from memory or an I/O port. On the other hand, logic 1 at W/$\overline{\text{R}}$ says that data are to be written into memory or an I/O device. For example, all bus cycles that read instruction code from memory are accompanied by logic 0 on the W/$\overline{\text{R}}$ line.

EXAMPLE 9.5

If the bus cycle indication code M/$\overline{\text{IO}}$ D/$\overline{\text{C}}$ W/$\overline{\text{R}}$ equals 010, what type of bus cycle is taking place?

Solution

Looking at the table in Fig. 9.7, we see that bus cycle indication code 010 identifies an I/O read (input) bus cycle.

Three bus cycle–control signals are produced directly by the 80386DX. They are identified in Figs. 9.3 and 9.4 as *address status* ($\overline{\text{ADS}}$), *transfer acknowledge* ($\overline{\text{READY}}$), and *next-address request* ($\overline{\text{NA}}$). The $\overline{\text{ADS}}$ output is switched to logic 0 to indicate that the bus cycle identification code (M/$\overline{\text{IO}}$D/$\overline{\text{C}}$W/$\overline{\text{R}}$), byte-enable code ($\overline{\text{BE}}_3\overline{\text{BE}}_2\overline{\text{BE}}_1\overline{\text{BE}}_0$), and address ($A_2$ through A_{31}) signals are all stable. Therefore, it is normally applied to an input of the external bus-control logic circuit and tells it that a valid bus cycle indication code and address are available. In Fig. 9.7 the bus cycle indication code M/$\overline{\text{IO}}$ D/$\overline{\text{C}}$ W/$\overline{\text{R}}$ = 001 is identified as *idle*. That is, it is the code that is output whenever no bus cycle is being performed.

$\overline{\text{READY}}$ can be used to insert wait states into the current bus cycle such that it is extended by a number of clock periods. In Fig. 9.4, we find that this signal is an input to the 80386DX. Normally, it is produced by the microcomputer's memory or input/output subsystem and supplied to the 80386DX by way of external bus control logic circuitry. By switching $\overline{\text{READY}}$ to logic 0, slow memory or I/O devices can tell the 80386DX when they are ready to permit a data transfer to be completed.

Earlier we pointed out that the 80386DX supports address pipelining at its bus interface. By address pipelining, we mean that the address and bus cycle indication code for the next bus cycle is output before $\overline{\text{READY}}$ becomes active to signal that the prior bus cycle can be completed. This mode of operation is optional. The external bus-control logic circuitry activates pipelining by switching the next-address request ($\overline{\text{NA}}$) input to logic 0. By using pipelining, delays introduced by the decode logic can be made transparent and the address to data access time is increased. In this way, the same level of performance can be obtained with slower, lower-cost memory devices.

One other bus interface control output that is supplied by the 80386DX is *bus lock indication* ($\overline{\text{LOCK}}$). This signal is needed to support multiple processor architectures. In multiprocessor systems that employ shared resources, such as global memory, this signal can be employed to assure that the 80386DX has uninterrupted control of the system bus and the shared resource. That is, by switching its $\overline{\text{LOCK}}$ output to logic 0, the MPU can lock up the shared resource for exclusive use.

The 80386SX has the same bus cycle indication and control signals as the 80386DX. The signals that identify the type of bus cycle are write/read indication (W/$\overline{\text{R}}$), data/control indication (D/$\overline{\text{C}}$), and memory/input-output indication (M/$\overline{\text{IO}}$), whereas those that control the bus cycle are address status ($\overline{\text{ADS}}$), transfer acknowledge ($\overline{\text{READY}}$), next address request ($\overline{\text{NA}}$) and bus lock indication ($\overline{\text{LOCK}}$). Each of these signals serve the exact same function as they do for the 80386DX MPU.

Interrupt Interface

Looking at Figs. 9.3 and 9.4, we find that the key interrupt interface signals are *interrupt request* (INTR), *nonmaskable interrupt request* (NMI), and *system reset* (RESET). INTR is an input to the 80386DX that can be used by external devices to signal that they need to be serviced. The 80386DX samples this input at the beginning of each instruction. Logic 1 on INTR represents an active interrupt request.

When an active interrupt request has been recognized by the 80386DX, it signals this fact to external circuitry and initiates an interrupt-acknowledge bus cycle sequence. In Fig. 9.7 we see that the occurrence of an interrupt-acknowledge bus cycle is signaled to external circuitry with the bus cycle definition M/$\overline{\text{IO}}$ D/$\overline{\text{C}}$ W/$\overline{\text{R}}$ equal to 000. This bus cycle indication code can be decoded in the external bus control logic circuitry to produce an interrupt acknowledge signal. With this interrupt acknowledge signal, the 80386DX tells the external device that its request for service has been granted. This completes the interrupt request/acknowledge handshake. At this point, program control is passed to the interrupt's service routine.

The INTR input is maskable. That is, its operation can be enabled or disabled with the interrupt flag (IF) within the 80386DX's flag register. On the other hand, the NMI input, as its name implies, is a nonmaskable interrupt input. On any 0-to-1 transition of NMI, a request for service is latched within the 80386DX. Independent of the setting of the IF flag, control is passed to the beginning of the nonmaskable interrupt service routine at the completion of execution of the current instruction.

Finally, the RESET input provides a method of implementing a hardware reset for the 80386DX microprocessor. For instance, the microcomputer system can be reset at power-on by using this input. Switching RESET to logic 1 initializes the internal registers of the 80386DX. When it is returned to logic 0, program control is passed to the beginning of a reset service routine. This routine is used to initialize the rest of the system's resources, such as I/O ports, the interrupt flag, and data memory. A diagnostic routine that tests the 80386DX microprocessor can also be initiated as part of the reset sequence. This assures an orderly startup of the microcomputer system.

The 80386SX's interrupt interface is exactly the same as that of the 80386DX. It is implemented with the same signals, interrupt request (INTR), nonmaskable interrupt request (NMI), and system reset (RESET) and is identified to external circuitry with the same bus cycle code.

DMA Interface

Now that we have examined the signals of the 80386DX's interrupt interface, let us turn our interest to the *direct memory access* (DMA) interface. From Figs. 9.3 and 9.4 we find that the DMA interface is implemented with just two signals: *bus hold request* (HOLD) and *bus hold acknowledge* (HLDA). When an external device, such as a *DMA controller*, wants to take over control of the local address and data buses, it signals this fact to the 80386DX by switching the HOLD input to logic 1. At completion of the current bus cycle, the 80386DX enters the hold state. When in the hold state, its local bus signals are in the high-impedance state. Next, the 80386DX signals external devices that it has given up control of the bus by switching its HLDA output to the 1 logic level. This completes the hold/hold acknowledge handshake sequence. The 80386DX remains in this state until the hold request is removed. The 80386SX's DMA interface is exactly the same as that of the 80386DX.

Coprocessor Interface

In Fig. 9.3 we find that a coprocessor interface is provided on the 80386DX microprocessor to permit it to easily interface to the *80387DX numeric coprocessor*. The 80387DX cannot perform transfers over the data bus by itself. Whenever the 80387DX needs to read or write operands from memory, it must signal the 80386DX to initiate the data transfers. The 80387DX does this by switching the *coprocessor request* (PEREQ) input of the 80386DX to logic 1.

The other two signals included in the external coprocessor interface are $\overline{\text{BUSY}}$ and $\overline{\text{ERROR}}$. *Coprocessor busy* ($\overline{\text{BUSY}}$) is an input of the 80386DX. Whenever the 80387DX is executing a numeric instruction, it signals this fact to the 80386DX by switching the $\overline{\text{BUSY}}$ input to logic 0. In this way, the 80386DX knows not to request the numeric coprocessor to perform another calculation until $\overline{\text{BUSY}}$ returns to 1. Moreover, if an error occurs in a calculation performed by the numeric coprocessor, this condition is signaled to the 80386DX by switching the *coprocessor error* ($\overline{\text{ERROR}}$) input to the 0 logic level. This interface is implemented the same way on the 80386SX MPU.

▲ 9.4 SYSTEM CLOCK

The time base for synchronization of the internal and external operations of the 80386DX micropocessor is provided by the *clock* (CLK2) input signal. At present, the 80386DX has been available with four different clock speeds. The original 80386DX-16 MPU operates at 16 MHz and its three faster versions, the 80386DX-20, 80386DX-25, and 80386DX-33, operate at 20, 25, and 33 MHz, respectively. The clock signal applied to the CLK2 input of the 80386DX is twice the frequency rating of the microprocessor. Therefore, CLK2 of an 80386DX-16 is driven by a 32-MHz signal. This signal must be generated in external circuitry. The 80386SX is also avalable in each of these four speeds.

The waveform of the CLK2 input of the 80386DX is given in Fig. 9.8. This signal is specified at CMOS-compatible voltage levels and not TTL levels. Its minimum and maximum low logic levels are $V_{\text{ILCmin}} = -0.3$ V and $V_{\text{ILCmax}} = 0.8$ V, respectively. Moreover, the

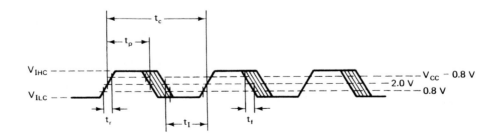

Figure 9.8   System clock (CLK2) waveform

minimum and maximum high logic levels are $V_{IHCmin} = V_{CC} - 0.8$ V and $V_{IHCmax} = V_{CC} + 0.3$ V, respectively. The minimum period of the 16-MHz clock signal is $t_{Cmin} = 31$ ns (measured at the 2.0-V level); its minimum high time t_{Pmin} and low time t_{lmin} (measured at the 2.0-V level) are both equal to 9 ns; and the maximum rise time t_{rmax} and fall time t_{fmax} of its edges (measured between the $V_{CC} - 0.8$-V and 0.8-V levels) are equal to 8 ns.

▲ 9.5 BUS STATES AND PIPELINED AND NONPIPELINED BUS CYCLES

Before looking at the bus cycles of the 80386DX, let us first examine the relationship between the timing of the 80386DX's CLK2 input and its bus cycle states. The *internal processor clock* (PCLK) signal is at half the frequency of the external clock input signal. Therefore, as shown in Fig. 9.9, one processor clock cycle corresponds to two CLK2 cycles. Notice that these CLK2 cycles are labeled *phase 1* (ϕ_1) and *phase 2* (ϕ_2). In a 20-MHz 80386DX microprocessor, CLK2 equals 40 MHz and each clock cycle has a duration of 25 ns. In Fig. 9.9, we see that the two phases ($\phi_1 + \phi_2$) of a processor cycle are identified as one processor clock period. A processor clock period is also called a *T state*. Therefore, the minimum length of an internal processor clock cycle is 50 ns.

Nonpipelined and Pipelined Bus Cycles

A *bus cycle* is the activity performed whenever a microprocessor accesses information in program memory, data memory, or an input/output device. The 80386DX can perform bus cycles with either of two types of timing: *nonpipelined* and *pipelined*. Here we will examine the difference between these two types of bus cycles.

Figure 9.10 shows a typical nonpipelined microprocessor bus cycle. Notice that the bus cycle contains two T states and that they are called T_1 and T_2. During the T_1 part of the bus cycle, the 80386DX outputs the address of the storage location that is to be accessed on the address bus, a bus cycle indication code, and control signals. In the case of a write cycle, write data are also output on the data bus during T_1. The second state, T_2, is the part of the bus cycle during which external devices are to accept write data from the data bus, or in the case of a read cycle, put data on the data bus.

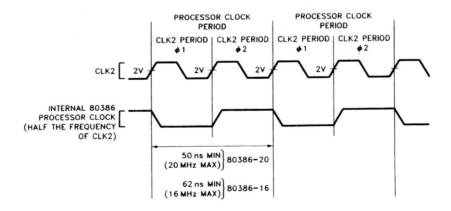

Figure 9.9 Processor clock cycles. (Reprinted by permission of Intel Corp. Copyright Intel/Corp. 1987)

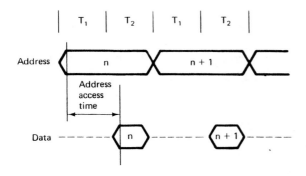

Figure 9.10 Typical read/write bus cycles.

For instance in Fig. 9.10, we see that the sequence of events starts with an address, denoted as n, being output on the address bus in clock state T_1. Later in the bus cycle, while the address is still available on the address bus, a read or write data transfer takes place over the data bus. Notice that the data transfer for address n is shown to occur in clock state T_2. Since each bus cycle has a minimum of two T states (4 CLK2 cycles), the minimum bus cycle duration for an 80386DX-20 is 100 ns.

Let us now look at a microprocessor bus cycle that employs *pipelining*. By pipelining we mean that addressing for the next bus cycle is overlapped with the data transfer of the prior bus cycle. When address pipelining is in use, the address, bus cycle indication code, and control signals for the next bus cycle are output during T_2 of the prior cycle, instead of the T_1 that follows.

In Fig. 9.11 we see that address n becomes valid in the T_2 state of the prior bus cycle, and then the data transfer for address n takes place in the next T_2 state. Moreover, notice that at the same time that data transfer n occurs, address $n + 1$ is output on the address bus. In this way we see that the microprocessor begins addressing the next storage location that it is to accessed while it is still performing the read or write of data for the previously addressed storage location. Due to the address/data pipelining, the memory or I/O subsystem actually has 5 CLK2 cycles (125 ns for an 80386DX-20 running at full speed) to perform the data transfer, even though the duration of every bus cycle is just four clock cycles (100 ns).

The interval of time denoted as *address-access time* in Fig. 9.10 represents the amount of time that the address is stable prior to the read or write of data actually taking place. Notice that this duration is less than the 4 CLK2 cycles in a nonpipelined bus cycle. Figure 9.11 shows that in a pipelined bus cycle the *effective address-access time* equals the duration

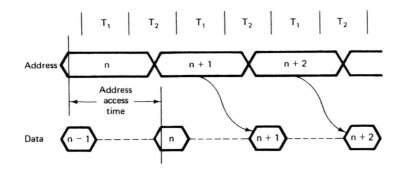

Figure 9.11 Pipelined bus cycle. (Reprinted by permission of Intel Corp. Copyright/Intel Corp. 1987)

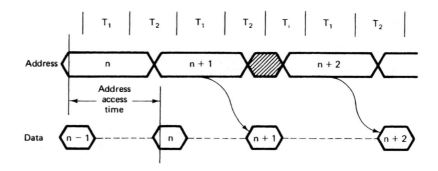

Figure 9.12 Idle states in bus activity.

of a complete bus cycle. This leads us to the benefit of the 80386DX's pipelined mode of bus operation over the nonpipelined mode of operation. It is that, for a fixed address-access time (equal speed memory design), the 80386DX pipelined bus cycle will have a shorter duration than that of its nonpipelined bus cycle. This results in improved bus performance.

Another way of looking at this is to say that when using equal-speed memory designs, an 80386DX that uses a pipelined bus can be operated at a higher clock rate than a design that executes a nonpipelined bus cycle. Once again the result is higher system performance.

In Fig. 9.11 we find that at completion of the bus cycle for address n, another bus cycle is initiated immediately for address $n + 1$. Sometimes another bus cycle will not be initiated immediately. For instance, if the 80386DX's prefetch queue is already full and the instruction that is currently being executed does not need to access operands in memory, no bus activity will take place. In this case the bus goes into a mode of operation known as an *idle state*. Figure 9.12 shows a sequence of bus activity in which an idle state exists between the bus cycles for addresses $n + 1$ and $n + 2$. The duration of a single idle state is equal to two CLK2 cycles.

Wait states can be inserted to extend the duration of the 80386DX's bus cycle. This is done in response to a request by an event in external hardware instead of an internal event such as a full queue. In fact, the $\overline{\text{READY}}$ input of the 80386DX is provided specifically for this purpose. This input is sampled in the later part of the T_2 state of every bus cycle to determine if the data transfer should be completed. Figure 9.13 shows that logic 1 at this

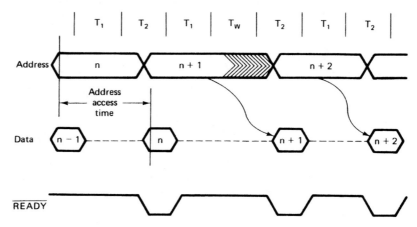

Figure 9.13 Bus cycle with wait states.

input indicates that the current bus cycle should not be completed. As long as $\overline{\text{READY}}$ is held at the 1 level, the read or write data transfer does not take place and the current T_2 state becomes a wait states (T_w) to extend the bus cycle. The bus cycle is not completed until external hardware returns $\overline{\text{READY}}$ back to logic 0. This ability to extend the duration of a bus cycle permits the use of slow memory or I/O devices in the microcomputer system.

▲ 9.6 READ AND WRITE BUS CYCLE TIMING

In the preceding sections we introduced the 80386DX's memory interface signals, bus cycle states, and the concepts of nonpipelined and pipelined bus cycles. We found that the 80386DX's bus can be dynamically configured to operate in either the 32-bit (standard) mode or 16-bit mode and that data transfers can be performed using either nonpipelined or pipelined bus cycles. In this section we continue by studying in detail the sequence of events that take place during the 80386DX's memory read and write bus cycles. The bus cycles of the 80386SX are identical to those we will describe for the 80386DX, but with two small exceptions. Both of them are due to the fact that the 80386SX's external data bus is 16 bits wide, not 32 bits. First, there is no $\overline{\text{BS 16}}$ signal, and second, a maximum of 16 bits of data can be transferred during a bus cycle.

Nonpipelined Read Cycle Timing

The memory interface signals that occur when the 80386DX reads data from memory are shown in Fig. 9.14. This diagram shows two separate nonpipelined read cycles. They are *cycle 1*, which is performed without wait states, and *cycle 2*, which includes one wait state. Let us now trace through the events that take place in cycle 1 as data or instructions are read from memory.

The occurrence of all the signals in the read bus cycle timing diagram are illustrated relative to the two timing states, T_1 and T_2, of the 80386DX's bus cycle. The read operation starts at the beginning of phase 1 (ϕ_1) in the T_1 state of the bus cycle. At this moment, the 80386DX outputs the address of the double-word memory location to be accessed on address bus lines A_2 through A_{31}, outputs the byte-enable signals $\overline{\text{BE}_0}$ through $\overline{\text{BE}_3}$ that identify the bytes of the double word that are to be fetched, and switches address strobe ($\overline{\text{ADS}}$) to logic 0 to signal that a valid address is on the address bus. Looking at Fig. 9.14, we see that the address and bus cycle indication signals are maintained stable during the complete bus cycle; however, they must be latched into the external bus control logic circuitry synchronously with the pulse to logic 0 on $\overline{\text{ADS}}$. At the end of ϕ_2 of T_1, $\overline{\text{ADS}}$ is returned to its inactive 1 logic level.

Notice in Fig. 9.14 that the bus cycle indication signals, $M/\overline{\text{IO}}$, $D/\overline{\text{C}}$, and $W/\overline{\text{R}}$, are also made valid at the beginning of ϕ_1 of state T_1. The bus cycle indication codes that apply to a memory read cycle are highlighted in Fig. 9.15. Here we see that, if code is being read from memory, $M/\overline{\text{IO}}D/\overline{\text{C}}W/\overline{\text{R}}$ equals 100. That is, signal $M/\overline{\text{IO}}$ is set to logic 1 to indicate to the circuitry in the memory interface that a memory bus cycle is in progress, $D/\overline{\text{C}}$ is set to 0 to indicate that code memory is to be accessed, and $W/\overline{\text{R}}$ is set to 0 to indicate that data are being read from memory.

At the beginning of ϕ_1 in T_2 of the read cycle, external circuitry must signal the 80386DX whether the bus is to operate in the 16- or 32-bit mode. In Fig. 9.14 we see that it does this with the $\overline{\text{BS 16}}$ signal. The 80386DX samples this input in the middle of the T_2 bus cycle state. The 1 logic level shown in the timing diagram indicates that a 32-bit data transfer is to take place.

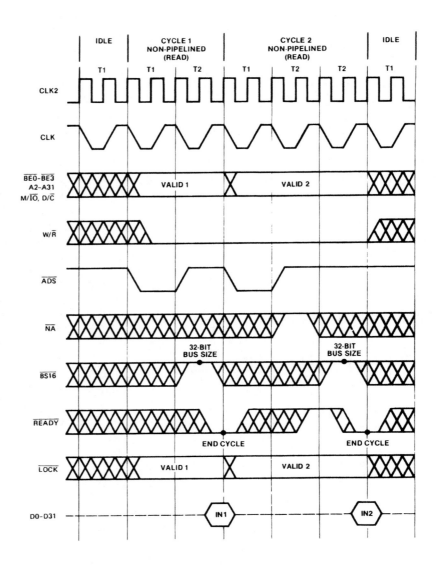

Figure 9.14 Nonpipelined read-cycle timing. (Reprinted by permission of Intel Corp. Copyright/Intel Corp. 1987)

M/$\overline{\text{IO}}$	D/$\overline{\text{C}}$	W/$\overline{\text{R}}$	Type of Bus Cycle
0	0	0	Interrupt acknowledge
0	0	1	Idle
0	1	0	I/O data read
0	1	1	I/O data write
1	0	0	Memory code read
1	0	1	Halt/shutdown
1	1	0	Memory data read
1	1	1	Memory data write

Figure 9.15 Memory-read bus-cycle indication codes.

Notice in Fig. 9.14 that at the end of T_2 the $\overline{\text{READY}}$ input is tested by the 80386DX. The logic level at this input signals whether the current bus cycle is to be completed or extended with wait states. The logic 0 shown at this input means that the bus cycle is to run to completion. For this reason we see that data available on data bus lines D_0 through D_{31} are read into the 80386DX at the end of T_2.

Nonpipelined Write Cycle Timing

The nonpipelined write bus cycle timing diagram, shown in Fig. 9.16, is similar to that given for a nonpipelined read cycle in Fig. 9.14. It includes waveforms for both a no-wait-state write operation (cycle 1) and a one-wait-state write operation (cycle 2). Looking at

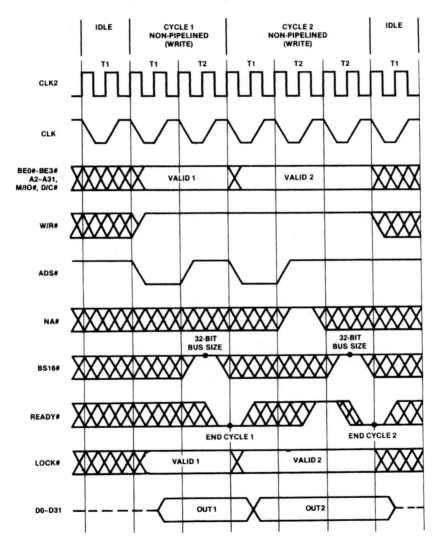

Figure 9.16 Nonpipelined write cycle timing. (Reprinted by permission of Intel Corp. Copyright/Intel Corp. 1987)

M/$\overline{\text{IO}}$	D/$\overline{\text{C}}$	W/$\overline{\text{R}}$	Type of Bus Cycle
0	0	0	Interrupt acknowledge
0	0	1	Idle
0	1	0	I/O data read
0	1	1	I/O data write
1	0	0	Memory code read
1	0	1	Halt/shutdown
1	1	0	Memory data read
1	1	1	Memory data write

Figure 9.17 Memory write-bus cycle-indication code.

the write cycle waveforms, we find that the address, byte enable, and bus cycle indication signals are output at the beginning of ϕ_1 of the T_1 state. All these signals are to be latched in external circuitry with the pulse at $\overline{\text{ADS}}$. The one difference here is that W/$\overline{\text{R}}$ is at the 1 logic level instead of 0. In fact, as shown in Fig. 9.17, the bus cycle indication code for a memory data write is M/$\overline{\text{IO}}$ D/$\overline{\text{C}}$ W/$\overline{\text{R}}$ equals 111; therefore, M/$\overline{\text{IO}}$ and D/$\overline{\text{C}}$ are also at the 1 logic level.

Let us now look at what happens on the data bus during a write bus cycle. Notice in Fig. 9.16 that the 80386DX outputs the data that are to be written to memory onto the data bus at the beginning of ϕ_2 in the T_1 state. These data are maintained valid until the end of the bus cycle. In the middle of the T_2 state, the logic level of the $\overline{\text{BS16}}$ input is tested by the 80386DX and indicates that the bus is to be used in the 32-bit mode. Finally, at the end of T_2, $\overline{\text{READY}}$ is tested and found to be at its active 0 logic level. Since the memory subsystem has made $\overline{\text{READY}}$ logic 0, the write cycle is complete and the buses and control signal lines are prepared for the next write cycle.

Wait States in a Nonpipelined Memory Bus Cycle

In the previous section we showed how wait states are used to lengthen the duration of the memory bus cycle of a microprocessor. Wait states are inserted with the $\overline{\text{READY}}$ input signal. Upon request from an event in external hardware, for instance, slow memory, the $\overline{\text{READY}}$ input is switched to logic 1. This signals the 80386DX that the memory subsystem is not ready and that the current bus cycle should not be completed. Instead, it is extended by repeating the T_2 state. Therefore, the duration of one wait state ($T_w = T_2$) equals 50 ns for 20-MHz clock operation.

Cycle 2 in Fig. 9.14 shows a read cycle extended by one wait state. Notice that the address, byte-enable, and bus cycle indication signals are maintained throughout the wait-state period. In this way, the read cycle is not completed until $\overline{\text{READY}}$ is switched to logic 0 in the second T_2 state.

EXAMPLE 9.6

If cycle 2 in Fig. 9.16 is for an 80386DX-20 running at full speed, what is the duration of the bus cycle?

Solution

Each T state in the bus cycle of an 80386DX running at 20 MHz is 50 ns. Since the write cycle is extended by one wait state, the write cycle takes 150 ns.

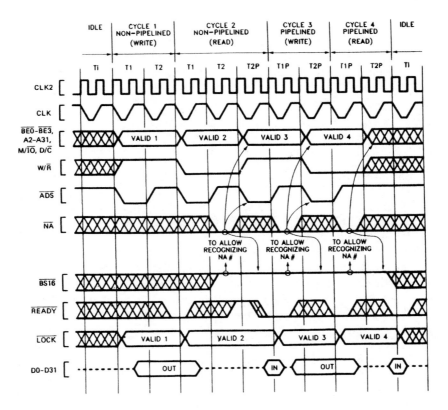

Figure 9.18 Pipelined read- and write-cycle timing (Reprinted by permission of Intel Corp. Copyright/Intel Corp. 1987)

Pipelined Read/Write Cycle Timing

Timing diagrams for both nonpipelined and pipelined read and write bus cycles are shown in Fig. 9.18. Here we find that the cycle identified as *cycle 3* is an example of a pipelined write bus cycle. Let us now look more closely at this bus cycle.

Remember that when pipelined addressing is in use, the 80386DX outputs the address information for the next bus cycle during the T_2 state of the current cycle. The signal next address ($\overline{NA}$) is used to signal the 80386DX that a pipelined bus cycle is to be initiated. This input is sampled by the 80386DX during any bus state when $\overline{ADS}$ is not active. In Fig. 9.18 we see that ($\overline{NA}$) is first tested as 0 (active) during T_2 of cycle 2. This nonpipelined read cycle is also extended with period T_{2P} because $\overline{READY}$ is not active. Notice that the address, byte enable, and bus cycle indication signals for cycle 3 become valid (identified as VALID 3 in Fig. 9.18) during this period and a pulse is produced at $\overline{ADS}$. This information is latched externally synchronously with $\overline{ADS}$ and decoded to produce bus enable and control signals. In this way, the memory access time for a zero-wait-state memory cycle has been increased.

Bus cycle 3 represents a pipelined write cycle. The data to be written to memory are output on D_0 through D_{31} at ϕ_2 of T_{1P} and remain valid for the rest of the cycle. Logic 0 on $\overline{READY}$ at the end of T_{2P} indicates that the write cycle is completed without wait states.

Looking at Fig. 9.18, we find that $\overline{NA}$ is also active during T_{1P} of cycle 3. This means that cycle 4 will be performed with pipelined timing. Cycle 4 is an example of a zero-wait-state pipelined read cycle. In this case, the address information, bus cycle indication, and address strobe are output during T_{2P} of cycle 3 (the previous cycle), and memory data are read into the MPU at the end of T_{2P} of cycle 4.

▲ 9.7 HARDWARE ORGANIZATION OF THE MEMORY ADDRESS SPACE

Earlier we indicated that in protected mode the 32-bit address bus of the 80386DX results in a 4G-byte physical memory address space. As shown in Fig. 9.19, from a software point of view, this memory is organized as individual bytes over the address range from 00000000_{16} through $FFFFFFFF_{16}$. The 80386DX can also access data in this memory as words or double words.

From a hardware point of view, the physical address space is implemented as four independent byte-wide banks, and each of these banks is 1G byte in size. In Fig. 9.20 we find that the banks are identified as *bank 0, bank 1, bank 2,* and *bank 3.* Notice that they correspond to addresses that produce byte-enable signals $\overline{BE}_0$, $\overline{BE}_1$, $\overline{BE}_2$, and $\overline{BE}_3$, respectively. Logic 0 at a byte-enable input selects the bank for operation. Looking at Fig. 9.20, we see that address bits A_2 through A_{31} are applied to all four banks in parallel. On the other hand, each memory bank supplies just eight lines of the 80386DX's 32-bit data bus. For example, byte data transfers for bank 0 take place over data bus lines D_0 through D_7, whereas byte data transfers for bank 3 are carried over bus lines D_{24} through D_{31}.

When the 80386DX is operated in real mode, only the value on address lines A_2 through A_{19} and the $\overline{BE}$ signals are used to select the storage location that is to be accessed. For this reason, the physical address space is 1M byte in length, not 4G bytes. The memory subsystem is once again partitioned into four banks, as shown in Fig. 9.20, but this time each bank is 256K bytes in size.

Figure 9.20 shows that in hardware the memory address space is physically organized as a sequence of double words. The address on lines A_2 through A_{31} selects the double-word storage location. Therefore, each aligned double word starts at a physical address that

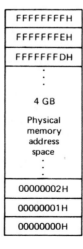

Figure 9.19 Physical address space.

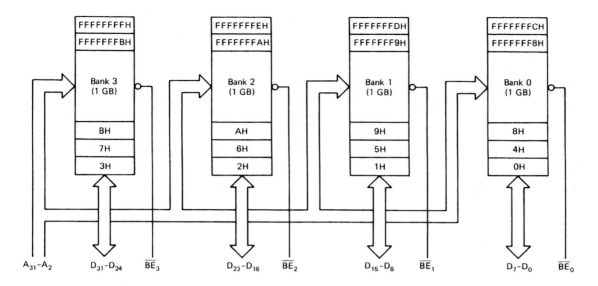

Figure 9.20 Hardware organization of the physical address space.

is a multiple of 4. For instance, in Fig. 9.20, we see that aligned double words start at address 00000000_{16}, 00000004_{16}, 00000008_{16}, up through $FFFFFFFC_{16}$.

Each of the 4 bytes of a double word corresponds to one of the byte-enable signals. For this reason, they are each stored in a different bank of memory. In Fig. 9.20, we have identified the range of byte addresses that correspond to the storage locations in each bank of memory. For example, byte data accesses to addresses such as 00000000_{16}, 00000004_{16}, and 00000008_{16} all produce $\overline{BE}_0$, which enables memory bank 0, and the read or write data transfer takes place over data bus lines D_0 through D_7. Figure 9.21(a) illustrates how the byte at double-word aligned memory address X is accessed.

On the other hand, in Fig. 9.20 we see that byte addresses 00000001_{16}, 00000005_{16}, and 00000009_{16} correspond to data held in memory bank 1. Figure 9.21(b) shows how the byte of data at address X + 1 is accessed. Notice that $\overline{BE}_1$ is made active to enable bank 1 of memory.

Most memory accesses produce more than one byte-enable signal. For instance, if the word of data beginning at aligned address X is read from memory, both $\overline{BE}_0$ and $\overline{BE}_1$ are generated. In this way, bank 0 and bank 1 of memory are enabled for operation. As shown in Fig. 9.21(c), the word of data is transferred to the MPU over data bus lines D_0 through D_{15}.

Let us now look at what happens when a double word of data is written to aligned double-word address X. As shown in Fig. 9.21(d), $\overline{BE}_0$, $\overline{BE}_1$, $\overline{BE}_2$, and $\overline{BE}_3$ are made 0 to enable all four banks of memory, and the MPU writes the data to memory over the complete data bus, D_0 through D_{31}.

All the data transfers we have described so far have been what are called *double-word aligned data*. For each of the pieces of data, all the bytes existed within the same double word, that is, a double word that is on an address boundary equal to a multiple of 4. The diagram in Fig. 9.22 illustrates a number of aligned words and double words of data. Byte, aligned word, and aligned double-word data transfers are all performed by the 80386DX in a single bus cycle.

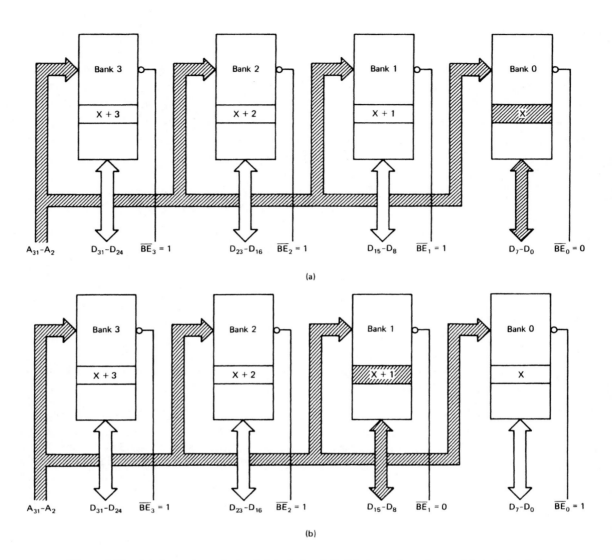

Figure 9.21 (a) Accessing a byte of data in bank 0. (b) Accessing a byte of data in bank 1.

It is not always possible to have all words or double words of data aligned at double-word boundaries. Figure 9.23 shows some examples of misaligned words and double words of data that can be accessed by the 80386DX. Notice that word 3 consists of byte 3 that is in aligned double word 0 and byte 4 that is in aligned double word 4. Let us now look at how misaligned data are transferrred over the bus.

The diagram in Fig 9.24 illustrates a misaligned double-word data transfer. Here the double word of data starting at address X + 2 is to be accessed. However, this word consists of bytes X + 2 and X + 3 of the aligned double word at physical address X and bytes Y and Y + 1 of the aligned double word at physical address Y. Looking at the diagram, we see that $\overline{BE}_0$ and $\overline{BE}_1$ are active during the first bus cycle, and the word at address Y is transferred over D_0 through D_{15}. A second bus cycle automatically follows in which $\overline{BE}_3$ and $\overline{BE}_4$ are active, address X is put on the address bus, and the second word of data, X + 2

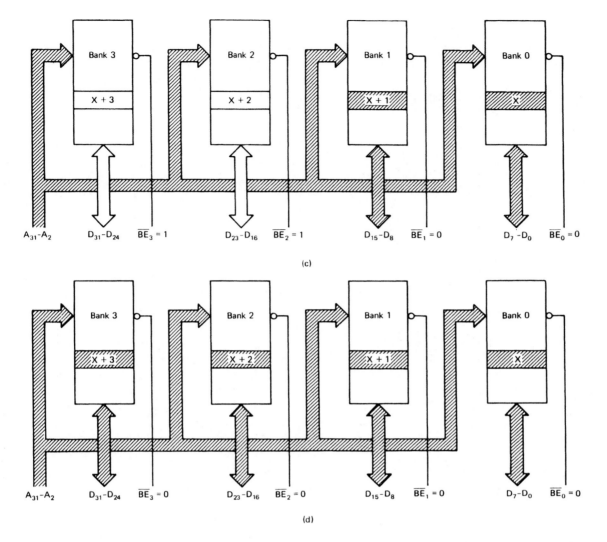

Figure 9.21 (c) Accessing an aligned word of data in memory. (d) Accessing an aligned double word in memory.

and $X + 3$, is carried over D_{16} through D_{31}. In this way we see that data transfers of misaligned words or double words take two bus cycles.

EXAMPLE 9.7

Is the word at address $0000123E_{16}$ aligned or misaligned? How many bus cycles are required to read it from memory?

Solution

The first byte of the word is the fourth byte at aligned double-word address $0000123C_{16}$ and the second byte of the word is the first byte of the aligned double word at address

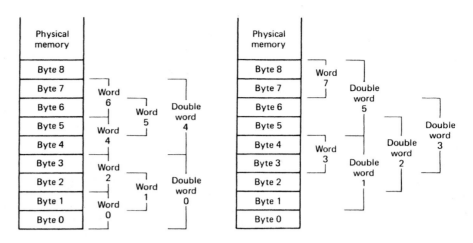

Figure 9.22 Examples of aligned data words and double words.

Figure 9.23 Examples of misaligned data words and double words.

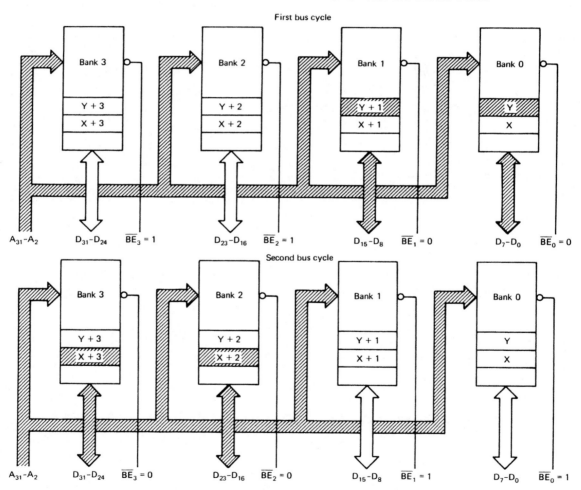

Figure 9.24 Misaligned double-word data transfer.

00001240_{16}. Therefore, the word is misaligned and requires two bus cycles to be read from memory.

▲ 9.8 MEMORY INTERFACE CIRCUITRY

A memory interface diagram for a protected-mode 80386DX-based microcomputer system is shown in Fig. 9.25. Here we find that the interface includes bus control logic, address bus latches and an address decoder, data bus transceiver/buffers, and bank write control logic. The bus cycle indication signals, M/$\overline{\text{IO}}$, D/$\overline{\text{C}}$, and W/$\overline{\text{R}}$, which are output by the 80386DX, are supplied directly to the bus control logic. Here they are decoded to produce the command and control signals needed to control data transfers over the bus. In Figs. 9.15 and 9.17, the status codes that relate to the memory interface are highlighted. For example, the code M/$\overline{\text{IO}}$ D/$\overline{\text{C}}$ W/$\overline{\text{R}}$ equal 110 indicates that a data memory-read bus cycle is in progress. This code makes the $\overline{\text{MRDC}}$ command output of the bus control logic switch to logic 0. Notice in Fig. 9.25 that $\overline{\text{MRDC}}$ is applied directly to the $\overline{\text{OE}}$ input of the memory subsystem.

Next let us look at how the address bus is decoded, buffered, and latched. Looking at Fig. 9.25, we see that address lines A_{29} through A_{31} are decoded to produce chip-enable outputs $\overline{\text{CE}}_0$ through $\overline{\text{CE}}_7$. These chip-enable signals are latched along with address bits A_2 through A_{28} and byte-enable lines $\overline{\text{BE}}_0$ through $\overline{\text{BE}}_3$ into the address bus latches. Notice that the bus-control logic receives $\overline{\text{ADS}}$ and the bus cycle indication code as inputs and produces the address latch enable (ALE), memory read command ($\overline{\text{MRDC}}$), and memory write command ($\overline{\text{MWTC}}$) signals as its output. ALE is applied to the CLK input of the latches and strobes the bits of the address, byte-enable, and chip-enable signals into the address latches. These signals are buffered by the address latch devices; the address and chip enables are output directly to the memory subsystem, and the byte enables are supplied as inputs to the bank write control logic circuit.

This part of the memory interface demonstrates one of the benefits of the 80386DX's pipelined bus mode. When working in the pipelined mode, the 80386DX actually outputs the address in the T_2 state of the prior bus cycle. Therefore, by putting the address decoder before the address latches instead of after, the code at address lines A_{28} through A_{31} can be fully decoded and stable prior to the T_1 state of the next bus cycle. In this way, the access time of the memory subsystem is reduced.

During read bus cycles, the $\overline{\text{MRDC}}$ output of the bus control logic enables the data at the outputs of the memory subsystem onto data bus lines D_0 through D_{31}. The 80386DX will read the appropriate byte, word, or double word of data. On the other hand, during write operations to memory, the bank write control logic determines into which of the four memory banks the data are written. This depends on whether a byte, word, or double-word data transfer is taking place over the bus.

Notice in Fig. 9.25 that the latched byte-enable signals $\overline{\text{BE}}_0$ through $\overline{\text{BE}}_3$ are gated with the memory-write command signal $\overline{\text{MWTC}}$ to produce a separate write-enable signal for each of the four banks of memory. These signals are denoted as $\overline{\text{WEB}}_0$ through $\overline{\text{WEB}}_3$ in Fig. 9.25. For example, if a word of data is to be written to memory over data bus lines D_0 through D_{15}, $\overline{\text{WEB}}_0$ and $\overline{\text{WEB}}_1$ are switched to their active 0 logic level.

The bus transceivers control the direction of data transfer between the MPU and memory subsystem. In Fig. 9.25 we see that the operation of the transceivers is controlled by the data-transmit/receive (DT/$\overline{\text{R}}$) and data bus–enable ($\overline{\text{DEN}}$) outputs of the bus control logic. $\overline{\text{DEN}}$ is applied to the enable ($\overline{\text{EN}}$) input of the transceivers and enables them for operation. This happens during all read and write bus cycles. DT/$\overline{\text{R}}$ selects the direction

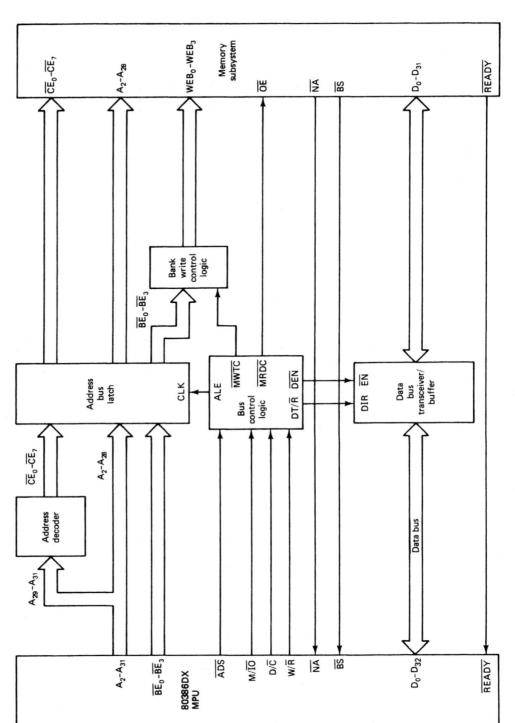

Figure 9.25 Memory interface block diagram.

of data transfer through the transceivers. Notice that it is supplied to the DIR input of the devices. When a read cycle is in process, DT/$\overline{\text{R}}$ is set to 0 and data are passed from the memory subsystem to the MPU. On the other hand, when a write cycle is taking place, DT/$\overline{\text{R}}$ is switched to logic 1 and data are carried from the MPU to the memory subsystem.

Address Latches and Buffers

The 74F373 is an example of an octal latch device that can be used to implement the *address latch* section of the 80386DX's memory interface circuit. A block diagram of this device is shown in Fig. 9.26(a), and its internal circuitry is shown in Fig. 9.26(b). Notice that it accepts eight inputs, 1D through 8D. As long as the clock (C) input is at logic 1, the outputs of the D-type flip-flops follow the logic level of the data applied to their corresponding inputs. When C is switched to logic 0, the current contents of the D-type flip-flops are latched. The latched information in the flip-flops are not output at data outputs 1Q through 8Q unless the output-enable ($\overline{\text{OC}}$) input is at logic 0. If $\overline{\text{OC}}$ is at logic 1, the outputs are in the high-impedance state. Figure 9.26(c) summarizes this operation.

In the 80386DX microcomputer system, the 30 address lines A_2 through A_{31} and four byte-enable signals $\overline{\text{BE}}_0$ through $\overline{\text{BE}}_3$ are normally latched in the address latch. The circuit configuration shown in Fig. 9.27 can be used to latch these signals. Notice that the latched outputs AL_2 through AL_{31} and $\overline{\text{BEL}}_0$ through $\overline{\text{BEL}}_3$ are permanently enabled by fixing $\overline{\text{OC}}$ at the 0 logic level. Moreover, the address information is latched at the outputs as the ALE signal from the bus control logic returns to logic 0—that is, when the C input of all devices is switched to logic 0.

In general, it is important to minimize the propagation delay of the address signals as they go through the bus interface circuit. The switching property of the 74F373 latches that determines this delay for the circuit of Fig. 9.27 is called *enable-to-output propagation delay* and has a maximum value of 13 ns. By selecting fast latches, that is, latches with a short propagation delay time, a maximum amount of the 80386DX's bus cycle time is preserved for the access time of the memory devices. In this way, slower, lower-cost memory ICs can be used. These latches also provide buffering for the 80386DX's address lines, which can sink only 4 mA. The outputs of the 74F373 latch can sink a maximum of 24 mA.

The 74F374 is another IC that is frequently used as an address latch in microcomputer systems. The circuit within this device is shown in Fig. 9.28(a). This circuit is similar to the 74F373 we just introduced, in that it is an octal latch device. However, the flip-flops used to implement the latches are edge triggered instead of transparent. Notice in Fig. 9.28(b) that when $\overline{\text{OC}}$ is logic 0, the data outputs become equal to the value of the data inputs synchronous with a low-to-high transition at the CLK input.

In some applications, additional buffering is required on the latched address lines. For example, the diagram in Fig. 9.29(a) shows that some of the address lines may be buffered to provide an independent I/O address bus. In this case a simple octal buffer/line driver device, such as the 74F244, can be used as the I/O bus buffer. Figure 9.29(b) shows the buffer circuitry provided by the 74F244. The outputs of this device can sink a maximum of 64 mA.

Data Bus Transceivers

The *data bus transceiver* block of the bus interface circuit can be implemented with 74F245 octal bus transceivers ICs. Figure 9.30(a) shows a block diagram of this device. Notice that its bidirectional input/output lines are called A_1 through A_8 and B_1 through B_8.

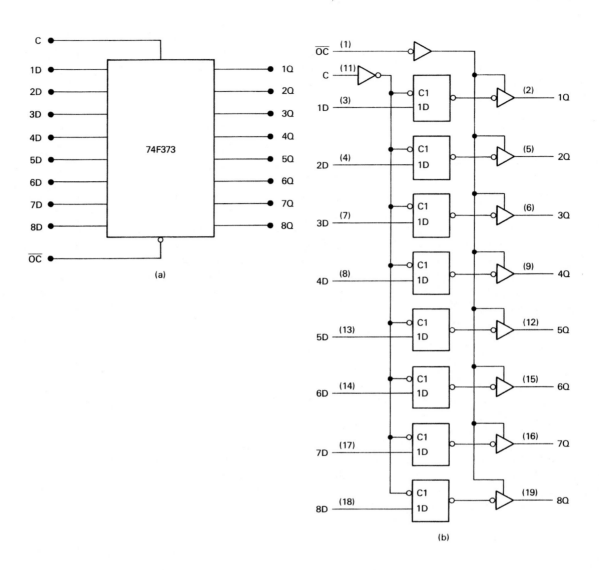

Figure 9.26 (a) Block diagram of an octal D-type latch. (b) Circuit diagram of the 74F373. (Courtesy of Texas Instruments Incorporated) (c) Operation of the 74F373. (Courtesy of Texas Instruments Incorporated)

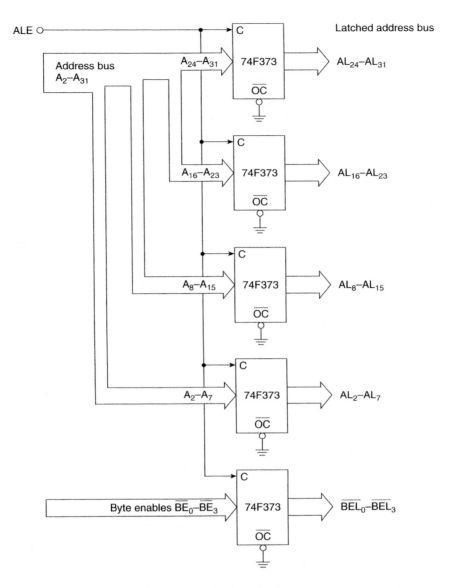

Figure 9.27 Address-latch circuit.

Looking at the circuit diagram in Fig. 9.30(b), we see that the $\overline{G}$ input is used to enable the buffer for operation. On the other hand, the logic level at the direction (DIR) input selects the direction in which data are transferred through the device. For instance, logic 0 at this input sets the transceiver to pass data from the B lines to the A lines. Switching DIR to logic 1 reverses the direction of data transfer.

Figure 9.31 shows a circuit that implements the data bus transceiver block of the bus interface circuit using the 74F245. For the 32-bit data bus of the 80386DX microcomputer, four devices are required. Here the DIR input is driven by the signal data transmit/receive

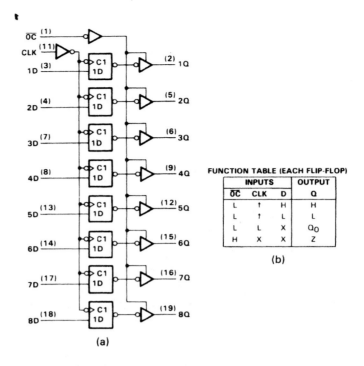

FUNCTION TABLE (EACH FLIP-FLOP)

INPUTS			OUTPUT
$\overline{OC}$	CLK	D	Q
L	↑	H	H
L	↑	L	L
L	L	X	Q_0
H	X	X	Z

(b)

(a)

Figure 9.28 (a) Circuit diagram of the 74F374. (Courtesy of Texas Instruments Incorporated) (b) Operation of the 74F374. (Courtesy of Texas Instruments Incorporated)

$(DT/\overline{R})$ and $\overline{G}$ is supplied by data bus enable $(\overline{DEN})$. These signals are outputs of the bus-control logic.

Another key function of the data bus transceiver circuit is to buffer the data bus lines. This capability is defined by how much current the devices can sink at their outputs. The I_{OL} rating of the 74F245 is 64 mA. The data lines of the 80386DX are rated for a maximum of 4 mA.

The 74F646 device is an octal bus transceiver with registers. This device is more versatile than the 74F245 we just described. It can be configured to operate either as a simple bus transceiver or as a registered bus transceiver. Figure 9.32(a) shows the operations that can occur when used as a simple transceiver. Notice that logic 0 at the SAB or SBA input selects the direction of data transfer through the device. Figure 9.32(b) shows that when configured for the registered mode of operation, data do not directly transfer between the A and B buses. Instead, data are passed from the bus to an internal register with one control signal sequence and from the internal register to the other bus with another control signal. Notice that lines CAB and CBA control the storage of the data from the A or B bus into the register and SAB and SBA control the transfer of stored data from the registers to the A or B bus. The table in Fig. 9.32(c) summarizes all the operations performed by the 74F646.

Address Decoders

As shown in Figs. 9.33(a) and (b), the *address decoder* in the 80386DX microcomputer system can be located on either side of the address latch. A typical device that is used to

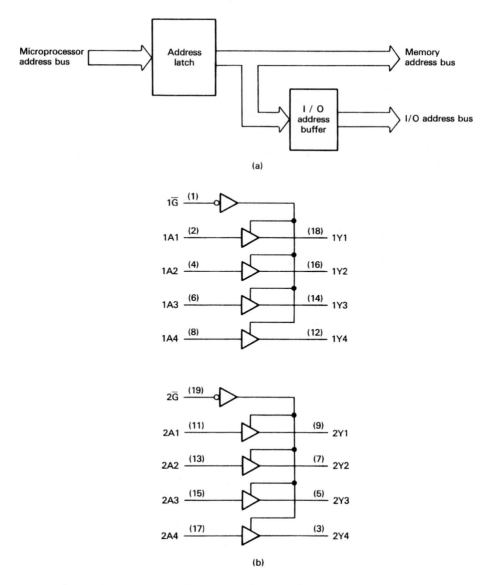

Figure 9.29 (a) Buffering the I/O address. (b) 74F244 circuit diagram. (Courtesy of Texas Instruments Incorporated)

perform this decode function is the 74F139 dual 2-line-to-4-line decoder. Figures 9.34(a) and 9.34(b) show a block diagram and circuit diagram for this device, respectively. When the enable ($\overline{G}$) input is at its active 0 logic level, the output corresponding to the code at the BA inputs switches to the 0 logic level. For instance, when BA = 01, output Y_1 is logic 0. The operation of the 74F139 is summarized in the table of Fig. 9.35.

The circuit in Fig. 9.36 employs the address decoder configuration of Fig. 9.33(a). Notice that address lines A30 and A31 are applied to the A and B inputs of the 74F139 decoder, respectively. Here they are decoded to produce chip enable outputs $\overline{CE}_0$ through

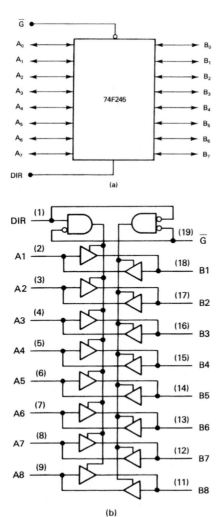

Figure 9.30 (a) Block diagram of the 74F245 octal bidirectional bus transceiver. (b) Circuit diagram of the 74F245. (Courtesy of Texas Instruments Incorporated)

$\overline{CE}_3$. The advantage of this configuration is that the propagation delay involved in decoding the address becomes transparent. That is, since the propagation delay of the decoder takes place prior to the trailing edge of $\overline{ALE}$, it does not contribute to the total access time of the memory interface. The chip-enable outputs $\overline{CE}_0$ through $\overline{CE}_3$ of the decoder are loaded into the address latch along with address signals A_2 through A_{31} and byte-enable signals $\overline{BE}_0$ through $\overline{BE}_3$. In some applications, independent decoders are used for memory and I/O. Typically, independent decoders are used for memory and I/O. In this case, one of the decoders in the 74F139 can be used to produce four memory-chip enables and the other decoder to provide four I/O device-chip selects.

The block diagram of another commonly used decoder, 74F138, is shown in Fig. 9.37(a). The 74F138 is similar to the 74F139, except that it is a single 3-line-to-8-line decoder. The circuit used in this device is shown in Fig. 9.37(b). Notice that it can be used to produce

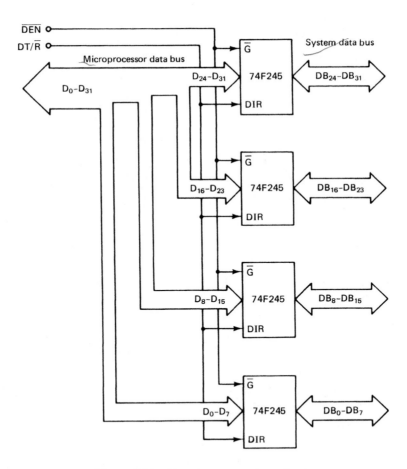

Figure 9.31 Data bus transceiver circuit.

eight $\overline{CE}$ outputs. The operation of the 74F138 is described by the table in Fig. 9.37(c). Here we find that when enabled, only the output that corresponds to the code at the CBA inputs switches to the active 0 logic level.

▲ 9.9 PROGRAMMABLE LOGIC ARRAYS: BUS-CONTROL LOGIC

In the preceding section we found that basic logic devices such as latches, transceivers, and decoders are required in the bus interface section of the 80386DX microcomputer system. We showed that these functions were performed with standard logic devices such as the 74F373 octal transparent latch, 74F245 octal bus transceiver, and 74F139 2-line-to-4-line decoder, respectively. Today *programmable logic array* (PLA) devices are becoming very important in the design of high-performance microcomputer systems. For example, the bus-control logic section of the memory interface in Fig. 9.25 is normally implemented with

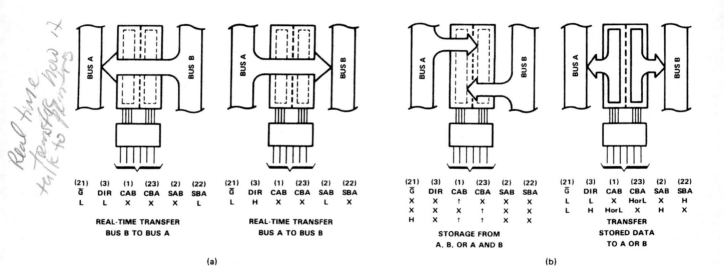

Figure 9.32 (a) Transceiver-mode data transfers of the 74F646. (Courtesy of Texas Instruments Incorporated) (b) Register-mode data transfers. (Courtesy of Texas Instruments Incorporated) (c) Control signals and data-transfer operations. (Courtesy of Texas Instruments Incorporated)

PLAs instead of with separate logic ICs. Unlike the earlier mentioned devices, PLAs do not implement a specific logic function. Instead, they are general-purpose logic devices that have the ability to perform a wide variety of specialized logic functions. A PLA contains a general-purpose AND-OR-NOT array of logic gate circuits. The user has the ability to interconnect the inputs to the AND gates of this array. The definition of these inputs determines the logic function that is implemented. The process used to connect or disconnect inputs of the AND gate array is known as *programming*, which leads to the name programmable logic array.

PLAs, GALs, and EPLDs

A variety of different types of PLA devices are available. Early devices were all manufactured with the bipolar semiconductor process. These devices are referred to as *PALs*™ and remain in wide use today. Bipolar devices are programmed with an interconnect

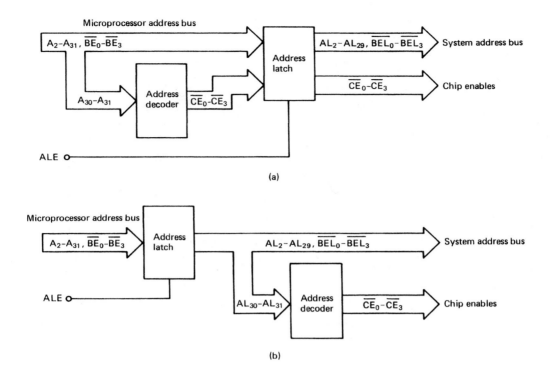

Figure 9.33 Bus configuration with address decoding before the address latch. (b) Bus configuration with address decoding after the address latch.

pattern by burning out fuse links within the device. In the initial state, all these fuse links are intact. During programming, unwanted links are open-circuited by injecting a current through the fuse to burn it out. For this reason, once a device is programmed it cannot be reused. If a design modification is required in the pattern, a new device must be programmed and substituted for the original device. Since PALs are made with an older bipolar technology, they are limited to simpler functions and characterized by slower operating speeds and high power consumption.

Newer PLA devices are manufactured with the CMOS process. With this process, very complex, high-speed, low-power devices can be made. Two kinds of CMOS PLAs are in wide use today: the GAL^{TM} and the $EPLD^{TM}$. These devices differ in the type of CMOS

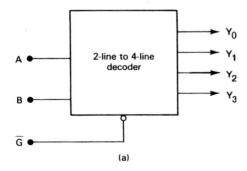

Figure 9.34(a) Block diagram of the 74F139 2-line-to-4-line decoder/demultiplexer.

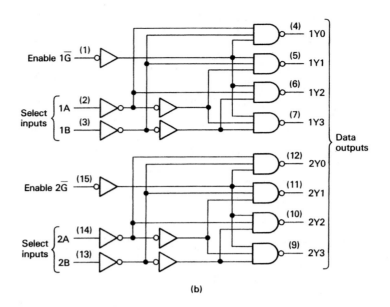

(b)

Figure 9.34(b) Circuit diagram of the 74F139. (Courtesy of Texas Instruments Incorporated)

technology used in their design. GALs are designed using *electrically erasable read-only memory* (E²ROM) technology. The input/output operation of this device is determined by the programming of cells. These electrically programmable cells are also electrically eraseable. For this reason, a GAL can be used for one application, erased, and then reprogrammed for another application. EPLDs are similar to GALs in that they can be programmed, erased, and reused; however, the erase mechanism is different. They are manufactured with *electrically programmable read only memory* (EPROM) technology. That is, they employ EPROM cells; instead of E²ROM cells. Therefore, to be erased an EPLD must be exposed to ultraviolet light. GALs and EPLDs are currently the most rapidly growing segments of the PLA marketplace.

Block Diagram of a PLA

The block diagram in Fig. 9.38 represents a typical PLA. Looking at this diagram, we see that it has 16 input leads, marked I_0 through I_{15}. There are eight output leads. These leads are labeled F_0 through F_7. This PLA is equipped with three-state outputs. For this reason, it has a chip-enable control lead. In the block diagram this control input is marked $\overline{CE}$. The logic level of $\overline{CE}$ determines if the outputs are enabled or disabled.

INPUTS			OUTPUTS			
ENABLE	SELECT					
$\overline{G}$	B	A	Y0	Y1	Y2	Y3
H	X	X	H	H	H	H
L	L	L	L	H	H	H
L	L	H	H	L	H	H
L	H	L	H	H	L	H
L	H	H	H	H	H	L

Figure 9.35 Operation of the 74F139 decoder. (Courtesy of Texas Instruments Incorporated)

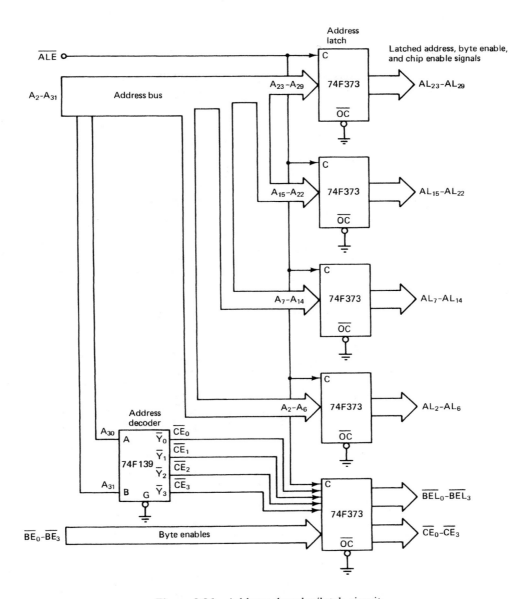

Figure 9.36 Address decoder/latch circuit.

When a PLA is used to implement random logic functions, the inputs represent Boolean variables, and the outputs are used to provide eight separate random logic functions. The internal AND-OR-NOT array is programmed to define a sum-of-product equation for each of these outputs in terms of the inputs and their complements. In this way we see that the logic levels applied at inputs I_0 through I_{15} and the programming of the AND array determine what logic levels are produced at outputs F_0 through F_7. Therefore, the capacity of a PLA is measured by three properties: the number of inputs, the number of outputs, and the number of product terms (P-terms).

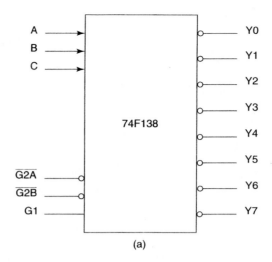

(a)

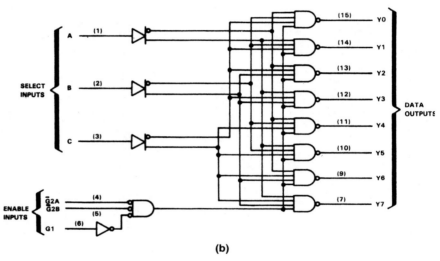

(b)

ENABLE INPUTS			SELECT INPUTS			OUTPUTS							
G1	G̅2̅A̅	G̅2̅B̅	C	B	A	Y0	Y1	Y2	Y3	Y4	Y5	Y6	Y7
X	H	X	X	X	X	H	H	H	H	H	H	H	H
X	X	H	X	X	X	H·	H	H	H	H	H	H	H
L	X	X	X	X	X	H	H	H	H	H	H	H	H
H	L	L	L	L	L	L	H	H	H	H	H	H	H
H	L	L	L	L	H	H	L	H	H	H	H	H	H
H	L	L	L	H	L	H	H	L	H	H	H	H	H
H	L	L	L	H	H	H	H	H	L	H	H	H	H
H	L	L	H	L	L	H	H	H	H	L	H	H	H
H	L	L	H	L	H	H	H	H	H	H	L	H	H
H	L	L	H	H	L	H	H	H	H	H	H	L	H
H	L	L	H	H	H	H	H	H	H	H	H	H	L

(c)

Figure 9.37 (a) Block diagram. (b) 74F138 circuit diagram. (Courtesy of Texas Instruments Incorporated) (c) Operation of the 74F138. (Courtesy of Texas Instruments Incorporated)

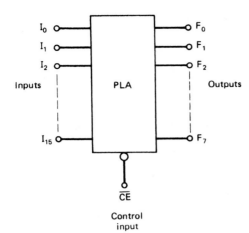

Figure 9.38 Block diagram of a PLA. (Courtesy of Walter A. Triebel)

Architecture of a PLA

We just pointed out that the circuitry of a PLA is a general-purpose AND-OR-NOT array. Figure 9.39(a) shows this architecture. Here we see that the input buffers supply input signals A and B and their complements $\overline{A}$ and $\overline{B}$. Programmable connections in the AND array permit any combination of these inputs to be combined to form a product term (P-term). The product term outputs of the AND array are supplied to fixed inputs of the OR array. The output of the OR gate produces a sum-of-products function. Finally, the inverter complements this function.

The circuit of Fig. 9.39(b) shows how the function $F = (\overline{A\overline{B} + \overline{A}B})$ is implemented with the AND-OR-NOT array. Notice that an X marked into the AND array means that the fuse is left intact, and no marking means that it has been blown to form an open circuit. For this reason, the upper AND gate is connected to A and $\overline{B}$ and produces the product term $A\overline{B}$. The second AND gate from the top connects to $\overline{A}$ and B to produce $\overline{A}B$. The bottom AND gate is marked with an X to indicate that it is not in use. Gates like this that are not to be active should have all of their input fuse links left intact.

In Fig. 9.40(a), we show the circuit structure that is most widely used in PLAs. It differs from the circuit shown in Fig. 9.39(a) in two ways. First, the inverter has a programmable three-state control and can be used to isolate the logic function from the output. Second, the buffered output is fed back to form another set of inputs to the AND array. This new output configuration permits the output pin to be programmed to work as a *standard output, standard input*, or *logic-controlled input/output*. For instance, if the upper AND gate, which is the control gate for the output buffer, is set up permanently to enable the inverter, and the fuse links for its inputs that are fed back from the outputs are all blown open, the output functions as a standard output.

PLAs are also available in which the outputs are latched with registers. A circuit for this type of device is shown in Fig. 9.40(b). Here we see that the output of the OR gate is applied to the D input of a clocked D-type flip-flop. In this way, the logic level produced by the AND-OR array is not presented at the output until a pulse is first applied at the CLOCK input. Furthermore, the feedback input is produced from the complemented output of the flip-flop, not the output of the inverter. This configuration is known as a *PLA with registered outputs* and is designed to simplify implementation of *state machines* designs.

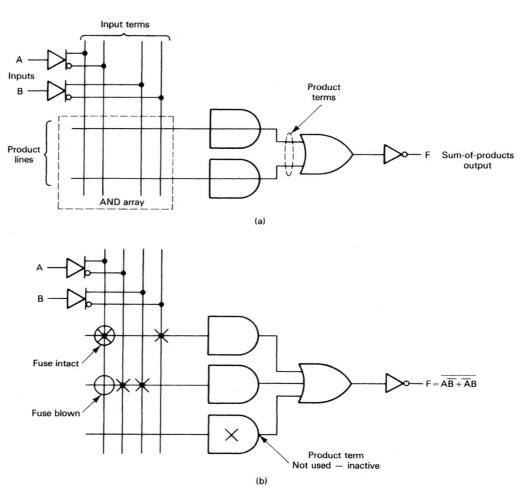

Figure 9.39 (a) Basic PLA architecture. (b) Implementing the logic function $F = (A\overline{B} + \overline{A}B)$.

Standard PLA Devices

Now that we have introduced the types of PLAs, block diagram of the PLA, and internal architecture of the PLA, let us continue by examining a few of the widely used PAL devices. A PAL, or a programmable array logic, is a PLA in which the OR array is fixed; only the AND array is programmable.

The 16L8 is one of the more widely used PAL ICs. Its internal circuitry and pin numbering are shown in Fig. 9.41(a). This device is housed in a 20-pin package, as shown in Fig. 9.41(b). Looking at this diagram, we see that it employs the PLA architecture illustrated in Fig. 9.40(a). Notice that it has 10 dedicated input pins. All these pins are labeled *I*. There are also two dedicated outputs, which are labeled with the letter *O*, and six programmable I/O lines, which are labeled *I/O*. Using the programmable I/O lines, the number of input lines can be expanded to as many as 16 inputs or the number of outputs can be increased to as many as 8 lines.

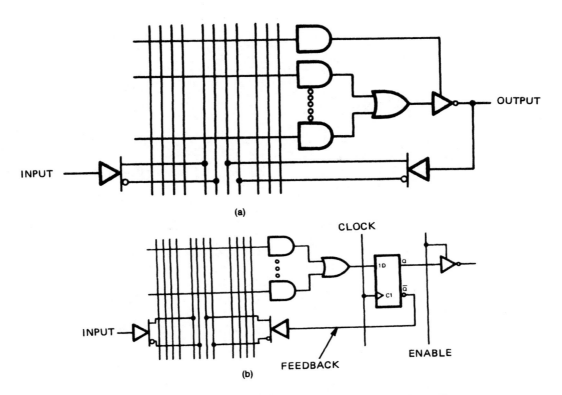

Figure 9.40 (a) Typical PLA architecture. (Courtesy of Texas Instruments Incorporated) (b) PLA with output latch. (Courtesy of Texas Instruments Incorporated)

All the 16L8's inputs are buffered and produce both the original form of the signal and its complement. The outputs of the buffer are applied to the inputs of the AND array. This array is capable of producing 64 product terms. Notice that the AND gates are arranged into 8 groups of 8. The outputs of 7 gates in each of these groups are used as inputs to an OR gate and the eighth output is used to produce an enable signal for the corresponding three-state output buffer. In this way we see that the 16L8 is capable of producing up to 7 product terms for each output, and the product terms can be formed using any combination of the 16 inputs.

The 16L8 is manufactured with bipolar technology. It operates from a $+5$-V $\pm$ 10% dc power supply and draws a maximum of 180 mA. Moreover, all its inputs and outputs are at TTL-compatible voltage levels. This device exhibits high-speed input/output propagation delays. In fact, the maximum I-to-O propagation delay is rated as 7 ns.

Another widely used PAL is the 20L8 device. Looking at the circuitry of this device in Fig. 9.42(a), we see that it is similar to that of the 16L8 just described. However, the 20L8 has a maximum of 20 inputs, 8 outputs, and 64 P terms. The device's 24-pin package is shown in Fig. 9.42(b).

The 16R8 is also a popular 20-pin PAL. The circuit diagram and pin layout for this device are shown in Fig. 9.43(a) and (b), respectively. From Fig. 9.43(a) we find that its eight fixed I inputs and AND-OR array are essentially the same as those of the 16L8. There is one change. The outputs of eight AND gates, instead of seven, are supplied to the inputs of each OR gate.

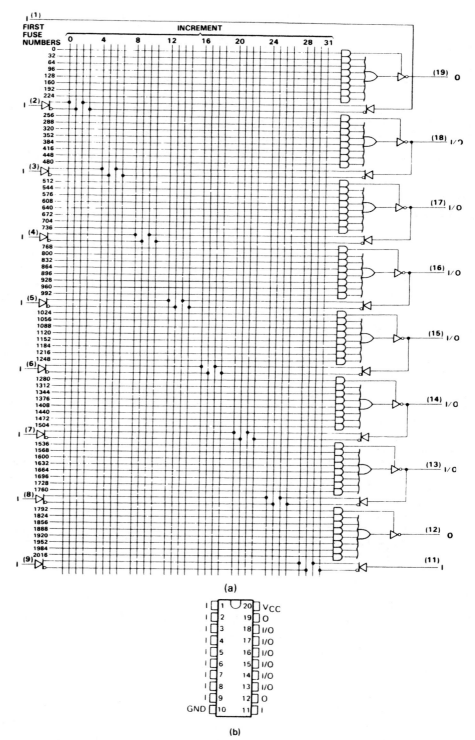

Figure 9.41 (a) 16L8 circuit diagram. (Courtesy of Texas Instruments Incorporated) (b) 16L8 pin layout. (Courtesy of Texas Instruments Incorporated)

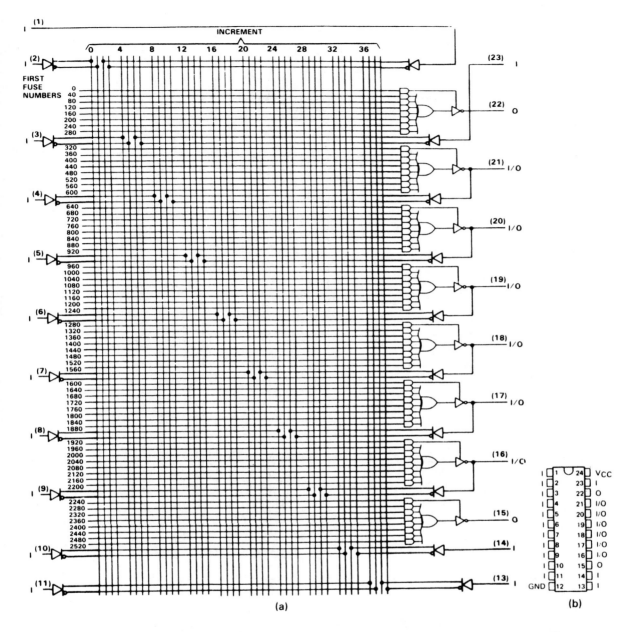

Figure 9.42 (a) 20L8 circuit diagram. (Courtesy of Texas Instruments Incorporated) (b) 20L8 pin layout. (Courtesy of Texas Instruments Incorporated)

A number of changes have been made at the output side of the 16R8. Notice that the outputs of the OR gates are first latched in D-type flip-flops with the CLK signal. They are then buffered and supplied to the eight Q outputs. Another change is that the enable signals for the output inverters are no longer programmable. Now all three-state outputs are enabled by the logic level of the $\overline{\text{OE}}$ control input.

Sec. 9.9   Programmable Logic Arrays: Bus-Control Logic   **415**

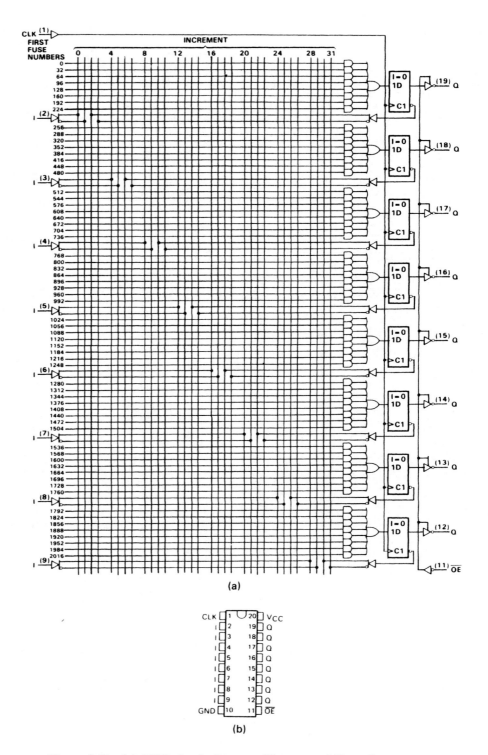

(a)

(b)

Figure 9.43 (a) 16R8 circuit diagram. (Courtesy of Texas Instruments Incorporated) (b) 16R8 pin layout. (Courtesy of Texas Instruments Incorporated)

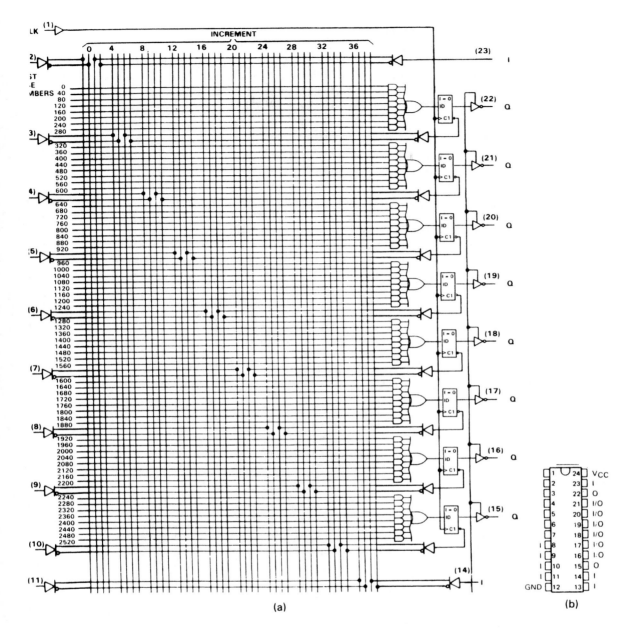

Figure 9.44 (a) 20R8 circuit diagram. (Courtesy of Texas Instruments Incorporated) (b) 20R8 pin layout. (Courtesy of Texas Instruments Incorporated)

The last change is in the part of the circuit that produces the feedback inputs. In the 16R8, these eight input signals are derived from the complementary output of the corresponding latch instead of the output of the buffer. For this reason the output leads can no longer be programmed to work as direct inputs.

The 20R8 is the registered output version of the 20L8 PAL. Its circuit diagram and pin layout are given in Fig. 9.44(a) and (b), respectively.

Expanding PLA Capacity

Some applications have requirements that exceed the capacity of a single PLA IC. For instance, a 16L8 device has the ability to supply a maximum of 16 inputs, 8 outputs, and 64 product terms. Capacity can be expanded by connecting several devices together. Let us now look at the way in which PLAs are interconnected to expand the number of inputs and outputs.

If a single PLA does not have enough outputs, two or more devices can be connected together into the configuration of Fig. 9.45(a). Here we see that the inputs I_0 through I_{15} on the two devices are individually connected in parallel. This connection does not change the number of inputs.

On the other hand, the 8 outputs of the two PLAs are separately used to form the upper and lower bytes of a 16-bit output word. The bits of this word are denoted O_0 through O_{15}. So with this connection, we have doubled the number of outputs.

When data are applied to the inputs, PLA 1 outputs the 8 least significant bits of data. At the same instant PLA 2 outputs the 8 most significant bits. These outputs can be used to represent individual logic functions.

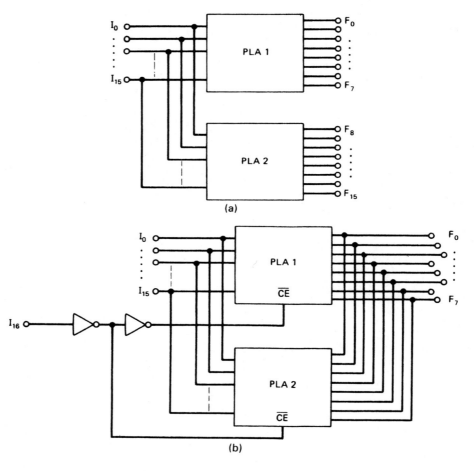

Figure 9.45 (a) Expanding output-word length. (b) Expanding input-word length. (Courtesy of Walter A. Triebel)

Another limitation on the application of PLAs is the number of inputs. The maximum number of inputs on a single 16L8 is 16. However, additional ICs can be connected to expand the capacity of inputs. Figure 9.45(b) shows how one additional input is added. This permits a 17-bit input denoted as I_0 through I_{16}. The new bit I_{16} is supplied through inverters to the $\overline{CE}$ inputs on the two PLAs. At the output side of the PLAs, outputs O_0 through O_7 of the two devices are individually connected in parallel. To involve this connection, PLA devices with open-collector or three-state outputs must be used.

When I_{16} is logic 0, $\overline{CE}$ on PLA 1 is logic 0. This enables the device for operation, and the output functions coded for input I_0 through I_{15} are output at O_0 through O_7. At the same instant, $\overline{CE}$ on PLA 2 is logic 1 and it remains disabled. Making the logic level of I_{16} equal 1 disables PLA 1 and enables PLA 2. Now the input at I_0 through I_{15} causes the output function defined by PLA 2 to be output at O_0 through O_7. Actually, this connection doubles the number of product terms as well as increases the number of inputs.

▲ 9.10 TYPES OF INPUT/OUTPUT

The 80386DX microcomputer can employ two different types of input/output (I/O). They are known as *isolated I/O* and *memory-mapped I/O*. These I/O methods differ in how I/O ports are mapped into the 80386DX's address spaces. Practical microcomputer systems usually employ both kinds of I/O. That is, some peripherals are treated as isolated I/O devices and others as memory-mapped I/O devices. Let us now look at each of these types of I/O.

Isolated Input/Output

When using isolated input/output in a microcomputer system, the I/O devices are treated separate from memory. This is achieved because the software and hardware architectures of the 80386DX support separate memory and I/O address spaces. Figure 9.46 illustrates the 80386DX's real-mode address spaces.

In our study of 80386DX software architecture in Chapter 2, we examined these address spaces from a software point of view. We found that information in memory or at I/O ports is organized as bytes of data; that the real-mode memory address space contains 1M consecutive byte addresses in the range 00000_{16} through $FFFFF_{16}$; and that the I/O address space contains 64K consecutive byte addresses in the range 0000_{16} through $FFFF_{16}$.

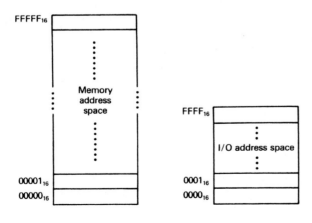

Figure 9.46 Isolated I/O real-mode memory and I/O address spaces.

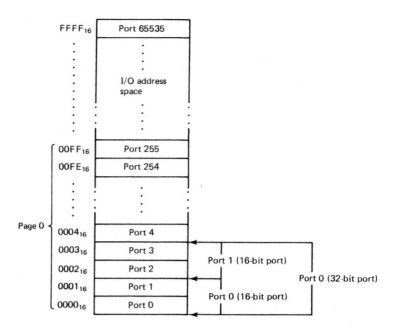

Figure 9.47 Byte-wide, word-wide, and double-word-wide I/O ports.

Figure 9.47 shows a more detailed map of this I/O address space. Here we find that the bytes of data in two consecutive memory or I/O addresses could be accessed as word-wide data and the contents of four consecutive byte addresses represent a double word of data. For instance, I/O addresses 0000_{16}, 0001_{16}, 0002_{16}, and 0003_{16} can be treated as independent byte-wide I/O ports, ports 0, 1, 2, and 3; ports 0 and 1 may be considered together as word-wide port 0; and ports 0 through 3 may be accessed as a double-word port.

Notice that the part of the I/O address map in Fig. 9.47 from address 0000_{16} through $00FF_{16}$ is referred to as *page 0*. Certain I/O instructions can only perform operations to ports in this part of the address range. Other I/O instructions can input or output data for ports anywhere in the I/O address space.

The isolated method of input/output offers some advantages. First, the complete 1M byte memory address space is available for use with memory. Second, special instructions have been provided in the instruction set of the 80386DX to perform isolated I/O input and output operations. These instructions have been tailored to maximize I/O performance. A disadvantage of this type of I/O is that all input and output data transfers must take place between the AL, AX, or EAX register and the I/O port.

Memory-Mapped Input/Output

I/O devices can be placed in the memory address space of the microcomputer as well as in the independent I/O address space. In this case the MPU looks at the I/O port as though it is a storage location in memory. For this reason, the method is known as *memory-mapped I/O*.

In a microcomputer system with memory-mapped I/O, some of the memory address space is dedicated to I/O ports. For example, in Fig. 9.48 the 4095 memory addresses in the range from $E0000_{16}$ through $E0FFF_{16}$ are assigned to I/O devices. Here the contents of

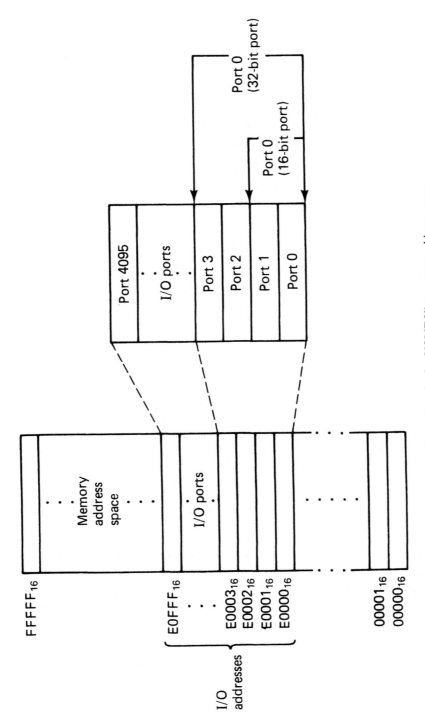

Figure 9.48 Memory-mapped I/O devices in the 80386DX's memory address space.

address $E0000_{16}$ represents byte-wide I/O port 0; the contents of addresses $E0000_0$ and $E0001_{16}$ correspond to word-wide port 0; and the contents of addresses $E0000_{16}$ through $E0003_{16}$ are double-word-wide port 0.

When I/O is configured in this way, instructions that affect data in memory are used instead of the special input/output instructions. This is an advantage in that many more instructions and addressing modes are available to perform I/O operations. For instance, the contents of a memory-mapped I/O port can be directly ANDed with a value in an internal register. In addition, I/O transfers can now take place between an I/O port and an internal register other than just AL, AX, or EAX. However, this also leads to a disadvantage. That is, the memory instructions tend to execute slower than those specifically designed for isolated I/O. Therefore, a memory mapped I/O routine may take longer to execute than an equivalent program using the input/output instructions.

Another disadvantage of using this method is that part of the memory address space is lost. For instance, in Fig. 9.48 addresses in the range from $E0000_{16}$ through $E0FFF_{16}$ cannot be used to implement memory.

▲ 9.11 THE ISOLATED INPUT/OUTPUT INTERFACE

The isolated input/output interface of the 80386-based microcomputer permits it to communicate with the outside world. The way in which the 80386DX and 80386SX MPU's deal with input/output circuitry is similar to the way in which they interface with memory circuitry. That is, input/output data transfers also take place over the data bus. This parallel bus permits easy interface to LSI peripheral devices such as parallel I/O expanders, interval timers, and serial communication controllers. Let us continue by looking at how the 80386DX interfaces to an isolated I/O subsystem.

Figure 9.49 shows a typical isolated I/O interface. Here we find that I/O devices 0 through N connect to the 80386DX through an I/O interface circuit. These blocks can represent input devices such as a keyboard, output devices such as a printer, or input/output devices such as an asynchronous serial communications port. An example of a typical I/O device used in the I/O subsystem is a programmable peripheral interface (PPI) IC, such as the 82C55A. This type of device is used to implement parallel input and output ports. Let us now look at the function of each of the blocks in this circuit more closely.

Notice that the interface between the microprocessor and I/O subsystem includes the bus controller logic, an I/O address decoder, I/O address latches, I/O data bus transceiver/buffers, and I/O bank write control logic. As in the memory interface, bus control logic is needed to produce the control signals for the I/O interface. In Fig. 9.49 we see that the inputs of the bus control logic are the bus cycle indication signals that are output by the 80386DX on $M/\overline{IO}$, $D/\overline{C}$, and $W/\overline{R}$. These 3-bit codes tell which type of bus cycle is in progress. In the table in Fig. 9.50, the bus cycle indication codes that correspond to input/output bus cycles are highlighted. Notice that the code $M/\overline{IO}$ $D/\overline{C}$ $W/\overline{R}$ equals 010 identifies an I/O read bus cycle, and $M/\overline{IO}$ $D/\overline{C}$ $W/\overline{R}$ equals 011 an I/O write bus cycle. In response to these inputs, the bus-control logic section produces bus control signals, such as ALE, $IODT/\overline{R}$, and $\overline{IODEN}$ at its output. These signals are needed to latch the address and set up the data bus for an input or output data transfer.

The bus-control logic must also generate the I/O read command ($\overline{IORC}$) and I/O write command ($\overline{IOWC}$) control signals. $\overline{IORC}$ is applied directly to the input/output read ($\overline{IORD}$) input of the I/O devices and tells them when data are to be input to the MPU. In this case 32-bit data are always put on the data bus. However, the 80386DX inputs only the appropriate byte, word, or double word. On the other hand, $\overline{IOWC}$ is gated with the

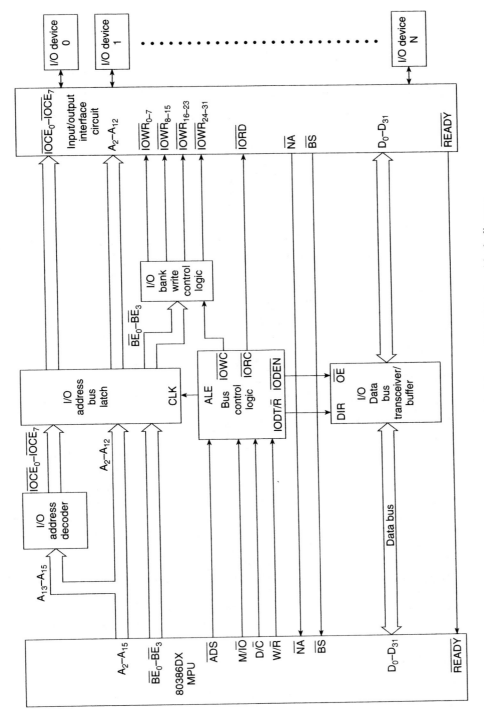

Figure 9.49 Byte, word, and double-word I/O interface block diagram.

423

M/$\overline{\text{IO}}$	D/$\overline{\text{C}}$	W/$\overline{\text{R}}$	Type of bus cycle
0	0	0	Interrupt acknowledge
0	0	1	Idle
0	1	0	I/O data read
0	1	1	I/O data write
1	0	0	Memory code read
1	0	1	Halt/shutdown
1	1	0	Memory data read
1	1	1	Memory data write

Figure 9.50 Input/output bus-cycle indication codes.

$\overline{\text{BE}}$ signals to produce a separate write-enable signal for each byte of the data bus. They are labeled $\overline{\text{IOWR}}_{0-7}$, $\overline{\text{IOWR}}_{8-15}$, $\overline{\text{IOWR}}_{16-23}$, and $\overline{\text{IOWR}}_{24-31}$. These signals are needed to support writing of 8-bit, 16-bit, or 32-bit data through the interface.

The I/O device that is accessed for input or output of data is selected by an *I/O address*. This address is specified as part of the instruction that performs the I/O operation. In the 80386DX architecture, all isolated I/O addresses are 16 bits in length and support 64K independent I/O ports. As shown in Fig. 9.49, they are output to the I/O interface over address bus lines A_2 through A_{15} and byte-enable lines $\overline{\text{BE}}_0$ through $\overline{\text{BE}}_3$. The more significant address bits, A_{16} through A_{31}, are held at the 0 logic level during the address period of all I/O bus cycles. The 80386DX signals external circuitry that an I/O address is on the bus by switching its M/$\overline{\text{IO}}$ output to logic 0.

The I/O interface shown in Fig. 9.51 is designed to support 8-bit, 16-bit, and 32-bit I/O data transfers. The address on lines A_2 through A_{15} is used to specify the double-word I/O port that is to be accessed. When data are output to output ports, the logic levels of $\overline{\text{BE}}_0$ through $\overline{\text{BE}}_3$ determine which byte-wide port or ports are enabled for operation.

Notice in the circuit diagram that part of the I/O address that is output on address lines A_2 through A_{15} of the 80386DX is decoded by the I/O address decoder. The bits of the address that are decoded produce I/O chip enable signals for the individual I/O devices. For instance, Fig. 9.51 shows that with three address bits, A_{13} through A_{15}, enough chip-enable outputs are produced to select up to eight I/O devices. Notice that the outputs of the I/O address decoder are labeled $\overline{\text{IOCE}}_0$ through $\overline{\text{IOCE}}_7$. The I/O chip-enable signals are latched along with the address and byte-enable signals in the I/O address latches. Latching of this information is achieved with a pulse at the ALE output of the bus control logic. If a microcomputer employs a very simple I/O subsystem, it may be possible to eliminate the address decoder and simply use some of the latched high-order address bits as I/O enable signals.

In Fig. 9.51 all the low-order address bits, A_2 through A_{12}, are shown to be latched and sent directly to the I/O devices. Typically, these address bits are used to select the register within the peripheral device that is to be accessed. For example, with just four of these address lines, we can select any one of 16 registers.

EXAMPLE 9.8

If address bits A_7 through A_{15} are used directly as chip-enable signals and address lines A_2 through A_6 are used as register-select inputs for the I/O devices, how many I/O devices can be used, and what is the maximum number of registers that each device can contain?

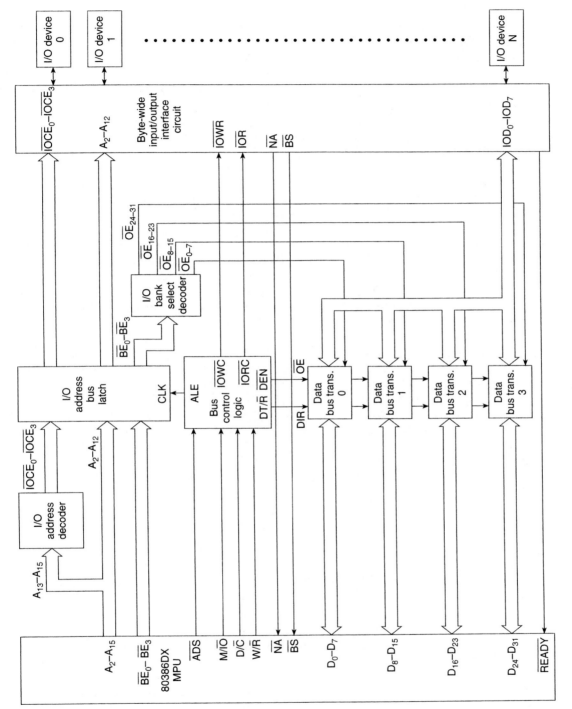

Figure 9.51 Byte-wide I/O interface block diagram.

425

Solution

When latched into the address latch, the nine address lines produce the nine I/O chip-enable signals, $\overline{IOCE_0}$ through $\overline{IOCE_8}$, for I/O devices 0 through 8. The lower 5 address bits are able to select between $2^5 = 32$ registers for each peripheral IC.

Earlier we indicated that all data transfers between the 80386DX and I/O interface take place over data bus lines D_0 through D_{31}. Data written by the MPU to an I/O device are referred to as *output data* and data read in from an I/O device are called *input data*. Earlier we pointed out that the 80386DX can input or output data in byte-wide, word-wide, or double-word-wide format. However, many LSI peripherals used as I/O controllers are designed to interface to an 8-bit bus. For this reason I/O operations frequently involve byte-wide data transfers. Just as for the memory interface, the signals $\overline{BE_0}$ through $\overline{BE_3}$ are used to signal which byte or bytes of data are being transferred over the bus. Again logic 0 at $\overline{BE_0}$ identifies that a byte of data is input or output over data bus lines D_0 through D_7. On the other hand, logic 0 at $\overline{BE_3}$ means that a byte-data transfer is taking place over bus lines D_{24} through D_{31}. All byte-wide I/O data transfers are performed in just one bus cycle.

During input and output bus cycles, data are passed between the selected register in the enabled I/O device and the 80386DX over the data bus lines D_0 through D_{31}. We just mentioned that many of the peripheral ICs that are used in the 80386DX microcomputer have a byte-wide data bus. For this reason they are normally attached to the lower part of the data bus. That is, they are connected to data bus lines D_0 through D_7. If this is done in the circuit of Fig. 9.51, all I/O addresses must be scaled by four. This is because the first byte I/O address that corresponds to a byte transfer across the lower eight data bus lines is 0000_{16}, the next byte address, which represents a byte transfer over D_0 through D_7, is 0004_{16}, the third I/O address is 0008_{16}, and so on. In fact, if only 8-bit peripherals are used in the 80386DX microcomputer system and they are all attached to I/O data bus lines D_0 through D_7, the byte-enable signals are not needed in the I/O interface. In this case, address bit A_2 is used as the least significant bit of the I/O address and A_{15} the most significant bit. Therefore, from a hardware point of view, the I/O address space appears as 16K contiguous byte-wide storage locations over the address range from

$$A_{15} \ldots \ldots A_3A_2 = 0000000000000_2$$

to

$$A_{15} \ldots \ldots A_3A_2 = 1111111111111_2$$

This puts the burden on software to assure that bytes of data are input or output only for addresses that are a multiple of 4 and correspond to a data transfer over data bus lines D_0 through D_7.

Earlier we found that the bus-control logic section produced the ALE signal that is used to latch the address. It also supplies the control signals needed to set up the data bus for input and output data transfers. The data bus transceiver/buffers control the direction of data transfers between the 80386DX and I/O devices. They are enabled for operation when their output-enable ($\overline{OE}$) inputs are switched to logic 0. Notice that the signal *I/O data bus enable* ($\overline{IODEN}$) is applied to the $\overline{OE}$ inputs.

The direction in which data are passed through the transceivers is determined by the logic level of the DIR input. This input is supplied by the *I/O data-transmit/receive* (IODT/$\overline{\text{R}}$) output of the bus-control logic. During all input cycles, IODT/$\overline{\text{R}}$ is logic 0 and the transceivers are set to pass data from the selected I/O device to the 80386DX. On the other hand, during output cycles, IODT/$\overline{\text{R}}$ is switched to logic 1 and data passes from the 80386DX to the I/O device.

Another input/output interface diagram is shown in Fig. 9.51. This circuit includes an I/O bank select decoder in the data bus interface. The decoder is used to multiplex the 32-bit data bus of the 80386DX to an 8-bit I/O data bus for connection to 8-bit peripheral devices. By using this circuit configuration, data can be input from or output to all 64K contiguous byte addresses in the I/O address space. In this case, hardware, instead of software, assures that byte data transfers to consecutive byte I/O addresses are performed to contiguous byte-wide I/O ports. The I/O bank-select decoder circuit maps bytes of data from the 32-bit data bus to the 8-bit I/O data bus. It does this by assuring that only one byte-enable ($\overline{\text{BE}}$) output of the 80386DX is active. That is, it checks to assure that a byte input or output operation is in progress. If more than one of the $\overline{\text{BE}}$ inputs of the decoder is active, none of the $\overline{\text{OE}}$ outputs of the decoder is produced, and the data transfer does not take place. Now the I/O address 0000_{16} corresponds to an I/O cycle over data bus lines D_0 through D_7 to an 8-bit peripheral attached to I/O data bus lines, IOD_0 through IOD_7, 0001_{16} corresponds to an input or output of a byte of data for the peripheral over lines D_8 through D_{15}, 0002_{16} represents a byte I/O transfer over D_{16} through D_{23}, and, finally, 0003_{16} accompanies a byte transfer over data lines D_{24} through D_{31}. That is, even though the byte of data for addresses 0000_{16} through 0003_{16} are output by the 80386DX on different parts of its data bus, they are all multiplexed in external hardware to the same 8-bit I/O data bus, IOD_0 through IOD_7. In this way the addresses of the peripheral's registers no longer need to be scaled by 4 in software.

▲ 9.12 INPUT AND OUTPUT BUS CYCLE TIMING

In Section 9.11 we found that the isolated I/O interface signals of the 80386DX microcomputer are essentially the same as those involved in the memory interface. In fact, the function, logic levels, and timing of all signals other than M/$\overline{\text{IO}}$ are identical to those already described for the memory interface earlier in Section 9.6.

The timing diagram in Fig. 9.52 shows some *nonpipelined input* and *output bus cycles*. Looking at the waveforms for the first input/output bus cycle, which is called cycle 1, we see that it represents a zero-wait-state input bus cycle. Notice that the byte-enable signals $\overline{\text{BE}}_0$ through $\overline{\text{BE}}_3$, the address lines A_2 through A_{15}, the bus cycle indication signals M/$\overline{\text{IO}}$, D/$\overline{\text{C}}$, and W/$\overline{\text{R}}$, and address status ($\overline{\text{ADS}}$) signal are all output at the beginning of the T_1 state. This time the 80386DX switches M/$\overline{\text{IO}}$ to logic 0, D/$\overline{\text{C}}$ to 1, and W/$\overline{\text{R}}$ to logic 0 to signal external circuitry that an I/O data input bus cycle is in progress.

As shown in the block diagram of Fig. 9.49, the bus cycle indication code is input to the bus-control logic. An input of M/$\overline{\text{IO}}$ D/$\overline{\text{C}}$ W/$\overline{\text{R}}$ equals 010 initiates an I/O input bus-control sequence. Let us continue with the sequence of events that takes place in external circuitry during the input cycle. First, the bus-control logic outputs a pulse to the 1 logic level on ALE. As shown in the circuit of Fig. 9.49, this pulse is used to latch the address information into the I/O address latch devices. The decoded part of the latched address ($\overline{\text{IOCE}}_0$ through $\overline{\text{IOCE}}_7$) selects the I/O device to be accessed, and the code on the lower address lines selects the register that is to be accessed. Later in the bus cycle, $\overline{\text{IORC}}$ is

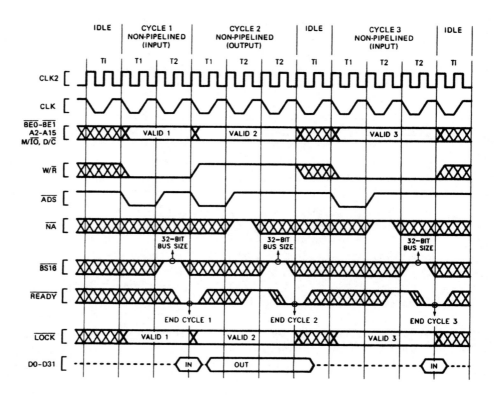

Figure 9.52 I/O read and I/O write-bus cycles. (Reprinted by permission of Intel Corp. Copyright/Intel Corp. 1987)

switched to logic 0 to signal the enabled I/O device that data are to be input to the MPU. In response to $\overline{\text{IORC}}$, the enabled input device puts the data from the addressed register onto the data bus. A short time later, IODT/$\overline{\text{R}}$ is switched to logic 0 to set the data bus transceivers to the input direction, and then the transceivers are enabled as $\overline{\text{IODEN}}$ is switched to logic 0. At this point the data from the I/O device are available on the 80386DX's data bus.

In the waveforms of Fig. 9.52 we see that at the end of the T_2 state the 80386DX tests the logic level at its ready input to determine if the I/O bus cycle should be completed or extended with wait states. As shown in Fig. 9.52, $\overline{\text{READY}}$ is at its active 0 logic level when sampled. Therefore, the 80386DX inputs the data off the bus. Finally, the bus-control logic returns $\overline{\text{IORC}}$, $\overline{\text{IODEN}}$, and IODT/$\overline{\text{R}}$ to their inactive logic levels, and the input bus cycle is finished.

Cycle 3 in Fig. 9.52 is also an input bus cycle. However, looking at the $\overline{\text{READY}}$ waveform, we find that this time it is not logic 0 at the end of the first T_2 state. Therefore, the input cycle is extended with a second T_2 state (wait state). Since some of the peripheral devices used with the 80386DX are older, slower devices, it is common to have several wait states in I/O bus cycles.

EXAMPLE 9.9

If the 80386DX that is executing cycle 3 in Fig. 9.52 is running at 20 MHz, what is the duration of this input cycle?

Solution

An 80386DX that is running at 20 MHz has a T state equal to 50 ns. Since the input bus cycle takes 3 T states, its duration is 150 ns.

Looking at the output bus cycle, cycle 2 in the timing diagram of Fig. 9.52, we see that the 80386DX puts the data to be output onto the data bus at the beginning of ϕ_2 in the T_1 state. This time the bus-control logic switches $\overline{IODEN}$ to logic 0 and maintains $IODT/\overline{R}$ at the 1 level for transmit mode. From Fig. 9.49 we find that since $\overline{IODEN}$ is logic 0 and $IODT/\overline{R}$ is 1, the transceivers are enabled and set up to pass data from the 80386DX to the I/O devices. Therefore, the data outputs on the bus are available on the data inputs of the enabled I/O device. Finally, the signal $\overline{IOWC}$ is switched to logic 0. It is gated with $\overline{BE_0}$ through $\overline{BE_3}$ in the I/O bank write-control logic to produce the needed bank write-enable signals. These signals tell the I/O device that valid output data are on the bus. Now the I/O device must read the data off the bus before the bus control logic terminates the bus cycle. If the device cannot read data at this rate, it can hold $\overline{READY}$ at the 1 logic level to extend the bus cycle.

EXAMPLE 9.10

If the output bus cycles performed to byte-wide ports by an 80386DX running at 20 MHz are to be completed in a minimum of 250 ns, how many wait states are needed?

Solution

Since each T state is 50 ns in duration, the bus cycle must last at least

$$\text{number of T states} = 250 \text{ ns}/50 \text{ ns} = 5 \text{ ns}$$

A zero-wait-state output cycle lasts just two T states; therefore, all output cycles must include three wait states.

Similar to memory, I/O bus cycle requirements exist for data transfers for aligned and unaligned I/O ports. That is, all word and double-word data transfers to aligned port addresses take place in just one bus cycle. However, two bus cycles are required to perform data transfers for unaligned 16-bit or 32-bit I/O ports.

▲ 9.13 INPUT/OUTPUT INSTRUCTIONS

Input/output operations are performed by the 80386DX microcomputer that employs isolated I/O using special input and output instructions together with the I/O port addressing modes. The input and output instructions are listed in Fig. 9.53. This table provides the mnemonic, name, and a brief description of operation for each instruction. Let us begin by looking at the *input* (IN) and *output* (OUT) instructions in more detail.

Notice that there are two different forms of IN and OUT instructions: the *direct I/O instruction* and *variable I/O instruction*. Either of these two types of instructions can be used to transfer a byte, a word, or a double word of data. All data transfers take place between an I/O device and the MPU's accumulator register. For this reason, this method

Mnemonic	Meaning	Format	Operation
IN	Input direct	IN Acc, Port	$(Acc) \leftarrow (Port)$ Acc = AL, AX, or EAX
	Input indirect (variable)	IN Acc, DX	$(Acc) \leftarrow ((DX))$
OUT	Output direct	OUT Port, Acc	$(Acc) \rightarrow (Port)$
	Output indirect (variable)	OUT DX, Acc	$(Acc) \rightarrow ((DX))$
INS	Input string byte	INSB	$(ES:DI) \leftarrow ((DX))$
			$E(DI) \leftarrow E(DI) \pm 1$
	Input string word	INSW	$(ES:DI) \leftarrow ((DX))$
			$E(DI) \leftarrow E(DI) \pm 2$
	Input string double	INSD	$(ES:DI) \leftarrow ((DX))$
			$E(DI) \leftarrow E(DI) \pm 4$
OUTS	Output string byte	OUTSB	$(ES:SI) \rightarrow ((DX))$
			$E(SI) \pm 1 \rightarrow E(SI)$
	Output string word	OUTSW	$(ES:SI) \rightarrow ((DX))$
			$E(SI) \pm 2 \rightarrow E(SI)$
	Output string double	OUTSD	$(ES:SI) \rightarrow ((DX))$
			$E(SI) \pm 4 \rightarrow E(SI)$

Figure 9.53   Input/output instructions.

of performing I/O is known as *accumulator I/O*. Byte transfers involve the AL register; word transfers, the AX register; and double-word transfers, the EAX register. In fact, specifying AL as the source or destination register in an I/O instruction indicates that it corresponds to a byte data transfer. That is, byte-wide, word-wide, or double-word-wide input/output is selected by specifying the accumulator (Acc) in the instruction as AL, AX, or EAX, respectively.

In a direct I/O instruction, the address of the I/O port is specified as part of the instruction. Eight bits are provided for this direct address. For this reason, its value is limited to the address range from 0_{10} equal 0000_{16} to 255_{10} equal $00FF_{16}$. This range is identified as page 0 in the I/O address space of Fig. 9.47.

EXAMPLE 9.11

How many aligned double-word addresses exist in page 0?

Solution

The number of aligned double-word addresses in page zero is found as

$$\text{number of aligned double-word addresses} = (\text{number of bytes in page 0})/4$$

$$= 256/4$$

$$= 64$$

An example is the instruction

```
IN   AL,0FEH
```

As shown in Fig. 9.53, execution of this instruction causes the contents of the byte-wide I/O port at address FE_{16} of the I/O address space to be input to the AL register. This data transfer takes place in one input bus cycle.

EXAMPLE 9.12

Write a sequence of instructions that will output the data FF_{16} to a byte-wide output port at address AB_{16} of the 80386DX's I/O address space.

Solution

First, the AL register is loaded with FF_{16} as an immediate operand in the instruction

```
MOV   AL, 0FFH
```

Now the data in AL can be output to the byte-wide output port with the instruction

```
OUT   0ABH, AL
```

The difference between the direct and variable I/O instructions lies in the way in which the address of the I/O port is specified. We just saw that for direct I/O instructions an 8-bit address is specified as part of the instruction. On the other hand, the variable I/O instructions use a 16-bit address that resides in the DX register within the MPU. The value in DX is not an offset. It is the actual address that is to be output on the address bus during the I/O bus cycle. Since this address is a full 16 bits in length, variable I/O instructions can access ports located anywhere in the 64K-byte I/O address space.

When using either type of I/O instruction, the data must be loaded into or removed from the AL, AX, or EAX register before another input or output operation is performed. In the case of variable I/O instructions, the DX register must be loaded with an address. This requires execution of additional instructions. For instance, the instruction sequence

```
MOV   DX,0A000H
IN    AL,DX
MOV   BL,AL
```

inputs the contents of the byte-wide input port at $A000_{16}$ of the I/O address space into AL and then saves it in BL.

EXAMPLE 9.13

Write a series of instructions that will output FF_{16} to an output port located at address $B000_{16}$ of the I/O address space.

Solution

The DX register must first be loaded with the address of the output port. This is done with the instruction

```
MOV   DX,0B000H
```

Next the data that is to be output must be loaded into AL.

```
MOV   AL,0FFH
```

Finally, the data are output with the instruction

```
OUT   DX,   AL
```

EXAMPLE 9.14

Data are to be read in from two byte-wide input ports at addresses AA_{16} and $A9_{16}$, respectively, and then output as a word to a word-wide output port at address $B000_{16}$. Write a sequence of instructions to perform this input/output operation.

Solution

We can first read in the byte from the port at address AA_{16} into AL and move it to AH. This is done with the instructions

```
IN    AL,0AAH
MOV   AH,AL
```

Now the other byte, which is at port $A9_{16}$, can be read into AL by the instruction

```
IN    AL,0A9H
```

The word is now held in AX. To write out the word of data, we load DX with the address $B000_{16}$ and use a variable output instruction. This leads to the following:

```
MOV   DX,0B000H
OUT   DX,AX
```

In Fig. 9.53 we find that input and output string instructions also exist in the instruction set of the 80386DX. These two instruction forms are not supported in the 8086/8088 software architecture. Using these string instructions, a programmer can either input data from an input port to a storage location directly in memory or output data from a memory location to an output port.

The first instruction, which is called *input string*, can be denoted in three ways, INSB, INSW, or INSD. INSB stands for *input string byte*; INSW means *input string word*; and INSD represents *input string double word*. Let us now look at the operation performed by the INSB instruction. INSB assumes that the address of the input port that is to be accessed is in the DX register. This value must be loaded prior to executing the instruction. Moreover, the address of the storage location into which the byte of data is input is identified by the values in ES and DI; that is, when executed, the input operation performed is

$$(ES:DI) \leftarrow ((DX))$$

In the protected mode, DI is replaced by the extended destination index register EDI.

Just as for the other string instructions, the value in DI (or EDI) is either incremented or decremented by 1 after the data transfer takes place.

$$DI \leftarrow DI \pm 1$$

In this way it points to the next byte-wide storage location in memory. Whether the value in DI is incremented or decremented depends on the setting of the DF flag. Notice in Fig 9.53 that INSW and INSD perform the same data-transfer operation except that since the word or double-word contents of the I/O port are stored in memory, the value in DI (EDI) is incremented or decremented by 2 or 4, respectively.

The INSB instruction performs the operation we just described on one data element, not an array of elements. This basic operation can be repeated to handle a block input operation. Block operations are done by inserting a repeat (REP) prefix in front of the string instruction. The operation performed by REP is described in Fig. 6.43. For example, the instruction

```
REPINSW
```

when executed in real mode will cause the contents of the word-wide port pointed to by the I/O address in DX to be input and saved in the memory location at address ES:DI. Then the value in DX is incremented by 2 (assuming that DF equals 0), the count in CX is decremented by 1, and the value in CX is tested to determine if it is 0. As long as the value in CX is not 0, the input operation is repeated. When CX equals 0, all elements of the array have been input and the input string operation is complete. Remember that the count of the number of times the string operation is to be repeated must be loaded into the CX register prior to executing the repeat input string instruction.

In Fig. 9.53 we see that OUTSB, OUTSW, and OUTSD are the three forms of the *output string* instruction. These instructions operate in a similar way to the input string instructions; however, they perform an output operation. For instance, executing OUTSW causes the operation that follows:

$$(ES:SI) \rightarrow ((DX))$$

$$SI \pm 2 \rightarrow SI$$

That is, the word of data held at the memory location pointed to by address ES:SI (ES:ESI in protected mode) is output to the word-wide port pointed to by the I/O address in DX. After the output data transfer is complete, the value in SI (ESI) is either incremented or decremented by two.

An example of an output string instruction that can be used to output an array of data is

```
REPOUTSB
```

When executed in the real mode, this instruction causes the data elements of the array of data in memory pointed to by ES:SI to be output one after the other to the output port located at the I/O address in DX. Again, the count in CX defines the size of the array.

When the 80386DX is in the protected-address mode, the input/output instructions can be executed only if the current privilege level is greater than or equal to the I/O privilege level (IOPL). That is, the numerical value of CPL must be lower than or equal to the numerical value of IOPL. Remember that IOPL is defined by the code in bits 12 and 13 of the flags register. If the current privilege level is less than IOPL, the instruction is not executed; instead, a general protection fault occurs. The general protection fault is an example of an 80386DX exception and is examined in more detail in Chapter 12.

In Chapter 8 we indicated that the task state segment (TSS) of a task includes a section known as the *I/O permission bit map*. This I/O permission bit map provides a second

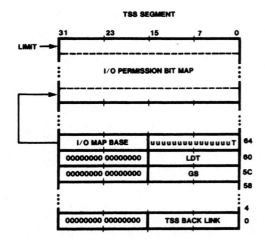

TSS SEGMENT

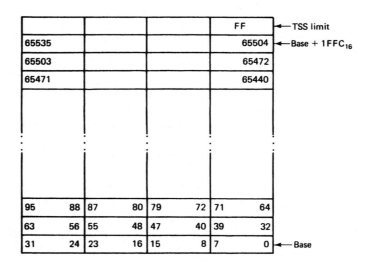

Figure 9.54 Location of the I/O permission bit map in the TSS. (Reprinted by permission of Intel Corp. Copyright/Intel Corp. 1986)

protection mechanism for the protected-mode I/O address space. Remember that the size of the TSS segment is variable. Its size is specified by the limit in the TSS descriptor. Figure 9.54 shows a typical task state segment. Here we see that the 16-bit *I/O map base* offset, which is held at word offset 66_{16} in the TSS, identifies the beginning of the I/O permission bit map. The upper end of the bit map is set by the limit field in the descriptor for the TSS. Let us now look at what the bits in the I/O permission bit map stand for.

Figure 9.55 shows a more detailed representation of the I/O permissions bit map. Notice that it contains one bit position for each of the 65,536 byte-wide I/O ports in the 80386DX's I/O address space. In the bit map we find that the bit position that corresponds to I/O port 0 (I/O address 0000_{16}) is the least significant bit at the address defined with the I/O bit map base offset. The rest of the bits in this first double word in the map represent I/O ports 1 through 31. Finally, the last bit in the table, which corresponds to port 65,535 and I/O address $FFFF_{16}$, is the most significant bit in the double word located at an offset

Figure 9.55 Contents of the I/O permission bit map.

of $1FFC_{16}$ from the I/O bit map base. In Fig. 9.55 we see that the byte address that follows the map must always contain FF_{16}. This is the least significant byte in the last double word of the TSS. The value of the I/O map base offset must be less than $DFFF_{16}$; otherwise, the complete map may not fit within the TSS.

Using this bit map, restrictions can be put on input/output operations to each of the 80386DX's 65,536 I/O port addresses. In protected mode, the bit for the I/O port in the I/O permission map is checked only if the CPL, when the I/O instruction is executed, is less privileged than the IOPL. If logic 0 is found in a bit position, it means that I/O operation can be performed to the port address. On the other hand, logic 1 inhibits the I/O operation. Any attempt to input or output data for an I/O address marked with a 1 in the I/O permission bit map by code with a CPL that is less privileged than the IOPL results in a general protection exception. In this way an operating system can detect attempts to access certain I/O devices and trap to special service routines for the devices through the general protection exception. In virtual 8086 mode, all I/O accesses reference the I/O permisson bit map.

The I/O permission configuration defined by a bit map applies only to the task that uses the TSS. For this reason, many different I/O configurations can exist within a protected-mode software system. Actually, a different bit map could be defined for every task.

In practical applications, most tasks would use the same I/O permission bit map configuration. In fact, in some applications not all I/O addresses need to be protected with the I/O permission bit map. It turns out that any bit map position that is located beyond the limit of the TSS is interpreted as containing a 1. Therefore, all accesses to an I/O address that corresponds to a bit position beyond the limit of the TSS will produce a general protection exception. For instance, a protected-mode I/O address space may be set up with a small block of I/O address to which access is permitted at the low end of the I/O address space and with access to the rest of the I/O address space restricted. A smaller table can be set up to specify this configuration. By setting the values of the bit map base and TSS limit such that the bit positions for all the restricted addresses fall beyond the end of the TSS segment, they are caused to result in an exception. On the other hand, the bit positions that are located within the table are all made 0 to permit I/O accesses to their corresponding ports. Moreover, if the complete I/O address space is to be restricted for a task, the I/O permission map base address can simply be set to a value greater than the TSS limit.

ASSIGNMENTS

Section 9.2

1. Name the technology used to fabricate the 80386DX microprocessor.

2. What is the transistor count of the 80386DX?

3. Which signal is located at pin B7?

Section 9.3

4. How large is the real-address mode address and physical address space of the 80386DX? How large is the protected-address mode address and physical address space? How large is the protected-mode virtual address space?

5. If the byte-enable code output during a data-write bus cycle is $\overline{BE_3}\,\overline{BE_2}\,\overline{BE_1}\,\overline{BE_0} = 1110_2$, is a byte, word, or double-word data transfer taking place? Over which data bus lines are the data transferred? Does data duplication occur?

6. For which byte-enable codes does data duplication take place?

7. What type of bus cycle is in progress when the bus status code M/$\overline{\text{IO}}$ D/$\overline{\text{C}}$ W/$\overline{\text{R}}$ equals 010?

8. Which signals implement the DMA interface?

9. What processor is most frequently attached to the processor extension interface?

Section 9.4

10. What are the speeds of the 80386DX ICs available from Intel Corporation? How are these speeds denoted in the part number?

11. At what pin is the CLK2 input applied?

12. What frequency clock signal must be applied to the CLK2 input of an 80386DX-25 if it is run at full speed?

Section 9.5

13. What is the duration of PCLK for an 80386DX that is driven by CLK2 equals 50 MHz?

14. What two types of bus cycles can be performed by the 80386DX?

15. How many CLK2 cycles are in an 80386DX bus cycle that has no wait states? How many T states are in this bus cycle? What would be the duration of this bus cycle if the 80386DX is operating at CLK2 equals 50 MHz?

16. What does T_1 stand for? What happens in this part of the bus cycle?

17. Explain what is meant by *pipelining* of the 80386DX's bus.

18. What is an idle state?

19. What is a wait state?

Section 9.6

20. If an 80386DX-25 is executing a nonpipelined write bus cycle that has no wait states, what would be the duration of this bus cycle if the 80386DX is operating at full speed?

21. If an 80386DX-25 that is running at full speed performs a read bus cycle with two wait states, what is the duration of the bus cycle?

Section 9.7

22. How is memory organized from a hardware point of view in a protected-mode 80386DX microcomputer system? In a real-mode 80386DX microcomputer system?

23. What are the five types of data transfers that can take place over the data bus? How many bus cycles are required for each type of data transfer?

24. If an 80386DX-25 is running at full speed and all memory accesses involve one wait state, how long will it take to fetch the word of data starting at address 0FF1A$_{16}$? The word at address 0FF1F$_{16}$?

25. During a bus cycle that involves a misaligned word transfer, which byte of data is transferred over the bus during the first bus cycle?

Section 9.8

26. Give an overview of the function of each block in the memory interface diagram of Fig. 9.25.

27. When the instruction PUSH AX is executed, what bus status code is output by the 80386DX, which byte enable signals are active, and what read/write control signal is produced by the bus control logic?

28. What type of basic logic function is provided by the 74F373 and 74F374 ICs?

29. What is the key difference between the 74F373 and 74F374?

30. Make a drawing like that in Fig. 9.27 for the real-mode 80386DX address bus.

31. Make a drawing to show how the I/O address buffer in Fig. 9.29(a) can be constructed with 74F244 ICs. Assume that the buffer circuits will be permanently enabled.

32. What logic function is provided by the 74F245 IC?

33. In the circuit of Fig. 9.31, what logic levels must be applied to the $\overline{\text{DEN}}$ and DT/$\overline{\text{R}}$ inputs to cause data on the system data bus to be transferred to the microprocessor data bus?

34. What are the logic levels of the $\overline{\text{G}}$, DIR, CAB, CBA, SAB, and SBA inputs of the 74F646 when stored data in the A register are transferred to the B bus?

35. Name an IC that implements a 2-line-to-4-line decoder logic function.

36. If the inputs to a 74F138 decoder are $G_1 = 1$, $\overline{G}_{2A} = 0$, $\overline{G}_{2B} = 0$, and CBA = 101, which output is active?

37. Make a drawing like that in Fig. 9.36 for a real-mode 80386DX address bus for which a 74F138 decoder is used to decode address line A_{17} through A_{19} into memory-chip selects.

Section 9.9

38. What does PLA stand for?

39. List three properties that measure the capacity of a PLA.

40. What does PAL stand for? Give two key differences between a PAL and a PLA. What do we call the programming mechanism used in a PAL?

41. Give two key differences between a PAL and GAL.

42. What type of cell technology is used to manufacture EPLDs?

43. Redraw the circuit in Fig. 9.39(b) to show how it can implement the logic function $F = (\overline{A}\,\overline{B} + AB)$.

44. How many dedicated inputs, dedicated outputs, programmable input/outputs, and product terms are supported on the 16L8 PAL?

45. What is the maximum number of inputs on a 20L8 PAL? The maximum number of outputs?

46. How do the outputs of the 16R8 differ from those of the 16L8?

Section 9.10

47. Name the two types of input/output.

48. What type of I/O is in use when peripheral devices are mapped into the 80386DX's I/O address space?

49. Which type of I/O has the disadvantage that part of the address space must be given up to implement I/O ports?

50. Which type of I/O has the disadvantage that all I/O data transfers must take place through the AL, AX, and EAX register?

51. How many byte-wide I/O ports can exist in the 80386DX's I/O address space? Aligned word-wide ports? Aligned double-word-wide ports?

52. What is the address range of page 0?

Section 9.11

53. What are the functions of the 80386DX's address and data bus lines relative to input/output operation?

54. Which signal indicates to external circuitry that the current bus cycle is for the I/O interface and not for the memory interface?

55. If the $\overline{BE_3}\overline{BE_2}\overline{BE_1}\overline{BE_0}$ = 1100 during an output bus cycle, is a byte, word, or double word of data being transferred? Over which data bus lines is it transferred?

56. What bus cycle indication code is output by the 80386DX during output bus cycles?

57. If the byte enable code $\overline{BE_3}\overline{BE_2}\overline{BE_1}\overline{BE_0}$ in the circuit of Fig. 9.49 is 0011, what are the $\overline{IOWR}$ signals for the current I/O write cycle?

58. If address lines $A_{15}A_{14}A_{13}$ = 100, which $\overline{IOCE}$ output is produced in the circuit of Fig. 9.49?

59. If in Fig. 9.49, just address lines A_2 through A_6 are used to select registers, what is the maximum number of registers that can be accessed?

60. Which signal in the circuit of Fig. 9.49 is used to enable the I/O data bus transceiver/buffer?

61. Describe briefly the function of each block in the I/O interface circuit in Fig. 9.51.

62. For the circuit in Fig. 9.51, if the address during an input bus cycle is A_{15} through $A_2 = 01100000000011_2$ and $\overline{BE_3}$ through $\overline{BE_0}$ = 1101, which chip-enable output is active, which register is accessed in the enabled I/O device, and over which of the 80386DX's data bus lines is the data input?

Section 9.12

63. What is the minimum duration of I/O bus cycles for an 80386DX running at 25 MHz?

64. If an 80386DX-25 running at full speed inserts two wait states into all I/O bus cycles, what is the duration of a nonpipelined bus cycle in which a byte of data is being output?

65. If the input cycles from byte-wide input ports of an 80386DX microcomputer running at 25 MHz are to be performed in a minimum of 250 ns, how many wait states are needed?

66. If the 80386DX in Problem 64 were outputting a word of data to a word-wide port at I/O address $1A3_{16}$, what would be the duration of the bus cycle?

Section 9.13

67. Describe the operation performed by the instruction IN AX,1AH.

68. Write an instruction sequence to perform the same operation as that of the instruction in Problem 67, but this time use variable or indirect I/O.

69. Describe the operation performed by the instruction OUT 2AH,AL.

70. Write an instruction sequence that will output the byte of data $0F_{16}$ to an output port at address 1000_{16}.

71. Write an instructions sequence that will input the byte of data from input ports at I/O addresses $A000_{16}$ and $B000_{16}$, add these values together, and save the sum in memory location IO_SUM.

72. Write a sequence of instructions that will input the contents of the port at address $F004_{16}$, mask off all but the lower 4 bits, and output this value to the port at address $F000_{16}$.

73. Write a sequence of instructions that will input the contents of the input port at I/O address $B0_{16}$ and jump to the beginning of a service routine identifed by the label ACTIVE_INPUT if the least significant bit of the data is 1.

74. Describe what happens when the instruction INSD is executed.

75. What operation is performed by the instruction sequence

```
MOV  DX, 0A000H
MOV  DI, 1001H
MOV  CX, 0FH
CDF
REPINSB
```

76. Write a sequence of instructions that will output the byte contents of memory addresses $ES:1001_{16}$ through $ES:100F_{16}$ to an output port at I/O address A010H.

77. What parameters identify the beginning of the I/O permission bit map in a TSS? At what address of the TSS is this parameter held?

78. At what double-word address in the I/O permission bit map is the bit for I/O port 64 held? Which bit of this double word corresponds to port 64?

79. To what logic level should the bit in Problem 78 be set if I/O operations are to be inhibited to the port in protected mode?

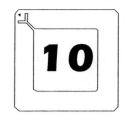

Memory Devices, Circuits, and Subsystem Design

▲ 10.1 INTRODUCTION

In the last chapter, we began our study of the hardware architecture of the 80386DX-based microcomputer system. This included a study of the MPU's memory interface and the circuits needed to connect to a memory subsystem. Topics covered were the memory interface signals, read and write bus cycles, hardware organization of the memory address space and memory interface circuits. In this chapter, we continue our study of microcomputer hardware by examining the devices, circuits, and techniques used in the design of memory subsystem. For this purpose, we have included the following topics in the chapter:

1. Program and data-storage memory
2. Read-only memory
3. Random access read/write memories
4. Parity, parity bit, and parity-checker/generator circuit
5. FLASH memory
6. Wait-state circuitry
7. 80386DX/SX microcomputer system memory interface circuitry
8. Cache memory
9. The 82385DX cache controller and the cache memory subsystem

▲ 10.2 PROGRAM AND DATA-STORAGE MEMORY

Memory provides the ability to store and retrieve digital information and is one of the key elements of a microcomputer system. By digital information, we mean that instructions and data are encoded with 0s and 1s and then saved in memory. The ability to store information

440

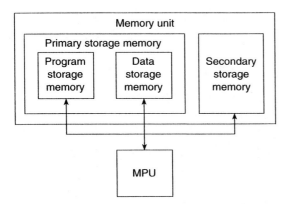

Figure 10.1 Partitioning of the microcomputer's memory unit.

is made possible by the part of the microcomputer system known as the *memory unit*. In Chapter 1 we indicated that memory unit of the microcomputer is partitioned into a *primary storage section* and *secondary storage section*. This subdivision of the memory unit is illustrated in Fig. 10.1.

Secondary storage memory is used for storage of data, information, and programs that are not in use. This part of the memory unit can be slow speed but requires very large storage capacity. For this reason, it is normally implemented with magnetic storage devices, such as the floppy disk and hard disk drive. Hard disk drives used in today's personal computers have the ability to store several gigabytes (G bytes) of information.

The other part, primary storage memory, is used for working information, such as the instructions of the program currently being executed and data that it is processing. This section normally requires high-speed operation but does not normally require very large storage capacity. Therefore, it is implemented with semiconductor memory devices. Most modern personal computers have 16 megabytes (16M bytes) of primary storage memory.

In Fig. 10.1 we see that the primary storage memory is further partitioned into *program-storage memory* and *data-storage memory*. The program-storage part of the memory subsystem is used to hold information such as the instructions of the program. That is, when a program is executed by the microcomputer, it is read one instruction at a time from the program-storage part of the memory subsystem. These programs can be either permanently stored in memory, which makes them always available for execution, or temporarily loaded into memory before execution. The program storage memory section does not normally just contain instructions. It can also store other fixed information such as constant data and look-up tables.

The program-storage memory in a personal computer is implemented exactly this way. It has a fixed part of program memory that contains the *basic input/output system* (BIOS). These programs are permanently held in a read-only memory device mounted on the main processor board. Programs held this way in ROM are called *firmware* because of their permanent nature. The typical size of a BIOS ROM used in a PC today is 2 megabits (MB) or 256K bytes.

The much larger part of the program storage memory in a PC is built with random access read/write memory devices. The DRAMs may be either mounted on the main processor board or on an add-in memory module or board. Use of DRAMs allows this part of the program-storage memory to be either read from or written into. Its purpose is again to store programs that are to be executed, but in this case they are loaded into memory only when needed. Programs are normally read in from the secondary storage device, stored in the program-storage part of memory, and then run. When the program is terminated, the part of the program memory where it resides is given back to the operating system for reuse. Moreover, if power is turned off, the contents of the RAM-based part of the program storage memory are lost. Due to the temporary nature of these programs, they are referred to as software.

Earlier we indicated that the primary storage memory of a microcomputer is typically 16M bytes. This number just represented the total of the DRAM part of the memory subsystem and was given as the size of memory because the ROM BIOS piece is almost negligible when compared to the amount of DRAM. In the PC, a good part of these 16M bytes is available for use as program-storage memory.

In other microcomputer applications, such as an electronic game or telephone, the complete program-storage memory is implemented with ROM devices.

Information that frequently changes is stored in the data-storage part of the microcomputer's memory subsystem. For instance, the data that is to be processed by the microcomputer is held in the data-storage part of the primary storage memory. When a program is run by the microcomputer, the values of the data can change repeatedly. For this reason, data-storage memory must be implemented with RAM. In a PC, the data does not automatically reside in the data-storage part of memory. Just like software, it is read into memory from a secondary storage device, such as the hard disk. Any part of the PC's 16M bytes of DRAM can be assigned for data storage. This is done by the operating system software.

When a program is run, data are modified while in DRAM and the new values are saved by writing them to the disk. Data do not have to be numeric in form; they can also be alphanumeric characters, codes, and graphical patterns. For instance, when running a word processor application, the data are alphanumeric and graphical information.

▲ 10.3 READ-ONLY MEMORY

We begin our study of semiconductor memory devices with the *read-only memory* (ROM). ROM is one type of semiconductor memory device. It is most widely used in microcomputer systems for storage of the program that determines overall system operation. The information stored within a ROM integrated circuit is permanent—or *nonvolatile*. This means that when the power supply of the device is turned off, the stored information is not lost.

ROM, PROM, and EPROM

For some ROM devices, information (the microcomputer program) must be built in during manufacturing, and for others the data must be electrically entered. The process of entering the data into a ROM is called *programming*. As the name ROM implies, once entered into the device this information can be read only. For this reason, these devices are used primarily in applications where the stored information would not change frequently.

Three types of ROM devices are in wide use today. They are known as the *mask-programmable read-only memory* (ROM), the *one-time-programmable read-only memory* (PROM), and the *erasable programmable read-only memory* (EPROM). Let us continue by looking more closely into the first type of device, the mask-programmable read-only memory. This device has its data pattern programmed as part of the manufacturing process. This is known as *mask programming*. Once the device is programmed, its contents can never be changed. Because of this fact and the cost for making the programming masks, ROMs are used mainly in high-volume applications where the data will not change.

The other two types of read-only memories, the PROM and EPROM, differ from the ROM in that the bit patterns for the data are electrically entered by the user. Programming is usually done with an instrument known as an *EPROM programmer*. Both the PROM and EPROM are programmed in the same way. Once a PROM is programmed, its contents cannot be changed. This is the reason they are sometimes called one-time-programmable PROMs. On the other hand, the contents of an EPROM can be erased by exposing it to ultraviolet light. That is, the programmed bit pattern is cleared out to restore the device to its unprogrammed state. In this way, the device can be used over and over again simply by erasing and reprogramming. PROMs and EPROMs are most often used during the design of a product, for early production, when the code of the microcomputer may need to be changed frequently, and for production in low-volume applications that do not warrant making a mask programmed device.

Figure 10.2(a) shows a typical EPROM programmer unit. Programming units like this usually have the ability to verify that an EPROM is erased, program it with new data, verify

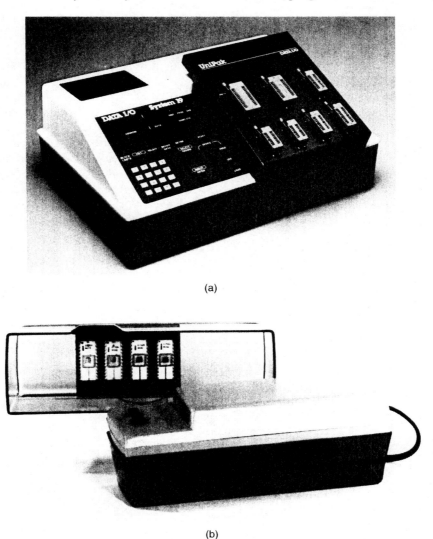

(a)

(b)

Figure 10.2 (a) EPROM programming unit. (Data I/O, Inc.) (b) EPROM erasing unit. (Ultra-Violet Products, Inc.)

correct programming, and read the information out of a programmed EPROM. An erasing unit such as that shown in Fig. 10.2(b) can be used to erase a number of EPROM ICs at a time.

Block Diagram of a Read-Only Memory

A block diagram of a typical read-only memory is shown in Fig. 10.3. Here we see that the device has three sets of signal lines: the address inputs, data outputs, and control inputs. This block diagram is valid for a ROM, PROM, or EPROM. Let us now look at the function of each of these sets of signal lines.

The address bus is used to input the signals that select between the storage locations within the ROM device. In Fig. 10.3, we find that this bus consists of 11 address lines, A_0 through A_{10}. The bits in the address are arranged so that A_{10} is the MSB and A_0 is the LSB. With an 11-bit address, the memory device has $2^{11} = 2048$ unique byte-storage locations. The individual storage locations correspond to consecutive addresses over the range $00000000000_2 = 000_{16}$ through $11111111111_2 = 7FF_{16}$.

Earlier we pointed out that information is stored inside a ROM, PROM, or EPROM as either a binary 0 or binary 1. Actually, 8 bits of data are stored at every address. Therefore, the organization of the ROM is described as 2048×8. The total storage capacity of the ROM is identified as the number of bits of information it can hold. We know 2048 bytes corresponds to 16,384 bits; that is, the device we are describing is actually a 16-KB ROM.

By applying the address of a storage location to the address inputs of the ROM, the byte of data held at the addressed location is read out onto the data lines. In the block diagram of Fig. 10.3, we see that the data bus consists of eight lines labeled as O_0 through O_7. Here O_7 represents the MSB and O_0 the LSB. For instance, applying the address $A_{10} \ldots A_1 A_0 = 10000000000_2 = 700_{16}$ will cause the byte of data held in this storage location to be output as $O_7 O_6 O_5 O_4 O_3 O_2 O_1 O_0$.

EXAMPLE 10.1

Suppose the block diagram in Fig. 10.3 has 15 address lines and 8 data lines. How many bytes of information can be stored in the ROM? What is its total storage capacity?

Solution

The number of bytes are

$$2^{15} = 32,768 \text{ bytes}$$

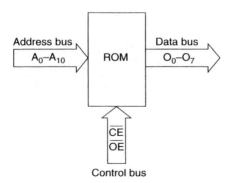

Figure 10.3 Block diagram of a ROM.

This gives a total of

$$32{,}768 \times 8 = 262{,}144 \text{ bits}$$

The control bus represents the control signals that are required to enable or disable the ROM, PROM, or the EPROM device. In the block diagram of Fig. 10.3, two control inputs, output enable ($\overline{OE}$) and chip enable ($\overline{CE}$), are identified. For example, logic 0 at $\overline{OE}$ enables the three state outputs, D_0 through D_7, of the device. If $\overline{OE}$ is switched to the 1 logic level, these outputs are disabled (put in the high-Z state). Moreover, $\overline{CE}$ must be at logic 0 for the device to be active. Logic 1 at $\overline{CE}$ puts the device in a low-power standby mode. When in this state, the data outputs are in the high-Z state independent of the logic level of $\overline{OE}$. In this way we see that both $\overline{OE}$ and $\overline{CE}$ must be at their active 0 logic levels for the device to be ready for operation.

Read Operation

It is the role of the MPU and its memory interface circuitry to provide the address and control input signals and to read the output data at the appropriate times during the memory-read bus cycle. The block diagram in Fig. 10.4 shows a typical read-only memory interface. For a microprocessor to read a byte of data from the device, it must apply a binary address to inputs A_0 through A_{10} of the EPROM. This address gets decoded inside the device to select the storage location of the byte of data that is to be read. Remember that the microprocessor must switch $\overline{CE}$ and $\overline{OE}$ to logic 0 to enable the device and its outputs. Now the byte of data is available at O_0 through O_7 and the microprocessor can read the data over its data bus.

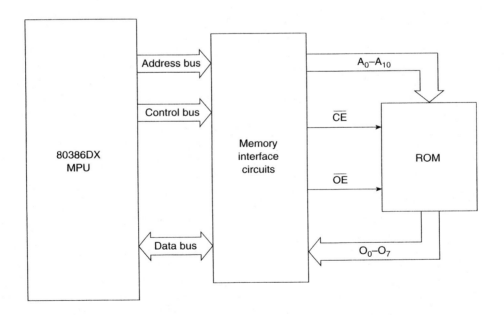

Figure 10.4 Read-only memory interface.

EPROM	Density (bits)	Capacity (bytes)
2716	16K	2K × 8
2732	32K	4K × 8
27C64	64K	8K × 8
27C128	128K	16K × 8
27C256	256K	32K × 8
27C512	512K	64K × 8
27C010	1M	128K × 8
27C020	2M	256K × 8
27C040	4M	512K × 8

Figure 10.5 Standard EPROM devices.

Standard EPROM ICs

A large number of standard EPROM ICs are available today. Figure 10.5 lists the part numbers, bit densities, and byte capacities of seven popular devices. They range in size from the 16-KB density (2K × 8) 2716 device to the 2M-byte (256K × 8) 27C020 device. Higher-density devices, such as the 27C256 through 27C020, are most popular for system designs. In fact, some of the older devices, such as the 2716 and 2732, have already been discontinued by many manufacturers. Let us now look at some of these EPROMs in more detail.

The 27C256 is an EPROM IC manufactured with the CMOS technology. Looking at Fig. 10.5, we find that it is a 256-KB device, and its storage array is organized as 32K × 8 bits. Figure 10.6 shows the pin layout of the 27C256. Here we see that it has 15 address inputs, labeled A_0 through A_{14}, eight data outputs, identified as O_0 through O_7, and two control signals, $\overline{CE}$ and $\overline{OE}$.

From our description of the read operation, it appears that after the inputs of the EPROM are set up, the output appears immediately; however, in practice this is not true. A short delay exists between address inputs and data outputs. This leads us to three important timing properties defined for the read cycle of an EPROM. They are called *access time* (t_{ACC}), *chip-enable time*, (t_{CE}), and *chip-deselect time* (t_{DF}). The values of these timing properties are provided in the read-cycle switching characteristics of Fig. 10.7(a) and identified in the waveforms shown in Fig. 10.7(b).

Access time tells us how long it takes to access data stored in an EPROM. Here we assume that both $\overline{CE}$ and $\overline{OE}$ are already at their active 0 levels, and then the address is applied to the inputs of the EPROM. In this case, the delay t_{ACC} occurs before the data stored at the addressed location are stable at the outputs. The microprocessor must wait at least this long before reading the data; otherwise, invalid results may be obtained. Figure 10.7(a) shows the standard EPROMs that are available with a variety of access times ratings. The maximum values of access time are given as 170 ns, 200 ns, and 250 ns. The speed of the device is selected to match that of the MPU. If the access time of the fastest standard device is too long for the MPU, wait-state circuitry needs to be added to the interface. In this way, wait states can be inserted to slow down the memory bus cycle.

Chip-enable time is similar to access time. In fact, for most EPROMs they are equal in value. They differ in how the device is set up initially. This time the address is applied and $\overline{OE}$ is switched to 0; then the read operation is initiated by making $\overline{CE}$ active. Therefore, t_{CE} represents the chip-enable-to-output delay instead of the address-to-output delay. Looking at Fig. 10.7(a), we see that the maximum values of t_{CE} are also 170 ns, 200 ns, and 250 ns.

Left table (pins 1–14):

27C512	27C128	27C64	2732A	2716	Pin
A_{15}	V_{PP}	V_{PP}			1
A_{12}	A_{12}	A_{12}			2
A_7	A_7	A_7	A_7	A_7	3
A_6	A_6	A_6	A_6	A_6	4
A_5	A_5	A_5	A_5	A_5	5
A_4	A_4	A_4	A_4	A_4	6
A_3	A_3	A_3	A_3	A_3	7
A_2	A_2	A_2	A_2	A_2	8
A_1	A_1	A_1	A_1	A_1	9
A_0	A_0	A_0	A_0	A_0	10
O_0	O_0	O_0	O_0	O_0	11
O_1	O_1	O_1	O_1	O_1	12
O_2	O_2	O_2	O_2	O_2	13
Gnd	Gnd	Gnd	Gnd	Gnd	14

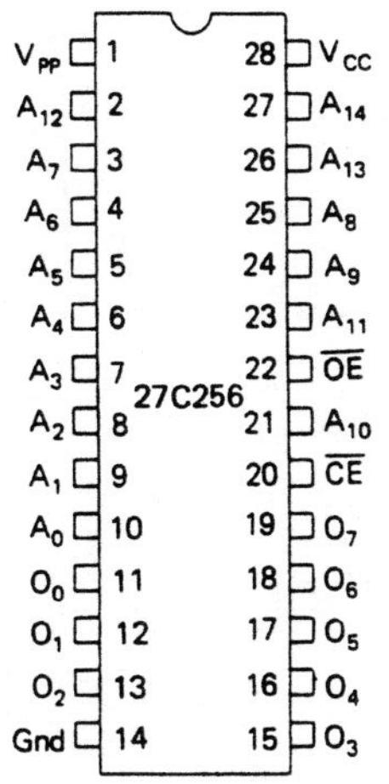

27C256 (28-pin DIP):

Pin	Signal	Signal	Pin
1	V_{PP}	V_{CC}	28
2	A_{12}	A_{14}	27
3	A_7	A_{13}	26
4	A_6	A_8	25
5	A_5	A_9	24
6	A_4	A_{11}	23
7	A_3	$\overline{OE}$	22
8	A_2	A_{10}	21
9	A_1	$\overline{CE}$	20
10	A_0	O_7	19
11	O_0	O_6	18
12	O_1	O_5	17
13	O_2	O_4	16
14	Gnd	O_3	15

Right table (pins 28–15):

Pin	2716	2732A	27C64	27C128	27C512
28			V_{CC}	V_{CC}	V_{CC}
27			$\overline{PGM}$	$\overline{PGM}$	A_{14}
26	V_{CC}	V_{CC}	N.C.	A_{13}	A_{13}
25	A_8	A_8	A_8	A_8	A_8
24	A_9	A_9	A_9	A_9	A_9
23	V_{PP}	A_{11}	A_{11}	A_{11}	A_{11}
22	$\overline{OE}$	$\overline{OE}/V_{PP}$	$\overline{OE}$	$\overline{OE}$	$\overline{OE}/V_{PP}$
21	A_{10}	A_{10}	A_{10}	A_{10}	A_{10}
20	$\overline{CE}$	$\overline{CE}$	$\overline{CE}$	$\overline{CE}$	$\overline{CE}$
19	O_7	O_7	O_7	O_7	O_7
18	O_6	O_6	O_6	O_6	O_6
17	O_5	O_5	O_5	O_5	O_5
16	O_4	O_4	O_4	O_4	O_4
15	O_3	O_3	O_3	O_3	O_3

Figure 10.6 Pin layouts of standard EPROMs.

	Versions	$V_{CC} \pm 5\%$		$V_{CC} \pm 10\%$											
		27C256-120V05		27C256-135V05 / 27C256-135V10		27C256-150V05 / 27C256-150V10		27C256-1 P27C256-1 N27C256-1		27C256-2 P27C256-2 N27C256-2 / 27C256-20 P27C256-20 N27C256-20		27C256 P27C256 N27C256 / 27C256-25 P27C256-25 N27C256-25		Unit
Symbol	Parameter	Min	Max	Min	Max	Min	Max	Min	Max	Min	Max	Min	Max	
t_{ACC}	Address to output delay		120		135		150		170		200		250	ns
t_{CE}	$\overline{CE}$ to output delay		120		135		150		170		200		250	ns
t_{OE}	$\overline{OE}$ to output delay		60		65		70		75		75		100	ns
t_{DF}[2]	$\overline{OE}$ high to output high-Z		30		35		45		55		55		60	ns
t_{OH}[2]	Output hold from addresses, $\overline{CE}$ or $\overline{OE}$ change—whichever is first	0		0		0		0		0		0		ns

Notes:

1. A.C. characteristics tested at $V_{IH} = 2.4$ V and $V_{IL} = 0.45$ V.
   Timing measurements made at $V_{OL} = 0.8$ V and $V_{OH} = 2.0$ V.
2. Guaranteed and sampled.
3. Package Prefixes: No Prefix = CERDIP; N = PLCC; P = Plastic DIP.

(a)

Figure 10.7(a) EPROM device timing characteristics. (Reprinted by permission of Intel Corp. Copyright/Intel Corp. 1989)

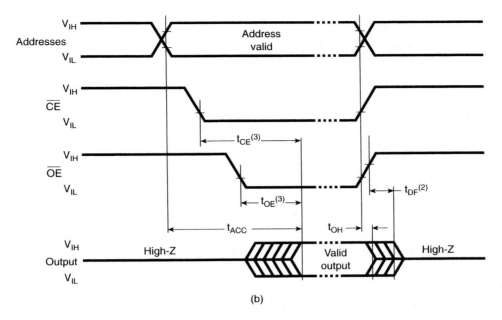

Figure 10.7(b) EPROM switching waveforms. (Reprinted by permission of Intel Corp. Copyright/Intel Corp. 1989)

Chip-deselect time is the opposite of access or chip-enable time. It represents the amount of time the device takes for the data outputs to return to the high-Z state after $\overline{OE}$ becomes inactive—that is, the recovery time of the outputs. In Fig. 10.7(a), we find that the maximum values for this timing property are 55 ns, 55 ns, and 60 ns.

In an erased EPROM, all storage cells hold logic 1. The device is put into the programming mode by switching on the V_{pp} power supply. Once in this mode, the address of the storage location that is to be programmed is applied to the address inputs, and the byte of data that is to be loaded into this location is supplied as inputs to the data leads. Note that the data outputs act as inputs when the EPROM is set up for the programming mode of operation. Next the $\overline{CE}$ input is pulsed to load the data. Actually, a complex series of program and verify operations are performed to program each storage location in an EPROM. The two programming sequences in wide use today are the *Quick-Pulse Programming Algorithm*™ and the *Intelligent Programming Algorithm*™. Flowcharts for these programming algorithms are given in Fig. 10.8(a) and (b), respectively.

Another group of important electrical characteristics for the 27C256 EPROM are given in Fig. 10.9. They are the device's dc electrical operating characteristics. CMOS EPROMs are designed to provide TTL-compatible input and output logic levels. Here we find the output logic level ratings are $V_{OHmin} = 3.5$ V and $V_{OLmax} = 0.45$ V. Also provided is the operating current rating of the device, which is identified as $I_{CC} = 30$ mA. This shows that if the device is operating at 5 V, it will consume 150 mW of power.

Figure 10.6 also shows the pin layouts for the 2716 through 27C512 EPROM devices. In this diagram, we find that both the 27C256 and 27C512 are available in a 28-pin package. A comparison of the pin configuration of the 27C512 with that of the 27C256 shows that the only differences between the two pinouts are that pin 1 on the 27C512 becomes the

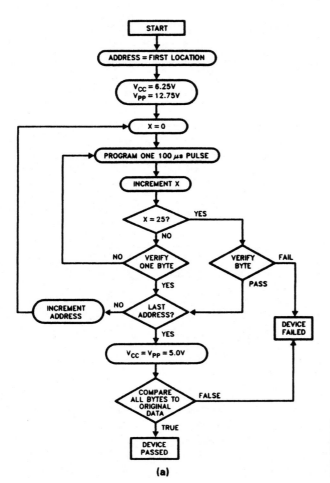

Figure 10.8(a) Quick-Pulse Programming™ algorithm flowchart. (Reprinted by permission of Intel Corp. Copyright/Intel Corp. 1989)

new address input A_{15}, and V_{pp}, which was at pin 1 on the 27C256, becomes a second function performed by pin 22 on the 27C512.

Expanding EPROM Word Length and Word Capacity

In many applications, the microcomputer system requirements for EPROM are greater than what is available in a single device. There are two basic reasons for expanding EPROM capacity: first, the byte-wide length is not large enough; and second, the total storage capacity is not enough bytes. Both of these expansion needs can be satisfied by interconnecting a number of ICs.

For example, the 80386DX microprocessor has a 32-bit data bus. Therefore, its program memory subsystem needs to be implemented with four 27C256 EPROMs connected as shown in Fig. 10.10(a). Notice that the individual address inputs, chip-enable lines, and output-enable lines on the two devices are connected in parallel. On the other hand, the eight data outputs of each device are used to supply eight lines of the MPU's 32-bit data

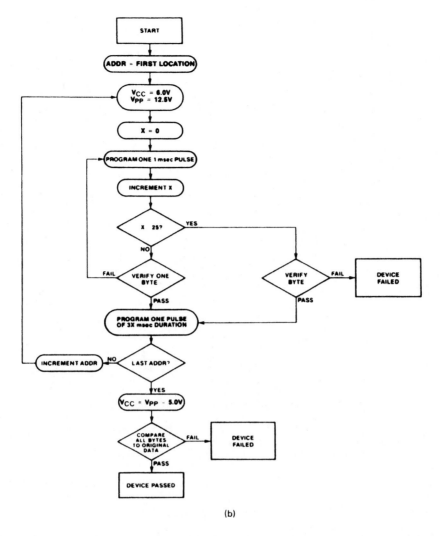

START

ADDR = FIRST LOCATION

V_{CC} = 6.0V
V_{PP} = 12.5V

X = 0

PROGRAM ONE 1 msec PULSE

INCREMENT X

X 25? — YES

NO

VERIFY ONE BYTE — FAIL

PASS

PROGRAM ONE PULSE OF 3X msec DURATION

VERIFY BYTE — FAIL — DEVICE FAILED

PASS

LAST ADDR? — NO — INCREMENT ADDR

YES

V_{CC} = V_{PP} = 5.0V

COMPARE ALL BYTES TO ORIGINAL DATA — FAIL — DEVICE FAILED

PASS

DEVICE PASSED

(b)

Figure 10.8(b) Intelligent Programming™ algorithm flowchart. (Reprinted by permission of Intel Corp. Copyright/Intel Corp. 1989)

bus. This circuit configuration has a total storage capacity equal to 32K 32-bit words, which equals 1 MB.

Figure 10.10(b) shows how two 27C256s can be interconnected to expand the number of bytes of storage. Here the individual address inputs, data outputs, and output-enable lines of the two devices are connected in parallel. However, the $\overline{CE}$ inputs of the individual devices remain independent and can be supplied by different chip-enable outputs, identified as $\overline{CE}_0$ and $\overline{CE}_1$, of an address decoder circuit. In this way, only one of the two devices will be enabled at a time. This configuration results in a total storage capacity of 64K bytes, or 512 KB. To double the word capacity of the circuit in Fig. 10.10(a), this same connection must be made for each of the EPROMs.

Symbol	Parameter		Notes	Min	Typ[3]	Max	Unit	Test Conditions
I_{LI}	Input load current				0.01	1.0	μA	$V_{IN} = 0V$ to V_{CC}
I_{LO}	Output leakage current					± 10	μA	$V_{OUT} = 0V$ to V_{CC}
I_{PP_1}	V_{PP} read current		5			200	μA	$V_{PP} = V_{CC}$
I_{SB_1}	V_{CC} current standby	TTL	8			1.0	mA	$\overline{CE} = V_{IH}$
I_{SB_2}		CMOS	4			100	μA	$\overline{CE} = V_{CC}$
I_{CC_1}	V_{CC} current active		5, 8			30	mA	$\overline{CE} = V_{IL}$ $f = 5$ MHz
V_{IL}	Input low voltage (± 10% supply) (TTL)			−0.5		0.8	V	
	Input low voltage (CMOS)			−0.2		0.8		
V_{IH}	Input high voltage (± 10% supply) (TTL)			2.0		$V_{CC} + 0.5$	V	
	Input high voltage (CMOS)			$0.7\,V_{CC}$		$V_{CC} + 0.2$		
V_{OL}	Output low voltage					0.45	V	$I_{OL} = 2.1$ mA
V_{OH}	Output high voltage			3.5			V	$I_{OH} = -2.5$ mA
I_{OS}	Output short circuit current		6			100	mA	
V_{PP}	V_{PP} read voltage		7	$V_{CC} - 0.7$		V_{CC}	V	

Notes:

1. Minimum D.C. input voltage is −0.5V. During transitions, the inputs may undershoot to −2.0V for periods less than 20 ns. Maximum D.C. voltage on output pins is $V_{CC} + 0.5$V which may overshoot to $V_{CC} + 2$V for periods less than 20 ns.
2. Operating temperature is for commercial product defined by this specification. Extended temperature options are available in EXPRESS and Military version.
3. Typical limits are at $V_{CC} = 5$V, $T_A = +25°C$.
4. $\overline{CE}$ is $V_{CC} \pm 0.2$V. All other inputs can have any value within spec.

5. Maximum Active power usage is the sum $I_{PP} + I_{CC}$. The maximum current value is with outputs O_0 to O_7 unloaded.
6. Output shorted for no more than one second. No more than one output shorted at a time. I_{OS} is sampled but not 100% tested.
7. V_{PP} may be one diode voltage drop below V_{CC}. It may be connected directly to V_{CC}. Also, V_{CC} must be applied simultaneously or before V_{PP} and removed simultaneously or after V_{PP}.
8. V_{IL}, V_{IH} levels at TTL inputs.

Figure 10.9 DC electrical characteristics of the 27C256. (Reprinted by permission of Intel Corp. Copyright/Intel Corp. 1989)

▲ 10.4 RANDOM ACCESS READ/WRITE MEMORIES

The memory section of a microcomputer system is normally formed from both read-only memories (ROM) and *random access read/write memories* (RAM). Earlier we pointed out that the ROM is used to store permanent information such as the microcomputer's hardware-control program. RAM is similar to ROM in that its storage location can be accessed in a random order, but it is different from ROM in two important ways. First, data stored in RAM are not permanent in nature; that is, they can be altered. RAM can be used to save data by writing to it, and later the data can be read back for additional processing. Because of its read and write features, RAM finds wide use where data and programs need to be

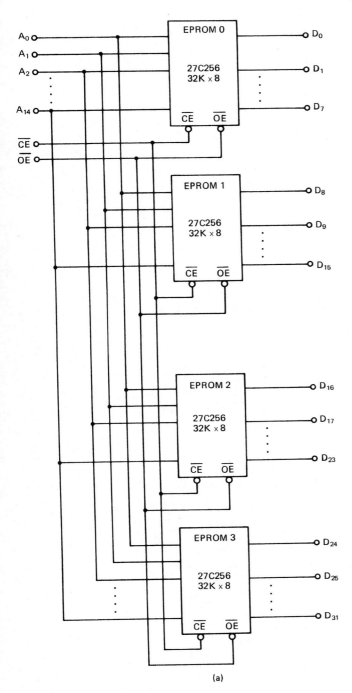

Figure 10.10(a) Expanding word length.

(a)

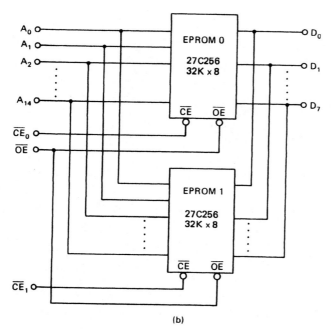

Figure 10.10(b) Expanding word capacity.

(b)

placed in memory only temporarily. For this reason, it is normally used to store data and application programs for execution. The second difference is that RAM is *volatile*; that is, if power is removed from RAM, the stored data are lost.

Static and Dynamic RAMs

There are two types of RAMs in wide use today, the *static RAM* (SRAM) and *dynamic RAM* (DRAM). For a static RAM, data, once entered, remain valid as long as the power supply is not turned off. On the other hand, to retain data in a DRAM, it is not sufficient just to maintain the power supply. For this type of device, we must both keep the power supply turned on and periodically restore the data in each storage location. This added requirement is necessary because the storage elements in a DRAM are capacitive nodes. If the storage nodes are not recharged within a specific interval of time, data are lost. This recharging process is known as *refreshing* the DRAM.

Block Diagram of a Static RAM

A block diagram of a typical static RAM IC is shown in Fig. 10.11. By comparing this diagram with the one shown for a ROM in Fig. 10.3, we see that they are similar in many ways. For example, they both have address lines, data lines, and control lines. These signal buses perform similar functions when the RAM is operated. Because of the RAMs read/ write capability, data lines, however, act as both inputs and outputs. For this reason, they are identified as a *bidirectional bus*.

A variety of static RAM ICs are currently available. They differ both in density and organization. The most commonly used densities in system designs are the 64- and 256-KB devices. The structure of the data bus determines the organization of the SRAMs storage

Memory Devices, Circuits, and Subsystem Design   Chap. 10

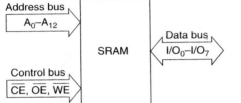

Figure 10.11 Block diagram of a static RAM.

array. In Fig. 10.11, an 8-bit data bus is shown. This type of organization is known as a *byte-wide* SRAM. Devices are also manufactured with by 1 and by 4 data I/O organizations. At the 64-KB density, this results in three standard device organizations: 64K × 1, 16K × 4, and 8K × 8.

The address bus on the SRAM in Fig. 10.11 consists of the lines labeled A_0 through A_{12}. This 13-bit address is what is needed to select between the 8K individual storage locations in an 8K × 8-bit SRAM IC. The 16K × 4 and 64K × 1 devices require a 14-bit and 16-bit address, respectively.

To either read from or write to SRAM, the device must first be chip-enabled. Just like for a ROM, this is done by switching the $\overline{CE}$ input of the SRAM to logic 0. Earlier we indicated that data lines I/O_0 through I/O_7 in Fig. 10.11 are bidirectional. This means that they will act as inputs when writing data into the SRAM or as outputs when reading data from the SRAM. The setting of a new control signal, the *write enable* ($\overline{WE}$) input, determines how the data lines operate. During all write operations to a storage location within the SRAM, the appropriate $\overline{WE}$ inputs must be switched to the 0 logic level. This configures the data lines as inputs. On the other hand, if data are to be read from a storage location, $\overline{WE}$ is left at the 1 logic level. When reading data from the SRAM, output enable ($\overline{OE}$) must be active. Applying the active memory signal, logic 0, at this input, enables the device's three-state outputs.

2 *milsec*

Three-state data bus lines allow for the parallel busing needed to expand data memory by interconnecting multiple devices. For example, in Fig. 10.12 we see how four 8K × 8-bit SRAMs are interconnected to form an 8K × 32-bit memory circuit. In this circuit, the separate $\overline{CE}$ inputs of the SRAM ICs are wired together to form a single chip-enable input. Normally this input is activated by an SRAM-select output of the address decoder circuit. The $\overline{OE}$ input of the individual SRAMs are also directly connected in parallel. The output-enable input that results is normally driven by the $\overline{MRDC}$ output of the bus-control logic and enables the outputs of all SRAMs during all memory-read bus cycles. Notice that an individual write-enable is provided for each SRAM. For instance, $\overline{WE}_0$ is applied to SRAM 0 and $\overline{WE}_1$ is applied to SRAM 1. They are required to permit byte and word write operations. For example, if a word of data is to be written to the SRAM array over data bus lines D_0 through D_{15}, just $\overline{WE}_0$ and $\overline{WE}_1$ are at their active 0 logic level. The $\overline{MWTC}$ signal that is an output of the bus control logic is used to derive the individual device write-enable signals.

Standard Static RAM ICs

Figure 10.13 is a list of a number of standard static RAM ICs. Here we find their part numbers, densities, and organizations. For example, the 4361, 4363, and 4364 are all 64-KB density devices; however, they are each organized differently. The 4361 is a 64K × 1-bit device, the 4363 is a 16K × 4-bit device, and the 4364 is an 8K × 8-bit device.

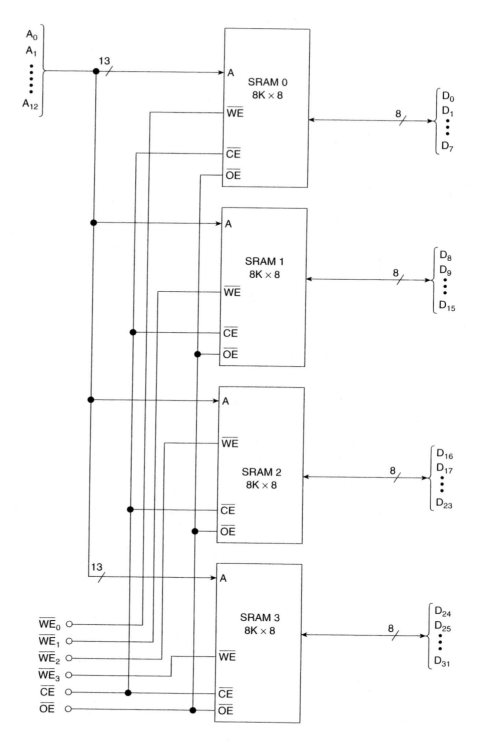

Figure 10.12 8K × 32-bit SRAM circuit.

SRAM	Density (bits)	Organization
4361	64K	$64K \times 1$
4363	64K	$16K \times 4$
4364	64K	$8K \times 8$
43254	256K	$64K \times 4$
43256A	256K	$32K \times 8$
431000A	1M	$128K \times 8$

Figure 10.13 Standard SRAM devices.

The pin layouts of the 4364 and 43256A ICs are given in Figs. 10.14(a) and (b), respectively. Looking at the 4364 we see that it is almost identical to the block diagram shown in Fig. 10.11. The one difference is that it has two chip-enable lines instead of one. They are labeled $\overline{CE}_1$ and CE_2. Notice that one is activated by logic 0 and the other by logic 1. Both of these chip-enable inputs must be at their active logic level to enable the device for operation.

EXAMPLE 10.2

How does the 43256A SRAM differ from the block diagram in Fig. 10.11?

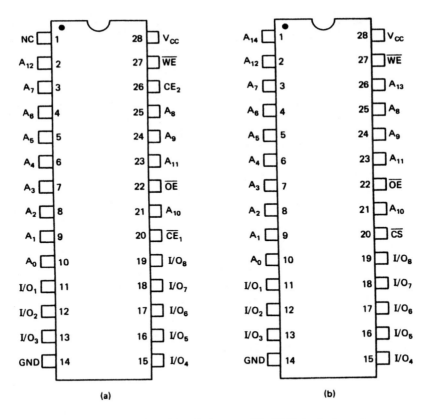

(a)   (b)

Figure 10.14   (a) 4364 pin layout. (b) 43256A pin layout.

Sec. 10.4   Random Access Read/Write Memories

Part number	Read/write cycle time
4364-10	100 ns
4364-12	120 ns
4364-15	150 ns
4364-20	200 ns

Figure 10.15 Speed selections for the 4364.

Solution

It has two additional address inputs A_{13} and A_{14} and the chip-enable input is labeled $\overline{CS}$ instead of $\overline{CE}$.

As shown in Fig. 10.15, the 4364 is available in four speeds. For example, the minimum read cycle and write cycle time for the 4364-10 is 100 ns.

Figure 10.16 is a list of the 4364's dc electrical characteristics. Note that it draws a maximum of 10 mA of dc current when operating. Therefore, it can consume as much as 55 mW when operating at its maximum power supply voltage of 5.5 V.

SRAM Read and Write Cycle Operation

The waveforms for a typical write cycle are illustrated in Fig. 10.17. Let us trace the events that take place during the write cycle. Here we see that all critical timing is referenced to the point at which the address becomes valid. Notice that the minimum duration of the write cycle is identified as t_{WC}. This is the 100-ns *write cycle time* of the 4364-10. The address must remain stable for this complete interval of time.

Next $\overline{CE}_1$ and CE_2 become active and must remain active until the end of the write cycle. The duration of these pulses are identified as $\overline{CE}_1$ *to end of write* (t_{CW1}) *time* and CE_2 *to end of write* (t_{CW2}) *time*. As shown in the waveforms, we are assuming here that they begin at any time after the occurrence of the address but before the leading edge of $\overline{WE}$. The minimum value for both of these times is 80 ns. On the other hand, $\overline{WE}$ is shown to not occur until the interval t_{AS} elapses. This is the *address-setup time* and represents the minimum amount of time the address inputs must be stable before $\overline{WE}$ can be switched to logic 0. For the 4364, however, this parameter is equal to 0 ns. The width of the write enable pulse is identified as t_{WP}, and its minimum value equals 60 ns.

Data applied to the D_{IN} data inputs are written into the device synchronous with the trailing edge of $\overline{WE}$. Notice that the data must be valid for an interval equal to t_{DW} before this edge. This interval, which is called *data valid to end of write*, has a minimum value of 40 ns for the 4364-10. Moreover, it is shown to remain valid for an interval of time equal to t_{DH} after this edge. This *data-hold time*, however, just like address-setup time, equals 0 ns for the 4364. Finally, a short recovery period takes place after $\overline{WE}$ returns to logic 1 before the write cycle is complete. This interval is identified as t_{WR} in the waveforms, and its minimum value equals 5 ns.

The read cycle of a static RAM, such as the 4364, is similar to that of a ROM. Waveforms of a read operation are given in Fig. 10.18.

Standard Dynamic RAM ICs

Dynamic RAMs are available in higher densities than static RAMs. Currently, the most widely used DRAMs are the 64-KB, 256-KB, 1-MB, and 4-MB devices. Figure 10.19

Parameter	Symbol	Min	Typ	Max	Unit	Test Conditions
Input leakage current	I_{LI}			1	μA	$V_{IN} = 0$ V to V_{CC}
Output leakage current	I_{LO}			1	μA	$V_{I/O} = 0$ V to V_{CC} $\overline{CE}_1 = V_{IH}$ or $CE_2 = V_{IL}$ or $\overline{OE} = V_{IH}$ or $\overline{WE} = V_{IL}$
Operating supply current	I_{CCA1}			(1)	mA	$\overline{CE}_1 = V_{IL}$, $CE_2 = V_{IH}$, $I_{I/O} = 0$, Min cycle
	I_{CCA2}		5	10	mA	$\overline{CE}_1 = V_{IL}$, $CE_2 = V_{IH}$, $I_{I/O} = 0$, DC current
	I_{CCA3}		3	5	mA	$\overline{CE}_1 \leq 0.2$ V, $CE_2 \geq V_{CC} - 0.2$ V, $V_{IL} \leq 0.2$ V, $V_{IH} \geq V_{CC} - 0.2$ V, $f = 1$ MHz, $I_{I/O} = 0$
Standby supply current	I_{SB}			(2)	mA	$\overline{CE}_1 \geq V_{IH}$ or $CE_2 = V_{IL}$
	I_{SB1}			(3)	mA	$\overline{CE}_1 \geq V_{CC}-0.2$ V $CE_2 \geq V_{CC}-0.2$ V
	I_{SB2}			(3)	mA	$CE_2 \leq 0.2$ V
Output voltage, low	V_{OL}			0.4	V	$I_{OL} = 2.1$ mA
Output voltage, high	V_{OH}	2.4			V	$I_{OH} = -1.0$ mA

Notes:

(1) μPD4364-10/10L: 45 mA max
  μPD4364-12/12L/12LL: 40 mA max
  μPD4364-15/15L/15LL: 40 mA max
  μPD4364-20/20L/20LL: 35 mA max

(2) μPD4364-xx: 5 mA max
  μPD4364-xxL: 3 mA max
  μPD4364-xxLL: 3 mA max

(3) μPD4364-xx: 2 mA max
  μPD4364-xxL: 100 μA max
  μPD4364-xxLL: 50 μA max

Figure 10.16 DC electrical characteristics of the 4364.

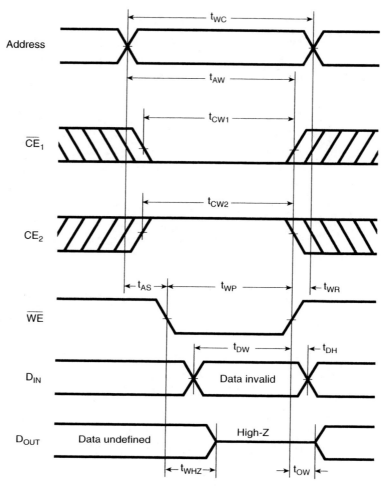

Notes:
1. A write occurs during the overlap of a low $\overline{CE}_1$ and a high CE_2 and a low $\overline{WE}$.
2. $\overline{CE}_1$ or $\overline{WE}$ [or CE_2] must be high [low] during any address transaction.
3. If $\overline{OE}$ is high the I/O pins remain in a high-impedance state.

Figure 10.17 Write-cycle timing diagram.

is a list of a number of popular DRAM ICs. Here we find the 2164B, which is organized as 64K × 1 bit, the 21256, which is organized as 256K × 1 bit, the 21464, which is organized as 64K × 4 bits, the 421000, which is organized as 1M × 1 bit, and the 424256, which is organized as 256K × 4 bits. Pin layouts for the 2164B, 21256, and 421000 are shown in Figs. 10.20(a), (b), and (c), respectively.

Some other benefits of using DRAMs over SRAMs are that they cost less, they consume less power, and their 16- and 18-pin packages take up less space. For these reasons, DRAMs are normally used in applications that require a large amount of memory. For example, most systems that support at least 1M byte of data memory are designed using DRAMs.

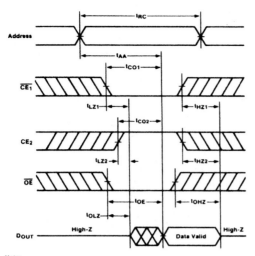

Figure 10.18 Read-cycle timing diagram.

The 2164B is one of the older NMOS DRAM devices. A block diagram of the device is shown in Fig. 10.21. Looking at the block diagram we find that it has eight address inputs, A_0 through A_7, a data input and a data output marked D and Q, respectively, and three control inputs, *row-address strobe* ($\overline{RAS}$), *column-address strobe* ($\overline{CAS}$), and *read/write* ($\overline{W}$).

The storage array within the 2164B is capable of storing 65,536 (64K) individual bits of data. To address this many storage locations, we need a 16-bit address; however, this device's package has just 16 pins. For this reason, the 16-bit address is divided into two separate parts: an 8-bit *row address* and an 8-bit *column address*. These two parts of the address are time-multiplexed into the device over a single set of address lines, A_0 through A_7. First the row address is applied to A_0 through A_7. Then $\overline{RAS}$ is pulsed to logic 0 to latch it into the device. Next, the column address is applied and $\overline{CAS}$ is strobed to logic 0. This 16-bit address selects which one of the 64K storage locations is to be accessed.

Data are either written into or read from the addressed storage location in the DRAMs. Write data are applied to the D input and read data are output at Q. The logic levels of

DRAM	Density (bits)	Organization
2164B	64K	64K × 1
21256	256K	256K × 1
21464	256K	64K × 4
421000	1M	1M × 1
424256	1M	256K × 4
44100	4M	4M × 1
44400	4M	1M × 4
44160	4M	256K × 16
416800	16M	8M × 2
416400	16M	4M × 4
416160	16M	1M × 16

Figure 10.19 Standard DRAM devices.

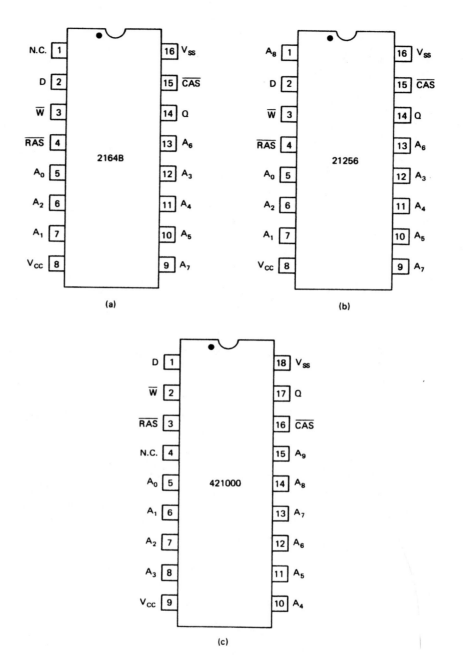

Figure 10.20 (a) 2164B pin layout. (b) 21256 pin layout. (c) 421000 pin layout.

control signals $\overline{W}$, $\overline{RAS}$, and $\overline{CAS}$ tell the DRAM whether a read or write data transfer is taking place and control the three-state outputs. For example, during a write operation, the logic level at D is latched into the addressed storage location at the falling edge of either $\overline{CAS}$ or $\overline{W}$. If $\overline{W}$ is switched to logic 0 by an active $\overline{MWTC}$ signal before $\overline{CAS}$, an early write cycle is performed. During this type of write cycle, the outputs are maintained

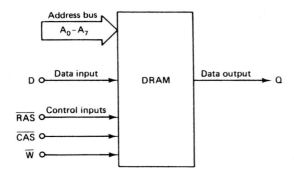

Figure 10.21 Block diagram of the 2164B DRAM.

in the high-Z state throughout the complete bus cycle. The fact that the output is put in the high-Z state during the write operation allows the D input and Q output of the DRAM to be tied together. The Q output is also in the high-Z state whenever $\overline{CAS}$ is logic 1. This is the connection and mode of operation normally used when attaching DRAMs to the bidirectional data bus of a microprocessor. Figure 10.22 shows how thirty-two 2164B devices are connected to make up a 64K × 32-bit DRAM array.

The 2164B also has the ability to perform what are called *page-mode* accesses. If $\overline{RAS}$ is left at logic 0 after the row address is latched inside the device, the address is maintained within the device. Then data cells along the selected row can be accessed by simply supplying successive column addresses. This permits faster access of memory by eliminating the time needed to set up and strobe additional row addresses.

Earlier we pointed out that the key difference between the DRAM and SRAM is that the storage cells in the DRAM need to be periodically refreshed; otherwise, they lose their data. To maintain the integrity of the data in a DRAM, each of the rows of the storage array must typically be refreshed periodically, such as every 2 ms. All the storage cells in an array are refreshed by simply cycling through the row addresses. As long as $\overline{CAS}$ is held at logic 1 during the refresh cycle, no data are output.

External circuitry is required to perform the address multiplexing, $\overline{RAS}/\overline{CAS}$ generation, and refresh operations for a DRAM subsystem. *DRAM-refresh controller* ICs are available to permit easy implementation of these functions.

Battery Backup for the RAM Subsystem

Even though RAM ICs are volatile, in some equipment it is necessary to make all or part of the RAM memory subsystem nonvolatile. An example is an electronic cash register. In this application, a power failure could result in the loss of irreplaceable information about the operation of the business.

To satisfy the nonvolatile requirement, additional circuitry can be included in the RAM subsystem. These circuits must sense the occurrence of a power failure and automatically switch the memory subsystem over to a backup battery. An orderly transition must take place from system power to battery power. Therefore, when a loss of power is detected, the power-fail circuit must permit the completion of any read or write cycle that is in progress and then lock the memory against the occurrence of additional read/write operations. The memory subsystem remains in this state until power is restored. In this way, the system can be made at least temporarily nonvolatile.

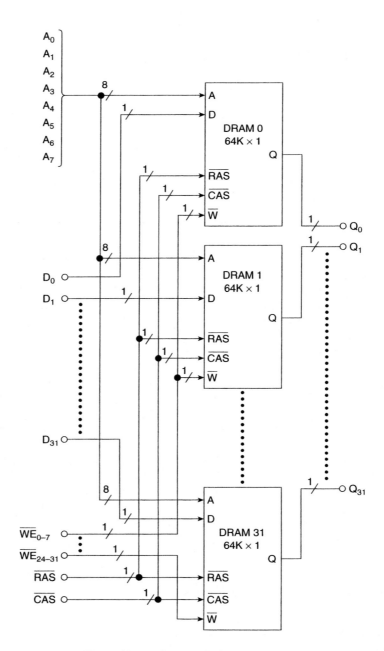

Figure 10.22 64K × 32-bit DRAM circuit.

▲ 10.5 PARITY, PARITY BIT, AND PARITY-CHECKER/ GENERATOR CIRCUIT

In microcomputer systems, the data exchanges that take place between the MPU and the memory must be done without error. However, problems such as noise, transient signals, or even bad memory bits can produce errors in the transfer of data and instructions. For instance, the storage location for one bit in a large DRAM array may be bad and stuck at the 0 logic level. This will not present a problem if the logic level of the data written to the storage location is 0, but if it is 1 the value will always be read as 0. To improve the reliability of information transfer between the MPU and memory, a *parity bit* can be added to each byte of data. To implement data transfers with parity, a *parity-checker/generator* circuit is required.

Figure 10.23 shows a parity-checker/generator circuit added to the memory interface of a microcomputer system. Notice that the data passed between the MPU and memory subsystem is applied in parallel to the parity-checker/generator circuit. Assuming that the microprocessor has an 8-bit data bus, data words read from or written to memory by the MPU over the data bus are still byte wide, but the data stored in memory are 9 bits long. The data in memory consist of 8 bits of data and 1 parity bit. Assuming that the memory array is constructed with 256K $\times$ 1-bit DRAMs, then, the memory array for a memory subsystem with parity would have nine DRAM ICs instead of eight. The extra DRAM is needed for storage of the parity bit for each byte of data stored in the other eight DRAM devices.

The parity-checker/generator circuit can be set up to produce either *even parity* or *odd parity*. The 9-bit word of data stored in memory has even parity if it contains an even number of bits that are at the 1 logic level and odd parity if the number of bits at logic 1 is odd.

Let us assume that the circuit in Fig. 10.23 is used to generate and check for even parity. If the byte of data written to memory over the MPU's data bus is FFH, the binary data word is 11111111_2. This byte has 8 bits at logic 1, that is, it already has even parity. Therefore, the parity-checker/generator circuit, which operates in the parity generate mode,

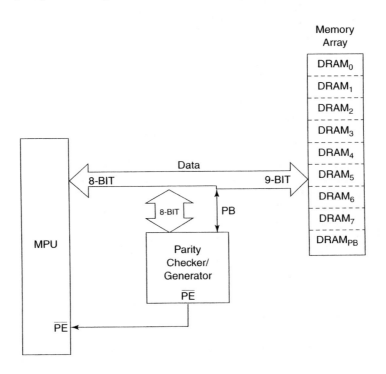

Figure 10.23 Data-storage memory interface with parity-checking/generation.

outputs logic 0 on the parity bit line (PB), and the 9 bits of data stored in memory is 011111111_2. On the other hand, if the byte written to memory is 7FH, the binary word is 01111111_2. Since only 7 bits are at logic 1, parity is odd. In this case, the parity-checker/generator circuit makes the parity bit logic 1, and the 9 bits of data saved in memory is 101111111_2. Notice that the data held in memory has even parity. In this way, we see that during all data memory write cycles, the parity-checker/generator circuit simply checks the data that are to be stored in memory and generates a parity bit. The parity bit is attached to the original 8 bits of data to make it 9 bits. The 9 bits of data stored in memory have even parity.

The parity-checker/generator works differently when data are read from memory. Now the circuit must perform its parity-check function. Notice that the 8 bits of data from the addressed storage location in memory are sent directly to the MPU. However, at the same time, this byte and the parity bit are applied to the inputs of the parity-checker/generator circuit. This circuit checks to determine whether there is an even or odd number of logic 1s in the word with parity. Again we will assume that the circuit is set up to check for even parity. If the 9 bits of data read from memory are found to have an even number of bits at the 1 logic level, parity is correct. The parity checker/generator signals this fact to the MPU by making the parity error ($\overline{PE}$) output inactive logic 1. This signal is normally sent to the MPU to identify whether or not a memory *parity error* has occurred. If an odd number of bits are found to be logic 1, a parity error has been detected and $\overline{PE}$ is set to 0 to tell the MPU of the error condition. Once alerted to the error, the MPU can do any one of a number of things under software control to recover. For instance, it could simply repeat the memory-read cycle to see if it takes place correctly the next time.

The 74AS280 device implements a parity-checker/generator function similar to that we just described. In Fig. 10.24(a) we have shown a block diagram of the device. Notice that it has nine data-input lines, which are labeled A through I. In the memory interface, lines A through H would be attached to data bus lines D_0 through D_7, respectively, and during a read operation the parity bit output of the memory array, D_{PB} would be applied to the I input.

The operation of the 74AS280 is described by the function table in Fig. 10.24(b). It shows how the Σ_{EVEN} and Σ_{ODD} outputs respond to an even or odd number of data inputs at logic 1. Notice that if there are 0, 2, 4, or 8 inputs at logic 1, the Σ_{EVEN} output switches to logic 1 and Σ_{ODD} to logic 0. This output response signals the even parity condition.

In practical applications, the Σ_{EVEN} and Σ_{ODD} outputs are used to produce the parity bit and parity error signal lines. Figure 10.24(c) is an even parity-checker/generator configuration. Notice that Σ_{ODD} is used as the parity bit (D_{PB}) output that gets applied to the data input of the parity bit DRAM in the memory array. During a write operation $\overline{MEMR}$ is 0, which makes the I input 0, and therefore the parity of the byte depends only on data bits D_0 through D_7, which are applied to the A through H inputs of the 74AS280. As long as the input at A through H has an even number of bits at logic 1 during a memory write cycle, Σ_{ODD}, which is D_{PB}, is at logic 0 and the 9 bits of data written to memory retain an even number of bits that are 1, that is, even parity. On the other hand, if the incoming byte at A through H has an odd number of bits that are logic 1, Σ_{ODD} switches to logic 1. The logic 1 at D_{PB} along with the odd number of 1s in the original byte again give the 9 bits of data stored in memory an even parity.

Let us next look at what happens in the parity-checker/generator circuit during a memory-read cycle for the data-storage memory subsystem. When the MPU is reading a byte of data from memory, the 74AS280 performs the parity-check operation. In response to the MPU's read request, 9 bits of data are output by the memory array. They are applied to inputs A through I of the parity-checker/generator circuit. The 74AS280 checks the

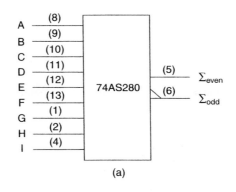

(a)

| NUMBER OF INPUTS A | OUTPUTS | |
THRU I THAT ARE HIGH	Σ EVEN	Σ ODD
0,2,4,6,8	H	L
1,3,5,7,9	L	H

(b)

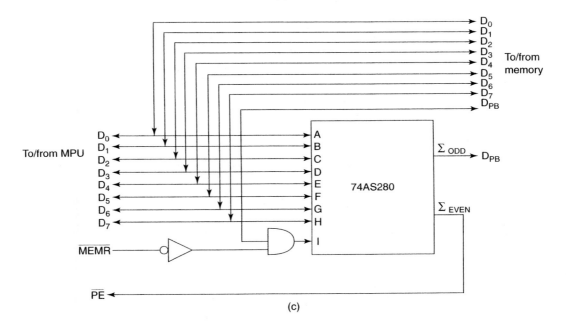

(c)

Figure 10.24 (a) Block diagram of the 74AS280. (Texas Instruments Incorporated) (b) Function table. (Texas Instruments Incorporated) (c) Even-parity checker/generator connection.

parity and adjusts the logic levels of Σ_{EVEN} and Σ_{ODD} to represent this parity. If parity is even as expected, Σ_{EVEN}, which represents the parity error ($\overline{PE}$) signal, is at logic 1. This tells the MPU that a valid data transfer has taken place. However, if the data at A through I has an odd number of bits at logic 1, Σ_{EVEN} switches to logic 0 and informs the MPU that a parity error has occurred.

In a 16-bit microcomputer system, such as that built with the 80386SX MPU, there

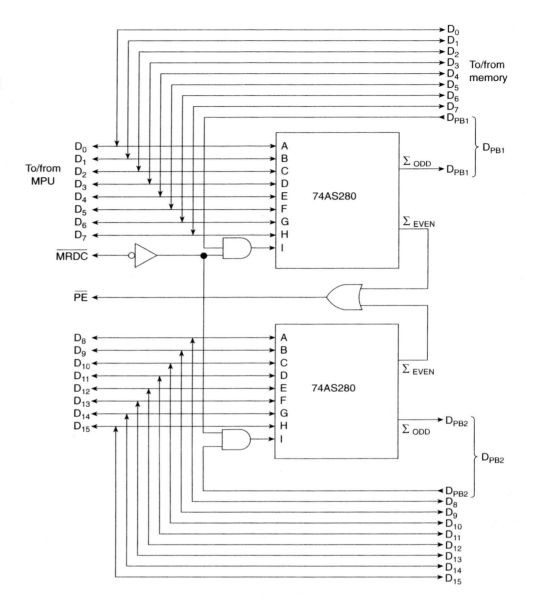

Figure 10.25 Parity checking/generation circuits for 16-bit bus memory interface.

are normally two 8-bit banks of DRAM ICs in the data-storage memory array. The circuit diagram in Fig. 10.25 shows this type of data-storage memory interface. Notice that a parity bit DRAM is added to each bank. Therefore, parity is implemented separately for each of the 2 bytes of a data word stored in memory. This is important because the 80386SX can read either bytes or words of data from memory. For this reason, two parity-checker/ generator circuits are also required, one for the upper eight lines of the data bus and one for the lower eight lines. The parity error outputs of the two circuits are combined by gating them together and the resulting parity error signal is supplied to the MPU. In this way, the

MPU is notified of a parity error if it occurs in an even-addressed byte data transfer, odd-addressed byte data transfer, or in either or both bytes of a 16-bit data transfer.

▲ 10.6 FLASH MEMORY

Another memory technology important to the study of microcomputer systems is what is known as *FLASH memory*. FLASH memory devices are similar to EPROMs in many ways but are different in several very important ways. In fact, FLASH memories act just like EPROMs. They are nonvolatile, are read just like an EPROM, and program with an EPROM-like algorithm.

The key difference between a FLASH memory and an EPROM is that its memory cells are erased electrically, instead of by exposure to ultraviolet light. That is, the storage array of a FLASH memory can be both electrically erased and reprogrammed with new data. Unlike RAMs, they are not byte erasable and writeable. When an erase operation is performed on a FLASH memory, either the complete memory array or a large block of storage locations, not just 1 byte, is erased. Moreover, the erase process is complex and can take as long as several seconds. This erase operation can be followed by a write operation, which is actually a programming cycle, that loads new data into the storage location. This write operation also take a long time when compared to the write cycle times of a RAM.

Even though FLASH memories are writeable, they find their widest use in microcomputer systems for storage of firmware, just like EPROMs. However, their limited erase/rewrite capability enables them to be used in a number of other important applications where data must be rewritten, but not frequently. Some examples are to implement a nonvolatile writeable lookup table, in-system programming for code updates, and solid state drives.

Block Diagram of a FLASH Memory

Earlier we pointed out that the FLASH memories operates in a way very similar to an EPROM. Figure 10.26 shows a block diagram of a typical FLASH memory device. Let us compare this block diagram to that of the ROM in Fig. 10.3. Address lines A_0 through A_{17}, chip enable ($\overline{CE}$), and output enable ($\overline{OE}$) serve the exact same function for both devices. That is, the address picks the storage location that is to be accessed, $\overline{CE}$ enables the device for operation, and $\overline{OE}$ enables the data to the outputs during read cycles.

We also see that they differ in two ways. First, the data bus is identified as bidirectional. This is because the FLASH memory can be used in an application where it is written into as well as read from. Second, another control input, write enable ($\overline{WE}$), is provided. This

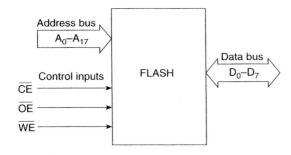

Figure 10.26 Block diagram of a FLASH memory.

signal must be at its active 0 logic level during all write operations to the device. In fact, this block diagram is exactly the same as that given for an SRAM in Fig. 10.11.

Bulk-Erase, Boot Block, and FlashFile™ FLASH Memories

FLASH memory devices are available with several different memory array architectures. These architectures relate to how the device is organized for the purpose of erasing. Earlier we pointed out that when an erase operation is performed to a FLASH memory device, either all or a large block of memory storage locations get erased. The three standard FLASH memory array architectures, *bulk-erase, boot block, and FlashFile™*, are shown in Fig. 10.27. In a bulk-erase device, the complete storage array is arranged as a single block. Whenever an erase operation is performed, the contents of all storage locations are cleared. This is the architecture used in the design of the earliest FLASH memory devices.

More modern FLASH memory devices employ either the boot block or FlashFile architecture for their memory array. They add granularity to the programming process. Now the complete memory array does not have to be erased. Instead, each of the independent blocks of storage locations erases separately. Notice that the blocks on a boot block device are asymmetrical is size. There is one small block known as the *boot block*. This block is intended for storage of the boot code for the system. It is followed by two small blocks that are called *parameter blocks*. Their intended use is for storage of certain system parameters, for instance, a system configuration table or lookup time. Finally, there are a number of much larger blocks of memory identified as *main blocks*. These are the blocks where the code of the firmware is stored.

Boot block devices are intended for use in a variety of applications that require smaller memory capacity and benefit from the asymmetrical blocking. One such application is known as *in-system programming*. In this type of application, the boot code used to start

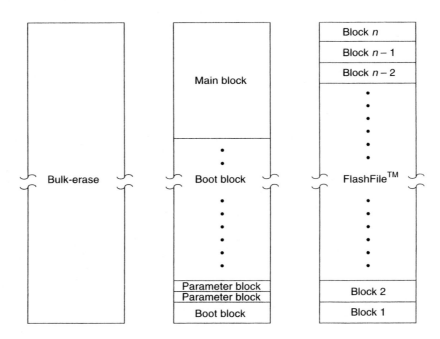

Figure 10.27 FLASH memory array architectures.

up the microcomputer is held in the boot block part of the FLASH memory. When the system is powered on, a memory-loading program is copied from the boot area of FLASH into RAM. Then, program execution is transferred to this program; the firmware that is to be loaded into the FLASH memory is downloaded from a communication line or external storage device such as a drive; the firmware is written into the main blocks of the FLASH memory devices; and finally the program is executed out of FLASH. In this way, we see that the FLASH memory devices are not loaded with the microcomputer's program in advance; instead, they are programmed while in the system.

FlashFile architecture FLASH memory devices differ from boot block devices in that the memory array is organized into equal-sized blocks. For this reason, it is said to be symmetrically blocked. This type of organization is primarily used in the design of high density devices. High-density flash devices are used in applications that require a large amount of code or data to be stored. An example is a FLASH memory drive.

Standard Bulk-Erase FLASH Memories

Bulk-erase FLASH memories are the oldest type of FLASH devices and are available in densities similar to those of EPROMs. Figure 10.28 lists the part number, bit density, and storage capacity of some of the popular devices. Notice that the part numbers of FLASH devices are similar to those used for the EPROMs earlier described. The differences are that the 7 in the EPROM part number is replaced by an 8, representing FLASH, and instead of a C, which is used to identify that the circuitry of the EPROM is made with a CMOS process, an F identifies FLASH technology. Remember the 2-MB EPROM was labeled 27C020; therefore, the 2-MB FLASH memory is the 28F020. This device is organized as 256K byte-wide storage locations.

Since FLASH memories are electrically erased, they do not need to be manufactured in a windowed package. For this reason, and the trend toward the use of surface-mount packaging, the most popular package for housing FLASH memory ICs is the plastic leaded chip carrier, or PLCC, as it is known. Figure 10.29(a) shows the PLCC pin layout of the 28F020. All the devices listed in Fig. 10.28 are manufactured in this same-size package and with compatible pin layouts.

Looking at the signals identified in the pin layout, we find that the device is exactly the same as the block diagram in Fig. 10.26. To select between its 256K byte-wide storage locations, it has 18 address inputs, A_0 through A_{17}, and to support byte-wide data-read and -write transfers, it has an 8-bit data bus, DQ_0 through DQ_7. Finally to enable the chip and its outputs and to distinguish between read- and write-data transfers, it has control lines $\overline{CE}$, $\overline{OE}$, and $\overline{WE}$, respectively. As shown in Fig. 10.29(b), the device is available with read access times ranging from 70 ns for the 28F020-70 to 150 ns for the 28F020-150.

The power supply voltage and current requirements depend on whether the FLASH memory is performing a read, erase, or write operation. During read mode of operation, the 28F020 is powered by 5 V $\pm$ 10% between the V_{cc} and V_{ss} pins and it draws a maximum

FLASH	Density (bits)	Capacity (bytes)
28F256	256K	32K × 8
28F512	512K	64K × 8
28F010	1M	128K × 8
28F020	2M	256K × 8

Figure 10.28 Standard bulk-erase FLASH memory devices.

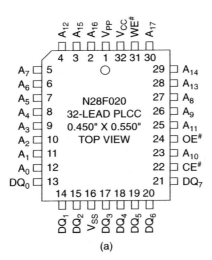

(a)

Part number	Access time
28F020-70	70 ns
28F020-90	90 ns
28F020-120	120 ns
28F020-150	150 ns

(b)

Figure 10.29 (a) Pin layout of the 28F020. (Reprinted by permission of Intel Corp. Copyright/Intel Corp. 1995) (b) Standard speed selections for the 28F020.

current of 30 mA. On the other hand, when either an erase or write cycle is taking place, 12 V ± 5% must also be applied to the V_{pp} power supply input.

The 28F256, 28F512, 28F010, and 28F020 employ a bulk erase storage array. For this reason, when an erase operation is performed to the device, all bytes in the storage array are restored to FF_{16}, which represents the erased state. The method employed to erase the 28F020 FLASH memory IC is known as the *quick-erase algorithm*. A flowchart that outlines the sequence of events that must take place to erase a 28F020 is given in Fig. 10.30. This programming sequence can be performed either with a FLASH memory-programming instrument or by the software of the microprocessor to which the FLASH device is attached. Let us next look more closely at how a 28F020 is erased.

To change the contents of a memory array, that is, either erase the storage array or write bytes of data into the array, commands must be written to the FLASH memory device. Unlike an EPROM, a FLASH memory has an internal command registers. Figure 10.31 lists the commands that can be issued to the 28F020. Notice that they include *read* (read memory), *set up and erase* (set up erase/erase), and *erase verification* (erase verify) commands. These three are used as part of the quick-erase algorithm process. The command register can be accessed only when +12 V is applied to the V_{pp} pin of the 28F020.

Figure 10.30 also includes a table of the bus operation and command activity that takes place during an erase operation of the 28F020. From the bus operation and command columns, we see that as part of the erase process, the microprocessor (or programming instrument) must issue a set up erase/erase command, followed by an erase verify command, and then a read command to the FLASH device. It does this by executing a FLASH memory programming control program that causes the write of the command to the FLASH memory

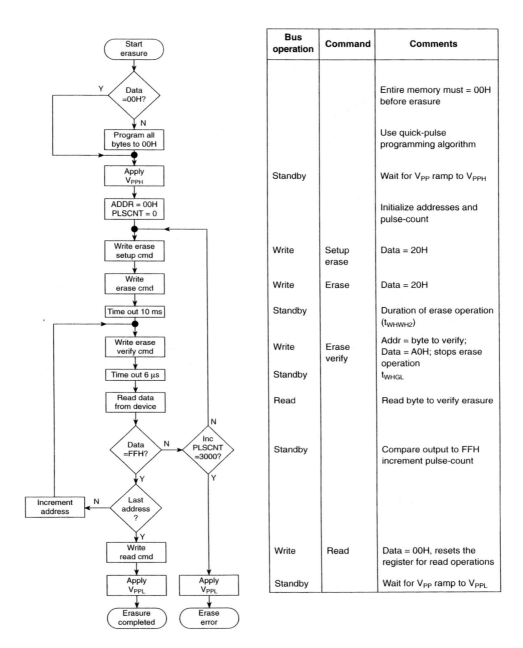

Bus operation	Command	Comments
		Entire memory must = 00H before erasure
		Use quick-pulse programming algorithm
Standby		Wait for V_{PP} ramp to V_{PPH}
		Initialize addresses and pulse-count
Write	Setup erase	Data = 20H
Write	Erase	Data = 20H
Standby		Duration of erase operation (t_{WHWH2})
Write	Erase verify	Addr = byte to verify; Data = A0H; stops erase operation
Standby		t_{WHGL}
Read		Read byte to verify erasure
Standby		Compare output to FFH increment pulse-count
Write	Read	Data = 00H, resets the register for read operations
Standby		Wait for V_{PP} ramp to V_{PPL}

Figure 10.30 Quick-erase algorithm of the 28F020. (Reprinted by permission of Intel Corp. Copyright/Intel Corp. 1995)

device at the appropriate time. Actually, all storage locations in the memory array must always be programmed with 00_{16} before initiating the erase process.

In Fig. 10.30, we see that two consecutive set up erase/erase commands are used in the quick-erase sequence. The first one prepares the 28F020 to be erased and the second initiates the erase process. Figure 10.31 shows that this sequence is identified as two write cycles. During both of these bus cycles, a value of data equal to 20_{16} is written to any address

Command	Bus Cycles Req'd	First Bus Cycle			Second Bus Cycle		
		Operation	Address	Data	Operation	Address	Data
Read memory	1	Write	X	00H			
Read intelligent identifier codes	3	Write	X	90H	Read		
Setup erase/erase	2	Write	X	20H	Write	X	20H
Erase verify	2	Write	EA	A0H	Read	X	EVD
Setup program/program	2	Write	X	40H	Write	PA	PD
Program verify	2	Write	X	C0H	Read	X	PVD
Reset	2	Write	X	FFH	Write	X	FFH

Figure 10.31 28F020 command definitions. (Reprinted by permission of Intel Corp. Copyright/Intel Corp. 1995)

in the address range of the FLASH device being erased. Once these commands have been issued, a state-machine within the device automatically initiates and directs the erase process through completion.

The next step in the quick-erase process is to determine whether or not the device has erased completely. This is done with the erase verify command. In Fig. 10.31, we see that this operation requires a write cycle followed by a read cycle. During the write cycle, the data bus carries the erase verify command, which is $A0_{16}$, and the address bus carries the address of the storage location that is to be tested, EA. The read cycle that follows is used to transfer the data from the storage location corresponding to EA to the MPU. This data is identified as EVD in Fig. 10.31. The flowchart shows that the MPU must verify that the value of data read out of FLASH is FF_{16}. This erase verify step is repeated for every storage location in the 28F020. If any storage location does not verify erasure by reading back FF_{16}, the complete erase process is immediately repeated.

After complete erasure has been verified, the software must issue a read command to the device. From Fig. 10.31, we find that it requires a single write bus cycle and is accompanied by any address that corresponds to the FLASH device being erased and a command data value of 00_{16}. Issuing this command puts the device into the read mode and readies it for read operation.

The *quick-pulse programming algorithm* of the 28F020 is outlined in Fig. 10.32. This process is similar to that just described for the erasing device; however, it uses the *set up and program* (set up program/program), *program verification* (program verify), and read (read memory) commands.

Standard Boot Block FLASH Memories

Earlier we pointed out that boot block FLASH memories are designed for use in embedded microprocessor applications. These newer devices are available in higher densities than bulk erase devices. In Fig. 10.33, the pin layouts for three compatible standard densities, the 2-MB, 4-MB, and 8-MB devices, are shown. Notice that the corresponding devices are identified with the part numbers 28F002, 28F004, and 28F008, respectively. These devices have different densities but have a common set of operating features and capabilities. The pinout information given is for a 40-pin *thin small outline package* (TSOP).

These devices offer a number of new architectural features when compared to the

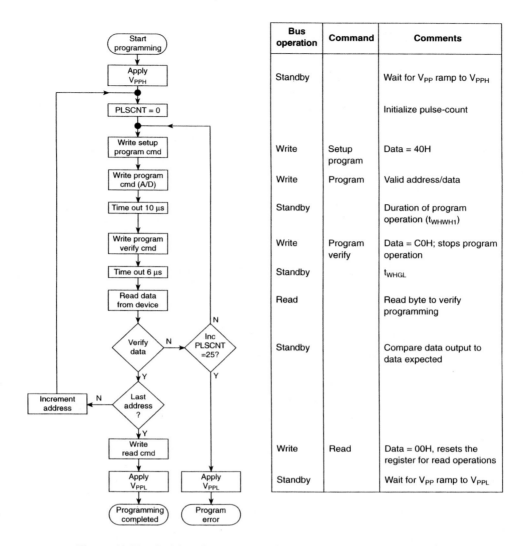

Bus operation	Command	Comments
Standby		Wait for V_{PP} ramp to V_{PPH}
		Initialize pulse-count
Write	Setup program	Data = 40H
Write	Program	Valid address/data
Standby		Duration of program operation (t_{WHWH1})
Write	Program verify	Data = C0H; stops program operation
Standby		t_{WHGL}
Read		Read byte to verify programming
Standby		Compare data output to data expected
Write	Read	Data = 00H, resets the register for read operations
Standby		Wait for V_{PP} ramp to V_{PPL}

Figure 10.32 Quick-pulse programming algorithm of the 28F020. (Reprinted by permission of Intel Corp. Copyright/Intel Corp. 1995)

bulk erase devices just described. One of the most important of these new features is what is known as *SmartVoltage*™. This capability enables the device to be programmed with either a 5-V or 12-V value of V_{pp}. In fact, the device can be installed into a circuit using either value of V_{pp}. This is because the device has the ability to automatically detect and adjust its programming operation to the value of the programming supply voltage in use. The SmartVoltage function not only allows the identical device to operate with either of two values of V_{pp}, it also permits the read voltage supply, V_{cc}, to be either 5 V or 3.3 V.

A second important difference is that devices are available at each of these three densities that can be organized with either an 8-bit or 16-bit bus. A block diagram of such a device is shown in Fig. 10.34(a). This device is identified as a 28F004/28F400. The 28F004 device is the one that is available in the 40-pin TSOP package and it operates only in the 8-bit data bus mode. On the other hand, the 28F400 device has 16-bit data lines, D_0 through

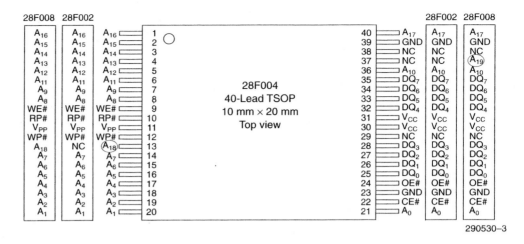

Figure 10.33 Pin-layout comparison of the TSOP 28F002, 28F004, and 28F008 ICs.

D_{15}. Actually, the 28F400 can be configured to operate with either an 8-bit or 16-bit data bus. This is done with the $\overline{\text{BYTE}}$ input. Logic 0 at $\overline{\text{BYTE}}$ selects byte-wide mode of operation and logic 1 chooses word-wide operation. To permit the extra data lines, the 28F400 is housed in a 56-lead TSOP.

Remember that the storage array of a boot block device is arranged as multiple asymmetrically sized, independently erasable blocks. In fact, the 28F004 has one 16K-byte boot block, two 8K-byte parameter blocks, three 128K-byte main blocks, and a fourth main block that is just 96K bytes in size. There are two different organizations of these blocks available, as shown in Fig. 10.34(b). The configuration on the left is known as the *top boot* (T) and that on the right as the *bottom boot* (B). Notice that they differ in how the blocks are assigned to the address space. That is, the T version has the 16K-byte boot block at the top (highest address) of the address space, and it is followed by the parameter blocks and then the main blocks. The address space of the B version is a mirror image; therefore, the 16K-byte boot block starts at the bottom (lowest address) of the address space.

Another new feature introduced with the boot block architecture is that of a hardware-lockable block. In the 28F004/28F400 the 16K-byte boot block section can be locked. If external hardware applies logic 0 to the write protect ($\overline{\text{WP}}$) input, the boot block is locked. Any attempt to erase or program this block when it is locked results in an error condition. Therefore, we say that the boot block is *write protected*. In an in-system programming application, the boot block part of the storage array would typically contain the part of the microcomputer program (boot program) that is used to load the system software into FLASH memory. For this reason, it would be locked and should remain that way.

Looking at Fig. 10.34(a), we find one more new input on the 28F004/28F400. This is the reset/deep power-down ($\overline{\text{RP}}$) input. This input must be at logic 1 to enable normal read, erase, and program operations. During read operations, the device can draw as much as 60 mA of current. If the device is not in use, it can be put into the deep power-down mode to conserve power. To do this, external circuitry must switch $\overline{\text{RP}}$ to logic 0. In this mode, it draws just .2 μA.

The last difference we will describe is that the 28F004/28F400 is equipped with what is known as *automatic erase and write*. No longer do we need to implement the complex

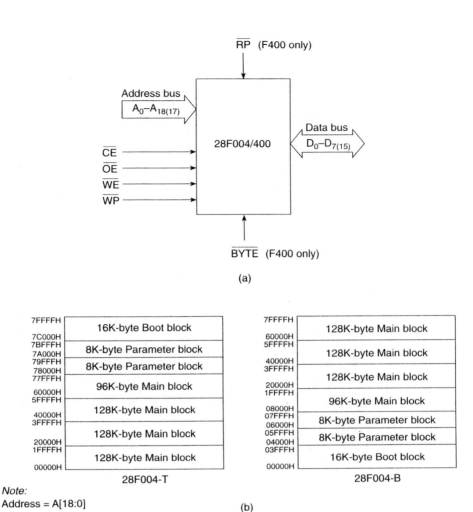

Figure 10.34 (a) Block diagram of the 28F004/400. (Reprinted by permission of Intel Corp. Copyright/Intel Corp. 1995) (b) Top and bottom boot block organization of the 28F004. (Reprinted by permission of Intel Corp. Copyright/Intel Corp. 1995)

quick-erase and quick-pulse programming algorithms in software as done for the 28F020. Instead, the 28F004/28F400 uses a *command user interface* (CUI), *status register*, and *write-state machine* to initiate an internally implemented, highly automated, method of erasing and programming the blocks of the storage array.

Let us now look briefly at how an erase operation is performed. The command bus definitions of the 28F004/28F400 are shown in Fig. 10.35(a), the bit definitions of its status register are given in Fig. 10.35 (b), and its erase cycle flowchart is shown in Fig. 10.35(c). Here we see that now all that needs to be done to initiate an erase operation is to write a command bus definition that includes an erase setup command and an erase confirm command to the device. These commands contain an address that identifies the block that is to be erased. In response to these commands, the write state machine drives a sequence that automatically programs all of the bits in this block to logic 0, verifies that they have

Command	Notes	First Bus Cycle			Second Bus Cycle		
		Oper	Addr	Data	Oper	Addr	Data
Read array	8	Write	X	FFH			
Intelligent identifier	1	Write	X	90H	Read	IA	IID
Read status register	2,4	Write	X	70H	Read	X	SRD
Clear status register	3	Write	X	50H			
Word/byte write		Write	WA	40H	Write	WA	WD
Alternate word/byte write	6,7	Write	WA	10H	Write	WA	WD
Block erase/confirm	6,7	Write	BA	20H	Write	BA	D0H
Erase suspend/resume	5	Write	X	B0H	Write	X	D0H

Address
BA = Block Address
IA = Identifier Address
WA = Write Address
X = Don't Care

Data
SRD = Status Register Data
IID = Identifier Data
WD = Write Data

Notes:
1. Bus operations are defined in Tables 4 and 5.
2. IA = Identifier Address: A0 = 0 for manufacturer code, A0 = 1 for device code.
3. SRD—Data read from Status Register.
4. IID = Intelligent Identifier Data. Following the Intelligent Identifier command, two Read operations access manufacturer and device codes.
5. BA = Address within the block being erased.
6. WA = Address to be written. WD = Data to be written at location WD.
7. Either 40H or 10H commands is valid.
8. When writing commands to the device, the upper data bus $[DQ_8-DQ_{15}] = X$ (28F400 only) which is either V_{IL} or V_{IH}, to minimize current draw.

(a)

Figure 10.35(a) 28F004 command bus definitions. (Reprinted by permission of Intel Corp. Copyright/Intel Corp. 1995)

programmed, erases all the bits in the block, and then verifies that each bit in the block has been erased. While it is performing this process, the WSM's bit of the status register is reset to 0 to say that the device is busy. The microcomputer's software can simply poll this bit to see if it is still busy. When WSM is read as logic 1 (ready), the erase operation is complete and all the bits in the erased block are at the 1 logic level. In this way, we see that the new programming software just has to initiate the automatic erase process and then poll the status register to determine when the erase operation is finished.

Standard FlashFile FLASH Memories

The highest-density FLASH memories available today are 8-MB and 16-MB devices. As pointed out earlier, they use the symmetrically sized, independently erasable, FlashFile organization for blocking their storage array. Two popular devices are the 28F008SA and the 28F016SA/SV. These devices are intended for use in large-code storage applications and to implement solid-state mass-storage devices such as the FLASH card and FLASH drive.

A block diagram of the 28F016SA/SV FLASH memory device is shown in Fig. 10.36(a)

WSMS	ESS	ES	DWS	VPPS	R	R	R
7	6	5	4	3	2	1	0

Notes:

SR.7 = WRITE STATE MACHINE STATUS (WSMS)
    1 = Ready
    0 = Busy

Write State Machine bit must first be checked to determine Byte/Word program or Block Erase completion before the Program or Erase Status bits are checked for success.

SR.6 = ERASE-SUSPEND STATUS (ESS)
    1 = Erase Suspended
    0 = Erase in Progress/Completed

When Erase Suspend is issued, WSM halts execution and sets both WSMS and ESS bits to "1." ESS bit remains set to "1" until an Erase Resume command is issued.

SR.5 = ERASE STATUS
    1 = Error in Block Erasure
    0 = Successful Block Erase

When this bit is set to "1," WSM has applied the maximum number of erase pulses to the block and is still unable to successfully verify block erasure.

SR.4 = PROGRAM STATUS
    1 = Error in Byte/Word Program
    0 = Successful Byte/Word Program

When this bit is set to "1," WSM has attempted but failed to program a byte or word.

SR.3 = V_{PP} STATUS
    1 = V_{PP} Low Detect, Operation Abort
    0 = V_{PP} OK

The V_{PP} Status bit, unlike an A/D converter, does not provide continuous indication of V_{PP} level. The WSM interrogates V_{PP} level only after the Byte Write or Erase command sequences have been entered, and informs the system if V_{PP} has not been switched on. The V_{PP} Status bit is not guaranteed to report accurate feedback bewteen V_{PPLK} and V_{PPH}.

SR.2–SR.0 = RESERVED FOR FUTURE ENHANCEMENTS

These bits are reserved for future use and should be masked out when polling the Status Register.

(b)

Figure 10.35(b) Status register bit definitions (Reprinted by permission of Intel Corp. Copyright/Intel Corp. 1995)

and its pin layout for a *shrink small outline package* (SSOP) is given in Fig. 10.36(b). Comparing this device to the 28F004/28F400 in Fig. 10.34(a), we find many similarities. For instance, both devices have address bus, data bus, and control signals $\overline{OE}$, $\overline{WE}$, $\overline{WP}$, $\overline{RP}$, and $\overline{BYTE}$, and they serve similar functions relative to device operation. One difference is that there are now two chip-enable inputs, which are labeled $\overline{CE}_0$ and $\overline{CE}_1$, instead of just one. Both of these inputs must be at logic 0 to enable the device for operation.

    Another change found on the 28F016SA/SV is the addition of the ready/busy (RY/$\overline{BY}$) output. This output has been provided to further reduce the dependency of the erase and programming processes on MPU software. When this output is 0, it signals that the on-chip write-state machine of the FLASH memory is busy performing an operation. Logic 1 means that it is ready to start a new operation. For the boot block devices we introduced earlier, the busy condition had to be determined through software by polling the WSM's bit of the status register. One approach for the 28F016SA/SV is that software could poll RY/$\overline{BY}$ as an input waiting for the FLASH device to be ready. On the other hand, this signal could be used as an interrupt input to the MPU. In this way, no software and MPU

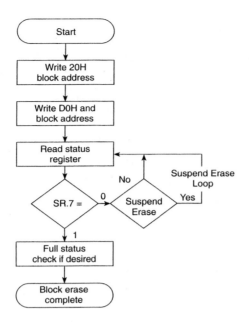

Bus Operation	Command	Comments
Write	Erase setup	Data = 20H Addr = Within block to be erased
Write	Erase confirm	Data = D0H Addr = Within block to be erased
Read		Status register data toggle CE# or OE# to update status register
Standby		Check SR.7 1 = WSM ready 0 = WSM busy

Repeat for subsequent block erasures.
Full Status Check can be done after each block erase,
   or after a sequence of block erasures.
Write FFH after the last operation to reset device to read
   array mode.

(c)

Figure 10.35(c) Erase operation flowchart and bus activity. (Reprinted by permission of Intel Corp. Copyright/Intel Corp. 1995)

overhead is needed to recognize when the FLASH memory is ready to perform another operation. This is the default mode of operation and is known as *level mode.*

The function of the $RY/\overline{BY}$ output can be configured for a number of different modes of operation under software control. This is done by writing a device configuration code to the 28F016SA/SV. For instance, it can be set to produce a pulse on write or a pulse on erase or even be disabled.

The last signal line in the block diagram that is new is the $3/\overline{5}$ input. Notice that this input is implemented only on the 28F016SA IC. Logic 1 at this input selects a 3.3 V operation for V_{cc} and logic 0 means that a 5-V supply is in use. Since the 28F016SV is a SmartVoltage device, this input is not needed.

The blocking of the 28F016SA/SV configured for the byte-wide mode of operation is illustrated in Fig. 10.36(c). Here we find that the 16-MB address space is partitioned into 32 independent 64-KB blocks. Notice that block 0 is in the address range from 000000_{16} through $00FFFF_{16}$ and block 31 corresponds to the range $1F0000_{16}$ through $1FFFFF_{16}$. If the device is strapped for word-mode operation with logic 1 at the $\overline{BYTE}$ input, there are still 32 blocks, but they are now 32K words in size.

Just as for the 28F004/28F400, the 28F016SA/SV supports block locking. However, in these devices, the 32 blocks are independently programmable as locked or unlocked. In fact, there is a separate block status register for each of the 32 blocks. The status register contains both control and status bits related to corresponding block. The *block-lock status* (BLS) bit, which is bit 6 in this register, is an example of a control bit. When it is set to logic 1 under software control, the corresponding block is configured as unlocked and write and erase operations are permitted. Changing it to logic 0 locks the block so that it cannot be written into or erased. Bit 7, *block status* (BS), is an example of a status bit that can be

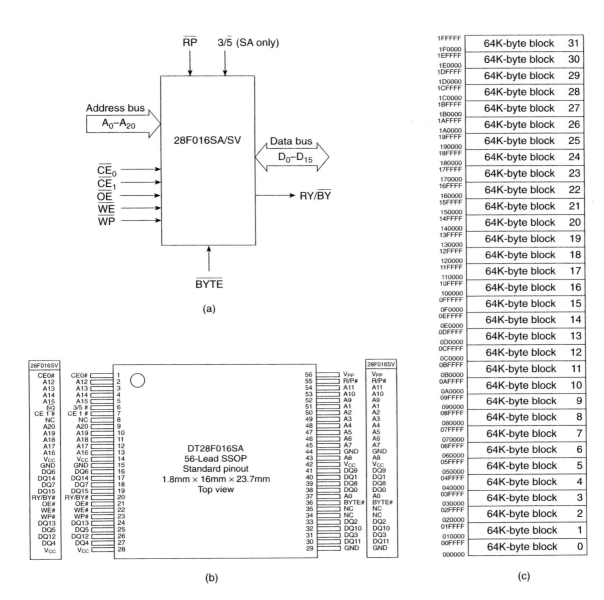

Figure 10.36 (a) Block diagram of the 28F016SA/SV FlashFile™ memory. (b) Pin layout. (Reprinted by permission of Intel Corp. Copyright/Intel Corp. 1995) (c) Byte-wide mode memory map. (Reprinted by permission of Intel Corp. Copyright/Intel Corp. 1995)

read by the MPU. Logic 1 in this bit means that the block is ready and logic 0 signals that it is busy. When the write-protect ($\overline{\text{WP}}$) input is active (logic 0), write and erase operations are not permitted to those blocks that are marked as locked with a 0 in the BLS bit in their corresponding block status register.

Finally, the internal algorithms and hardware of the 28F016SA/SV have been expanded to improve programming performance. For instance, two 256-byte (128-word) write buffers have been added into the architecture to enable paged data writes. Moreover, the program algorithm has been enhanced to support queuing of commands and overlapping of erase and write operations. Therefore, additional commands can be sent to a device while it is still executing a prior command. They are held in the queue until processed. The overlapping write/erase capability enables the devices to erase one block while writing data to another. All these features result in easier and faster programming for the 28F016SA/SV.

▲ 10.7 WAIT-STATE CIRCUITRY

Depending upon the access time of the memory device used and the clock rate of the MPU, a number of wait states may need to be inserted into external memory read and write operations. In our study of 80386DX bus cycles in Chapter 9, we found that the memory subsystem signals the 80386DX whether or not wait states are needed in a bus cycle with the logic level applied to its $\overline{\text{READY}}$ input.

The circuit that implements this function for a microcomputer system is known as a *wait-state generator*. A block diagram of this type circuit is shown in Fig. 10.37(a). Notice that the circuit has six inputs and just one output. The two inputs located at the top of the circuit, $\overline{\text{CS}}_0$ and $\overline{\text{CS}}_1$, are outputs of the memory chip-select logic. They could represent the chip selects for the program storage and data-storage memory subsystems, respectively, and tell the circuit whether or not these parts of the memory subsystem are being accessed.

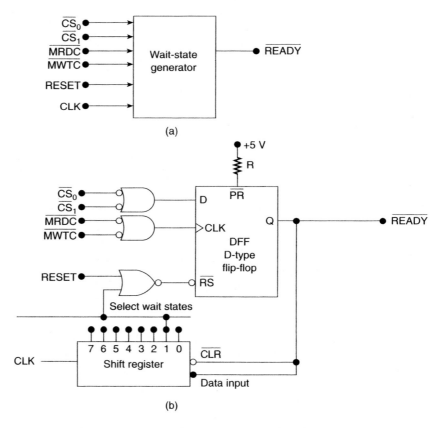

Figure 10.37 (a) Wait-state generator circuit block diagram. (b) Typical wait-state generator circuit.

The two middle inputs are the memory read command ($\overline{\text{MRDC}}$) and memory write command ($\overline{\text{MWTC}}$) outputs of the bus-control logic. They indicate that a read or write operation is taking place to the memory subsystem. The last two inputs are the system reset (RESET) signal and system clock (CLK) signals. The output is $\overline{\text{READY}}$.

The circuit in Fig. 10.37(b) can be used to implement a wait-state generator for an 80386DX-based microcomputer system. To design this circuit, we must select the appropriate D-type flip-flop, shift register, and gates.

Let us look briefly at how this wait-state generator circuit works. The $\overline{\text{READY}}$ output is returned directly to the $\overline{\text{READY}}$ input of the MPU. Logic 0 at this output tells the MPU that the current read/write operation is to be completed. On the other hand, logic 1 means that the memory bus cycle must be extended by inserting wait states.

Whenever an external memory bus cycle is initiated, the D-type flip-flop is used to start the wait-state generation circuit. Before either $\overline{\text{CS}}_0$, $\overline{\text{CS}}_1$, or RESET becomes active, the Q output of the flip-flop is held at logic 0 and signals that wait states are not needed. The Q output is also applied to the $\overline{\text{CLR}}$ input of the shift register. Logic 0 at $\overline{\text{CLR}}$ holds it in the reset state and holds outputs 0 through 7 all at logic 0.

Whenever a read or write operation takes place to a storage location in the external memory's address range, a logic 0 is produced at either $\overline{\text{CS}}_0$ or $\overline{\text{CS}}_1$ and a logic 0 is produced at either $\overline{\text{MRDC}}$ or $\overline{\text{MWTC}}$. The active chip-select input makes the D input of the flip-flop logic 1 and the transition to logic 0 by the read/write command signal causes the Q output to switch to logic 1. Now $\overline{\text{READY}}$ tells the MPU to start inserting wait states into the memory bus cycle.

The Q output also makes both the $\overline{\text{CLR}}$ and data input of the shift register logic 1. Therefore, it is released and the logic 1 at the data input shifts up through the register synchronous with clock pulses from the MPU's system clock. When the select wait-state output becomes logic 1, it makes the $\overline{\text{RS}}$ input of the flip-flop active, thereby resetting the Q output to logic 0. Thus, the $\overline{\text{READY}}$ output returns to logic 0 and terminates the insertion of wait states, and the MPU completes the bus cycle. The number of wait states inserted depends upon how many clock periods $\overline{\text{READY}}$ remains at logic 0. This can be changed by simply attaching the select wait-state line to a different output of the shift register. For instance, the connection shown in Fig. 10.37(b) represents two wait-state operations.

A zero wait-state memory read/write cycle of the 80386DX has at most two clock cycles available for the data-transfer operation to be completed. For an MPU running at 33 MHz, this means that memory devices are needed with access times less than 67 ns. To use 100-ns devices, we must insert at least two wait states into the memory bus cycles to give the memory subsystem enough time to respond.

▲ 10.8 80386DX/SX MICROCOMPUTER SYSTEM MEMORY INTERFACE CIRCUITRY

In Chapter 9 we introduced the bus cycles, hardware organization of the memory address space, and memory interface circuits of the 80386DX-based microcomputer system. The earlier sections of this chapter covered the types of memory devices that are used in the memory subsystem. Here we show how the memory interface circuits and memory subsystem are interconnected to implement a simple microcomputer system. This microcomputer system is given in Fig. 10.38. Notice that the circuitry of the memory interface and memory array have been highlighted in the circuit diagram.

Let us begin by examining the bus interface circuitry. Looking at the circuit diagram, we see that address line A_{31} and control signal M/$\overline{\text{IO}}$ are decoded by the upper 74F139

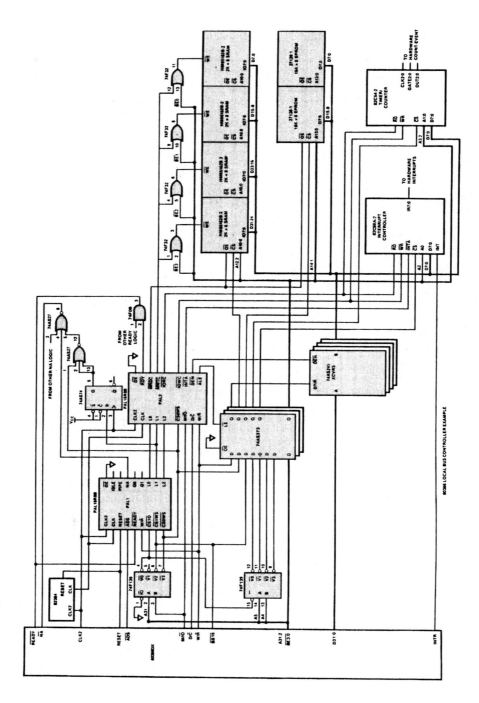

Figure 10.38 Memory circuit for a typical 80386DX microcomputer.

2-line-to-4-line decoder. This device represents the memory address decoder. When M/$\overline{\text{IO}}$ equals 1 and A$_{31}$ equals 0, the $\overline{\text{Y}}_2$ output of the decoder is at its active 0 logic level. This signal indicates that a memory address is on the bus. Notice that it is input to PAL 1 of the bus-control logic as chip-select 1 wait state ($\overline{\text{CS1WS}}$). $\overline{\text{CS1WS}}$ is also latched into the 74LS373 address latch. Here it is used as the chip-select signal for the 27C128 EPROMs. On the other hand, an input of 11 to the decoder produces chip select 0 wait state ($\overline{\text{CS0WS}}$). This signal is also used as an input to PAL 1. Moreover, it is latched for use as the chip-select signal for the static RAMs.

The address latch is implemented with three 74AS373 octal latch devices. Notice that the outputs of the latches are permanently enabled by applying logic 0 to the $\overline{\text{OE}}$ input. The address signals, byte-enable signals, chip-select outputs of the memory address decoder, and W/$\overline{\text{R}}$ are latched into the address latch synchronously with the $\overline{\text{ALE}}$ output from PAL 2 of the bus-control logic. Actually, just address lines A$_2$ through A$_{14}$ are used in the memory interface.

The bus-control logic section is formed from the devices labeled PAL 1 and PAL 2. In the circuit diagram, we find that the inputs to this section include the bus cycle definition signals M/$\overline{\text{IO}}$, D/$\overline{\text{C}}$, and W/$\overline{\text{R}}$, control signal $\overline{\text{ADS}}$, and chip selects $\overline{\text{CSIO}}$, $\overline{\text{CS1WS}}$, and $\overline{\text{CS0WS}}$. At the output side the bus-control logic produces the control signals $\overline{\text{MRDC}}$, $\overline{\text{MWTC}}$, $\overline{\text{DEN}}$, and $\overline{\text{ALE}}$ for the memory interface, along with $\overline{\text{NA}}$ for activating pipelining and $\overline{\text{RDY}}$ for introducing wait states. $\overline{\text{NA}}$ and $\overline{\text{RDY}}$ are supplied through control logic to inputs of the 80386DX.

The bank write-control logic section is implemented with the gates of a 74F32 quad OR gate IC. The memory write command ($\overline{\text{MWTC}}$) signal is applied to one input of each of the four two-input OR gates. The other input on each of the four gates is supplied by one of the latched byte-enable signals. For instance, $\overline{\text{BE}}_0$ enables $\overline{\text{MWTC}}$ to the $\overline{\text{WE}}$ input of the rightmost static RAM. The signals produced at the outputs are used to activate the write-enable ($\overline{\text{WE}}$) input of the static RAMs in banks 0 through 3.

The final section in the bus interface section consists of the data bus transceivers/buffers. In the circuit diagram, we see that the data bus lines are buffered with four 74AS245 transceivers. The $\overline{\text{DEN}}$ output of the bus-control logic enables the transceivers for operation, and the latched W/$\overline{\text{R}}$ output of the 80386DX is used to control the direction in which data are passed through them.

Let us now look at the program-storage memory section. It is implemented with two 27C128 EPROMs. The data outputs of these devices attach to the lower 16 data bus lines, D$_0$ through D$_{15}$, and together they provide 16K $\times$ 16 bits for program storage. Notice that program memory is 16 bits wide, not 32 bits wide. The storage location to be accessed is selected by the 14-bit address that is applied to the A$_0$ through A$_{13}$ inputs of both EPROMs. Chip-select signal $\overline{\text{CS1WS}}$ is used to enable the EPROMs. $\overline{\text{CS1WS}}$ is also returned to the bus size 16 ($\overline{\text{BS16}}$) input of the 80386DX and tells it to perform all read cycles from the program memory address space as 16-bit bus cycles. Finally, memory-read command ($\overline{\text{MRDC}}$) supplies the $\overline{\text{OE}}$ input of the EPROMs and enables the data held at the addressed storage location onto the data bus whenever a memory-read bus cycle is in progress.

In Fig. 10.38 we see that the data-storage memory section is constructed with four 2K $\times$ 8-bit static RAMs. Each of the SRAMs represents a bank and attaches to eight of the data bus lines. For example, the leftmost device is bank 3 and connects to data bus lines D$_{24}$ through D$_{31}$. This memory subsystem gives a total data storage capacity of 8K bytes, which is organized 2K $\times$ 32 bits. The storage location to be accessed is selected by the 11-bit address A$_2$ through A$_{12}$, which is applied to address inputs A$_0$ through A$_{10}$ of all four SRAMs in parallel. During bus cycles to data memory, all four SRAMs are enabled by the chip-select signal $\overline{\text{CS0WS}}$. For this reason, data read and write bus cycles involve no wait states. When data are read from memory, the signal $\overline{\text{MRDC}}$, which is applied to the $\overline{\text{CS}}$ input of all four SRAMs in parallel, enables their outputs. Therefore, the 32-bit word held at the addressed storage location is output onto D_0 through D_{31} and the 80386DX reads a byte, word, or double word of data off the bus.

On the other hand, during write bus cycles to SRAM, the 80386DX does not always output a 32-bit word; instead, it outputs the appropriate-size element of data: a byte, word, or double word. At the same time, it outputs a code on $\overline{\text{BE}}_0$ through $\overline{\text{BE}}_3$ to identify which part of the bus carries the data. The byte-enable signals gate $\overline{\text{MWTC}}$ to the $\overline{\text{WE}}$ input of the appropriate SRAM banks. In this way, we see that data are written only into the appropriate banks of memory, not to all four of them.

EXAMPLE 10.3

Design a memory subsystem for an 80386DX MPU that consists of 64K bytes of R/W memory and 64K bytes of ROM memory. Use 4363 SRAM devices to implement R/W memory and 27C128 EPROM devices to implement ROM memory. Block diagrams of these two memory devices are shown in Fig. 10.39(a). R/W memory is to reside over the address range 00000_{16} through $0FFFF_{16}$ and the address range of ROM memory is to be $F0000_{16}$ through $FFFFF_{16}$. Assume that the microprocessor system bus signals that follow are available for use in the circuit design: A_2 through A_{19}, D_0 through D_{31}, $\overline{\text{BE}}_0$ through $\overline{\text{BE}}_3$, $\overline{\text{MRDC}}$, and $\overline{\text{MWTC}}$.

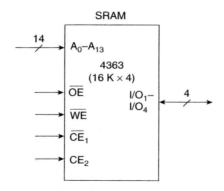

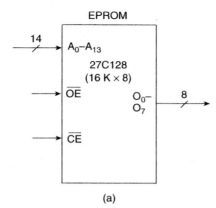

(a)

Figure 10.39(a) Devices to be used in the design of Example 10.3.

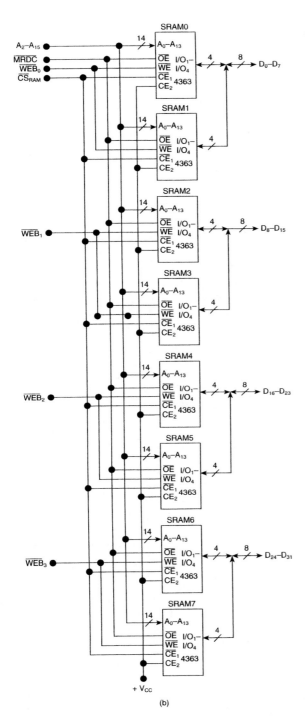

Figure 10.39(b) SRAM bank.

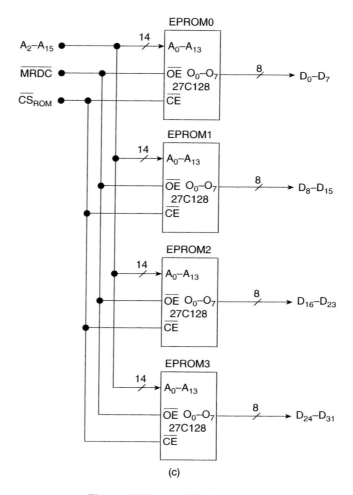

Figure 10.39(c) EPROM bank.

Solution

First let us determine the number of SRAM devices that are needed to implement the R/W memory bank. Since each 4363 provides $2^{14} \times 4$, or $16K \times 4$, of storage, the number of SRAM devices needed to implement 64K bytes of storage is

$$\text{no. SRAM} = 64K \text{ byte}/(16K \times 4) = 8$$

To attach to a 32-bit data bus, all eight SRAMs, $SRAM_0$ through $SRAM_7$, must be connected together to form a $16K \times 32$-bit R/W memory bank, as shown in Fig. 10.39(b). The memory map in Fig. 10.39(d) shows that this group of devices implement the address range 00000_{16} through $0FFFF_{16}$.

Next let us determine the number of EPROM devices that are needed to implement the bank of ROM memory. In this case, each 27C128 provides $2^{14} \times 8$, or 16K bytes, of

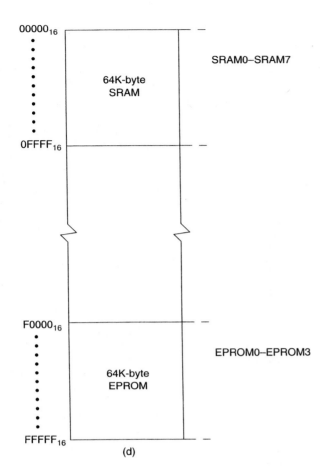

Figure 10.39(d) Memory map of the memory subsystem to be designed.

A_{19} •••••••••••••••••••• A_0

$00000_{16} = 0000\ 0000\ 0000\ 0000\ 0000$

⋮ ⋮

$0FFFF_{16} = 0000\ 1111\ 1111\ 1111\ 1111$

$\underbrace{\qquad}$
$\overline{CS}_{RAM}$

$F0000_{16} = 1111\ 0000\ 0000\ 0000\ 0000$

⋮ ⋮

$FFFFF_{16} = 1111\ 1111\ 1111\ 1111\ 1111$

$\underbrace{\qquad}$
$\overline{CS}_{ROM}$

(e)

Figure 10.39(e) Address range analysis for the design of the chip-select logic.

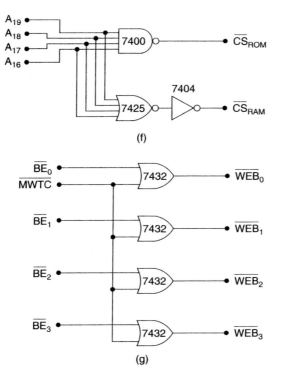

(f)

(g)

Figure 10.39 (f) Chip-select logic circuit. (g) Write-control logic circuit.

storage. To implement 64K bytes of storage, the number of EPROM devices needed is

$$\text{no. EPROM} = 64\text{K byte}/16\text{K byte} = 4$$

These four devices must be connected together as shown in Fig. 10.39(c) and implement the ROM address range $0F000_{16}$ through $0FFFF_{16}$ as shown in Fig. 10.39(d).

The two chip-select signals, $\overline{\text{CS}}_{\text{RAM}}$ and $\overline{\text{CS}}_{\text{ROM}}$, that are used in the circuit need to be produced for the appropriate address range. To design a circuit for generating these chip-select signals, we first analyze the addresses for each of these ranges. Figure 10.39(e) shows them expressed in binary form. Here we see that $A_{19}A_{18}A_{17}A_{16} = 0000_2$ must be decoded to produce $\overline{\text{CS}}_{\text{RAM}}$. Similarly the other address ranges tell us that $A_{19}A_{18}A_{17}A_{16} = 1111_2$ is decoded to produce $\overline{\text{CS}}_{\text{ROM}}$. This information is used in Fig. 10.39(f) to design the chip-select logic circuit with a 7400 NAND gate, 7425 NOR gate, and 7404 inverter.

Finally, the signals $\overline{\text{MWTC}}$ and $\overline{\text{BE}}_0$ through $\overline{\text{BE}}_3$ must be combined in the write-control logic to produce a separate write enable signal for each of the bytes of the 32-bit data word when writing the R/W memory bank. This is done with 7432 OR gates, as shown in Fig. 10.39(g).

▲ 10.9 CACHE MEMORY

When an 80386DX microcomputer system employs a large main memory subsystem of several megabytes, it is normally made with high-capacity but relative slow-speed dynamic RAMs and EPROMs. Even though DRAMs are available with access times as short as

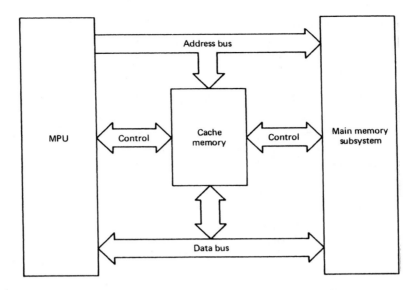

Figure 10.40 Microcomputer system with a cache memory.

60 ns and EPROMs as fast as 70 ns, these high-speed versions of the devices are expensive and still too slow to work in an 80386DX system that is running with zero wait states. For example, an 80386DX microprocessor running at 25 MHz would require DRAMs with a 40-ns access time to implement a zero-wait state memory design. For this reason, wait states are introduced in all bus cycles to data and program memory. These waits states degrade the overall performance of the microcomputer system.

Addition of a cache memory subsystem to the 80386DX microcomputer provides a means for improving overall system performance while still permitting the use of low-cost, slow-speed memory devices in main memory. In a microcomputer system with cache, a second smaller, but very fast, memory section is added between the MPU and main memory subsystem. Figure 10.40 illustrates this type of system architecture. This small, high-speed memory section is known as the *cache memory*. The cache is designed with fast, more expensive static RAMs and can be accessed without wait states. During system operation, the cache memory contains recently used instructions and data. The objective is that the MPU accesses code and data in the cache most of the time, instead of from main memory. This results in close to a zero-wait-state memory system operation and higher performance for the microcomputer even though accesses of the main memory require one or more wait states.

Cache memories are widely used in high-performance microcomputer systems today. Notice in Fig. 10.40 that on one side the cache memory subsystem attaches to the local bus of the 80386DX, and at the other side it drives the system bus of the microcomputer's main memory subsystem. Caches typically range in size from 16K bytes to 256K bytes and can be used to cache both data and code.

Let us continue our examination of cache for the 80386DX microcomputer system by looking at how it affects the execution of a program. The first time the 80386DX executes a segment of program, one instruction after the other is fetched from main memory and executed. The most recently fetched instructions are automatically saved in the cache memory. That is, a copy of these instructions is held within the cache. For example, a segment of program that implements a loop operation could be fetched, executed, and placed in the cache. In this way, we find that the cache always holds some of the most recently executed instructions.

Now that we know what the cache holds, let us look at how the cached instructions are used during program execution. Many software operations involve repeated execution of the same sequence of instructions. A loop is a good example of this type of program structure. In Fig. 10.41 we find that the first execution of the loop references code held in the slow main program memory. During this access, the routine is copied into the cache part of memory. When the instructions of the loop are repeated, the MPU reaccesses the routine by using the instructions held in the cache instead of refetching them from main memory. Accesses to code in the cache are performed with no wait states, whereas those of code in main memory normally require wait states. In this way we see that the use of cache has reduced the number of accesses made from the slower main memory. The more frequent instructions held in the cache are used; the closer to zero-wait-state operation is achieved, the more the overall execution time of the program is decreased, and the higher is the performance of the microcomputer system.

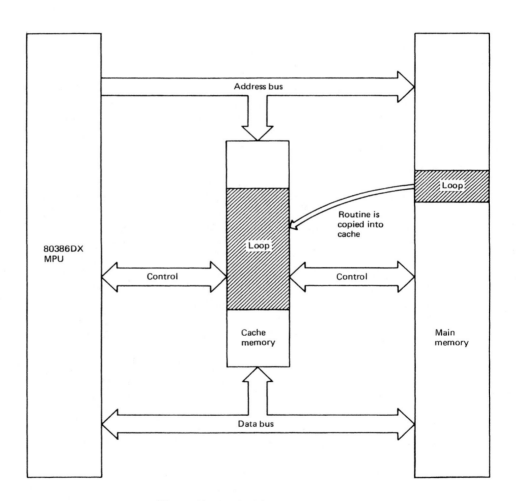

Figure 10.41 Caching a loop routine.

During execution of the loop routine, data operands that are accessed can also be cached in the cache memory. If these operands are reaccessed during the repeated execution of the loop, they are read from the cache instead of from main data memory. This further reduces execution time of the segment of program.

We just found that the concept behind cache memory is that it stores recently used code and data and that if this information is to be reaccessed, it may be read from the cache with zero-wait-state bus cycles rather than from main memory. When the address of a code or data-storage location that is to be read is output on the local bus, the cache subsystem must determine whether or not the information to be accessed resides in both main memory and cache memory. If it does, the memory cycle is considered a *cache hit* condition. In this case a bus cycle is not initiated to the main memory subsystem; instead, the copy of the information in the cache is accessed.

On the other hand, if the address output on the local bus does not correspond to information that is already cached, the condition represents what is called a *cache miss*. This time the MPU reads the code or data from main memory and writes it into a corresponding location in cache.

Hit rate is a measure of how effective the cache subsystem operates. Hit rate is defined as the ratio of the number of cache hits to the total number of memory accesses, expressed as a percentage. That is, hit rate equals

$$\text{hit rate} = \frac{\text{number of hits}}{\text{number of bus cycles}} \times 100\%$$

The higher the value of hit rate, the better the cache memory design. For instance, a cache may have a hit rate of 85%. This means that the MPU reads data from the cache memory for 85% of its memory bus cycles. In other words, just 15% of the memory accesses are from the main memory subsystem. Hit rate is not a fixed value for a cache design. It depends on the code being executed. That is, hit rate may be one value for a specific application program and a totally different value for another.

A number of features of the cache design also affects the hit rate. For instance, the size, organization, and update method of the cache memory subsystem all determine the maximum hit rate that may be achieved by a cache. Earlier we pointed out that practical cache memories for microcomputer systems range in size from as small as 16K bytes to as large as 256K bytes. In general, the larger the size of the cache, the higher the hit rate. This is because a larger cache can contain more data and code, which yields a greater chance that the information to be accessed resides in the cache. However, due to the fact that the improvement in hit rate decreases with increasing cache size and that the cost of the cache subsystem increases substantially with increasing size, most caches used with the 80386DX are either 32K bytes or 64K bytes.

The two most widely used cache memory organizations are those known as the *direct-mapped cache* and the *two-way set associative cache*. The direct-mapped cache is also called a *one-way set associative cache*. The organization of a 64K-byte direct-mapped cache memory is illustrated in Fig. 10.42. Notice that the cache memory array is arranged as a single 64K-byte bank of memory, and the main memory is viewed as a series of 64K-byte pages, denoted page 0 through page *n*. Notice that the data-storage location at the same offset (X) in all pages of main memory, these storage locations are identified as X(0) through X(n) in Fig. 10.42, map to a single storage location, marked X, in the cache memory array. That is, each location in a 64K-byte page of main memory maps to a different location in the cache memory array.

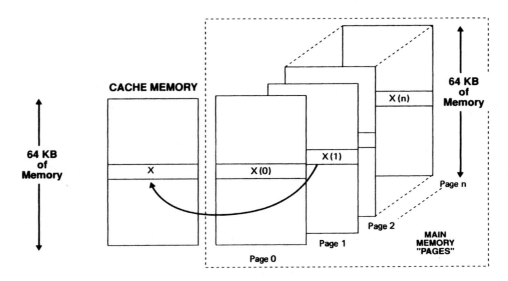

Figure 10.42 Organization of a direct-mapped memory subsystem. (Reprinted by permission of Intel Corp. Copyright/Intel Corp. 1990)

On the other hand, the 64K-byte memory array of a two-way set associative cache memory is organized into two 32K-byte banks. That is, the cache array is divided two ways: *BANK A* and *BANK B*. This cache memory subsystem configuration is shown in Fig. 10.43. Again main memory is mapped into pages equal to the size of a bank in the cache array. But because a bank is now 32K bytes, there are twice as many main-memory pages as in direct-mapped organizations. In this case the storage location at a specific offset in every page of main memory can map to the same storage location in either the A or B bank. For

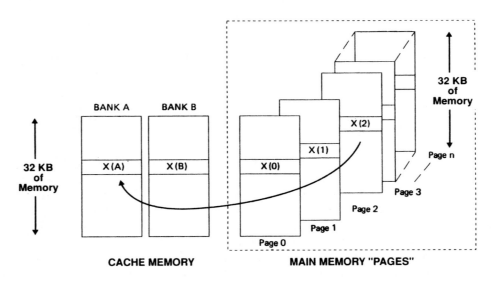

Figure 10.43 Organization of a two-way set associative memory subsystem. (Reprinted by permission of Intel Corp. Copyright/Intel Corp. 1990)

example, the contents of storage location X(2) can be cached into either X(A) or X(B). The two-way set associative organization results in higher hit rate operation.

One example of a memory update method that affects hit rate is the information replacement algorithm. Replacement methods are based on the fact that there is a higher chance that more recently used information will be reused. For instance, the two-way set associative cache organization permits the use of a *least recently used* (LRU) replacement algorithm. In this method, the cache subsystem hardware keeps track of whether information X(A) in the BANK A storage location or X(B) in BANK B is most recently used. For example, let us assume that the value at storage location X(A) in BANK A of Fig. 10.43 is X(0) from page 0 and that it was just loaded into the cache. On the other hand, X(B) in BANK B is from page 1 and has not been accessed for a long time. Therefore, the value of X(B) in BANK B is tagged as the least recently used information. When a new value of code or data, for instance, from offset X(3) in page 3 is accessed, it must replace the value of X(A) in BANK A or X(B) in BANK B. Therefore, the cache replacement algorithm automatically selects the cache storage location corresponding to the least recently accessed bank for storage of X(3). For our example this would be storage location X(B) in BANK B. This shows that the replacement algorithm maintains more recently used information in the cache memory array. The results are a higher maximum hit rate and a higher level of performance for the microcomputer system.

We found earlier that use of a cache reduced the number of accesses of main memory over the system bus. In our example of a loop routine we saw that repeated accesses of the instructions that perform the loop operation were made from the cache, not from main memory. Therefore, fewer code and data accesses are performed across the system bus. That is, availability of the bus has been increased for external devices. This is another advantage of using cache in a microcomputer system. The freed-up bus bandwidth is available to other bus masters, such as DMA controllers or other processors in a multiple-processor system.

Posted write-through is an example of a cache memory update method that affects microcomputer system performance but not hit rate. In this method, all write operations that are used to update addresses of storage locations in main memory go through (*write through*) the cache. When a write bus cycle is initiated, the cache memory checks to determine if the information is cached. If a miss occurs, the write is performed only to the storage location in main memory. However, if the information for the addressed location is cached (a cache hit), the corresponding entry in the cache memory array is also updated. *Posted write* capability allows write operations to main memory to be completed by the cache memory subsystem rather than by the MPU. For this reason, the MPU can terminate the write cycle without incurring any wait states. This capability increases the bus bandwidth of the local bus.

▲ 10.10 THE 82385DX CACHE CONTROLLER AND THE CACHE MEMORY SUBSYSTEM

The 82385DX is a VLSI device that is designed to implement the control function for the cache memory subsystem in an 80386DX-based microcomputer system. Use of the 82385DX results in a high-performance, versatile, compact cache design. It is designed to attach directly to the 80386DX microprocessor. For this reason the amount of circuitry needed to construct the cache memory is minimized. In terms of versatility, the 82385DX controller can be used to implement several different cache configurations. For instance, it can work

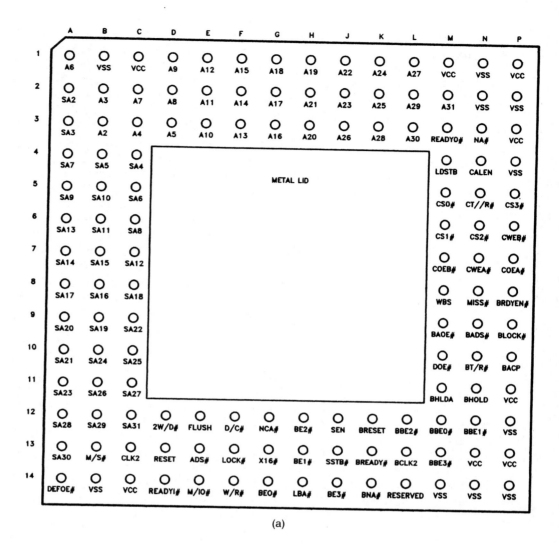

(a)

Figure 10.44(a) Pin layout of the PGA 82385DX cache controller. (Reprinted by permission of Intel Corp. Copyright/Intel Corp. 1990)

as either a master or a slave. Master mode of operation is selected when used in a single-processor microcomputer application. The 82385DX supports a 32K-byte cache memory array. This memory array can be set up to operate as either a direct-mapped cache or a two-way set associative cache. When configured for two-way set associative organization, it uses the least recently used replacement algorithm. Finally, memory write cycles are performed in a posted write-through manner. Depending on the application, an 82385DX-based cache memory subsystem can achieve hit rates as high as 99%.

The 82385DX IC is housed in a 132-lead pin grid array package. The pin layout of the 82385DX is shown in Fig. 10.44(a) and Fig. 10.44(b) lists the signal at each lead. For example, address line A_{31} is at pin M2 and byte enable input $\overline{BE}_0$ is at pin G14.

Pin	Signal	Pin	Signal	Pin	Signal	Pin	Signal
M2	A31	C12	SA31	—	V_{CC}	B1	V_{SS}
L3	A30	A13	SA30	C1	V_{CC}	B14	V_{SS}
L2	A29	B12	SA29	C14	V_{CC}	M14	V_{SS}
K3	A28	A12	SA28	M1	V_{CC}	N1	V_{SS}
L1	A27	C11	SA27	N13	V_{CC}	N2	V_{SS}
J3	A26	B11	SA26	P1	V_{CC}	N14	V_{SS}
K2	A25	C10	SA25	P3	V_{CC}	P2	V_{SS}
K1	A24	B10	SA24	P11	V_{CC}	P4	V_{SS}
J2	A23	A11	SA23	P13	V_{CC}	P12	V_{SS}
J1	A22	C9	SA22	E13	$\overline{ADS}$	P14	V_{SS}
H3	A20	A10	SA21	F14	$W/\overline{R}$	N9	$\overline{BADS}$
H1	A19	A9	SA20	F12	$D/\overline{C}$	M12	$\overline{BBE0}$
G1	A18	B9	SA19	E14	$M/\overline{IO}$	N12	$\overline{BBE1}$
G2	A17	C8	SA18	F13	$\overline{LOCK}$	L12	$\overline{BBE2}$
G3	A16	A8	SA17	N3	$\overline{NA}$	M13	$\overline{BBE3}$
F1	A15	B8	SA16	G13	$\overline{X16}$	P9	$\overline{BLOCK}$
F2	A14	B7	SA15	G12	$\overline{NCA}$	K14	$\overline{BNA}$
F3	A13	A7	SA14	H14	$\overline{LBA}$	N4	CALEN
E1	A12	A6	SA13	D14	$\overline{READYI}$	P7	$\overline{COEA}$
E2	A11	C7	SA12	M3	$\overline{READYO}$	M7	$\overline{COEB}$
E3	A10	B6	SA11	E12	FLUSH	N7	$\overline{CWEA}$
D1	A9	B5	SA10	M8	WBS	P6	$\overline{CWEB}$
D2	A8	A5	SA9	N8	$\overline{MISS}$	M5	$\overline{CS0}$
C2	A7	C6	SA8	A14	$\overline{DEFOE}$	M6	$\overline{CS1}$
A1	A6	A4	SA7	D12	$2W/\overline{D}$	N6	$\overline{CS2}$
D3	A5	C5	SA6	B13	$M/\overline{S}$	P5	$\overline{CS3}$
C3	A4	B4	SA5	M10	$\overline{DOE}$	N5	$CT/\overline{R}$
B2	A3	C4	SA4	M4	LDSTB	P8	$\overline{BRDYEN}$
B3	A2	A3	SA3	N11	BHOLD	K13	$\overline{BREADY}$
G14	$\overline{BE0}$	A2	SA2	M11	BHLDA	P10	BACP
H13	$\overline{BE1}$	J12	SEN			M9	$\overline{BAOE}$
H12	$\overline{BE2}$	J13	$\overline{SSTB}$			N10	$BT/\overline{R}$
J14	$\overline{BE3}$	L14	RESERVED				
C13	CLK2						
D13	RESET						
K12	BRESET						
L13	BCLK2						

(b)

Figure 10.44(b) Signal mnemonics and pin numbers. (Reprinted by permission of Intel Corp. Copyright/Intel Corp. 1990)

Architecture of an 80386DX Microcomputer with an82385DX-Based Cache Memory

Let us now look at the architecture of an 80386DX microcomputer that employs an 82385DX-based cache memory. A block diagram of such a microcomputer system is shown in Fig. 10.45. Notice that the 82385DX cache controller and cache memory both connect onto the 80386DX's memory address bus, control bus, and data bus in parallel with the system bus to the main memory subsystem. For this reason, the diagram shows that the cache controller and cache memory attach to the 80386's *local bus*. On the other hand, the memory control signals for the main memory subsystem are produced by the 82385DX. These signals, together with the buffered address and data buses, are called the *82385DX local bus*. Therefore, the main memory subsystem, which is attached to the *system bus*, is supplied by the 82385DX's local bus.

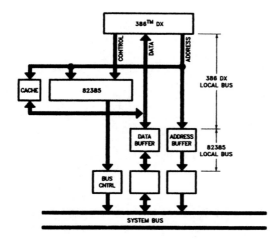

Figure 10.45 82385DX cache-memory subsystem architecture. (Reprinted by permission of Intel Corp. Copyright/Intel Corp. 1990)

Signal Interfaces of the 82385DX

A block diagram of the 82385DX is shown in Fig. 10.46. Here we see that it has *80386DX/82385DX interface signals, 80386DX local bus decode inputs, bus watching support signals, 82385DX bus data transceiver and address latch control signals, bus arbitration signals, 82385DX local bus interface signals, status and control signals, cache memory control signals,* and *configuration inputs.* Let us now look briefly at the function of these signals relative to operation of the cache memory subsystem.

The configuration inputs are strapped to set the mode of operation of the 82385DX. Looking at Fig. 10.46, we find two configuration inputs, *master/slave select* (M/$\overline{\text{S}}$) and *two-way/direct mapped select* (2W/$\overline{\text{D}}$). To put the 82385DX in master mode, the M/$\overline{\text{S}}$ input is strapped to the 1 logic level. This is the normal mode of operation in single-processor microcomputer systems. A multiprocessor system would normally employ some cache-controller devices configured as masters and others as slaves. Next, the cache organization is picked with the 2W/$\overline{\text{D}}$ input. Wiring this input to logic 0 selects direct-mapped cache operation, and setting it to 1 selects two-way set associative operation.

The circuit diagram in Fig. 10.47 shows how the 82385DX connects to the 80386DX MPU, the signals of the 82385DX's local bus, which goes to the main memory subsystem, and the signal interface to the cache memory subsystem. Notice that most of the signals identified as 80386DX/82385DX interface signals in Fig. 10.46 connect directly to the corresponding pins of the 80386DX. For instance, the $\overline{\text{ADS}}$ output of the 80386DX is attached to the $\overline{\text{ADS}}$ input of the 82385DX, the CLK2 input of both devices are tied together and driven by the external clock oscillator, and the $\overline{\text{NA}}$ input of the 80386DX is driven by the $\overline{\text{NA}}$ output of the 82385DX.

Next we will examine the signals at the cache memory interface. In Fig. 10.47 we find that this interface includes data bus lines D_0 through D_{31}, address lines A_2 through A_{13} or A_2 through A_{14}, and a number of control signals. The address and data bus lines are supplied directly by the 80386DX. On the other hand, the control signals are generated by the 82385DX. Figure 10.46 identifies these control signals as the cache memory control signals.

An example of a direct-mapped cache memory array is shown in Fig. 10.48(a) and a two-way set associative cache array in Fig. 10.48(b). The first control signal, as shown in Fig. 10.46, is *cache address latch enable* (CALEN). This output is used to enable the address

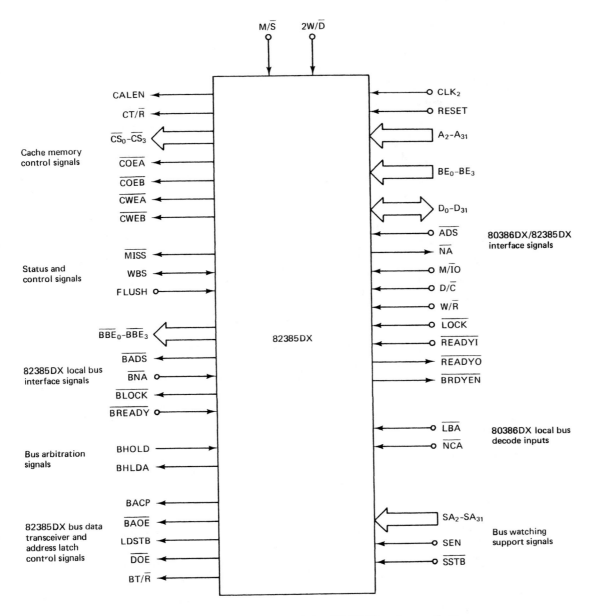

Figure 10.46 Block diagram of the 82385DX cache controller.

latch in the cache memory subsystem. In both circuits CALEN is applied to the enable (E) input of the 74F373 address latches. The next control signal, *cache transmit/receive* $(CT/\overline{R})$, is used to control the direction of data transfer through the cache data bus transceivers during read and write bus cycles. Notice in Fig. 6.48(a) and (b) that this signal is applied to the direction (DIR) inputs of the 74F245 data bus transceiver devices.

The 32K-byte memory array in the direct-mapped cache of Fig. 10.48(a) is implemented with four 8K × 8-bit SRAMs. Each RAM is attached to eight of the data bus lines. The

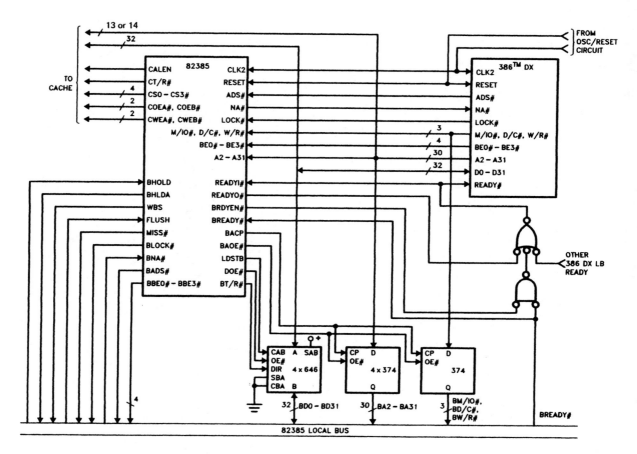

Figure 10.47 Connection of the 82385DX to the 80386DX MPU. (Reprinted by permission of Intel Corp. Copyright/Intel Corp. 1990)

cache chip select ($\overline{CS}_0$ through $\overline{CS}_3$) outputs of the 82385DX are applied to chip select inputs of the RAMs and are used to enable them for operation. For example, Fig. 10.49 shows that SRAM 0 supplies data bus lines D_0 through D_7 and is enabled by $\overline{CS}_0$. During byte, word, and double-word data transfers, the cache chip select signals are used to enable the appropriate static RAMs for operation.

In Fig. 10.46, we find two more groups of cache control signals, the *cache output enable* ($\overline{COEA}$ and $\overline{COEB}$) signals and *cache write enable* ($\overline{CWEA}$ and $\overline{CWEB}$) signals. They are used to enable the data bus transceivers and the outputs of the SRAMs, respectively. Looking at Fig. 10.48(a), we see that in a direct-mapped cache only $\overline{COEA}$ and $\overline{CWEA}$ are used. $\overline{COEA}$ is applied to the $\overline{OE}$ input of the four 74F245 data bus transceivers and enables them for operation during read and write bus cycles. Moreover, $\overline{CWEA}$ drives the $\overline{WE}$ inputs of the four SRAMs in parallel and signals them whether a read or write data transfer is taking place.

Earlier we found that the 80386DX microcomputer's system bus was formed from the 82385DX's local bus signals. Let us now look at the signals of the *82385DX local bus*.

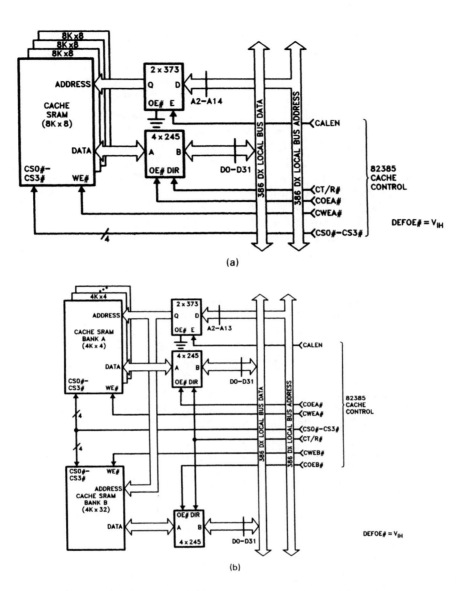

Figure 10.48 (a) Direct-mapped cache-memory array. (Reprinted by permission of Intel Corp. Copyright Intel Corp. 1990) (b) Two-way set associative cache-memory array. (Reprinted by permission of Intel Corp. Copyright/Intel Corp. 1990)

SRAM	Data bus lines	Cache chip select
0	D_7-D_0	CS_0
1	$D_{15}-D_8$	CS_1
2	$D_{23}-D_{16}$	CS_2
3	$D_{31}-D_{24}$	CS_3

Figure 10.49 Direct-mapped cache-memory SRAM data bus connections and the cache chip-select signals.

Notice in Fig. 10.47 that the data bus, address bus, and bus cycle indication signals of the 80386DX are used as part of the 82385DX's local bus. Buffered address lines BA_2 through BA_{31} and buffered bus cycle indication signals $BM/\overline{IO}$, $BD/\overline{C}$, and $BW/\overline{R}$ are formed by latching the corresponding signal of the 80386DX into 74F374 latches. The address and bus cycle indication signals are latched into the octal latches with a pulse from the *bus address clock pulse* (BCAP) output of the 82385DX and the buffer is enabled by the *bus address output enable* ($\overline{BAOE}$) output. Data bus lines BD_0 through BD_{31} are buffered by 74F646 registered data bus transceivers. Signals *bus transmit/receive* ($BT/\overline{R}$), *data output enable* ($\overline{DOE}$), and *local data strobe* (LDSTB) are used to set the direction of data transfer through the transceivers, enable the transceivers for operation, and clock the data into the register within the 74F646s during write cycles, respectively. These outputs are identified in Fig. 10.46 as the 82385DX bus data transceiver and address latch–control signals.

At the other side of the 82385DX in Fig. 10.46, we find the control signals of the 82385DX local bus. The *bus byte enable* ($\overline{BBE}_0$ through $\overline{BBE}_3$), *bus address status* ($\overline{BADS}$), *bus next address request* ($\overline{BNA}$), *bus lock* ($\overline{BLOCK}$), *bus hold acknowledge* (BHLDA), and *bus hold* (BHOLD) signals perform similar functions as their corresponding signals of the 80386DX. For instance, the $\overline{BADS}$ output indicates that a valid address and bus cycle indication code are available. There are also a few new signals that are unique to the 82385DX local bus. The *cache flush* (FLUSH) input can be used by external circuitry to initiate clearing of the cache memory. If FLUSH is switched to the 1 logic level for eight CLK_2 cycles, all of the cache directory valid bits are cleared. In this way, the contents of the cache SRAMs are flushed by invalidating them. *Cache miss* ($\overline{MISS}$) is another output. It signals external circuitry that the current address on the bus does not correspond to data that are already in the cache and that the information must be read from the main memory subsystem. Finally, *write buffer status* (WBS) is an output that signals that the registers within the 74F646 data bus transceivers contain write data that has not yet been written to the main memory subsystem. These control signals correspond to the 82385DX local bus interface signals, bus arbitration signals, and status and control signals groups in Fig. 10.46.

Direct-Mapped Cache Operation

We just examined the signal interfaces of the 82385DX cache controller and the hardware of its cache memory subsystem. Let us look next at the operation of an 82385DX-based direct-mapped cache memory subsystem.

We begin by describing how the 82385DX keeps track of which main memory locations are cached. The 82385DX accomplishes this through what is called a *cache directory*. The cache directory is located within the 82385DX and contains 1024 26-bit entries. Figure 10.50 illustrates the organization of the direct-mapped cache memory subsystem. Notice that the 80386DX's 4G-byte main memory physical address space is treated as $2^{17} - 1$ (131,071) 32K-byte pages. Since the 80386DX reads code and data as 32-bit double words, the organization of a page in main memory can also be viewed as 8192 (8K) double words.

As expected, the page size in main memory matches the size of the cache memory. The 32K-byte cache is organized as double-word storage locations. Figure 10.50 shows that this results in an 8K × 32-bit memory array. Each of these 32-bit double words is called a *line*. Code and data are read from and written into cache one line (a double word) at a time.

The cache directory within the 82385DX is organized in a different way. Within the cache controller, the status of cache memory locations are tracked with 26-bit cache directory entries. In Fig. 10.50 we see that the entries within the cache directory are called *SET 0* through *SET 1023*. Each of the 1024 SET entries corresponds to eight consecutive lines

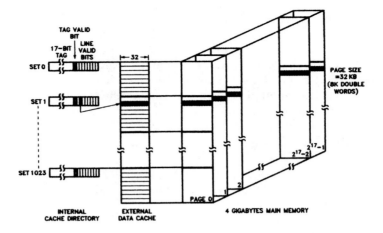

Figure 10.50 Direct-mapped cache organization. (Reprinted by permission of Intel Corp. Copyright/Intel Corp. 1990)

(double words) of information in the cache memory array. For instance, SET 0 corresponds to lines 0 through 7 of the array.

EXAMPLE 10.4

In which SET entry of the cache directory in Fig. 10.50 is the information for line 9 of the memory array stored?

Solution

Looking at Fig. 10.50, we find that line 9 of the cache memory array is in SET 1. In fact, line 9 is the 32-bit storage location that is highlighted in the external data cache section of the drawing.

The format of the cache directory SET entry used by an 82385DX configured for direct-mapped mode is given in Fig. 10.51. Notice that bits 0 through 7 are eight independent *line valid bits*, the eighth bit is the *tag valid bit*, and bits 9 through 25 are a 17-bit *tag*. The 17-bit tag identifies the page number of one of the 131,071 pages in main memory. The tag

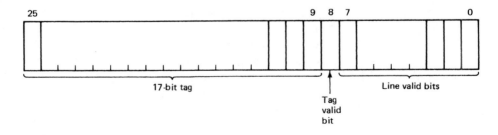

Figure 10.51 Direct-mapped cache directory SET entry format.

bit indicates whether the 17-bit tag is valid or invalid. If this bit is 0, all lines of information in the SET are invalid. However, logic 1 in the tag bit does not mean that they are all valid. When the tag bit is logic 1, the eight line valid bits identify whether the individual lines in the SET are valid or invalid. For example, logic 1 in line valid bit 0 of SET 0 means that line 0 in the cache contains valid information, logic 1 in bit 1 means that line 1 is valid, and so on. This information is updated automatically each time a noncached line of information is read from memory.

EXAMPLE 10.5

If the cache directory entry for SET 1 equals $00005FF_{16}$, from which page of main memory is the cached information? Is the cache entry valid? Which lines in SET 1 are valid?

Solution

Expressing the SET 1 entry in binary form gives

$$\text{SET } 1 = 00000000000000010111111111_2$$

From the binary form of SET 1, we find that

$$\text{TAG} = 00000000000000010_2 = 2_{10}$$

Therefore, the entry is from page 2 of the main memory array. Next, the valid bit is

$$\text{TAG VALID} = 1$$

and the tag entry is valid. Finally, the line valid bits are

$$\text{LINE VALID} = 11111111_2$$

This means that all lines of information in SET 1 are valid.

Now that we have looked at what information is used to track entries in the cache memory, we will continue by examining the events that take place when data are read from memory. Whenever the 80386DX initiates a read bus cycle, the address output on the address bus is applied in parallel to the 82385DX local bus address latch, the address inputs of the 82385DX cache controller, and the cache memory interface. The cache controller quickly makes a decision whether the information for this address should be read from the cache or from main memory. It does this by interpreting the address and then comparing this information to the corresponding cache directory SET entry.

Figure 10.52 shows how addresses are interpreted by the 82385DX cache controller. Notice that the 17 most significant bits, A_{15} through A_{31}, are the tag that identifies the page of main memory from which information is to be read. The next 10 address bits, A_5 through A_{14}, are called the SET address. This part of the address identifies which SET of storage locations in the cache memory array are to be accessed. Finally, the 3 least significant bits, A_2 through A_4, are a line-select code and select one of the eight lines in this SET.

Let us now look at what happens when an address is applied to the address inputs of the 82385DX. The SET part of the address is used to select one of the 1024 SET entries

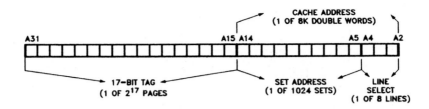

Figure 10.52   Address bit fields for direct-mapped cache. (Reprinted by permission of Intel Corp. Copyright/Intel Corp. 1990)

in the cache directory. Then, three checks are made to determine if the information is already in the cache. First, the 82385DX compares the tag field of the address to the value of the tag field in the selected SET entry to verify that they match. Second, the tag valid bit in the SET entry is examined to verify that it is set. Finally, the line valid bit corresponding to the line-select code part of the address is tested to verify that it is 1. If all three of these conditions are satisfied, the information held in the storage location to be read is already cached and is valid. That is, a cache hit has occurred. In this case, the 82385DX initiates the read cycle from the cache memory instead of main memory.

We just saw the result of a cache hit. Let us now determine what can cause a cache miss and what happens when one occurs. If during the address-interpretation process, the tag is found to match and the tag valid bit is 1, but the line valid bit is 0, a cache miss has occurred. This result presents a situation where the SET entry in the directory corresponds to the correct page of main memory, but the double word (line) to be read has not yet been cached. This is called a *line miss*. For this situation the cache controller initiates a read of the double word from main memory to the MPU. As part of the bus operation, the double word of data is written into the corresponding storage location in the cache memory array, and then the line valid bit in the SET entry in the cache directory is set to 1. That is, the information is copied into the cache and marked as valid.

If, during the entry-verification operation, either the tags do not match or the tag valid bit is found to be 0, a *tag miss* has occurred. This corresponds to an attempt to read information from either a page that has not already been cached or one that is cached but is invalid. When this type of cache miss occurs, the cache controller again initiates a read from main memory and the value transferred over the data bus is written into the cache memory. However, this time the tag in the SET entry of the directory is updated with the value from the tag part of the address, the tag valid bit is set, the line valid bit identified by the address is set, and all other line valid bits are cleared. In this way, a new valid page and line entry have been created and all other entries in the SET, which now correspond to information in another page of main memory, are invalidated.

Two-Way Set Associative Cache Operation

The operation of an 82385DX-based cache memory subsystem configured for two-way set associative mode is somewhat different from that we just described for the direct-mapped subsystem. Let us begin by examining the organization of the two-way set associative cache memory subsystem shown in Fig. 10.53. Here we find that the 82385DX's internal cache directory contains two SET entry directories, not one as in the direct-mapped configuration. These directories are identified as *DIRECTORY A* and *DIRECTORY B*, and each contains 512 SET entries. The external 32K-byte cache memory array is partitioned into

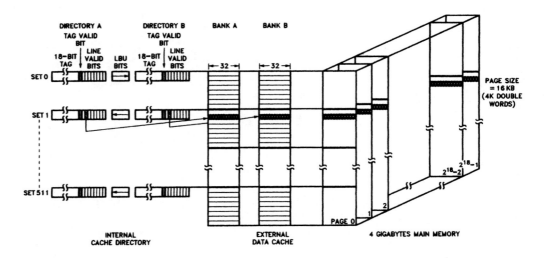

Figure 10.53 Two-way set associative cache organization. (Reprinted by permission of Intel Corp. Copyright/Intel Corp. 1990)

two separate 4K $\times$ 32-bit banks called *BANK A* and *BANK B*. This permits information from the same offset in two different pages of main memory to reside in the cache at the same time. The cache directory also includes 512 one-bit *least recently used (LRU) flags*. The settings of these flags keep track of whether BANK A or BANK B holds the least recently used information. These flags are checked by the least recently used replacement algorithm that is employed by the 82385DX. Finally, we find that the 80386DX's 4G-byte physical address space is divided into 2^{18} (262,144) 16K-byte (4K double-word) pages. This page size is half that used in the direct-mapped main memory subsystem.

The SET entries in DIRECTORY A and DIRECTORY B both have the same format. Figure 10.54 shows this format in more detail. Notice that they are 27 bits in length. The difference between this SET entry and that shown in Fig. 10.51 for a direct-mapped cache is that the tag is 18 bits long instead of 17 bits. The extra bit in the tag is needed because making the page size 16K bytes results in twice as many pages.

A read operation from a two-way set associative cache memory is similar to that just described for the direct-mapped cache. The 82385DX interprets the memory address as shown in Fig. 10.55. Notice that the SET address is 9 bit, not 10 bits. The reason the SET address has 1 less bit than that for a direct-mapped cache is that the cache directory contains 512 SET entries rather than 1024. First, the 9-bit SET address is used to select one of the

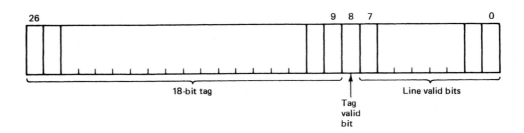

Figure 10.54 Two-way set associative cache directory SET entry format.

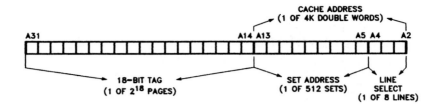

Figure 10.55 Address bit fields for two-way set associative cache. (Reprinted by permission of Intel Corp. Copyright/Intel Corp. 1990)

512 SET entries in both DIRECTORY A and DIRECTORY B. Then the 18-bit tag address is compared to the tag in each of these SET entries; there tag valid bits are checked; and the line valid bit corresponding to the line select code is checked in each SET entry. If these three conditions are satisfied for either of the two SET entries, a cache hit has occurred and the line of information is read from the corresponding bank of cache memory.

On the other hand, a miss is detected if there is no match of either the tags, if both valid bits are cleared, or if the line valid bit is not set in either entry. At this point, operation differs for the two-way set associative and the direct-mapped cache. This is because we are at the point where the least recently used (LRU) algorithm takes over. It checks the LRU flag for the SET selected by the SET address to determine whether the BANK A or BANK B cache entry was least recently used. Then the read operation is initiated from the main memory subsystem and the line of information is read by the MPU and written into the cache location of the least recently used bank. Now the SET entry is updated by replacing the tag; setting the tag valid bit and the line valid bit; and switching the LRU flag. The updated cache memory location now holds the most recently used information.

Cache Coherency and Bus Watching

The ability to maintain the information held at an address in main memory and its copy in a cache memory the same is called cache *coherency*. In multiprocessor applications, another bus master can take over control of the system bus. This bus master could write data into a main memory storage location whose data are already held in the cache of another processor. When this happens, the data in the cache no longer match those held in main memory. That is, the data are invalid. If this cached value of data were read, processed, and written back to memory, the contents of the main memory would be contaminated. To protect against this problem, the 82385DX is equipped with a feature known as *bus watching*.

The diagram in Fig. 10.56 shows the architecture of an 80386DX microcomputer system that employs an 82385DX-based cache memory with bus watching. Here we find that a *snoop bus* has been added between the microcomputer's system bus and the 82385DX. The snoop bus is implemented with the signals identified as the bus watching support signals in Fig. 10.46. They include the *snoop address bus* (SA_2 through SA_{31}) inputs, the *snoop enable* (SEN) input, and the *snoop strobe* ($\overline{SSTB}$) input.

Whenever a bus master is performing a write bus cycle to the main memory subsystem, the 82385DXs in the system examine the address to determine if they contain valid data for this storage location. The main memory subsystem must signal the 82385DXs that a write cycle is in progress by applying logic 1 at their SEN inputs. The address of the storage location being accessed is applied to the snoop address inputs of the 82385DXs and is

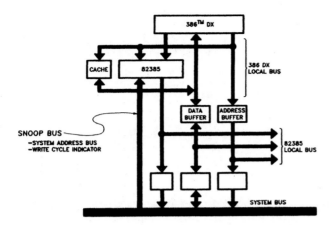

Figure 10.56 82385DX cache-memory subsystem architecture that employs bus watching. (Reprinted by permission of Intel Corp. Copyright/Intel Corp. 1990)

strobed into the device with logic 0 at $\overline{\text{SSTB}}$. If the cache directory entry corresponding to this address indicates that the cache contains valid data, a *snoop hit* has occurred. The 82385DX automatically clears the tag to invalidate this piece of data in the cache. In this way we see that the 82385DX has the ability to maintain cache coherency by watching what system bus activity takes place.

Noncacheable Memory Address Space

The 82385DX supports partitioning of the main memory subsystem's address space into a cacheable and a noncacheable address range. This feature is achieved with the *noncacheable access* ($\overline{\text{NCA}}$) input of the 82385DX. By decoding the address in external circuitry and returning logic 0 to the $\overline{\text{NCA}}$ input for those addresses in the noncacheable address range, bus cycles to these storage locations are made noncacheable. That is, the cache controller causes their bus cycles to be performed over the 82385DX local bus to the main memory subsystem.

ASSIGNMENTS

Section 10.2

1. Which part of the primary storage memory is used to store instructions of the program and fixed information such as constant data and look up tables? Data that changes frequently?
2. What does BIOS stand for?
3. What term is used to refer to programs stored in ROM?
4. Can DRAMs be used to construct a program storage memory?

Section 10.3

5. What is meant by the term *nonvolatile memory*?
6. What does PROM stand for? EPROM?
7. What must an EPROM be exposed to in order to erase its stored data?

8. If the block diagram of Fig. 10.3 has address lines A_0 through A_{16} and data lines D_0 through D_7, what are its bit density and byte capacity?

9. Summarize the read cycle of an EPROM. Assume that both $\overline{CE}$ and $\overline{OE}$ are active before the address is applied.

10. Which standard EPROM stores 64K 8-bit words?

11. What is the difference between a 27C64A and a 27C64A-1?

12. What are the values of V_{CC} and V_{pp} for the Intelligent Programming Algorithm?

13. What is the duration of the programming pulses used for the Intelligent Programming Algorithm?

Section 10.4

14. What do SRAM and DRAM stand for?

15. Are RAM ICs examples of nonvolatile or volatile memory devices?

16. What must be done to maintain the data in a DRAM valid?

17. Find the total storage capacity of the circuit in Fig. 10.12 if the memory devices are 43256As.

18. List the minimum values of each of the write cycle parameters that follow for the 4364-10 SRAM: t_{WC}, t_{CW1}, t_{CW2}, t_{WP}, t_{DW}, and t_{WR}.

19. Give two benefits of DRAMs over SRAMs.

20. Name the two parts of a DRAM address.

21. Show how the circuit in Fig. 10.22 can be expanded to 128K $\times$ 32 bits.

22. Give a disadvantage of the use of DRAMs in an application that does not require a large amount of memory.

Section 10.5

23. What type of circuit can be added to the data storage memory interface to improve the reliability of data transfers over the data bus?

24. If in Fig. 10.23, the datum read from memory is 100100100_2, what is its parity? Repeat if the datum is 011110000_2.

25. If the input to a 74AS280 parity-checker/generator circuit that is set up for odd parity checking and generation is I H ... A = 111111111_2, what are its outputs?

26. What changes must be made in the circuit of Fig. 10.24(c) to convert it to an odd parity configuration?

27. Make a drawing similar to that shown in Fig 10.25 that can be used as the parity-checker/generator in the data storage memory subsystem of an 80386DX microcomputer system. Assume that parity checking is performed independently for each byte-wide bank of the memory array and that the parity error outputs for the four banks are combined to form a single parity error signal.

Section 10.6

28. What is the key difference between a FLASH memory and an EPROM?

29. What is the key difference between the bulk-erase architecture and the boot block and FlashFile architectures?

30. What is the key difference between the boot block architecture and the FlashFile architecture?

31. What architecture is used in the 28F010 FLASH memory IC?

32. What power supply voltage must be applied to a 28F010 device when it is being erased or written into?

33. Give the names of two SmartVoltage boot block FLASH devices.

34. What value V_{cc} power supply voltages can be applied to a SmartVoltage boot block FLASH IC? What value V_{pp} power supply voltages?

35. Name the three types of blocks used in the storage array of the 28F004. How many of each is provided? What are their sizes?

36. Name two FlashFile FLASH memory devices.

37. What is the function of the $RY/\overline{BY}$ output of the 28F016SA/SV?

Section 10.7

38. What function is performed by a wait-state generator circuit?

39. What output signal is produced by the wait-state generator?

40. Does the circuit in Fig. 10.37(b) produce the same number of wait states for the memory subsystems corresponding to both chip select?

41. What is the maximum number of wait states that can be produced with the circuit in Fig. 10.37(b)?

Section 10.8

42. The output of what gate in the circuit of Fig. 10.38 supplies the $\overline{READY}$ input of the 80386DX?

43. In the microcomputer of Fig. 10.38, for what logic levels at $M/\overline{IO}$ and A_{31} will the $\overline{Y}_3$ output of the memory address decoder be active?

44. The output of which gate drives the $\overline{NA}$ input of the 80386DX in the circuit of Fig. 10.38? For what input conditions does it become active?

45. What type of PALs are used to implement the bus control logic in Fig. 10.38?

46. For the microcomputer in Fig. 10.38, what input conditions make the output of an OR gate in the bank write control logic active?

47. If we assume that the unused address bits in the circuits formed in Fig. 10.38 are all logic 0, what is the address range of the program memory subsystem?

48. How many SRAMs would be needed in the memory array of the circuit in Fig. 10.38 if the capacity of data storage memory is to be expanded to 128K bytes?

Section 10.9

49. What is a cache memory?

50. What is the result obtained by using a cache memory in a microcomputer system?

51. What is the range of cache memory sizes commonly used in microcomputer systems?

52. Define the term *cache hit*.

53. When an application program is tested on a microcomputer system with a code cache, it is found that 1340 instruction acquisition bus cycles are from the cache memory and 97 are from main memory. What is the hit rate?

54. If the cache memory in Problem 53 operates with zero wait states and main memory bus cycles are performed with three wait states, what is the average number of wait states experienced executing the application?

55. Name two widely used cache memory organizations.

56. Name the element into which main memory is organized for use with a cache.

57. How does the organization of a direct-mapped cache memory array differ from that of a two-way set associative cache memory?

58. What does LRU stand for?

59. What is posted write-through?

Section 10.10

60. At what pin of the 82385DX's package is the signal CALEN located?

61. Which signal is output at pin M10 of the 82385DX's package?

62. What must be done to configure an 82385DX cache controller for two-way set associative operation?

63. Make a list of the mnemonics and signal names for all of the control signals at the cache memory interface.

64. What are the functions of the signals BCAP and $\overline{BAOE}$?

65. Which signals indicate whether a byte, word, or double word of data is being transferred across the 82385DX's local data bus?

66. Which input, when activated, clears the cache memory?

67. Into what size pages does a direct-mapped 82385DX organize the memory of a real-mode 80386DX microcomputer system? Into how many pages is the real-mode address space organized?

68. What is the size of the cache memory array in an 82385DX-based direct-mapped cache memory subsystem? Express the answer in double words.

69. If the cache directory entry for SET 3 in a direct-mapped cache equals $0001F03_{16}$, from which page of main memory is the cached information? Is the cache entry valid? Which lines in the set are valid?

70. What is the physical address range of the lines of the SET 3 entry in Problem 69?

71. What are the sizes of BANK A and BANK B in the memory array of an 82385DX-based two-way set associative cache memory? Express the answer in kilobytes.

72. What is the line length in a two-way set associative cache memory that is designed with the 82385DX cache controller?

73. What function is performed by the LRU flag?

74. Which feature of the 82385DX is used to assure cache coherency?

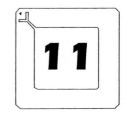

Input/Output Interface Circuits and LSI Peripheral Devices

▲ 11.1 INTRODUCTION

In Chapter 9 we introduced the input/output interface of the 80386DX microprocessor. At that time, we discussed the topics of isolated and memory-mapped I/O, isolated I/O interface circuits, I/O bus cycles, and I/O instructions. Here we continue our study of input/output by examining circuits and large-scale-integrated peripheral ICs that are used to implement input/output subsystems for the microcomputer systems. These are the topics in the order in which they are covered:

1. Core and special-purpose I/O interfaces
2. Byte-wide output ports using isolated I/O
3. Byte-wide input ports using isolated I/O
4. Input/output handshaking and a parallel printer interface
5. 82C55A programmable peripheral interface (PPI)
6. Implementing isolated I/O parallel input/output ports using the 82C55A
7. Implementing memory-mapped I/O parallel input/output ports using the 82C55A
8. 82C54 programmable interval timer
9. 82C37A programmable direct memory access controller
10. 80386DX microcomputer system I/O circuitry
11. Serial communications interface
12. Programmable communication interface controllers
13. Keyboard and display interface
14. 8279 programmable keyboard/display controller

▲ 11.2 CORE AND SPECIAL-PURPOSE I/O INTERFACES

In Chapter 1 we indicated that the input/output unit provides the microcomputer with the means for communicating with the outside world. For instance, the keyboard of a PC permits the user to input information such as programs or data for an application. The display outputs information about the program or application for the user to read. These examples represent one type of input/output function, which we will call *special-purpose I/O interfaces*. Other examples of special-purpose I/O interfaces are parallel printer interfaces, serial communication interfaces, and local area network interfaces. They are referred to as special-purpose interfaces because not all microcomputer systems employ each of these types.

Both the original PC and PC/AT are capable of supporting a wide variety of input/output interfaces. In fact all the interfaces we just mentioned are available for the PC. Since they are special-purpose interfaces, they are all implemented as add-on cards. That is, to support a keyboard and display interface on the PC, a special keyboard/display controller card is inserted into a slot of the PC and then the keyboard and display are attached to the card with cables connected at connectors at the back of the PC.

In microcomputer circuit design, a variety of other types of circuits are also classified as input/output circuitry. For instance, parallel input/output ports, interval timers, and direct memory access control are examples of interfaces that are also considered to be part of the I/O subsystem. These I/O functions are employed by most microcomputer systems. For this reason, we will refer to them as *core input/output interfaces*.

The core I/O functions are not as visible to the user of the microcomputer. However, they are just as important to overall microcomputer function. In fact, the circuitry of the original PC and PC/AT contains each of these core microcomputer functions. For example, parallel I/O is the method used to implement the reading settings of the DIP switches on the processor board. Also an interval time is used as part of the DRAM refresh process and to keep track of the time of day. The circuitry for these core I/O functions is built right on the PC's main processor board. They are also included as part of the MPU IC in some newer highly integrated versions, such as the 80386EX.

In the sections that follow, we explore the function, circuits, and operation of both the core and special-purpose input/output functions.

▲ 11.3 BYTE-WIDE OUTPUT PORTS USING ISOLATED I/O

Now we show circuits that can be used to implement parallel output ports in a microcomputer system employing isolated I/O. Figure 11.1(a) is such a circuit for an 80386DX-based microcomputer. It provides 8-byte-wide output ports that are implemented using 74F374 octal latches. In this circuit, the ports are labeled port 0 through port 7. These eight ports give a total of 64 parallel output lines, which are labeled O_0 through O_{63}.

Looking at the circuit, we see that it employs a byte-wide I/O interface similar to that of Fig. 9.51. The only difference between this and the earlier circuit is that here the I/O address is first latched and then decoded. The latched address outputs are labeled A_{2L} through A_{15L}. Remember that address lines A_{16} through A_{31} are not involved in the I/O interface. For this reason, they are not shown in the circuit diagram.

In the byte-wide I/O interface, the 32-bit data bus of the 80386DX is multiplexed to provide an 8-bit I/O data bus. These data bus lines, D_0 through D_7, are shown connecting to the input side of the output latches. It is over these lines that the 80386DX writes data into the output ports.

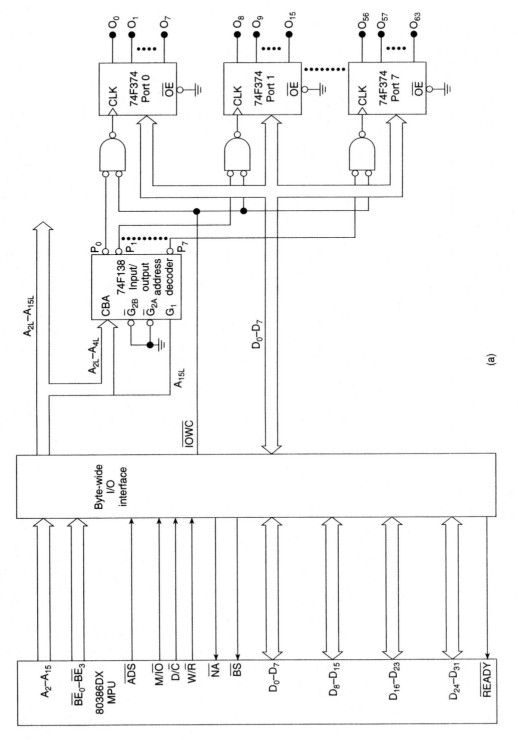

Figure 11.1(a) Sixty-four-line parallel output circuit for an 80386DX-based microcomputer.

(a)

I/O port	I/O address
Port 0	$1XXXXXXXXXX000XX_2$
Port 1	$1XXXXXXXXXX001XX_2$
Port 2	$1XXXXXXXXXX010XX_2$
Port 3	$1XXXXXXXXXX011XX_2$
Port 4	$1XXXXXXXXXX100XX_2$
Port 5	$1XXXXXXXXXX101XX_2$
Port 6	$1XXXXXXXXXX110XX_2$
Port 7	$1XXXXXXXXXX111XX_2$

(b)

Figure 11.1(b) I/O address decoding for ports 0 through 7.

Address line A_{15L} provides one of the three enable inputs of the 74F138 input/output address decoder. This signal is applied directly to enable inputs G_1. The decoder requires two more enable inputs, $\overline{G}_{2A}$ and $\overline{G}_{2B}$. They are both permanently enabled by connecting them to ground. The enable inputs must be $\overline{G}_{2B}\overline{G}_{2A}G_1 = 001$ to enable the decoder for operation. The first two inputs, $\overline{G}_{2A}$ and $\overline{G}_{2B}$, are fixed at the 0 logic level, and the third condition, $G_1 = 1$, is an additional requirement that A_{15L} be at logic 1 during all data transfers for this section of parallel output ports.

Notice that the three address lines, A_{2L}, A_{3L}, and A_{4L}, are applied to select inputs CBA of the 74F138 1-of-8 decoder. When the decoder is enabled, the P output corresponding to these select inputs switches to logic 0. Logic 0 at this output enables the input/output write command ($\overline{IOWC}$) signal to the clock (CLK) input of the corresponding output latch. In this way, just one of the eight ports is selected for operation.

When valid output data are on D_0 through D_7, the 80386DX switches $\overline{IOWC}$ to logic 0. This change in logic level causes the selected 74F374 device to latch in the data from the bus. The outputs of the latches are permanently enabled by the 0 logic level at their $\overline{OE}$ inputs. Therefore, the latched data appear at the appropriate port outputs.

Notice in Fig. 11.1(a) that not all address bits are used in the I/O address decoding. Here only latched address bits A_{2L}, A_{3L}, A_{4L}, and A_{15L} are decoded. Figure 11.1(b) shows the addresses that select each of the I/O ports. Unused bits are shown as don't cares. By assigning various logic combinations to the unused bit, the same port can be selected by different addresses. In this way, we see that many addresses will decode to select each of the I/O ports. For instance, if all of the don't-care address bits are made 0, the address of port 0 is

$$1000000000000000_2 = 8000_{16}$$

However, if these bits are all made equal to 1 instead of 0, the address is

$$1111111111100011_2 = FFE3_{16}$$

and it still decodes to enable PORT 0. In fact, every I/O address in the range from 8000_{16} through $FFE3_{16}$ that has address bits $A_{4L}A_{3L}A_{2L}$ equal to 000_2 decodes to enable port 0. Some other examples are 8003_{16} and $FFE0_{16}$.

EXAMPLE 11.1

To which output port in Fig. 11.1(a) are data written when the address used during the output bus cycle is 8004_{16}?

Solution

Expressing the address in binary form, we get

$$A_{15} \ldots A_0 = 1000000000000100_2$$

The important address bits are

$$A_{15L} = 1$$

and

$$A_{4L}A_{3L}A_{2L} = 001$$

Therefore, the enable inputs of the 74F138 decoder are

$$\overline{G}_{2A} = \overline{G}_{2B} = 0$$

$$G_1 = A_{15L} = 1$$

These inputs enable the decoder for operation. At the same time, its select inputs are supplied with the code 001. This input causes output P_1 to switch to logic 0.

$$P_1 = 0$$

The gate at the CLK input of port 1 has as its inputs P_1 and $\overline{IOWC}$. When valid output data are on the bus, $\overline{IOWC}$ switches to logic 0. Since P_1 is also 0, the CLK input of the 74F374 for port 1 switches to logic 0. At the end of the $\overline{IOWC}$ pulse, CLK switches from 0 to 1, a positive transition. This causes the data on D_0 through D_7 to be latched at output lines O_8 through O_{15} of port 1.

EXAMPLE 11.2

Write a series of instructions that will output the byte contents of the memory address DATA to output port 0 in the circuit of Fig. 11.1(a).

Solution

To write a byte to output port 0, the address that must be used in the out instruction is

$$A_{15}A_{14} \ldots A_0 = 1XXXXXXXXXX000XX_2$$

Assuming that the don't-care bits are all made logic 0, we get

$$A_{15}A_{14} \ldots A_0 = 1000000000000000_2$$

$$= 8000_{16}$$

Then the instruction sequence needed to output the contents of memory address DATA is

```
MOV    DX,8000H
MOV    AL,[DATA]
OUT    DX,AL
```

The Time-Delay Loop and Blinking an LED at an Output

The circuit in Fig. 11.2 has an LED attached to output O_7 of parallel port 0. This circuit is identical to that shown in Fig. 11.1(a). Therefore, the port address as found in Example 11.2 is 8000H, and the LED corresponds to bit 7 of the byte of data that is written to port 0. For the LED to turn on, O_7 must be switched to logic 0, and it will remain on until this output is switched back to 1. The 74374 is not an inverting latch; therefore, to make O_7 logic 0 we simply write 0 to that bit of the octal latch. To make the LED blink, we must write a program that first makes O_7 logic 0 to turn on the LED; delays for a short period of time; and then switches O_7 back to 1 to turn off the LED. This piece of program can be run as a loop to make the LED continuously blink.

Let us begin by writing the sequence of instructions needed to initialize O_7 to logic 0. This is done as follows:

```
         MOV    DX,8000H    ;Initialize address of port 0
         MOV    AL,00H      ;Load data with bit 7 as logic 0
ON_OFF:  OUT    DX,AL       ;Output the data to port 0
```

After the out operation is performed, the LED will be turned on.

Next we must delay for a short period of time. This can be done with a software loop. The following instruction sequence produces such a delay.

```
         MOV    CX,0FFFFH   ;Load delay count of FFFFH
HERE:    LOOP   HERE        ;Time delay loop
```

First the count register is loaded with $FFFF_{16}$. Then the loop instruction is repeatedly executed. With each occurrence of the loop, the count in CX is decremented by 1. After 65,335 repeats of the loop, the count in CX is 0000_{16} and the loop operation is complete. These executions perform no software function for the program other than to use time, which is the duration of the time delay. By using $FFFF_{16}$ as the count, the maximum delay is obtained. The duration of the delay can be shortened by simply loading a smaller number in CX.

Next the value in bit 7 of AL is complemented to 1 and then a jump performed to return to the output operation that writes the data to the output port.

```
         XOR    AL,80H    ;Complement bit 7 of AL
         JMP    ON_OFF    ;Return to output the new data
```

By performing an exclusive-OR operation on the value in AL with the value 80_{16}, the most significant bit is complemented to 1. The jump instruction returns control to the OUT instruction. Now the new value in AL, with MSB equal to 1, is output to port 0 and the LED turns off. After this, the time delay repeats; the value in AL is complemented back to 00_{16}; and the LED turns back on. In this way, we see that the LED blinks repeatedly with an equal period of on and off time that is set by the count in CX.

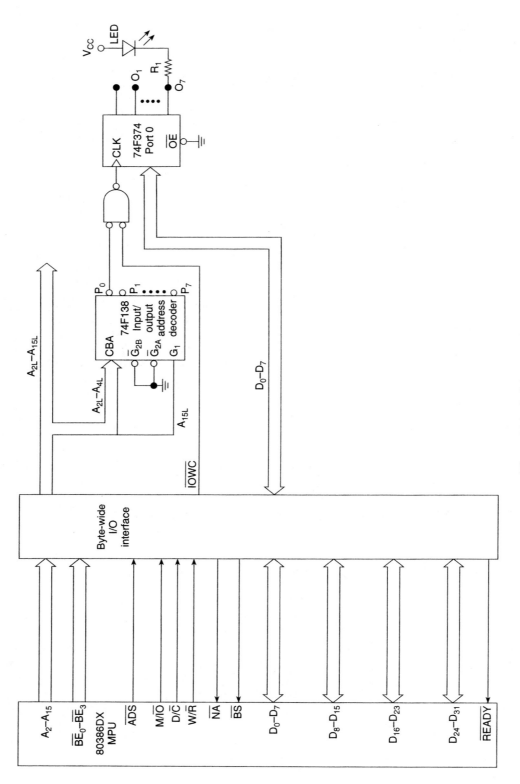

Figure 11.2 Driving an LED.

In Section 11.3, we showed a circuit that implemented eight byte-wide output ports for the 80386DX microcomputer system. These circuits used the 74F374 octal latch to provide the output ports. Here we examine a similar circuit that implements input ports for the microcomputer system.

The circuit in Fig. 11.3 employs the same byte-wide I/O interface as the output interface of Fig. 11.1(a), but it provides eight byte-wide input ports for an 80386DX-based microcomputer system employing isolated I/O. Just as in the output circuit, the ports are labeled port 0 through port 7; however, this time the 64 parallel port lines are inputs, I_0 through I_{63}. Notice that eight 74F244 octal buffers are used to implement the ports. The outputs of the buffer are applied to the data bus for input to the MPU. These buffers are equipped with three-state outputs.

When an input bus cycle is in progress, the I/O address output on A_{2L} through A_{15L} selects the port whose data are to be input. Notice that the $\overline{G}_{2A}$ and $\overline{G}_{2B}$ input of the I/O address decoder are again both fixed at logic 0. Therefore, if—during the bus cycle—address bit $A_{15L} = 1$, the address decoder is enabled for operation. Then the code $A_{4L}A_{3L}A_{2L}$ is decoded to produce an active logic level at one of the decoder's outputs. For instance, an input of $A_{4L}A_{3L}A_{2L} = 001$ switches the P_1 output to logic 0. P_1 is gated with $\overline{IORC}$ to produce the $\overline{G}$ enable input for the port 1 buffer. If both $\overline{IORC}$ and P_1 are logic 0, the $\overline{G}$ input for port 1 is switched to logic 0 and the outputs of the 74F244 are enabled. In this case, the logic levels at inputs I_8 through I_{15} are passed onto data bus lines D_0 through D_7, respectively. This byte of data is carried through the byte-wide I/O interface to the appropriate lines of the 80386DX's 32-bit data bus.

EXAMPLE 11.3

What is the I/O address of port 7 in the circuit of Fig. 11.3? Assume that all unused address bits are at logic 0.

Solution

For the I/O address decoder to be enabled, address bit A_{15} must be

$$A_{15} = 1$$

and to select port 7, the address applied to the CBA inputs of the decoder must be

$$A_{4L}A_{3L}A_{2L} = 111$$

Filling the unused bits with 0s gives the address

$$A_{15L} \ldots A_{1L}A_{0L} = 1000000000011100_2$$
$$= 801C_{16}$$

EXAMPLE 11.4

For the circuit of Fig. 11.3, write an instruction sequence that will input the byte contents of input port 7 to the memory location DATA_7.

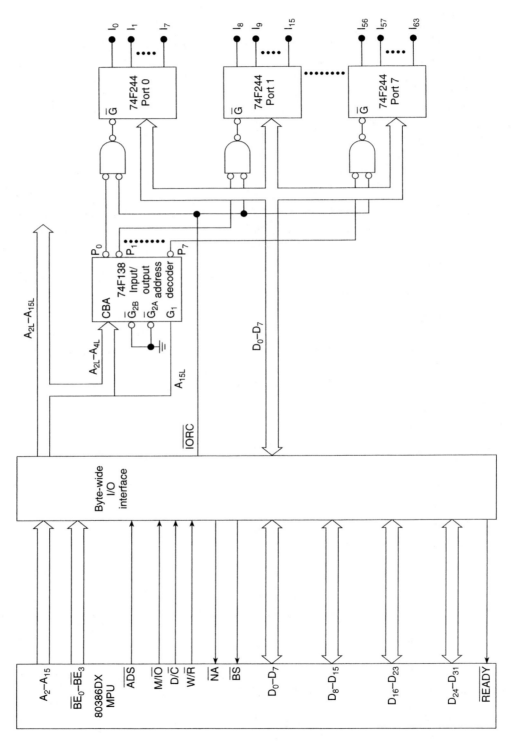

Figure 11.3 Sixty-four-line parallel input circuit for an 80386DX-based microcomputer.

Solution

In Example 11.3 we found that the address of port 7 is $801C_{16}$. This address is loaded into the DX register with the instruction

```
MOV   DX,801CH
```

Now the contents of this port are input to the AL register by executing the instruction

```
IN   AL,DX
```

Finally, the byte of data is copied to memory location DATA_7 with the instruction

```
MOV   DATA_7,AL
```

In practical applications, it is sometimes necessary within an I/O service routine to repeatedly read the value at an input line and test this value for a specific logic level. For instance, input I_3 at port 0 in Fig. 11.3 can be checked to determine if it is at the 1 logic level. Normally, the I/O routine does not continue until the input under test switches to the appropriate logic level. This mode of operation is known as *polling* an input. The polling technique can be used to synchronize the execution of an I/O routine to an event in external hardware.

Let us now look at how a polling software routine is written. The first step in the polling operation is to read the contents of the input port. For instance, the instructions needed to read the contents of port 0 in the circuit of Fig. 11.3 are

```
          MOV   DX,8000H
POLL_I3:  IN    AL,DX
```

A label has been added to identify the beginning of the polling routine. After executing these instructions, the byte contents of port 0 are held in the AL register. Let us assume that input I_3 at this port is the line that is being polled. Therefore, all other bits in AL are masked off with the instruction

```
AND AL,08H
```

After this instruction is executed, the contents of AL will be either 00H or 08H. Moreover the zero flag is 1 if AL contains 00H or else it is 0. The state of the zero flag can be tested with a jump-on-zero instruction

```
JZ POLL_I3
```

If zero flag is 1, a jump is initiated to POLL_I3, and the sequence repeats. On the other hand, if it is 0, the jump is not made; instead, the instruction following the jump instruction is executed. That is, the polling loop repeats until input I_3 is tested and found to be logic 1.

Polling the Setting of a Switch

Figure 11.4 shows a switch connected to input 7 of an input port similar to that shown as port 0 of Fig. 11.3. Notice that when the switch is open input I_7 is pulled to +5 V (logic

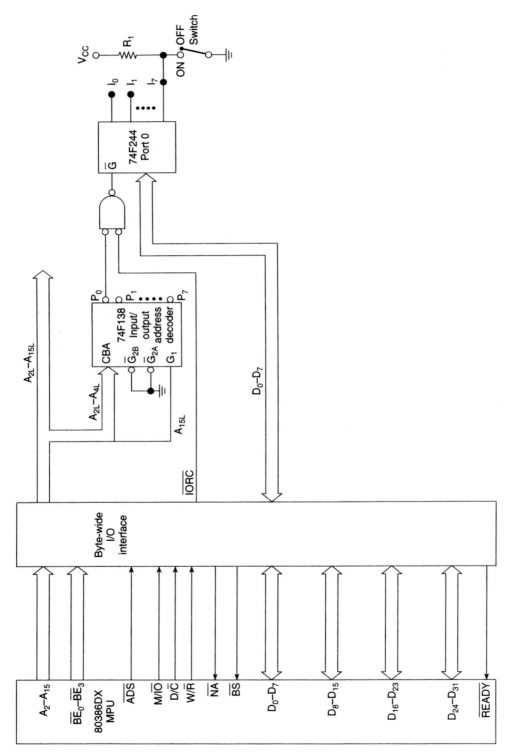

Figure 11.4 Reading the setting of a switch.

1) through pull-up resistor R_1. When the switch is closed, I_7 is connected to ground (logic 0). It is a common practice to poll a switch like this with software waiting for it to close. The instruction sequence that follows will poll the switch at I_7.

```
                      MOV   DX, 8000H
          POLL_I7:    IN    AL,DX
                      SHL   AL,1
                      JC    POLL_I7
          CONTINUE:
```

First, DX is loaded with the address of port 0. Then the contents of port 0 are input to the AL register. Since the logic level at I_7 is in bit 7 of the byte of data in AL, a shift left by one bit position will put this logic level into CF. Now a jump-on-carry instruction is executed to test CF. If CF is 1, the switch is not yet closed. In this case, control is returned to the IN instruction and the poll sequence repeats. On the other hand, if the switch is closed, bit 7 in AL is 0 and this value is shifted into CF. When the JC instruction detects this condition, the polling operation is complete, and the instruction following JC is executed.

▲ 11.5 INPUT/OUTPUT HANDSHAKING AND A PARALLEL PRINTER INTERFACE

In some applications, the microcomputer must synchronize the input or output of information to a peripheral device. Two examples of interfaces that may require a synchronized data transfer are a serial communications interface and a parallel printer interface. Sometimes it is necessary as part of the I/O synchronization process first to poll an input from an I/O device and, after receiving the appropriate level at the poll input, to acknowledge this fact to the device with an output. This type of synchronization is achieved by implementing what is known as *handshaking* as part of the input/output interface.

Figure 11.5(a) shows a conceptual view of the interface between the printer and a *parallel printer port*. There are three general types of signals at the *printer interface*: data, control, and status. The data lines are the parallel paths used to transfer data to the printer. Transfers of data over this bus are synchronized with an appropriate sequence of control signals. However, data transfers can take place only if the printer is ready to accept data. Printer readiness is indicated through the parallel interface by a set of signals called status lines. This interface handshake sequence is summarized by the flowchart of Fig. 11.5(b).

The printer is attached to the microcomputer system at a connector known as the *parallel printer port*. On the IBM PC, a 25-pin connector is used to attach the printer. The actual signals supplied at the pins of this connector are shown in Fig. 11.5(c). Notice that there are five status signals available at the interface, and they are called Ack, Busy, Paper Empty, Select, and Error. In a particular implementation only some of these signals may be used. For instance, to send a character to the printer, the software may test only the Busy signal. If it is inactive, it may be a sufficient indication to proceed with the transfer.

A detailed block diagram of a simple parallel printer interface is shown in Fig. 11.6(a). Here we find eight data-output lines, D_0 through D_7, control signal strobe ($\overline{STB}$), and status signal busy (BUSY). The MPU outputs data representing the character to be printed through the parallel printer interface. Character data are latched at the outputs of the parallel interface and are carried to the data inputs of the printer over data lines D_0 through D_7. The $\overline{STB}$ output of the parallel printer interface is used to signal the printer that new character data are available. Whenever the printer is already busy printing a character, it

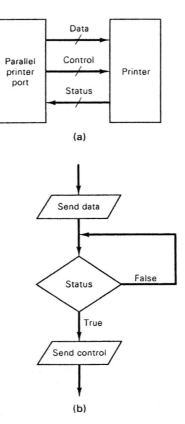

(a)

Send data

Status — False

True

Send control

(b)

Figure 11.5 (a) Parallel printer interface. (b) Flowchart showing the data transfer in a parallel printer interface.

signals this fact to the MPU with the BUSY input of the parallel printer interface. This handshake signal sequence is illustrated in Fig. 11.6(b).

Let us now look at the sequence of events that takes place at the parallel printer interface when data are output to the printer. Figure 11.6(c) is a flowchart of a subroutine that performs a parallel printer interface character-transfer operation. First the BUSY input of the parallel printer interface is tested. Notice that this is done with a polling operation. That is, the MPU tests the logic level of BUSY repeatedly until it is found to be at the not-busy logic level. *Busy* means that the printer is currently printing a character. On the other hand, *not busy* signals that the printer is ready to receive another character for printing. After finding a not-busy condition, a count of the number of characters in the printer buffer (microprocessor memory) is read; a byte of character data is read from the printer buffer; the character is output to the parallel interface; and then a pulse is produced at $\overline{STB}$. This pulse tells the printer to read the character off the data bus lines. The printer is again printing a character and signals this fact at BUSY. The handshake sequence is now complete. Now the count that represents the number of characters in the buffer is decremented and checked to see if the buffer is empty. If empty, the print operation is complete. Otherwise, the character transfer sequence is repeated for the next character.

The circuit in Fig. 11.6(d) implements the parallel printer interface of Fig. 11.6(a).

EXAMPLE 11.5 _____

What are the addresses of the ports that provide the data lines, strobe output, and busy input in the circuit of Fig. 11.6(d)? Assume that all unused address bits are 0.

Pin	Assignment
1	Strobe
2	Data 0
3	Data 1
4	Data 2
5	Data 3
6	Data 4
7	Data 5
8	Data 6
9	Data 7
10	Ack
11	Busy
12	Paper empty
13	Select
14	Auto foxt
15	Error
16	Initialize
17	Slctin
18	Ground
19	Ground
20	Ground
21	Ground
22	Ground
23	Ground
24	Ground
25	Ground

```
Data:          Data0, Data1, ........., Data7

Control:       Strobe
               Auto Foxt
               Initialize
               Slctin

Status:        Ack
               Busy
               Paper Empty
               Select
               Error
```

(c)

Figure 11.5(c) Parallel printer port pin assignments and types of interface signals.

Solution

The I/O addresses that enable port 0 for the data lines, port 1 for the strobe output, and
port 2 for the busy input are found as follows:

$$\text{Address of port 0} = 1000000000000000_2 = 8000_{16}$$

$$\text{Address of port 1} = 1000000000000010_2 = 8004_{16}$$

$$\text{Address of port 2} = 1000000000000100_2 = 8008_{16}$$

EXAMPLE 11.6 ⎯⎯⎯⎯⎯⎯⎯⎯⎯⎯⎯⎯⎯⎯⎯⎯⎯⎯⎯⎯⎯⎯⎯⎯⎯⎯⎯⎯⎯⎯

Write a program that will implement the sequence in Fig. 11.6(c) for the circuit in Fig.
11.6(d). Character data are held in memory starting at address PRNT_BUFF, and the

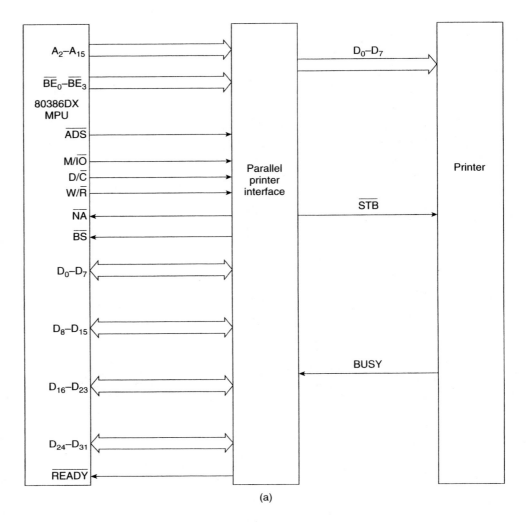

(a)

Figure 11.6(a) I/O interface that employs handshaking.

number of characters held in the buffer is identified by the count at address CHAR_COUNT. Use the port addresses from Example 11.5.

Solution

First the character counter and the character pointer are set up with the instructions

```
MOV  CL,[CHAR_COUNT]
MOV  SI,PRNT_BUFF
```

Next the BUSY input is checked with the instructions

```
POLL_BUSY:  MOV  DX,8008H
            IN   AL,DX
            AND  AL,01H
            JNZ  POLL_BUSY
```

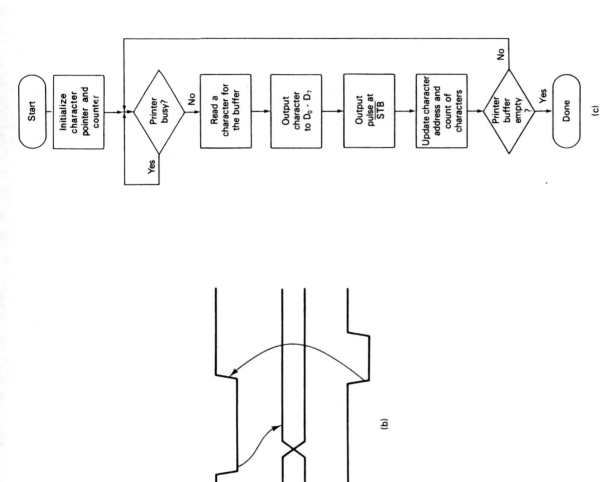

Figure 11.6 (b) Handshake signals. (c) Handshake sequence flow chart.

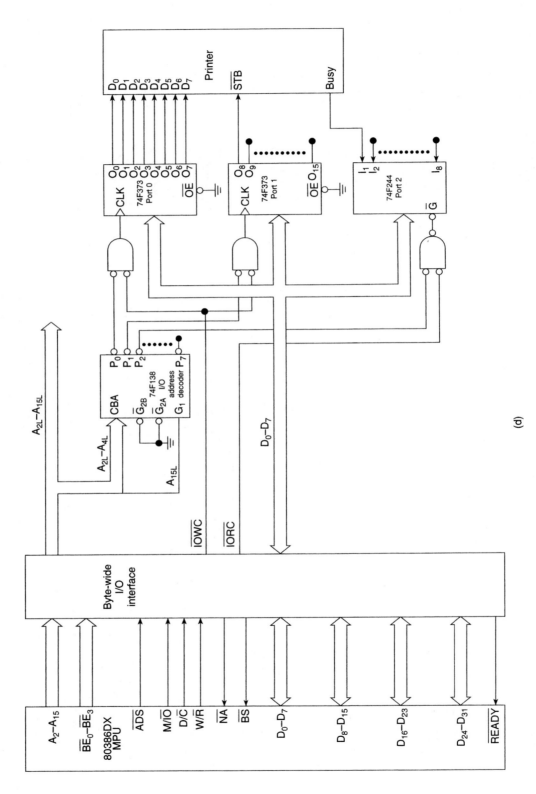

Figure 11.6(d) Handshaking printer interface circuit.

The character is copied into AL, and then it is output to port 0.

```
MOV  AL,[SI]
MOV  DX,8000H
OUT  DX,AL
```

Now a strobe pulse is generated at port 1 with the instructions

```
        MOV   AL,00H    ;STB = 0
        MOV   DX,8004H
        OUT   DX,AL
        MOV   BX,0FH    ;Delay for STB duration
STROBE: DEC   BX
        JNZ   STROBE
        MOV   AL,01H    ;STB = 1
        OUT   DX,AL
```

At this point, the value of PRNT_BUFF must be incremented, and the value of CHAR_COUNT must be decremented:

```
INC  SI
DEC  CL
```

Finally, a check is made to see if the printer buffer is empty. If it is not empty, we need to repeat the prior instruction sequence. To do this, we execute the instruction

```
       JNZ   POLL_BUSY
DONE:  -
```

The program comes to the DONE label after all characters are transferred to the printer.

▲ 11.6 82C55A PROGRAMMABLE PERIPHERAL INTERFACE (PPI)

The 82C55A is an LSI peripheral designed to permit easy implementation of *parallel I/O* in the 80386DX microcomputer systems. It provides a flexible parallel interface, which includes features such as single-bit, 4-bit, and byte-wide input and output ports; level-sensitive inputs; latched outputs; strobed inputs or outputs; and strobed bidirectional input/outputs. These features are selected under software control.

A block diagram of the 82C55A is shown in Fig. 11.7(a) and its pin layout appears in Fig. 11.7(b). The left side of the block represents the *microprocessor's interface*. It includes an *8-bit bidirectional data bus* D_0 through D_7. Over these lines, commands, status information, and data are transferred between the MPU and 82C55A. These data are transferred whenever the MPU performs an input or output bus cycle to an address of a register within the device. Timing of the data transfers to the 82C55A is controlled by the *read/write control* ($\overline{RD}$ and $\overline{WR}$) signals.

The source or destination register within the 82C55A is selected by a 2-bit *register-select code*. The MPU must apply this code to the *register-select inputs* A_0 and A_1 of the 82C55A. The *port A, port B,* and *port C registers* correspond to codes $A_1A_0 = 00$, $A_1A_0 = 01$, and $A_1A_0 = 10$, respectively.

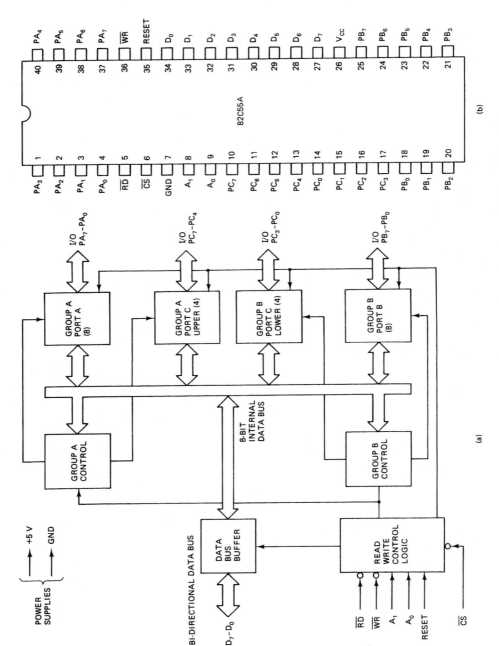

Figure 11.7 (a) Block diagram of the 82C55A. (Reprinted by permission of Intel Corp. Copyright/Intel Corp. 1980) (b) Pin layout. (Reprinted by permission of Intel Corp. Copyright/Intel Corp. 1980)

Two other signals are shown on the microprocessor interface side of the block diagram. They are the *reset* (RESET) and *chip-select* ($\overline{CS}$) inputs. $\overline{CS}$ must be logic 0 during all read or write operations to the 82C55A. It enables the 82C55A's microprocessor interface circuitry for an input or output operation.

On the other hand, RESET is used to initialize the device. Switching it to logic 0 at power-up causes the internal registers of the 82C55A to be cleared. *Initialization* configures all I/O ports for input mode of operation.

The other side of the block corresponds to three *byte-wide I/O ports*. They are called *port A, port B*, and *port C* and represent *I/O lines* PA_0 through PA_7, PB_0 through PB_7, and PC_0 through PC_7, respectively. These ports can be configured for input or output operation. This gives us a total of 24 I/O lines.

We already mentioned that the operating characteristics of the 82C55A can be configured under software control. It contains an 8-bit internal control register for this purpose. This register is represented by the *group A* and *group B control blocks* in Fig. 11.7(a). Logic 0 or 1 can be written to the bit positions in this register to configure the individual ports for input or output operation and to enable one of its three modes of operation. The control register is write only and its contents can be modified using microprocessor instructions. A write bus cycle to the 82C55A with register-select code $A_1A_0 = 11$ and an appropriate control word is used to modify the control registers.

The circuit in Fig. 11.8 is an example of how the 82C55A can be interfaced to a

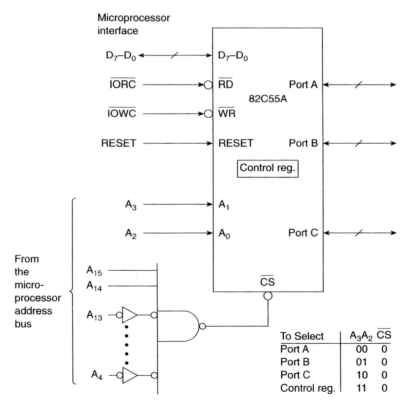

Figure 11.8 Addressing an 82C55A using the microprocessor interface signals.

microprocessor. Here we see that address lines A_0 and A_1 of the microprocessor drive the 82C55A's register-select inputs A_1 and A_0, respectively. The $\overline{CS}$ input of the 82C55A is supplied from the output of the address decoder circuit whose inputs are address lines A_4 through A_{15}. To access either a port or the control register of the 82C55A, $\overline{CS}$ must be active. Then the code A_3A_2 selects the port or control register to be accessed. The select codes are shown in Fig. 11.8.

For instance, to access port A, $A_3A_2 = 00$, $A_{15} = A_{14} = 1$, and $A_{13} = A_{12} = \cdots = A_4 = 0$, which gives the port A address as

$$1100 \cdots 00XX_2 = C000_{16}$$

Similarly, it can be determined that the address of port B equals $C004_{16}$, that of port C is $C008_{16}$, and the address of the control registers is $C00C_{16}$.

The bits of the control register and their control functions are shown in Fig. 11.9. Here we see that bits D_0 through D_2 correspond to the group B control block in the diagram of Fig. 11.7(a). Bit D_0 configures the lower four lines of port C for input or output operation. Notice that logic 1 at D_0 selects input operation, and logic 0 selects output operation. The next bit, D_1, configures port B as an 8-bit-wide input or output port. Again, logic 1 selects input operation and logic 0 selects output operation.

The D_2 bit is the mode-select bit for port B and the lower 4 bits of port C. It permits selection of one of two different modes of operation called *mode 0* and *mode 1*. Logic 0 in bit D_2 selects mode 0, whereas logic 1 selects mode 1. These modes are discussed in detail shortly.

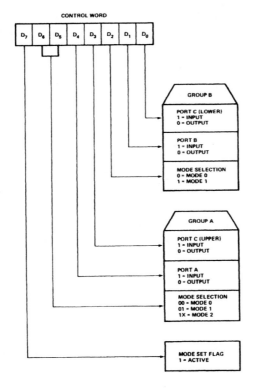

Figure 11.9 Control-word bit functions. (Reprinted by permission of Intel Corp. Copyright/Intel Corp. 1980)

The next 4 bits in the control register, D_3 through D_6, correspond to the group A control block in Fig. 11.7(a). Bits D_3 and D_4 of the control register are used to configure the operation of the upper half of port C and all of port A, respectively. These bits work in the same way as D_0 and D_1 to configure the lower half of port C and port B. However, there are now two-mode select bits, D_5 and D_6, instead of just one. They are used to select between three modes of operation, known as *mode 0, mode 1,* and *mode 2*.

The last control register bit, D_7, is the *mode-set flag*. It must be at logic 1 (active) whenever the mode of operation is to be changed.

Mode 0 selects what is called *simple I/O operation*. By simple I/O, we mean that the lines of the port can be configured as level-sensitive inputs or latched outputs. To set all ports for this mode of operation, load bit D_7 of the control register with logic 1, bits $D_6 D_5 = 00$, and $D_2 = 0$. Logic 1 at D_7 represents an active mode set flag. Now port A and port B can be configured as 8-bit input or output ports, and port C can be configured for operation as two independent 4-bit input or output ports. This is done by setting or resetting bits D_4, D_3, D_1, and D_0. Figure 11.10 summarizes the port pins and the functions they can perform in mode 0.

For example, if $80_{16} = 10000000_2$ is written to the control register, the 1 in D_7 activates the mode-set flag. Mode 0 operation is selected for all three ports because bits D_6, D_5, and D_2 are logic 0. At the same time, the zeros in D_4, D_3, D_1, and D_0 set up all port lines to work as outputs. This configuration is illustrated in Fig. 11.11(a).

By writing different binary combinations into bit locations D_4, D_3, D_1, and D_0, any one of 16 different mode 0 I/O configurations can be obtained. The control word and I/O setup for the rest of these combinations are shown in Fig. 11.11(b) through (p).

Pin	MODE 0	
	IN	OUT
PA_0	IN	OUT
PA_1	IN	OUT
PA_2	IN	OUT
PA_3	IN	OUT
PA_4	IN	OUT
PA_5	IN	OUT
PA_6	IN	OUT
PA_7	IN	OUT
PB_0	IN	OUT
PB_1	IN	OUT
PB_2	IN	OUT
PB_3	IN	OUT
PB_4	IN	OUT
PB_5	IN	OUT
PB_6	IN	OUT
PB_7	IN	OUT
PC_0	IN	OUT
PC_1	IN	OUT
PC_2	IN	OUT
PC_3	IN	OUT
PC_4	IN	OUT
PC_5	IN	OUT
PC_6	IN	OUT
PC_7	IN	OUT

Figure 11.10 Mode 0 port pin functions.

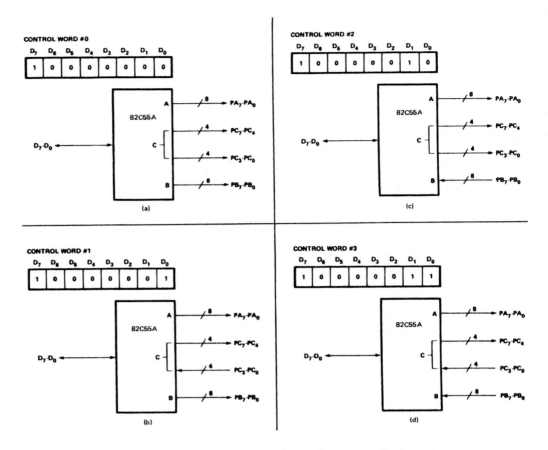

Figure 11.11(a–p) Mode 0 control words and corresponding input/output configurations. (Reprinted by permission of Intel Corp. Copyright/Intel Corp. 1980)

EXAMPLE 11.7

What are the mode and I/O configuration for ports A, B, and C of an 82C55A after its control register is loaded with 82_{16}?

Solution

Expressing the control register contents in binary form, we get

$$D_7D_6D_5D_4D_3D_2D_1D_0 = 10000010_2$$

Since D_7 is 1, the modes of operation of the ports are selected by the control word. The 3 least significant bits of the word configure port B and the lower 4 bits of port C. They give

$D_0 = 0$ Lower 4 bits of port C are outputs.
$D_1 = 1$ Port B are inputs.
$D_2 = 0$ Mode 0 operation for both port B and the lower four bits of port C.

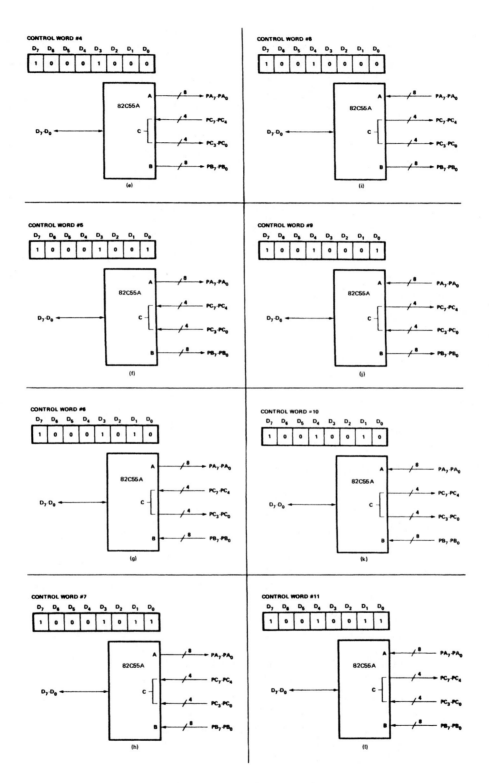

Figure 11.11 (Continued)

535

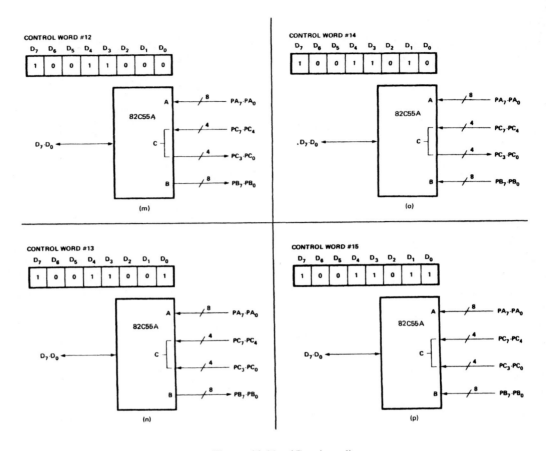

Figure 11.11 (Continued)

The next 4 bits configure the upper part of port C and port A.

$$D_3 = 0 \quad \text{Upper 4 bits of port C are outputs.}$$
$$D_4 = 0 \quad \text{Port A are outputs.}$$
$$D_6D_5 = 00 \quad \text{Mode 0 operation for both port A and the upper part of port C.}$$

This mode 0 I/O configuration is shown in Fig. 11.11(c).

Mode 1 operation represents what is known as *strobed I/O*. The ports of the 82C55A are put into this mode of operation by setting $D_7 = 1$ to activate the mode-set flag and setting $D_6D_5 = 01$ and $D_2 = 1$.

In this way, the A and B ports are configured as two independent *byte-wide I/O ports*, each of which has a *4-bit control/data port* associated with it. The control/data ports are formed from the lower and upper nibbles of port C, respectively. Figure 11.12 lists the mode 1 functions of each pin at ports A, B, and C.

When configured in this way, data applied to an input port must be strobed in with a signal produced in external hardware. An output port in mode 1 is provided with handshake

	MODE 1	
Pin	**IN**	**OUT**
PA_0	IN	OUT
PA_1	IN	OUT
PA_2	IN	OUT
PA_3	IN	OUT
PA_4	IN	OUT
PA_5	IN	OUT
PA_6	IN	OUT
PA_7	IN	OUT
PB_0	IN	OUT
PB_1	IN	OUT
PB_2	IN	OUT
PB_3	IN	OUT
PB_4	IN	OUT
PB_5	IN	OUT
PB_6	IN	OUT
PB_7	IN	OUT
PC_0	$INTR_B$	$INTR_B$
PC_1	IBF_B	$\overline{OBF_B}$
PC_2	$\overline{STB_B}$	$\overline{ACK_B}$
PC_3	$INTR_A$	$INTR_A$
PC_4	$\overline{STB_A}$	I/O
PC_5	IBF_A	I/O
PC_6	I/O	$\overline{ACK_A}$
PC_7	I/O	$\overline{OBF_A}$

Figure 11.12 Mode 1 port pin functions.

signals that indicate when new data are available at its outputs and when an external device has read these values.

As an example, let us assume for the moment that the control register of an 82C55A is loaded with $D_7D_6D_5D_4D_3D_2D_1D_0 = 10111XXX$. This configures port A as a mode 1 input port. Figure 11.13(a) shows the function of the signal lines for this example. Notice that PA_7 through PA_0 form an 8-bit input port. On the other hand, the function of the upper port C leads are reconfigured to provide the port A control/data lines. The PC_4 line becomes

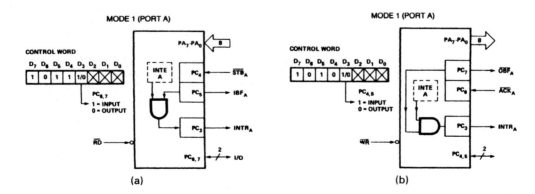

Figure 11.13 (a) Mode 1, port A input configuration. (Reprinted by permission of Intel Corp. Copyright/Intel Corp. 1980) (b) Mode 1, port A output configuration. (Reprinted by permission of Intel Corp. Copyright/Intel Corp. 1980)

strobe input ($\overline{\text{STB}}_A$), which is used to strobe data at PA_7 through PA_0 into the input latch. Moreover, PC_5 becomes *input buffer full* (IBF_A). Logic 1 at this output indicates to external circuitry that a word has already been strobed into the latch.

The third control signal is at PC_3 and is labeled *interrupt request* ($INTR_A$). It switches to logic 1 when $\overline{\text{STB}}_A = 1$ making $IBF_A = 1$, and an internal signal *interrupt enable* ($INTE_A$) = 1. $INTE_A$ is set to logic 0 or 1 under software control by using the bit set/reset feature of the 82C55A. This feature is discussed later. Looking at Fig. 11.13(a), we see that logic 1 in $INTE_A$ enables the logic level of IBF_A to the $INTR_A$ output. This signal can be applied to an interrupt input of the MPU to signal it that new data are available at the input port. The corresponding interrupt-service routine reads the data, which clears $INTR_A$ and IBF_A. The timing diagram in Fig. 11.14(a) summarizes these events for an input port configured in mode 1.

As another example, let us assume that the contents of the control register are changed to $D_7D_6D_5D_4D_3D_2D_1D_0 = 10100XXX$. This I/O configuration is shown in Fig. 11.13(b). Notice that port A is now configured for output operation instead of input operation. PA_7 through PA_0 make up the 8-bit output port. The control line at PC_7 is *output buffer full*

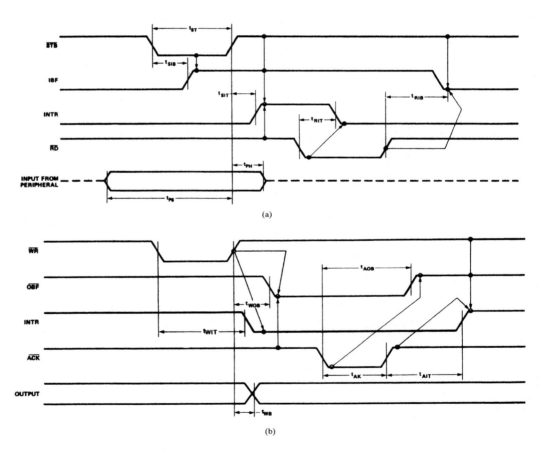

(a)

(b)

Figure 11.14   (a) Timing diagram for an input port in mode 1 configuration. (Reprinted by permission of Intel Corp. Copyright/Intel Corp. 1980) (b) Timing diagram for an output port in mode 1 configuration. (Reprinted by permission of Intel Corp. Copyright/Intel Corp. 1980)

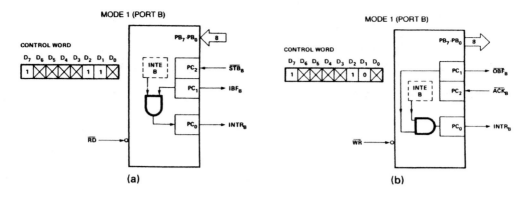

Figure 11.15 (a) Mode 1, port B input configuration. (Reprinted by permission of Intel Corp. Copyright/Intel Corp. 1980) (b) Mode 1, port B output configuration. (Reprinted by permission of Intel Corp. Copyright/Intel Corp. 1980)

($\overline{\text{OBF}}_A$). When data have been written into the output port, $\overline{\text{OBF}}_A$ switches to the 0 logic level. In this way, it signals external circuitry that new data are available at the port outputs.

Signal line PC_6 becomes *acknowledge* ($\overline{\text{ACK}}_A$), which is an input. An external device reads the data and signals the 82C55A that it has accepted the data provided at the output port by switching $\overline{\text{ACK}}_A$ to logic 0. When the $\overline{\text{ACK}}_A = 0$ is received by the 82C55A, it in turn deactivates the $\overline{\text{OBF}}_A$ output. The last signal at the control port is the *interrupt request* (INTR_A) output, which is produced at the PC_3 lead. This output is switched to logic 1 when the $\overline{\text{ACK}}_A$ input becomes inactive. It is used to signal the MPU with an interrupt that indicates that an external device has accepted the data from the outputs. To produce the INTR_A, INTE_A must equal 1. Again the interrupt enable (INTE_A) bit must be set to 1 using the bit set/reset feature. The timing diagram in Fig. 11.14(b) summarizes these events for an output port configured in mode 1.

EXAMPLE 11.8

Figures 11.15(a) and (b) show how port B can be configured for mode 1 operation. Describe what happens in Fig. 11.15(a) when the $\overline{\text{STB}}_B$ input is pulsed to logic 0. Assume that INTE_B is already set to 1.

Solution

As $\overline{\text{STB}}_B$ is pulsed, the byte of data at PB_7 through PB_0 is latched into the port B register. This causes the IBF_B output to switch to 1. Since INTE_B is 1, INTR_B switches to logic 1.

The last mode of operation, mode 2, represents what is known as *strobed bidirectional I/O*. The key difference is that now the port works as either inputs or outputs, and control signals are provided for both functions. Only port A can be configured to work in this way. The I/O port and control signal pins are shown in Fig. 11.16.

To set up this mode, the control register is set to $D_7D_6D_5D_4D_3D_2D_1D_0 = 11\text{XXXXXX}$. The I/O configuration that results is shown in Fig. 11.17. Here we find that PA_7 through PA_0 operate as an *8-bit bidirectional port* instead of a unidirectional port. Its control signals

| | MODE 2 |
Pin	GROUP A ONLY
PA_0	↔
PA_1	↔
PA_2	↔
PA_3	↔
PA_4	↔
PA_5	↔
PA_6	↔
PA_7	↔
PB_0	—
PB_1	—
PB_2	—
PB_3	—
PB_4	—
PB_5	—
PB_6	—
PB_7	—
PC_0	I/O or $\overline{INTR_B}$
PC_1	I/O or $\overline{ACK_B}$ or IBF_B
PC_2	I/O or $\overline{OBF_B}$ or $\overline{STB_A}$
PC_3	$INTR_A$
PC_4	$\overline{STB_A}$
PC_5	IBF_A
PC_6	$\overline{ACK_A}$
PC_7	$\overline{OBF_A}$

MODE 0 OR MODE 1 ONLY (applies to PB_0–PB_7)

Figure 11.16 Mode 2 port pin functions.

are $\overline{OBF_A}$ at PC_7, $\overline{ACK_A}$ at PC_6, $\overline{STB_A}$ at PC_4, IBF_A at PC_5, and $INTR_A$ at PC_3. Their functions are similar to those already discussed for mode 1. One difference is that $INTR_A$ is produced by either gating $\overline{OBF_A}$ with $INTE_1$ or IBF_A with $INTE_2$.

In our discussion of mode 1, we mentioned that the *bit set/reset* feature could be used to set the INTE bit to logic 0 or 1. This feature also allows the individual bits of port C to be set or reset. To do this, we write logic 0 to bit D_7 of the control register. This resets the bit set/reset flag. The logic level that is to be latched at a port C line is included as bit D_0 of the control word. This value is latched at the I/O line of port C, which corresponds to the 3-bit code at $D_3D_2D_1$.

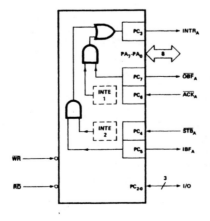

Figure 11.17 Mode 2 input/output configuration. (Reprinted by permission of Intel Corp. Copyright/Intel Corp. 1980)

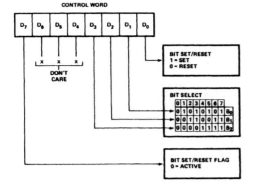

Figure 11.18 Bit set/reset format. (Reprinted by permission of Intel Corp. Copyright/Intel Corp. 1980)

The relationship between the set/reset control word and input/output lines is illustrated in Fig. 11.18. For instance, writing $D_7D_6D_5D_4D_3D_2D_1D_0 = 00001111_2$ into the control register of the 82C55A selects bit 7 and sets it to 1. Therefore, output PC_7 at port C is switched to the 1 logic level.

EXAMPLE 11.9

The interrupt-control flag $INTE_A$ for output port A in mode 1 is controlled by PC_6. Using the set/reset feature of the 82C55A, what command code must be written to the control register of the 82C55A to set it to enable the control flag?

Solution

To use the set/reset feature, D_7 must be logic 0. Moreover, $INTE_A$ is to be set; therefore, D_0 must be logic 1. Finally, to select PC_6, the code at bits $D_3D_2D_1$ must be 110. The rest of the bits are don't-care states. This gives us the control word

$$D_7D_6D_5D_4D_3D_2D_1D_0 = 0XXX1101_2$$

Replacing the don't-care states with the 0 logic level, we get

$$D_7D_6D_5D_4D_3D_2D_1D_0 = 00001101_2 = 0D_{16}$$

We have just described and given examples of each of the modes of operation that can be assigned to the ports of the 82C55A. It is also possible to configure the A and B ports with different modes. For example, Fig. 11.19(a) shows the control word and port configuration of an 82C55A set up for bidirectional mode 2 operation of port A and input mode 0 operation of port B. It should also be noted that in all modes, the unused leads of port C can still be used as general purpose inputs or outputs.

EXAMPLE 11.10

What control word must be written into the control register of the 82C55A such that port A is configured for bidirectional operation and port B is set up with mode 1 outputs?

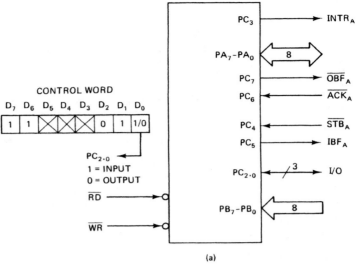

(a)

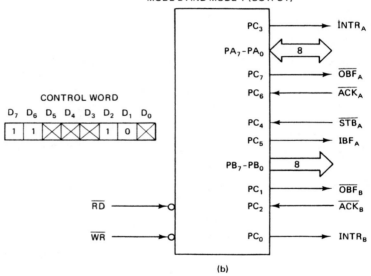

(b)

Figure 11.19 (a) Combined mode 2 and mode 0 (input) control-word and I/O configuration. (Reprinted by permission of Intel Corp. Copyright/Intel Corp. 1980) (b) Combined mode 2 and mode 1 (output) control-word and I/O configuration. (Reprinted by permission of Intel Corp. Copyright/Intel Corp. 1980)

Solution

To configure the operating mode of the ports of the 82C55A, D_7 must be 1:

$$D_7 = 1$$

Port A is set up for bidirectional operation by making D_6 logic 1. In this case, D_5 through D_3 are don't-care states.

$$D_6 = 1$$

$$D_5 D_4 D_3 = XXX$$

Mode 1 is selected for port B by logic 1 in bit D_2 and output operation by logic 0 in D_1. Since mode 1 operation has been selected, D_0 is a don't-care state.

$$D_2 = 1$$

$$D_1 = 0$$

$$D_0 = X$$

This gives the control word

$$D_7 D_6 D_5 D_4 D_3 D_2 D_1 D_0 = 11XXX10X_2$$

Assuming logic 0 for the don't-care states, we get

$$D_7 D_6 D_5 D_4 D_3 D_2 D_1 D_0 = 11000100_2 = C4_{16}$$

This configuration is shown in Figure 11.19(b).

EXAMPLE 11.11

Write the sequence of instructions needed to load the control register of an 82C55A with the control word formed in Example 11.10. Assume that the 82C55A resides at address $0F_{16}$ of the I/O address space and that the microcomputer uses the byte-wide I/O interface illustrated in Fig. 9.51.

Solution

First we must load AL with $C4_{16}$. This is the value of the control word that is to be written to the control register at address $0F_{16}$. The move instruction used to load AL is

```
MOV   AL,0C4H
```

These data are output to the control register with the OUT instruction

```
OUT   0FH,AL
```

In this case we have used direct I/O. This is because the I/O address of the control register is less than FF_{16}.

Input configuration

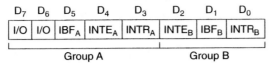

D$_7$	D$_6$	D$_5$	D$_4$	D$_3$	D$_2$	D$_1$	D$_0$
I/O	I/O	IBF$_A$	INTE$_A$	INTR$_A$	INTE$_B$	IBF$_B$	INTR$_B$

Group A   Group B

Output configurations

D$_7$	D$_6$	D$_5$	D$_4$	D$_3$	D$_2$	D$_1$	D$_0$
$\overline{OBF}_A$	INTE$_A$	I/O	I/O	INTR$_A$	INTE$_B$	$\overline{OBF}_B$	INTR$_B$

Group A   Group B

(a)

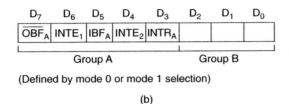

D$_7$	D$_6$	D$_5$	D$_4$	D$_3$	D$_2$	D$_1$	D$_0$
$\overline{OBF}_A$	INTE$_1$	IBF$_A$	INTE$_2$	INTR$_A$			

Group A   Group B

(Defined by mode 0 or mode 1 selection)

(b)

Figure 11.20 (a) Mode 1 status information for port C. (Reprinted by permission of Intel Corp. Copyright/Intel Corp. 1980) (b) Mode 2 status information for port C. (Reprinted by permission of Intel Corp. Copyright/Intel Corp. 1980)

In our descriptions of mode 1 and mode 2 operations, we found that when the 82C55A is configured for either of these modes, most of the pins of port C perform I/O control functions. For instance, Fig. 11.12 shows that in mode 1 PC$_3$ works as the INTR$_A$ output. Control information can be read from port C through software. This is known as reading the status of port C. The format of the status information input by reading port C of an 82C55A operating in mode 1 is shown in Fig. 11.20(a). Notice that if the ports are configured for input operation, the status byte contains the values of the IBF and INTR outputs and the INTE flag for both ports. Once read by the MPU, these bits can be tested with other software to control the flow of the program. On the other hand, if the ports are configured as outputs, the status byte contains the values of outputs $\overline{OBF}$ and INTR and the INTE flag for each port. Figure 11.20(b) shows the status byte format for an 82C55A configured for mode 2 operation.

▲ 11.7 IMPLEMENTING ISOLATED I/O PARALLEL INPUT/OUTPUT PORTS USING THE 82C55A

In Sections 11.3 and 11.4, we showed how parallel input and output ports can be implemented for the 80386DX microcomputer systems using logic devices such as the 74F244 octal buffer and 74F373 octal latch, respectively. Even though logic ICs can be used to implement parallel input and output ports, the 82C55A PPI, which was introduced in Section 11.6, can be used to design a more versatile I/O interface. This is because its ports can be configured either as inputs or outputs under software control. Here we show how the 82C55A is used to design isolated parallel I/O interfaces for 80386DX-based microcomputers.

The circuit in Fig. 11.21(a) shows how 82C55A devices can be connected to the bus of the 80386DX to implement isolated parallel input/output ports. Here we find that four

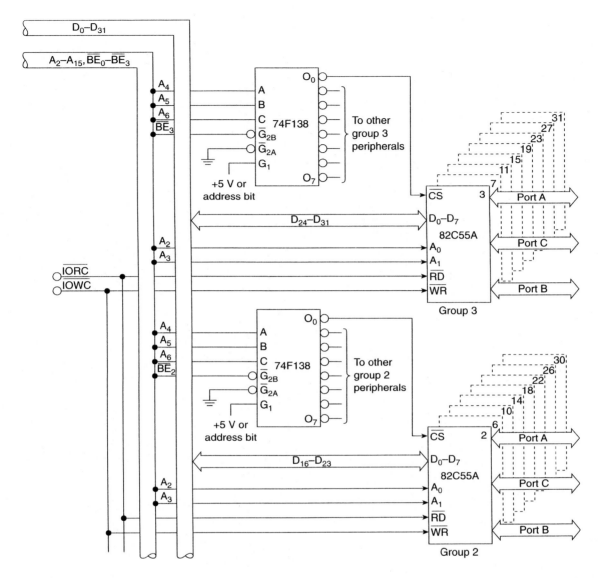

Figure 11.21(a) 32-bit-wide 82C55A isolated parallel I/O ports for an 80386DX-based microcomputer.

groups, each with up to eight 82C55As, are connected to the data bus. Notice that each group has its own address decoder. The output of the decoder selects one of the devices in the group at a time for input or output of data. Each of these PPI devices provides up to three byte-wide ports. In the circuit, they are labeled port A, port B, and port C. These ports can be individually configured as inputs or outputs through software. Therefore, this circuit is capable of implementing up to 192 I/O lines, for a total of 768 I/O lines.

Let us look more closely at the group 3 decoder circuit. Starting with the inputs of the 74F138 address decoder, we see that its enable inputs are $\overline{G}_{2B} = \overline{BE}_3$, $\overline{G}_{2A} = 0$, and

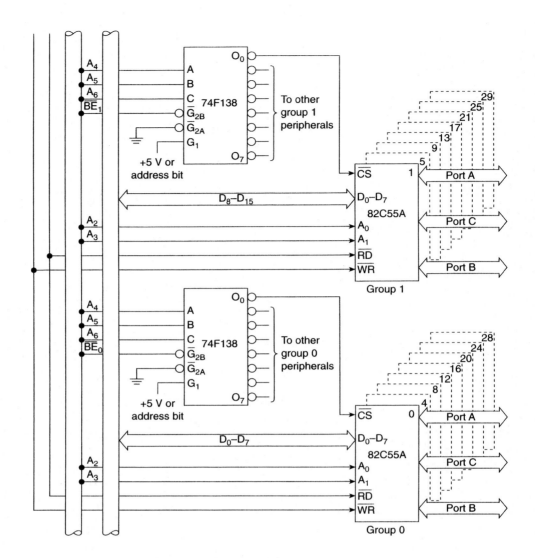

Figure 11.21(b)

$G_1 = 1$. Whenever the 80386DX outputs an I/O address on the bus with $\overline{BE}_3$ equal 0, the decoder is enabled for operation. Notice that the other three group decoders are connected in a similar way, except that their $\overline{G}_{2B}$ input is enabled by a different $\overline{BE}$ signal. For example, the $\overline{G}_{2B}$ input of the group 1 decoder is driven by $\overline{BE}_1$.

When this 74F138 decoder is enabled, the code at its CBA inputs causes one of the eight 82C55A PPIs attached to its outputs to get enabled for operation. Bits A_6 through A_4 of the I/O address are applied to these inputs of the decoder. It responds by switching the output corresponding to this 3-bit code to the 0 logic level. Decoder outputs O_0 through O_7 are applied to the chip select ($\overline{CS}$) inputs of the PPIs in group 3. For instance, $A_6A_5A_4 = 000$ switches output O_0 to logic 0. This enables the first 82C55A device, which is numbered 3 in Fig. 11.21(b).

At the same time that the PPI 3 is selected, the 2-bit code A_3A_2 at inputs A_1A_0 of the 82C55A selects the port for which data are input or output. For example, $A_3A_2 = 00$ indicates that Port A is to be accessed. The input or output data transfer to this port takes place over data bus lines D_{31} through D_{24}. The timing of these read or write transfers is controlled by signals $\overline{IORC}$ and $\overline{IOWC}$, respectively.

EXAMPLE 11.12

What must be the address bus inputs of the circuit in Fig. 11.21 if port C of PPI 14 is to be accessed?

Solution

To enable PPI 14, the group 2 74F138 must be enabled for operation and its O_3 output switched to logic 0. This requires enable input $\overline{G}_{2B} = 0$ and chip-select code CBA = 011. This in turn requires from the bus that

$$\overline{BE}_2 = 0 \qquad \text{Enables 74F138 for group 2}$$

$$A_6A_5A_4 = 011 \qquad \text{Selects PPI 14}$$

Port C of PPI 14 is selected with $A_1A_0 = 10$, which from the bus requires that

$$A_3A_2 = 10 \qquad \text{Accesses port C}$$

The rest of the address bits are don't-care states.

EXAMPLE 11.13

Assume that in the circuit of Fig. 11.21, PPI 14 is configured such that port A is an output port, ports B and C are both input ports, and all three ports are set up for mode 0 operation. Write a program that will input the data at ports B and C, find the difference C − B, and output this difference to port A.

Solution

From the circuit diagram in Fig. 11.21, we find that the addresses of the three I/O ports of PPI 14 are

$$\text{port A} = 0110010_2 = 32_{16}$$

$$\text{port B} = 0110110_2 = 36_{16}$$

$$\text{port C} = 0111010_2 = 3A_{16}$$

The data at ports B and C can be input with the instruction sequence

```
IN    AL,36H   ;Read port B
MOV   BL,AL    ;Save data from port B
IN    AL,3AH   ;Read port C
```

Now the data from port B are subtracted from the data at port C with the instruction

```
SUB   AL,BL   ;Subtract B from C
```

Finally, the difference is output to port A with the instruction

```
OUT   32H,AL   ;Write to port A
```

Notice in Fig. 11.21 that not all address bits are used in the I/O address decoding. Here only latched address bits A_4, A_5, and A_6 are decoded. Unused address bits are don't-care states.

$$XXXXXXXXXA_6A_5A_4A_3A_2$$

For this reason, many addresses decode to select each of the I/O ports. For instance, if all of the don't-care address bits are made 0, the address of the 82C55A labeled 0 is

$$A_{15} \dots A_4 = 00000000000_2$$
$$\overline{BE_3}\,\overline{BE_2}\,\overline{BE_1}\,\overline{BE_0} = 1110_2$$

and the code at bits A_3A_2 selects port A, B, or C. However, if the don't-care address bits are all made equal to 1 instead of 0, the address is

$$A_{15} \dots A_4 = 11111111000_2$$
$$\overline{BE_3}\,\overline{BE_2}\,\overline{BE_1}\,\overline{BE_0} = 1110_2$$

and it still decodes to enable 82C55A number 0. In fact, every I/O address that has $A_6A_5A_4 = 000_2$ and $\overline{BE_0} = 0$ decodes to enable this 82C55A device.

▲ 11.8 IMPLEMENTING MEMORY-MAPPED I/O PARALLEL INPUT/OUTPUT PORTS USING THE 82C55A

The *memory-mapped I/O interface* of a real-mode 80386DX microcomputer is essentially the same as that employed in the isolated I/O circuit of Fig. 11.21. Figure 11.22(a) shows the equivalent memory-mapped circuit. Ports are still selected by an address on the address bus, and byte, word, or double words of data are transferred between the MPU and I/O device over the data bus. One difference is that now the full 20-bit address is available for addressing I/O. Therefore, memory-mapped I/O devices can reside anywhere in the 1M-byte real-mode memory address space of the 80386DX.

Another difference is that during I/O operation, memory read and write bus cycles are initiated instead of I/O bus cycles. This is because we are using memory instructions, not input/output instructions, to perform the data transfer.

Since memory-mapped I/O devices reside in the memory address space and are accessed with read and write cycles, additional I/O address latch, address buffer, data bus transceiver, and address decoder circuitry are not needed. The circuitry provided for the memory interface can be used to access the memory-mapped ports.

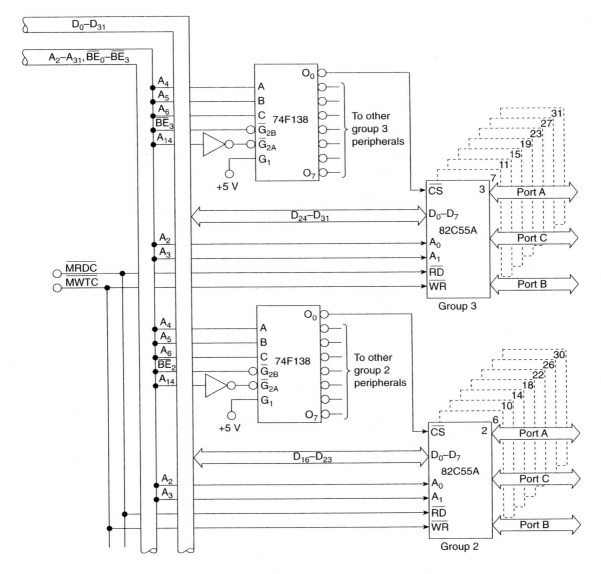

Figure 11.22(a) Memory-mapped 82C55A parallel I/O ports for an 80386DX-based micro-computer.

The key difference between the circuits in Fig. 11.21 and 11.22 is that $\overline{\text{MRDC}}$ and $\overline{\text{MWTC}}$ are now used as the $\overline{\text{RD}}$ and $\overline{\text{WR}}$ input, respectively. Therefore, the I/O circuits are accessed whenever a memory bus cycle takes place in which A_{14} is equal to logic 1.

EXAMPLE 11.14

Which I/O port in Fig. 11.22 is selected for operation when a byte access is performed to memory address 04002_{16}?

Sec. 11.8   Implementing Memory-Mapped I/O Parallel I/O Ports Using 82C55A   **549**

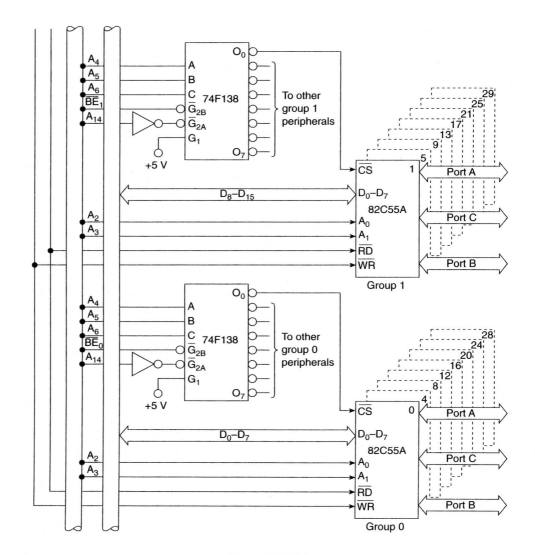

Figure 11.22(b)

Solution

We begin by converting the address to binary form, which gives

$$A_{19} \cdots A_2 A_1 = 0000010000000000000010_2$$

In this address, bits $A_{14} = 1$ and $\overline{BE}_2 = 0$. Therefore, the group 2 74F138 address decoder is enabled.

A memory-mapped I/O operation takes place to the port selected by $A_6 A_5 A_4 = 000$. This input code switches decoder output O_0 to logic 0 and chip selects PPI 2 for operation.

$$A_6 A_5 A_4 = 000 \text{ Selects} \qquad \text{PPI 2}$$

$$O_0 = 0$$

The port-select inputs of the PPI are $A_3A_2 = 00$. These inputs cause port A to be accessed.

$$A_3A_2 = 00 \qquad \text{Port A accessed}$$

Thus the address 00402_{16} selects port A on PPI 2 for memory-mapped I/O.

EXAMPLE 11.15

Write the sequence of instructions needed to initialize the control register of PPI 0 in the circuit of Fig. 11.22 such that port A is an output port, ports B and C are input ports, and all three ports are configured for mode 0 operation.

Solution

Referring to Fig. 11.9, we find that the control byte required to provide this configuration is

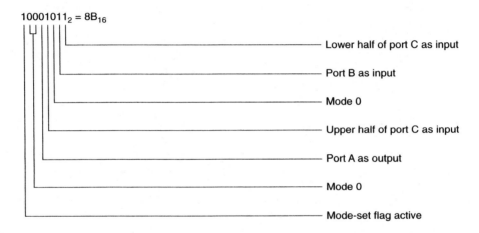

$10001011_2 = 8B_{16}$

- Lower half of port C as input
- Port B as input
- Mode 0
- Upper half of port C as input
- Port A as output
- Mode 0
- Mode-set flag active

From the circuit diagram, the memory address of the control register for PPI 0 is found to be

$$\text{CONTROL REGISTER} = 0000010000000001100_2 = 0400C_{16}$$

Since PPI 0 is memory-mapped, move instructions can be used to initialize the control register.

```
MOV   AX,0H        ;Create data segment at 00000₁₆
MOV   DS,AX
MOV   AL,8BH       ;Load AL with control byte
MOV   [400CH],AL   ;Write control byte to PPI 0 control register
```

EXAMPLE 11.16

Assume that PPI 0 in Fig. 11.22 is configured as described in Example 11.15. Write a program that will input the contents of ports B and C, AND them together, and output the results to port A.

Solution

From the circuit diagram, we find that the addresses of the three I/O ports on PPI 0 are

$$\text{port A address} = 04000_{16}$$

$$\text{port B address} = 04004_{16}$$

$$\text{port C address} = 04008_{16}$$

Now we set up a data segment at 00000_{16} and input the data from ports B and C.

```
MOV   AX,0H        ;Create data segment at 00000H
MOV   DS,AX
MOV   BL,[4004H]   ;Read port B
MOV   AL,[4008H]   ;Read port C
```

Next, the contents of AL and BL must be ANDed and the result output to port A. This can be done with the instructions

```
AND   AL,BL        ;AND data at ports B and C
MOV   [4000H],AL   ;Write to port A
```

▲ 11.9 82C54 PROGRAMMABLE INTERVAL TIMER

The 82C54 is an LSI peripheral designed to permit easy implementation of *timer* and *counter functions* in a microcomputer system. It contains three independent 16-bit counters that can be programmed to operate in a variety of ways to implement timing functions. For instance, they can be set up to work as a one-shot pulse generator, square-wave generator, or rate generator.

Block Diagram of the 82C54

Let us begin our study of the 82C54 by looking at the signal interfaces shown in the block diagram of Fig. 11.23(a). The actual pin location for each of these signals is given in Fig. 11.23(b). In a microcomputer system, the 82C54 is treated as a peripheral device. Moreover, it can be memory-mapped into the memory address space or I/O-mapped into the I/O address space. The microprocessor interface of the 82C54 allows the MPU to read from or write to its internal registers. In this way, it can be configured in various modes of operation.

Now let us look at the signals of the microprocessor interface. The microprocessor interface includes an 8-bit bidirectional data bus, D_0 through D_7. It is over these lines that data are transferred between the MPU and 82C54 . Register address inputs A_0 and A_1 are used to select the register to be accessed and control signals read ($\overline{RD}$) and write ($\overline{WR}$) indicate whether it is to be read from or written into, respectively. A chip-select ($\overline{CS}$) input is also provided to enable the 82C54's microprocessor interface. This input allows the designer to locate the device at a specific memory or I/O address.

At the other side of the block in Fig. 11.23(a), we find three signals for each counter. For instance, counter 0 has two inputs that are labeled CLK_0 and $GATE_0$. Pulses applied to the clock input are used to decrement counter 0. The gate input is used to enable or

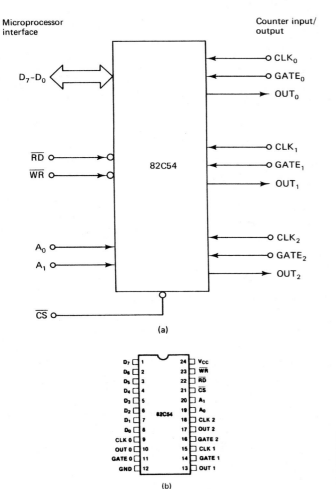

Microprocessor
interface

Counter input/
output

D_7-D_0

$\overline{RD}$

$\overline{WR}$

82C54

A_0

A_1

$\overline{CS}$

CLK_0
$GATE_0$
OUT_0

CLK_1
$GATE_1$
OUT_1

CLK_2
$GATE_2$
OUT_2

(a)

D_7 ⬚ 1		24 ⬚ V_{CC}
D_6 ⬚ 2		23 ⬚ $\overline{WR}$
D_5 ⬚ 3		22 ⬚ $\overline{RD}$
D_4 ⬚ 4		21 ⬚ $\overline{CS}$
D_3 ⬚ 5		20 ⬚ A_1
D_2 ⬚ 6	82C54	19 ⬚ A_0
D_1 ⬚ 7		18 ⬚ CLK 2
D_0 ⬚ 8		17 ⬚ OUT 2
CLK 0 ⬚ 9		16 ⬚ GATE 2
OUT 0 ⬚ 10		15 ⬚ CLK 1
GATE 0 ⬚ 11		14 ⬚ GATE 1
GND ⬚ 12		13 ⬚ OUT 1

(b)

Figure 11.23 (a) Block diagram of the 82C54 interval timer. (b) Pin layout. (Reprinted by permission of Intel Corp. Copyright/Intel Corp. 1987)

disable the counter. $GATE_0$ must be switched to logic 1 to enable counter 0 for operation. For example, in the square-wave mode of operation, the counter is to run continuously; therefore, $GATE_0$ is fixed at the 1 logic level and a continuous clock signal is applied to CLK_0. The 82C54 is rated for a maximum clock frequency of 10 MHz. Counter 0 also has an output line that is labeled OUT_0. The counter produces either a clock or a pulse at OUT_0, depending on the mode of operation selected. For instance, when configured for the square-wave mode of operation, this output is a clock signal.

Architecture of the 82C54

The internal architecture of the 82C54 is shown in Fig. 11.24. Here we find the *data bus buffer, read/write logic, control word register*, and three *counters*. The data bus buffer and read/write control logic represent the microprocessor interface we just described.

The control word register section actually contains three 8-bit registers that are used to configure the operation of counters 0, 1, and 2. The format of a *control word* is shown in Fig. 11.25. Here we find that the 2 most significant bits are a code that assigns the control

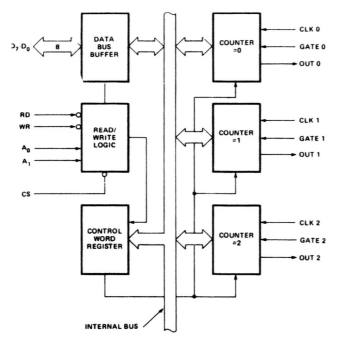

Figure 11.24 Internal architecture of the 82C54. (Reprinted by permission of Intel Corp. Copyright/Intel Corp. 1987)

word to a counter. For instance, making these bits 01 selects counter 1. Bits D_1 through D_3 are a 3-bit mode-select code, $M_2M_1M_0$, that selects one of six modes of counter operation. The least significant bit, D_0, is labeled BCD and selects either binary or BCD mode of counting. For instance, if this bit is set to logic 0, the counter acts as a 16-bit binary counter. Finally, the 2-bit code $RW/W_1RW/W_0$ is used to set the sequence in which bytes are read from or loaded into the 16-bit count registers.

EXAMPLE 11.17

An 82C54 receives the control word 10010000_2 over the bus. What configuration is set up for the timer?

Solution

Since the SC bits are 10, the rest of the bits are for setting up the configuration of counter 2. Following the format in Fig. 11.25, we find that 01 in the RW/W bits sets counter 2 for the read/write sequence identified as the least significant byte only. This means that the next write operation performed to counter 2 will load the data into the least significant byte of its count register. Next the mode code is 000, and this selects mode 0 operation for this counter. The last bit, BCD, is also set to 0 and selects binary counting.

The three counters shown in Fig. 11.24 are each 16 bits in length and operate as *down counters*. That is, when enabled by an active gate input, the clock decrements the count. Each counter contains a 16-bit *count register* that must be loaded as part of the initialization cycle. The value held in the count register can be read at any time through software.

Control word format

D7	D6	D5	D4	D3	D2	D1	D0
SC_1	SC_0	RW/W_1	RW/W_0	M_2	M_1	M_0	BCD

Definition of control

SC—select counter:

SC_1	SC_0	
0	0	Select counter 0
0	1	Select counter 1
1	0	Select counter 2
1	1	Read back command

RW—read/write:

RW/W_1	RW/W_0	
0	0	Counter latch command
1	0	Read/write most significant byte only
0	1	Read/write least significant byte only
1	1	Read/write least significant byte first, then most significant byte

M—mode:

M_2	M_1	M_0	
0	0	0	Mode 0
0	0	1	Mode 1
X	1	0	Mode 2
X	1	1	Mode 3
1	0	0	Mode 4
1	0	1	Mode 5

BCD:

0	Binary counter 16-bits
1	Binary coded decimal (BCD) counter (4 decades)

Figure 11.25 Control word format of the 82C54. (Reprinted by permission of Intel Corp. Copyright/Intel Corp. 1987)

$\overline{CS}$	$\overline{RD}$	$\overline{WR}$	A_1	A_0	
0	1	0	0	0	Write into Counter 0
0	1	0	0	1	Write into Counter 1
0	1	0	1	0	Write into Counter 2
0	1	0	1	1	Write Control Word
0	0	1	0	0	Read from Counter 0
0	0	1	0	1	Read from Counter 1
0	0	1	1	0	Read from Counter 2
0	0	1	1	1	No-Operation (3-State)
1	X	X	X	X	No-Operation (3-State)
0	1	1	X	X	No-Operation (3-State)

Figure 11.26 Accessing the registers of the 82C54. (Reprinted by permission of Intel Corp. Copyright/Intel Corp. 1987)

Sec. 11.9   82C54 Programmable Interval Timer

To read from or write to the counters of the 82C54 or load its control word register, the microprocessor needs to execute instructions. Figure 11.26 shows the bus-control information needed to access each register. For example, to write to the control register, the register address lines must be $A_1A_0 = 11$ and the control lines must be $\overline{WR} = 0$, $\overline{RD} = 1$, and $\overline{CS} = 0$.

EXAMPLE 11.18 _____

Write an instruction sequence to set up the three counters of the 82C54 in Fig. 11.27 as follows:

Counter 0:   Binary counter operating in mode 0 with an initial value of 1234H.
Counter 1:   BCD counter operating in mode 2 with an initial value of 100H.
Counter 2:   Binary counter operating in mode 4 with initial value of 1FFFH.

Assume that the device is attached to the I/O bus of the circuit in Fig. 9.51.

Solution

First we need to determine the base address of the 82C54. The base address, which is also the address of counter 0, is determined with A_3A_2 set to 00. In Fig. 11.27 we find that to

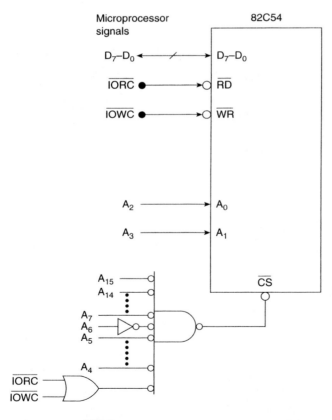

Figure 11.27   Microprocessor interface for the 82C54.

select the 82C54, $\overline{\text{CS}}$ must be logic 0. This requires that

$$A_{15}A_{14} \ldots A_6A_4 = 000000000100_2$$

Combining this part of the address with the 00 at A_1A_0 gives the base address as

$$0000000001000000_2 = 40\text{H}$$

Since the base address of the 82C54 is 40H and to select the mode register requires $A_3A_2 = 11$, its address is 4CH. Similarly, the three counters 0, 1, and 2 are at addresses 40H, 44H, and 48H, respectively. Let us first determine the mode words for the three counters. Following the bit definitions in Fig. 11.25, we get

$$\text{mode word for counter } 0 = 00110000_2 = 30_{16}$$

$$\text{mode word for counter } 1 = 01010101_2 = 55_{16}$$

$$\text{mode word for counter } 2 = 10111000_2 = \text{B8}_{16}$$

The following instruction sequence can be used to set up the 82C54 with the modes and counts:

```
MOV   AL,30H    ;Set up counter 0 mode
OUT   4CH,AL
MOV   AL,55H    ;Set up counter 1 mode
OUT   4CH,AL
MOV   AL,0B8H   ;Set up counter 2 mode
OUT   4CH,AL
MOV   AL,34H    ;Load counter 0
OUT   40H,AL
MOV   AL,12H
OUT   40H,AL
MOV   AL,00H    ;Load counter 1
OUT   44,AL
MOV   AL,01H
OUT   44,AL
MOV   AL,0FFH   ;Load counter 2
OUT   48,AL
MOV   AL,1FH
OUT   48,AL
```

Earlier we pointed out that the contents of a count register can be read at any time. Let us now look at how this is done in software. One approach is simply to read the contents of the corresponding register with an input instruction. In Fig. 11.26 we see that to read the contents of count register 0, the control inputs must be $\overline{\text{CS}} = 0$, $\overline{\text{RD}} = 0$, and $\overline{\text{WR}} = 1$, and the register address code must be $A_1A_0 = 00$. To ensure that a valid count is read out of count register 0, the counter must be inhibited before the read operation takes place. The easiest way to do this is to switch the GATE_0 input to logic 0 before performing the read operation. The count is read as two separate bytes, low byte first followed by the high byte.

The contents of the count registers can also be read without first inhibiting the counter. That is, the count can be read on the fly. To do this in software, a command must first be issued to the mode register to capture the current value of the counter into a temporary storage register. In Fig. 11.25, we find that setting bits D_5 and D_4 of the mode byte to 00 specifies the latch mode of operation. Once this mode byte has been written to the 82C54, the contents of the temporary storage register for the counter can be read just as before.

EXAMPLE 11.19

Write an instruction sequence to read the contents of counter 2 on the fly. The count is to be loaded into the AX register. Assume that the 82C54 is located at I/O address 40H and that the device is attached to the I/O bus of the circuit in Fig. 9.51.

Solution

First, we will latch the contents of counter 2, and then this value is read from the temporary storage register. This is done with the following sequence of instructions:

```
MOV   AL,1000XXXXB   ;Latch counter 2,
                     ;XXXX must be as per the mode and counter type
OUT   4CH,AL
IN    AL,48H         ;Read the low byte
MOV   BL,AL
IN    AL,48H         ;Read the high byte
MOV   AH,AL
MOV   AL,BL          ;(AX) = counter 2 value
```

Another mode of operation, called *read-back mode*, permits a programmer to capture the current count values and status information of all three counters with a single command. In Fig. 11.25 we see that a read-back command has bits D_6 and D_7 both set to 1. The read-back command format is shown in more detail in Fig. 11.28. Notice that bits D_1 (CNT 0), D_2 (CNT 1), and D_3 (CNT 2) are made logic 1 to select the counters, logic 0 in bit D_4 means that status information will be latched, and logic 0 in D_5 means that the counts will be latched. For instance, to capture the values in all three counters, the read-back command is $11011110_2 = DE_{16}$. This command must be written into the control word register of the 82C54. Figure 11.29 shows some other examples of read-back commands. Notice that both count and status information can be latched with a single command.

Our read-back command example, DE_{16}, latches only the values of the three counters.

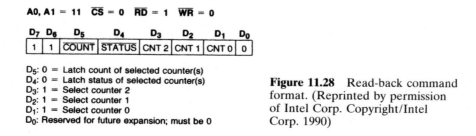

A0, A1 = 11 $\overline{CS}$ = 0 $\overline{RD}$ = 1 $\overline{WR}$ = 0

D_7	D_6	D_5	D_4	D_3	D_2	D_1	D_0
1	1	COUNT	STATUS	CNT 2	CNT 1	CNT 0	0

D_5: 0 = Latch count of selected counter(s)
D_4: 0 = Latch status of selected counter(s)
D_3: 1 = Select counter 2
D_2: 1 = Select counter 1
D_1: 1 = Select counter 0
D_0: Reserved for future expansion; must be 0

Figure 11.28 Read-back command format. (Reprinted by permission of Intel Corp. Copyright/Intel Corp. 1990)

Command								Description	Results
D_7	D_6	D_5	D_4	D_3	D_2	D_1	D_0		
1	1	0	0	0	0	1	0	Read back count and status of Counter 0	Count and status latched for Counter 0
1	1	1	0	0	1	0	0	Read back status of Counter 1	Status latched for Counter 1
1	1	1	0	1	1	0	0	Read back status of Counters 2, 1	Status latched for Counters 1 and 2
1	1	0	1	1	0	0	0	Read back count of Counter 2	Count latched for Counter 2
1	1	0	0	0	1	0	0	Read back count and status of Counter 1	Count and status latched for Counter 1
1	1	1	0	0	0	1	0	Read back status of Counter 0	Status latched for Counter 0

Figure 11.29 Read-back command examples. (Reprinted by permission of Intel Corp. Copyright/Intel Corp. 1990)

These values must next be read by the programmer by issuing read commands for the individual counters. Once the value of a counter or status is latched, it must be read before a new value can be captured.

An example of a command that latches just the status for counters 1 and 2 is given in Fig. 11.29. This command is coded as

$$11101100_2 = EC_{16}$$

The format of the status information latched with this command is shown in Fig. 11.30. Here we find that bits D_0 through D_5 contain the mode-control information that was written into the counter. These bits are identical to the 6 least significant bits of the control word in Fig. 11.25. In addition to this information, the status byte contains the logic state of the counters output pin in bit position D_7 and the value of the null count flip-flop in bit position D_6. Latched status information is also read by the programmer by issuing a read-counter command to the 82C54.

The first command in Fig. 11.29, which is $11000010_2 = C2_{16}$, captures both the count and status information for counter 0. When both count and status information are captured with a read-back command, two read-counter commands are required to return the information to the MPU. During the first read operation, the value of the count is read and the status information is transferred during the second read operation.

Operating Modes of 82C54 Counters

As indicated earlier, each of the 82C54's counters can be configured to operate in one of six modes. Figure 11.31 shows waveforms that summarize operation for each mode. Notice that mode 0 operation is known as interrupt on terminal count and mode 1 is

D_7	D_6	D_5	D_4	D_3	D_2	D_1	D_0
OUTPUT	NULL COUNT	RW1	RW0	M2	M1	M0	BCD

D_7 1 = Out Pin is 1
    0 = Out Pin is 0
D_6 1 = Null count
    0 = Count available for reading
D_5-D_0 Counter Programmed Mode

Figure 11.30 Status byte format. (Reprinted by permission of Intel Corp. Copyright/Intel Corp. 1990)

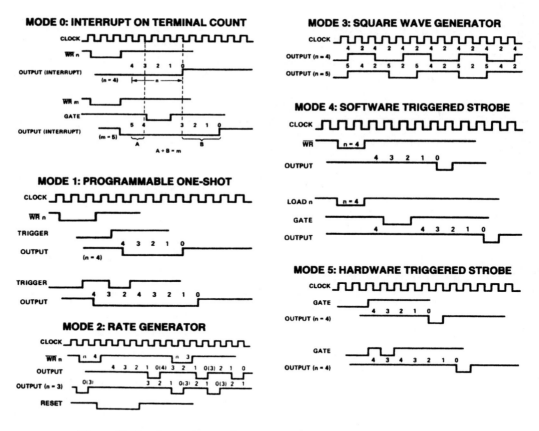

Figure 11.31 Operating modes of the 82C54. (Reprinted by permission of Intel Corp. Copyright/Intel Corp. 1987)

called programmable one-shot. The GATE input of a counter takes on different functions, depending on which mode of operation is selected. The effect of the gate input is summarized in Fig. 11.32. For instance, in mode 0, GATE disables counting when set to logic 0 and enables counting when set to 1. Let us now discuss each of these modes of operation in more detail.

The *interrupt on terminal count* mode of operation is used to generate an interrupt to the microprocessor after a certain interval of time has elapsed. As shown in the waveforms for mode 0 operation in Fig. 11.31, a count of $n = 4$ is written into the count register synchronously with the pulse at $\overline{\text{WR}}$. After the write operation is complete, the count is loaded into the counter on the next clock pulse and then the count is decremented by 1 for each clock pulse that follows. When the count reaches 0, the terminal count, a 0-to-1 transition occurs at OUTPUT. This occurs after $n + 1$ (five) clock pulses. This signal can be used as the interrupt input to the microprocessor.

Earlier we found in Fig. 11.32 that GATE must be at logic 1 to enable the counter for interrupt on terminal count mode of operation. Figure 11.31 also shows waveforms for the case in which GATE is switched to logic 0. Here we see that the value of the count is 4 when GATE is switched to logic 0. It holds at this value until GATE returns to 1.

Signal Status Modes	Low Or Going Low	Rising	High
0	Disables counting	—	Enables counting
1	—	1) Initiates counting 2) Resets output after next clock	—
2	1) Disables counting 2) Sets output immediately high	Initiates counting	Enables counting
3	1) Disables counting 2) Sets output immediately high	Initiates counting	Enables counting
4	Disables counting	—	Enables counting
5	—	Initiates counting	—

Figure 11.32 Effect of the GATE input for each mode. (Reprinted by permission of Intel Corp. Copyright/Intel Corp. 1987)

EXAMPLE 11.20

The counter of Fig. 11.33 is programmed to operate in mode 0. Assuming that the decimal value 100 is written into the counter, compute the time delay (T_D) that occurs until the positive transition takes place at the counter 0 output. The counter is configured for BCD counting. Assume the relationship between the GATE and the CLK signal as shown in the figure.

Solution

Once loaded, counter 0 needs to decrement down for 100 pulses at the clock input. During this period, the counter is disabled by logic 0 at the GATE input for two clock periods. Therefore, the time delay is calculated as

$$T_D = (n + 1 + d)(T_{CLK0})$$
$$= (100 + 1 + 2)(1/1.19318) \, \mu s$$
$$= 86.3 \, \mu s$$

Mode 1 operation implements what is known as a *programmable one-shot*. As shown in Fig. 11.31, when set for this mode of operation, the counter produces a single pulse at its output. The waveforms show that an initial count, which in this example is the number 4, is written into the counter synchronous with a pulse at $\overline{WR}$. When GATE, called TRIGGER in the waveshapes, switches from logic 0 to 1, OUTPUT switches to logic 0 on the next pulse at CLOCK and the count begins to decrement with each successive clock pulse. The pulse is completed as OUTPUT returns to logic 1 when the terminal count, which is

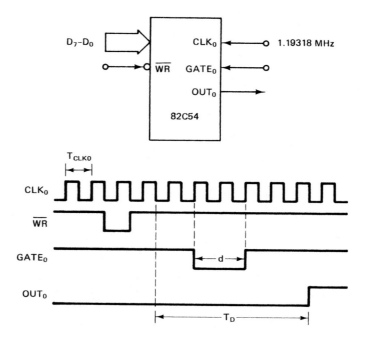

CLK₀ labeled; the following signal labels appear in the figure:

T_{CLK0}

CLK$_0$

$\overline{WR}$

GATE$_0$

OUT$_0$

T_D

d

D$_7$–D$_0$

$\overline{WR}$ GATE$_0$

CLK$_0$ ← 1.19318 MHz

OUT$_0$

82C54

Figure 11.33 Mode 0 configuration.

zero, is reached. In this way, we see that the duration of the pulse is determined by the value loaded into the counter.

The pulse generator produced with an 82C54 counter is what is called a *retriggerable one-shot*. By retriggerable we mean that if, after an output pulse has been started, another rising edge is experienced at TRIGGER, the count is reloaded and the pulse width is extended by restarting the count operation. The lower one-shot waveform in Fig. 11.31 shows this type of operation. Notice that after the count is decremented to 2, a second rising edge occurs at TRIGGER. On the next clock pulse, the value 4 is reloaded into the counter to extend the pulse width to 7 clock cycles.

EXAMPLE 11.21

Counter 1 of an 82C54 is programmed to operate in mode 1 and is loaded with the decimal value 10. The gate and clock inputs are as shown in Fig. 11.34. How long is the output pulse? Assume that the counter is configured for BCD counting.

Solution

The GATE input in Fig. 11.34 shows that the counter is operated as a nonretriggerable one-shot. Therefore, the pulse width is given by

$$T = (\text{counter contents}) \times (\text{clock period})$$
$$= (10)(1/1.19318) \text{ MHz}$$
$$= 8.38 \ \mu s$$

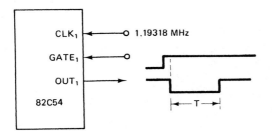

Figure 11.34 Mode 1 configuration.

When set for mode 2, *rate generator* operation, the counter within the 82C54 is set to operate as a divide-by-*N* counter. Here *N* stands for the value of the count loaded into the counter. Figure 11.35 shows counter 1 of an 82C54 set up in this way. Notice that the gate input is fixed at the 1 logic level. As shown in the table of Fig. 11.32, this enables counting operation. Looking at the waveforms for mode 2 operation in Fig. 11.31, we see that OUTPUT is at logic 1 until the count decrements to 1. Then the output switches to the active 0 logic level for just one clock pulse width. In this way, we see that there is one clock pulse at the output for every *N* clock pulses at the input. This is why it is called a divide-by-*N* counter.

EXAMPLE 11.22

Counter 1 of the 82C54, as shown in Fig. 11.35, is programmed to operate in mode 2 and is loaded with decimal number 18. Describe the signal produced at OUT_1. Assume that the counter is configured for BCD counting.

Solution

In mode 2 the output goes low for one period of the input clock after the counter contents decrement to zero. Therefore,

$$T_2 = 1/1.19318 \text{ MHz} = 838 \text{ ns}$$

and

$$T = 18 \times T_2 = 15.094 \ \mu s$$

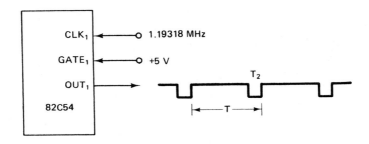

Figure 11.35 Mode 2 configuration.

Mode 3 sets the counter of the 82C54 to operate as a *square-wave rate generator*. In this mode, the output of the counter is a square wave with 50% duty cycle whenever the counter is loaded with an even number. That is, the output is at the 1 logic level for exactly the same amount of time that it is at the 0 logic level. As shown in Fig. 11.31, the count decrements by 2 with each pulse at the clock input. When the count reaches zero, the output switches logic levels, the original count ($n = 4$) is reloaded, and the count sequence repeats. Transitions of the output take place with respect to the negative edge of the input clock. The period of the symmetrical square wave at the output equals the number loaded into the counter multiplied by the period of the input clock.

If an odd number (N) is loaded into the counter instead of an even number, the time for which the output is high depends on $(N + 1)/2$, and the time for which the output is low depends on $(N - 1)/2$.

EXAMPLE 11.23

The counter in Fig. 11.36 is programmed to operate in mode 3 and is loaded with the decimal value 15. Determine the characteristics of the square wave at OUT_1. Assume that the counter is configured for BCD counting.

Solution

$$T_{CLK1} = 1/1.19318 \text{ MHz} = 838 \text{ ns}$$

$$T_1 = T_{CLK1}(N + 1)/2 = 838 \text{ ns} \times [(15 + 1)/2]$$

$$= 6.704 \ \mu s$$

$$T_2 = T_{CLK1}(N - 1)/2 = 838 \text{ ns} \times [(15 - 1)/2]$$

$$= 5.866 \ \mu s$$

$$T = T_1 + T_2 = 6.704 \ \mu s + 5.866 \ \mu s$$

$$= 12.57 \ \mu s$$

Selecting mode 4 operation for a counter configures the counter to work as a *software-triggered strobed counter*. When in this mode, the counter automatically begins to decrement one clock pulse after it is loaded with the initial value through software. Again, it decrements at a rate set by the clock input signal. At the moment the terminal count is reached, the counter generates a single strobe pulse with duration equal to one clock pulse at its output.

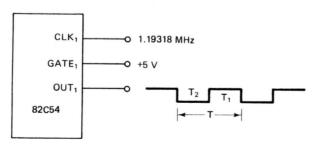

Figure 11.36 Mode 3 configuration.

That is, a strobe pulse is produced at the output after $n + 1$ clock pulses. Here n again stands for the value of the count loaded into the counter. This output pulse can be used to perform a timed operation. Figure 11.31 shows waveforms illustrating this mode of operation initiated by writing the value 4 into a counter. For instance, if CLOCK is 1.19318 MHz, the strobe occurs 4.19 μs after the count 4 is written into the counter. In the table of Fig. 11.32, we find that the gate input needs to be at logic 1 for the counter to operate.

This mode of operation can be used to implement a long-duration interval timer or a free-running timer. In either application, the strobe at the output can be used as an interrupt input to a microprocessor. In response to this pulse, an interrupt service routine can be used to reload the timer and restart the timing cycle. Frequently, the service routine also counts the strobes as they come in by decrementing the contents of a register. Software can test the value in this register to determine if the timer has timed out a certain number of times; for instance, to determine if the contents of the register have decremented to zero. When it reaches zero, a specific operation, such as a jump or call, can be initiated. In this way, we see that software has been used to extend the interval of time at which a function occurs beyond the maximum duration of the 16-bit counter within the 82C54.

EXAMPLE 11.24

Counter 1 of Fig. 11.37 is programmed to operate in mode 4. What value must be loaded into the counter to produce a strobe signal 10 μs after the counter is loaded?

Solution

The strobe pulse occurs after counting down the counter to zero. The number of input clock periods required for a period of 10 μs is given by

$$N = T/T_{CLK}$$
$$= 10\,\mu s/(1/1.19318\,\text{MHz})$$
$$= 12_{10} = C_{16} = 00001100_2$$

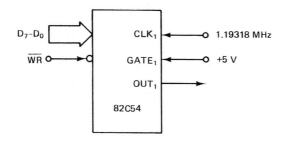

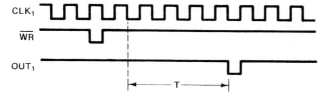

Figure 11.37 Mode 4 configuration.

Thus the counter should be loaded with the number $n = 0B_{16}$ to produce a strobe pulse 10 μs after loading.

The last mode of 82C54 counter operation, mode 5, is called the *hardware-triggered strobe*. This mode is similar to mode 4 except that now counting is initiated by a signal at the gate input. That is, it is hardware triggered instead of software triggered. As shown in the waveforms of Fig. 11.31 and the table of Fig. 11.32, a rising edge at GATE starts the countdown process. Just as for software-triggered strobed operation, the strobe pulse is output after the count is decremented to zero. But in this case, OUTPUT switches to logic 0 N clock pulses after GATE becomes active.

▲ 11.10 82C37A PROGRAMMABLE DIRECT MEMORY ACCESS CONTROLLER

The 82C37A is the LSI controller IC that is widely used to implement the *direct memory access* (DMA) function in 80386DX-based microcomputer systems. DMA capability permits devices, such as peripherals, to perform high-speed data transfers between either two sections of memory or between memory and an I/O device. In a microcomputer system, the memory or I/O bus cycles initiated as part of a DMA transfer are not performed by the MPU; instead, they are performed by a device known as a *DMA controller*, such as the 82C37A. DMA mode of operation is frequently used when blocks or packets of data are to be transferred. For instance, disk controllers, local area network controllers, and communication controllers are devices that normally process data as blocks or packets. A single 82C37A supports up to four peripheral devices for DMA operation.

Microprocessor Interface of the 82C37A

A block diagram that shows the interface signals of the 82C37A DMA controller is given in Fig. 11.38(a). The pin layout in Fig. 11.38(b) identifies the pins at which these signals are available. Let us now look briefly at the operation of the microprocessor interface of the 82C37A.

In a microcomputer system, the 82C37A acts as a peripheral controller device, and its operation must be initialized through software. This is done by reading from or writing to the bits of its internal registers. These data transfers take place through its microprocessor interface. Figure 11.39 shows how the 80386DX connects to the 82C37A's microprocessor interface.

Whenever the 82C37A is not in use by a peripheral device for DMA operation, it is in a state known as the *idle state*. When in this state, the microprocessor can issue commands to the DMA controller and read from or write to its internal registers. Data bus lines DB_0 through DB_7 are the path over which these data transfers take place. Which register is accessed is determined by a 4-bit register address that is applied to address inputs A_0 through A_3. As shown in Fig. 11.39, these inputs are supplied by address lines A_2 through A_5 of the microprocessor.

During the data-transfer bus cycle, other bits of the address are decoded in external circuitry to produce a chip-select ($\overline{CS}$) input for the 82C37A. When in the idle state, the 82C37A continuously samples this input, waiting for it to become active. Logic 0 at this input enables the microprocessor interface. The microprocessor tells the 82C37A whether

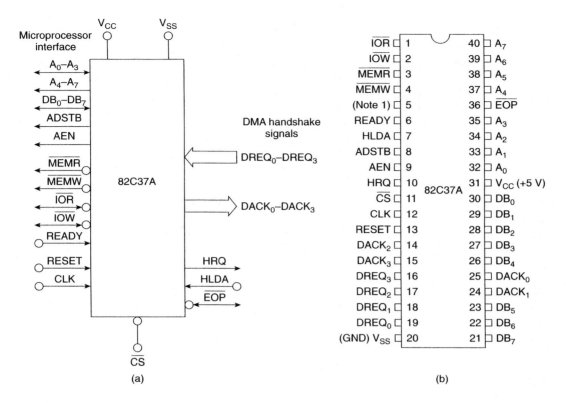

Figure 11.38 (a) Block diagram of the 82C37A DMA controller. (b) Pin layout. (Reprinted by permission of Intel Corp. Copyright/Intel Corp. 1987)

an input or output bus cycle is in progress with the signal $\overline{IOR}$ or $\overline{IOW}$, respectively. In this way, we see that the 82C37A is intended to be mapped into the I/O address space of the 80386DX microcomputer.

DMA Interface of the 82C37A

Now that we have described how a microprocessor talks to the registers of the 82C37A, let us continue by looking at how peripheral devices initiate DMA service. The 82C37A contains four independent DMA channels called channels 0 through 3. Typically, each of these channels is dedicated to a specific peripheral device. In Fig. 11.40, we see that the device has four DMA request inputs, denoted as $DREQ_0$ through $DREQ_3$. These DREQ inputs correspond to channels 0 through 3, respectively. In the idle state, the 82C37A continuously tests these inputs to see if one is active. When a peripheral device wants to perform DMA operations, it makes a request for service at its DREQ input by switching it to logic 1.

In response to the active DMA request, the DMA controller switches the hold request (HRQ) output to logic 1. Normally, this output is supplied to the HOLD input of the 80386DX and signals the microprocessor that the DMA controller needs to take control of the system bus. When the 80386DX is ready to give up control of the bus, it puts its bus

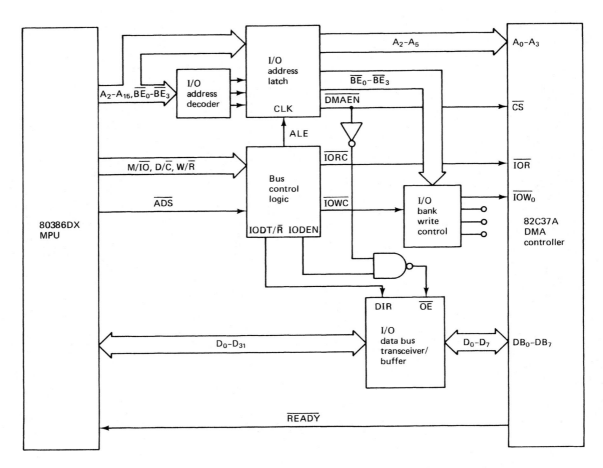

Figure 11.39 Microprocessor interface to the 80386DX.

signals into the high-impedance state and signals this fact to the 82C37A by switching the HLDA (hold acknowledge) output to logic 1. HLDA of the 80386DX is applied to the HLDA input of the 82C37A and signals that the system bus is now available for use by the DMA controller.

When the 82C37A has control of the system bus, it tells the requesting peripheral device that it is ready by outputting a DMA-acknowledge (DACK) signal. Notice in Fig. 11.40 that each of the four DMA request inputs, $DREQ_0$ through $DREQ_3$, has a corresponding DMA-acknowledge output, $DACK_0$ through $DACK_3$. Once this DMA-request/acknowledge handshake sequence is complete, the peripheral device gets direct access to the system bus and memory under control of the 82C37A.

During DMA bus cycles, the system bus is driven by the DMA controller, not the MPU. The 82C37A generates the address and all control signals needed to perform the memory or I/O data transfers. At the beginning of all DMA bus cycles, a 16-bit address is output on lines A_0 through A_7 and DB_0 through DB_7. The upper 8 bits of the address, which are available on the data bus lines, appear at the same time that address strobe (ADSTB) becomes active. Thus ADSTB is intended to be used to strobe the most significant byte of the address into an external address latch. This 16-bit address gives the 82C37A the ability to directly address up to 64K bytes of storage locations. The address enable (AEN) output

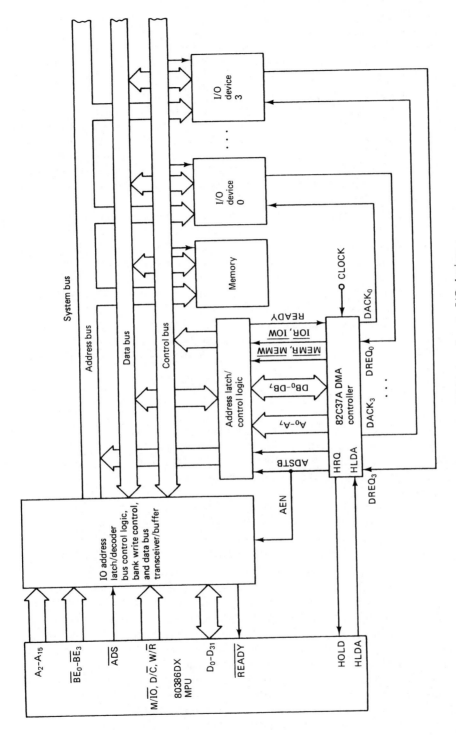

Figure 11.40 DMA interface to I/O devices.

signal is active during the complete DMA bus cycle and can be used to both enable the address latch and disable other devices connected to the bus.

Let us assume for now that an I/O peripheral device is to transfer data to memory. That is, the I/O device wants to write data to memory. In this case, the 82C37A uses the $\overline{\text{IOR}}$ output to signal the I/O device to put the data onto data bus lines DB_0 through DB_7. At the same time, it asserts $\overline{\text{MEMW}}$ to signal that the data available on the bus are to be written into memory. In this case, the data are transferred directly from the I/O device to memory and do not go through the 82C37A.

In a similar way, DMA transfers of data can take place from memory to an I/O device. In this case, the I/O device reads data from memory and outputs it to the peripheral. For this data transfer, the 82C37A activates the $\overline{\text{MEMR}}$ and $\overline{\text{IOW}}$ control signals.

The 82C37A performs both the memory-to-I/O and I/O-to-memory DMA bus cycles in just four clock periods. The duration of these clock periods is determined by the frequency of the clock signal applied to the CLOCK input. For instance, at 5 MHz the clock period is 200 ns and the bus cycle takes 800 ns.

The 82C37A is also capable of performing memory-to-memory DMA transfers. In such a data transfer, both the $\overline{\text{MEMR}}$ and $\overline{\text{MEMW}}$ signals are utilized. Unlike the I/O-to-memory operation, this memory-to-memory data transfer takes eight clock cycles. This is because it is actually performed as a separate four-clock read bus cycle from the source memory location to a temporary register within the 82C37A and then another four-clock write bus cycle from the temporary register to the destination memory location. At 5 MHz, a memory-to-memory DMA cycle takes 1.6 μs.

The READY input is used to accommodate for slow memory or I/O devices. READY must go active, logic 1, before the 82C37A will complete a memory or I/O bus cycle. As long as READY is at logic 0, wait states are inserted to extend the duration of the current bus cycle.

Internal Architecture of the 82C37A

Figure 11.41 is a block diagram of the internal architecture of the 82C37A DMA controller. Here we find the following functional blocks: the timing and control, the priority encoder and rotating priority logic, the command control, and 12 different types of registers. Let us now look briefly at the functions performed by each of these sections of circuitry and registers.

The timing and control part of the 82C37A generates the timing and control signals needed by the external bus interface. For instance, it accepts as inputs the READY and $\overline{\text{CS}}$ signals and produces outputs signals such as ADSTB and AEN. These signals are synchronized to the clock signal that is input to the controller. The highest-speed version of the 82C37A available today operates at a maximum clock rate of 5 MHz.

If multiple requests for DMA service are received by the 82C37A, they are accepted on a priority basis. One of two priority schemes can be selected for the 82C37A under software control. They are called *fixed priority* and *rotating priority*. The fixed-priority mode assigns priority to the channels in descending numeric order. That is, channel 0 has the highest priority and channel 3 has the lowest priority. Rotating priority starts with the priority levels initially the same way as in fixed priority. However, after a DMA request for a specific level gets serviced, priority is rotated such that the previously active channel is reassigned to the lowest priority level. For instance, assuming that channel 1, which was initially at priority level 1, was just serviced, then $DREQ_2$ is now at the highest priority level and $DREQ_1$ rotates to the lowest level. The priority logic circuitry shown in Fig. 11.41

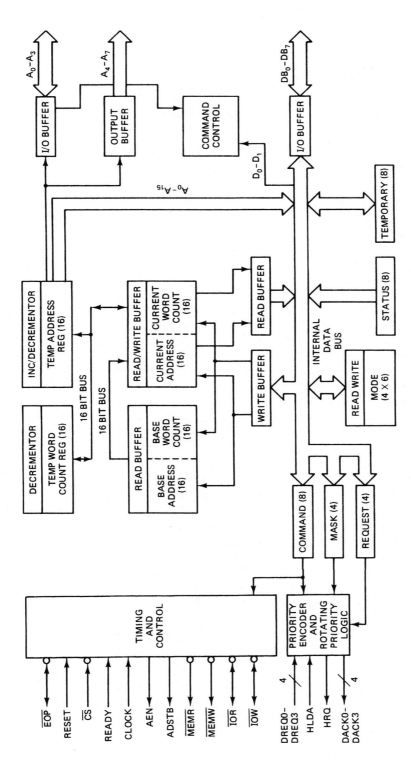

Figure 11.41 Internal architecture of the 82C37A. (Reprinted by permission of Intel Corp. Copyright/Intel Corp. 1987)

571

Name	Size	Number
Base Address Registers	16 bits	4
Base Word Count Registers	16 bits	4
Current Address Registers	16 bits	4
Current Word Count Registers	16 bits	4
Temporary Address Register	16 bits	1
Temporary Word Count Register	16 bits	1
Status Register	8 bits	1
Command Register	8 bits	1
Temporary Register	8 bits	1
Mode Registers	6 bits	4
Mask Register	4 bits	1
Request Register	4 bits	1

Figure 11.42 Internal registers of the 82C37A. (Reprinted by permission of Intel Corp. Copyright/Intel Corp. 1987)

resolves priority for simultaneous DMA requests from peripheral devices based on the programmed priority scheme.

The command control circuit decodes the register commands applied to the 82C37A through the microprocessor interface. In this way it determines which register is to be accessed and what type of operation is to be performed. Moreover, it is used to decode the programmed operating modes of the device during DMA operation.

Looking at the block diagram in Fig. 11.41, we find that the 82C37A has 12 different types of internal registers. Some examples are the current address register, current count register, command register, mask register, and status register. The names for all the internal registers are listed in Fig. 11.42, along with their size and how many are provided in the 82C37A. Note that there are actually four current address registers and they are all 16 bits long. That is, there is one current address register for each of the four DMA channels. We will now describe the function served by each of these registers in terms of overall operation of the 82C37A DMA controller. Addressing information for the internal registers is summarized in Fig. 11.43.

Each DMA channel has two address registers. They are called the *base address register* and the *current address register*. The base address register holds the starting address for the DMA operation, and the current address register contains the address of the next storage location to be accessed. Writing a value to the base address register automatically loads the same value into the current address register. In this way, we see that initially the current address register points to the starting I/O or memory address.

These registers must be loaded with appropriate values prior to initiating a DMA cycle. To load a new 16-bit address into the base register, we must write two separate bytes, one after the other, to the address of the register. The 82C37A has an internal flip-flop called the *first/last flip-flop*. This flip-flop identifies which byte of the address is being written into the register. As shown in the table of Fig. 11.43, if the beginning state of the internal flip-flop (FF) is logic 0, then software must write the low byte of the address word to the register. On the other hand, if it is logic 1, the high byte must be written to the register. For example, to write the address 1234_{16} into the base address register and the current address register for channel 0 of a DMA controller located at base I/O address DMA (where DMA $\leq$ F0H and it is decided by how the $\overline{\text{CS}}$ for the 82C37A is generated), the following instructions may be executed:

```
MOV    AL,34H        ;Write low byte
OUT    DMA + 0,AL
MOV    AL,12H        ;Write high byte
OUT    DMA + 0,AL
```

Channel(s)	Register	Operation	Register address	Internal FF	Data bus
0	Base and current address	Write	0_{16}	0 1	Low High
	Current address	Read	0_{16}	0 1	Low High
	Base and current count	Write	1_{16}	0 1	Low High
	Current count	Read	1_{16}	0 1	Low High
1	Base and current address	Write	2_{16}	0 1	Low High
	Current address	Read	2_{16}	0 1	Low High
	Base and current count	Write	3_{16}	0 1	Low High
	Current count	Read	3_{16}	0 1	Low High
2	Base and current address	Write	4_{16}	0 1	Low High
	Current address	Read	4_{16}	0 1	Low High
	Base and current count	Write	5_{16}	0 1	Low High
	Current count	Read	5_{16}	0 1	Low High
3	Base and current address	Write	6_{16}	0 1	Low High
	Current address	Read	6_{16}	0 1	Low High
	Base and current count	Write	7_{16}	0 1	Low High
	Current count	Read	7_{16}	0 1	Low High
All	Command register	Write	8_{16}	X	Low
All	Status register	Read	8_{16}	X	Low
All	Request register	Write	9_{16}	X	Low
All	Mask register	Write	A_{16}	X	Low
All	Mode register	Write	B_{16}	X	Low
All	Temporary register	Read	B_{16}	X	Low
All	Clear internal FF	Write	C_{16}	X	Low
All	Master clear	Write	D_{16}	X	Low
All	Clear mask register	Write	E_{16}	X	Low
All	Mask register	Write	F_{16}	X	Low

Figure 11.43   Accessing the registers of the 82C37A.

This routine assumes that the internal flip-flop was initially set to 0. Looking at Fig. 11.43, we find that a command can be issued to the 82C37A to clear the internal flip-flop. This is done by initiating an output bus cycle to address DMA + C_{16}.

If we read the contents of the register at address DMA + 0_{16}, the value obtained is the contents of the current address register for channel 0. Once loaded, the value in the base address register cannot be read out of the device. Assuming that the microcomputer

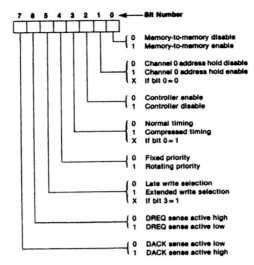

Figure 11.44 Command register format. (Reprinted by permission of Intel Corp. Copyright/Intel Corp. 1987)

uses the byte-wide I/O interface illustrated in Fig. 9.51, the address input A_0 must be driven by address line A_2. Therefore, the hardware address of the clear flip-flop would be DMA + 30_{16}.

The 82C37A also has two word-count registers for each of its DMA channels. They are called the *base count register* and the *current count register*. In Fig. 11.42, we find that these registers are also 16 bits in length, and Fig. 11.43 identifies their address as 1_{16}, relative to the base address DMA for channel 0. The number of bytes of data that are to be transferred during a DMA operation is specified by the value in the base word-count register. Actually, the number of bytes transferred is always 1 more than the value programmed into this register. This is because the end of a DMA cycle is detected by the rollover of the current word count from 0000_{16} to $FFFF_{16}$. At any time during the DMA cycle, the value in the current word count register tells how many bytes remain to be transferred.

The count registers are programmed in the same way as was just described for the address registers. For instance, to program a count of $0FFF_{16}$ into the base and current count registers for channel 1 of a DMA controller located at address DMA, (where DMA ≤ F0H), the instructions that follow can be executed:

```
MOV   AL,0FFH       ;Write low byte
OUT   DMA + 8,AL
MOV   AL,0FH        ;Write high byte
OUT   DMA + 18,AL
```

Again we have assumed that the internal flip-flop was initially cleared and that the microcomputer employs the byte-wide I/O interface of Fig. 9.51.

In Fig. 11.42, we find that the 82C37A has a single 8-bit command register. The bits in this register are used to control operating modes that apply to all channels of the DMA controller. Figure 11.44 identifies the function of each of its control bits. Notice that the settings of the bits are used to select or deselect operating features such as memory-to-memory, DMA transfer, and the priority scheme. For instance, when bit 0 is set to logic 1, the memory-to-memory mode of DMA transfer is enabled, and when it is logic 0, DMA transfers take place between I/O and memory. Moreover, setting bit 4 to logic 0 selects the fixed priority scheme for all four channels or logic 1 in this location selects rotating priority. Looking at Fig. 11.43, we see that the command register is loaded by outputting the command code to the register at address 8_{16}, relative to the base address for the 82C37A.

EXAMPLE 11.25

If the command register of an 82C37A is loaded with 01_{16}, how does the controller operate?

Solution

Representing the command word as a binary number, we get

$$01_{16} = 00000001_2$$

Referring to Fig. 11.44, we find that the DMA operation can be described as follows:

bit 0 = 1 = memory-to-memory transfers are disabled

bit 1 = 0 = channel 0 address increments/decrements normally

bit 2 = 0 = 82C37A is enabled

bit 3 = 0 = 82C37A operates with normal timing

bit 4 = 0 = channels have fixed priority, channel 0 having the highest priority and channel 3 the lowest priority

bit 5 = 0 = write operation occurs late in the DMA bus cycle

bit 6 = 0 = DREQ is an active high (logic 1) signal

bit 7 = 0 = DACK is an active low (logic 0) signal

The *mode registers* are also used to configure operational features of the 82C37A. In Fig. 11.42 we find that there is a separate mode register for each of the four DMA channels and that they are each 6 bits in length. Their bits are used to select various operational features for the individual DMA channels. A typical mode register command is shown in Fig. 11.45. As shown in the diagram, the 2 least significant bits are a 2-bit code that identifies

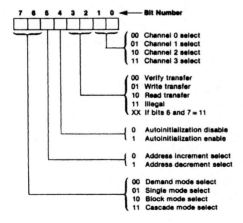

Figure 11.45 Mode register format. (Reprinted by permission of Intel Corp. Copyright/Intel Corp. 1987)

the channel to which the mode command byte applies. For instance, in a mode register command written for channel 1, these bits must be made 01. Bits 2 and 3 specify whether the channel is to perform data write, data read, or verify bus cycles. For example, if these bits are set to 01, the channel will perform only write-data transfers (DMA data transfers from an I/O device to memory).

The next 2 bits of the mode register affect how the values in the current address and current count registers are updated at the end of a DMA cycle and DMA data transfer, respectively. Bit 4 enables or disables the autoinitialization function. When autoinitialization is enabled, the current address and current count registers are automatically reloaded from the base address and base count registers, respectively, at the end of a DMA operation. In this way, the channel is prepared for the next DMA operation to begin. The setting of bit 5 determines whether the value in the current address register is automatically incremented or decremented at completion of each DMA data transfer.

The 2 most significant bits of the mode register select one of four possible modes of DMA operation for the channel. The four modes are called *demand mode, single mode, block mode,* and *cascade mode.* These modes allow for either 1 byte of data to be transferred at a time or a block of bytes. For example, when in the demand transfer mode, once the DMA cycle is initiated, bytes are continuously transferred as long as the DREQ signal remains active and the terminal count (TC) is not reached. By reaching the terminal count, we mean that the value in the current word count register, which automatically decrements after each data transfer, rolls over from 0000_{16} to $FFFF_{16}$.

Block-transfer mode is similar to demand transfer mode in that, once the DMA cycle is initiated, data are continuously transferred until the terminal count is reached. However, they differ in that, when in the demand mode, the return of DREQ to its inactive state halts the data transfer sequence. But, when in block-transfer mode, DREQ can be released at any time after the DMA cycle begins, and the block transfer will still run to completion.

In the single-transfer mode, the channel is set up such that it performs just one data transfer at a time. At the completion of the transfer, the current word count is decremented and the current address either incremented or decremented (based on the selected option). Moreover, an autoinitialization, if enabled, will not occur unless the terminal count has been reached at the completion of the current data transfer. If the DREQ input becomes inactive before the completion of the current data transfer, another data transfer will not take place until DREQ once more becomes active. On the other hand, if DREQ remains active during the complete data transfer cycle, the HRQ output of the 82C37A is switched to its inactive 0 logic level to allow the microprocessor to gain control of the system bus for one bus cycle before another single transfer takes place. This mode of operation is typically used when it is necessary to not lock the microprocessor off the bus for the complete duration of the DMA operation.

EXAMPLE 11.26

Specify the mode byte for DMA channel 2 if it is to transfer data from an input peripheral device to a memory buffer starting at address $A000_{16}$ and ending at $AFFF_{16}$. Ensure that the microprocessor is not completely locked off the bus during the DMA cycle. Moreover, at the end of each DMA cycle, the channel is to be reinitialized so that the same buffer is to be filled when the next DMA operation is initiated.

Solution

For DMA channel 2, bit 1 and bit 0 must be loaded with 10_2.

$$B_1B_0 = 10$$

Transfer of data from an I/O device to memory represents a write bus cycle. Therefore, bit 3 and bit 2 must be set to 01.

$$B_3B_2 = 01$$

Selecting autoinitialization will set up the channel to automatically reset so that it points to the beginning of the memory buffer at completion of the current DMA cycle. This feature is enabled by making bit 4 equal to 1.

$$B_4 = 1$$

The address that points to the memory buffer must increment after each data transfer. Therefore, bit 5 must be set to 0.

$$B_5 = 0$$

Finally, to ensure that the 80386DX is not locked off the bus during the complete DMA cycle, we will select the single-transfer mode of operation. This is done by making bits B_7 and B_6 equal to 01.

$$B_7B_6 = 01$$

Thus the mode register byte is

$$B_7B_6B_5B_4B_3B_2B_1B_0 = 01010110_2 = 56_{16}$$

Up to now, we have discussed how DMA cycles can be initiated by a hardware request at a DREQ input. However, the 82C37A is also able to respond to software-initiated requests for DMA service. The *request register* has been provided for this purpose. Figure 11.42 shows that the request register has just 4 bits, one for each of the DMA channels. When the request bit for a channel is set, DMA operation is started, and when reset, the DMA cycle is stopped. Any channel used for software-initiated DMA must be programmed for block-transfer mode of operation.

The bits in the request register can be set or reset by issuing software commands to the 82C37A. The format of a request register command is shown in Fig. 11.46. For instance, if a command is issued to the address of the request register with bits 0 and 1 equal to 01 and with bit 2 at logic 1, a block-mode DMA cycle is initiated for channel 1. In Fig. 11.43, we find that the request register is located at register address 9_{16}, relative to the base address for the 82C37A.

A 4-bit *mask register* is also provided within the 82C37A. One bit is provided in this register for each of the DMA channels. When a mask bit is set, the DREQ input for the

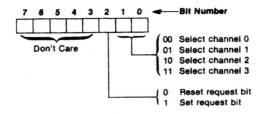

Figure 11.46 Request register format. (Reprinted by permission of Intel Corp. Copyright/Intel Corp. 1987)

corresponding channel is disabled. Therefore, hardware requests to the channel are ignored. That is, the channel is masked out. On the other hand, if the mask bit is cleared, the DREQ input is enabled, and its channel can be activated by an external device.

The format of a software command that can be used to set or reset a single bit in the mask register is shown in Fig. 11.47(a). For example, to enable the DREQ input for channel 2, the command is issued with bits 1 and 0 set to 10 to select channel 2 and with bit 2 equal to 0 to clear the mask bit. Therefore, the software command byte would be 02_{16}. The table in Fig. 11.43 shows that this command byte must be issued to the 82C37A with register address A_{16}, relative to the base address for the 82C37A.

A second mask register command is shown in Fig. 11.47(b). This command can be used to load all 4 bits of the register at once. In Fig. 11.43, we find that this command is issued to relative register address F_{16} instead of A_{16}. For instance, to mask out channel 2 while enabling channels 0, 1, and 3, the command code is 04_{16}. Either of these two methods can be used to mask or enable the DREQ input for a channel.

At system initialization, it is a common practice to clear the mask register. Looking at Fig. 11.43, we see that a special command is provided to perform this operation. The mask register can be cleared by executing an output cycle to the register with relative address E_{16}.

The 82C37A has a *status register* that contains information about the operating state of its four DMA channels. Figure 11.48 shows the bits of the status register and defines their functions. Here we find that the 4 least significant bits identify whether or not channels 0 through 3 have reached their terminal count. When the DMA cycle for a channel reaches the terminal count, this fact is recorded by setting the corresponding TC bit to the 1 logic

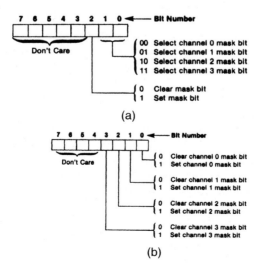

Figure 11.47 (a) Single-channel mask-register command format. (Reprinted by permission of Intel Corp. Copyright/Intel Corp. 1987.) (b) Four-channel mask-register command format. (Reprinted by permission of Intel Corp. Copyright/Intel Corp. 1987)

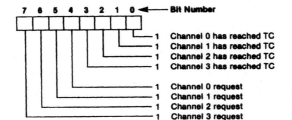

Figure 11.48 Status register. (Reprinted by permission of Intel Corp. Copyright/Intel Corp. 1987)

level. The 4 most significant bits of the register tell if a request is pending for the corresponding channel. For instance, if a DMA request has been issued for channel 0 either through hardware or software, bit 4 is set to 1. The 80386DX can read the contents of the status register through software. This is done by initiating an input bus cycle for register address 8_{16} relative to the base address for the 82C37A.

Earlier we pointed out that during memory-to-memory DMA transfers, the data read from the source address are held in a register known as the *temporary register*, and then a write cycle is initiated to write the data to the destination address. At the completion of the DMA cycle, this register contains the last byte that was transferred. The value in this register can be read by the microprocessor.

EXAMPLE 11.27

Write an instruction sequence to issue a master clear to the 82C37A and then enable all its DMA channels. Assume that the device is located at base I/O address DMA < F0H in a byte-wide I/O interface such as the one shown in Fig. 9.51.

Solution

In Fig. 11.43, we find that a special software command is provided to perform a master reset of the 82C37A's registers. Since the contents of the data bus are a don't-care state when executing the master clear command, it is performed by simply writing to the register at relative address D_{16} (hardware address 34H). For instance, the instruction

```
OUT   DMA + 34H,AL
```

can be used. To enable the DMA request inputs, all 4 bits of the mask register must be cleared. The clear-mask register command is issued by performing a write to the register at relative address E_{16} (hardware address 38H). Again, the data put on the bus during the write cycle are a don't-care state. Therefore, the command can be performed with the instruction

```
OUT   DMA + 38H,AL
```

▲ 11.11 80386DX MICROCOMPUTER SYSTEM I/O CIRCUITRY

In the last chapter, we examined the memory of the 80386DX-based microcomputer system shown in Fig. 11.49. This system includes an 82C54-2 timer/counter device in its I/O section.

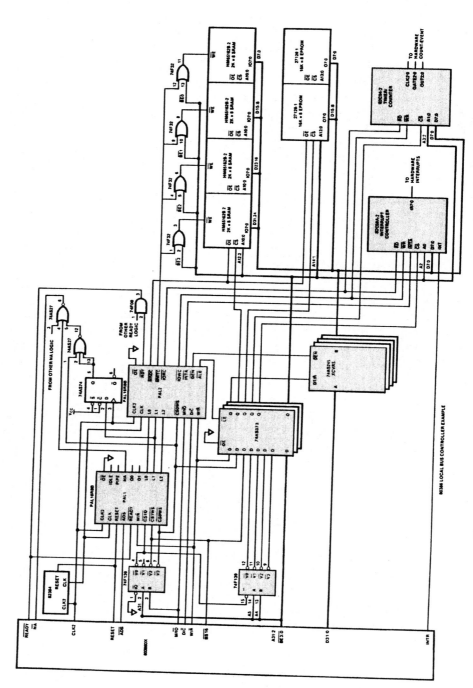

Figure 11.49 I/O interface of a typical 80386DX-based microcomputer.

Moreover, the internal registers of the 82C59A-2 interrupt controller are mapped into the I/O address space. Here we will continue by studying the operation of this microcomputer's I/O interface.

Let us begin with the bus interface circuitry. The 74F139 memory address decoder device accepts inputs A_{31} and $M/\overline{IO}$ and decodes them to produce chip-select outputs. When both $M/\overline{IO}$ and A_{31} equal 0, the $\overline{Y}_0$ output of the decoder is at its active 0 logic level. This signal indicates that an I/O address is on the bus. Notice that it is input to PAL 1 of the bus control logic as chip select I/O ($\overline{CSIO}$) and at the same time used to enable the I/O address decoder, which is another 74F139 device.

The I/O address decoder produces chip selects for the I/O peripherals. The control inputs of the decoder are address lines A_4 and A_5. When the decoder is enabled, these inputs are decoded to activate one of the four chip select signals, outputs $\overline{Y}_0$ through $\overline{Y}_3$. Tracing these signals in the circuit, we find that the $\overline{Y}_0$ output is supplied to the $\overline{CS}$ input of the 82C54-2 timer, and $\overline{Y}_1$ is sent to the $\overline{CS}$ input of the 82C59A-2. Logic 0 must be applied to the $\overline{CS}$ input of these peripherals to enable their microprocessor interface for operation.

PAL 1 and PAL 2, which form the bus control logic, produce the control signals for the I/O bus interface. Notice that the inputs to this section of the circuit include the bus cycle indication signals $M/\overline{IO}$, $D/\overline{C}$, and $W/\overline{R}$, control signal $\overline{ADS}$, and chip selects $\overline{CSIO}$, $\overline{CS1WS}$, and $\overline{CS0WS}$. At the outputs they produce the signals needed to control the transfer of data between the 80386DX and I/O devices. These signals are address latch enable ($\overline{ALE}$), I/O read command ($\overline{IORC}$), I/O write command ($\overline{IOWC}$), data bus enable ($\overline{DEN}$), and ready ($\overline{RDY}$).

The I/O address signals, byte-enable signals, chip-select outputs of the I/O address decoder, and $W/\overline{R}$ are all latched with three 74AS373 latch devices. These latches are permanently enabled by the logic 0 applied at the $\overline{OE}$ input. Information is latched into the address latches synchronously with the $\overline{ALE}$ output from PAL 2 of the bus control logic.

The 82C54-2 timer supplies three independent 16-bit interval timers for use by the microcomputer. The inputs to the timers are clocks CLK_0 through CLK_2 and gates $GATE_0$ through $GATE_2$. The timer's outputs are OUT_0 through OUT_2. Earlier we pointed out that both the 82C59A-2 and 82C54-2 are I/O mapped. For this reason, the read ($\overline{RD}$) and write ($\overline{WR}$) inputs on both devices are supplied by I/O read command ($\overline{IORC}$) and I/O write command ($\overline{IOWC}$), respectively, not $\overline{MRDC}$ and $\overline{MWTC}$. Moreover, the register to be accessed is selected by the codes on A_2 and A_3 for the 82C54-2 and by just A_2 on the 82C59-2. Input and output data transfers between the 80386DX and the I/O peripherals are byte wide and take place over data bus lines D_0 through D_7. In the circuit diagram, we see that the data bus lines are buffered with 74AS245 transceivers. The $\overline{DEN}$ output of the bus control logic enables the transceivers for operation whenever an I/O data transfer is in progress and the latched logic level of the $W/\overline{R}$ output of the 80386DX selects the direction in which data is transferred.

▲ 11.12 SERIAL COMMUNICATIONS INTERFACE

Another type of I/O interface that is widely used in microcomputer systems is known as a *serial communication port*. This is the type of interface that is commonly used to connect peripheral units, such as CRT terminals, modems, and printers, to a microcomputer. It permits data to be transferred between two units using just two data lines. One line is used for transmitting data and the other for receiving data. For instance, data input at the

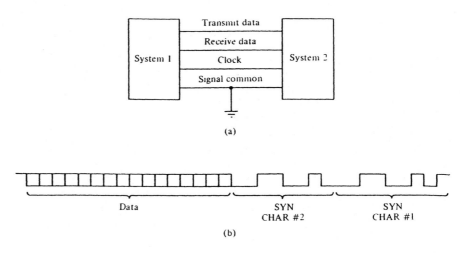

(a)

(b)

Figure 11.50 (a) Synchronous communications interface. (b) Synchronous data-transmission format.

keyboard of a terminal are passed to the MPU part of the microcomputer through this type of interface. Let us now look into the two different types of serial interfaces that are implemented in microcomputer systems.

Synchronous and Asynchronous Data Communications

Two types of *serial data communications* are widely used in microcomputer systems. They are called *asynchronous communication* and *synchronous communications*. By synchronous, we mean that the receiver and transmitter sections of the two pieces of equipment that are communicating with each other must run synchronously. For this reason, as shown in Fig. 11.50(a), the interface includes a Clock line as well as Transmit data, Receive data, and Signal common lines. It is the clock signal that synchronizes both the transmission and reception of data.

The format used for synchronous communication of data is shown in Fig. 11.50(b). To initiate synchronous transmission, the transmitter first sends out synchronization characters to the receiver. The receiver reads the synchronization bit pattern and compares it to a known sync pattern. Once they are identified as being the same, the receiver begins to read character data off the data line. Transfer of data continues until the complete block of data is received. If large blocks of data are being sent, the synchronization characters may be periodically resent to assure that synchronization is maintained. The synchronous type of communications is typically used in applications where high-speed data transfer is required.

The asynchronous method of communications eliminates the need for the Clock signal. As shown in Fig. 11.51(a), the simplest form of an asynchronous communication interface could consist of a Receive data, Transmit data, and Signal common communication lines. In this case, the data to be transmitted are sent out one character at a time, and at the receiver end of the communication line synchronization is performed by examining synchronization bits that are included at the beginning and end of each character.

The format of a typical asynchronous character is shown in Fig. 11.51(b). Here we see that the synchronization bit at the beginning of the character is called the *start bit* and that at the end of the character, the *stop bit*. Depending on the communications scheme,

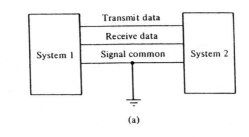

(a)

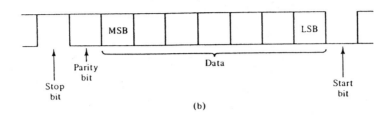

MSB　　　　　　　　　　　LSB

Parity
bit

Data

Stop
bit

Start
bit

(b)

Figure 11.51　(a) Asynchronous communications interface. (b) Asynchronous data transmission format.

1, 1½, or 2 stop bits can be used. The bits of the character are embedded between the start and stop bits. Notice that the start bit is either input or output first. It is followed in the serial bit stream by the LSB of the character, the rest of the character's bits, a parity bit, and the stop bits. For instance, 7-bit ASCII can be used and parity added as an eighth bit for higher reliability in transmission. The duration of each bit in the format is called a *bit time*.

The fact that a 0 or 1 logic level is being transferred over the communication line is identified by whether the voltage level on the line corresponds to that of a *mark* or a *space*. The start bit is always to the mark level. It synchronizes the receiver to the transmitter and signals that the unit receiving data should start assembling the character. Stop bits are to the space level. The nontransmitting line is always at the space logic level. This scheme assures that the receiving unit sees a transition of logic level at the start bit of the next character.

Simplex, Half-Duplex, and Full-Duplex Communication Links

Applications require different types of asynchronous links to be implemented. For instance, the communication link needed to connect a printer to a microcomputer just needs to support communications in one direction. That is, the printer is an output-only device; therefore, the MPU needs only to transmit data to the printer. Data are not transmitted back. In this case, as shown in Fig. 11.52(a), a single unidirectional communication line can be used to connect the printer and microcomputer together. This type of connection is known as a *simplex communication link*.

Other devices, such as the CRT terminal with keyboard shown in Fig. 11.52(b), need to both transmit data to and receive data from the MPU. That is, they must both input and output data. This requirement can also be satisfied with a single communication line by setting up a *half-duplex communication link*. In a half-duplex link, data are transmitted and received over the same line; therefore, transmission and reception of data cannot take place at the same time.

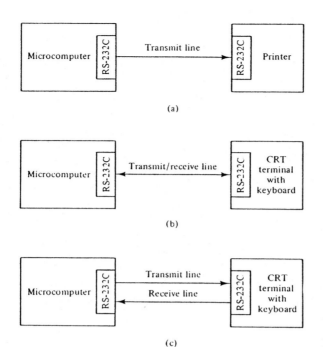

Figure 11.52 (a) Simplex communication link. (b) Half-duplex communications link. (c) Full-duplex communications link.

If higher-performance communication is required, separate transmit and receive lines can be used to connect the peripheral and microcomputer. When this is done, data can be transferred in both directions at the same time. This type of link is illustrated in Fig. 11.52(c). It is called a *full-duplex communication link.*

The USART and UART

Because serial communication interfaces are so widely used in modern electronic equipment, special LSI peripheral devices have been developed to permit easy implementation of these types of interfaces. Some of the names that these devices go by are UART (universal asynchronous receiver/transmitter) and USART (universal synchronous/asynchronous receiver/transmitter).

Both UARTs and USARTs have the ability to perform the parallel-to-serial conversions needed in the transmission of data and the serial-to-parallel conversions needed in the reception of data by a microprocessor. For data that are transmitted asynchronously, they also have the ability to frame the character automatically with a start bit, parity bit, and the appropriate stop bits.

For reception of data, UARTs and USARTs typically have the ability to check characters automatically as they are received for correct parity, and for two other errors, known as framing error and overrun error. A framing error means that after the detection of the beginning of a character with a start bit, the appropriate number of stop bits were not detected. This means that the character that was transmitted was not received correctly and should be resent. An overrun error means that the prior character that was received was not read out of the USART's receive data register by the microprocessor before another character was received. Therefore, the first character was lost and should be retransmitted.

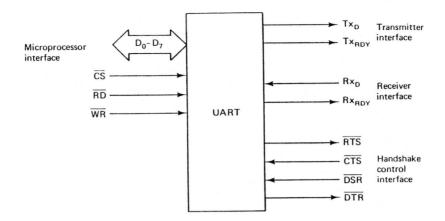

Figure 11.53 Block diagram of a UART.

A block diagram of a UART is shown in Fig. 11.53. Here we see that it has four key signal interfaces: the microprocessor interface, the transmitter interface, the receiver interface, and the handshake-control interface. Let us now look at each of these interfaces in more detail.

A UART cannot stand alone in a communication system. Its operation must typically be controlled by a microprocessor. The microprocessor interface is the interface that is used to connect the UART to an MPU. Looking at Fig. 11.53, we see that this interface consists of an 8-bit bidirectional data bus (D_0–D_7) and three control lines, $\overline{CS}$, $\overline{RD}$, and $\overline{WR}$.

All data transfers between the UART and MPU take place over the 8-bit data bus. Two uses of this bus are for the input of character data from the receiver of the UART and for the output of character data to its transmitter. Other types of information are also passed between the MPU and UART. Examples are mode control instructions, operation command instructions, and status.

LSI UARTs, just like the 82C55A LSI peripheral we discussed earlier in the chapter, can be configured for various modes of operation through software. Mode-control instructions are what must be issued to a UART to initialize its control registers for the desired mode of operation. For example, the format of the data frame used for transmitted or received data can be configured through software. Typical options are character length equal to between 5 and 8 bits; even, odd, or no parity; and 1, $1\frac{1}{2}$, or 2 stop bits.

Most UARTs have a status register that contains information related to its current state. For example, it may contain an Rx_{RDY} flag bit that represents the current state of receiver lines such as $\overline{RTS}$ and $\overline{DTR}$. This permits the MPU to examine the logic state of these lines through software and based on their settings take the necessary actions.

Besides information about transmission/receive status, the status register typically contains flag bits for error conditions such as parity error, overrun error, and framing error. After reception of a character, the MPU can first read these bits to assure that a valid character has been receive. On finding a valid character status, the character can be read from the receive data register within the UART.

In the block in Fig. 11.53, we find the transmitter and receiver interfaces. The transmitter interface has two signal lines: transmit data (Tx_D) and transmitter ready (Tx_{RDY}). Tx_D is the line over which the transmitter section of the UART outputs serial data. This output line is connected to the receive data (Rx_D) input of the receiver section in the system at the other end of the communication line.

Usually, the transmitter section of an LSI UART can hold only one character at a time. This character data is held in the transmit data register within the UART. Since only one character can be held within the UART, it must signal the MPU when it has completed transmission of this character. The Tx_{RDY} line is provided for this purpose. As soon as transmission of the character is complete, the transmitter switches Tx_{RDY} to its active logic level. This signal can be returned to an interrupt input of the MPU. In this way, its occurrence can cause program control to be passed quickly to a service routine that will output another character to the transmitter data register and then reinitiate transmission.

The receiver section is similar to the transmitter we just described. However, here the receive-data (Rx_D) line is the input that accepts bit-serial data that are transmitted from the other system's transmitter. The receive-data input connects to the transmit-data (Tx_D) output of the transmitter section in the system at the other end of the communications line. Here the receiver-ready (Rx_{RDY}) output can be used as an interrupt to the MPU, and it signals the MPU that a character has been received. The service routine that is initiated must first determine whether or not the character is valid; if it is, it can read the character out of the UART's receive-data register.

Using the handshake-control signals $\overline{RTS}$, $\overline{DSR}$, $\overline{DTR}$, and $\overline{CTS}$, different types of asynchronous communication protocols can be implemented through the serial I/O interface. By protocol we mean a handshake sequence by which two systems signal their status to each other during communications.

An example of an asynchronous communication interface that uses these control lines is shown in Fig. 11.54. When the system 1 UART wants to send data to the UART of system 2, it issues a request at its request to sent ($\overline{RTS}$) output. $\overline{RTS}$ of the system 1 UART is applied to the data set ready ($\overline{DSR}$) input of the system 2 UART. In this way, it tells the UART of system 2 that the UART of system 1 wants to transmit data to it.

When the UART of system 2 is ready to receive data, it acknowledges this fact to system 1 by activating the data terminal ready ($\overline{DTR}$) output of its UART. This signal is returned to the clear to send ($\overline{CTS}$) input of system 1's UART and tells it to output data on Tx_D. At the same time, the receiver section of the UART within system 2 begins to read data from its Rx_D input.

The implementation of this kind of interface using handshake signals requires programming of the UARTs and proper microprocessor interfaces to control the UARTs. In many practical cases, the handshake lines are not used in this way. Instead, the input lines on

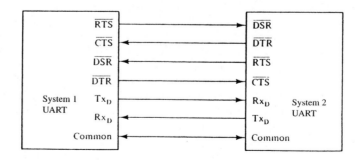

Figure 11.54 Asynchronous communications interface between two UARTs.

both UARTs are permanently connected to their enabled state by directly connecting them to logic 0 or 1. Moreover, the output handshake lines are simply ignored. This avoids the need for physical connections between the UARTs for handshake purpose.

Baud Rate and the Baud-Rate Generator

The rate at which data transfers take place over the receive and transmit lines is known as the *baud rate*. By baud rate we mean the number of bits of data transferred per second. For instance, some of the common data transfer rates are 300 baud, 1200 baud, and 9600 baud. They correspond to 300 bits/second (bps), 1200 bps, and 9600 bps, respectively.

The baud rate at which data are transferred determines the bit time. That is, the amount of time each bit of data is on the communication line. At 300 baud, the bit time is found to be

$$t_{BT} = 1/300 \text{ bps} = 3.33 \text{ ms}$$

EXAMPLE 11.28

The data transfer across an asynchronous serial data communications line is observed and the bit time of a character is measured as 0.833 ms. What is the baud rate?

Solution

Baud rate is calculated from the bit time as

$$\text{baud rate} = 1/t_{BT} = 1/0.833 \text{ ms} = 1200 \text{ bps}$$

Baud rate is set by a part of the serial communication interface called the *baud-rate generator*. This part of the interface generates the clock signal that is used to drive the receiver and transmitter parts of the UART. Some LSI UARTs have a built-in baud-rate generator; others need an external circuit to provide this function.

The RS-232C Interface

The *RS-232C interface* is a standard hardware interface for implementing asynchronous serial data communication ports on devices such as printers, CRT terminals, keyboards, and modems. The pin definitions and electrical characteristics of this interface are defined by the Electronic Industries Association (EIA). The aim behind publishing standards, such as the RS-232C, is to assure compatibility between equipment made by different manufacturers.

Peripherals that connect to a microcomputer can be located anywhere from several feet to many feet from the system. For instance, in large systems it is common to have the microcomputer part of the system in a separate room from the terminals and printers. This leads us to the main advantage of using a serial interface to connect peripherals to a microcomputer, which is that as few as three signal lines can be used to connect the peripheral to the MPU: a receive-data line, a transmit-data line, and signal common. This results in a large savings in wiring costs, and the small number of lines that need to be put in place also leads to higher reliability.

Pin	Signal
1	Protective ground
2	Transmitted data
3	Received data
4	Request to send
5	Clear to send
6	Data Set ready
7	Signal ground (common return)
8	Received line signal detector
9	Reserved for data set testing
10	Reserved for data set testing
11	Unassigned
12	Secondary received line signal detector
13	Secondary clear to send
14	Secondary transmitted data
15	Transmission signal element timing
16	Secondary received data
17	Receiver signal element timing
18	Unassigned
19	Secondary request to send
20	Data terminal ready
21	Signal quality detector
22	Ring indicator
23	Data signal rate selector
24	Transmit signal element timing
25	Unassigned

Figure 11.55 RS-232C interface pins and functions.

The RS-232C standard defines a 25-pin interface. Figure 11.55 lists each pin and its function. Note that the three signals that we mentioned earlier, transmit data, receive data, and signal ground, are located at pins 2, 3, and 7, respectively. Pins are also provided for additional control functions. For instance, pins 4 and 5 are the request to send and clear to send control signals.

The RS-232C interface is specified to operate correctly over a distance of up to 100 ft. To satisfy this distance specification, a bus driver is used on the transmit line and a bus receiver is used on the receive line. RS-232C drivers and receivers are available as standard ICs. These buffers do both the voltage-level translation needed to convert the TTL-compatible outputs of the UART to the mark (logic 1) and space (logic 0) voltage levels defined for the RS-232C interface. The voltage levels that are normally transmitted for a mark and a space are $+12$ V dc and -12 V dc, respectively. For the RS-232C interface, voltages from -3 V dc to -15 V dc are equal to a mark, and all voltages from $+3$ V dc to $+15$ V dc are considered a space.

The RS-232C interface is specified to support baud rates of up to 20,000 bps. In general, the receive and transmit baud rates do not have to be the same; however, in most simpler systems they are set to the same value.

The programmable communication interface controller is another important LSI peripheral for the 80386DX microcomputer system. It permits simple implementation of a serial data communications interface. For instance, it can be used to implement an RS-232C port. This is the type of interface that is used to connect a CRT terminal or a modem to a microcomputer. To support connection of these two peripheral devices, the microcomputer would need two independent RS-232C I/O ports. This function is normally implemented with a programmable communication controller known as a *universal synchronous/asynchronous receiver transmitter* (USART). As the name implies, a USART is capable of implementing either an asynchronous or synchronous communication interface. Here we will concentrate on its use in implementing an asynchronous communication interface.

The programmability of the USART provides for a very flexible asynchronous communication interface. Typically, they contains a full-duplex receiver and transmitter, which can be configured through software for communication of data using formats with character lengths between 5 and 8 characters, with even or odd parity, and with 1, $1\frac{1}{2}$, or 2 stop bits. Moreover, they have the ability to detect automatically the occurrence of parity, framing, and overrun errors during data reception.

The 8251A USART

A block diagram of the internal architecture of the 8251A is shown in Fig. 11.56(a), and its pin layout is in Fig. 11.56(b). From this diagram we find that it includes four key sections: the bus interface section, which consists of the data bus buffer and read/write-control logic blocks; the transmit section, which consists of the transmit buffer and transmit-control blocks; the receive section, which consists of the receive buffer and receive-control blocks; and the modem-control section. Let us now look at each of these sections in more detail.

The bus interface section is used to connect the 8251A to a microprocessor such as the 80386DX. Notice that the interface includes an 8-bit bidirectional data bus D_0 through D_7 that is driven by the data bus buffer. It is over these lines that the microprocessor transfers commands to the 8251A, reads its status register, and inputs or outputs character data.

Data transfers over the bus are controlled by the signals C/$\overline{D}$ (control/data), $\overline{RD}$ (read), $\overline{WR}$ (write), and $\overline{CS}$ (chip select), which are all inputs to the read/write control logic section. Typically, the 8251A is located at a specific address in the microcomputer's I/O or memory address space. When the microprocessor is to access registers within the 8251A, it puts this address on the address bus. The address is decoded by external circuitry and must produce logic 0 at the $\overline{CS}$ input for a read or write bus cycle to take place to the 8251A.

The other three control signals, C/$\overline{D}$, $\overline{RD}$, and $\overline{WR}$, tell the 8251A what type of data transfer is to take place over the bus. Figure 11.57 shows the various types of read/write operations that can occur. For example, the first state in the table, C/$\overline{D}$ = 0, $\overline{RD}$ = 0, and $\overline{WR}$ = 1, corresponds to a character data transfer from the 8251A to the microprocessor. Notice that in general, $\overline{RD}$ = 0 signals that the microprocessor is reading data from the 8251A, $\overline{WR}$ = 0 indicates that data are being written into the 8251A, and the logic level of C/$\overline{D}$ indicates whether character data, control information, or status information is on the data bus.

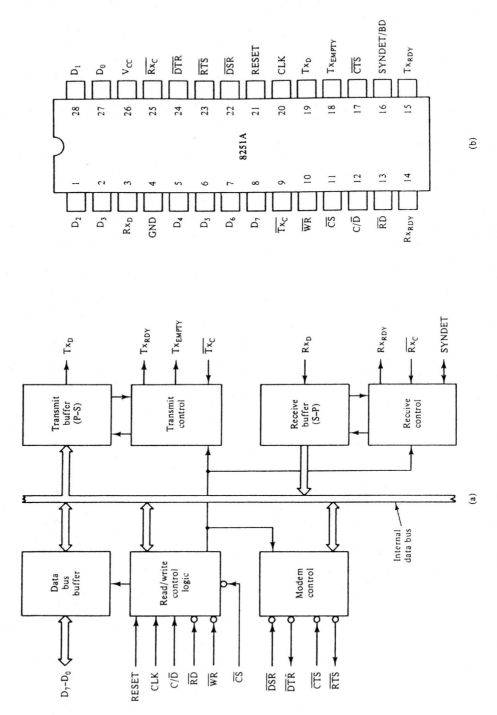

Figure 11.56 (a) Block diagram of the 8251A. (Reprinted by permission of Intel Corp. Copyright/Intel Corp. 1987) (b) Pin layout. (Reprinted by permission of Intel Corp. Copyright/Intel Corp. 1987)

C/$\overline{\text{D}}$	$\overline{\text{RD}}$	$\overline{\text{WR}}$	$\overline{\text{CS}}$	Operation
0	0	1	0	8251A Data → Data bus
0	1	0	0	Data bus → 8251A data
1	0	1	0	Status → Data bus
1	1	0	0	Data bus → Control
X	1	1	0	Data bus → 3-state
X	X	X	1	Data bus → 3-state

Figure 11.57 Read/write operations. (Reprinted by permission of Intel Corp. Copyright/Intel Corp. 1987)

EXAMPLE 11.29

What type of data transfer is taking place over the bus if the control signals are at $\overline{\text{CS}}$ = 0, C/$\overline{\text{D}}$ = 1, $\overline{\text{RD}}$ = 0, and $\overline{\text{WR}}$ = 1?

Solution

Looking at the table in Fig. 11.57, we see that $\overline{\text{CS}}$ = 0 means that the 8251A's data bus has been enabled for operation. Since C/$\overline{\text{D}}$ is 1 and $\overline{\text{RD}}$ is 0, status information is being read from the 8251A.

The receiver section is responsible for reading the serial bit stream of data at the Rx_D (receive-data) input and converting it to parallel form. When a mark voltage level is detected on this line, indicating a start bit, the receiver enables a counter. As the counter increments to a value equal to one-half a bit time, the logic level at the Rx_D line is sampled again. If it is still at the mark level, a valid start pulse has been detected. Then Rx_D is examined every time the counter increments through another bit time. This continues until a complete character is assembled and the stop bit is read. After this, the complete character is transferred into the receive-data register.

During reception of a character, the receiver automatically checks the character data for parity, framing, or overrun errors. If one of these conditions occurs, it is flagged by setting a bit in the status register. Then the Rx_{RDY} (receiver ready) output is switched to the 1 logic level. This signal is sent to the microprocessor to tell it that a character is available and should be read from the receive-data register. Rx_{RDY} is automatically reset to logic 0 when the MPU reads the contents of the receive-data register.

The 8251A does not have a built-in baud-rate generator. For this reason, the clock signal that is used to set the baud rate must be externally generated and applied to the Rx_C input of the receiver. Through software the 8251A can be set up to internally divide the clock signal input at Rx_C by 1, 16, or 64 to obtain the desired baud rate.

The transmitter does the opposite of the receiver section. It receives parallel character data from the MPU over the data bus. The character is then automatically framed with the start bit, appropriate parity bit, and the correct number of stop bits and put into the transmit data buffer register. Finally, it is shifted out of this register to produce a bit-serial output on the Tx_D line. When the transmit data buffer register is empty, the Tx_{RDY} output switches to logic 1. This signal can be returned to the MPU to tell it that another character should be output to the transmitter section. When the MPU writes another character out to the transmitter buffer register, the Tx_{RDY} output resets.

Data are output on the transmit line at the baud rate set by the external transmitter clock signal that is input at Tx_C. In most applications, the transmitter and receiver operate

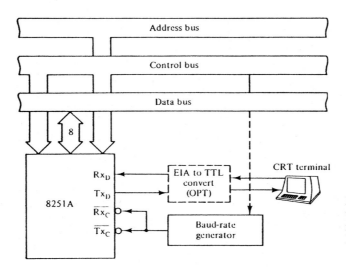

Figure 11.58 Receiver and transmitter driven at the same baud rate. (Reprinted by permission of Intel Corp. Copyright/Intel Corp. 1987)

at the same baud rate. Therefore, both Rx_C and Tx_C are supplied by the same baud-rate generator. The circuit in Fig. 11.58 shows this type of system configuration.

The operation of the 8251A is controlled through the setting of bits in three internal control registers: the mode-control register, command register, and the status register. For instance, the way in which the 8251A's receiver and transmitter operate is determined by the contents of the mode control register.

Figure 11.59 shows the organization of the mode control register and the function of each of its bits. Note that the 2 least significant bits B_1 and B_2 determine whether the device operates as an asynchronous or synchronous communication controller and in asynchronous mode how the external baud rate clock is divided within the 8251A. For example, if these 2 bits are 11, it is set for asynchronous operation with divide-by-64 for the baud-rate input. The 2 bits that follow these, L_1 and L_2, set the length of the character. For instance, when information is being transmitted and received as 7-bit ASCII characters, these bits should be loaded with 10.

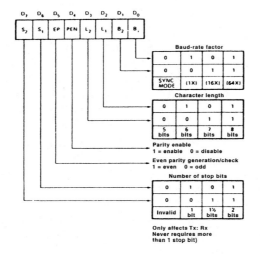

Figure 11.59 Mode instruction format. (Reprinted by permission of Intel Corp. Copyright/Intel Corp. 1987)

The next 2 bits, PEN and EP, determine whether parity is in use and if so, whether it is even parity or odd parity. Looking at Fig. 11.59, we see that PEN enables or disables parity. To enable parity, it is set to 1. Furthermore, when parity is enabled, logic 0 in EP selects odd parity or logic 1 in this position selects even parity. To disable parity, all we need to do is reset PEN.

We will assume that the 8251A is working in the asynchronous mode; therefore, bits S_1 and S_2 determine the number of stop bits. Note that if 11 is loaded into these bit positions, the character is transmitted with 2 stop bits.

EXAMPLE 11.30

What value must be written into the mode-control register in order to configure the 8251A such that it works as an asynchronous communications controller with the baud rate clock internally divided by 16? Character size is to be 8 bits; parity is odd; and 1 stop bit is used.

Solution

From Fig. 11.59, we find that B_2B_1 must be set to 10 in order to select asynchronous operation with divide by 16 for the external baud clock input.

$$B_2B_1 = 10$$

To select a character length of 8 bits, the next 2 bits are both made logic 1. This gives

$$L_2L_1 = 11$$

To set up odd parity, EP and PEN must be made equal to 0 and 1, respectively.

$$EP\ PEN = 01$$

Finally, S_2S_1 are set to 01 for 1 stop bit.

$$S_2S_1 = 01$$

Therefore, the complete control word is

$$D_7D_6 \ldots D_0 = 01011110_2$$
$$= 5E_{16}$$

Once the configuration for asynchronous communications has been set up in the mode-control register, the operation of the serial interface can be controlled by the microprocessor by issuing commands to the command register within the 8251A. The format of the command instruction byte and the function of each of its bits is shown in Fig. 11.60. Let us look at the function of just a few of its bits.

Tx_{EN} and Rx_{EN} are enable bits for the transmitter and receiver. Since both the receiver and transmitter can operate simultaneously, these two bits can both be set. Rx_{EN} is actually an enable signal to the Rx_{RDY} signal. It does not turn the receiver section on and off. The receiver runs at all times, but if Rx_{EN} is set to 0, the 8251A does not signal the MPU that

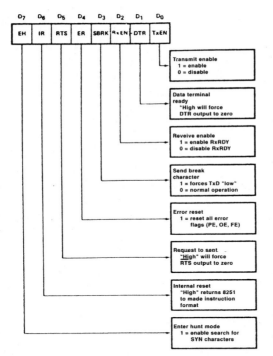

Figure 11.60 Command instruction format. (Reprinted by permission of Intel Corp. Copyright/Intel Corp. 1987)

a character has been received by switching Rx_{RDY} to logic 1. The same is true for Tx_{EN}. It enables the Tx_{RDY} signal.

The ER bit of the command register can be used to reset the error bits of the status register. The status register of the 8251A is shown in Fig. 11.61. Notice that bits PE, OE, and FE are error flags for the receiver. If the incoming character is found to have a parity

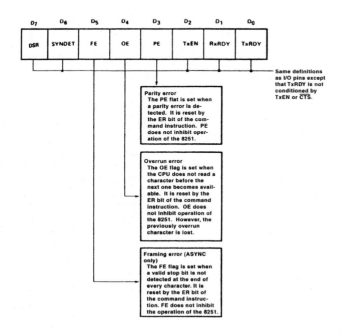

Figure 11.61 Status register. (Reprinted by permission of Intel Corp. Copyright/Intel Corp. 1987)

594   Input/Output Interface Circuits and LSI Peripheral Devices   Chap. 11

error, the PE (parity error) bit gets set. On the other hand, if an overrun or framing error condition occurs, the OE (overrun error) or FE (framing error) flag is set, respectively. The MPU should always examine these error bits before reading a character from the receive data register. If an error has occurred, a command can be issued to the command register to write a 1 into the ER bit. This causes all three of the error flags in the status register to be reset. Then a software routine can be initiated to cause the character to be retransmitted.

Let us look at just one more bit of the command register. The IR bit, which stands for internal reset, allows the 8251A to be initialized under software control. To initialize the device, the MPU simply writes a 1 into the IR bit.

Before the 8251A can be used to receive or transmit characters, its mode control and command registers must be initialized. The flowchart in Fig. 11.62 shows the sequence that must be followed when initializing the device. Let us just briefly trace through the sequence of events needed to set up the controller for asynchronous operation.

As the microcomputer powers up, it should issue a hardware reset to the 8251A. This is done by switching its RESET input to logic 1. After this, a load-mode instruction must be issued to write the new configuration byte into the mode-control register. Assuming that the 8251A is in the I/O address space of the 80386DX, the command byte formed in Example 11.32 can be written to the command register with the instruction sequence

```
MOV   DX,MODE_REG_ADDR
MOV   AL,5EH
OUT   DX,AL
```

where MODE_REG_ADDR is a variable equal to the address of the mode register of the 8251A.

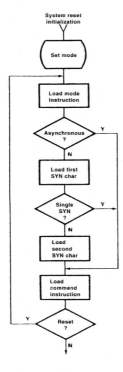

Figure 11.62 8251A initialization flowchart. (Reprinted by permission of Intel Corp. Copyright/Intel Corp. 1987)

Since bits B_2B_1 of this register are not 00, asynchronous mode of operation is selected. Therefore, we go down the branch in the flowchart to the load command instruction. Execution of another OUT instruction can load the command register with its initial value. For instance, this command could enable the transmitter and receiver by setting the Tx_{EN} and Rx_{EN} bits, respectively. During its operation the status register can be read by the microprocessor to determine if the device has received the next byte, if it is ready to send the next byte, or if any problem occurred in the transmission such as a parity error.

EXAMPLE 11.31

The circuit in Fig. 11.63(a) implements serial I/O for the 80386DX microprocessor using an 8251A. Write a program that continuously reads serial characters from the RS-232C interface, complements the received characters, and sends them back on the RS-232C interface. Each character is received and transmitted as an 8-bit character using 2 stop bits and no parity. Assume that the device is attached to the byte-wide I/O bus of the circuit in Fig. 9.51.

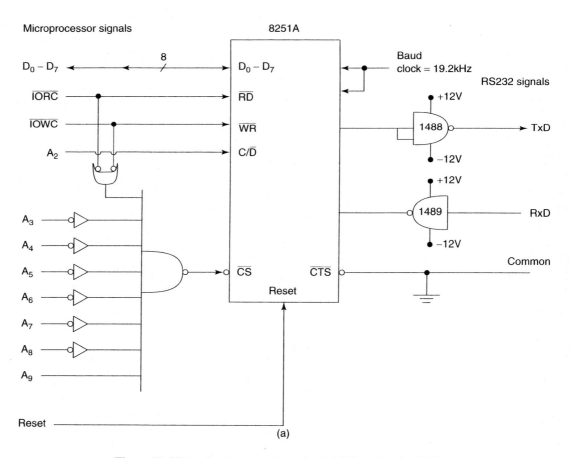

Figure 11.63(a) Implementation of serial I/O using the 8251A.

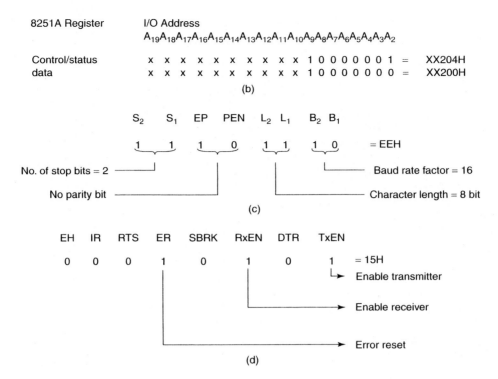

| 8251A Register | I/O Address |
| | $A_{19}A_{18}A_{17}A_{16}A_{15}A_{14}A_{13}A_{12}A_{11}A_{10}A_9A_8A_7A_6A_5A_4A_3A_2$ |

Control/status x x x x x x x x x x 1 0 0 0 0 0 0 1 = XX204H
data x x x x x x x x x x 1 0 0 0 0 0 0 0 = XX200H

(b)

S_2 S_1 EP PEN L_2 L_1 B_2 B_1

1 1 1 0 1 1 1 0 = EEH

No. of stop bits = 2
No parity bit
Baud rate factor = 16
Character length = 8 bit

(c)

EH IR RTS ER SBRK RxEN DTR TxEN

0 0 0 1 0 1 0 1 = 15H

Enable transmitter
Enable receiver
Error reset

(d)

Figure 11.63 (b) Addresses for the 8251A registers. (c) Mode word. (d) Command word.

Solution

We must first determine the addresses for the registers in the 8251A that can be accessed from the microprocessor interface. Chip select ($\overline{CS}$) is enabled for I/O read or write operations to addresses for which

$$A_9A_8A_7A_6A_5A_4A_3 = 1000000$$

Bit A_2 of the address bus is used to select between the data and control (or status) registers. As shown in Fig. 11.63(b), the addresses for the data and the control (or status) register are XX200H and XX204H, respectively.

Next we must determine the mode word to select and 8-bit character with 2 stop bits and no parity. As shown in Fig. 11.63(c), it is EEH. Here we have used a baud-rate factor of 16, which means that the baud rate is given as

$$\text{baud rate} = \text{baud-rate clock}/16 = 19,200/16 = 1200 \text{ bps}$$

To enable the transmitter as well as receiver operation of the 8251A, the required command word as shown in Fig. 11.63(d) is equal to 15H. Notice that error reset has also been implemented by making the ER bit equal 1.

The flowchart of Fig. 11.63(e) shows how we can write software to implement initialization, the receive operation, and transmit operation. The program written to perform this sequence is shown in Fig. 11.63(f).

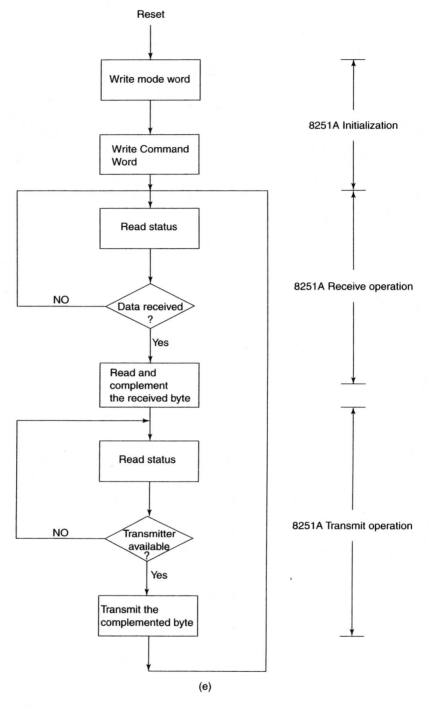

Figure 11.63(e) Flowchart for initialization, receive operation, and transmit operation.

```
INIT8251:    MOV   AL, 0 EEH    ; Write the mode word
             OUT   204H, AL
             MOV   AL, 15H      ; Write the command word
             OUT   204H, AL

CHKRX :      IN    AL, 204H     ; Check if a character is received
             ROR   AL, 1
             JNC   CHKRX
             IN    AL, 200H     ; Read the received character
             NOT   AL           ; Complement the received character
             MOV   BL, AL       ; Save it for later

CHKTX :      IN    AL, 204H     ; Check if transmitter is available
             ROR   AL, 1
             ROR   AL, 1
             JNC   CHKTX
             MOV   AL, BL       ; Transmit the complemented character
             OUT   200H, AL
             JMP   CHKRX        ; Repeat the process
```

(f)

Figure 11.63(f)   Program for the implementation of initialization, receive operation, and transmit operation.

Initialization involves writing the mode word followed by the command word to the control register of the 8251A. It is important to note that this is done after the device has been reset. Since the control register's I/O address is 204H, the two words are output to this address using appropriate instructions.

The receive operation starts by reading the contents of the status register at address 204H and checking if the LSB, which is Rx_{RDY}, is at logic 1. If it is not 1, the routine keeps reading and checking until it does become 1. Next we read the data register at 200H for the receive data. The byte of data received is complemented and then saved for transmission.

The transmit operation also starts by reading the status register at address 204H and checking if bit 1, which is Tx_{RDY}, is logic 1. If it is not, we again keep reading and checking until it becomes 1. Next the byte of data that was saved for transmission is written to the data register at address 200H. This causes it to be transmitted at the serial interface. The receive and transmit operations are repeated by jumping back to the point where the receive operation begins.

The 8250/16450 UART

The 8250 and 16450 are pin-for-pin and functionally equivalent universal asynchronous receiver transmitter ICs. These devices are newer than the 8251A USART and implements a more versatile serial I/O operation. For instance, they have a built-in programmable baud-rate generator, double buffering on communication data registers, and enhanced status and interrupt signaling. The common pin layout for these devices is shown in Fig. 11.64(a).

The connection of the 8250/16450 to implement a simple RS-232C serial communications interface is shown in Fig. 11.64(b). Looking at the microprocessor interface, we find chip-select inputs CS_0, CS_1, and $\overline{CS}_2$. To enable the interface, these inputs must be at logic 1, 1, and 0, respectively, at the same time that address strobe ($\overline{ADS}$) is logic 0. Therefore,

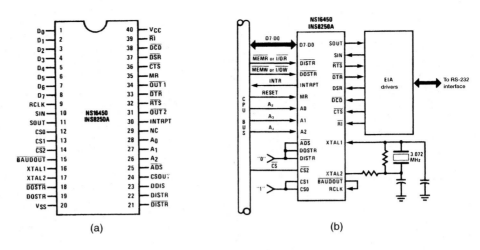

Figure 11.64 (a) Pin layout of the 8250/16450 UART. (National Semiconductor Corporation) (b) 8250/16450 RS-232C interface. (National Semiconductor Corporation)

the interface in Fig. 11.64(b) is enabled whenever logic 0 is applied to $\overline{CS}_2$ from the MPU's bus.

Let us next look at how data are read from or written into the registers of the 8250/16450. Data transfers between the MPU and communication controller take place over data bus lines D_0 through D_7. The MPU signals the peripheral whether a data input or output operation is to occur with the logic level at the data-input strobe ($\overline{DISTR}$) and data-output strobe ($\overline{DOSTR}$) inputs. Notice that when data are output during a memory write or output bus cycle, the MPU notifies the 8250/16450 with logic 0 on the $\overline{MEMW}$ or $\overline{I/OW}$ signal line, which is applied to the $\overline{DOSTR}$ input.

During the read or write bus cycle, the register that is accessed is determined by the code at register-select inputs A_0, A_1, and A_2. In Fig. 11.64(b), we find that these inputs are attached to address lines A_2 through A_4, respectively. The registers selected by the various register-select codes are shown in Fig. 11.65. Notice that the setting of the divisor latch bit

DLAB	A_2	A_1	A_0	Register
0	0	0	0	Receiver buffer (read), transmitter holding register (write)
0	0	0	1	Interrupt enable
X	0	1	0	Interrupt identification (read only)
X	0	1	1	Line control
X	1	0	0	MODEM control
X	1	0	1	Line status
X	1	1	0	MODEM status
X	1	1	1	Scratch
1	0	0	0	Divisor latch (least significant byte)
1	0	0	1	Divisor latch (most significant byte)

Figure 11.65 Register-select codes. (National Semiconductor Corporation)

(DLAB), which is in the line-control register, is also involved in the selection of the register. For example, to write to the line-control register, the code at $A_2A_1A_0$ must be 011_2. Moreover, to read the receive buffer register, the DLAB bit in the line-control register must first be set to 0 and then a read performed with register-select code $A_2A_1A_0$ equal to 000_2

The function of the various bits of the 8250/16450's registers are summarized in the table of Fig. 11.66(a). Notice that the receive buffer register (RBR) and transmitter hold register (THR) correspond to the read and write functions of register 0. However, as mentioned earlier to perform these read or write operations the divisor latch bit (DLAB), which is bit 7 of the line control register (LCR), must have already been set to 0. From the

Bit No.	Register Address										
	0 DLAB=0	0 DLAB=0	1 DLAB=0	2	3	4	5	6	7	0 DLAB=1	1 DLAB=1
	Receiver Buffer Register (Read Only)	Transmitter Holding Register (Write Only)	Interrupt Enable Register	Interrupt Ident. Register (Read Only)	Line Control Register	MODEM Control Register	Line Status Register	MODEM Status Register	Scratch Register	Divisor Latch (LS)	Latch (MS)
	RBR	THR	IER	IIR	LCR	MCR	LSR	MSR	SCR	DLL	DLM
0	Data Bit 0*	Data Bit 0	Enable Received Data Available Interrupt (ERBFI)	"0" if Interrupt Pending	Word Length Select Bit 0 (WLS0)	Data Terminal Ready (DTR)	Data Ready (DR)	Delta Clear to Send (DCTS)	Bit 0	Bit 0	Bit 8
1	Data Bit 1	Data Bit 1	Enable Transmitter Holding Register Empty Interrupt (ETBEI)	Interrupt ID Bit (0)	Word Length Select Bit 1 (WLS1)	Request to Send (RTS)	Overrun Error (OE)	Delta Data Set Ready (DDSR)	Bit 1	Bit 1	Bit 9
2	Data Bit 2	Data Bit 2	Enable Receiver Line Status Interrupt (ELSI)	Interrupt ID Bit (1)	Number of Stop Bits (STB)	Out 1	Parity Error (PE)	Trailing Edge Ring Indicator (TERI)	Bit 2	Bit 2	Bit 10
3	Data Bit 3	Data Bit 3	Enable MODEM Status Interrupt (EDSSI)	0	Parity Enable (PEN)	Out 2	Framing Error (FE)	Delta Data Carrier Detect (DDCD)	Bit 3	Bit 3	Bit 11
4	Data Bit 4	Data Bit 4	0	0	Even Parity Select (EPS)	Loop	Break Interrupt (BI)	Clear to Send (CTS)	Bit 4	Bit 4	Bit 12
5	Data Bit 5	Data Bit 5	0	0	Stick Parity	0	Transmitter Holding Register (THRE)	Data Set Ready (DSR)	Bit 5	Bit 5	Bit 13
6	Data Bit 6	Data Bit 6	0	0	Set Break	0	Transmitter Empty (TEMT)	Ring Indicator (RI)	Bit 6	Bit 6	Bit 14
7	Data Bit 7	Data Bit 7	0	0	Divisor Latch Access Bit (DLAB)	0	0	Data Carrier Detect (DCD)	Bit 7	Bit 7	Bit 15

*Bit 0 is the least significant bit. It is the first bit serially transmitted or received.

(a)

Figure 11.66(a)   Register bit functions. (National Semiconductor Corporation)

Bit 1	Bit 0	Word Length
0	0	5 Bits
0	1	6 Bits
1	0	7 Bits
1	1	8 Bits

(b)

Figure 11.66(b) Word-length select bits. (National Semiconductor Corporation)

table, we find that other bits of LCR are used to define the serial character data structure. For instance, Fig. 11.66(b) shows how the *word-length select bits*, bit 0 (WLS_0) and bit 1 (WLS_1) of LCR, select the number of bits in the serial character. Bit 2, *number of stop bits* (STB), selects the number of stop bits. If it is set to logic 0, 1 stop bit is generated for all transmitted data. On the other hand, if bit 2 is set to 1, $1\frac{1}{2}$ stop bits are produced if character length is set to 5 bits and 2 stop bits are supplied if character length is 6 or more bits. The next 2 bits, bit 3 *parity enable* (PEN) and bit 4 *even parity select* (EPS), are used to select parity. First parity is enabled by making bit 3 logic 1 and then even or odd parity is selected by making bit 4 logic 1 or 0, respectively. The LCR can be loaded with the appropriate configuration information under software control.

Looking at Fig. 11.64(b) we see that the baud-rate generator is operated off a 3.072 MHz crystal. This crystal frequency can be divided within the 8250/16450 to produce a variety of data communication baud rates. The divisor values required to produce standard baud rates are shown in Fig. 11.67. For example, to set the asynchronous data communication rate to 300 baud, a divisor equal to 640 must be used. The 16-bit divider must be loaded under software control into the divisor latch registers, DLL and DLM. Figure 11.66(a) shows that the 8 least significant bits of the divisor are in DLL and the 8 most significant bits are in DLM.

EXAMPLE 11.32

What count must be loaded into the divisor latch registers to set the data communication rate to 2400 baud? What register-select code must be applied to the 8250/16450 when writing the bytes of the divider count into the DLL and DLM registers?

Desired Baud Rate	Divisor Used to Generate 16 x Clock	Percent Error Difference Between Desired and Actual
50	3840	—
75	2560	—
110	1745	0.026
134.5	1428	0.034
150	1280	—
300	640	—
600	320	—
1200	160	—
1800	107	0.312
2000	96	—
2400	80	—
3600	53	0.628
4800	40	—
7200	27	1.23
9600	20	—
19200	10	—
38400	5	—

Figure 11.67 Baud rates and corresponding divisors. (National Semiconductor Corporation)

Solution

Looking at Fig. 11.67, we find that the divisor for 2400 baud is 80. When writing the byte into DLL, the address must make

$$A_2 A_1 A_0 = 000_2 \quad \text{with} \quad DLAB = 1$$

and the value that is written is

$$DLL = 80 = 50H$$

For DLM, the address must make

$$A_2 A_1 A_0 = 001_2 \quad \text{with} \quad DLAB = 1$$

and the value is

$$DLM = 0 = 00H$$

Let us now turn our attention to the right side of the 8250/16450 in Fig. 11.64(b). Here the RS-232C serial communication interface is implemented. We find that the transmit data are output in serial form over the serial output (S_{OUT}) line and receive data are input over the serial input (S_{IN}) line. Handshaking for the asynchronous serial interface are implemented with the request to send ($\overline{RTS}$) and data terminal ready ($\overline{DTR}$) outputs and the data set ready ($\overline{DSR}$), data carrier detect ($\overline{DCD}$), clear to send ($\overline{CTS}$), and ring indicator (RI) inputs.

The serial interface input/output signals are buffered by EIA drivers for compatibility with RS-232C voltage levels and drive currents. For example, an MC1488 driver IC can be used to buffer the output lines. It contains four TTL level to RS-232C drivers, each of which is actually a NAND gate. The MC1488 requires $+12$ V, -12 V, and ground supply connections to provide the mark and space transmission-voltage levels. The input lines of the interface can be buffered by the gates of an MC1489 RS-232C to TTL level driver. This IC contains four inverting buffers with tristate outputs and is operated from a single $+5$ V supply. Figure 11.68 shows an RS-232C interface including the EIA driver circuitry.

▲ 11.14 KEYBOARD AND DISPLAY INTERFACE

The keyboard and display are important input and output devices in microcomputer systems, such as the PC. Different types of keyboards and displays are used in many other types of digital electronic systems. For instance, all calculators have both a keyboard and a display, and many electronic test instruments have a display.

The circuit diagram in Fig. 11.69 shows how a keyboard is most frequently interfaced to a microcomputer. Note that the switches in the keyboard are arranged in an *array*. The size of the array is described in terms of the number of rows and the number of columns. In our example, the keyboard array has four rows, which are labeled R_0 through R_3, and four columns, which are labeled C_0 through C_3. The location of the switch for any key in the array is uniquely defined by a row and a column. For instance, the 0 key is located at the junction of R_0 and C_0, whereas the 1 key is located at R_0 and C_1.

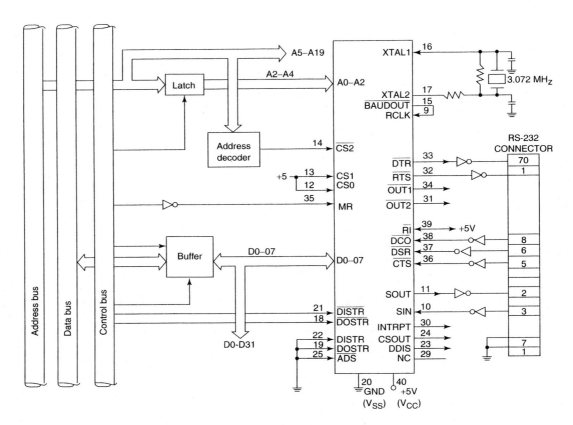

Figure 11.68 RS-232C interface with EIA drivers.

Now that we know how the keys of the keyboard are arranged, let us look at how the microcomputer services them. In most applications, the microcomputer scans the keyboard array. That is, it strobes one row of the keyboard after the other by sending out a short-duration pulse, to the 0 logic level, on the row line. During each row strobe, all column lines are examined by reading them in parallel. Typically, the column lines are pulled up to the 1 logic level; therefore, if a switch is closed, a logic 0 will be read on the corresponding column line. If no switches are closed, all 1s will be read when the column lines are examined.

For instance, if the 2 key is depressed when the microcomputer is scanning R_0, the column code read-back will be $C_3C_2C_1C_0 = 1011$. Since the microcomputer knows which row it is scanning (R_0) and which column the strobe was returned on (C_2), it can determine that the number 2 key was depressed. The microcomputer does not necessarily store the row and column codes in the form that we have shown. It is more common just to maintain the binary equivalent of the row or column. In our example, the microcomputer would internally store the row number as $R_0 = 00$ and the column number as $C_2 = 10$. This is a more compact representation of the row and column information.

Several other issues arise when designing keyboards for microcomputer systems. One is that when a key in the keyboard is depressed, its contacts bounce for a short period of time. Depending on the keyboard sampling method, this could result in incorrect reading of the keyboard input. This problem is overcome by a technique known as *keyboard debouncing*. Debouncing is achieved by resampling the column lines a second time, about

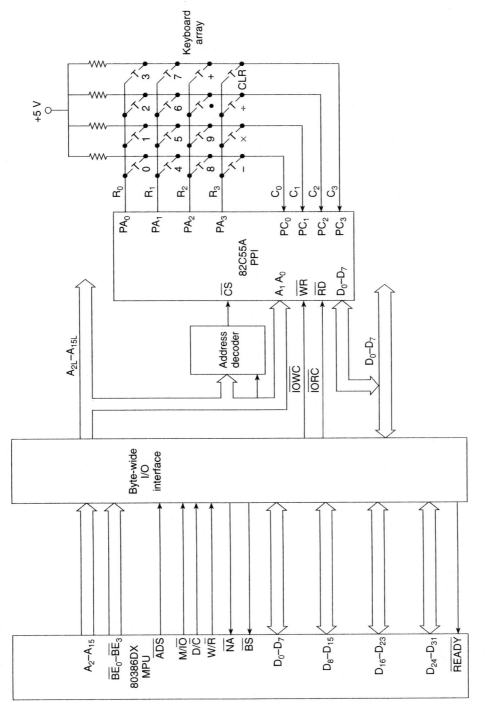

Figure 11.69 Keyboard interfaced to a microcomputer.

605

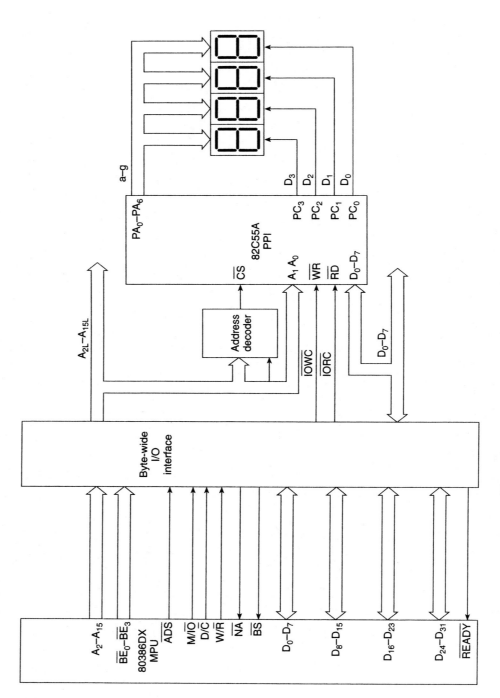

Figure 11.70 Display interfaced to a microcomputer.

10 ms later, to assure that the same column line is at the 0 logic level. If so, it is then accepted as a valid input. This technique can be implemented either in hardware or software.

Another problem occurs in keyboard sampling when more than one key is depressed at a time. In this case, the column code read by the microcomputer would have more than 1 bit that is logic 0. For instance, if the 0 and 2 keys were depressed, the column code read back during the scan of R_0 would be $C_3C_2C_1C_0 = 1010$. Typically, two keys are not actually depressed at the same time. It is more common that the second key is depressed while the first one is still being held down and that the column code showing two key closures would show up in the second test, which is made for debouncing.

Several different techniques are used to overcome this problem. One is called *two-key lockout*. With this method, the occurrence of a second key during the debounce scan causes both keys to be locked out, and neither is accepted by the microcomputer. If the second key that was depressed is released before the first key is released, the first key entry is accepted and the second key is ignored. On the other hand, if the first key is released before the second key, only the second key is accepted.

A second method of solving this problem is that known as *N-key rollover*. In this case, more than one key can be depressed at a time and be accepted by the microcomputer. The microcomputer keeps track of the order in which they are depressed and as long as the switch closures are still present at another keyboard scan 10 ms later, they are accepted. That is, in the case of multiple key depressions, the key entries are accepted in the order in which their switches are closed.

The display interface used in most microcomputer systems is shown in Fig. 11.70. Here we are using a 4-digit, 7-segment numeric display. Notice that segment lines a through g of all digits of the display are driven in parallel by outputs of the microcomputer. It is over these lines that the microcomputer outputs signals to tell the display which segments are to be lighted to form the numbers in its digits. The way in which the segments of a 7-segment display digit are labeled is shown in Fig. 11.71. For instance, to form the number 1, a code is output to light just segments b and c.

The other set of lines in the display interface correspond to the digits of the display. These lines, which are labeled D_0 through D_3, correspond to digits 0, 1, 2, and 3, respectively. It is with these signals that the microcomputer tells the display in which digit the number corresponding to the code on lines a through g should be displayed.

The way in which the display is driven by the microcomputer is said to be *multiplexed*. That is, data are not permanently displayed; instead, they are output to one digit after the other in time. This scanning sequence is repeated frequently such that the fact that the display is not permanently lighted cannot be recognized by the user.

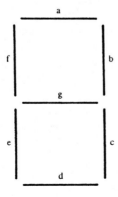

Figure 11.71 Seven-segment display labeling.

The scanning of the digits of the display is similar to the scanning we have just described for the rows of the keyboard. A digit-drive signal is output to one digit of the display after the other in time, and during each digit-drive pulse the 7-segment code for the number that is to be displayed in that digit is output on segment lines a through g. In fact, in most systems the digit-drive signals for the display and row-drive signals of the keyboard are supplied by the same set of outputs.

▲ 11.15 8279 PROGRAMMABLE KEYBOARD/DISPLAY CONTROLLER

Here we will introduce an LSI device, the *8279 programmable keyboard/display interface,* that can be used to implement a keyboard and display interfaces similar to those described in the previous section. Use of the 8279 makes implementation of a keyboard/display interface quite simple. This device can drive an 8×8 keyboard switch array and a 16-digit, 8-segment display. Moreover, it can be configured through software to support key debouncing, two-key lockout, or *N*-key rollover modes of operation and either left or right data entry to the display.

A block diagram of the device is shown in Fig. 11.72(a) and its pin layout in Fig. 11.72(b). From this diagram we see that there are four signal sections: The MPU interface, the key data inputs, the display data outputs, and the scan lines that are used by both the keyboard and display. Let us first look at the function of each of these interfaces.

The bus interface of the 8279 is similar to that found on the other peripherals that we have considered up to this point. It consists of the eight data bus lines DB_0 through DB_7. These are the lines over which the MPU outputs data to the display, inputs key codes, issues commands to the controller, and reads status information from the controller. Other signals found at the interface are the read ($\overline{RD}$), write ($\overline{WR}$), chip-select ($\overline{CS}$), and address buffer (A_0) control signals. They are the signals that control the data bus transfers that take place between the microprocessor and 8279.

A new signal introduced with this interface is *interrupt request* (IRQ). This is an output that gets returned to an interrupt input of the microcomputer. This signal is provided so that the 8279 can tell the MPU that it contains key codes that should be read.

The scan lines are used as row-drive signals for the keyboard and digit-drive signals for the display. There are just four of these lines, SL_0 through SL_3. However, they can be configured for two different modes of operation through software. In applications that require a small keyboard and display (four or fewer rows and digits), they can be used in what is known as the *decoded mode.* Scan output waveforms for this mode of operation are shown in Fig. 11.73(a). Notice that a pulse to the 0 logic level is produced at one output after the other in time.

The second mode of operation, which is called *encoded mode*, allows use of a keyboard matrix with up to 8 rows and a display with up to 16 digits. When this mode of operation is enabled through software, the binary-coded waveforms shown in Fig. 11.73(b) are output on the SL lines. These signals must be decoded with an external decoder circuit to produce the digit and column drive signals.

Even though 16-digit drive signals are produced, only 8-row drive signals can be used for the keyboard. This is because the key code that is stored when a key depression has been sensed has just 3 bits allocated to identify the row. Figure 11.74 shows this kind of circuit configuration.

The key data lines include the eight return lines RL_0 through RL_7. These lines receive inputs from the column outputs of the keyboard array. They are not tested all at once as

Input/Output Interface Circuits and LSI Peripheral Devices   Chap. 11

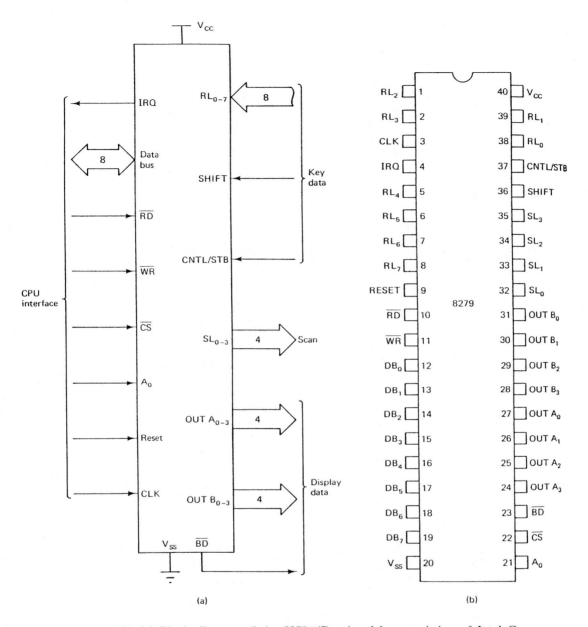

Figure 11.72 (a) Block diagram of the 8279. (Reprinted by permission of Intel Corp. Copyright/Intel Corp. 1987) (b) Pin layout. (Reprinted by permission of Intel Corp. Copyright/Intel Corp. 1987)

we described earlier. Looking at the waveforms in Fig. 11.75, we see that the RL lines are examined one after the other during each 640-μs row pulse.

If logic 0 is detected at a return line, the number of the column is coded as a 3-bit binary number and combined with the 3-bit row number to make a 6-bit key code. This key code input is first debounced and then loaded into an 8×8 key code FIFO within the 8279. Once the FIFO contains a key code, the IRQ output is set to logic 1. This signal can

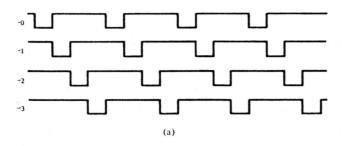

(a)

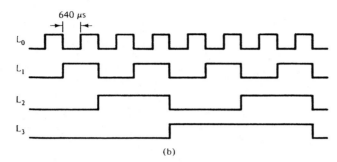

640 μs

(b)

Figure 11.73 (a) Decoded-mode scan-line signals. (Reprinted by permission of Intel Corp. Copyright/Intel Corp. 1987) (b) Encoded-mode scan-line signals. (Reprinted by permission of Intel Corp. Copyright/Intel Corp. 1987)

be used to tell the MPU that a keyboard input should be read from the 8279. There are two other signal inputs in this section. They are shift (SHIFT) and control/strobed (CNTL/ STB). The logic levels at these two inputs are also stored as part of the key code when a switch closure is detected. The format of the complete key code byte that is stored in the FIFO is shown in Fig. 11.76

A status register is provided within the 8279 that contains flags that indicate the status of the *key code FIFO*. The bits of the status register and their meanings are shown in Fig. 11.77. Notice that the 3 least significant bits, which are labeled NNN, identify the number of key codes that are currently held in the FIFO. The next bit, F, indicates whether or not the FIFO is full. The two bits that follow it, U and O, represent two FIFO error conditions. O, which stands for overrun, indicates that an attempt was made to enter another key code into the FIFO when it was already full. This condition could occur if the microprocessor does not respond quickly enough to the IRQ signal by reading key codes out of the FIFO. The other error condition, underrun (U), means that the microprocessor attempted to read the contents of the FIFO when it was empty. The microprocessor can read the contents of the status register under software control.

The display data lines include two 4-bit output ports, OUT A_0 through OUT A_3 and OUT B_0 through OUT B_3, that are used as display segment drive lines. Segment data that are output on these lines are held in a dedicated display RAM area within the 8279. This RAM is organized 16×8 and must be loaded with segment data by the microprocessor. In Fig. 11.75 we see that during each 640-μs scan time the segment data for one of the digits are output at the OUT A and OUT B ports.

The operation of the 8279 must be configured through software. Eight command words are provided for this purpose. These control words are loaded into the device by performing write (output) operations to the device with buffer address bit A_0 set to logic 1. Let us now look briefly at the function of each of these control words.

The first command (*command word 0*) is used to set the mode of operation for the keyboard and display. The general format of this word is shown in Fig. 11.78(a). Here we

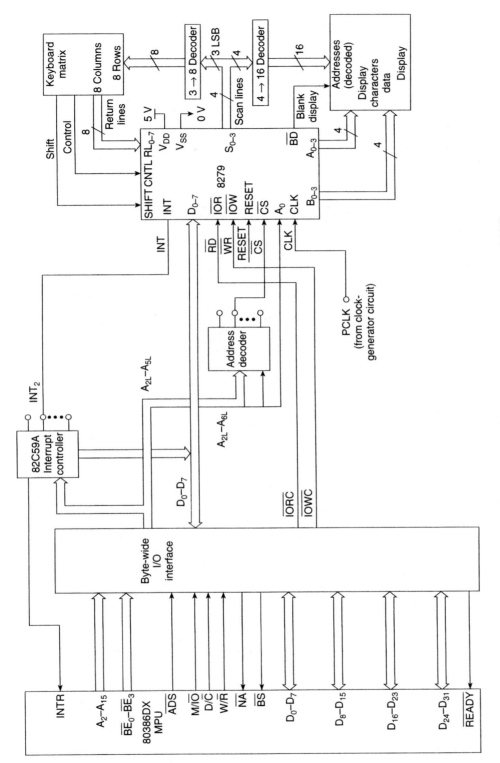

Figure 11.74 System configuration using the 80386DX and 8279.

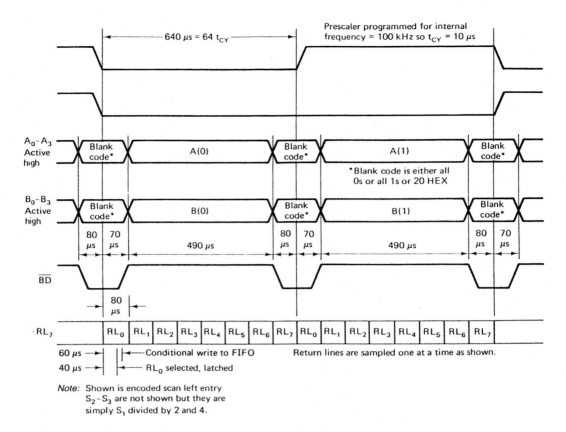

Figure 11.75 Keyboard and display signal timing. (Reprinted by permission of Intel Corp. Copyright/Intel Corp. 1987)

see that the 3 most significant bits are always reset. These 3 bits are a code by which the 8279 identifies which command is being issued by the microprocessor. The next 2 bits, which are labeled DD, are used to set the mode of operation for the display. The table in Fig. 11.78(b) shows the options that are available. After power-up reset, these bits are set to 01. From the table we see that this configures the display for 16 digits with left entry. By left entry we mean that characters are entered into the display starting from the left.

The 3 least significant bits of the command word (KKK) set the mode of operation of the display. They are used to configure the operation of the keyboard according to the table in Fig. 11.78(c). The default code at power-up is 000 and selects encoded scan operation with two-key lockout.

EXAMPLE 11.33

What should be the value of command word 0 if the display is to be set for eight 8-segment digits with right entry and the keyboard for decoded scan with *N*-key rollover?

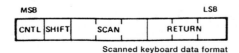

Figure 11.76 Key code byte format. (Reprinted by permission of Intel Corp. Copyright/Intel Corp. 1987)

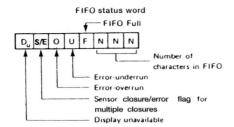

Figure 11.77 Status register. (Reprinted by permission of Intel Corp. Copyright/Intel Corp. 1987)

Solution

The 3 MSBs of the command word are always 0. The next 2 bits, DD, must be set to 10 for eight 8-segment digits with right entry. Finally, the 3 LSBs are set to 011 for decoded keyboard scan with N-key rollover. This gives

$$\text{command word } 0 = 000DDKKK$$
$$= 00010011_2$$
$$= 13_{16}$$

MSB							LSB
0	0	0	D	D	K	K	K

(a)

D	D	Display operation
0	0	8 8-bit character display – left entry
0	1	16 8-bit character display – left entry
1	0	8 8-bit character display – right entry
1	1	16 8-bit character display – right entry

(b)

K	K	K	Keyboard operation
0	0	0	Encoded scan keyboard – 2-key lockout
0	0	1	Decoded scan keyboard – 2-key lockout
0	1	0	Encoded scan keyboard – N-key rollover
0	1	1	Decoded scan keyboard – N-key rollover
1	0	0	Encoded scan sensor matrix
1	0	1	Decoded scan sensor matrix
1	1	0	Strobed input, encoded display scan
1	1	1	Strobed input, encoded display scan

(c)

Figure 11.78 (a) Command word 0 format. (Reprinted by permission of Intel Corp. Copyright/Intel Corp. 1987) (b) Display mode-select codes. (Reprinted by permission of Intel Corp. Copyright/Intel Corp. 1987) (c) Keyboard-select codes. (Reprinted by permission of Intel Corp. Copyright/Intel Corp. 1987)

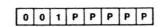

Figure 11.79 Command word 1 format. (Reprinted by permission of Intel Corp. Copyright/Intel Corp. 1987)

Command word 1 is used to set the frequency of operation of the 8279. It is designed to run at 100 kHz; however, in most applications a much higher frequency signal is available to supply its CLK input. For this reason, a *5-bit programmable prescaler* is provided within the 8279 to divide down the input frequency. The format of this command word is shown in Fig. 11.79.

Let us skip now to *command word 6* because it is also used for initialization of the 8279. It can be used to initialize the complete display memory, the FIFO status, and the interrupt-request output line. The format of this word is given in Fig. 11.80(a). The three C_D bits are used to control initialization of the display RAM. Figure 11.80(b) shows what values can be used in these locations. The C_F bit is provided for clearing the FIFO status and resetting the IRQ line. To perform the reset operation, a 1 must be written to C_F. The last bit, clear all (C_A), can be used to initiate both the C_D and C_F functions.

EXAMPLE 11.34

What clear operations are performed if the value of command word 6 written to the 8279 is $D2_{16}$?

Solution

First we need to express the command word in binary form. This gives

$$\text{command word } 6 = D2_{16} = 11010010_2$$

Note that the most significant C_D bit is set and the C_D bit that follows it is reset. This combination causes display memory to be cleared with all 0s. The C_F bit is also set, and this causes the FIFO status and IRQ output to be reset.

As shown in Fig. 11.81, only one bit of *command word 7* is functional. This bit is labeled E and is an enable signal for what is called the special error mode. When this mode is enabled and the keyboard has *N*-key rollover selected, a multiple-key depression causes

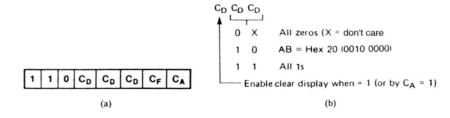

(a)   (b)

Figure 11.80 (a) Command word 6 format. (Reprinted by permission of Intel Corp. Copyright/Intel Corp. 1987) (b) C_D coding. (Reprinted by permission of Intel Corp. Copyright/Intel Corp. 1987)

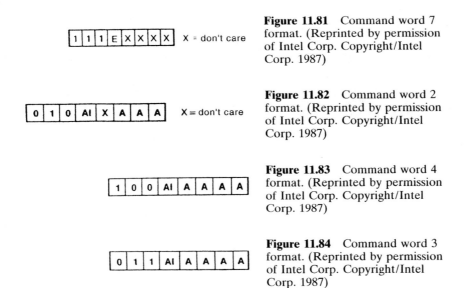

| 1 | 1 | 1 | E | X | X | X | X |

X = don't care

Figure 11.81 Command word 7 format. (Reprinted by permission of Intel Corp. Copyright/Intel Corp. 1987)

| 0 | 1 | 0 | AI | X | A | A | A |

X = don't care

Figure 11.82 Command word 2 format. (Reprinted by permission of Intel Corp. Copyright/Intel Corp. 1987)

| 1 | 0 | 0 | AI | A | A | A | A |

Figure 11.83 Command word 4 format. (Reprinted by permission of Intel Corp. Copyright/Intel Corp. 1987)

| 0 | 1 | 1 | AI | A | A | A | A |

Figure 11.84 Command word 3 format. (Reprinted by permission of Intel Corp. Copyright/Intel Corp. 1987)

the S/E flag of the FIFO status register to be set. This flag can be read by the microprocessor through software.

The rest of the command codes are related to accessing the key code FIFO and display RAM. The key code FIFO is read only. However, before the microprocessor can access it, a read FIFO command must be issued to the 8279. This is *command word 2* and has the format shown in Fig. 11.82. When the 8279 is set up for keyboard scanning, the AI and AAA bits are don't-care states. Then all that needs to be done is issue the command $01000000_2 = 40_{16}$ to the 8279 and initiate read (input) cycles to the address of the 8279. For each read bus cycle, the key code at the top of the FIFO is read into the MPU.

The display RAM can be both read from or written into by the MPU. Just like for the FIFO, a command must be sent to the 8279 before reading or writing can be initiated. For instance, when the microprocessor wants to send new data to the display, it must first issue *command word 4*. This command has the format shown in Fig. 11.83. Here the AAAA in the 4 least significant bit locations is the address of the first location to be accessed. For instance, if 0000_2 is put into these bits of the command, the first write operation will be to the first location in display RAM. Moreover, if the AI bit is set in the command, autoincrement addressing is enabled. In this way, the display RAM address pointer is automatically incremented after the write operation is complete and a write cycle can be initiated to address 0001_2 of display RAM without first issuing another write command.

The MPU can also read the contents of the display RAM in a similar way. This requires that *command word 3* be issued to the 8279. Figure 11.84 shows the format of this read display RAM command.

ASSIGNMENTS

Section 11.2

1. Give three examples of special-purpose input/output interfaces of a microcomputer.

2. List three core input/output interfaces commonly used in microcomputer systems.

Section 11.3

3. What is the address of port 7 in the circuit of Fig. 11.1(a)?

4. What are the inputs of the I/O address decoder in Fig. 11.1(a) when the I/O address on the bus is 8014_{16}? Which output is active? Which output port does this enable?

5. What operation does the instruction sequence

```
MOV    AL,0FFH
MOV    DX,8008H
OUT    DX,AL
```

perform to the circuit in Fig. 11.1(a)?

6. Write a sequence of instructions to output the word contents of the memory location called DATA to output ports 0 and 1 in the circuit of Fig. 11.1(a).

Section 11.4

7. Which input port in the circuit of Fig. 11.3 is selected for operation if the I/O address output on the bus is 8010_{16}?

8. What operation is performed to the circuit in Fig. 11.3 when the instruction sequence

```
MOV    DX,8000H
IN     AL,DX
AND    AL,0FH
MOV    [LOW_NIBBLE],AL
```

is executed?

9. Write a sequence of instructions to read in the contents of ports 1 and 2 in the circuit of Fig. 11.3 and save them at consecutive memory addresses $A0000_{16}$ and $A0001_{16}$ in memory.

10. Write an instruction sequence that will poll input I_{63} in the circuit of Fig. 11.3, checking for it to switch to logic 0.

Section 11.5

11. Name a method that can be used to synchronize the input or output of information to a peripheral device.

12. List the control signals in the parallel printer interface circuit of Fig. 11.6(a). Identify whether they are an input or output of the printer and briefly describe their functions.

13. What type device provides the data lines for the printer interface circuit of Fig. 11.6(d)?

14. Overview what happens when a write bus cycle of byte-wide data is performed to I/O address 8000_{16}.

15. Show what push and pop instructions are needed in the program written in Example 11.6 to preserve the contents of registers used by it so that it can be used as a subroutine.

Section 11.6

16. What kind of input/output interface is a PPI used to implement?

17. How many I/O lines are available on the 82C55A?

18. What are the signal names of the I/O port lines of the 82C55A?

19. Describe the mode 0, mode 1, and mode 2 I/O operations of the 82C55A.

20. What function can be served by the port B lines of the 82C55A when port A is configured for mode 2 operation?

21. How is an 82C55A configured if its control register contains 9BH?

22. If the value $A4_{16}$ is written to the control register of an 82C55A, what is the mode and I/O configuration of port A? port B?

23. What should be the control word if ports A, B, and C of an 82C55A are to be configured for mode 0 operation? Moreover, ports A and B are to be used as inputs and C as an output.

24. What value must be written to the control register of the 82C55A to configure the device such that both port A and port B are to be configured for mode 1 input operation?

25. If the control register of the 82C55A in Problem 23 is at I/O address 1000_{16}, write an instruction sequence that will load the control word.

26. Assume that the control register of an 82C55A resides at memory address 00100_{16}. Write an instruction sequence to load it with the control word formed in Problem 23.

27. What control word must be written to the control register of an 82C55A shown in Fig. 11.15(a) to enable the $INTR_B$ output? $INTE_B$ corresponds to bit PC_4 of port C.

28. If the value 03_{16} is written to the control register of an 82C55A set for mode 2 operation, what bit at port C is affected by the bit set/reset operation? Is it set to 1 or cleared to 0?

29. Assume that the control register of an 82C55A is at I/O address 0100_{16}. Write an instruction sequence that will load it with the bit set/reset value given in Problem 28.

Section 11.7

30. If I/O address $003D_{16}$ is applied to the circuit in Fig. 11.21 during a byte-write cycle and the data output on the bus is 98_{16}, which 82C55A is being accessed? Are data being written into port A, port B, port C, or the control register of this device?

31. If the instruction

```
IN  AL, 08H
```

is executed to the I/O interface circuit in Fig. 11.21, what operation is performed?

32. What are the addresses of the A, B, and C ports of PPI 2 in the circuit of Fig. 11.21?

33. Assume that PPI 2 in Fig. 11.21 is configured as defined in Problem 23. Write a program that will input the data at ports A and B, add these values together, and output the sum to port C.

Section 11.8

34. Distinguish between memory-mapped I/O and isolated I/O.

35. What address inputs must be applied to the circuit in Fig. 11.22 in order to access port B of device 4? Assuming that all unused bits are 0, what would be the memory address?

36. Write an instruction that will load the control register of the port identified in Problem 35 with the value 98_{16}.

37. Repeat Problem 33 for the circuit in Fig. 11.22.

Section 11.9

38. What are the inputs and outputs of counter 2 of an 82C54?

39. Write a control word for counter 1 that selects the following options: load least significant byte only, mode 5 of operation, and binary counting.

40. What are the logic levels of inputs $\overline{CS}$, $\overline{RD}$, $\overline{WR}$, A_1, and A_0 when the byte in Problem 39 is written to an 82C54?

41. Write an instruction sequence that will load the control word in Problem 39 into an 82C54 that is located starting at address 01000_{16} of the memory address space. Assume that the device is attached to the I/O interface circuit in Fig. 9.51 and that address inputs A_0 and A_1 are supplied by address bits A_2 and A_3, respectively.

42. Write an instruction sequence that will write the value 12_{16} into the least significant byte of the count register for counter 2 of an 82C54 located starting at memory address 01000_{16}. Assume that the registers are attached as in Problem 41.

43. Repeat Example 11.19 for the 82C54 located at memory address 01000_{16}, but this time just read the least significant byte of the counter.

44. What is the maximum time delay that can be generated with the timer in Fig. 11.33? What would be the maximum time delay if the clock frequency is increased to 2 MHz? Assume that it is configured for binary counting.

45. What is the resolution of pulses generated with the 82C54 in Fig. 11.33? What will be the resolution if the clock frequency is increased to 2 MHz?

46. Find the pulse width of the one-shot in Fig. 11.34 if the counter is loaded with the value 1000_{16}. Assume that the counter is configured for binary count operation.

47. What count must be loaded into the square-wave generator of Fig. 11.36 in order to produce a 25-kHz output?

48. If the counter in Fig. 11.37 is loaded with the value 120_{16}, how long of a delay occurs before the strobe pulse is output?

Section 11.10

49. Are signal lines $\overline{MEMR}$ and $\overline{MEMW}$ of the 82C37A used in the microprocessor interface?

50. Summarize the 82C37A's DMA request/acknowledge handshake sequence.

51. What is the total number of user-accessible registers in the 82C37A?

52. Write an instruction sequence that will read the value of the address from the current address register for channel 0 into the AX register. Assume that the 82C37A has the base address 10H.

53. Assuming that an 82C37A is located at I/O address 1000H, write an instruction sequence to perform a master clear operation.

54. Write an instruction sequence that will write the command word 00_{16} into the command register of an 82C37A that is located at address 2000H in the I/O address space.

55. Write an instruction sequence that will load the mode register for channel 2 with the mode byte obtained in Example 11.26. Assume that the 82C37A is located at I/O address F0H.

56. What must be output to the mask register in order to disable all of the DRQ inputs?

57. Write an instruction sequence that will read the contents of the status register into the AL register. Assume that the 82C37A is located at I/O address 5000H.

Section 11.11

58. Which Y output of the memory address decoder in the circuit of Fig. 11.49 is used as the signal $\overline{\text{CSIO}}$?

59. What code at M/$\overline{\text{IO}}A_{31}A_5A_4$ will make the Y output for the 82C54-2 in the circuit of Fig. 11.49 become active?

60. What signal is applied to the $\overline{\text{RD}}$ inputs of the 82C59A-2 and 82C54-2 in the circuit of Fig. 11.49?

61. Which data bus lines are used in the I/O interface in the circuit of Fig. 11.49?

62. Which address lines are used in the circuit of Fig. 11.49 to select the registers within the 82C54-2 timer IC?

Section 11.12

63. Name a signal line that distinguishes an asynchronous communication interface from that of a synchronous communication interface.

64. Describe the sequence of signals that become active in Fig. 11.54 when system 2 transfers a character to system 1.

65. Define a simplex, a half-duplex, and a full-duplex communication link.

Section 11.13

66. To write a byte of data to the 8251A, what logic levels must the microprocessor apply to control inputs C/$\overline{\text{D}}$, $\overline{\text{RD}}$, $\overline{\text{WR}}$, and $\overline{\text{CS}}$?

67. The mode-control register of an 8251A contains 11111111_2. What are the asynchronous character length, type of parity, and the number of STOP bits?

68. Write an instruction sequence to write the control word given in Problem 67 to a memory-mapped 8251A that is located at address MODE.

69. Describe the difference between a mode instruction and a command instruction used in 8251A initialization.

Section 11.14

70. Referring to Fig. 11.69, what is the maximum number of keys that can be supported using all 24 I/O lines of an appropriately configured 82C55A?

71. In the circuit of Fig. 11.69, what row and column code would identify the 9 key?

72. What codes would need to be output on the digital and segment lines of the circuit in Fig. 11.70 to display the number 7 in digit 1?

Section 11.15

73. Specify the mode of operation for the keyboard and display when an 8279 is configured with command word 0 equal to $3F_{16}$.

74. Determine the clock frequency applied to the input of an 8279 if it needs command word 1 equal to $1E_{16}$ to operate.

75. Summarize the function of each command word of the 8279.

Interrupt and Exception Processing of the 80386DX Microprocessor

▲ 12.1 INTRODUCTION

In Chapters 10 and 11 we covered the input/output interface of the 80386DX-based microcomputer system. Here we continue with a special input interface, the *interrupt interface,* and the *exception processing* capability of the 80386 microprocessor family. A list of the topics, in the order they are presented in this chapter, is as follows:

1. Types of interrupts and exceptions
2. Interrupt vector and descriptor tables
3. Interrupt instructions
4. Enabling/disabling of interrupts
5. External hardware-interrupt interface
6. External hardware-interrupt sequence
7. 82C59A programmable interrupt controller
8. Interrupt interface circuits using the 82C59A
9. Software interrupts
10. Nonmaskable interrupt
11. Reset
12. Internal interrupts and exception functions

▲ 12.2 TYPES OF INTERRUPTS AND EXCEPTIONS

Interrupts provide a mechanism for quickly changing program environments. Transfer of program control is initiated by the occurrence of either an event internal to the microprocessor or an event in its external hardware. For instance, when an interrupt signal occurs in

621

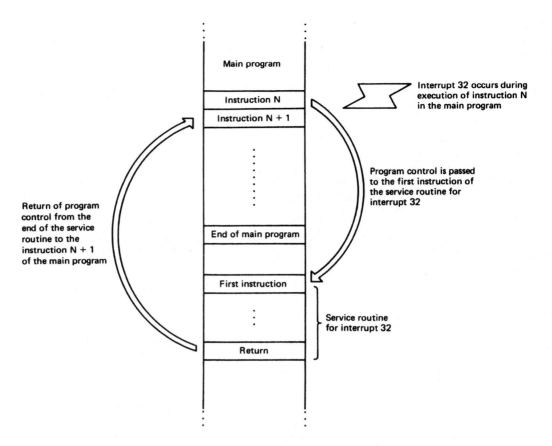

Figure 12.1 Interrupt program context switch mechanism.

external hardware indicating that an external device, such as a printer, requires service, the MPU must suspend what it is doing in the main part of the program and pass control to a special routine that performs the function required by the device. The section of program to which control is passed is called the *interrupt-service routine.* In the case of our example of a printer, the routine is usually called the *printer driver,* which is the piece of software that, when executed, drives the printer output interface.

As shown in Fig. 12.1, interrupts supply a well-defined context-switching mechanism for changing program environments. Here we see that interrupt 32 occurs as instruction N of the program is being executed. When the 80386DX terminates execution of the main program in response to interrupt 32, it first saves information that identifies the instruction following the one where the interrupt occurred, which is instruction $N + 1$, and then picks up execution with the first instruction in the service routine. After this routine has run to completion, program control is returned to the point where the MPU originally left the main program, instruction $N + 1$, and then execution resumes.

The 80386DX is capable of implementing any combination of up to 256 interrupts. As shown in Fig. 12.2, they are divided into five groups: *external hardware interrupts, nonmaskable interrupt, software interrupts, internal interrupts and exceptions,* and *reset.* The function of the external hardware, software, and nonmaskable interrupts are defined by the user. For instance, hardware interrupts are often assigned to devices such as the keyboard,

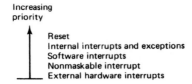

Increasing
priority

Reset
Internal interrupts and exceptions
Software interrupts
Nonmaskable interrupt
External hardware interrupts

Figure 12.2 Types of interrupts and their priority.

printer, and timers. On the other hand, the functions of the internal interrupts and exceptions and reset are not user-defined. They perform dedicated system functions.

Hardware, software, and internal interrupts and exceptions are serviced on a *priority* basis. Priority is achieved in two ways. First, the interrupt-processing sequence implemented in the 80386DX tests for the occurrence of the various groups based on the hierarchy shown in Fig. 12.2. Thus we see that internal interrupts and exceptions are the highest-priority group, and the hardware interrupts are the lowest-priority group.

Second, each of the interrupts and exceptions is given a different priority level by assigning it a *type number*. Type 0 identifies the highest-priority interrupt, and type 255 identifies the lowest-priority interrupt. Actually, a few of the type numbers are not available for use with software or hardware interrupts. This is because they are reserved for special interrupt functions of the 80386DX, such as the internal interrupts and exceptions. For instance, within the internal interrupt and exception group, the exception known as *divide error* is assigned to type number 0. Therefore, it has the highest priority of the exception functions. Another exception, called *general protection*, is assigned the type number 13.

The importance of priority lies in the fact that, if an interrupt-service routine has been initiated to perform the function assigned to a specific priority level, only devices with higher priority are allowed to interrupt the active service routine. Lower-priority devices will have to wait until the current routine is completed before their request for service can be acknowledged. For hardware interrupts, this priority scheme is implemented in external hardware. For this reason, the user normally assigns tasks that must not be interrupted frequently to higher-priority levels and those that can be interrupted to lower-priority levels.

An example of a high-priority service routine that should not be interrupted is that for a power failure. Once initiated, this routine should be quickly run to completion to assure that the microcomputer goes through an orderly power-down. A keyboard should also be assigned to a high-priority interrupt. This will assure that the keyboard buffer does not get full and lock out additional entries. On the other hand, devices such as the floppy disk or hard disk controller are typically assigned to a lower priority level.

We just pointed out that once an interrupt service routine is initiated, it can be interrupted only by a function that corresponds to a higher-priority level. For example, if a type 50 external hardware interrupt is in progress, it can be interrupted by the nonmaskable interrupt, all internal interrupts and exceptions, software interrupts, or any external hardware interrupt with type number less than 50. That is, external hardware interrupts with priority levels equal to 50 or greater are *masked out*.

▲ 12.3 INTERRUPT VECTOR AND INTERRUPT DESCRIPTOR TABLES

An address pointer table is used to link the interrupt type numbers to the location of their service routines in program-storage memory. In a real-mode 80386DX-based microcomputer system, this table is called the *interrupt-vector table*. On the other hand, in a protected-mode system, the table is referred to as the *interrupt-descriptor table*. Figure 12.3 shows a

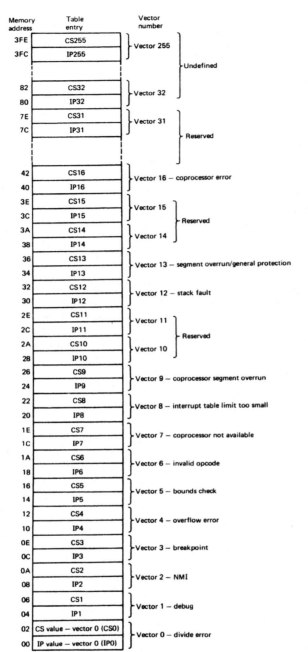

Figure 12.3 Real-mode interrupt-vector table.

map of the interrupt-vector table in the memory of a real-mode 80386DX microcomputer. Looking at the table, we see that it contains 256 *address pointers (vectors)*, which are identified as *vector 0* through *vector 255*. That is, one pointer corresponds to each of the interrupt types 0 through 255. These address pointers identify the starting locations of their service routines in program memory. The contents of these tables may be either held as firmware in EPROMs or loaded into RAM as part of the system initialization routine.

Notice in Fig. 12.3 that the interrupt-vector table is located at the low-address end of the memory address space. It starts at address 00000_{16} and ends at $003FE_{16}$. This represents the first 1K bytes of memory. Actually, the interrupt-vector table or interrupt-descriptor table can be located anywhere in the memory address space. Its starting location and size are defined by the contents of a register within the 80386DX called the *interrupt-descriptor table register* (IDTR). When the 80386DX is reset at power on, it comes up in the real mode with the bits of the base address in IDTR all equal to zero and the limit set to $03FF_{16}$. This positions the interrupt-vector table, as shown in Fig. 12.3. Moreover, when in the real mode, the value in IDTR is normally left at this initial value to maintain compatibility with 8088/8088-based microcomputer software.

Each of the 256 vectors requires 2 words (1 double word) of memory. These words are always stored at a double-word aligned address boundary. The higher-addressed word of the two-word vector is called the *base address*. It identifies the program memory segment in which the service routine resides. For this reason, it is loaded into the code segment (CS) register within the MPU. The lower-addressed word of the vector is the *offset* of the first instruction of the service routine from the beginning of the code segment defined by the base address loaded into CS. This offset is loaded into the instruction pointer (IP) register. For example, the offset and base address for type number 255, IP_{255} and CS_{255}, are stored at word addresses $003FC_{16}$ and $003FE_{16}$, respectively. When loaded into the MPU, it points to the instruction at $CS_{255}{:}IP_{255}$.

Looking more closely at the table in Fig. 12.3, we find that the first 31 vectors either have dedicated functions or are reserved. For instance, pointers 0, 1, 3, and 4 are used by the 80386DX's internal interrupts or exceptions: *divide error, debug exception, breakpoint,* and *overflow error*, respectively. The remainder of the table, the 224 vectors in the address range 00080_{16} through $003FF_{16}$, is available to the user for storage of software or hardware-interrupt vectors. These pointers correspond to type numbers 32 through 255. In the case of external hardware interrupts, each type number (priority level) is associated with an interrupt input in the external interrupt interface circuitry.

EXAMPLE 12.1

At what address should vector 50, CS_{50} and IP_{50}, be stored in memory?

Solution

Each vector requires 4 consecutive bytes of memory for storage. Therefore, its address can be found by multiplying the type number by 4. Since CS_{50} and IP_{50} represent the words of the type 50 interrupt pointer, we get

$$\text{address} = 4 \times 50 = 200$$

Converting to binary form gives

$$\text{address} = 11001000_2$$

and expressing it as a hexadecimal number results in

$$\text{address} = C8_{16}$$

Therefore, IP_{50} is stored at $000C8_{16}$ and CS_{50} at $000CA_{16}$.

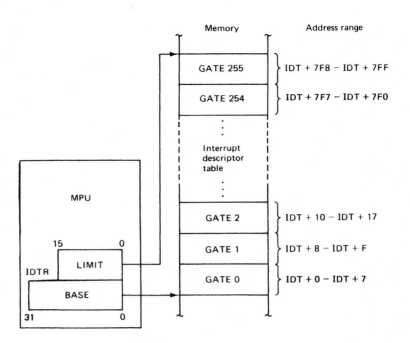

Figure 12.4 Accessing a gate in the protected-mode interrupt-descriptor table.

The protected-mode interrupt-descriptor table can reside anywhere in the 80386DX's physical address space. The location and size of this table is again defined by the contents of the IDTR. Figure 12.4 shows that IDTR contains a 32-bit *base address* and a 16-bit *limit*. The base address identifies the starting point of the table in memory. On the other hand, the limit determines the number of bytes in the table.

The interrupt-descriptor table contains gate descriptors, not vectors. In Fig. 12.4 we find that the table contains a maximum of 256 gate descriptors. These descriptors are identified as *gate 0* through *gate 255*. Each gate descriptor can be defined as a *trap gate, interrupt gate*, or *task gate*. Interrupt and trap gates permit control to be passed to a service routine that is located within the current task. On the other hand, the task gate permits program control to be passed to a different task.

Just as a real-mode interrupt vector, a protected-mode gate acts as a pointer that is used to direct program execution to the starting point of a service routine. However, unlike an interrupt vector, a gate descriptor takes up 8 bytes of memory. For instance, in Fig. 12.4 we see that gate 0 is located at addresses IDT + 0H through IDT + 7H and gate 255 is at addresses IDT + 7F8H through IDT + 7FFH. If all 256 gates are not needed for an application, limit in the IDTR can be set to a value lower than $07FF_{16}$ to minimize the amount of memory reserved for the table.

Figure 12.5 illustrates the format of a typical interrupt or trap gate descriptor. Here we see that the two lower-addressed words, 0 and 1, are the interrupt's *code offset 0 through 15* and *segment selector*, respectively. The highest-addressed word, word 3, is the interrupt's *code offset 16 through 31*. These three words identify the starting point of the service routine. The upper byte of word 2 of the descriptor is called the *access rights byte*. The settings of the bits in this byte identify whether or not this gate descriptor is valid, the privilege level of the service routine, and the type of gate. For example, the *present bit* (P) needs to be

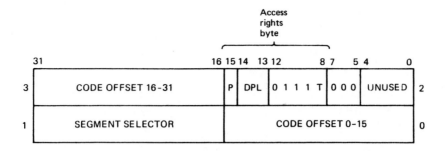

Figure 12.5 Format of a trap or interrupt-gate descriptor.

set to logic 1 if the gate descriptor is to be active. The next 2 bits, identified as DPL in Fig. 12.5, are used to assign a privilege level to the service routine. If these bits are made 00, level 0, which is the most privileged level, is assigned to the gate. Finally, the setting of the *type bit* (T) determines if the descriptor works as a trap gate or an interrupt gate. T equal to 0 selects interrupt-gate mode of operation. The only difference between the operation of these two types of gates is that, when a trap gate context switch is performed, IF is not cleared to disable external hardware interrupts.

Normally, external hardware interrupts are configured with interrupt-gate descriptors. Once an interrupt request has been acknowledged for service, the external hardware-interrupt interface is disabled with IF. In this way, additional external interrupts cannot be accepted unless the interface is reenabled under software control. On the other hand, internal interrupts, such as software interrupts, usually use trap gate descriptors. In this case, the hardware-interrupt interface in not affected when the service routine for the software interrupt is initiated. Sometimes low-priority hardware interrupts are assigned trap gates instead of an interrupt gate. This will permit higher-priority external events to easily interrupt their service routine.

▲ 12.4 INTERRUPT INSTRUCTIONS

A number of instructions are provided in the instruction set of the 80386DX microprocessor for use with interrupt and exception processing. These instructions are listed, with brief descriptions of their functions, in Fig. 12.6.

For instance, the first two instructions, which are CLI and STI, permit manipulation of the 80386DX's interrupt flag through software. STI stands for *set interrupt flag*. Execution of this instruction enables the external interrupt request (INTR) input for operation. That is, it sets interrupt flag (IF). On the other hand, execution of CLI (*clear interrupt flag*) disables the external interrupt input. It does this by resetting IF. When STI or CLI is executed in protected mode, the current privilege level is compared to the I/O privilege level (IOPL) in the flag register. If the current level is less than IOPL, the instruction is not executed; instead, a general protection exception results. General protection exceptions are explained later in the chapter.

Earlier we pointed out that the contents of IDTR determines the location and size of the interrupt-descriptor table and that after reset the base address is initialized to 00000000_{16} and the limit, to $000003FF_{16}$. The next two instructions in Fig. 12.6 let us modify or examine the contents of this register.

Mnemonic	Meaning	Format	Operation	Flags affected
CLI	Clear interrupt flag	CLI	$0 \rightarrow (IF)$	IF
STI	Set interrupt flag	STI	$1 \rightarrow (IF)$	IF
LIDT	Load interrupt descriptor table register	LIDT EA	$(EA) \rightarrow (LIMIT_{0-15})$ $(EA + 2) \rightarrow (BASE_{0-15})$ $(EA + 4) \rightarrow (BASE_{16-32})$	None
SIDT	Store interrupt descriptor table register	SIDT EA	$(LIMIT_{0-15}) \rightarrow (EA)$ $(BASE_{0-15}) \rightarrow (EA + 2)$ $(BASE_{16-32}) \rightarrow (EA + 4)$	None
INT n	Type n software interrupt	INT n	[Real Mode] $(Flags) \rightarrow ((SP) - 2)$ $0 \rightarrow TF, IF$ $(CS) \rightarrow ((SP) - 4)$ $(2 + 4 \cdot n) \rightarrow (CS)$ $(IP) \rightarrow ((SP) - 6)$ $(4 \cdot n) \rightarrow (IP)$	TF, IF
IRET	Interrupt return	IRET	[real mode] $((SP)) \rightarrow (IP)$ $((SP) + 2) \rightarrow (CS)$ $((SP) + 4) \rightarrow (Flags)$ $(SP) + 6 \rightarrow (SP)$	All
INTO	Interrupt on overflow	INTO	INT 4 steps	TF, IF
BOUND	Check array index against bounds	BOUND D, S	$(D) < (S) \rightarrow INT5$ or $(D) > (S + 2) \rightarrow INT5$	None
HLT	Halt	HLT	Wait for an external interrupt or reset to occur	None
WAIT	Wait	WAIT	Wait for $\overline{BUSY}$ to go inactive	None

Figure 12.6 Interrupt instructions.

As its name implies, the *load interrupt-descriptor table register* (LIDT) is the instruction used to modify the contents of IDTR. The general form of the instruction is given in Fig. 12.6 as

```
LIDT    EA
```

Here EA stands for the effective address of the operand in memory. This operand is 3 words in length and contains the values of the base address and limit that are to be loaded into IDTR. Figure 12.7 shows the format of these data. Notice that the lowest addressed word is the 16-bit limit, and the next 2 words are the 32-bit base address. For instance, executing the instruction

```
LIDT    IDT_TABLE
```

causes IDTR to be loaded with the limit held at address IDT_TABLE and the base address held at IDT_TABLE+2 and IDT_TABLE+4. These values must be stored in memory prior to execution of the LIDT instruction. In the protected mode, the LIDT instruction can be executed only when the 80386DX is operating at privilege level 0.

The instruction *store interrupt-descriptor table register* (SIDT) can be used to examine the contents of IDTR. As shown in Fig. 12.6, its format is

```
SIDT    EA
```

When executed it saves the current contents of the IDTR in the format shown in Fig. 12.6, starting in memory at the storage location pointed to by effective address EA. Once the value is stored in memory, it can be examined with additional software.

The next instruction listed in Fig. 12.6 is the *software-interrupt* instruction INT n. It is used to initiate a vectored call of a subroutine. Executing the instruction causes program control to be transferred to the subroutine pointed to by the vector or gate for the number n specified in the instruction.

The operation outlined in Fig. 12.6 describes the effect of executing the INT instruction in the real mode. For example, execution of the instruction INT 50 initiates execution of a subroutine whose starting point is identified by vector 50 in the pointer table of Fig. 12.3. First, the 80386DX saves the old flags on the stack, clears TF and IF, and saves the old program context, CS and I'', on the stack. Then it reads the values of IP_{50} and CS_{50} from addresses $000C8_{16}$ and $000CA_{16}$, respectively, in memory, loads them into the IP and CS registers, calculates the physical address $CS_{50}:IP_{50}$, and starts to fetch instructions from this new location in program memory.

An *interrupt-return* (IRET) instruction must be included at the end of each interrupt-service routine. It is required to pass control back to the point in the program where execution was terminated due to the occurrence of the interrupt. As shown in Fig. 12.6, when executed in real mode, IRET causes the old values of IP, CS, and flags to be popped

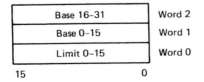

Base 16–31	Word 2
Base 0–15	Word 1
Limit 0–15	Word 0

15   0

Figure 12.7 LIDT instruction data type.

from the stack back into the internal registers of the MPU. This restores the original program environment.

INTO is the *interrupt-on-overflow* instruction. This instruction must be included after arithmetic instructions that can result in an overflow condition, such as divide. It tests the overflow flag, and if the flag is found to be set, a type 4 internal interrupt is initiated. In the real mode, this condition causes program control to be passed to an overflow service routine that is located at the starting address identified by the vector IP_4 at 00010_{16} and CS_4 at 00012_{16} of the pointer table in Fig. 12.3.

As its name implies, the *check array index against bounds* (BOUND) instruction can determine if the contents of a register, called the *array index*, lies within a set of minimum/maximum values, called the *upper bound* and *lower bound*. This type of operation is important when accessing elements of an array of data in memory. The format of the BOUND instruction is given in Fig. 12.6.

An example is the instruction

```
BOUND   SI,LIMITS
```

Notice that the instruction contains two operands. The first operand represents the register whose word contents are to be tested to verify whether or not it lies within the boundaries. In our example, this is the source index register (SI). The second operand is the effective relative address of the first of two word-storage locations in memory that contain the values of the lower and upper boundaries. In the example the word of data starting at address LIMITS is the value of the lower boundary and that at address LIMITS+2 is the value of the upper boundary.

When this BOUND instruction is executed, the contents of SI are compared to both the value of the lower bound at LIMITS and upper bound at LIMITS+2. If it is found to be either less than the lower bound or more than the upper bound, an exception occurs and control is passed to a service routine through the vector or gate for type number 5. Otherwise, the next sequential instruction is performed. The operands can also be 32 bits in length.

EXAMPLE 12.2

For the instruction,

```
BOUNDS   EDI,LIMITS
```

where LIMITS equals 101000_{16}, what are the addresses of the values of the upper and lower bounds for the value in EDI?

Solution

The lower boundary is the 32-bit word starting at address 101000_{16}, and the upper boundary is the double word at address 101004_{16}.

The last two instructions associated with the interrupt interface are *halt* (HLT) and *wait* (WAIT). They produce similar responses by the 80386DX and permit the operation of the MPU to be synchronized to an event in external hardware. For instance, when HLT is executed, the MPU suspends operation and enters the idle state. It no longer executes instructions; instead, it remains idle waiting for the occurrence of an external hardware interrupt, nonmaskable interrupt, or reset. With the occurrence of any of these events, the MPU resumes execution with the corresponding service routine. HLT is a privileged instruction and can only be executed at privilege level 0 in a protected-mode system.

If the WAIT instruction is used instead of the HLT instruction, the 80386DX checks the logic level of the $\overline{BUSY}$ input prior to going into the idle state. Only if $\overline{BUSY}$ is tested and found to be active, logic 0, will the MPU go into the idle state. While in the idle state, the MPU continues to check the logic level at $\overline{BUSY}$, looking for a transition back to its inactive level, logic 1. As $\overline{BUSY}$ switches back to 1, execution resumes with the next sequential instruction in the program. For example, the $\overline{BUSY}$ input is normally used by the 80387DX numerics coprocessor and is switched to logic 0 whenever a numeric operation is in progress. Therefore, WAIT can be used to determine when a numerics operation is completed.

▲ 12.5 ENABLING/DISABLING OF INTERRUPTS

An *interrupt-enable flag* bit is provided within the 80386DX MPU. Earlier we found that it is identified as IF. It affects only the external hardware-interrupt interface, not software interrupts, the nonmaskable interrupt, or internal interrupts or exceptions. The ability to initiate an external hardware interrupt at the INTR input is enabled by setting IF or masked output by resetting it. Through software, this can be done by executing the STI instruction or the CLI instruction, respectively.

During the initiation sequence of a service routine for an external hardware interrupt, the 80386DX automatically clears IF. This masks out the occurrence of any additional external hardware interrupts. In some applications, it may be necessary to permit other higher-priority external hardware interrupts to interrupt the active service routine. If this is the case, the interrupt flag bit can be set with an STI instruction located at the beginning of the service routine to reenable the INTR input. Otherwise, at the end of the service routine, the external hardware-interrupt interface is reenabled by the IRET instruction.

▲ 12.6 EXTERNAL HARDWARE-INTERRUPT INTERFACE

Up to this point, we have introduced the types of interrupts supported by the 80386DX, its interrupt vector and descriptor tables, interrupt instructions, and masking of interrupts. Earlier we pointed out that type numbers 32 through 255 can be used by external hardware interrupts. Let us now look at the *external hardware-interrupt interface* of the 80386DX microcomputer.

A general interrupt interface for an 80386DX-based microcomputer system is illustrated in Fig. 12.8. Here we see that it includes the address and data buses, byte-enable signals, bus cycle indication signals, lock output, and the ready and interrupt-request inputs. Moreover, external circuitry is required to interface the interrupt inputs, INT_{32} through INT_{255}, to the 80386DX's interrupt interface. This interface circuit must identify which of the pending active interrupts has the highest priority, perform an interrupt-request/ acknowledge handshake, and then set up the bus to pass an interrupt-type number to the MPU.

In this circuit we see that the key interrupt interface signals are *interrupt request* (INTR) and *interrupt acknowledge* ($\overline{INTA}$). The logic-level input at the INTR line signals the 80386DX that an external device is requesting service. The 80386DX samples this input at the beginning of each instruction execution cycle, that is, at instruction boundaries.

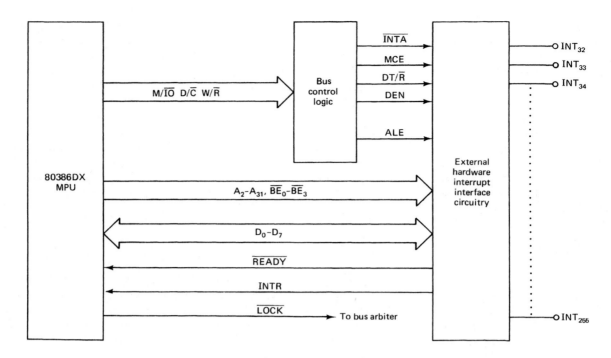

Figure 12.8   80386DX microcomputer system external hardware-interrupt interface.

Logic 1 at INTR represents an active interrupt request. INTR is *level triggered*; therefore, the external hardware must maintain the active level until it is tested by the MPU. If it is not maintained, the request for service may not be recognized. For this reason, inputs INT_{32} through INT_{255} are normally latched. Moreover, the 1 at INTR must be removed before the service routine runs to completion; otherwise, the same interrupt may get acknowledged a second time.

When an interrupt request has been recognized by the 80386DX, it signals this fact to external circuitry by outputting the interrupt-acknowledge bus cycle indication code on $M/\overline{IO}$ $C/\overline{D}$ $W/\overline{R}$. This code, which equals 000_2, is highlighted in Fig. 12.9. Notice in Fig. 12.8 that this code is input to the bus controller logic, where it is decoded to produce a pulse to logic 0 at the $\overline{INTA}$ output. Actually, there are two pulses produced at $\overline{INTA}$ during the *interrupt-acknowledge bus cycle* sequence. The first pulse, which is output during cycle 1, signals external circuitry that the interrupt request has been acknowledged and to prepare to send its type number to the MPU. The second pulse, which occurs during cycle 2, tells

$M/\overline{IO}$	$D/\overline{C}$	$W/\overline{R}$	Type of Bus Cycle
0	0	0	Interrupt acknowledge
0	0	1	Idle
0	1	0	I/O data read
0	1	1	I/O data write
1	0	0	Memory code read
1	0	1	Halt/shutdown
1	1	0	Memory data read
1	1	1	Memory data write

Figure 12.9   Interrupt-acknowledge bus cycle indication code.

the external circuitry to put the type number on the data bus. The ready ($\overline{\text{READY}}$) input can be used to insert wait states into these bus cycles.

Notice that the lower 8 lines of the data bus, D_0 through D_7, are also part of the interrupt interface. During the second cycle in the interrupt-acknowledge bus cycle, external circuitry must put the 8-bit type number of the highest-priority active interrupt-request input onto this part of the data bus. The 80386DX reads the type number off the bus to identify which external device is requesting service. It uses the type number to generate the address of the interrupt's vector or gate in the interrupt vector or descriptor table, respectively, and to read the new values of IP and CS into the corresponding internal registers. IP and CS values from the interrupt vector table are transferred to the MPU over the data bus. Before loading IP and CS with new values, their old values and the values of the internal flags are automatically written to the stack part of memory.

Address lines A_2 through A_{31} and byte-enable lines $\overline{\text{BE}}_0$ through $\overline{\text{BE}}_3$ are also shown in the interrupt interface circuit of Fig. 12.8. This is because LSI interrupt-controller devices are typically used to implement most of the external circuitry. When a read or write bus cycle is performed to the controller, for example, to initialize its internal registers after system reset, some of the address bits are decoded to produce a chip select to enable the controller device, and other address bits are used to select the internal register that is to be accessed. The interrupt controller could be I/O mapped instead of memory mapped; in this case only address lines A_2 through A_{15} are used in the interface. Addresses are also output on A_2 through A_{31} during the write cycles that save the old program context in the stack and the read cycles that are used to load the new program context from the vector table in program memory.

Another signal shown in the interrupt interface of Fig. 12.8 is the *bus lock indication* ($\overline{\text{LOCK}}$) output of the 80386DX. $\overline{\text{LOCK}}$ is used as an input to the bus arbiter circuit in multiprocessor systems. The 80386DX switches this output to its active 0 logic level and maintains it at this level throughout the complete interrupt acknowledge bus cycle. In response to this signal, the arbitration logic assures that no other devices can take over control of the system bus until the interrupt acknowledge bus cycle sequence is completed.

▲ 12.7 EXTERNAL HARDWARE-INTERRUPT SEQUENCE

In the preceding section we showed the interrupt interface for external hardware interrupts in an 80386DX-based microcomputer system. We will continue by describing in detail the events that take place during the interrupt request, interrupt-acknowledge bus cycle, and device service routine.

The interrupt sequence begins when an external device requests service by activating one of the interrupt inputs, INT_{32} through INT_{255}, of the external interrupt interface circuit in Fig. 12.8. For example, if the INT_{50} input is switched to the 1 logic level, it signals the microprocessor that the device associated with priority level 50 wants to be serviced.

The external circuitry evaluates the priority of this input. If there is no other interrupt already in progress or if the new interrupt is of higher priority than the one presently active, the external circuitry issues a request for service to the MPU.

Let us assume that INT_{50} is the only active interrupt input. In this case, the external circuitry switches INTR to logic 1. This tells the 80386DX that an interrupt is pending for service. To ensure that it is recognized, the external circuitry must maintain INTR active until an interrupt-acknowledge bus cycle indication code is output by the 80386DX.

Figure 12.10 is a flow diagram that outlines the events that take place when an 80386DX that is configured for real-mode operation processes an interrupt. The 80386DX tests for an active interrupt request at the end of the current instruction. Note that it tests first for the occurrence of an internal interrupt or exception, then the occurrence of the nonmaskable interrupt, and finally checks the logic level of INTR to determine if an external hardware interrupt has occurred.

If INTR is logic 1, a request for service is recognized. Before the 80386DX initiates the interrupt-acknowledge sequence, it checks the setting of IF (interrupt flag). If IF is logic 0, external interrupts are masked out and the request is ignored. In this case, the next

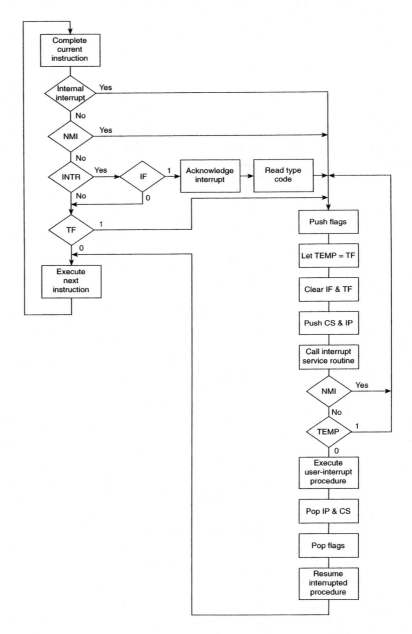

Figure 12.10 Real-mode interrupt processing sequence. (Reprinted by permission of Intel Corp. Copyright/Intel Corp. 1979)

sequential instruction is executed. On the other hand, if IF is at logic 1, external hardware interrupts are enabled and the service routine is to be initiated.

Let us assume that IF is set to permit interrupts to occur when INTR is tested as 1. The 80386DX responds by initiating the interrupt-acknowledge bus cycles. This bus sequence is illustrated in Fig. 12.11. Here we see that at the beginning of T_1 of the first bus cycle (interrupt-acknowledge cycle 1 in Fig. 12.11) the MPU switches $\overline{LOCK}$ to its active 0 logic level and holds it at this value for the complete bus cycle sequence. This locks the bus for uninterrupted use by the 80386DX. At the same time, address lines A_3 through A_{31} and $\overline{BE}_0$ are set to logic 0, whereas A_2 and $\overline{BE}_1$ through $\overline{BE}_3$ are set to 1. Moreover, the bus cycle indication code $M/\overline{IO}\ D/\overline{C}\ W/\overline{R} = 000$ is output to the bus control logic. These signals are latched into external circuitry with the pulse at $\overline{ADS}$. The code 000 is decoded by the bus-control logic to produce a pulse at $\overline{INTA}$.

In the waveforms of Fig. 12.11, $\overline{READY}$ is shown to be logic 1 at the end of the first T_2 state of cycle 1. This signals not ready to the MPU. The response to this condition is that another T_2 state occurs. This T_2 state acts as a wait state to extend the current bus cycle. Notice that in the second T_2 state $\overline{READY}$ is logic 0 and interrupt-acknowledge cycle 1 is complete. The data bus lines of the 80386DX are in the high-Z state during this cycle; therefore, any data on the bus are ignored.

If a single-interrupt controller is used in the interrupt interface circuit of an 80386DX-based microcomputer system, the MPU uses the first interrupt-acknowledge bus cycle to acknowledge to external circuitry that an interrupt request has been accepted. On the other hand, systems that employ a master-slave interrupt-controller configuration use this first interrupt-acknowledge cycle both to signal external circuitry that an interrupt request has been acknowledged and to tell the master controller to tell the slave controllers which

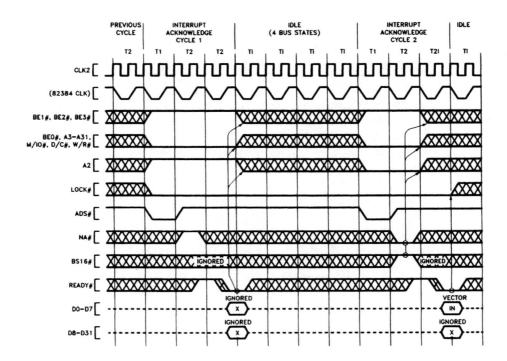

Figure 12.11 Interrupt-acknowledge bus cycle. (Reprinted by permission of Intel Corp. Copyright/Intel Corp. 1979)

Sec. 12.7   External Hardware-Interrupt Sequence

635

interrupt request has been accepted. The master interrupt controller outputs what is called the *cascade address* to all slave controllers in parallel. This address identifies the interrupt controller that is to supply the type number to the MPU.

At the completion of the first interrupt-acknowledge bus cycle, the bus controller automatically inserts four idle states (T_I) before initiating the second interrupt-acknowledge cycle. Looking at Fig. 12.11, we find that during interrupt-acknowledge cycle 2, the levels and timing of all signals except A_2 are essentially the same as those for the first cycle. It is during this second bus cycle that the interrupt controller passes one of the interrupt-type numbers, $32 = 20_{16}$ through $255 = FF_{16}$, from the interrupt interface circuit to the 80386DX. The type number, identified as vector in Fig. 12.11, is supplied to the MPU over data bus lines D_0 through D_7. For the case of INT_{50}, the code would be $00110010_2 = 32_{16}$. This completes the interrupt-request/acknowledge handshake.

Looking at Fig. 12.10, we see that the 80386DX next saves the contents of the flag register by pushing it to the stack. This requires one write cycle. Then the TF and IF flags are cleared. This disables the single-step mode of operation if it happens to be active and masks out additional external hardware interrupts. Now the MPU automatically pushes the contents of CS and IP onto the stack. This requires two more write cycles to take place over the system bus. The current value of the stack pointer is decremented by four as each of these values is put onto the top of the stack.

Now the 80386DX knows the type number associated with the external device that is requesting service. It must next call the service routine by fetching the interrupt vector that defines its starting point in the memory. The type number is internally multiplied by four, and this result is used as the address of the first word of the interrupt vector in the vector table. A read operation is performed to read the two-word vector from the memory. The lower-addressed word is loaded into IP, and the higher-addressed word is loaded into CS. For instance, the words of the vector for INT_{50} would be read as a double word from address $000C8_{16}$.

The service routine is now initiated. That is, execution resumes with the first instruction of the service routine. It is located at the address generated from the new values in CS and IP. Figure 12.12 shows the structure of a typical interrupt-service routine. The service routine includes PUSH instructions to save the contents of those internal registers that it will use. In this way, their original contents are saved in the stack during execution of the routine. The contents of all registers can be saved simply by including a single PUSHA instruction instead of a separate PUSH instruction for each register.

At the end of the service routine, the original program environment must be restored. This is done by first popping the contents of the appropriate registers from the stack by executing POP instructions (or POPA). An IRET instruction must be executed as the last instruction of the service routine. This instruction reenables the interrupt interface and causes the old contents of the flags, CS and IP, to be popped from the stack back into the

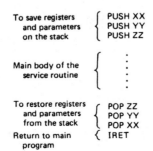

Figure 12.12 Structure of an interrupt-service routine.

internal registers of the MPU. The original program environment has now been completely restored and execution resumes at the point in the program where it was interrupted.

If an 80386DX-based microcomputer is configured for protected mode of operation, the interrupt-processing sequence is different than we just described. Actually, the interrupt-request/acknowledge handshake sequence appears to take place exactly the same way in the external hardware; however, a number of changes do occur in the internal interrupt-processing sequence of the MPU. Let us now look at how the protected mode 80386DX reacts to an interrupt request.

When processing interrupts in protected mode, the general protection mechanism of the 80386DX comes into play. The general protection rules dictate that program control can be directly passed only to a service routine that is in a segment with equal or higher privilege—that is, a segment with an equal or lower-numbered descriptor privilege level. Any attempt to transfer program control to a routine in a segment with lower privilege (higher-numbered descriptor privilege level) results in an exception unless the transition is made through a gate.

Typically, interrupt drivers are in code segments at a high privilege level, possibly level 0. Moreover, interrupts occur randomly; therefore, there is a good chance that the microprocessor will be executing application code that is at a low privilege level. In the case of interrupts, the current privilege level (CPL) is the privilege level assigned by the descriptor of the software that was executing when the interrupt occurred. This could be any of the 80386DX's valid privilege levels. The privilege level of the service routine is that defined in the interrupt or trap gate descriptor for the type number. That is, it is the descriptor privilege level (DPL).

When a service routine is initiated, the current privilege level may change. This depends on whether the software that was interrupted was in a code segment that was configured as *conforming* or *noncomforming*. If the interrupted code is in a conforming code segment, CPL does not change when the service routine is initiated. In this case, the contents of the stack after the context switch is as illustrated in Fig. 12.13(a). Since the privilege level does

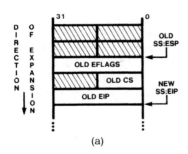

(a)

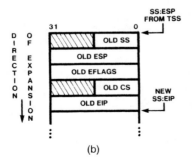

(b)

Figure 12.13 (a) Stack after context switch with no privilege-level transition. (Reprinted by permission of Intel Corp. Copyright/Intel Corp. 1986.) (b) Stack after context switch with a privilege-level transition. (Reprinted by permission of Intel Corp. Copyright/Intel Corp. 1986)

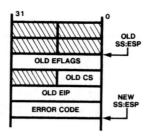

Figure 12.14 Stack contents after interrupt with an error. (Reprinted by permission of Intel Corp. Copyright/Intel Corp. 1986)

not change, the current stack (OLD SS:ESP) is used. Notice that as part of the interrupt initiation sequence, the OLD EFLAGS, OLD CS, and OLD EIP are automatically saved on the stack. Actually, the *requested privilege level* (RPL) code is also saved on the stack. This is because it is part of OLD CS. RPL identifies the protection level of the interrupted routine.

However, if the segment is nonconforming, the value of DPL is assigned to CPL as long as the service routine is active. As shown in Fig. 12.13(b), this time the stack is changed to that for the new privilege level. The MPU is loaded with a new SS and new ESP from TSS, and then the old stack pointer, OLD SS and OLD SP, are saved on the stack, followed by the OLD EFLAGS, OLD CS, and OLD EIP. Remember that for an interrupt gate, IF is cleared as part of the context switch, but for a trap gate, IF remains unchanged. In both cases, the TF flag is reset after the contents of the flag register are pushed to the stack.

Figure 12.14 shows the stack as it exists after an attempt to initiate an interrupt-service routine that did not involve a privilege-level transition has failed. Notice that the context switch to the exception service routine caused an *error code* to be pushed onto the stack following the values of OLD EFLAGS, OLD CS, and OLD EIP.

One format of the error code is given in Fig. 12.15. This type error code is known as an *IDT error code*. Here we see that the least significant bit, which is labeled EXT, indicates whether the error was for an externally or an internally initiated interrupt. For external interrupts, such as the hardware interrupts, the EXT bit is always set to logic 1. The next bit, which is labeled IDT, is set to 1 if the error is produced as the result of an interrupt. That is, it is the result of a reference to a descriptor in the IDT. If IDT is not set, the third bit indicates whether the descriptor is in the GDT (TI = 0) or the LDT (TI = 1). The next 14 bits contain the segment selector that produced the error condition. With this information available on the stack, the exception service routine can determine which interrupt attempt had failed and whether it was internally or externally initiated. A second format is used for errors that result from a protected-mode page fault. Figure 12.16 illustrates this error code and the function of its bits.

Just as in real mode, the IRET instruction is used to return from a protected-mode interrupt-service routine. For service routines using an interrupt gate or trap gate, IRET is restricted to the return from a higher privilege level to a lower privilege level, for instance, from level 1 to level 3. Once the flags, OLD CS and OLD EIP, are returned to the 80386DX, the RPL bits of OLD CS are tested to see if they equal CPL. If RPL = CPL, an intralevel

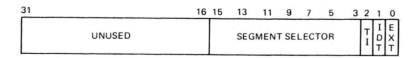

Figure 12.15 IDT error-code format.

Field	Value	Description
U/S	0	The access causing the fault originated when the processor was executing in supervisor mode.
	1	The access causing the fault originated when the processor was executing in user mode.
W/R	0	The access causing the fault was a read.
	1	The access causing the fault was a write.
P	0	The fault was caused by a not-present page.
	1	The fault was caused by a page-level protection violation.

Figure 12.16 Page-fault error-code format and bit functions. (Reprinted by permission of Intel Corp. Copyright/Intel Corp. 1986)

return is in progress. In this case the return is complete and program execution resumes at the point in the program where execution had stopped.

If RPL is greater than CPL, an interlevel return is taking place, not an intralevel return. During an interlevel return, checks are performed to determine if a protection violation will occur due to the protection-level transition. Assuming that no violations occur, the OLD SS and OLD ESP are popped from the stack into the MPU and then program execution resumes.

EXAMPLE 12.3

The interrupting device in Fig. 12.17(a) interrupts the microprocessor each time the Interrupt Request input signal has a transition from 0 to 1. The corresponding interrupt type number generated by the 74LS244 in response to $\overline{\text{INTA}}$ is 60H.

    a. Describe the operation of the hardware for an active request at the Interrupt Request input.

    b. What is the value of the type number sent to the microprocessor?

    c. Assume that the original values in the segment registers are CS = DS = 1000H and SS = 4000H; the main program is located at offsets of 200H from the beginning of the original code segment; the count is held at an offset of 100H from the beginning of the current data segment; the interrupt-service routine starts at offset 1000H from the beginning of another code segment that begins at address 2000H:0000H; and the stack starts at an offset of 500H from the beginning of the current stack segment. Make a map showing the organization of the memory address space.

    d. Write the main program and the service routine for the circuit so that the positive transitions at INTR are counted as a decimal number.

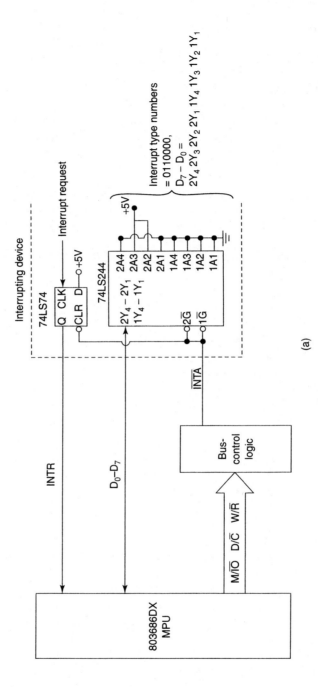

Figure 12.17(a) Circuit for Example 12.3.

(a)

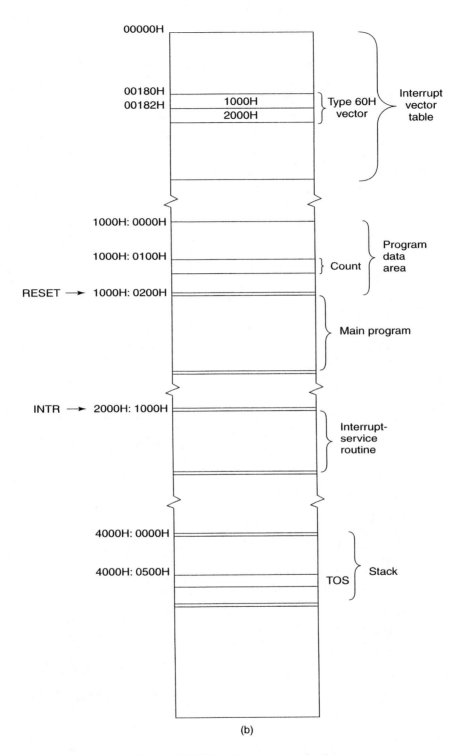

Figure 12.17(b) Memory organization.

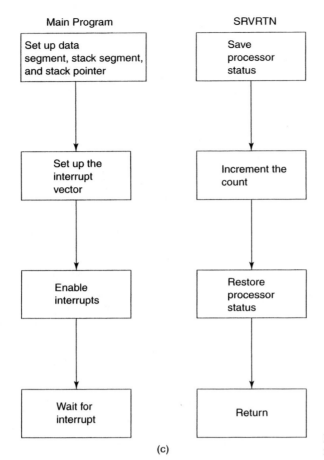

Main Program

| Set up data segment, stack segment, and stack pointer |

↓

| Set up the interrupt vector |

↓

| Enable interrupts |

↓

| Wait for interrupt |

(c)

SRVRTN

| Save processor status |

↓

| Increment the count |

↓

| Restore processor status |

↓

| Return |

Figure 12.17(c) Flowcharts for the main program and service routine.

Solution

a. Analysis of the circuit in Fig. 12.17(a) shows that a positive transition at the CLK input of the flip-flop (Interrupt Request) makes the Q output of the flip-flop logic 1 and presents a positive level signal at the INTR input of 80386DX. When the MPU recognizes this as an interrupt request, it responds by outputting the interrupt-acknowledge bus cycle indication code. This code is decoded by the bus-control logic to produce the $\overline{\text{INTA}}$ control signal. The logic 0 output on this line clears the flip-flop and enables the 74LS244 buffer to present the type number to the 80386DX. This number is read off the data bus by the MPU and is used to initiate the interrupt-service routine.

b. From the inputs and outputs of the 74LS244, we see that the type number is

$$D_7 . . . D_1 D_0 = 2Y_4 2Y_3 2Y_2 2Y_1 1Y_4 1Y_3 1Y_2 1Y_1 = 01100000_2$$

$$D_7 . . . D_1 D_0 = 60H$$

c. The memory organization in Fig. 12.17(b) shows where the various pieces of program and data are located. Here we see that the type 60H vector is located in the interrupt-vector table at address 60H $\times$ 4 = 180H. Notice that the byte-wide memory location

```
START:      MOV  AX,1000H           ;Setup data segment at 1000H:0000H
            MOV  DS,AX
            MOV  AX,4000H           ;Setup stack segment at 4000H:0000H
            MOV  SS,AX
            MOV  SP,0500H           ;TOS is at  4000H:0500H
            MOV  AX,0000H           ;Segment for interrupt vector table
            MOV  ES,AX
            MOV  AX,1000H           ;Service routine offset
            MOV  [ES:180H],AX
            MOV  AX,2000H           ;Service routine segment
            MOV  [ES:182H],AX
            STI                     ;Enable interrupts
HERE:       JMP  HERE               ;Wait for interrupt

; Interrupt Service Routine, SRVRTN = 2000H:1000H

SRVRTN:     PUSH AX                 ;Save register to be used
            MOV  AL,[0100H]         ;Get the count
            INC  AL                 ;Increment the count
            DAA                     ;Decimal asdjust the count
            MOV  [0100H],AL         ;Save the updated count
            POP  AX                 ;Restore the register used
(d)         IRET                    ;Return from the interrupt
```

Figure 12.17(d)   Main program and service routine.

used for count is at address 1000H:0100H. This part of the memory address space is identified as the program data area in the memory map. The main part of the program, which is entered after reset, starts at address 2000H:1000H. On the other hand, the service routine is located at address 2000H:1000H in a separate code segment. For this reason, the vector held at 180H of the interrupt-vector table is (CS) = 2000H and (IP) = 1000H. Finally, the stack begins at 4000H:0000H with the current top of the stack located at 4000H:0500H.

d. The flowcharts in Fig. 12.17(c) show how the main program and interrupt-service routines are to function. The corresponding software is given in Fig. 12.17(d).

▲ 12.8 82C59A PROGRAMMABLE INTERRUPT CONTROLLER

The 82C59A is an LSI peripheral IC that is designed to simplify the implementation of the interrupt interface in the 80386DX-based microcomputer systems. This device is known as a *programmable interrupt controller*, or *PIC*. It is manufactured using the CMOS technology.

The operation of the PIC is programmable under software control, and it can be configured for a wide variety of applications. Some of its programmable features are the ability to accept level-sensitive or edge-triggered inputs, the ability to be easily cascaded to expand from 8 to 64 interrupt inputs, and its ability to be configured to implement a wide variety of priority schemes.

Block Diagram of the 82C59A

Let us begin our study of the PIC with its block diagram in Fig. 12.18(a). We just mentioned that the 82C59A is treated as a peripheral in the microcomputer. Therefore, its operation must be initialized by the microprocessor. The *host processor interface* is provided for this purpose. This interface consists of 8 *data bus lines* D_0 through D_7 and control signals *read* ($\overline{RD}$), *write* ($\overline{WR}$), and *chip select* ($\overline{CS}$). The data bus is the path over which data are transferred between the MPU and 82C59A. These data can be command words, status information, or interrupt-type numbers. Control input $\overline{CS}$ must be at logic 0 to enable the host processor interface. Moreover, $\overline{WR}$ and $\overline{RD}$ signal the 82C59A whether data are to be written into or read from its internal registers. They also control the timing of these data transfers.

Two other signals, INT and $\overline{INTA}$, are identified as part of the host processor interface. Together, these two signals provide the handshake mechanism by which the 82C59A can signal the MPU of a request for service and receive an acknowledgment that the request has been accepted. INT is the interrupt request output of the 82C59A. It is applied directly to the INTR input of the 80386DX. Logic 1 is produced at this output whenever the interrupt controller receives a valid request from an interrupting device.

On the other hand, $\overline{INTA}$ is an input of the 82C59A. It is connected to the $\overline{INTA}$ output of the 80386DX's bus-control logic. This input of the 82C59A is pulsed to logic 0 twice during the interrupt-acknowledge bus cycle, thereby signaling the 82C59A that the interrupt request has been acknowledged and that it should output the type number of the highest-priority active interrupt on data bus lines D_0 through D_7 such that it can be read

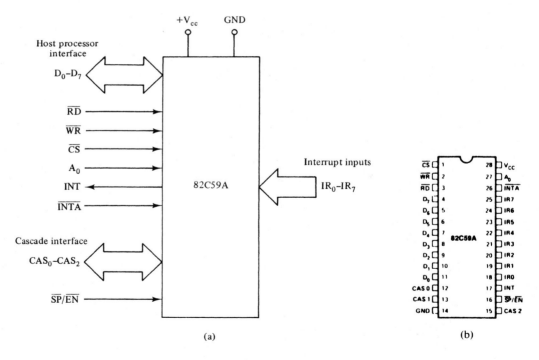

(a)   (b)

Figure 12.18   (a) Block diagram of the 82C59A. (b) Pin layout. (Reprinted by permission of Intel Corp. Copyright/Intel Corp. 1979)

by the MPU. The last signal line involved in the host processor interface is the A_0 input. This input is normally supplied by an address line of the microprocessor such as A_2. The logic level at this input is involved in the selection of the internal register that is accessed during read and write operations.

At the other side of the block in Fig. 12.18(a), we find the 8 *interrupt inputs* of the PIC. They are labeled IR_0 through IR_7. It is through these inputs that external devices issue a request for service. One of the software options of the 82C59A permits these inputs to be configured for *level-sensitive* or *edge-triggered operations*. When configured for level-sensitive operation, logic 1 is the active level of the IR inputs. In this case, the request for service must be removed before the service routine runs to completion. Otherwise, the interrupt will be requested a second time and the service routine initiated again. Moreover, if the input returns to logic 0 before it is acknowledged by the MPU, the request for service will be missed.

Some external devices produce a short-duration pulse instead of a fixed logic level for use as an interrupt-request signal. If the MPU is busy servicing a higher-priority interrupt when the pulse is produced, the request for service could be completely missed if the 82C59A is in level-sensitive mode. To overcome this problem, the edge-triggered mode of operation is used.

Inputs of the 82C59A that are set up for edge-triggered operation become active on the transition from the inactive 0 logic level to the active 1 logic level. This represents what is known as a *positive edge-triggered input*. The fact that this transition has occurred at an IR line is latched internal to the 82C59A. If the IR input remains at the 1 logic level even after the service routine is completed, the interrupt is not reinitiated. Instead, it is locked out. To be recognized a second time, the input must first return to the 0 logic level and then be switched back to 1. The advantage of edge-triggered operation is that if the request at the IR input is removed before the MPU acknowledges service of the interrupt, its request is kept latched internal to the 82C59A until it can be serviced.

The last group of signals on the PIC implement what is known as the *cascade interface*. As shown in Fig. 12.18(a), it includes bidirectional *cascading bus lines* CAS_0 through CAS_2 and a multifunction control line labeled $\overline{SP/EN}$. The primary use of these signals is in cascaded systems where a number of 82C59A ICs are interconnected in a *master/slave configuration* to expand the number of IR inputs from 8 to as high as 64. One of these 82C59A devices is configured as the *master* and all others are set up as *slaves*.

In a cascaded system, the CAS lines of all 82C59As are connected together to provide a private bus between the master and slave devices. In response to the first $\overline{INTA}$ pulse during the interrupt-acknowledge bus cycle, the master PIC outputs a 3-bit code on the CAS lines. This code identifies the highest-priority slave that is to be serviced. It is this device that is to be acknowledged for service. All slaves read this code off the *private cascading bus* and compare it to their internal ID code. A match condition at one slave tells the PIC that it has the highest-priority input. In response, it must put the type number of its highest-priority active input on the data bus during the second interrupt-acknowledge bus cycle.

When the PIC is configured through software for the cascaded mode, the $\overline{SP/EN}$ line is used as an input. This corresponds to its $\overline{SP}$ (*slave program*) function. The logic level applied at $\overline{SP}$ tells the device whether it is to operate as a master or slave. Logic 1 at this input designates master mode and logic 0 designates slave mode.

If the PIC is configured for single mode instead of cascade mode, $\overline{SP/EN}$ takes on another function. In this case, it becomes an enable output that can be used to control the direction of data transfer through the bus transceiver that buffers the data bus.

A pin layout of the 82C59A is given in Fig. 12.18(b).

Internal Architecture of the 82C59A

Now that we have introduced the input/output signals of the 82C59A, let us look at its internal architecture. Figure 12.19 is a block diagram of the PIC's internal circuitry. Here we find eight functional parts: the *data bus buffer, read/write logic, control logic, in-service register, interrupt-request register, priority resolver, interrupt mask register,* and *cascade buffer/comparator.*

We will begin with the function of the data bus buffer and read/write logic sections. It is these parts of the 82C59A that let the MPU have access to the internal registers. Moreover, they provide the path over which interrupt-type numbers are passed to the microprocessor. The data bus buffer is an 8-bit bidirectional three-state buffer that interfaces the internal circuitry of the 82C59A to the data bus of the MPU. The direction, timing, and source or destination for data transfers through the buffer are under control of the outputs of the read/write logic block. These outputs are generated in response to control inputs $\overline{RD}$, $\overline{WR}$, A_0, and $\overline{CS}$.

The interrupt-request register, in-service register, priority resolver, and interrupt mask register are the key internal blocks of the 82C59A. The interrupt mask register (IMR) can be used to enable or mask out individually the interrupt request inputs. It contains 8 bits identified by M_0 through M_7. These bits correspond to interrupt-request inputs IR_0 through IR_7, respectively. Logic 0 in a mask register bit position enables the corresponding interrupt input and logic 1 masks it out. The register can be read from or written into through software control.

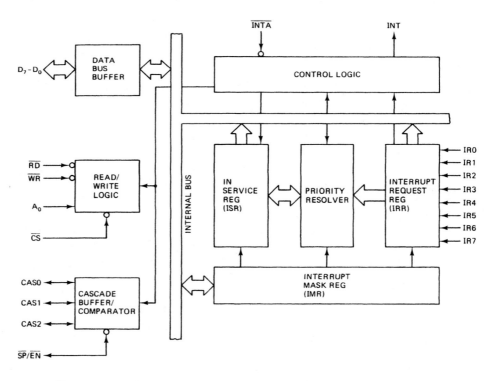

Figure 12.19 Internal architecture of the 82C59A. (Reprinted by permission of Intel Corp. Copyright/Intel Corp. 1979)

On the other hand, the interrupt-request register (IRR) stores the current status of the interrupt-request inputs. It also contains one bit position for each of the IR inputs. The values in these bit positions reflect whether the interrupt inputs are active or inactive.

Which of the active interrupt inputs is identified as having the highest priority is determined by the priority resolver. This section can be configured to work using a number of different priority schemes through software. Following the selected scheme, it identifies which of the active interrupts has the highest priority and signals the control logic that an interrupt is active. In response, the control logic causes the INT signal to be issued to the microprocessor.

The in-service register differs in that it stores the interrupt level that is presently being serviced. During the first $\overline{\text{INTA}}$ pulse of an interrupt-acknowledge bus cycle, the level of the highest active interrupt is strobed into ISR. Loading of ISR occurs in response to output signals of the control logic section. This register cannot be written into by the microprocessor; however, its contents may be read as status.

The cascade buffer/comparator section provides the interface between master and slave 82C59As. As we mentioned earlier, this interface permits easy expansion of the interrupt interface using a master/slave configuration. Each slave has an *ID code* that is stored in this section.

Programming the 82C59A

The way in which the 82C59A operates is determined by how the device is programmed. Two types of command words are provided for this purpose. They are the *initialization command words* (ICW) and the *operational command words* (OCW). ICW commands are used to load the internal control registers of the 82C59A to define the basic configuration or mode in which it is used. There are four such command words, and they are identified as ICW_1, ICW_2, ICW_3, and ICW_4. On the other hand, the three OCW commands permit the 80386DX to initiate variations in the basic operating modes defined by the ICW commands. These three commands are called OCW_1, OCW_2, and OCW_3.

Depending on whether the 82C59A is I/O-mapped or memory-mapped, the MPU issues commands to the 82C59A by initiating output or write cycles. This can be done by executing either the OUT instruction or MOV instruction, respectively. The address put on the system bus during the output bus cycle must be decoded with external circuitry to chip select the peripheral. When an address assigned to the 82C59A is on the bus, the output of the decoder must produce logic 0 at the $\overline{\text{CS}}$ input. This signal enables the read/write logic within the PIC, and data applied at D_0 through D_7 are written into the command register within the control logic section synchronously with a write strobe at $\overline{\text{WR}}$.

The interrupt-request input (INTR) of the 80386DX must be disabled whenever commands are being issued to the 82C59A. This can be done by clearing the interrupt-enable flag by executing the CLI (clear interrupt flag) instruction. After completion of the command sequence, the interrupt input must be reenabled. To do this, the microprocessor must execute the STI (set interrupt flag) instruction.

The flow diagram in Fig. 12.20 shows the sequence of events that must take place to initialize the 82C59A with ICW commands. The cycle begins with the MPU outputting initialization command word ICW_1 to the address of the 82C59A.

The moment that ICW_1 is written into the control logic section of the 82C59A, certain internal setup conditions automatically occur. First, the internal sequence logic is set up such that the 82C59A will accept the remaining ICWs as designated by ICW_1. It turns out that if the least significant bit of ICW1 is logic 1, command word ICW_4 is required in the

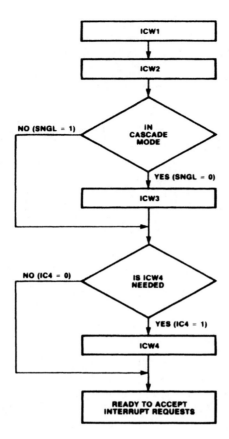

Figure 12.20 Initialization sequence of the 82C59A. (Reprinted by permission of Intel Corp. Copyright/Intel Corp. 1979)

initialization sequence. Moreover, if the next least significant bit of ICW_1 is logic 0, the command word ICW_3 is also required.

In addition to this, writing ICW_1 to the 82C59A clears ISR and IMR. Also three operation command word bits, *special mask mode* (SMM) in OCW_3, *interrupt-request register* (IRR) in OCW_3, and *end of interrupt* (EOI) in OCW_2, are cleared to logic 0. Furthermore, the *fully nested masked mode* of interrupt operation is entered with an initial priority assignment such that IR_0 is the highest-priority input and IR_7 is the lowest-priority input. Finally, the edge-sensitive latches associated with the IR inputs are all cleared.

If the LSB of ICW_1 is initialized to logic 0, one additional event occurs: all bits of the control register associated with ICW_4 are cleared.

In Fig. 12.20 we see that once the MPU starts initialization of the 82C59A by writing ICW_1 into the control register, it must continue the sequence by writing ICW_2 and then, optionally, ICW_3 and ICW_4 in that order. Notice that it is not possible to modify just one of the initialization command registers. Instead, all words that are required to define the device's operating mode must be written into the 82C59A once again.

We found that all four words need not always be used to initialize the 82C59A. However, for its use in an 80386DX microcomputer system, words ICW_1, ICW_2, and ICW_4 are always required. ICW_3 is optional and is needed only if the 82C59A is to function in the cascade mode.

Initialization Command Words

Now that we have introduced the initialization sequence of the 82C59A, let us look more closely at the functions controlled by each of the initialization command words. We will begin with ICW_1. Its format and bit functions are identified in Fig. 12.21(a). Notice that address input A_0 is included as a ninth bit, and it must be logic 0.

Here we find that the logic level of the LSB D_0 of the initialization word indicates to the 82C59A whether or not ICW_4 will be included in the programming sequence. As we mentioned earlier, logic 1 at D_0 (IC_4) specifies that it is needed. The next bit, D_1 (SNGL), selects between *single device* or *multidevice cascaded mode* of operation. When D_1 is set to logic 0, the internal circuitry of the 82C59A is configured for cascaded mode. Selecting this state also sets up the initialization sequence such that ICW_3 must be issued as part of the initialization cycle. Bit D_2 has functions specified for it in Fig. 12.21(a); however, it can be ignored when the 82C59A is being connected to the 80386DX and is a don't-care state. D_3, which is labeled LTIM, defines whether the eight IR inputs operate in the level-sensitive or edge-triggered mode. Logic 1 in D_3 selects level-triggered operation, and logic 0 selects edge-triggered operation. Finally, bit D_4 is fixed at the 1 logic level and the three MSBs, D_5 through D_7, are not required in 80386DX-based systems.

EXAMPLE 12.4

What value should be written into ICW_1 in order to configure the 82C59A such that ICW_4 is needed in the initialization sequence, the system is going to use multiple 82C59As, and its inputs are to be level-sensitive? Assume that all unused bits are to be logic 0. Give the result in both binary and hexadecimal form.

Solution

Since ICW_4 is to be initialized, D_0 must be logic 1.

$$D_0 = 1$$

For cascaded mode of operation, D_1 must be 0.

$$D_1 = 0$$

And for level-sensitive inputs, D_3 must be 1.

$$D_3 = 1$$

Bits D_2 and D_5 through D_7 are don't-care states and are all made logic 0.

$$D_2 = D_5 = D_6 = D_7 = 0$$

Moreover, D_4 must be fixed at the 1 logic level.

$$D_4 = 1$$

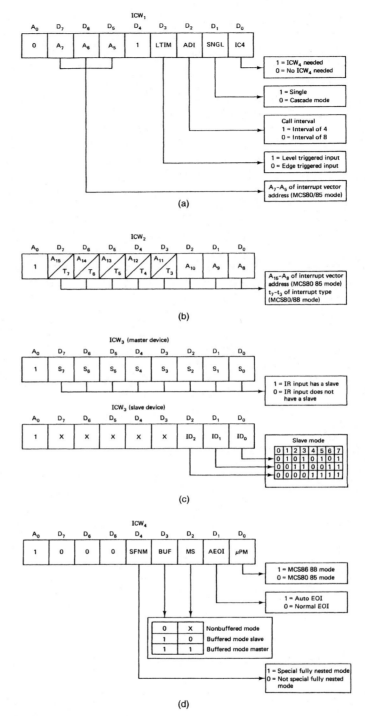

Figure 12.21 (a) ICW₁ format. (Reprinted by permission of Intel Corp. Copyright/Intel Corp. 1979) (b) ICW₂ format. (Reprinted by permission of Intel Corp. Copyright/Intel Corp. 1979) (c) ICW₃ format. (Reprinted by permission of Intel Corp. Copyright/Intel Corp. 1979) (d) ICW₄ format. (Reprinted by permission of Intel Corp. Copyright/Intel Corp. 1979)

This gives the complete command word

$$D_7D_6D_5D_4D_3D_2D_1D_0 = 00011001_2 = 19_{16}$$

The second initialization word, ICW$_2$, has a single function in the 80386DX microcomputer. As shown in Fig. 12.21(b), its 5 most significant bits, D$_7$ through D$_3$, define a fixed binary code, T$_7$ through T$_3$, that is used as the most significant bits of its type number. Whenever the 82C59A puts the 3-bit interrupt type number corresponding to its active input onto the bus, it is automatically combined with the values T$_7$ through T$_3$ to form an 8-bit type number. The 3 least significant bits of ICW$_2$ are not used. Notice that logic 1 must be applied to the A$_0$ input when this command word is put on the bus.

EXAMPLE 12.5

What should be programmed into register ICW$_2$ if the type numbers output on the bus by the device are to range from F0$_{16}$ through F7$_{16}$?

Solution

To set the 82C59A up such that type numbers are in the range of F0$_{16}$ through F7$_{16}$, its device code bits must be

$$D_7D_6D_5D_4D_3 = 11110_2$$

The lower 3 bits are don't-care states and all can be 0s. This gives the command word

$$D_7D_6D_5D_4D_3D_2D_1D_0 = 11110000_2 = F0_{16}$$

The information of initialization word ICW$_3$ is required by only those 82C59As that are configured for the cascaded mode of operation. Figure 12.21(c) shows its bits. Notice that ICW$_3$ is used for different functions, depending on whether the device is a master or slave. In the case of a master, bits D$_0$ through D$_7$ of the word are labeled S$_0$ through S$_7$. These bits correspond to IR inputs IR$_0$ through IR$_7$, respectively. They identify whether or not the corresponding IR input is supplied by either the INT output of a slave or directly by an external device. Logic 1 loaded in an S position indicates that the corresponding IR input is supplied by a slave.

On the other hand, ICW$_3$ for a slave is used to load the device with a 3-bit identification code ID$_2$ID$_1$ID$_0$. This number must correspond to the IR input of the master to which the slave's INT output is wired. The ID code is required within the slave so that it can be compared to the cascading code output by the master on CAS$_0$ through CAS$_2$.

EXAMPLE 12.6

Assume that a master PIC is to be configured such that its IR$_0$ through IR$_3$ inputs are to accept inputs directly from external devices, but IR$_4$ through IR$_7$ are to be supplied by the INT outputs of slaves. What code should be used for the initialization command word ICW$_3$?

Solution

For IR_0 through IR_3 to be configured to allow direct inputs from external devices, bits D_0 through D_3 of ICW_3 must be logic 0.

$$D_3D_2D_1D_0 = 0000_2$$

The other IR inputs of the master are to be supplied by INT outputs of slaves. Therefore, their control bits must be all 1.

$$D_7D_6D_5D_4 = 1111_2$$

This gives the complete command word

$$D_7D_6D_5D_4D_3D_2D_1D_0 = 11110000_2 = F0_{16}$$

The fourth control word, ICW_4, which is shown in Fig. 12.21(d), is used to configure the device for use with the 80386DX and selects various features that are available in its operation. The LSB D_0, which is called microprocessor mode (μPM), must be set to logic 1 whenever the device is connected to the 80386DX. The next bit, D_1, is labeled AEOI for *automatic end of interrupt*. If this mode is enabled by writing logic 1 into the bit location, the EOI (*end of interrupt*) command does not have to be issued as part of the service routine.

Of the next two bits in ICW_4, BUF is used to specify whether or not the 82C59A is to be used in a system where the data bus is buffered with a bidirectional bus transceiver. When buffered mode is selected, the $\overline{SP}/\overline{EN}$ line is configured as $\overline{EN}$. As indicated earlier, $\overline{EN}$ is a control output that can be used to control the direction of data transfer through the bus transceiver. It switches to logic 0 whenever data are transferred from the 82C59A to the MPU.

If buffered mode is not selected, the $\overline{SP}/\overline{EN}$ line is configured to work as the master/slave mode select input. In this case, logic 1 at the $\overline{SP}$ input selects master mode operation and logic 0 selects slave mode.

Assume that the buffered mode was selected; then the $\overline{SP}$ input is no longer available to select between the master and slave modes of operation. Instead, the MS bit of ICW_4 defines whether the 82C59A is a master or slave device.

Bit D_4 is used to enable or disable another operational option of the 82C59A. This option is known as the *special fully nested mode*. This function is only used in conjunction with the cascaded mode. Moreover, it is enabled only for the master 82C59A, not for the slaves. This is done by setting the SFNM bit to logic 1.

The 82C59A is put into the fully nested mode of operation as command word ICW_1 is loaded. When an interrupt is initiated in a cascaded system that is configured in this way, the occurrence of another interrupt at the slave corresponding to the original interrupt is masked out even if it is of higher priority. This is because the bit in ISR of the master 82C59A that corresponds to the slave is already set; therefore, the master 82C59A ignores all interrupts of equal or lower priority.

This problem is overcome by enabling special fully nested mode of operation at the master. In this mode, the master will respond to those interrupts that are at lower or higher priority than the active level.

The last three bits of ICW_4, D_5 through D_7, must be logic 0.

Operational Command Words

Once the appropriate ICW commands have been issued to the 82C59A, it is ready to operate in the fully nested mode. Three operational command words are also provided for controlling the operation of the 82C59A. These commands permit further modifications to be made to the operation of the interrupt interface after it has been initialized. Unlike the initialization sequence, which requires that the ICWs be output in a special sequence after power-up, the OCWs can be issued under program control whenever needed and in any order.

The first operational command word, OCW_1, is used to access the contents of the interrupt mask register (IMR). A read operation can be performed to the register to determine the present setting of the mask. Moreover, write operations can be performed to set or reset its bits. This permits selective masking of the interrupt inputs. Notice in Fig. 12.22(a) that bits D_0 through D_7 of command word OCW_1 are identified as mask bits M_0 through M_7, respectively. In hardware, these bits correspond to interrupt inputs IR_0 through IR_7, respectively. Setting a bit to logic 1 masks out the associated interrupt input. On the other hand, clearing it to logic 0 enables the interrupt input.

For instance, writing $F0_{16} = 11110000_2$ into the register causes inputs IR_0 through IR_3 to be unmasked and IR_4 through IR_7 to be masked. Input A_0 must be logic 1 whenever the OCW_1 command is issued.

EXAMPLE 12.7

What should be the OCW_1 code if interrupt inputs IR_0 through IR_3 are to be masked and IR_4 through IR_7 are to be unmasked?

Solution

For IR_0 through IR_3 to be masked, their corresponding bits in the mask register must be made logic 1.

$$D_3D_2D_1D_0 = 1111_2$$

On the other hand, for IR_4 through IR_7 to be unmasked, D_4 through D_7 must be logic 0.

$$D_7D_6D_5D_4 = 0000_2$$

Therefore, the complete word for OCW_1 is

$$D_7D_6D_5D_4D_3D_2D_1D_0 = 00001111_2 = 0F_{16}$$

The second operational command word, OCW_2, selects the appropriate priority scheme and assigns an IR level for those schemes that require a specific interrupt level. The format of OCW_2 is given in Fig. 12.22(b). Here we see that the three LSBs define the interrupt level. For example, using $L_2L_1L_0 = 000_2$ in these locations specifies interrupt level 0, which corresponds to input IR_0.

The other three active bits of the word D_7, D_6, and D_5 are called *rotation* (R), *specific level* (SL), and *end of interrupt* (EOI), respectively. They are used to select a priority scheme according to the table in Fig. 12.22(b). For instance, if these bits are all logic 1, the priority

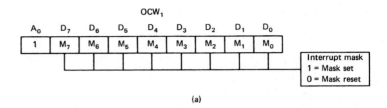

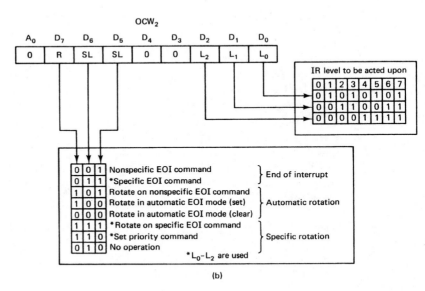

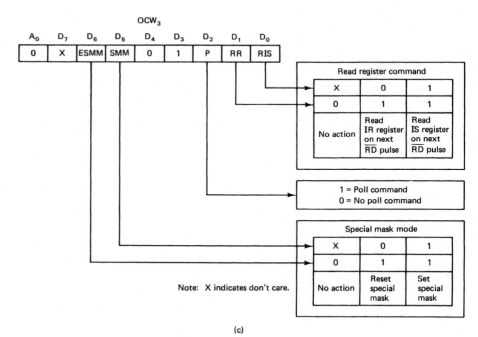

Figure 12.22 (a) OCW₁ format. (Reprinted by permission of Intel Corp. Copyright/Intel Corp. 1979) (b) OCW₂ format. (Reprinted by permission of Intel Corp. Copyright/Intel Corp. 1979) (c) OCW₃ format. (Reprinted by permission of Intel Corp. Copyright/Intel Corp. 1979)

scheme known as *rotate on specific EOI command* is enabled. Since this scheme requires a specific interrupt, its value must be included in $L_2L_1L_0$. Input A_0 must be logic 0 whenever this command is issued to the 82C59A.

EXAMPLE 12.8 _____

What OCW_2 must be issued to the 82C59A if the priority scheme rotate on nonspecific EOI command is to be selected?

Solution

To enable the rotate on nonspecific EOI command priority scheme, bits D_7 through D_5 must be set to 101. Since a specific level does not have to be specified, the rest of the bits in the command word can be 0. This gives OCW_2 as

$$D_7D_6D_5D_4D_3D_2D_1D_0 = 10100000_2 = A0_{16}$$

_____

The last control word OCW_3, which is shown in Fig. 12.22(c), permits reading of the contents of the ISR or IRR registers through software, issue of the poll command, and enable/disable of the special mask mode. Bit D_1, which is called *read register* (RR), is set to 1 to initiate reading of either the in-service register (ISR) or interrupt-request register (IRR). At the same time, bit D_0, which is labeled RIS, selects between ISR and IRR. Logic 0 in RIS selects IRR and logic 1 selects IRS. In response to this command, the 82C59A makes the contents of the selected register available on the data bus so that they can be read by the MPU.

If the next bit, D_2, in OCW_3 is logic 1, a *poll command* is issued to the 82C59A. The result of issuing a poll command is that the next $\overline{RD}$ pulse to the 82C59A is interpreted as an interrupt acknowledge. In turn, the 82C59A causes the ISR register to be loaded with the value of the highest-priority active interrupt. After this, a *poll word* is automatically put on the data bus. The MPU must read it off the bus.

Figure 12.23 illustrates the format of the poll word. Looking at this word, we see that the MSB is labeled I for interrupt. The logic level of this bit indicates to the MPU whether or not an interrupt input was active. Logic 1 indicates that an interrupt is active. The three LSBs, W_2, W_1 and W_0, identify the priority level of the highest-priority active interrupt input. This poll word can be decoded through software, and when an interrupt is found to be active, a branch is initiated to the starting point of its service routine. The poll command represents a software method of identifying whether or not an interrupt has occurred; therefore, the INTR input of the 80386DX should be disabled.

D_5 and D_6 are the remaining bits of OCW_3 for which functions are defined. They are used to enable or disable the special mask mode. ESMM (*enable special mask mode*) must be logic 1 to permit changing of the status of the special mask mode with the SMM (*special mask mode*) bit. Logic 1 at SMM enables the special mask mode of operation. If the 82C59A is initially configured for the fully nested mode of operation, only interrupts of higher priority are allowed to interrupt an active service routine. However, by enabling the special

Figure 12.23 Poll word format. (Reprinted by permission of Intel Corp. Copyright/Intel Corp. 1979)

mask mode, interrupts of higher or lower priority are enabled, but those of equal priority remain masked out.

EXAMPLE 12.9

Write a program that will initialize an 82C59A with the initialization command words ICW_1, ICW_2, and ICW_3 derived in Examples 12.4, 12.5, and 12.6, respectively. Moreover, ICW_4 is to be equal to $1F_{16}$. Assume that the 82C59A resides at address $A000_{16}$ in the memory address space. Assume that address bit A_2 of the 80386DX is applied to the A_0 input and that the device is attached to data bus lines D_0 through D_7.

Solution

Since the 82C59A resides in the memory address space, we can use a series of move instructions to write the initialization command words into its registers. Notice that the memory address for an ICW is $A000_{16}$ if $A_2 = 0$ and it is $A004_{16}$ if $A_2 = 1$. However, before doing this, we must first disable interrupts. This is done with the instruction

```
CLI   ;Disable interrupts
```

Next we will set up a data segment starting at address 00000_{16}.

```
MOV   AX,0H   ;Create a data segment at 00000₁₆
MOV   DS,AX
```

Now we are ready to write the command words to the 82C59A.

```
MOV   AL,19H     ;Load ICW1
MOV   0A000H,AL  ;Write ICW1 to 82C59A
MOV   AL,0F0H    ;Load ICW2
MOV   0A004H,AL  ;Write ICW2 to 82C59A
MOV   AL,0F0H    ;Load ICW3
MOV   0A004H,AL  ;Write ICW3 to 82C59A
MOV   AL,1FH     ;Load ICW4
MOV   0A004H,AL  ;Write ICW4 to 82C59A
```

Initialization is now complete and the interrupts can be enabled with the interrupt instruction

```
STI   ;Enable interrupts
```

▲ 12.9 INTERRUPT INTERFACE CIRCUITS USING THE 82C59A

Now that we have introduced the 82C59A programmable interrupt controller, let us look at how it is used to implement the interrupt interface in an 80386DX-based microcomputer system.

Figure 12.24(a) includes an interrupt interface circuit that uses a single 82C59A-2 programmable interrupt controller. Let us begin by looking at how the 82C59A's microprocessor interface is attached to the 80386DX. Notice that data bus lines D_0 through D_7 of

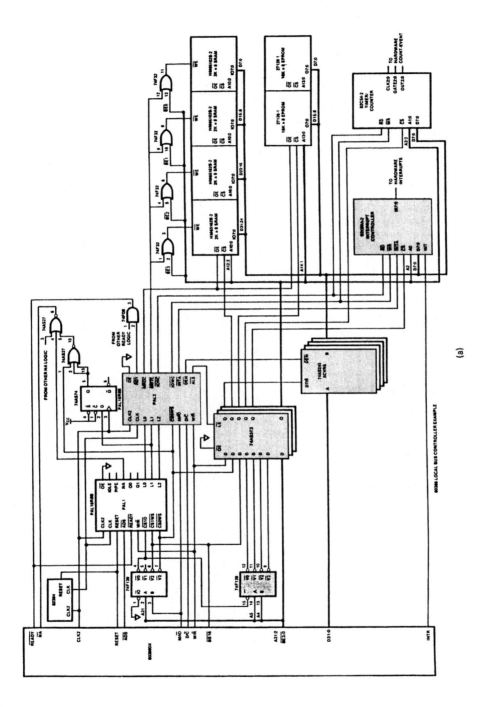

Figure 12.24(a) Interrupt interface of the 80386DX-based microcomputer system. (Reprinted by permission of Intel Corp. Copyright/Intel Corp. 1990)

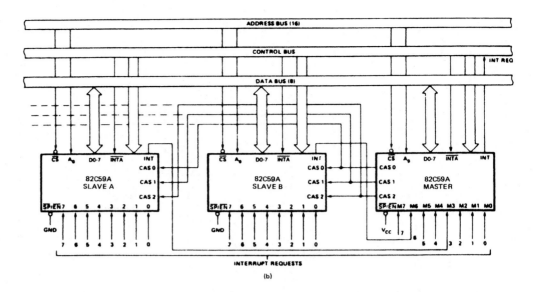

Figure 12.24(b) Cascading 82C59As. (Reprinted by permission of Intel Corp. Copyright/Intel Corp. 1979)

the 82C59A are connected through the 74AS245 transceiver to the 80386DX's data bus. It is over these lines that the MPU initializes the internal registers of the 82C59A, reads the contents of these registers, and reads the type number of the active interrupt input during the interrupt-acknowledge bus cycle.

In this circuit the registers of the 82C59A are assigned to unique I/O addresses. During input or output bus cycles to one of these addresses, bits A_4 and A_5 of the address are decoded by the 74F139 I/O address decode logic to produce the $\overline{\text{CS}}$ input for the 82C59A. When this input is at its active, logic 0, level, the 82C59A's microprocessor interface is enabled for operation. At the same time, the A_2 output of the 74AS373 address latches is applied to the A_0 input of the interrupt controller. It is this signal that selects the register that is to be accessed.

The bus-control logic produces the signals that identify whether an input or output data transfer is to take place. Notice that the I/O read ($\overline{\text{IORC}}$) and I/O write ($\overline{\text{IOWC}}$) control outputs from PAL2 of the control logic and are supplied to the $\overline{\text{RD}}$ and $\overline{\text{WR}}$ inputs of the 82C59A, respectively. Logic 0 at these outputs tells the 82C59A whether data are to be input or output over the bus, respectively.

Next we will trace the sequence of events that takes place as a device requests service through the interrupt interface circuit. The external interrupt request inputs are identified as IR_0 through IR_7 in the circuit of Fig. 12.24(a). Whenever an interrupt input becomes active, and either no other interrupt is active or the priority level of the new interrupt is higher than that of the already active interrupt, the 82C59A switches its INT output to logic 1. This output is returned to the INTR input of the 80386DX. In this way it signals that an external device needs to be serviced.

As long as the interrupt flag within the 80386DX is set to 1, the interrupt interface is enabled. Assuming that IF is 1, the interrupt request is accepted and an interrupt-acknowledge bus cycle sequence is initiated. During the first interrupt-acknowledge bus cycle, the 80386DX outputs the status code $M/\overline{\text{IO}}$ $D/\overline{\text{C}}$ $W/\overline{\text{R}} = 000_2$ to the bus control logic. This input causes the $\overline{\text{INTA}}$ output of PAL2 to be pulsed to logic 0. $\overline{\text{INTA}}$ is applied to the

$\overline{\text{INTA}}$ input of the 82C59A and when logic 0, it signals that the active interrupt request will be serviced.

As the second interrupt-acknowledge bus cycle is executed, the status code at the inputs M/$\overline{\text{IO}}$D/$\overline{\text{C}}$W/$\overline{\text{R}}$ of PAL2 in the bus control logic is again 000_2, and another pulse is output at $\overline{\text{INTA}}$. This pulse signals the 82C59A to output the type number of its highest-priority active interrupt at D_0 through D_7. Then the signal $\overline{\text{DEN}}$ is switched to 0 to enable the 74AS245 bus transceivers for operation. At the same time, the W/$\overline{\text{R}}$ output of the 80386DX switches to 0. This signal is latched in the address latch and applied to the DT/$\overline{\text{R}}$ input of the transceiver. The transceivers are now set to pass the type number from data outputs D_0 through D_7 of the interrupt controller onto the data bus lines of the 80386DX. The 80386DX reads this number off the bus and initiates a vectored transfer of program control to the starting point of the corresponding service routine in program memory.

For applications that require more than eight interrupt-request inputs, 82C59A devices are cascaded into a master-slave configuration. Figure 12.24(b) shows such a circuit. Here we find that the rightmost device is identified as the master and the devices to the left as slave A and slave B. Notice that the bus connections are similar to those explained for the circuit in Fig. 12.24(a).

At the interrupt request side of the devices, we find that slaves A and B are cascaded to the master 82C59A by attaching their INT outputs to the M_3 (IR$_3$) and M_6 (IR$_6$) inputs, respectively. Moreover, the CAS lines on all three PICs are tied in parallel. It is over these CAS lines that the master signals the slaves whether or not their interrupt request has been acknowledged. As the first pulse is output at $\overline{\text{INTA}}$, the master PIC is signaled to output the 3-bit cascade code of the device whose interrupt request is being acknowledged on the CAS bus. The slaves read this code and then compare it to their internal code. In this way the slave corresponding to the code is signaled to output the type number of its highest-priority active interrupt onto the data bus during the second interrupt-acknowledge bus cycle.

EXAMPLE 12.10

Analyze the circuit in Fig. 12.25(a) and write an appropriate main program and a service routine that counts as a decimal number the positive edges of the clock signal applied to IR$_0$ input of the 82C59A.

Solution

The microprocessor addresses to which the 82C59A in the circuit of Fig. 12.25(a) responds depend on how the $\overline{\text{CS}}$ signal for the 82C59A is generated as well as the logic level of address line A_2 that is connected to its A_0 input. Thus the 82C59A responds to

$$A_{15}\, A_{14}\, A_{13}\, A_{12}\, A_{11}\, A_{10}\, A_9\, A_8\, A_7\, A_6\, A_5\, A_4\, A_3\, A_2$$
$$= 11111111000000_2 \qquad \text{for } A_2 = 0,\, M/\overline{\text{IO}} = 0 \quad \text{and}$$
$$= 11111111000011_2 \qquad \text{for } A_2 = 1,\, M/\overline{\text{IO}} = 0$$

These two are I/O addresses FF00H and FF04H, respectively. The address FF00H is for the ICW$_1$ and FF04H is for the ICW$_2$, ICW$_3$, ICW$_4$, and OCW$_1$ command words. Let us now determine the ICWs and OCWs for the 82C59A.

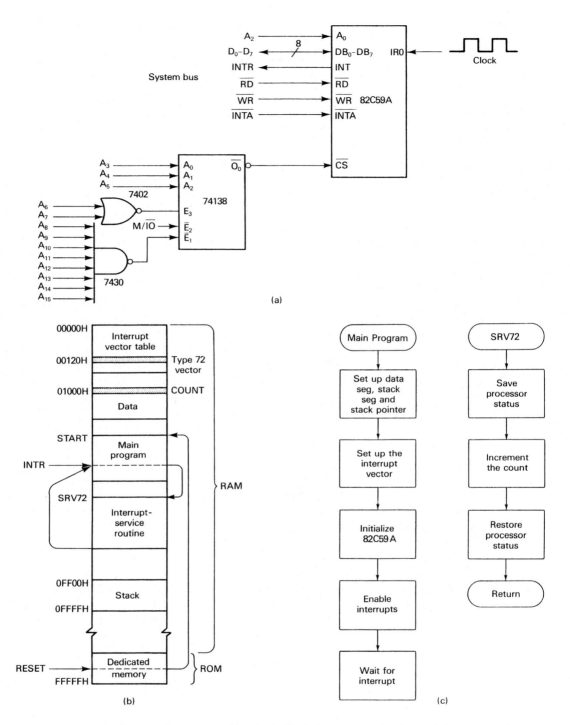

Figure 12.25 (a) Circuit for Example 12.10. (b) Software organization. (c) Flowcharts for the main program and service routine.

Because the 82C59A interface is for the 80386DX microprocessor; there is only one 82C59A in the system; and the interrupt input is an edge, we are led to the following ICW_1:

$$ICW_1 = 00010011_2 = 13H$$

Let us assume that we will use interrupt type 72 to service an interrupt generated by an edge presented to the IR_0. This leads to the following ICW_2:

$$ICW_2 = 01001000_2 = 48H$$

For a single 82C59A, ICW_3 is not needed. To determine ICW_4 let us assume that we will use auto EOI and nonbuffered mode of operation. This leads to the following ICW_4:

$$ICW_4 = 00000011_2 = 03H$$

For OCWs we will use only OCW_1 to mask all other interrupts but IR_0. This gives OCW_1 as

$$OCW_1 = 11111110_2 = FEH$$

Figure 12.25(b) shows the memory organization for the software. Let us understand the information presented in this memory organization. In the interrupt-vector table we need to set up the type 72 vector. The type 72 vector is located at $4 \times 72 = 288 = 120H$. At address 120H we need to place the offset of the service routine and at address 122H the code segment value of the service routine.

In the data area we need a location to keep a decimal count of the edges of the input clock. Let us assume that it is location 01000H. The stack starts at 0FF00H and ends at 0FFFFH. The start address of the main program is denoted as START, and that of the service routine as SRV72.

The flowcharts in Fig. 12.25(c) are for the main program and the service routine. The main program initializes the microprocessor and the 82C59A. First of all we establish various segments for data and stack. This can be done using the following instructions:

```
;MAIN    PROGRAM
         CLI                ;Start with interrupts disabled
START:   MOV   AX,0         ;Extra segment at 00000H
         MOV   ES,AX
         MOV   AX,100H      ;Data segment at 01000H
         MOV   DS,AX
         MOV   AX,0FF0H     ;Stack segment at 0FF00H
         MOV   SS,AX
         MOV   SP,1000H     ;Stack end at 10000H
```

Next we can set up the IP and CS for the type 72 vector in the interrupt-vector table. This can be accomplished using the following instructions:

```
MOV   AX,OFFSET SRV72    ;Get offset for the service routine
MOV   [ES:120H],AX       ;Setup the IP
MOV   AX,SEGMENT SRV72   ;Get code seg for the service routine
MOV   [ES:122H],AX       ;Setup the CS
```

Having set up the interrupt-type vector, let us proceed now to initialize the 82C59A. Using the analyzed information, the following instructions can be executed to initialize the 82C59A:

```
MOV   DX,0FF00H   ;ICW1 address
MOV   AL,13H      ;Edge trig input, single 82C59A
OUT   DX,AL
MOV   DX,0FF04H   ;ICW2, ICW4, OCW1 address
MOV   AL,48H      ;ICW2, type 72
OUT   DX,AL
MOV   AL,03H      ;ICW4, AEOI, nonbuff mode
OUT   DX,AL
MOV   AL,0FEH     ;OCW1, mask all but IR0
OUT   DX,AL
STI               ;Enable the interrupts
```

Now the processor is ready to accept interrupts. We can write an endless loop to wait for the interrupt to occur. In a real situation we may be doing some other operation in which the interrupt will be received and serviced. For simplicity let us use the following instruction to wait for the interrupt:

```
HERE:  JMP  HERE  ;Wait for interrupt
```

Figure 12.25(c) shows the flowchart for the interrupt-service routine as well. The operations shown in the flowchart can be implemented using the following instructions:

```
SRV72:  PUSH  AX            ;Save register to be used
        MOV   AL,[COUNT]     ;Get the count
        INC   AL            ;Increment the count
        DAA                 ;Decimal adjust the count
        MOV   [COUNT],AL     ;Save the new count
        POP   AX            ;Restore the register used
        IRET                ;Return from interrupt
```

▲ 12.10 SOFTWARE INTERRUPTS

The 80386DX microcomputer system is capable of implementing up to 256 *software interrupts*. They differ from the external hardware interrupts in that their service routines are initiated in response to the execution of a software interrupt instruction, not an event in external hardware.

The INT n instruction is used to initiate a software interrupt. Earlier in this chapter we indicated that n represents the type number associated with the service routine. The software interrupt service routines are vectored to, using pointers from the same memory locations as the corresponding external hardware interrupts. These locations are shown in the real-mode vector table of Fig. 12.3 and the protected-mode gate table of Fig. 12.4. Our earlier example was INT 50. It has a type number of 50, and in the real mode causes a vector in program control to the service routine whose starting address is defined by the values IP_{50} and CS_{50} stored at addresses $000C8_{16}$ and $000CA_{16}$, respectively.

The mechanism by which a real-mode or protected-mode software interrupt is initiated is similar to that described for the external hardware interrupts. However, no external interrupt-acknowledge bus cycles are initiated. Instead, control is passed to the start of the service routine immediately upon completion of execution of the interrupt instruction. In real mode, first the old flags are automatically saved on the stack; then IF and TF are cleared; next the old CS and old IP are pushed onto the stack; now the new CS and new IP are read from memory and loaded into the 80386DX's registers; and finally program execution resumes at $CS_{NEW}:IP_{NEW}$. On the other hand, when in protected mode the events that take place during the context switch depend on whether an interrupt gate, trap gate, or task gate is specified for the type number.

If necessary, the contents of other internal registers can be saved on the stack by including the appropriate PUSH instructions at the beginning of the service routine. Toward the end of the service routine, POP instructions are inserted to restore these registers. Finally, an IRET instruction is used at the end of the routine to return to the original program environment. In this way, we see that the structure of a software-interrupt routine is identical to that shown in Fig. 12.12.

Software interrupts are of higher priority than the external interrupts and are not masked out by IF. They actually work like *vectored subroutine calls*. A common use of these software routines is as *emulation routines* for more complex functions. For instance, INT 50 could define a floating-point addition instruction and INT 51, a floating-point subtraction instruction. These emulation routines are written using assembly language instructions, are assembled into machine code, and then are stored in the main memory of the microcomputer system. Other examples of their use are for *supervisor calls* from an operating system and for testing of external hardware interrupt service routines.

▲ 12.11 NONMASKABLE INTERRUPT

The *nonmaskable interrupt* (NMI) is another interrupt that is initiated from external hardware. However, it differs from the other external hardware interrupts in several ways. First, as its name implies, it cannot be masked out with the interrupt flag. Second, requests for service by this interrupt are signaled to the 80386DX by applying logic 1 at the NMI input, not the INTR input. Third, the NMI input is positive edge-triggered. Therefore, a request for service is automatically latched internal to the MPU.

On the 0 to 1 transition of the NMI input, the NMI flip-flop within the 80386DX is set. If the contents of this latch remains active for four consecutive internal clock cycles (eight CLK_2 cycles), it is recognized and at completion of the current instruction the nonmaskable interrupt transition sequence is initiated. Just as with the other interrupts we have studied, initiation of NMI causes the current flags, current CS, and current IP to be pushed onto the stack. Moreover, the interrupt-enable flag is cleared to disable all external hardware interrupts, and the trap flag is cleared to disable the single-step mode of operation. Next the 80386DX fetches the words of the NMI vector from memory and loads them into IP and CS. Finally, execution resumes with the first instruction of the NMI service routine.

Once the NMI input has returned to the inactive 0 logic level, it must remain at this level for at least 4 internal clock cycles (eight CLK_2 cycles) before returning to the active 1 level. Otherwise, it will not be recognized by the 80386DX.

As shown in Fig. 12.3, NMI has a dedicated type number. It automatically vectors from the type 2 vector location in the interrupt vector table. In a real-mode 80386DX microcomputer system, this vector is stored in memory at word addresses 00008_{16} and

$0000A_{16}$. In protected mode, the NMI service routine is entered through either an interrupt gate or task gate. No interrupt-acknowledge bus cycles are performed during the NMI initiation sequence.

Typically, NMI is assigned to hardware events that must be responded to immediately. Two examples are the detection of a power failure and detection of a memory-read error.

▲ 12.12 RESET

The RESET input of the 80386DX provides a hardware means for initializing the microcomputer. This is typically done at power-on to provide an orderly startup of the system. However, some systems, such as a personal computer, also allow for a *warm start*—that is, a software-initiated reset.

Figure 12.26 shows a typical reset interface circuit for the 80386DX microcomputer system. This circuit is used to detect an active reset input and synchronize the application and removal of the RESET signal with the clock. Notice that the $\overline{RES}$ input of the circuit is attached to an RC circuit. At power-on, the $\overline{RES}$ input is shorted to ground through the capacitor. This represents logic 0 at the input of the reset control logic and causes the RESET output to switch to its active 1 logic level. This output is supplied to the RESET input at pin C9 of the 80386DX and initiates an orderly reset of the MPU. Actually, the 0-to-1 transition at RESET is synchronized to CLK_2 by the circuit. As long as the voltage across the capacitor is below the 1-logic-level threshold of the $\overline{RES}$ input, the RESET output stays at logic 1.

RESET can also be applied in parallel to the reset inputs of other LSI and VLSI

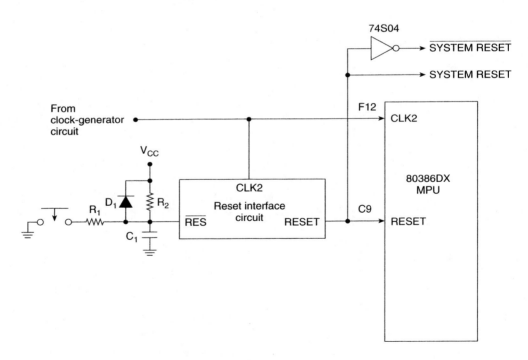

Figure 12.26 Reset interface of the 80386DX.

peripheral devices in the microcomputer system. In this way, they are also initialized at power-on. In Fig. 12.26 the signals $\overline{\text{SYSTEM RESET}}$ and SYSTEM RESET are provided for this purpose.

The RESET input of the 80386DX must be held at the active 1 logic level for a minimum of 15 CLK$_2$ cycles; otherwise, it may not be recognized. When RESET is recognized, the 80386DX terminates operation; puts the data bus lines in the high-Z state; and forces the address lines to the 1 logic level. Moreover, the bus cycle definition and control outputs are all switched to specific logic levels. For instance, W/$\overline{\text{R}}$ and M/$\overline{\text{IO}}$ are both switched to logic 0, while $\overline{\text{ADS}}$ and D/$\overline{\text{C}}$ are set to 1. The 80386DX's signal levels during reset are summarized in Fig. 12.27(a). The timing and transition sequence to these levels are illustrated in the waveforms of Fig. 12.27(b).

The return of the 80386DX's RESET input to logic 0 is also synchronized to CLK$_2$ by the reset control logic in Fig. 12.26. When RESET returns to logic 0, the MPU initiates its internal initialization routine and then comes up in the real-address mode. At completion of initialization, the flag register is set to UUUU0002$_{16}$ (U stands for undefined); the instruction pointer is set to 0000FFF0$_{16}$; the CS register is set to F000$_{16}$; the DS, SS, ES, FS, and GS registers are set to 0000$_{16}$; and the instruction queue is emptied. The table in Fig. 12.28 summarizes this initial state.

Since the interrupt flag is cleared as part of initialization, external hardware interrupts are disabled. Moreover, the code segment register contains F000$_{16}$, the instruction pointer contains 0000FFF0$_{16}$, and address lines A$_{20}$ through A$_{31}$ are all forced to logic 1. Therefore, real-mode program execution begins after reset at address FFFFFFF0$_{16}$. This storage location can contain an instruction that will cause a jump to the start-up program that is used to initialize the rest of the microcomputer system's resources, such as I/O ports, the interrupt flag, and data memory. This start-up routine is also known as a *boot-strap program*. The 80386DX could also be switched to the protected-address mode at this time. After system-level initialization is complete, another jump can be performed to the starting point of the microcomputer's operating system or application program. If the 80386DX is to be used only in real mode, the upper address bits can be ignored. In this case, after reset, execution would start at address FFFF0$_{16}$, just as in an 8086 or 80286 microcomputer system.

Looking at Fig. 12.27(b), we find that an automatic diagnostic routine, called *self-test*, can be initiated as part of the 80386DX's internal initialization sequence. At the 1-to-0 transition of RESET, the state of the $\overline{\text{BUSY}}$ pin is tested by the 80386DX. If this input is logic 0, a built-in self-test diagnostic routine is automatically run on the 80386DX. If the diagnostic passes, the EAX register is cleared to zero. On the other hand, if the test fails, a nonzero value is loaded into EAX. The value in EAX can be examined by system-level diagnostic routines that are normally run as part of system initialization software to verify that the 80386DX is fully functional. If the diagnostic is to be used as part of the reset procedure, RESET must be held at its active 1 logic level for at least 78 CLK$_2$ periods instead of just 15.

One other 80386DX register is nonzero after reset, the EDX register. After reset it contains a component identifier. The more significant byte, DH, will contain the value 3 to identify that the chip is an 80386DX microprocessor and the lower byte DL contains a number that determines the revision level of the device. That is, whether it is a B-step or D-step device. For instance, a value of 3 in DL indicates that a B-step 80386DX is in use. This information can also be examined by system-level initialization software. For example, the operating system could use this information to install a set of software patches for the B-step errata of the 80386DX.

Pin Name	Signal Level During Reset
ADS#	High
D0–D31	High impedance
BE0#–BE3#	Low
A2–A31	High
W/R#	Low
D/C#	High
M/IO#	Low
LOCK#	High
HLDA	Low

(a)

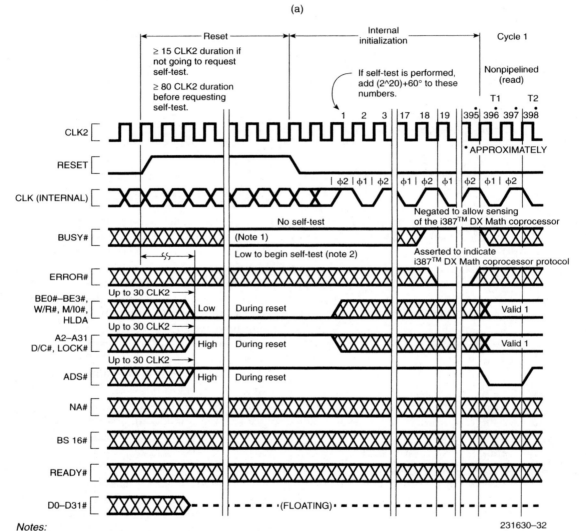

Notes:

1. BUSY # should be held stable for 8 CLK2 periods before and after the CLK2 period in which RESET falling edge occurs.

2. If self-test is requested, the Intel386 DX outputs remain in their reset state as shown here and in Table 5–3.

(b)

231630–32

Figure 12.27 (a) Signal levels during system reset. (b) Reset sequence waveforms. (Reprinted by permission of Intel Corp. Copyright/Intel Corp. 1990)

Flag word	UUUU0002H
Machine status word (CR0)	UUUUUUU0H
Instruction pointer	0000FFF0H
Code segment	F000H
Data segment	0000H
Stack segment	0000H
Extra segment (ES)	0000H
Extra segment (FS)	0000H
Extra segment (GS)	0000H
DX register	Component and stepping ID
All other registers	Undefined

Figure 12.28 Internal state of the 80386DX after reset. (Reprinted by permission of Intel Corp. Copyright/Intel Corp. 1979)

▲ 12.13 INTERNAL INTERRUPT AND EXCEPTION FUNCTIONS

Earlier we indicated that some of the 256 interrupt vectors of the 80386DX are dedicated to internal interrupt and exception functions. Internal interrupts and exceptions differ from external hardware interrupts in that they occur due to the result of executing an instruction, not an event that takes place in external hardware. That is, an internal interrupt or exception is initiated because an error condition was detected before, during, or after execution of an instruction. In this case, a routine must be initiated to service the internal condition before resuming execution of the same or next instruction of the program.

Looking at Fig. 12.10, we find that internal interrupts and exceptions are not masked out with the interrupt enable flag. For this reason, occurrence of any one of these internal conditions is automatically detected by the 80386DX and causes an interrupt of program execution and a vectored transfer of control to a corresponding service routine. During the control-transfer sequence, no interrupt-acknowledge bus cycles are produced.

Figure 12.29 identifies the internal interrupts and exceptions that are active in real mode. Here we find internal interrupts such as breakpoint and exception functions such as divide error, overflow error, and bounds check. Each of these functions is assigned a unique type number. Notice that the 32 highest-priority locations in the interrupt-vector table, vectors 0 through 31, are reserved for internal functions.

Internal interrupts and exceptions are further categorized as a *fault*, *trap*, or *abort* based on how the failing function is reported. In the case of an exception that causes a fault, the values of CS and IP saved on the stack point to the instruction that resulted in the fault. Therefore, after servicing the exception, the faulting instruction can be reexecuted. On the other hand, for those exceptions that result in a trap, the values of CS and IP are pushed to the stack point to the next instruction that is to be executed, instead of the instruction that caused the trap. Therefore, upon completion of the service routine, program execution resumes with the instruction that follows the instruction that produced the trap. Finally, exceptions that produce an abort do not preserve any information that identifies the location that caused the error. In this case the system may need to be restarted. Let us now look at each of the real-mode internal interrupts and exceptions in more detail.

Divide Error Exception

The *divide error exception* represents an error condition that can occur in the execution of a division instruction. If the quotient that results from a DIV (divide) instruction or an

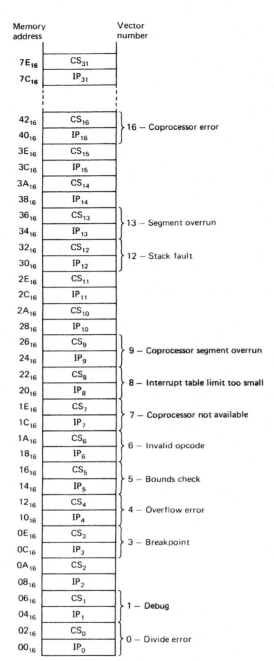

Memory address	Vector number
$7E_{16}$	CS_{31}
$7C_{16}$	IP_{31}
42_{16}	CS_{16}
40_{16}	IP_{16}
$3E_{16}$	CS_{15}
$3C_{16}$	IP_{15}
$3A_{16}$	CS_{14}
38_{16}	IP_{14}
36_{16}	CS_{13}
34_{16}	IP_{13}
32_{16}	CS_{12}
30_{16}	IP_{12}
$2E_{16}$	CS_{11}
$2C_{16}$	IP_{11}
$2A_{16}$	CS_{10}
28_{16}	IP_{10}
26_{16}	CS_9
24_{16}	IP_9
22_{16}	CS_8
20_{16}	IP_8
$1E_{16}$	CS_7
$1C_{16}$	IP_7
$1A_{16}$	CS_6
18_{16}	IP_6
16_{16}	CS_5
14_{16}	IP_5
12_{16}	CS_4
10_{16}	IP_4
$0E_{16}$	CS_3
$0C_{16}$	IP_3
$0A_{16}$	CS_2
08_{16}	IP_2
06_{16}	CS_1
04_{16}	IP_1
02_{16}	CS_0
00_{16}	IP_0

16 — Coprocessor error

13 — Segment overrun

12 — Stack fault

9 — Coprocessor segment overrun

8 — Interrupt table limit too small

7 — Coprocessor not available

6 — Invalid opcode

5 — Bounds check

4 — Overflow error

3 — Breakpoint

1 — Debug

0 — Divide error

Figure 12.29 Real-mode internal interrupt and exception-vector locations.

IDIV (integer divide) instruction is larger than the specified destination, a divide error has occurred. This condition causes automatic initiation of a type 0 interrupt and passes control to a service routine whose starting point is defined by the values of IP_0 and CS_0 at addresses 00000_{16} and 00002_{16}, respectively, in the pointer table. Divide error produces a fault; therefore, the CS and IP pushed to the stack are those for the divide instruction.

Debug Exception

The *debug exception* relates to the debug mode of operation of the 80386DX. The 80386DX has a set of eight on-chip debug registers. Using these registers, the programmer can specify up to four breakpoint addresses and specify conditions under which they are to be active. For instance, the activating condition could be an instruction fetch from the address, a data write to the address, or either a data read or write for the address, but not an instruction fetch. Moreover, for data accesses, the size of the data element can be specified as a byte, word, or double word. Finally, the individual addresses can be locally or globally enabled or disabled. If an access that matches any of these debug conditions is attempted, a debug exception occurs and control is passed to the service routine defined by IP_1 and CS_1 at word addresses 00004_{16} and 00006_{16}, respectively. The service routine could include a mechanism that allows the programmer to view the contents of the 80386DX's internal registers and its external memory.

If the trap flag (TF) bit in the flags register is set, the single-step mode of operation is enabled. This flag bit can be set or reset under software control. When TF is set, the 80386DX initiates a type 1 interrupt to the service routine defined by IP_1 and CS_1 at address 00004_{16} and 00006_{16}, respectively, at the completion of execution of every instruction. This permits implementation of the single-step mode of operation so that the program can be executed one instruction at a time.

Breakpoint Interrupt

The breakpoint function can also be used to implement a software diagnostic tool. A *breakpoint interrupt* is initiated by execution of the breakpoint instruction (1-byte instruction with code $= 11001100_2 = CC_{16}$). This instruction can be inserted at strategic points in a program that is being debugged to cause execution to be stopped automatically. Breakpoint interrupt can be used in a way similar to that of the debug exceptions when debugging programs. The breakpoint service routine can stop execution of the main program, permit the programmer to examine the contents of registers or memory, and allow for the resumption of execution of the program down to the next breakpoint. Breakpoint is an example of a trap. That is, upon execution of the breakpoint instruction, the CS and IP of the next instruction to be executed are saved on the stack.

Overflow Error Exception

The *overflow error exception* is an error condition similar to that of divide error. However, it can result from the execution of any arithmetic instruction. Whenever an overflow occurs, the overflow flag is set. Unlike divide error, the transfer of program control to a service routine is not automatic at occurrence of the overflow condition. Instead, the INTO (interrupt on overflow) instruction must be executed to test the overflow flag (OF) and determine if the overflow service routine should be initiated. If OF is tested and found to be set, a type 4 service routine is initiated. Its vector consists of IP_4 and CS_4, which are stored at 00010_{16} and 00012_{16}, respectively, in memory. The routine pointed to by this vector can be written to service the overflow condition. For instance, it could cause a message to be displayed to identify that an overflow has occurred.

Bounds Check Exception

Earlier we pointed out that the BOUND (check array index against bounds) instruction can be used to test an operand that is used as the index into an array to verify that it is within a predefined range. If the index is less than the lower bound (minimum value) or greater than the upper bound (maximum value), a *bounds check exception* has occurred and control is passed to the exception handler pointed to by $CS_5:IP_5$. The exception produced by the BOUND instruction is an example of a fault. Therefore, the values of CS and IP pushed to the stack represent the address of the instruction that produced the exception.

Invalid Opcode Exception

The exception-processing capability of the 80386DX permits detection of undefined opcodes. This feature of the 80386DX allows it to detect automatically whether or not the opcode to be executed as an instruction corresponds to one of the instructions in the instruction set. If it does not, execution is not attempted; instead, the opcode is identified as being undefined and the *invalid opcode exception* is initiated. In turn, control is passed to the exception handler identified by IP_6 and CS_6. This *undefined opcode-detection mechanism* permits the 80386DX to detect errors in its instruction stream. Invalid opcode is an example of an exception that produces a fault.

Coprocessor Extension Not Available Exception

When the 80386DX comes up in the real mode, both the EM (emulate coprocessor) and MP (math present) bits of its machine status word are reset. This mode of operation corresponds to that of the 8088 or 8086 microprocessor. When set in this way, the *coprocessor not available exception* cannot occur. However, if the EM bit has been set to 1 under software control (do not monitor coprocessor) and the 80386DX executes an ESC (escape) instruction for the math coprocessor, a processor extension not present exception is initiated through the vector at $CS_7:IP_7$. This service routine could pass control to a software emulation routine for the floating-point arithmetic operation. Moreover, if the MP and TS bits are set (meaning that a math coprocessor is available in the system and a task is in progress), when an ESC or WAIT instruction is executed, an exception also takes place.

Interrupt Table Limit Too Small Exception

Earlier we pointed out that the LIDT instruction can be used to relocate or change the limit of the interrupt vector table in memory. If the real-mode table has been changed, for example, its limit is set lower than address $003FF_{16}$ and an interrupt is invoked that attempts to access a vector stored at an address higher than the new limit, the *interrupt table limit too small exception* occurs. In this case, control is passed to the service routine by the vector $CS_8:IP_8$. This exception is a fault; therefore, the address of the instruction that exceeded the limit is saved on the stack.

Coprocessor Segment-Overrun Exception

The *coprocessor segment-overrun exception* signals that the 80387 numeric coprocessor has overrun the limit of a segment while attempting to read or write its operand. This event is detected by the coprocessor data channel within the 80386DX and passes control to the service routine through interrupt vector 9. This exception handler can clear the exception,

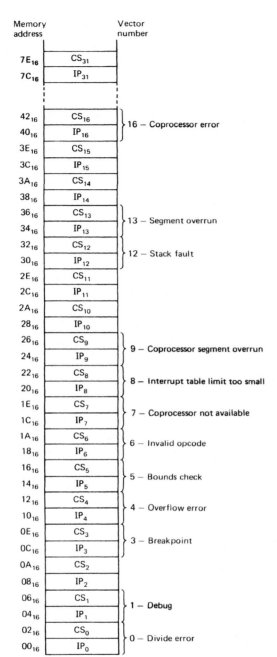

Memory address		Vector number
$7E_{16}$	CS_{31}	
$7C_{16}$	IP_{31}	
42_{16}	CS_{16}	16 – Coprocessor error
40_{16}	IP_{16}	
$3E_{16}$	CS_{15}	
$3C_{16}$	IP_{15}	
$3A_{16}$	CS_{14}	
38_{16}	IP_{14}	
36_{16}	CS_{13}	13 – Segment overrun
34_{16}	IP_{13}	
32_{16}	CS_{12}	12 – Stack fault
30_{16}	IP_{12}	
$2E_{16}$	CS_{11}	
$2C_{16}$	IP_{11}	
$2A_{16}$	CS_{10}	
28_{16}	IP_{10}	
26_{16}	CS_9	9 – Coprocessor segment overrun
24_{16}	IP_9	
22_{16}	CS_8	8 – Interrupt table limit too small
20_{16}	IP_8	
$1E_{16}$	CS_7	7 – Coprocessor not available
$1C_{16}$	IP_7	
$1A_{16}$	CS_6	6 – Invalid opcode
18_{16}	IP_6	
16_{16}	CS_5	5 – Bounds check
14_{16}	IP_5	
12_{16}	CS_4	4 – Overflow error
10_{16}	IP_4	
$0E_{16}$	CS_3	3 – Breakpoint
$0C_{16}$	IP_3	
$0A_{16}$	CS_2	
08_{16}	IP_2	
06_{16}	CS_1	1 – Debug
04_{16}	IP_1	
02_{16}	CS_0	0 – Divide error
00_{16}	IP_0	

Figure 12.30 Protected-mode internal exception gate locations.

reset the 80387, determine the cause of the exception by examining the registers within the 80387, and then initiate a corrective action.

Stack Fault Exception

In the real mode, if the address of an operand access for the stack segment crosses the boundaries of the stack, a stack fault exception is produced. This causes control to be transferred to the service routine defined by CS_{12} and IP_{12}.

Segment Overrun Exception

This exception occurs in the real mode if an instruction attempts to access an operand that extends beyond the end of a segment. For instance, if a word access is made to the address CS:FFFFH, DS:FFFFH, or ES:FFFFH, a fault occurs to the *segment overrun exception* service routine.

Coprocessor Error Exception

As part of the handshake sequence between the 80386DX microprocessor and 80387 numeric coprocessor, the 80386DX checks the status of its $\overline{ERROR}$ input. If the 80387 encounters a problem performing a numeric operation, it signals this fact to the 80386DX by switching its $\overline{ERROR}$ output to logic 0. This signal is normally applied directly to the $\overline{ERROR}$ input of the 80386DX and signals that an error condition has occurred. Logic 0 at this input causes a *processor extension exception* through vector 16.

Protected-Mode Internal Interrupts and Exceptions

In protected mode, more internal conditions can initiate an internal interrupt or exception. Figure 12.30 identifies each of these functions and its corresponding type number.

ASSIGNMENTS

Section 12.2

1. What are the five groups of interrupts supported on the 80386DX MPU?
2. What name is given to the special software routine to which control is passed when an interrupt occurs?
3. List in order the interrupt groups; start with the lowest priority and end with the highest priority.
4. What is the range of type numbers assigned to the interrupts in a real-mode 80386DX microcomputer system?
5. Is the interrupt assigned to type 21 at a higher or lower priority than the interrupt assigned to type 35?

Section 12.3

6. What is the real-mode interrupt address pointer table called? Protected-mode address pointer table?
7. What is the size of a real-mode interrupt vector? Protected-mode gate?

8. The contents of which register determines the location of the interrupt address pointer table? To what value is this register initialized at reset?

9. What two elements make up a real-mode interrupt vector?

10. The breakpoint routine in a real-mode 80386DX microcomputer system starts at address $AA000_{16}$ in the code segment located at address $A0000_{16}$. Specify how the breakpoint vector will be stored in the interrupt-vector table.

11. At what addresses is the real-mode interrupt vector table are CS and IP for type number 40 stored?

12. At what addresses is the protected-mode gate for type number 20 stored in memory? Assume that the table starts at address 00000_{16}.

13. Assume that gate 3 consists of the four words that follow:

$$(IDT + 8H) = 1000_{16}$$

$$(IDT + AH) = B000_{16}$$

$$(IDT + CH) = AE00_{16}$$

$$(IDT + EH) = 0000_{16}$$

  (a) Is the gate descriptor active?
  (b) What is the privilege level?
  (c) Is the gate a trap gate or an interrupt gate?
  (d) What is the starting address of the service routine?

Section 12.4

14. What does STI stand for?

15. Values stored in memory locations are as follows:

$$(IDT\_TABLE) = 01FF_{16}$$

$$(IDT\_TABLE + 2H) = 0000_{16}$$

$$(IDT\_TABLE + 4H) = 0001_{16}$$

What address is loaded into the interrupt-descriptor table register when the instruction LIDT [IDT_TABLE] is executed? What is the maximum size of the table? How many gates are provided for in this table?

16. Explain the operation performed by the following sequence of instructions:

```
INIT_IDTR   MOV    [IDT_TABLE],ILIMIT
            MOV    [IDT_TABLE+2],IBASE_LOW
            MOV    [IDT_TABLE+4],IBASE_HIGH
            LIDT   IDT_TABLE
            SIDT   IDT_COPY
            CMP    IDT_TABLE,IDT_COPY
            JNZ    INIT_IDTR
```

17. At what addresses is the gate for INT 50 stored in a protected-mode 80386DX microcomputer system?

18. Which type of instruction does INTO normally follow? Which flag does it test?

19. Explain the function of the bound instruction in the sequence

```
SCAN    DEC    DI
        BOUND  DI,LIMITS
          .
          .
          .
        JNZ    SCAN
```

if the contents of memory locations LIMITS and LIMITS + 2 are 0000_{16} and $00FF_{16}$.

20. What happens when the instruction HLT is executed?

Section 12.5

21. Explain how the CLI and STI instructions can be used to mask out external hardware interrupts during the execution of an uninterruptible subroutine.

22. How can the interrupt interface be reenabled during the execution of an interrupt-service routine?

Section 12.6

23. What does $\overline{\text{INTA}}$ stand for?

24. Is the INTR input of the 80386DX edge triggered or level triggered?

25. Explain the function of the INTR and $\overline{\text{INTA}}$ signals in the circuit diagram of Fig. 12.8.

26. What bus cycle identification code must be decoded to produce the $\overline{\text{INTA}}$ signal?

27. During the interrupt-acknowledge bus cycle, what does the first $\overline{\text{INTA}}$ pulse tell the external circuitry? The second pulse?

28. If the interface diagram in Fig. 12.8 represents an I/O-mapped interrupt-controller interface, which address lines are part of the interface?

29. Over which lines does external circuitry send the type number of the active interrupt to the 80386DX?

Section 12.7

30. Give an overview of the events in the order they take place during the interrupt-request, interrupt-acknowledge, and interrupt-vector-fetch cycles of a real-mode 80386DX microcomputer system.

31. If a real-mode 80386DX microcomputer system is running at 20 MHz with no wait states in all I/O bus cycle, how long does it take to perform the interrupt-acknowledge bus cycle sequence?

32. How long does it take the 80386DX in Problem 31 to push the values of the old flags, old CS, and old IP to the stack? How much stack space do these information take up?

33. How long does it take the 80386DX in Problem 31 to fetch the vector $\text{CS}_{\text{NEW}}:\text{IP}_{\text{NEW}}$ from memory?

34. What is the primary difference between the real-mode and protected-mode interrupt-request/acknowledge handshake sequence for the 80386DX microprocessor?

35. If a program that is interrupted is in a conforming code segment, what is the new privilege level equal to? What stack is used? What information is saved on the stack?

36. If the program transition identified in Problem 35 fails, what additional information is pushed to the stack?

37. If an IDT error code has the value $0000F307_{16}$, what are the values of the EXT, IDT, and TI bits, and what do they stand for?

38. What is the relationship between CPL and RPL in an intralevel interrupt return? An interlevel interrupt return?

Section 12.8

39. Specify the value of ICW_1 needed to configure an 82C59A as follows: ICW_4 not needed, single-device interface, and edge-triggered inputs.

40. Specify the value of ICW_2 if the type numbers produced by the 82C59A are to be in the range 70_{16} through 77_{16}.

41. Specify the value of ICW_4 such that the 82C59A is configured for use in an 80386DX system, with normal EOI, buffered-mode master, and special fully nested-mode disabled.

42. Write a program that will initialize an 82C59A with the initialization command words derived in Problems 39, 40, and 41. Assume that the 82C59A resides at address $0A000_{16}$ in the memory address space and that the contents of DS are 0000_{16}.

43. Write an instruction that, when executed, will read the contents of OCW_1 and place it in the AL register. Assume that the 82C59A has been configured by the software of Problem 42.

44. What priority scheme is enabled if OCW_2 equals 67_{16}?

45. Write an instruction sequence that when executed will toggle the state of the read register bit in OCW_3. Assume that the 82C59A is located at memory address $0A000_{16}$ and that the contents of DS are 0000_{16}.

Section 12.9

46. How many interrupt inputs can be directly accepted by the circuit in Fig. 12.24(a)?

47. How many interrupt inputs can be directly accepted by the circuit in Fig. 12.24(b)?

48. Summarize the interrupt-request/acknowledge handshake sequence for an interrupt initiated at an input to slave B in the circuit of Fig. 12.24(b).

49. What is the maximum number of interrupt inputs that can be achieved by expanding the number of slaves in the master-slave configuration of Fig. 12.24(b)?

Section 12.10

50. Give another name for a software interrupt.

51. If the instruction INT 80 is to pass control to a subroutine at address $A0100_{16}$ in the code segment starting at address $A0000_{16}$, what vector should be loaded into the interrupt vector table?

52. At what address would the real-mode vector for the instruction INT 80 be stored in memory?

Section 12.11

53. What type number and interrupt vector table addresses are assigned to NMI?
54. What are the key differences between NMI and the other external-hardware-initiated interrupts?
55. Give a very common use of the NMI input.

Section 12.12

56. What is the active logic level of the RESET input of the 80386DX?
57. To which signal must the application of the RESET input be synchronized?
58. List the states of the address bus lines, byte-enable signals, data bus lines, and control signals $\overline{ADS}$, W/$\overline{R}$, D/$\overline{C}$, and M/$\overline{IO}$ when reset is at its active level.
59. To what value is the flag register initialized after reset?
60. To what address does a reset operation pass control in a real-mode microcomputer? A protected-mode microcomputer?
61. Write a reset subroutine that initializes the block of memory locations from address $0A000_{16}$ to $0A0FF_{16}$ to zero. The initialization routine is at address 01000_{16}.
62. How is the built-in self-test of the 80386DX initiated?

Section 12.13

63. List the real-mode internal interrupts serviced by the 80386DX.
64. Internal interrupts and exceptions are categorized into groups based on how the failing function is reported. List the three groups.
65. Which real-mode vector numbers are allocated to internal interrupts and exceptions?
66. Into which reporting group is the invalid opcode exception classified?
67. What is the cause of a stack fault exception?
68. Which exceptions take on a new meaning or are active only in the protected mode?

80386DX PC/AT
Microcomputer System
Hardware

▲ 13.1 INTRODUCTION

Having learned about the 80386DX microprocessor and its memory, input/output, and interrupt interfaces, we now turn our attention to a microcomputer system design using this hardware. The microcomputer we study in this chapter is one that implements an 80386DX-based PC/AT (personal computer/advanced technology). The material covered in this chapter is organized as follows:

1. Architecture of the system processor board in the original IBM PC/AT
2. High-integration PC/AT-compatible peripheral ICs
3. Core 80386DX microcomputer
4. 82345 data buffer
5. 82346 system controller
6. 82344 ISA controller
7. 82341 high-integration peripheral combo
8. 82077AA floppy disk controller

▲ 13.2 ARCHITECTURE OF THE SYSTEM PROCESSOR
BOARD IN THE ORIGINAL IBM PC/AT

Before we begin our study of an 80386DX-based PC/AT-compatible microcomputer system, we will introduce the architecture implemented in IBM's original 80286-based PC/AT. The IBM PC/AT is a practical implementation of the 80286 microprocessor and its peripheral chip set as a general-purpose microcomputer. A block diagram of the system processor board (main circuit board) of the PC/AT is shown in Fig. 13.1(a). This diagram identifies

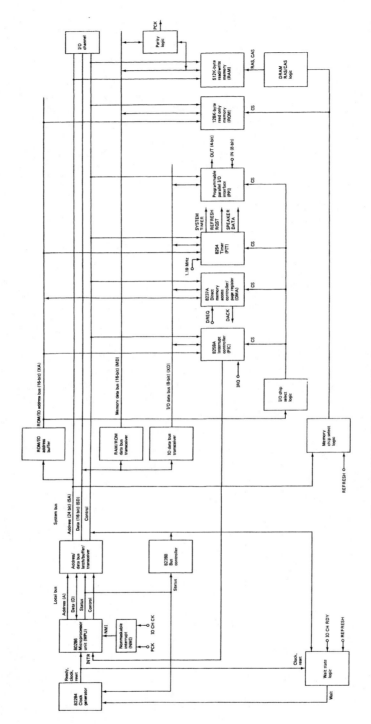

Figure 13.1(a) IBM PC/AT microcomputer block diagram.

Address	Name	Function
000000H to 07FFFFH	512KB system board	System board memory
080000H to 09FFFFH	128KB	I/O channel memory — IBM personal computer AT 128KB memory expansion option
0A0000H to 0BFFFFH	128KB video RAM	Reserved for graphics display buffer
0C0000H to 0DFFFFH	128KB I/O expansion ROM	Reserved for ROM on I/O adapters
0E0000H to 0EFFFFH	64KB reserved on system board	Duplicated code assignment at address FE0000H
0F0000H to 0FFFFFH	64KB ROM on the system board	Duplicated code assignment at address FF0000H
100000H to FDFFFFH	Maximum memory 15MB	I/O channel memory — IBM personal computer AT 512KB memory expansion option
FE0000H to FEFFFFH	64KB reserved on system board	Duplicated code assignment at address 0E0000H
FF0000H to FFFFFFH	64KB ROM on the system board	Duplicated code assignment at address 0F0000H

(b)

Address	Device
000H–01FH	DMA controller 1, 8237A-5
020H–03FH	Interrupt controller 1, 8259A, master
040H–05FH	Timer, 8254-2
060H–06FH	8042 (Keyboard)
070H–07FH	Real-time clock, NMI (nonmaskable interrupt) mask
080H–09FH	DMA page register, 74LS612
0A0H–0BFH	Interrupt controller 2, 8259A
0C0H–0DFH	DMA controller 2, 8237A-5
0F0H	Clear math coprocessor busy
0F1H	Reset math coprocessor
0F8H–0FFH	Math coprocessor
1F0H–1F8H	Fixed disk
200H–207H	Game I/O
278H–27FH	Parallel printer port 2
2F8H–2FFH	Serial port 2
300H–31FH	Prototype card
360H–36FH	Reserved
378H–37FH	Parallel printer port 1
380H–38FH	SDLC, bisynchronous 2
3A0H–3AFH	Bisynchronous 1
3B0H–3BFH	Monochrome display and printer adapter
3C0H–3CFH	Reserved
3D0H–3DFH	Color/graphics monitor adapter
3F0H–3F7H	Diskette controller
3F8H–3FFH	Serial port 1

(c)

Figure 13.1 (b) Memory Map. (c) Input/output address map.

the major functional elements of the PC/AT: MPU, PIC, DMA, PIT, parallel I/O, ROM, and RAM. Here we describe briefly the architecture of the PC/AT's microcomputer system.

The heart of the PC/AT's system processor board is the 80286 microprocessor unit (MPU). It is here that instructions of the program are fetched and executed. To interface to the peripherals and other circuitry such as memory, the 80286 microprocessor generates address, data, status, and control signals. Together, these signals form what is called the

Input/output	Signal	Function
OUT_0	TIM 2 GATE SPK	Speaker enable
OUT_1	SPKR DATA	Speaker data
OUT_2	ENB RAM PCK	RAM parity check enable
OUT_3	ENA IO CK	Enable I/O channel check
IN_0	TIM 2 GATE SPK	Speaker enable
IN_1	SPKR DATA	Speaker data
IN_2	ENB RAM PCK	RAM parity check enable
IN_3	ENA IO CK	Enable I/O channel check
IN_4	REF DET	Refresh detect
IN_5	OUT 2	Timer 2 output
IN_6	IO CH CK	I/O channel check
IN_7	PCK	Parity check

(d)

Priority level	Input		Function
	Microprocessor NMI		Parity or I/O channel check
	Interrupt controllers MASTER	SLAVE	
0	IRQ 0		Timer output 0
1	IRQ 1		Keyboard (output buffer full)
2	IRQ 2 ◄		Interrupt from CTLR 2
		IRQ 8	Real-time clock interrupt
		IRQ 9	Software redirected to INT 0AH (IRQ 2)
		IRQ 10	Reserved
		IRQ 11	Reserved
		IRQ 12	Reserved
		IRQ 13	Coprocessor
		IRQ 14	Fixed disk controller
		IRQ 15	Reserved
3	IRQ 3		Serial port 2
4	IRQ 4		Serial port 1
5	IRQ 5		Parallel port 2
6	IRQ 6		Diskette controller
7	IRQ 7		Parallel port 1

(e)

Figure 13.1 (d) Input/output functions. (e) Interrupt levels and functions. (Parts b, c, and e, courtesy of International Business Machines Corporation)

local bus in Fig. 13.1(a). Notice that the local address (A) lines are latched and buffered to form a 24-bit *system address bus* (SA) and the local data (D) lines are buffered to provide a 16-bit *system data bus* (SD).

Notice in Fig. 13.1(a) that the system address lines are further buffered by the *ROM/IO address buffer*. This *extended address bus* (XA) is distributed to ROM memory and the I/O peripherals. The system data bus is buffered by the *RAM/ROM data bus transceiver* to give the *memory data bus* (MD). Looking at Fig. 13.1(a), we see that the MD lines are the data path for both the ROM and RAM banks. In a similar way, the lower eight lines of the system data bus are buffered by the *I/O data bus transceiver* to give an 8-bit *extended data bus* (XD). The extended data bus is the data path for all of the I/O devices.

At the same time, the status and control lines of the local bus are decoded by the 82288 bus controller to generate the *system-control bus*. This control bus consists of memory and I/O read and write control signals. The bus controller also produces the data bus control signals. For instance, it provides a signal that determines the direction of data transfer through the data bus transceivers; that is, the signal needed to make the data bus lines work as inputs to the microprocessor during memory and I/O read operations and as outputs during write operations.

The operation of the microprocessor and other devices in a microcomputer system must be synchronized. The circuitry of the clock generator block generates clock signals for this purpose. The clock-generator section also produces a power-on reset signal that is needed to initiate initialization of the microprocessor and peripherals at power-up.

The clock generator section also works in conjunction with the wait-state logic to synchronize the MPU to slow peripheral devices. In Fig. 13.1(a) we see that the wait-state logic circuitry monitors the system control bus and the signals IO CH RDY and REFRESH. Depending on the state of these inputs, it generates a wait signal for input to the 82284 clock generator. In turn, the clock generator synchronizes this wait input with the system clock to produce a ready signal at its output. READY is input to the 80286 MPU and provides the ability to extend automatically bus cycles that are performed to slow devices by inserting wait states. For instance, all bus cycles to on-board memory take three clock periods. Therefore, they include one wait state. On the other hand, I/O bus cycles to the 8-bit LSI peripherals have four wait states and take a total of six clock periods.

The memory subsystem of the PC/AT system processor board has 256K bytes of dynamic R/W memory (RAM). This on-board RAM is implemented using 128K $\times$ 1-bit dynamic RAMs and can be expanded to 512K bytes by replacing the ICs with 256-KB DRAMs. A memory map for the PC/AT's memory is shown in Fig. 13.1(b). From the map we find that the 512K bytes of system board RAM resides in the address range from 000000_{16} through $07FFFF_{16}$. In the microcomputer system, RAM is used to store operating system routines, application programs, and data that are to be processed. These programs and data are typically loaded into RAM from a mass storage device such as a floppy diskette or hard disk.

Looking at Fig. 13.1(a), we see that the RAM subsystem includes parity logic. This circuit generates and adds a parity bit to all data written to RAM. Moreover, whenever data are read from RAM, parity is checked, and if an error occurs, it signals this fact with the PCK signal.

The memory subsystem also contains 64K bytes of read-only memory (ROM). ROM is implemented with 32K $\times$ 8 EPROMs and is expandable to 128K bytes. The memory map shows that on-board ROM is located in two separate 64K-byte address ranges. The first 64K bytes are located from $0F0000_{16}$ to $0FFFFF_{16}$. Expansion is allowed by the second 64K-byte block of addresses from $FF0000_{16}$ to $FFFFFF_{16}$. This part of the memory subsystem contains the system ROM of the PC/AT. Included in these ROMs are fixed programs such as the *BASIC interpreter, power-on system procedures,* and *I/O device drivers* or *BIOS* as they are better known.

The memory and I/O chip-select logic sections, which are shown in the block diagram of Fig. 13.1(a), are used to select and enable the appropriate memory or peripheral device whenever a bus cycle takes place over the system bus. Here we see that they accept address information at their inputs and produce chip-select signals at their outputs. To select a device in the I/O address space, such as the DMA controller, the I/O chip-select logic decodes the extended address XA on the ROM/IO address bus to generate a chip-select (CS) output for the corresponding I/O device. This chip-select signal is applied to the I/O device and enables its microprocessor interface for operation.

The ROM and RAM chip selects are produced in a similar way. But this time they are produced by decoding the address on the system address (SA) bus. Notice that the RAM chip selects are converted to row-address-select (RAS) and column-address-select (CAS) signals before they are applied to the RAM array.

The LSI peripheral devices included on the PC/AT system processor board are the 8237A direct memory access (DMA) controller, 8254 programmable interval timer (PIT), and 8259A programmable interrupt controller (PIC). Notice that each of these devices is identified with a separate block in Fig. 13.1(a). These peripherals are all located in the 80286's I/O address space, and their registers are accessed through software using the address ranges given in Fig. 13.1(c). For instance, the four registers within the PIT are located at addresses 0040_{16}, 0041_{16}, 0042_{16}, and 0043_{16}.

To support high-speed memory and I/O data transfers, two 8237A direct memory-access controllers are provided on the PC/AT system board. DMA controller 1 contains four DMA channels, *DMA channel 0* through *DMA channel 3*. One channel, DMA 1, is dedicated to support an SDLC synchronous communications channel and the other three DMA channels are available at the I/O channel [industry standard architecture (ISA) bus] for use with peripheral devices. For instance, DMA 2 is used to support floppy disk drive data transfers over the ISA bus. DMA controller 2 provides *DMA channels* 4 through 7. Here channel 4 is used as a cascade input for DMA controller 1 and channels 5 through 7 are also available at the ISA bus.

The 8254-based timer circuitry is used to generate time-related functions and signals in the PC/AT. There are three 16-bit counters in the 8254 PIT and they are all driven by a 1.19-MHz clock input signal. Timer 0 is used to generate an interrupt to the microprocessor. This timing function is known as the *system timer*. On the other hand, timer 1 is used to produce a request to initiate refresh of the dynamic RAM every 15 μs. The last timer is used to generate programmable tones for driving the speaker.

The parallel I/O section of the PC/AT microcomputer in Fig. 13.1(a) provides both an 8-bit input port and a 4-bit output port. The signal and function for each of the I/O lines is identified in Fig. 13.1(d). Here we see that the I/O lines are used to output tones to the speaker (SPKR DATA), enable or disable I/O channel check (ENA IO CK) and RAM parity check (ENB RAM PCK), and input the state of signals such as PCK, IO CH CK, and OUT 2.

The circuitry in the nonmaskable interrupt (NMI) logic block allows a nonmaskable interrupt request derived from two sources to be applied to the microprocessor. As shown in Fig. 13.1(a), these interrupt sources are *R/W memory parity check* (PCK) and *I/O channel check* (IO CH CK). If either of these inputs is active, the NMI logic outputs a request for service to the 80286 over the NMI signal line. The 80286 can determine the cause of the NMI request by reading the state of PCK and IO CH CK through the parallel I/O interface.

In addition to the nonmaskable interrupt, the PC/AT architecture provides for requests for service to the MPU by interrupts at another interrupt input called *interrupt request* (INTR). Notice in Fig. 13.1(a) that this signal is supplied to the 80286 by the output of the interrupt controller (PIC) block. Two 8259A LSI interrupt controllers are used in the PC/AT. Each provides for eight prioritized interrupt inputs. The inputs of the interrupt controller are supplied by peripherals such as the timer, keyboard, diskette drive, printer, and communication devices. Interrupt priority assignments for these devices are listed in Fig. 13.1(e).

For example, the timer (actually just timer 0 of the 8254) is at the highest-priority level, which is IRQ_0.

I/O channel (the ISA bus), which is a collection of address, data, control, and power lines, is provided to support expansion of the PC/AT system. The chassis of the PC/AT has eight expansion slots. Six slots have both a 36-pin and a 62-pin card edge socket. The other two slots have just the 62-pin socket. In this way the system configuration can be expanded by adding special function adapter cards, such as boards to control a monochrome or color display, floppy disk drives, a hard disk drive, or expanded memory or to attach a printer. In Fig. 13.1(b) we see that I/O channel expanded RAM (memory on the ISA bus) resides in the part of the memory address space from 100000_{16} through $FDFFFF_{16}$.

▲ 13.3 HIGH-INTEGRATION PC/AT-COMPATIBLE PERIPHERAL ICS

In the preceding section we provided an overview of the architecture of the main processor board of the original IBM PC/AT personal computer. The circuit implementation used on this board has become an architectural standard. PC/AT compatibles, whether they use the 80286, 80386DX, 80486DX, or Pentium® processor, must provide equivalent circuitry to assure 100% hardware and 100% software compatibility. Here we will begin our study of the circuit implementation techniques used in the design of an 80386DX-based PC/AT-compatible microcomputer systems.

Since the introduction of the original IBM PC/AT, much work has been done to develop special-purpose ICs that simplify the design of PC/AT-compatible personal computers. The cornerstone of this effort has been the development of high-integration peripheral ICs. The objective of these peripherals are to reduce the complexity and cost of the circuitry on the main processor board while increasing the functionality. In this section we begin a study of one of the more popular PC/AT-compatible peripheral IC chip sets and its use in the design of a modern PC/AT-compatible main processor board.

The *82340 PC/AT Chip Set* offers a four-chip high-integration solution for implementing 80386DX-based PC/AT-compatible personal computers. These devices implement most of the circuitry needed on the main processor board. That is, the 82340 chips, along with a few other ICs, produce a complete solution for the main processor board of a PC/AT-compatible microcomputer. In fact, they implement more functions than are provided on the original 80286-based PC/AT main processor board. Examples of additional functions included are two serial communication channels, a parallel printer port, and a hard disk drive interface.

Figure 13.2 is a block diagram of the architecture of a main processor board that uses the 82340 chip set. Here we find that the core of the microcomputer are the 80386DX MPU, the 82387DX numerics coprocessor, and 82385DX cache controller. Attached to the local bus of the MPU, we find three of the 82340 peripheral ICs: the 82345 data buffer, the 82346 system controller, and 82344 ISA controller. At the other side, these peripherals connect to the ISA expansion bus. Notice that the 82341 peripheral combo does not connect directly to the 80386DX MPU; instead, it is tied to the ISA bus.

The 82340 chip set is very versatile. With its programmable features, the microcomputer design can be optimized for low cost, maximum performance, minimum physical size, or low power. The circuit diagram in Fig. 13.3 shows a typical design of a 33-MHz 80386DX PC/AT main processor board. Notice that some additional devices such as PALs, octal

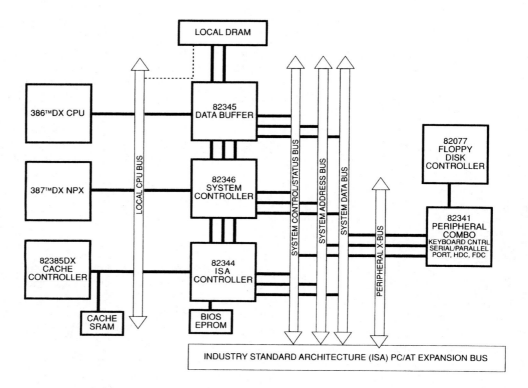

Figure 13.2 Block diagram of a PC/AT main processor board using the 82340 chip set. (Reprinted by permission of Intel Corp. Copyright/Intel Corp. 1989)

latches, octal transceivers, octal buffers, and logic gates are needed along with the 82340 chips to complete the design.

EXAMPLE 13.1

What function is performed by the 82077AA device?

Solution

In Fig. 13.2 we find that the 82077AA is the floppy disk controller for the microcomputer system.

▲ 13.4 CORE 80386DX MICROCOMPUTER

Let us begin our study of the PC/AT main processor board in Fig. 13.3 with the circuitry of the core 80386DX microcomputer. This part of the microcomputer is shown on sheet 1 of the circuit diagram. Notice that it consists of the 80386DX MPU (IC U_{12}), the 80387DX numerics coprocessor (U_{10}), the 82385DX cache controller (U_{15}), and the cache memory array (U_{16} and U_{17}). The 80386DX, 80387DX, and 82385DX are designed to attach to each

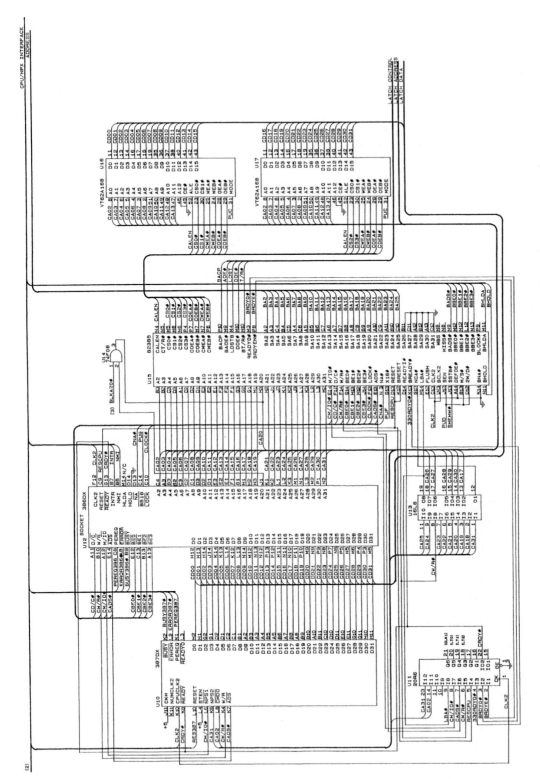

Figure 13.3 Circuit diagram for an 80386DX-based PC/AT compatible microcomputer main processor board. (Sheet 1 of 8) (Reprinted by permission of Intel Corp. Copyright/Intel Corp. 1989)

685

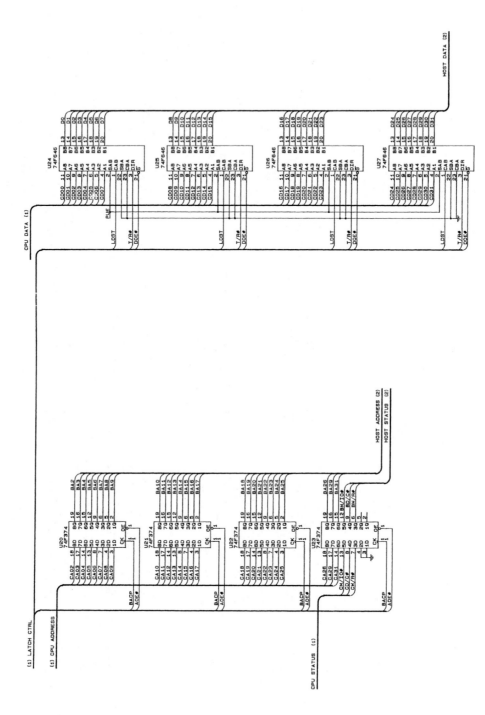

Figure 13.3 (Continued) Sheet 2 of 8.

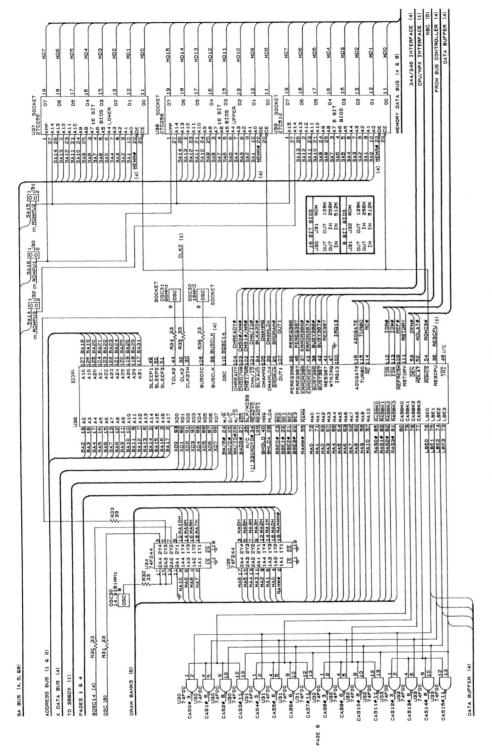

Figure 13.3 (Continued) Sheet 3 of 8.

687

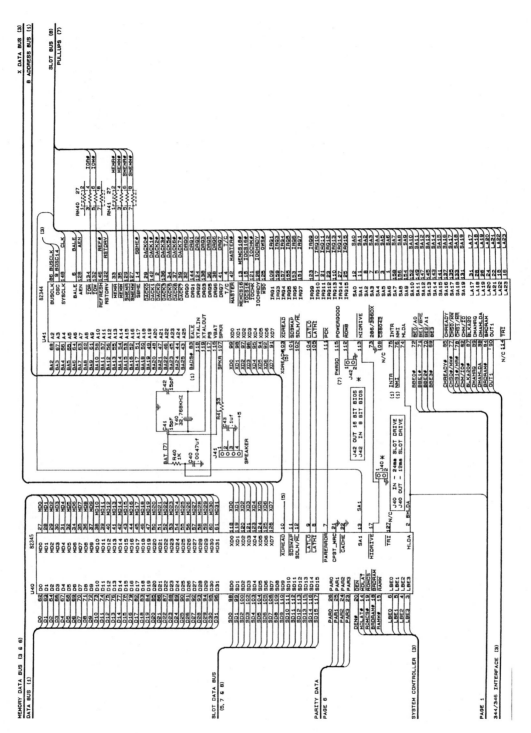

Figure 13.3 (Continued) Sheet 4 of 8.

688

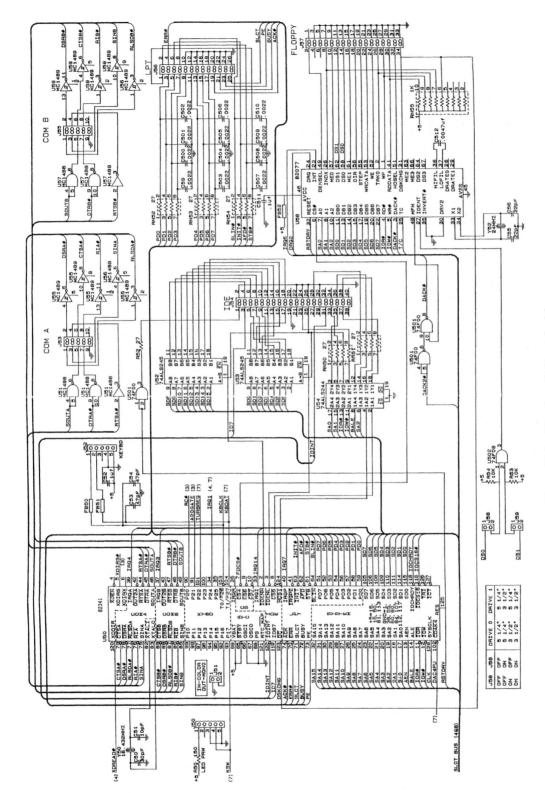

Figure 13.3 (Continued) Sheet 5 of 8.

689

Figure 13.3 (Continued) Sheet 6 of 8.

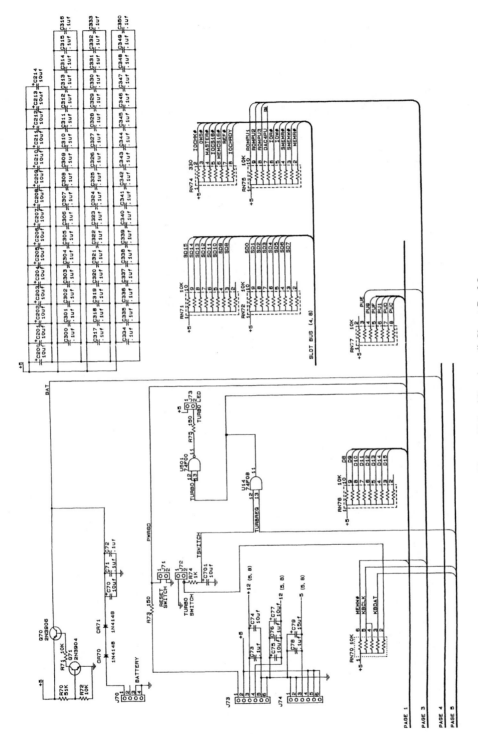

Figure 13.3 (Continued) Sheet 7 of 8.

691

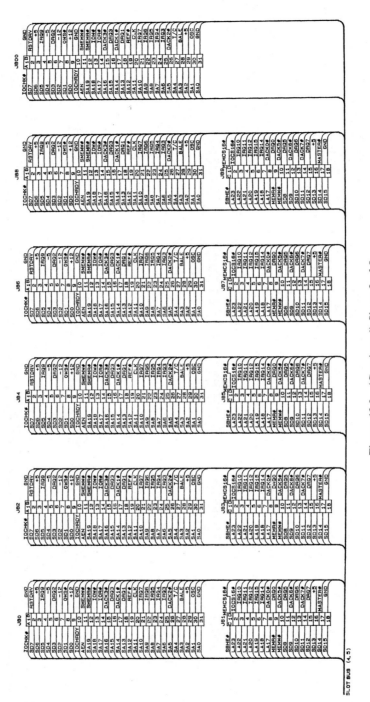

Figure 13.3 (Continued) Sheet 8 of 8.

SLOT BUS (4, 5)

692

other directly. Looking at their interconnection in Fig. 13.3, we see that address lines A_2 through A_{31} of the 80386DX attach directly to the corresponding address line of the 82385DX and data bus lines D_0 through D_{31} of the 80386DX connect directly to the corresponding data line on the 80387DX. Moreover, notice that the 80386DX's bus cycle definition outputs, $D/\overline{C}$, $W/\overline{R}$, and $M/\overline{IO}$, are applied directly to the $D/\overline{C}$, $W/\overline{R}$, and $M/\overline{IO}$ inputs of the 82385DX.

EXAMPLE 13.2

What signals supply the PEREQ, $\overline{ERROR}$, and $\overline{BUSY}$ inputs of the 80386DX? What are the sources of these signals?

Solution

From sheet 1 of Fig. 13.3, we find that the PEREQ, $\overline{ERROR}$, and $\overline{BUSY}$ inputs of the 80386DX are supplied by the signals PEREQ386, $\overline{ERROR386}$, and $\overline{BUSY386}$, respectively. These signals are produced at outputs of the 82346 system controller (sheet 3 of Fig. 13.3).

The cache memory array is implemented with two VT62A16B 8K $\times$ 16-bit static RAMs. These SRAMs, which are labeled U_{16} and U_{17} on sheet 1 of Fig. 13.3, are specifically designed for use in an 82385DX-based cache memory subsystem. For this reason no external address latches or data bus transceivers are required in the cache memory interface. They are built within the SRAM ICs. Notice that cache memory-control outputs CALEN, $\overline{COEA}$, $\overline{COEB}$, $\overline{CWEA}$, and $\overline{CWEB}$ are applied to equivalent inputs on both VT62A16Bs.

EXAMPLE 13.3

Is the 82385DX on sheet 1 in Fig. 13.3 configured to operate in direct-mapped or two-way set associative mode?

Solution

On sheet 1 of Fig. 13.3 we find that the $2W/\overline{D}$ input is supplied by the signal PUD. Tracing this signal back to sheet 7, we see that it supplies +5 V to the $2W/\overline{D}$ input through a 10-kΩ pull-up resistor. This is a logic 1 at the $2W/\overline{D}$ input and selects two-way set associative operation.

EXAMPLE 13.4

Which SRAM in the cache memory array is selected with chip-select outputs $\overline{CS}_2$ and $\overline{CS}_3$ of the 82385DX cache controller?

Solution

Looking at sheet 1 in Fig. 13.3, we find that the $\overline{CS}_2$ and $\overline{CS}_3$ outputs of the 82385DX are applied to the $\overline{CS}_0$ and $\overline{CS}_1$ inputs, respectively, of SRAM U_{17}. Therefore, $\overline{CS}_2$ and $\overline{CS}_3$ enable the lower of the two SRAM ICs.

The ready/wait logic is implemented with a 20R6 PAL. This is device U_{11} on sheet 1 of Fig. 13.3. Notice that it accepts signals such as CPU read/write (CW/$\overline{R}$), CPU address strobe ($\overline{CADS}$), CPU memory/IO (CM/$\overline{IO}$), and 82385DX bus ready ($\overline{BREADY}$) as inputs and decodes them to produce the appropriately timed CPU ready output ($\overline{CRDY}$). $\overline{CRDY}$ is applied in parallel to the $\overline{READY}$ input at pin G13 of the 80386DX, the $\overline{READY}$ input at pin K8 of the 80387DX, and $\overline{READYI}$ input at pin D14 of the 82385DX. Based on the input conditions, $\overline{CRDY}$ inserts the appropriate number of wait states into the bus cycle.

EXAMPLE 13.5

What is the source of the $\overline{LBA}$ input of the ready/wait state PAL? To what other device is this signal also supplied?

Solution

Tracing the $\overline{LBA}$ line at pin 9 of U_{11}, we find that it is supplied by output O_8 at pin 19 of PAL U_{13}. O_8 is also connected to the $\overline{LBA}$ input at pin 14 of the 82385DX.

One other circuit is shown on sheet 1 of the circuit diagram in Fig. 13.3. This circuit, which is implemented with the 16L8 PAL U_{13}, is used to map the 80386DX's address space into cacheable and noncacheable regions. Notice that the noncacheable access ($\overline{NCA}$) input of the 82385DX is driving by output O_1 at pin 12 of U_{13}. Remember that if $\overline{NCA}$ is logic 0 during a bus cycle, the bus cycle is performed to main memory, not to cache memory; that is, a noncacheable bus cycle is in progress. The inputs to this PAL are address lines CA_{17} through CA_{30} and CW/$\overline{R}$.

The local address and data bus lines of the 80386DX are latched and buffered by the address latch and data bus transceiver circuits. On sheet 2 of Fig. 13.3 we find that four 74F374 octal latches are used to latch the address and bus cycle definition signals of the 80386DX. These devices are the ICs labeled U_{20} through U_{23}. For example, U_{20} latches *CPU address* inputs CA_{02} through CA_{09} to produce *buffered address* outputs BA_2 through BA_9. Address information is clocked into the latches with the bus address clock pulse (BACP) signal and enabled to the outputs with address output enable ($\overline{AOE}$). Both of these signals are produced by the 82385DX. The address output on BA_2 through BA_{31} is identified on sheet 2 of Fig. 13.3 as the *host address bus*.

EXAMPLE 13.6

Which 74F374 IC is used to latch the CPU's bus cycle definition signals? How are the buffered bus cycle definition signals labeled? What is the output status bus called?

Solution

Looking at sheet 2 in Fig. 13.3, we see that *CPU status* signals CM/$\overline{IO}$, CD/$\overline{C}$, and CW/$\overline{R}$ are inputs of the 74F374 U_{23} and are latched to produce the buffered bus cycle definition signals BM/$\overline{IO}$, BD/$\overline{C}$, and BW/$\overline{R}$. These signals produce what is called the *host status bus*.

The data bus buffer is formed from four 74F646 registered transceiver devices. These ICs are labeled U_{24} through U_{27} on sheet 2 of Fig. 13.3. The A bus side of the transceivers are attached to the *CPU data bus* lines and their B sides supply the *host data bus*.

Let us continue by looking at the connection of the control signals of the data bus transceivers. In Fig. 13.3 we find that control inputs S_{BA} and C_{BA} are both wired to ground (logic 0) and S_{AB} is supplied by the signal PUE. PUE is traced to sheet 7 of the circuitry in Fig. 13.3. Here it is found to supply +5 V through a 10-kΩ pull-up resistor. Therefore, S_{AB} is at logic 1. The other three control signals of the latches are supplied by the 82385DX. For example, the C_{AB} clock inputs are driven by the *load data strobe* (LDSTB) output of the 82385DX.

EXAMPLE 13.7

What signal supplies the direction input of the 74F646 transceivers? What is the source of this signal?

Solution

Sheet 2 of Fig. 13.3 shows that the DIR input of all four 74F646 transceivers is driven by the signal T/$\overline{R}$. This signal is traced back to pin N10 (BT/$\overline{R}$) of the 82385DX on sheet 1.

Next we examine how the transceivers operate during read bus cycles from main memory. Since S_{BA} is at logic 0, the 74F646s are set for real-time transfers of data at bus B to bus A. That is, when $\overline{DOE}$ and T/$\overline{R}$ are logic 0, data supplied to the host data bus by memory or an I/O device are immediately passed through the transceivers to the CPU data bus.

When data are written to memory or an I/O device, the data bus transceivers work differently. In this case, $\overline{DOE}$ is 0, T/$\overline{R}$ is 1, and the write data are strobed from the A side of the transceivers to the B sides with a pulse on the LDST line.

▲ 13.5 82345 DATA BUFFER

The 82345 data buffer is a VLSI device that is used to perform data bus buffering and bus switching for the PC/AT microcomputer system. That is, it is responsible for automatic routing of data between the MPU's data bus and the microcomputer's, BIOS ROM/DRAM data bus, I/O data bus, and system data bus. The 82345 also implements some other system functions. For instance, it contains both the parity-generation and parity-checker logic. The 82345 is packaged in a high-density 128-pin plastic quad flat package (PQFP).

Block Diagram of the 82345

Figure 13.4 is a block diagram of the circuits contained within the 82345 data buffer. Notice that it includes several data bus transceivers, multiplexers, and control logic. For example, there are separate transceivers and multiplexers in the write data path for the XD, SD, and MD data buses. These multiplexers are used to route data between the buses. For instance, the XD bus multiplexer accepts SD_0 through SD_{15} at its A port, D_0 through D_{31} at its B port, and MD_0 through MD_{31} at its C port. The control logic determines which

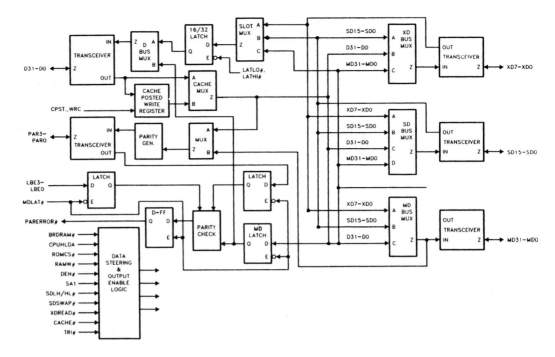

Figure 13.4 Block diagram of the 82345 data buffer. (Reprinted by permission of Intel Corp. Copyright/Intel Corp. 1989)

of these inputs are passed to the IN inputs of the XD bus transceiver and onto the XD data bus.

EXAMPLE 13.8

Which multiplexers and latches are in the read-data path?

Solution

Read data are passed to the MPU's data bus (D_0 through D_{31}) from the IN inputs of the D bus transceiver. The multiplexers and latches in the data paths to this input are the D bus multiplexer, the 16/32 latch, the slot multiplexer, and the MD latch.

EXAMPLE 13.9

What are the sources of input data for the slot multiplexer?

Solution

Looking at the slot multiplexer block in Fig. 13.4, we find that the source of read/input data for the A, B, and C inputs are the XD bus, SD bus, and MD bus, respectively.

The 82345 contains both a parity-generator circuit and parity-checker circuit. Parity generation and checking applies only to memory accesses that take place over the memory data (MD) bus. When a read bus cycle is in progress, the parity information held in the DRAM memory array is input to the 82345 over the *parity bit* (PAR$_0$ through PAR$_3$) lines. Notice that the outputs (OUT) of the parity transceiver are latched and applied as inputs to the parity checker circuit. At the same time, the *latched byte-enable* (LBE$_0$ through LBE$_3$) signals are latched within the 82345 and passed to another set of inputs on the parity checker block. Finally, memory data from MD$_0$ through MD$_{31}$ are both multiplexed to the CPU's data bus and latched into the MD latch. The outputs of the MD latch are supplied to a third set of inputs on the parity checker. After these three set of inputs are present, parity-check operations take place on each byte of the 32-bit word that is selected by the byte-enable code. If a parity error occurs in any of these bytes, the parity error D-type flip-flop is reset. In this way, a parity error condition is signaled at the $\overline{\text{PARERROR}}$ output.

Parity generation takes place whenever data are written into memory attached to the MD bus of the 82345. If a write operation is taking place, write data from the MPU's data bus (D$_0$–D$_{31}$) are multiplexed to the input (IN) port of the MD bus transceiver. In Fig. 13.4, we find that these data are also applied to the B input port of the multiplexer for the parity generator circuit. When data are applied to the input of the parity generator, a 4-bit parity code (one parity bit for each byte of the 32-bit word) is generated and sent to the input (IN) port of the parity bit transceiver. Parity code PAR$_3$PAR$_2$PAR$_1$PAR$_0$ is output along with the word of data on MD$_0$ through MD$_{31}$, and both are written into memory.

Inputs and Outputs of the 82345

The input and output signals of the 82345 are listed in Fig. 13.5. This table gives the name, pin number, type, and a brief description of function for each signal. Notice that the signals are grouped into six categories: CPU interface, cache interface, system-controller interface, bus-controller interface, buffer interface, and test mode pin. Let us next just briefly look at a few of these signals.

In the system-controller interface signal group, we find a signal called *RAM write* ($\overline{\text{RAMW}}$). This input is supplied by the 82346 bus controller IC and tells the 82345 that a bus cycle is in progress in which data are being written into DRAM on the main processor board. Logic 0 at this input activates the parity generation logic and memory data bus circuitry. *ROM chip select* ($\overline{\text{ROMCS}}$) is another input of the 82345 that is supplied by the 82346. Logic 0 at this input means that data are being read from ROM. Finally, the signal *board memory selected* ($\overline{\text{BRDRAM}}$) is at its active 0 logic level whenever a data storage location in on-board DRAM is being accessed.

EXAMPLE 13.10

At which pins on the package of the 82345 are the signals $\overline{\text{LATHI}}$ and $\overline{\text{LATHO}}$ input?

Solution

From Fig. 13.5 we find that $\overline{\text{LATHI}}$ is input at pin 8, and $\overline{\text{LATHO}}$ at pin 9 of the 82345's package.

Using the 82345 in the PC/AT Microcomputer

In the block diagram of Fig. 13.2 we find that the 82345 data buffer resides between the core microcomputer's local CPU bus and the ISA expansion bus. This data buffer

SIGNAL DESCRIPTIONS

Signal Name	Pin Number	Signal Type	Signal Description
CPU INTERFACE			
HLDA	2	I-TTL	**CPU HOLD ACKNOWLEDGE, ACTIVE HIGH:** This is the hold acknowledge pin directly from the CPU. It indicates the CPU has given up the bus for either a DMA master or a slot bus master. It is used in the steering logic to determine data routing.
D31–D0	96–82, 79–66, 64–62	I/O-TTL	**CPU DATA BUS:** This is the data bus directly connected to the CPU. It is also referred to as the local data bus. This bus is output enabled by the DEN # signal.
CACHE INTERFACE			
CPST__WRC	21	I-TPU	**POSTED CACHE WRITE CLOCK:** This clock signal is driven by the cache controller and is needed to latch the write data during a posted cache write cycle. The data is latched on the rising edge of this signal. The latch inside of the Data Buffer is bypassed if the CACHE # input is high. Also, when CACHE # is high, the state of CPST__WRC determines on which bus (D or MD) system DRAM is accessed. When high, DRAM is accessed on the D bus. When low, DRAM is accessed on the MD bus. This pin is pulled up internally.
CACHE #	22	I-TPU	**CACHE ENABLE, ACTIVE LOW:** This signal is used to enable the cache posted write register. When there is not a cache in the system, data bypasses the register. When CACHE # is inactive (high) the state of the CPST__WRC pin determines whether the system DRAM is on the CPU's D bus or on the MD bus. This pin is pulled up internally.
SYSTEM CONTROLLER INTERFACE			
MDLAT #	14	I-TTL	**MEMORY DATA LATCH:** This latching signal serves two purposes simultaneously and is only activated during on-board memory read and write cycles. As a memory data latch, this transparent low signal allows read data to flow through to the CPU's local bus. It follows CAS # on early CAS # high read cycles and on the positive going edge, latches the memory data and holds it for the CPU to sample. As a parity clock, it clocks out PARERROR # on its falling edge an on the rising edge it latches the parity bits (PAR3–PAR0), the byte enables (LBE3–LBE0) and the memory data for parity error processing. Any parity errors will be reported on the next read cycle. It is the negative NOR of all CAS # signals gated by W/R #.
RAMW #	15	I-TTL	**RAM WRITE, ACTIVE LOW:** This signal is supplied by the System Controller to indicate to the Bus Controller that an on-board memory write cycle is occurring. It is used internally to direct the parity logic and to enable the MD bus outputs.
ROMCS #	19	I-TTL	**ROM CHIP SELECT:** This signal tells the Data Buffer when the ROM is to be accessed so that it can latch the data and convert it from 16 or 8 bits to 32 bits. This signal is driven by the System Controller.
BRDRAM #	18	I-TTL	**BOARD MEMORY SELECTED, ACTIVE LOW:** This signal is driven by the System Controller and indicates when on-board DRAM is being accessed.

Figure 13.5 82345 input/output signal descriptions. (Reprinted by permission of Intel Corp. Copyright/Intel Corp. 1989)

performs the data bus buffer, multiplexer, and transceiver functions needed to implement the PC/AT's local DRAM data bus, peripheral X-data bus, and system (slot) data bus. The 82345 device is IC U_{40} on sheet 4 of the circuit diagram in Fig. 13.3. Notice that the interface between the 82345 and the core microcomputer is the 32 host data bus lines D_0 through D_{31}. The *memory data bus* to the local DRAM memory array and BIOS ROM is also 32

Signal Name	Pin Number	Signal Type	Signal Description
SYSTEM CONTROLLER INTERFACE (Continued)			
DEN#	20	I-TTL	**DATA ENABLE, ACTIVE LOW:** This is a control signal generated by the System Controller. It is used to enable data transfers on the local data bus and as an output enable for the D bus.
LBE3–LBE0	3–6	I-TTL	**LATCH BYTE ENABLES 3 THROUGH 0:** These signals are driven by the System Controller. They are used internally to enable the appropriate bytes (in a 4 byte wide memory configuration) for parity generation and checking.
PARERROR#	7	O	**PARITY ERROR, ACTIVE LOW:** This signal is the result of a parity check on the appropriate bytes being read from memory. It is generated on the falling edge of MDLAT#.
BUS CONTROLLER INTERFACE			
SA1	13	I-TTL	**SYSTEM ADDRESS BUS BIT 1:** This input will be driven by the Bus Controller or by the Controlling DMA or bus master. This signal is used for 16- to 32-bit conversion. When low, this signal indicates the low word is to be used.
SDLH/HL#	12	I-TTL	**SYSTEM DATA BUS LOW TO HIGH/HIGH TO LOW SWAP:** This signal is driven by the Bus Controller. It is used to establish the direction of byte swaps. (Similar to DIR245 in the existing PC/AT-type chip sets).
SDSWAP#	11	I-TTL	**SYSTEM DATA BUS BYTE SWAP ENABLE, ACTIVE LOW:** This signal is driven by the Bus Controller. It is the qualifying signal needed for SDLH/HL#. (It was formerly named GATE245 on the existing PC/AT-type chip sets).
XDREAD#	10	I-TTL	**PERIPHERAL DATA BUS (XD BUS) READ, ACTIVE LOW:** This signal is driven by the Bus Controller and it determines the direction of the XD bus data flow. (It is analogous to the XDATADIR control pin on the existing PC/AT-type chip sets). When this signal is high, the XD Bus is output enabled.
LATHI#	8	I-TTL	**SD BUS HIGH BYTE LATCH:** This signal is needed to latch the SD bus' high byte to the local data bus until the CPU is ready to sample the bus. When SA1 is low, the high byte is latched into both the one byte and the three byte of the 16/32 latch. When SA1 is high, the high byte is only latched into the three byte. This signal is driven by the Bus Controller.
LATLO#	9	I-TTL	**SD BUS LOW BYTE LATCH:** This signal is needed to latch the SD bus' low byte to the local data bus until the CPU is ready to sample the bus. When SA1 is low, the low byte is latched into both the zero byte and the two byte of the 16/32 latch. When SA1 is high, the high byte is only latched into the two byte. This signal is driven by the Bus Controller.
BUFFER INTERFACE			
MD31–MD0	61–50, 47–34, 32–27	I/O-TTL	**MEMORY DATA BUS:** This bus connects to the on-board DRAM and BIOS ROM. It is used to transfer data to/from memory during memory write/read bus cycles.
SD15–SD0	117–115, 103–109, 106–102, 100–98	I/O-TTL	**SYSTEM DATA BUS:** This bus connects directly to the slots. It is used to transfer data to/from local and system devices.

Figure 13.5   (Continued)

SIGNAL DESCRIPTIONS (Continued)

Signal Name	Pin Number	Signal Type	Signal Description
BUFFER INTERFACE (Continued)			
XD7–XD0	126–123 121–118	I/O-TTL	**PERIPHERAL DATA BUS:** This bus is connected to the Bus Controller and the System Controller. These I/O's are used to read and write to on-board 8-bit peripherals.
PAR3–PAR0	23–26	I/O-TTL	**PARITY BIT BYTES 3 THROUGH 0:** These bits are generated by the parity generation circuitry located on the Data Buffer chip. They are written to memory along with their corresponding bytes during memory write operations. During memory read operations, these bits become inputs and are used along with their respective data bytes to determine if a parity error has occurred. The generation and check of each bit is enabled only when their respective LBE3–LBE0 bits are active.
HIDRIVE#	17	I-TPU	**HIGH DRIVE ENABLE:** This pin is intended to be a wire option. When this pin is low, all bus drivers defined with an I_{OL} of 24 mA will sink the full 24 mA of current. When the input is high, all pins defined as 24 mA will have the output low drive capability cut in half to 12 mA. Note that al A.C. specifications are done with the outputs in the high drive mode and a 200 pF capacitive load. HIDRIVE# has an internal pull-up and can be left unconnected in 12 mA drive if desired. It should be tied low if 24 mA drive is desired.
TEST MODE PIN			
TRI#	127	I-TPU	**THREE-STATE:** This pin is used to drive all outputs to a high impedance state. When TRI# is low, all outputs and bidirectional pins are three-stated. This pin should be pulled up via a 10 kΩ pull-up resistor in a standard system configuration.

SIGNAL TYPE LEGEND

Signal Code	Signal Type
I-TTL	TTL Level Input
I-TPD	Input with 30 kΩ Pull-Down Resistor
I-TPU	Input with 30 kΩ Pull-Up Resistor
I-TSPU	Schmitt-Trigger Input with 30 kΩ Pull-Up Resistor
I-CMOS	CMOS Level Input
I/O-TTL	TTL Level Input/Output
IT-OD	TTL Level Input/Open Drain Output
I/O-OD	Input or Open Drain, Slow Turn On
O	CMOS and TTL Level Compatible Output
O-TTL	TTL Level Output
O-TS	Three-State Level Output
I1	Input Used for Testing Purposes
GND	Ground
PWR	Power

Figure 13.5 (Continued)

bits in length. These lines are labeled MD_0 through MD_{31}. At the ISA bus, the PC/AT microcomputer system provides just 16 data bus lines. These *system data bus* lines are identified as SD_0 through SD_{15}. They are called the *slot data bus*. Finally, the peripheral X-data bus, which is the I/O data bus part of the PC/AT's ISA expansion bus, is labeled XD_0 through XD_7.

EXAMPLE 13.11

At what pins of the 82345 are the eight peripheral X-bus lines located?

Solution

On sheet 4 of Fig. 13.3, we find that XD_0 through XD_3 are at pins 118 through 121 and XD_4 through XD_7 are at pins 123 through 126.

A number of signals are input to the 82345 from the 82346 system controller. They are identified as the system-controller interface signals in Fig. 13.5. On sheet 4 of Fig. 13.3, the system controller interface is located at the lower left corner of the 82345 IC. Here we find signals such as *memory data latch* ($\overline{MDLAT}$) and *data enable* ($\overline{DEN}$). A brief description of each of these signals is supplied in Fig. 13.5. For example, the $\overline{MDLAT}$ input is at its active 0 logic level during all read and write bus cycles to memory on the main processor board. During read cycles, logic 0 at this input enables the read data to the MPU's data bus. This signal is also used to clock in the *parity bits* (PAR_0 through PAR_3), *latched byte enables* (LBE_0 through LBE_3), and clock out the *parity error* ($\overline{PARERROR}$) signal.

The cache controller interface of the 82345 consists of just two signals: *cache enable* ($\overline{CACHE}$) and *posted cache write clock* (CPST_WRC). In Figs. 13.4 and 13.5 we see that $\overline{CACHE}$ is a control input that is used to enable or disable operation of the cache posted write register within the 82345. From Fig. 13.3 we find that $\overline{CACHE}$ is fixed at the 0 logic level. According to the description of $\overline{CACHE}$ in the table of Fig. 13.5, this logic level enables the cache posted write register. Looking at the block diagram in Fig. 13.4, we see that write data available on D_0 through D_{31} are output at OUT of the CPU data bus transceiver. These outputs directly feed the A port of the cache multiplexer and the inputs of the cache posted write register. When posted write operation is enabled, write data are clocked into the cache posted write register with the signal CPST_WRC, and the cache multiplexer is set up to select the input at the B port for output to memory. Notice that data at the output of the cache multiplexer are sent as D_0 through D_{31} to the XD, SD, and MD bus multiplexers and also to the A input port of the multiplexer to the parity generation circuit. If the data are written into the on-board DRAM, parity information is also saved in memory.

EXAMPLE 13.12

If $\overline{CACHE}$ is logic 1, what is the input path of write data from the D bus of the CPU?

Solution

Looking at the block diagram in Fig 13.4, we see that data from the MPU's data bus are passed to OUT of the transceiver. When a posted write operation is disabled, the data applied to the A port of the cache multiplexer are selected. These data are not latched as in a posted write operation. Instead, they are passed directly to the inputs of the parity generation circuit and the XD, SD, and MD multiplexers in parallel.

The last group of interface signals on the 82345 we will consider are those called the bus-controller interface signals. In the table of Fig. 13.5, we find that this interface consists of six input signals: SA_1, SDLH/$\overline{HL}$, $\overline{SDSWAP}$, $\overline{XDREAD}$, $\overline{LATHI}$, and $\overline{LATLO}$. The source for each of these signals is a corresponding output on the 82344 ISA bus controller. For example, Fig. 13.3 shows that the $\overline{SDSWAP}$ input at pin 11 of the 82345 is attached to the $\overline{SDSWAP}$ output at pin 101 of the 82344 (U_{41}). Another example is $\overline{XDREAD}$, which is a control signal that determines the direction of data flow on the XD data bus. This input is supplied by the $\overline{XDREAD}$ output at pin 103 of the ISA controller.

▲ 13.6 82346 SYSTEM CONTROLLER

The second VLSI device that we consider from the 82340 PC/AT chip set is the 82346 system controller. In the block diagram of the PC/AT microcomputer system in Fig. 13.2, we see that the system controller is the heart of the 82340 chip set. It attaches to the local bus of the 80386DX, the ISA expansion bus, the 82345 data buffer, and the 82344 ISA controller. The 82346 is responsible for performing a number of functions for the microcomputer system. For instance, it generates the clock signals, produces addresses for the DRAM array, control signals for all on-board memory, creates the reset signal for the MPU, and interfaces numeric coprocessors to the 80386DX. Just like the 82345, the 82346 is packaged in a 128-pin plastic PQFP.

Block Diagram of the 82346

A block diagram of the circuits within the 82346 system controller is given in Fig. 13.6. Here we find that it contains the *clock-control logic, hold/hold-acknowledge arbitration logic, bus controller, address decoder, DRAM controller, remap logic, ready-control logic, numeric coprocessor interface, reset-control logic,* and *configuration-control registers.* Let us next look briefly at the function of some of these blocks.

We begin with the clock-control logic block. This section of circuitry is used to produce the CPU clock and bus clock signals for the 80386DX-based microcomputer system. The clock signal that is applied to the $TCLK_2$ input must be generated in external circuitry. Within the 82346, this signal is divided by 2 and buffered to CMOS levels to produce system clock output CLK_2. For instance, in a 33-MHz 80386DX microcomputer, $TCLK_2$ must be driven by a 66-MHz clock. The 33-MHz CLK_2 output is used to drive the 80386DX MPU.

The TURBO input can be used to scale the operating frequency of the MPU. When TURBO is active (logic 1), the MPU runs at full speed. That is, CLK_2 equals $\frac{1}{2}TCLK_2$. However, if TURBO is switched to logic 0, CLK_2 is decreased to a lower frequency. This is done by dividing it by a scale factor that has been loaded into one of the internal configuration registers of the 82346. In this way the microcomputer can be set up in a low-power, low-performance mode waiting for an entry to be made at the keyboard. When an entry is detected, TURBO can be switched to 1, and the microcomputer runs at full speed (high performance mode).

A second clock input, bus oscillator (BUSOSC), is used to produce the bus clock (BUSCLK) signal that is needed for the ISA bus. For this reason the BUSCLK output is supplied to the 82344 ISA bus controller device.

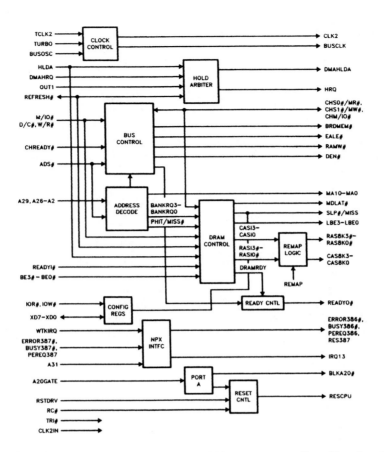

Figure 13.6 Block diagram of the 82346 system controller. (Reprinted by permission of Intel Corp. Copyright/Intel Corp. 1989)

Next we examine the function of the bus-controller, address decoder, DRAM controller, remap logic, and ready-control blocks of the 82346. The circuitry within these blocks are used to produce control signals for on-board memory, RAS and CAS address and strobe signals for the on-board DRAM array, and a control bus interface for communicating with the 82344 ISA bus controller. Notice in Fig. 13.6 that the 80386DX's bus cycle definition signals M/$\overline{\text{IO}}$, D/$\overline{\text{C}}$, and W/$\overline{\text{R}}$ are inputs to the bus controller. Another input is channel ready ($\overline{\text{CHREADY}}$) from the 82344 ISA bus controller. These signals are strobed into the bus-control logic with the address strobe ($\overline{\text{ADS}}$) signal. The bus controller converts them into appropriately timed memory-control outputs. For example, board memory selected ($\overline{\text{BRDMEM}}$) and RAM write ($\overline{\text{RAMW}}$) are outputs of the bus controller. These memory-control signals are used to drive the 82345 data buffer.

At the same time, $\overline{\text{ADS}}$ strobes address information into the address decoder. Here the address is converted into appropriately timed row and column addresses. The RAS and CAS addresses are output on memory address lines MA_0 through MA_{10}. This address is supplied to the DRAM array.

Looking at Fig. 13.6, we see that the DRAM controller accepts the bank-request outputs of the address decoder, byte-enable signals from the 80386DX or 82385DX, and a number of control signals as inputs. At its output it generates row-address strobes

$\overline{RASI}_0$ through $\overline{RASI}_3$ and column-address strobes $CASI_0$ through $CASI_3$. These RAS/CAS signals are mapped to the RAS and CAS bank outputs ($\overline{RASBK}_0$–$\overline{RASBK}_3$ and $CASBK_0$–$CASBK_3$) by the remap logic circuit. The RAS/CAS bank outputs are used to select DRAMs in the memory array.

Finally, the reset-control logic produces the CPU reset signal for the microcomputer. It accepts the various signals that can be used to initiate a reset of the microcomputer at its inputs. In Fig. 13.6, we find that these signals are reset control ($\overline{RC}$), reset drive (RSTDRV), and A20GATE (which is supplied through the port A circuit). These inputs initiate a hardware reset of the 80386DX MPU by activating the reset CPU (RESCPU) output.

Inputs and Outputs of the 82346 and Their Use in the PC/AT Microcomputer

Now that we have introduced the function and block diagram of the 82346, let us turn our attention to the inputs and outputs of the device and their connection in the 80386DX microcomputer system.

Figure 13.7 lists the input and output signals of the 82346. In this table the signals are arranged into seven groups. They are called the *CPU interface signals, on-board memory system interface signals, coprocessor signals, bus-control signals, peripheral interface signals, bus-interface signals,* and *test mode pin.* Again, the table includes the mnemonic, pin number, type, name, and a brief description of each signal. We will begin by examining signals from the CPU interface.

Looking at the list of CPU interface signals in Fig. 13.7, we find that address lines A_2 through A_{26}, A_{29}, and A_{31}, byte enables $\overline{BE}_0$ through $\overline{BE}_3$, and control signals W/$\overline{R}$, D/$\overline{C}$, M/$\overline{IO}$, $\overline{ADS}$, $\overline{READYI}$, and $\overline{HLDA}$ are all inputs of the 82346. These signals are supplied by either the 80386DX MPU or 82385DX cache controller. For example, the address inputs of the 82346 on sheet 3 of the circuit diagram in Fig. 13.3 are supplied by the outputs of the 80386DX's address bus latch on sheet 2. On the other hand, the byte-enable inputs are traced back to the 82385DX bus byte-enable outputs of the cache controller on sheet 1.

EXAMPLE 13.13

What is the function of the $\overline{ADS}$ input of the 82346? Is this signal supplied by the 80386DX or by the 82385DX in the microcomputer of Fig. 13.3?

Solution

In the table of Fig. 13.7 we find that logic 0 at this input tells the 82346 that the information at the address and control inputs of the CPU interface is valid. The signal applied to this input in Fig. 13.3 (sheet 3) is $\overline{BADS}$. On sheet 1 of Fig. 13.3, $\overline{BADS}$ is found to be produced at pin N9 of the 82385DX cache controller.

The CPU interface also includes the CLK_{2IN} and $TCLK_2$ clock inputs and the CLK_2 output. In the block diagram of Fig. 13.8 we see that $TCLK_2$ is one input of the clock-control logic. Earlier we pointed out that this input is supplied by an external crystal oscillator and that the frequency of this oscillator must be twice that of the microcomputer system clock, CLK_2. From the circuits on sheet 3 of Fig. 13.3, we find that the $TCLK_2$ input of the 82346 is driven by a 66-MHz oscillator (OSC_{31}). This input produces a 33-MHz clock at the CLK_2 output.

SIGNAL DESCRIPTIONS

Signal Name	Pin Number	Signal Type	Signal Description
CPU INTERFACE SIGNALS			
A31, A29, A26	119–117	I-TPU	A26 is used to prevent aliasing above 64 MByte. A29 is used to separate upper BIOS accesses from Weitek 3167 accesses. A 31 is used to determine 387DX accesses. These address lines are driven only by the CPU. When HLDA is active, these signals are held low internally. Since externally these pins are three-stated, this is required in order to prevent errant Bus Master and DMA accesses to on-board memory.
A25–2	120–127, 2–7, 9–13, 15–19	I-TTL	Address bits driven by the CPU when it is Bus Master. They are driven by the 82344 Bus Controller whenever HLDA is active. These bits allow direct access of up to 64 Mbytes of memory.
BE3# – BE0#	20–23	I-TTL	**BYTE ENABLES 3 THROUGH 0, ACTIVE LOW:** These signals are driven by the CPU or the 82344.
W/R#	24	I-TPU	WRITE or active low READ enable driven by the CPU. W/R# is decoded with the remaining CPU control signals to indicate the type of bus cycle requested. The bus cycle types include: Interrupt Acknowledge, Halt, Shutdown, I/O Reads and Writes, Memory Data Reads and Writes, and Memory Code reads. WR# is internally pulled up.
D/C#	25	I-TPU	DATA or active low CODE enable driven by the CPU. D/C# is decoded with the remaining CPU control signals to indicate the type of bus cycle requested. See W/R# definition for bus cycle types. D/C# is internally pulled up.
M/IO#	26	I-TPU	MEMORY or active low I/O enable driven by the CPU. M/IO# is decoded with the remaining CPU control signals to indicate the type of bus cycle requested. See W/R# definition for bus cycle types. M/IO# is internally pulled up.
ADS#	27	I-TPU	**ADDRESS STROBE, ACTIVE LOW:** Driven by the CPU as an indicator that the address and control signals currently supplied by the CPU are valid. Ths signal is used internally to indicate that the data and command are valid and to determine the beginning of a memory cycle. ADS# is internally pulled up.
CLK2IN	30	I-CMOS	This is the main clock input to the System Controller and is connected to the CLK2 signal that is output by the System Controller. This signal is used internally to clock the System Controller's logic.
TCLK2	44	I-TTL	This input is connected to a crystal ocscillator whose frequency is equal to two times the system frequency. The TTL level oscillator output is converted internally to CMOS levels and sent to the CLK2 output.
CLK2	32	O	This output signal is a CMOS level converted TCLK2 signal. It is output to the CPU and other on-board logic for synchronization.
SLP#/MISS	35	IO-od	As a "power on reset" default, this bit is an output that reflects the inverse state of the SLEEP[7] configuration register bit. It is active low when sleep mode is active. Sleep mode is activated by setting SLEEP[7] = 1. When configuration register CTRL1(0) = 1, this pin becomes a MISS input for use with a future 82340 compatible product.

Figure 13.7   82346 input/output signal descriptions. (Reprinted by permission of Intel Corp. Copyright/Intel Corp. 1989)

Signal Name	Pin Number	Signal Type	Signal Description
CPU INTERFACE SIGNALS (Continued)			
READYO#	34	O	**READY OUT, ACTIVE LOW:** This signal is an indication that the current memory or I/O bus cycle is complete. It is generated from the internal DRAM controller or the synchronized version of CHREADY# for slot bus accesses. Outside the chip it is ORed with any other local bus I/O or master such as a coprocessor or cache controller. The culmination of these ORed READY signals is sent to the 386DX and is also connected to the System Controller's ReadyL# input.
READYI#	29	I-TTL	**READY INPUT, ACTIVE LOW:** This signal is the ORed READY signals from the coprocessor, cache controller, or other optional add-in device. See the READYO# description for more details on how the signal is used inside the System Controller.
HLDA	28	I-TTL	**HOLD ACKNOWLEDGE, ACTIVE HIGH:** This signal is issued by the CPU in response to the HRQ driven by the System Controller. It indicates that the CPU is floating its outputs to the high impedance state so that another master can take control of the bus. When HLDA is active, the memory control is generated from CHS1#/MR# and CHS0#/MW# rather than CPU status signals.
HRQ	40	O	**HOLD REQUEST, ACTIVE HIGH:** Driven by the System Controller to the CPU, this output indicates that a bus master, such as a DMA or AT channel master, is requesting control of the bus. HRQ is a result of the DMAHRQ input or a coupled refresh cycle. It is synchronized to CLK2.
RESCPU	36	O	**RESET CPU, ACTIVE HIGH:** This signal is sent to the CPU by the System Controller. It is issued in response to the control bit for software reset located in the Port A register or a dummy read to I/O port EFh. It is also issued in response to signals on the RSTDRV or RC inputs and in response to System Controller detection of a shutdown command. In all cases, it is synchronized to CLK2.
ERROR386#	37	O	**ERROR 386, ACTIVE LOW:** This signal is sent to the 386DX. On any CPU reset it is pulled low to set the 386DX to 32-bit coprocessor interface mode.
BUSY386#	38	O	**BUSY 386, ACTIVE LOW:** This signal is sent to the 386DX. The state of BUSY387# is always passed through to BUSY387# indicating that the 387DX is processing a command. On occurrence of an ERROR387# signal, it is latched and held active until an occurrence of a write to ports F0h, F1h, or RES387. The former case is the normal mechanism used to reset the active latched signal. The latter two are resets. Since ERROR387# generates IRQ13 for PC/AT-compatibility, BUSY386# is held active to prevent software access of the 387DX until the interrupt service routine writes F0h. The System Controller also activates BUSY386# for 16 CLK2 cycles when no 387DX is connected and I/O writes to the coprocessor space are detected.

Figure 13.7 (Continued)

Signal Name	Pin Number	Signal Type	Signal Description
CPU INTERFACE SIGNALS (Continued)			
PEREQ386	39	O	**PROCESSOR EXTENSION REQUEST 386, ACTIVE HIGH:** Sent to the CPU in response to a PEREQ387, which is issued by the coprocessor to the System Controller. It indicates to the CPU that the coprocessor is requesting a data operand to be sent to or from memory by the CPU. For PC/AT-compatiblity, PEREQ386 is returned active on occurrence of ERROR387 # after BUSY387 # has gone inactive. A write to F0h by the interrupt 13 handler returns control of the PEREQ386 signal to directly follow the PEREQ387 input.
ON-BOARD MEMORY SYSTEM INTERFACE SIGNALS			
RAMW #	55	O	**RAM ACTIVE LOW WRITE OR ACTIVE HIGH READ:** Output to the 82345 Data Buffer and DRAM memory to control the direction of data flow of the on-board memory. It is a result of the address and bus control decode. It is active during memory write cycles and is high at all other times.
MA10–MA0	57, 58, 60, 62–64, 66, 67, 69, 71, 72	O	**MEMORY ADDRESSES 10 THROUGH 0:** These address bits are the row and column addresses sent to on-board memory. They are buffered and multiplexed versions of the bus master addresses. Along with LBE3–LBE0 they allow addressing of up to 16 MBytes per bank.
RASBK3 # – RASBK0 #	81–83, 85	O	**ROW ADDRESS STROBE BANK 0 THROUGH 3, ACTIVE LOW:** These signals are sent to their respective RAM banks to strobe in the row address during on-board memory bus cycles. The active period for this signal is completely programmable.
CASBK3– CASBK0	77–80	O	**COLUMN ADDRESS STROBE BANK 0 THROUGH 3:** These signals are the respective column address strobes for each of the banks. These signals are externally gated (NAND) with the LBE signals to generate the CAS # strobes for each byte of a DRAM memory bank.
LBE3–LBE0	73–76	O	**LATCHED BYTE ENABLE 0 THROUGH 3, ACTIVE HIGH:** These signals select one of four banks to access memory data from when an on-board memory access is activated. They are the latched version of the CPU's BE3 # –BE0 # signals when the CPU is bus master or is the latched version of SA1, SA0, and BHE # when the master or DMA is in control.
REFRESH #	109	I-CMOS/ O-OD	**REFRESH SIGNAL, ACTIVE LOW:** This output is used by the System Controller to initiate an off-board DRAM refresh operation in coupled refresh mode. In decoupled mode, the Bus Controller drives refresh active to indicate to the System Controller that it has decoded a refresh request command and is initiating an off-board refresh cycle.

Figure 13.7 (Continued)

Signal Name	Pin Number	Signal Type	Signal Description
ON-BOARD MEMORY SYSTEM INTERFCE SIGNALS (Continued)			
ROMCS#	54	O	**ROM CHIP SELECT:** This output is active in CPU mode only (CPUHLDA is negated). It is active anytime the address on the A bus selects the address range between AFFE0000–AFFFFFFF, BFFE0000-BFFFFFFF, EFFE0000–EFFFFFFF, or FFFE0000h–FFFFFFFFh. It is also active during a memory read of 000E0000h–000FFFFFh when RAM-MAP[7] = 1. On reset, it also decodes the middle BIOS space between 00FE0000h-00FFFFFFh. However, this decode space can be changed via internal configuration register to System Board DRAM space after RESET if desired. NOTE: The lower ROM are from 00FE0000h–00FFFFFFh is impacted by shadow and/or EMS. Any 16k segments for which EMS is active (00EXXXXh only) or for which the shadow code had been changed from its 00b default are mapped out of the -ROMCS space.
COPROCESSOR SIGNALS			
PEREQ387	46	I-TPD	**COPROCESSOR EXTENSION REQUEST, ACTIVE HIGH:** this input signal is driven by the coprocessor and indicates that it needs transfer of data operands to or from memory. For PC/AT-compatibility, this signal is gated with the internal ERROR/BUSY control logic before being output to the CPU as PEREQ386.
ERROR387#	43	I-TPU	This is an active low numerics signal which is driven by the coprocessor to indicate that an error has occurred in the previous instruction. This signal is decoded internally with BUSY387# to produce IRQ13.
BUSY387#	42	I-TPU	This is an active low numerics input signal which is driven by the coprocessor to indicate that it is currently executing a previous instruction and is not ready to accept another. This signal is decoded internally to produce IRQ13 and to control PEREQ386 and BUSY386#.
RES387	41	O	**RESET 387, ACTIVE HIGH:** This output is connected to the 387DX reset input. It is triggered through an internally generated system reset or via a write to port F1h. In the case of a system reset, the CPURESET signal is also activated. A write to port F1h only resets the coprocessor. A software FNINT signal must occur after an F1h generated reset before the coprocessor is reset to the same internal state that a 287 is put into by a hardware reset alone. For, compatibility, the F1h reset may be disabled by setting bit 6 of MISCSET to 1.
WTKIRQ	47	I-TPD	**WEITEK 3167 INTERRUPT REQUEST, ACTIVE HIGH:** An input from the Weitek 3167 coprocessor.
IRQ13	100	O	**INTERRUPT REQUEST 13, ACTIVE HIGH:** This signal is driven to the Bus Controller to indicate than an error has occurred within the coprocessor. This signal is a decode of the BUSY387# and ERROR387# inputs ORed with the WTKIRQ input.
BUS CONTROL SIGNALS			
CHREADY#	104	I-CMOS	**CHANNEL READY, ACTIVE LOW:** This signal is issued by the Bus Controller as an indication that the current channel bus cycle is complete. This signal is synchronized internally then combined with ready signals from the coprocessor and DRAM controller to form the final version of READYO# which is sent to the CPU.

Figure 13.7 (Continued)

Signal Name	Pin Number	Signal Type	Signal Description
BUS CONTROL SIGNALS (Continued)			
CHS0#/MW#	103	IO-TTL	**CHANNEL SELECT 0/MEMORY WRITE, ACTIVE LOW:** This signal is a decode of the 386DX's bus control signals and is sent to the Bus Controller. When combined with CHS1# and CHM/IO# and decoded, the bus cycle type is defined for the Bus Controller. Activation of CPUHLDA reverses this signal to become an input from the Bus Controller. It is then a MEMW# signal for DMA or bus master access to system memory.
CHS1#/MR#	102	IO-TTL	**CHANNEL SELECT 1/MEMORY READ, ACTIVE LOW:** This signal is a decode of the 386DX's bus control signals and is sent to the Bus Controller. When combined with CHS0# and CHM/IO# and decoded, the bus cycle type is defined for the Bus Controller. Activation of CPUHLDA reverses this signal to become an input from the Bus Controller. It is the a MEMR# signal for DMA or bus master access to system memory.
CHM/-IO	101	O	**CHANNEL MEMORY/ACTIVE LOW IO:** A decode of the M/IO# signal sent by the CPU to the System Controller. It is an indicator that the current bus cycle is a channel access. When combined with CHS0#, and CHM/IO# and decoded, the bus cycle type is defined for the Bus Controller.
BLKA20#	94	O	**BLOCK A20, ACTIVE LOW:** Driven to the Bus Controller to deactivate address bit 20. It is a decode of the A20GATE signal and Port A bit 1 indicating the dividing line of the 1 MByte memory boundary. Port A bit 1 may be directly written or set by a dummy read of I/O port EEh. BLKA20# is forced high when HLDA is active. (Refer to the "Sleep Mode Control Subsystem" section.)
BUSOSC	106	I-TTL	**BUS OSCILLATOR:** This signal is supplied from an external oscillator. It is supplied to the Bus Controller when the System Controller's internal configuration registers are set for asynchronous slot bus mode. This signal is two times the AT bus clock speed (SYSCLK).
BUSCLK	98	O-TTL	**BUS CLOCK:** This is the source clock used by the Bus Controller to drive the slot bus. It is two times the AT bus clock (SYSCLK). It is a programmable division from CLK2 or BUSOSC when in a synchronous bus mode.
DMAHRQ	105	I-CMOS	**DMA HOLD REQUEST, ACTIVE HIGH:** This input is sent by the Bus Controller, it is internally synchronized by the System Controller before it is sent out to the CPU as the HRQ signal. It is the indicator of the DMA controller or an other bus masters' desire to control the bus.
DMAHLDA	99	O	**DMA HOLD ACKNOWLEDGE, ACTIVE HIGH:** This output to the Bus controller indicates that the current hold acknowledge state is for the DMA controller or an other bus master.
BRDRAM#	95	O	**BOARD DRAM, ACTIVE LOW:** An output to Bus Controller and Data Buffer to indicate that on-board DRAM is being addressed.
OUT1	107	I-CMOS	Indicate a refresh request from the Bus Controller.

Figure 13.7 (Continued)

Signal Name	Pin Number	Signal Type	Signal Description
PERIPHERAL INTERFACE SIGNALS			
A20GATE	116	I-TTL	**ADDRESS BIT 20 ENABLE:** This is an input from the keyboard controller and is used internally along with Port A bit 1 to determine if address bit 20 from the CPU is true or gated low. It also determines the state of BLKA20#.
TURBO	115	I-TTL	**TURBO, ACTIVE HIGH:** This input to the System Controller determines the speed at which the system board operates. It is normally the externally ANDed signal from the keyboard controller and a turbo switch. It is internally ANDed with a software settable latch. When high, operation is full speed. When low, CLK2 is divided by the value coded in configuration register MISCSET. A range is provided that allows slow operation at or below 8 MHz for any valid CPU speed. Slow speed takes precedence. When any one request for slow mode is present, slow mode is active. Turbo mode is active only when all TURBO requests are active.
RC#	114	I-TTL	**RESET CONTROL, ACTIVE LOW:** The falling edge of this signal causes a RESCPU signal. RC# is generated by the keyboard controller and its inverse is ORed with Port A bit 0 to form RESCPU.
SLEEP1	49	O-OD	**SLEEP SIGNAL 1, ACTIVE HIGH:** This pin is the logical OR of the enable and external control bits (bits 1 and 7) of the sleep indexed configuration register. It can be used with external interface logic to control external devices. The pin is always active while in sleep mode but can also be controlled via software when sleep mode is inactive. It is pulled low when inactive and three-states when active. An external pull-up is required. This allows an external interface to control logic operation at voltages different than VDD.
SLEEP2	50	O-OD	**SLEEP SIGNAL 2, ACTIVE HIGH:** This pin is the logical OR of the enable and external control bits (bits 2 and 7) of the sleep indexed configuration register. It can be used with external interface logic to control external devices. The pin is always active while in sleep mode but can also be controlled via software when sleep mode is inactive. It is pulled low when inactive and three-states when active. An external pull-up is required. This allows an external interface to control logic operation at voltages different than VDD.
SLEEP3	51	O-OD	SLEEP SIGNAL 3, ACTIVE HIGH: This pin is the logical OR of the enable and external control bits (bits 3 and 7) of the Sleep indexed configuration register. It can be used with external interface logic to control external devices. The pin is always active while in sleep mode but can also be controlled via software when sleep mode is inactive. It is pulled low when inactive and three-states when active. An external pull-up is required. This allows an external interface to control logic operation at voltages different than VDD.

Figure 13.7   (Continued)

SIGNAL DESCRIPTIONS (Continued)

Signal Name	Pin Number	Signal Type	Signal Description
BUS INTERFACE SIGNALS			
XD7–XD0	86–93	IO–TTL	**PERIPHERAL DATA BUS:** This bus is used to read and write the internal configuration registers.
DEN#	53	O	**DATA ENABLE, ACTIVE LOW:** This signal is an output to the 82345 Data Buffer to enable data transfers on the local bus. This signal is low during any CPU read cycles or INTA cycles.
IOR#	112	I-TTL	**I/O READ CYCLE, ACTIVE LOW:** Driven by the Bus Controller to indicate to the 82346 that an I/O read cycle is occurring on the bus. Whenever an I/O cycle occurs, the memory interface signals are inactive.
IOW#	113	I-TTL	**I/O WRITE CYCLE, ACTIVE LOW:** Driven by the Bus Controller to indicate to the 82346 that an I/O write cycle is occurring on the bus. Whenever an I/O cycle occurs, the memory interface signals are inactive.
RSTDRV	111	I-TTL	**RESET DRIVE, ACTIVE HIGH:** This reset signal is output by the Bus Controller. It indicates that a hardware reset signal has been activated. This is the same signal which is output to the channel. This signal is used to reset internal logic and to derive the RESCPU which is output by the System Controller.
MDLAT#	52	O	**MEMORY DATA BUS LATCH:** This is an output signal to the Data Buffer. On the rising edge, the Data Buffer latches the memory data bus. MDLAT# is low anytime one of the CASBK signals is high. When low, the Data Buffer latches are transparent.
OSC	110	I-TTL	**OSCILLATOR:** This is the buffered input of the external 14.318 MHz oscillator.
TEST MODE PIN			
TRI#	48	I1	**THREE-STATE:** This pin is used to drive all outputs to a high impedance state. When TRI# is low, all outputs and bidirectional pins are three-stated.

Signal Type Legend

Signal Code	Signal Type
I-TTL	TTL Level Input
I-TPD	Input with 30 kΩ Pull-Down Resistor
I-TPU	Input with 30 kΩ Pull-Up Resistor
I-TSPU	Schmitt-Trigger Input with 30 kΩ Pull-Up Resistor
I-CMOS	CMOS Level Input
IO-TTL	TTL Level Input/Output
IT-OD	TTL Level Input/Open

Signal Code	Signal Type
IO-OD	Input or Open Drain, Slow Turn On
O	CMOS and TTL Level Compatible Output
O-TTL	TTL Level Output
O-TS	Three-State Level Output
I1	Input used for Testing Purposes
GND	Ground
PWR	Power

Figure 13.7 (Continued)

EXAMPLE 13.14

Where is the CLK_2 output of the 82346 system controller on sheet 3 of Fig. 13.3 distributed to?

Solution

Clock output CLK_2 is supplied to the CLK_2 input at pin F12 of the 80386DX on sheet 1 of Fig. 13.3.

Most of the signals in the on-board memory system interface group are used to drive the DRAM array on the main processor board. Figure 13.7 indicates that memory address lines MA_0 through MA_{10} of the 82346 carry the row and column addresses to the on-board DRAMs. Looking at sheet 3 of Fig. 13.3, we find that these signals and buffered versions MA_{0H} through MA_{10H} are output to the DRAM array on sheet 6. Notice on sheet 6 that MA_0 through MA_{10} are applied through resistor networks RN_{67} through RN_{69} to address inputs A_0 through A_{10} of all the DRAM SIM modules in banks 0 and 1.

EXAMPLE 13.15

What is the destination of the $\overline{RAS}_0$ row-address strobe bank output of the 82346 on sheet 3 in Fig. 13.3?

Solution

In Fig. 13.3 we find that $\overline{RAS}_0$ is sent to the DRAM array. In the array on sheet 6, $\overline{RAS}_0$ is passed through resistor network RN_{600} to the $\overline{RAS}$ input of each DRAM SIM in bank 0. They are the four DRAM SIMs labeled U_{60}, U_{61}, U_{62}, and U_{63}.

The column-address strobe bank ($CASBK_0$ through $CASBK_3$) outputs and latched byte-enable (LBE_0 through LBE_3) outputs of the 82346 are not supplied directly to the DRAM array. Instead, they are decoded to produce a separate column-address strobe signal for each of the DRAM SIMs in the on-board RAM array. This decoder circuit is formed with NAND gates U_{30}, U_{31}, U_{32}, and U_{33} on sheet 3 of Fig. 13.3. For example, the LBE_0 output of the 82346 is applied to the input at pin 2 of each of these four NAND gates. On the other hand, the $CASBK_0$ output is applied to one input of each gate on NAND gate U_{30}. If LBE_0 and $CASBK_0$ are both logic 1 and all the other byte-enable and column-address strobe signals are logic 0, the only column-address strobe output of the decoder that is active is $\overline{CAS}_0$. Notice in the DRAM array on sheet 6 of Fig. 13.3 that $\overline{CAS}_0$ is applied as $\overline{CAS}_{0R}$ to the $\overline{CAS}$ and $\overline{CAS}_9$ inputs of DRAM SIM U_{63}. If this type of CAS strobe follows a $\overline{RAS}_0$ strobe, it represents a byte data access from bank 0 DRAM SIM U_{63}. This data transfer takes place over memory data bus lines MD_0 through MD_7. Remember that the memory data bus is interfaced to the 80386DX's local bus by the 82345 data buffer. The logic level of the $\overline{RAMW}$ output of the 82346 signals the DRAM array whether a read or write data transfer is in progress.

EXAMPLE 13.16

Assume that during a memory bus cycle to the on-board DRAM array, the $CASBK_1$ and all four latched byte-enable outputs of the 82346 are at logic 1. Which outputs of the CAS decoder circuit are active? To which DRAMs are they applied? Over which data bus lines are data carried? What size data transfer is taking place?

Solution

Looking at the CAS decoder circuit on sheet 3 of Fig. 13.3, we find that $CASBK_1$ is applied to one input of all four NAND gates in IC U_{31}. Since LBE_0, LBE_1, LBE_2, and LBE_3 are all equal to 1, the four outputs of U_{31}, $\overline{CAS_4}$, $\overline{CAS_5}$, $\overline{CAS_6}$, and $\overline{CAS_7}$, are all at their active 0 logic level.

On sheet 6 of Fig. 13.3, these CAS lines are found to be applied to DRAM SIMs U_{67}, U_{66}, U_{65}, and U_{64}, respectively. Therefore, all four DRAM SIMs in bank 1 of the DRAM array are accessed, and the data transfer takes place over the complete memory data bus, MD_0 through MD_{31}. This transfer represents a double-word data access.

A few of the 80387DX's handshake signals are interfaced to the 80386DX MPU through the 82346 system controller. In Fig. 13.7 the coprocessor signals are found to include four input signals, $PEREQ_{387}$, $\overline{ERROR_{387}}$, $\overline{BUSY_{387}}$, and WTKIRQ. The 80387DX numeric coprocessor uses the first three signals to tell the 80386DX MPU that it needs to perform data operations, that an error has occurred in a numerical calculation, and that it is busy making a numerical calculation, respectively. These conditions are signaled to the 80386DX through the $PEREQ_{386}$, $\overline{ERROR_{386}}$, and $\overline{BUSY_{386}}$ outputs of the 82346's CPU interface.

EXAMPLE 13.17

What signals are listed as coprocessor outputs of the 82346 in the table of Fig. 13.7?

Solution

The coprocessor output signals are reset 387 (RES_{387}) and interrupt request 13 (IRQ_{13}).

▲ 13.7 82344 ISA CONTROLLER

The 82344 ISA controller is the last of the three 82340 PC/AT chip-set devices that attach between the local bus and ISA expansion bus. The 82344 is another VLSI device and is manufactured in a 160-pin PQFP. In Fig. 13.3 (sheet 4) we find that the 82344 connects to the buffered address outputs of the 80386DX core microcomputer on the local bus side and the 82346 system bus controller over the 344/346 communication interface lines. The primary function of the 82344 is to supply most of the signals of the ISA PC/AT expansion bus. It is also used to select between an 8- or 16-bit data bus for the BIOS EPROMs and to drive the speaker.

Block Diagram of the 82344

Figure 13.8 is a block diagram of the circuits within the 82344 IC. From the block diagram we find that it contains four main sections, called the *data-conversion and wait-state control, 284/288 ready and bus control, peripheral control,* and *address decoder.* Let us next look briefly at the functions of these blocks.

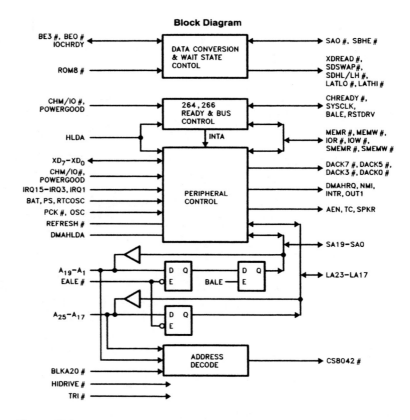

Block Diagram

Figure 13.8 Block diagram of the 82344 ISA bus controller. (Reprinted by permission of Intel Corp. Copyright/Intel Corp. 1989)

One function that is performed by the data-conversion and wait-state control block is to translate the byte-enable signals of the 80386DX to the 80286-compatible odd- and even-byte-select signals needed for the ISA expansion bus. In Fig. 13.8 we see that it accepts as inputs byte-enable signals $\overline{BE}_0$ through $\overline{BE}_3$ from the MPU, the $\overline{ROM}_8$ strap input, and IOCHRDY from the ISA bus. In response to these signals, the bus controller produces ISA odd- and even-byte-select signals $\overline{SA}_0$ and $\overline{SBHE}$. Notice that the byte enables and byte selects are shown as bidirectional lines. This is because they act as inputs when supplied by the MPU and as outputs when driven by the 82344's on-chip DMA controllers.

The 284/288 ready and bus-control logic section produces more control signals for the ISA expansion bus. For example, its SYSCLK, BALE, and RSTDRV outputs are distributed to the expansion bus. These two blocks also produce data buffer interface signals, such as $\overline{XDREAD}$, $\overline{SDSWAP}$, and $\overline{CHREADY}$. They are used for communication between the 82344, 82345, and 82346 ICs.

The peripheral-control section represents much of the circuitry within the 82344. This block contains the LSI peripheral devices that are used to implement the microcomputer in the PC/AT. Figure 13.9 shows the circuitry within the peripheral control block in more detail. Here we find that it contains the *82C59A interrupt controllers, 82C37A DMA controllers, 74LS612 page address register, 82C54 programmable timer/counter, speaker driver, parallel I/O port B, real-time clock,* and *refresh counter.*

Finally, the address-decoder block decodes address inputs to produce chip-select

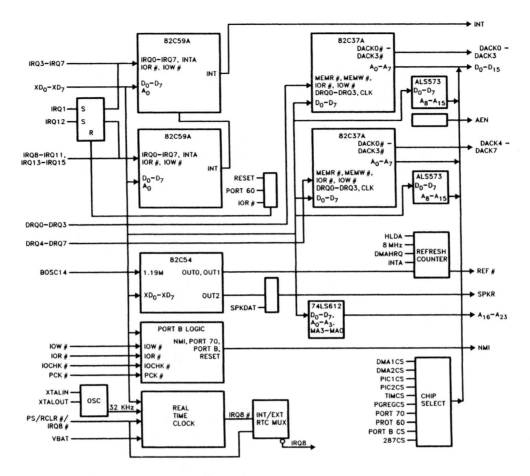

Figure 13.9 Block diagram of the peripheral control block of the 82344.
(Reprinted by permission of Intel Corp. Copyright/Intel Corp. 1989)

signals $\overline{\text{CS8042}}$ and $\overline{\text{ROMCS}}$. The $\overline{\text{CS8042}}$ output is used to enable the 8042 keyboard-controller device. This keyboard controller is optional in an 82340-based PC/AT system. On the other hand, $\overline{\text{ROMCS}}$ is used to enable the BIOS EPROMs, which are located on the main processor board.

Inputs and Outputs of the 82344 and Their Use in the PC/AT Microcomputer

Now that we have introduced the 82344 ISA bus controller and its block diagram, let us continue by examining its signal interfaces. Figure 13.10 lists the signals at each of the 82344's interfaces. Notice that the signals are divided into seven groups: the *CPU interface, system-controller interface, ROM interface, bus interface, peripheral interface, data buffer interface,* and *test-mode pin.* We continue by examining some of these interface signals and their connections in the PC/AT microcomputer system.

The 82344 is designed for use in either an 80286- or 80386DX-based PC/AT microcomputer system. For this reason, the device can be set up to operate with CPU interface signals that are either 80286 or 80386DX compatible. The mode of operation is selected with the

SIGNAL DESCRIPTIONS

Name	Pin Number	Type	Description
CPU INTERFACE			
A25, A24	43–44	O-TS	Address Bus—These pins are outputs during DMA, master, or standard refresh modes. They are high impedance at all other times. A25 and A24 are driven from the alternate 612 registers during DMA and refresh cycles and are driven low during master cycles.
A23–A2	45–58, 61–68	IO-TTL	Address Bus—These pins are outputs during DMA, master, or standard refresh modes. They are inputs at all other times. As inputs, they are passed to the SA and LA buses and A15–A2 are used to address I/O registers internal to the bus control chip. As outputs, they are driven from different sources depending on which mode the Bus Controller is in. While in refresh mode, these pins are driven from the 612 and refresh address counter. While in DMA mode, they are driven from the 612 and DMA controller subsection. If the Bus Controller is in master mode, the pins A23–A17 are driven from the inputs LA23–LA17 and the pins A16–A2 are driven from the inputs SA16–SA2.
—BE3	69	IO-TTL	Byte Enable 3, active low—This pin is an output during DMA, master, or standard refresh modes. It is an input at all other times. As an input in 386DX mode, it is decoded along with the other byte enable signals to generate SA1, SA0 and —SBHE. As an output in 386DX mode SA1, SA0, and —SBHE are used to determine the value of —BE3. This pin should be left unconnected when using this part in 286 mode. The pin has an internal pull-up.
—BE2/A1	70	IO-TTL	Byte Enable 2, active low, or A1—This pin has a dual function depending on the state of the 286/–386DX input. If 286/–386DX is high (286 mode), then the pin is treated as address bit 1. If 286/–386DX is low (386DX mode), the pin is treated as —BE2. This pin is an output during DMA, master, or standard refresh modes. It is an input at all other times. As an input in 386DX mode, it is decoded along with the other byte enable signals to generate SA1, SA0, and —SBHE. As an output in 386DX mode. SA1, SA0, and —SBHE are used to determine the value of —BE2. When in 286 mode, it is interpreted as address A1 and passed to SA1. As an output in 286 mode it is driven from the SA1 input.
—BE1/—BHE	71	IO-TTL	Byte Enable 1 or Byte High Enable, active low—This pin has a dual function depending on the state of the 286/—386DX input. If 286/—386DX is high (286 mode), then the pin is treated as —BHE. If 286/—386 is low (386 mode), the pin is treated as —BE1. This pin is an output during DMA, master, or standard refresh modes. It is an input at all other times. As an input in 386 mode, it is decoded along with the other byte enable signals to generate SA1, SA0, and —SBHE. As an output in 386 mode, SA1, SA0, and —SBHE are used to determine the value of —BE1. When in 286 mode, it is interpreted as —BHE and passed to —SBHE. As an output in 286 mode, it is driven from the —SBHE input.

Figure 13.10 82344 input/output signal descriptions. (Reprinted by permission of Intel Corp. Copyright/Intel Corp. 1989)

Name	Pin Number	Type	Description
CPU INTERFACE (Continued)			
—BE0/A0	72	IO-TTL	Byte Enable 0, active low, or A0—This pin has a dual function depending on the state of the 286/—386DX input. If 286/—386DX is high (286 mode), then the pin is treated as address bit 0. If 286/—386DX is low (386 mode), the pin is treated as —BE0. This pin is an output during DMA, master, or standard refresh modes. It is an input at all other times. As an input in 386 mode, it is decoded along with the other Byte Enable signals to generate SA1, SA0, and —SBHE. As an output in 386 mode, SA1, SA0, and —SBHE are used to determine the value of —BE0. When in 286 mode, it is interpreted as A0 and passed to SA0. As an output in 286 mode, it is driven from the SA0 input.
286/—386DX	73	I-TPU	CPU is 286 or 386DX—This pin defines the type of address bus to which the bus controller chip is interfaced. If the pin is tied high, the address bus is assumed to be emulating 286 signals. In this mode, A25, A24, and —BE3 would be left unconnected. The pins —BE2/A1, —BE1/—BHE and —BE0/A0 would take on the 286 functions. If the pin is tied low, A25, A24 can be used to generate up to 64 Mbyte addressing for DMA, and the byte enable pins will take on the normal 386DX addressing functions. This pin has an internal pull-up to cause the chip to default to 286 mode if left unconnected. This pin is a hard wiring option and must not be changed dynamically during operation. When strapped for 286 mode, the Bus Controller is assumed to be interfaced to the 82343 System Controller which in turn may be strapped for 286 or 386SX operation. The 82344 is strapped for 286 operation when used with the 82343 strapped for 386SX operation.
HLDA	74	I-TTL	Hold Acknowledge—This is the hold acknowledge pin directly from the CPU. It is used to control direction on address and command pins. When HLDA is low, the Bus Controller is defined as being in the CPU mode. In the CPU mode, the local address bus (A bus) pins are inputs. The system address bus (SA and LA) pins along with the command pins (—MEMR, —MEMW, —IOR and —IOW) are outputs. When HLDA is high, the Bus Controller can be in DMA, refresh, or master modes. In both DMA and refresh modes, the commands and all address buses (A, SA and LA) are outputs. In master mode, the commands and system address bus (SA and LA) pins are inputs and the local address bus (A bus) pins are outputs. The SA bus is passed directly to the A bus except bits 17, 18, and 19 are ignored. LA23–LA17 is passed directly to A23–A17.
INTR	75	O	Interrupt Request—INTR is used to interrupt the CPU and is generated by the 8259 megacells any time a valid interrupt request input is received.
NMI	76	O	Non-Maskable Interrupt—This output is used to drive the NMI input to the CPU. This signal is asserted by either a parity error (indicated by —PCK being asserted after the ENPARCK bit in Port B has been asserted), or an I/O channel error (indicated by —IOCHCK being asserted after the ENIOCK bit in Port B has been asserted). The NMI output is enabled by writing a 0 to bit D7 of I/O port 70h. NMI is disabled on reset.

Figure 13.10 (Continued)

Name	Pin Number	Type	Description
SYSTEM CONTROLLER INTERFACE			
—CHS0/—MW	77	IO-TTL	Channel Status 0 or active low Memory Write—This input is used along with —CHS1 and CHM/—IO to determine what type of bus cycle the Bus Controller is to perform. This input has the same meaning and timing requirements as the S0 signal for a 286 microprocessor. —CHS0 going active indicates a write cycle unless —CHS1 is also active. When both status inputs are active it indicates an interrupt acknowledge cycle. This input is synchronized to the BUSCLK input. Activation of CPUHLDA reverses this signal to become an output to the System Controller. It is then a —MEMW signal for DMA or bus master access to system memory.
—CHS1/—MR	78	IO-TTL	Channel Status 1 or active low Memory Read—This input is used along with —CHS0 and CHM/—IO to determine the bus cycle type. This input has the same meaning and timing requirements as the S1 signal for a 286 microprocessor —CHS1 going active indicates a read cycle unless —CHS0 is also active. When both status inputs are active it indicates an interrupt acknowledge cycle. This input is synchronized to the BUSCLK input. Activation of CPUHLDA reverses this signal to become an output to the System Controller. It is then a —MEMR signal for DMA or bus master access to system memory.
CHM/—IO	82	I-TTL	Channel Memory or active low I/O select—This input is used along with —CHS0 and —CHS1 to determine the bus cycle type. This input has the same meaning and timing requirements as the M/—IO signal for a 286 microprocessor. CHM/—IO is sampled anytime —CHS0 or —CHS1 is active. If sampled high, it indicates a memory read or write cycle. If sampled low, an I/O read or write cycle should be executed. This input is synchronized to the BUSCLK input.
—EALE	83	I-TTL	Early Address Latch Enable, active low—This input is used to latch the A25–A2 and Byte Enable signals. The latches are open when —EALE is low and hold their value when —EALE is high. The latched addresses are fed directly to the LA23–LA17 bus to provide more address setup time on the bus before a command goes active. The lower latched addresses are latched again with an internal ALE signal as soon as —CHS0 or —CHS1 is sampled active and fed to the SA19–SA0 and —SBHE outputs. In a 386DX system, this input is connected directly to the —ADS output from the CPU. In a 286 system, this input is connected to the —EALE output from the 82343 System Controller.
—BRDRAM	84	I-TTL	On-board DRAM, active low—An input from the System Controller indicating that the on-board DRAM is being addressed.

Figure 13.10   (Continued)

Name	Pin Number	Type	Description
SYSTEM CONTROLLER INTERFACE (Continued)			
—CHREADY	85	O	Channel Ready, active low—This output is maintained in the active state when no bus accesses are active. This indicates that the Bus Controller is ready to accept a new command. During normal bus accesses, —CHREADY is negated as soon as a valid bus requested is sampled on the —CHS0 and —CHS1 inputs. It is asserted again to indicate that the Bus Controller is ready to complete the current cycle. The bus command signals are then terminated on the next falling edge of the BUSCLK input.
BUSCLK	86	I-CMOS	Bus Clock—This is the main clock input for the Bus Controller. It runs at twice the frequency desired for the SYSCLK output. All inputs are synchronous with the falling edge of this input.
—BLKA20	87	I-TTL	Block A20, active low—This input is used while CPUHLDA is low to force the LA20 and SA20 outputs low anytime it is active. When —BLKA20 is negated LA20 and SA20 are generated from A20.
DMAHRQ	89	O	Hold Request—This output is generated by the DMA controller any time a valid DMA request is received. It is connected to the DMAHRQ pin on the System Controller.
DMAHLDA	88	I-TTL	DMA Hold Acknowledge—An input from the System Controller which indicates that the current hold acknowledge state is for the DMA controller or other bus master.
OUT1	90	O	Output 1—Indicates a refresh request to the System Controller. This is the 15 μs output of timer channel 1.
ROM INTERFACE			
—ROM8	112	I-TPU	8/16 bit ROM select—This input indicates the width of the ROM BIOS. If —ROM8 is low, the Bus Controller chip generates 8- to 16-bit conversions for ROM accesses. Data buffer controls are generated assuming the ROM is on the MD bus. If —ROM8 is high, data buffer controls are generated assuming 16-bit wide ROMs are on the MD bus.
BUS INTERFACE			
—IOR	134	IO-TTL	I/O Read, active low—This signal is an input when CPUHLDA is high and —MASTER is low. It is an output at all other times. When CPUHLDA is low, —IOR is driven from the 288 bus controller megacell. When CPUHLDA is high and —MASTER is high, it is driven by the 8237 DMA controller megacells. This pin requires an external 10 KΩ pull-up resistor.
—IOW	132	IO-TTL	I/O Write, active low—This signal is an input when CPUHLDA is high and —MASTER is low. It is an output at all other times. When CPUHLDA is low, —IOW is driven from the 288 bus controller megacell. When CPUHLDA is high and —MASTER is high, it is driven by the 8237 DMA controller megacells. This pin requires an external 10 KΩ pull-up resistor.

Figure 13.10 (Continued)

Name	Pin Number	Type	Description
BUS INTERFACE (Continued)			
—MEMR	33	IO-TTL	Memory Read, active low—This signal is an input when CPUHLDA is high and —MASTER is low. It is an output at all other times. When CPUHLDA is low, —MEMR is driven from the 288 bus controller megacell. When CPUHLDA is high and —MASTER is high, it is driven by the 8237 DMA controller megacells. This signal does not pulse low for DMA addresses above 16 Mbytes. DMA above 16 Mbytes is only performed to the system board, never to the slot bus. This pin requires an external 10 KΩ pull-up resistor.
—MEMW	35	IO-TTL	Memory Write, active low—This signal is an input when CPUHLDA is high and —MASTER is low. It is an output at all other times. When CPUHLDA is low, —MEMW is driven from the 288 bus controller megacell. When CPUHLDA is high and —MASTER is high, it is driven by the 8237 DMA controller megacells. This pin requires an external 10 KΩ pull-up resistor.
—SMEMR	129	IO-TTL	Memory Read, active low—This signal is an input when CPUHLDA is high and —MASTER is low. It is an output at all other times. When CPUHLDA is low, —MEMR is driven from the 288 bus controller megacell. When CPUHLDA is high and —MASTER is high, it is driven by the 8237 DMA controller megacells. —SMEMR is active on memory read cycles to addresses below 1 Mbyte. This pin requires an external 10 KΩ pull-up resistor.
—SMEMW	127	IO-TTL	Memory Write, active low—This signal is an input when CPUHLDA is high and —MASTER is low. It is an output at all other times. When CPUHLDA is low, —MEMW is driven from the 288 bus controller megacell. When CPUHLDA is high and —MASTER is high, it is driven by the 8237 DMA controller megacells. —SMEMW is active on memory write cycles to addresses below 1 Mbyte. This pin requires an external 10 KΩ pull-up resistor.
LA23–LA17	16, 18, 22, 24, 26, 28, 31	IO-TTL	Latchable Address bus—This bus in an input when CPUHLDA is high and —MASTER is low. It is an output bus at all other times. When CPUHLDA is low, the LA bus is driven by the latched values for the A bus. When CPUHLDA is high and —MASTER is high, the SA bus is driven by the 612 memory mapper for DMA cycles and normal refresh. The LA bus is latched internally with the —EALE input.
SA19–SA0	131, 133, 135, 137, 141, 143, 145, 147, 149, 152, 154, 156, 158, 1, 3, 5, 7, 8, 11, 12	IO-TTL	System Address bus—This bus is an input when CPUHLDA is high and —MASTER is low. It is an output bus at all other times. When CPUHLDA is low, the SA bus is driven by the latched values from the A bus. When CPUHLDA is high and —MASTER is high, the SA bus is driven by the 8237 DMA controller megacells or refresh address generator. The SA bus will become valid in the middle of the status cycle generated by the —CHS0 and —CHS1 inputs. They are latched with an internally generated ALE signal.

Figure 13.10 (Continued)

Name	Pin Number	Type	Description
BUS INTERFACE (Continued)			
—SBHE	14	IO-TTL	System Byte High Enable, active low—This pin is controlled the same way as the SA bus. It is generated from a decode of the —BE inputs in CPU mode. It is forced low for 16-bit DMA cycles and forced to the opposite value of SA0 for 8-bit DMA cycles.
—REFRESH	146	IT-OD	Refresh signal, active low—This I/O signal is pulled low whenever a decoupled refresh command is received from the System Controller. It is used as an input to sense refresh requests from external sources such as the System Controller for coupled refresh cycles or bus masters. It is used internally to clock the refresh address counter and select a location in the memory mapper which drives A23–A17. —REFRESH is an open drain output capable of sinking 24 mA and requires an external pull-up resistor.
SYSCLK	148	O	System Clock—This output is half the frequency of the BUSCLK input. The bus control outputs BALE and the —IOR, —IOW, —MEMR and —MEMW are synchronized to SYSCLK.
OSC	9	I-TTL	Oscillator—This is the buffered input of the external 14.318 MHz oscillator.
RSTDRV	122	O	Reset Drive, active high—This output is a system reset generated from the POWERGOOD input. RSTDRV is synchronized to the BUSCLK input.
BALE	6	O	Buffered Address Latch Enable, active high—A pulse which is generated at the beginning of any bus cycle initiated from the CPU. BALE is forced high anytime CPUHLDA is high.
AEN	128	O	Address Enable—This output goes high anytime the inputs CPUHLDA and —MASTER are both high.
T/C	4	O	Terminal Count—This output indicates that one of the DMA channels terminal count has been reached. This signal directly drives the system bus.
—DACK7- —DACK5, —DACK3- —DACK0	39, 37, 34, 136, 2, 142, 29	O	DMA Acknowledge, active low—These outputs are the acknowledge signals for the corresponding DMA requests. The active polarity of these lines is set active low on reset. Since the 8237 megacells are internally cascaded together, the polarity of the —DACK signals must not be changed. This signal directly drives the system bus.
DRQ7–DRQ5 DRQ3–DRQ0	41, 38, 36, 138, 124, 144, 32	I-TSPU	DMA Request—These asynchronous inputs are used by an external device to indicate when they need service from the internal DMA controllers. DRQ0–DRQ3 are used for transfers from 8-bit I/O adapters to/from system memory. DRQ5–DRQ7 are used for transfers from 16-bit I/O adapters to/from system memory. DRQ4 is not available externally as it is used to cascade the two DMA controllers together. All DRQ pins have internal pull-ups.

Figure 13.10 (Continued)

SIGNAL DESCRIPTIONS (Continued)

Name	Pin Number	Type	Description
BUS INTERFACE (Continued)			
IRQ15–IRQ9, IRQ7–IRQ3, IRQ1	25, 27, 110, 23, 21, 17, 123, 151, 153, 155, 157, 159, 109	I-TPSU	Internal Request—These are the asynchronous interrupt request inputs for the 8259 megacells. IRQ0, IRQ2, and IRQ8 are not available as external inputs to the chip, but are used internally. IRQ0 is connected to the output of the 8254 counter 0. IRQ2 is used to cascade the two 8259 megacells together. IRQ8 is output from the RTC megacell to the 8259 megacell. All IRQ input pins are active high and have internal pull-ups
—MASTER	42	I-TTL	Master, active low—This input is used by an external device to disable the internal DMA controllers and get access to the system bus. When asserted it indicates that an external bus master has control of the bus.
—MEMCS16	13	I-TTL	Memory Chip Select 16-bit—This input is used to determine when a 16-bit to 8-bit conversion is needed for CPU accesses. A 16 to 8 conversion is done anytime the System Controller requests a 16-bit memory cycle and —MEMCS16 is sampled high.
—IOCS16	15	I-TTL	I/O Chip Select 16-bit—This input is used to determine when a 16-bit to 8-bit conversion is needed for CPU accesses. A 16 to 8 conversion is done anytime the System Controller requests a 16-bit I/O cycle and —IOCS16 is sampled high.
—IOCHK	121	I-TTL	I/O Channel Check, active low—This input is used to indicate that an error has taken place on the I/O bus. If I/O checking is enabled, an —IOCHK assertion by a peripheral device generates an NMI to the processor. The state of the —IOCHK signal is read as data bit D6 of the Port B register
IOCHRDY	126	I-TTL	I/O Channel Ready—This input is pulled low in order to extend the read or write cycles of any bus access when required. The cycle can be initiated by the CPU, DMA controllers or refresh controller. The default number of wait states for cycles intiated by the CPU are four wait states for 8-bit peripherals, one wait state for 16-bit peripherals and three wait states for ROM cycles. One DMA wait state is inserted as the default for all DMA cycles. Any peripheral that cannot present read data, or strobe-in write data in this amount of time must use —IOCHRDY to extend these cycles
—WS0	125	I-TTL	Wait State 0, active low—This input is pulled low by a peripheral on the S bus to terminate a CPU controlled bus cycle earlier than the default values defined internally on the chip.
POWERGOOD	115	I-TSPU	System power on reset—This input signals that power to the board is stable. A Schmitt-trigger input is used. This allows the input to be connected directly to an RC network.
PERIPHERAL INTERFACE			
—CS8042	108	O	Chip select for 8042. This output is active any time an SA address is decoded at 60h or 64h. It is intended to be connected to the chip select of the keyboard controller. If BUSCTL[6] = 1, this pin is also active for RTC accesses at 70h and 71h. This is for use when the internal RTC is disabled and an external RTC is used.

Figure 13.10   (Continued)

Name	Pin Number	Type	Description
PERIPHERAL INTERFACE (Continued)			
XTALIN	118	I-CMOS	Crystal Input—An internal oscillator input for the real time clock crystal. It requires a 32.768 KHz external crystal or stand-alone oscillator.
XTALOUT	119	O	Crystal Output—An internal oscillator output for the real time clock crystal. See XTALIN. This pin is a no connect when an external oscillator is used.
PS/—RCLR/ IRQ8	117	I-TSPU	The Power Sense input (active high) is used to reset the status of the Valid RAM and Time (VRT) bit. This bit is used to indicate that the power has failed, and that the contents of the RTC may not be valid. This pin is connected to an external RC network. When BUSCTL[6] = 1, this pin becomes —IRQ8 input for use with an external RTC.
VBAT	116	I	Voltage Battery—Connected to the RTC hold-up battery between 2.4 and 5V.
SPKR	107	O	Speaker—This output drives an externally buffered speaker. This signal is created by gating the output of timer 2. Bit 1 of Port B, 61H, is used to enable the speaker output, and bit 0 is used to gate the output timer.
DATA BUFFER INTERFACE			
XD7–XD0	91, 92, 94–99	IO-TTL	Peripheral data bus—The bidirectional X data bus outputs data on an INTA cycle or I/O read cycle to any valid address within the Bus Controller. It is configured as an input at all other times.
—SDSWAP	101	O	System Data Swap, active low during some 8-bit accesses—It indicates that the data on the SD bus must be swapped from low byte to high byte or vice versa depending on the state of the SDLH/—HL pin. —SDSWAP is active for 8-bit DMA cycles when an odd address access occurs for data more than one byte wide. For non-DMA accesses. —SDSWAP is active for any bus cycle to an 8-bit peripheral that is addressing the odd byte.
SDLH/—HL	102	O	System Data Low to High, or High to Low—This signal is used to determine which direction data bytes must be swapped when —SDSWAP is active. When SDLH/—HL is high, it indicates that data on the low byte must be transferred to the high byte. When SDLH/—HL is low, it indicates that data on the high byte must be transferred to the low byte. SDLH/—HL is low for 8-bit DMA memory read cycles. For non-DMA accesses, SDLH/—HL is low for any memory write or I/O write when —SBHE is low. SDLH/—HL is high at all other times.
—XDREAD	103	O	Peripheral Data Read—This output is active low any time an INTA cycle occurs or an I/O read occurs to the address space from 0000h to 00FFh, which is defined as being resident on the peripheral bus.

Figure 13.10   (Continued)

Name	Pin Number	Type	Description
—LATLO	104	O	Latch Low byte—This output is generated for all I/O read and memory read bus accesses to the low byte. It is active with the same timing as the read command and returns high at the same time as the read command. This signal latches the data into the data buffer chip so that it can be presented to the CPU at a later time. This step is required due to the asynchronous interface between the System Controller and Bus Controller.
—LATHI	105	O	Latch High byte—This output is generated for all I/O read and memory read bus accesses to the high byte. It is active with the same timing as the read command and returns high at the same time as the read command. This signal latches the data into the data buffer chip so that it can be presented to the CPU at a later time. This step is required due to the asynchronous interface between the System Controller and Bus Controller.
—PCK	111	I-TPU	Party Check input, active low with pull-up—Indicates that a parity error has occurred in the on-board memory array. Assertion of this signal (if enabled) generates an NMI to the processor. The state of the —PCK signal is read as data bit D7 of the Port B register.
—HIDRIVE	113	I-TPU	High Drive Enable—This pin is a wire strap option. When this input is low, all bus drivers defined with an IOL spec of 24 mA will sink the full 24 mA of current. When this input is high, all pins defined as 24 mA have the output low drive capability cut in half to 12 mA. Note that all AC specifications are done with the outputs in the high drive mode and a 200 pF capacitive load —HIDRIVE has an internal pull-up and can be left unconnected if 12 mA drive is desired. It is tied low if 24 mA drive is desired.

TEST MODE PIN

Name	Pin Number	Type	Description
—TRI	114	I-TPU	Three-state—This pin is used to control the three-state drive of all outputs and bidirectional pins on the chip. If this pin is pulled low, all pins on the chip except XTALOUT are in a high impedance mode. This is useful during system test when test equipment or other chips drive the signals or for hardware fault tolerant applications. —TRI has an internal pull-up.

Figure 13.10 (Continued)

CPU 286/386 (C286/$\overline{386DX}$) input. Looking at Fig. 13.10, we find that applying logic 0 at this input enables 80386DX mode of operation.

On sheet 4 of Fig. 13.3, C286/$\overline{386DX}$, which is pin 73, is connected to ground to select 80386DX-compatible interface signals. This means that signal lines $\overline{BE}_2/A_1$, $\overline{BE}_1/\overline{BHE}$, and $\overline{BE}_0/A_0$ are configured to act like byte-enable lines. In the circuit diagram of Fig. 13.3, the address lines of the 82344 are traced back to the outputs of the address buffers on sheet 2 and the byte-enable lines are traced back to the bus byte-enable lines of the 82385DX cache controller.

EXAMPLE 13.18

In the microcomputer of Fig. 13.3, what are the destinations of the INTR and NMI outputs of the 82344?

Signal Type Legend

Signal Code	Signal Type
I-TTL	TTL Level Input
I-TPD	Input with 30 KΩ Pull-Down Resistor
I-TPU	Input with 30 KΩ Pull-Up Resistor
I-TSPU	Schmitt-Trigger Input with 30 KΩ Pull-Up Resistor
I-CMOS	CMOS Level Input
IO-TTL	TTL Level Input/Output
IT-OD	TTL Level Input/Open Drain Output
IO-OD	Input or Open Drain, Slow Turn On
O	CMOS and TTL Level Compatible Output
O-TTL	TTL Level Output
O-TS	Three-State Level Output
I1	Input used for Testing Purposes
GND	Ground
PWR	Power

Figure 13.10 (Continued)

Solution

In the circuit diagram of Fig. 13.3, the INTR and NMI CPU interface outputs of the 82344 are found to be applied directly to the corresponding input of the 80386DX MPU.

Let us next look at the ROM interface and BIOS EPROMs. $\overline{ROM_8}$ is an option-select input of the 82344. Notice on sheet 4 of Fig. 13.3 that jumper J_{42} is used to fix the setting of this input. Figure 13.10 identifies that if this input is made logic 0, the microcomputer is configured for 8-bit-wide BIOS memory. This is done by connecting jumper J_{42}. In this case, a single 27C512 EPROM is used to hold the BIOS software for the microcomputer. Sockets for both 8- and 16-bit BIOS are illustrated on sheet 3 of the circuit diagram in Fig. 13.3. U_{39} is populated with the 27C512 EPROM when the 8-bit BIOS mode is selected.

Now we look more closely at the connection of the 27C512 BIOS EPROM in the PC/AT microcomputer system. Storage locations in BIOS EPROM U_{39} are addressed by system address bus signals SA_0 through SA_{15}. These address lines are outputs of the 82344 ISA bus controller. The occurrence of a read bus cycle to the BIOS EPROM is signaled to the EPROM by another output of the 82344. Notice that the signal $\overline{MEMR}$ is applied to the $\overline{OE}$ input at pin 22 of the 27C512. Logic 0 at $\overline{OE}$ enables the outputs of the EPROM for operation. At the same time, the $\overline{ROMCS}$ output of the 82346 is supplied to the $\overline{CE}$ input of the EPROM and selects U_{39} for operation. Finally, information held at the addressed storage location of the EPROM are output on memory data bus lines MD_0 through MD_7. This byte of data is passed through the 82345 data buffer to data bus lines D_0 through D_7 of the MPU.

EXAMPLE 13.19

A chart for the settings of BIOS EPROM address jumpers J_{30}, J_{31}, and J_{32} is given on sheet 3 of the circuit diagram in Fig. 13.3. What should be the settings of the jumpers if the 8-bit mode is selected and a single 27C512 EPROM is used?

Solution

The chart indicates that the jumper setting should be

$$J_{30} = \text{don't care}$$
$$J_{31} = \text{IN}$$
$$J_{32} = \text{IN}$$

The ISA expansion bus of the PC/AT is a collection of address, data, control, and power lines that are provided to support expansion of the microcomputer system. The main processor board design of Fig. 13.3 includes connectors for six expansion slots. The pin layout of these sockets are illustrated on sheet 8 of the circuit diagram. Notice that five slots have both a 62-pin and 36-pin card edge connector. They are full 16-bit data bus expansion slots. An example is the slot made by the connectors labeled J_{80} and J_{81}. The sixth slot, which has just a 62-pin connector (J_{90}), supports only an 8-bit data bus. Using these slots, the microcomputer system is expanded by plugging in special function adapter cards, such as boards to control a monochrome or color display, expanded memory, or communication ports.

The pins of these connectors are attached to inputs or outputs of the 82345 data buffer, 82344 ISA controller, or power supply. Let us now examine the source of some of these signals. Sheet 4 of Fig. 13.3 shows that system data bus lines SD_0 through SD_{15} at J_{80} and J_{81} are tied to the system data bus lines of the 82345 data buffer. On the other hand, the address and control signals of the ISA bus are all supplied by the 82344 ISA controller. For instance, address signal SA_{19} at pin A12 of connector J_{80} is attached to pin 131 of the 82344, which is also labeled SA_{19}. Another example is control signal $\overline{\text{MEMR}}$ at pin C9 of J_{81}, which connects to $\overline{\text{MEMR}}$ at pin 33 of the 82344. The signal names for each of the signals at the ISA expansion bus are listed in Fig. 13.11.

EXAMPLE 13.20

What power supply voltage is available at pin B5 of J_{80}?

Solution

From sheet 8 of the circuit diagram, we find that -5 V dc is supplied from pin B5.

EXAMPLE 13.21

What does the signal mnemonic RSTDRV stand for? Is it active when at the 0 or 1 logic level?

Mnemonic	Name	Function
AEN	Address enable	O
BALE	Buffered address latch enable	O
CLK	Clock	O
$\overline{DACK}_0$-$\overline{DACK}_3$	DMA acknowledge 0-3 and 5-7	O
$\overline{DACK}_5$-$\overline{DACK}_7$		
DRQ_0-DRQ_3	DMA request 1-3 and 5-7	I
DRQ_5-DRQ_7		
$\overline{I/O\ CH\ CK}$	I/O channel check	I
I/O CH RDY	I/O channel ready	I
I/O CS16	I/O 16-bit chip select	I
$\overline{IOR}$	I/O read command	I/O
$\overline{IOW}$	I/O write command	I/O
IRQ_3-IRQ_7,	Interrupt request 3-7,	I
IRQ_9-IRQ_{12}	9-12, and 14-15	
and IRQ_{14}-IRQ_{15}		
LA_{17}-LA_{23}	System address lines 17-23	I/O
$\overline{MASTER}$	Master	I
$\overline{MEM\ CS16}$	Memory 16 chip select	I
$\overline{MEMR}$	Memory read command	I/O
$\overline{MEMW}$	Memory write command	I/O
OSC	Oscillator	O
$\overline{REFRESH}$	Refresh	I/O
RESET DRV	Reset drive	O
SA_0-SA_{19}	System address lines 0-19	I/O
SBHE	System high byte enable	O
SD_0-SD_{15}	Data lines 0-15	I/O
$\overline{SMEMR}$	System memory read command	O
$\overline{SMEMW}$	System memory write command	O
T/C	Terminal count	O
0WS	Zero wait state	I

Figure 13.11 ISA bus signals.

Solution

In Fig. 13.10 we see that RSTDRV means reset drive and is active when at the 1 logic level.

The peripheral interface corresponds to a number of special-purpose inputs and outputs of the 82344. Earlier we identified one of these signals, $\overline{CS8042}$, as an output that is used to enable the 8042 keyboard controller. Three other peripheral interface signals are PS/$\overline{RCLR}$/IRQ_8, VBAT, and SPKR. In the circuitry of sheet 4 of Fig. 13.3, we find that SPKR is output to pin 1 of connector J_{41}. J_{41} is used to connect the speaker to the microcomputer. VBAT and PS/$\overline{RCLR}$/IRQ_8 are inputs of the 82344. For instance, in the circuit diagram, power sense (PS) is found to be the voltage across capacitor C_{40}.

Earlier we found that a special interface was provided on the 82345 data buffer for communication with the 82344 ISA controller. Figure 13.10 shows that this data buffer interface includes peripheral data bus lines XD_0 through XD_7. This part of the interface is used to input data from or output data to the registers of the peripherals within the 82344 and to transfer type numbers from the interrupt controller to the MPU during interrupt-acknowledge bus cycles.

This interface also includes a number of signals that tell the 82345 data buffer how to set its multiplexers to switch data that are being transferred over the system data bus. For instance, the $\overline{SDSWAP}$ and SDLH/$\overline{HL}$ signals are used together to indicate that byte data on the bus must be swapped and whether the data must be swapped from the low-byte data bus lines to the high-byte lines, or vice versus.

EXAMPLE 13.22

To what logic level is SDLH/$\overline{\text{HL}}$ set to signal the 82345 that a byte of data on the system data bus must be swapped from the low-byte lines to the high-byte lines?

Solution

Figure 13.10 indicates that SDLH/$\overline{\text{HL}}$ must be set to 1 when data are to be transferred from the low-byte lines to the high-byte lines.

A special interface is also provided on the 82344 for communication with the 82346 system controller. It includes signals that identify the type of bus cycle that is taking place, on-board DRAM is being accessed, a request and acknowledgment for DMA, and a request for refresh. On sheet 4 of Fig. 13.3 we find that these signals are called the 344/346 interface; however, in Fig. 13.10, they are identified as those for the system-controller interface. The function of each of these signals is briefly described in Fig. 13.10. Let us trace the DMA handshake-signal sequence that takes place between the 82344 and 82346. Notice that the signal DMAHRQ at pin 89 is the DMA hold request output of the 82344. Logic 1 at this output signals that the DMA controllers within the 82344 want access to the system bus. This signal is input to the 82346 system-controller at pin 105 (DMAHRQ). When the bus is available to the DMA controller, the 82346 signals this fact to the 82344 by switching its DMA hold-acknowledge (DMAHLDA) output at pin 99 to logic 1. DMAHLDA is input to the 82344 at pin 88. Now a DMA controller within the 82344 has control of the system bus and can initiate data transfer.

▲ 13.8 82341 HIGH-INTEGRATION PERIPHERAL COMBO

The 82341 IC implements a number of the peripheral functions that are needed in the microcomputer of the PC/AT. It contains circuitry for two *16C450 compatible asynchronous serial communication ports, a parallel printer port, a real-time clock, a scratchpad RAM, keyboard and mouse controllers, an integrated drive electronics (IDE) hard disk interface,* and *programmable chip selects.* For this reason it is known as the *PC/AT peripheral combo chip.* This device is packaged in a 128-pin PQFP.

Some of the functions provided in the 82341 did not reside on the main processor board of IBM's original 80286-based PC/AT microcomputer. Instead, they were provided with add-on cards that connect to the ISA bus. This is the reason that the 82341 peripheral combo is shown attached to the ISA expansion bus in the 82340 PC/AT block diagram of Fig. 13.2.

Block Diagram of the 82341

The circuitry within the 82341 is illustrated with the block diagram in Fig. 13.12. Here we see that the device acts like a group of programmable peripheral ICs attached to the system address and data buses. They include an 82C042 keyboard/mouse controller, 146818A real-time clock, 16C452 dual UART and parallel printer port, and IDE hard disk interface. The operation of the various peripheral functions implemented within the 82341 are set up under software control. Addresses applied to the SA_0 through SA_{15} inputs are decoded by the port address decoder to select a control register within the device. Then programming information or data are written into the selected register over system data bus lines SD_0

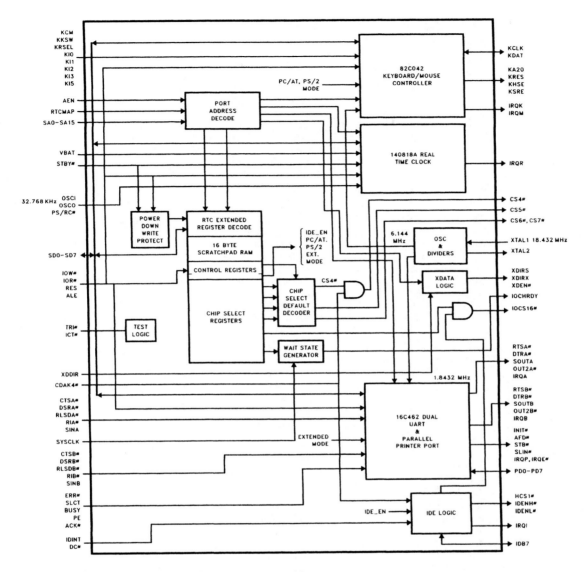

Figure 13.12 Block diagram of the 82341 PC/AT peripheral combo chip. (Reprinted by permission of Intel Corp. Copyright/Intel Corp. 1989)

through SD₇. Status, configuration information, and data can also be read from a register within the device.

Inputs and Outputs of the 82341 and Their Connection in the PC/AT Microcomputer System

Now that we have introduced the peripheral functions implemented with the 82341 peripheral combo, let us look at its signal interfaces and how they are used in a PC/AT microcomputer system. The signals of the 82341 are listed in Fig. 13.13. Earlier we pointed out that the operation of the peripheral functions of the 82341 are programmed over the

SIGNAL DESCRIPTIONS

Signal Name	Pin Number	Signal Type	Signal Description
COMMUNICATIONS PORT A			
RTSA #	44	O1	Request to Send, Port A
DTRA #	45	O1	Data Terminal Ready, Port A
SOUTA	46	O1	Serial Data Output, Port A
CTSA #	79	I4	Clear to Send, Port A
DSRA #	78	I4	Data Set Ready, Port A
RLSDA #	77	I4	Receive Line Signal Detect, Port A
RIA #	76	I4	Ring Indicator, Port A
SINA	75	I4	Serial Input, Port A
IRQA	39	O6	Interrupt Request, Port A
OUT2A #	42	O1	Output 2, Port A
COMMUNICATIONS PORT B			
RTSB #	47	O1	Request to Send, Port B
DTRB #	48	O1	Data Termina Ready, Port B
SOUTB	49	O1	Serial Data Output, Port B
CTSB #	89	I4	Clear to Send, Port B
DSRB #	88	I4	Data Set Ready, Port B
RLSDB #	87	I4	Receive Line Signal Detect, Port B
RIB #	86	I4	Ring Indicator, Port B
SINB	85	I4	Serial Input, Port B
IRQB	37	O6	Interrupt Request, Port B
OUT2B #	43	O1	Output 2, Port B
PARALLEL PRINTER PORT			
PD0	59	IO5	Printer Data Port, Bit 0
PD1	58	IO5	Printer Data Port, Bit 1
PD2	57	IO5	Printer Data Port, Bit 2
PD3	56	IO5	Printer Data Port, Bit 3
PD4	54	IO5	Printer Data Port, Bit 4
PD5	53	IO5	Printer Data Port, Bit 5
PD6	52	IO5	Printer Data Port, Bit 6
PD7	51	IO5	Printer Data Port, Bit 7
INIT #	63	O4	Initialize Printer Signal
AFD #	62	O4	Autofeed Printer Signal
STB #	61	O4	Data Strobe to Printer
SLIN #	64	O4	Select Signal from Printer
ERR #	70	I4	Error Signal from Priner
SLCT	71	I4	Select Signal from Printer

Figure 13.13   82341 input/output signal descriptions. (Reprinted by permission of Intel Corp. Copyright/Intel Corp. 1989)

system data bus. The system address bus and data bus signals are listed in the *common bus I/O* group in Fig. 13.13. Here we find that I/O bus cycles are used to read from or write to the 82341. The control signals for these data transfers are AEN, ALE, $\overline{\text{IOR}}$, and $\overline{\text{IOW}}$. Looking at sheet 5 of Fig. 13.3, we find that these control inputs are supplied by ISA bus signals AEN, BALE, $\overline{\text{IOR}}$, and $\overline{\text{IOW}}$, respectively. These four signals are produced by the 82344 ISA bus controller.

EXAMPLE 13.23

Is the IOCHRDY signal an input or output of the 82341? Where is this signal connected in the PC/AT microcomputer of Fig. 13.3?

SIGNAL DESCRIPTIONS (Continued)

Signal Name	Pin Number	Signal Type	Signal Description
PARALLEL PRINTER PORT (Continued)			
BUSY	72	I4	Busy Signal from Printer
PE	73	I4	Paper Error Signal from Printer
ACK#	74	I4	Acknowledge Signal from Printer
IRQP	40	O6	Printer Interrupt Request Output
IRQE#	41	O1	Printer Interrupt Request Enable Signal
REAL TIME CLOCK PORT			
VBAT	69	NA	Standby Power—Normally 3V to 5V, battery backed
STBY#	65	I5	Power Down Control
OSCI	66	NA	Crystal Connection Input—32 kHz
OSCO	67	NA	Crystal Connection Output—32 kHz
PS/RC#	68	I5	Power Sense/RAM Clear Input
IRQR	36	O1	Real Time Clock Interrupt Request Output
RTCMAP	121	I4	High—RTC is mapped to 70H and 71H, Low—RTC is mapped to 170H and 171H
KEYBOARD CONTROLLER PORT			
KCLK	103	IO4	Keyboard Clock
KDAT	104	IO4	Keyboard Data
KCM	92	I4	General Purpose Input, Normally Color/Monochrome
KKSW	93	I4	General Purpose Input, Normally Keyboard Switch
KA20	91	O1	General Purpose Output, Normally A20 Gate
KRES	90	O1	General Purpose Output, Normally Reset
KHSE	101	O1/IO4	General Purpose Input, Normally Speed Select
KSRE	100	O1/IO4	General Purpose Output, Normally Shadow RAM Enable
IRQK	34	O1	Keyboard Interrupt Request
IRQM	35	O1	Mouse Interrupt Request
KRSEL	94	I4	General Purpose Input, Normally RAM Select
K10	99	I4	General Purpose Input, Bit 0
K11	98	I4	General Purpose Input, Bit 1
K12	97	I4	General Purpose Input, Bit 2
K13	96	I4	General Purpose Input, Bit 3
K15	95	I4	General Purpose Input, Bit 5
IDE BUS I/O			
IDENH#	2	O1	IDE Bus Transceiver High Byte Enable
IDENL#	3	O1	IDE Bus Transceiver Low Byte Enable
IDINT	122	I4	IDE Bus Interrupt Request Input
IDB7	119	IO6	IDE Bus Data Bit 7
DC#	123	I4	Floppy Disk Change Signal
HCS1#	124	O1	IDE Host Chip Select 1
IRQI#	33	O6	IDE Interrupt Request Output

Figure 13.13 (Continued)

Solution

Figure 13.13 identifies IOCHRDY as an output of the 82341. In the circuits on sheet 4 of Fig. 13.3, we see that IOCHRDY is input to the microcomputer at pin 126 of the 82344 ISA bus controller.

We will continue with the asynchronous communication interfaces that are implemented with the UARTs of the 82341. Notice on sheet 5 of Fig. 13.3 that the signals of the

SIGNAL DESCRIPTIONS (Continued)

Signal Name	Pin Number	Signal Type	Signal Description
COMMON BUS I/O			
SD0	115	IO2	System Bus Data, Bit 0
SD1	114	IO2	System Bus Data, Bit 1
SD2	111	IO2	System Bus Data, Bit 2
SD3	110	IO2	System Bus Data, Bit 3
SD4	109	IO2	System Bus Data, Bit 4
SD5	108	IO2	System Bus Data, Bit 5
SD6	106	IO2	System Bus Data, Bit 6
SD7	105	IO2	System Bus Data, Bit 7
SA0	17	I1	System Bus Address, Bit 0
SA1	18	I1	System Bus Address, Bit 1
SA2	19	I1	System Bus Address, Bit 2
SA3	20	I1	System Bus Address, Bit 3
SA4	21	I1	System Bus Address, Bit 4
SA5	22	I1	System Bus Address, Bit 5
SA6	23	I1	System Bus Address, Bit 6
SA7	24	I1	System Bus Address, Bit 7
SA8	25	I1	System Bus Address, Bit 8
SA9	26	I1	System Bus Address, Bit 9
SA10	27	I1	System Bus Address, Bit 10
SA11	28	I1	System Bus Address, Bit 11
SA12	29	I1	System Bus Address, Bit 12
SA13	30	I1	System Bus Address, Bit 13
SA14	31	I1	System Bus Address, Bit 14
SA15	32	I1	System Bus Address, Bit 15
XTAL1	82	NA	Crystal Clock Input—18.432 MHz
XTAL2	83	NA	Crystal Clock Output—18.432 MHz
IOR #	11	I1	System Bus I/O Read
IOW #	12	I1	System Bus I/O Write
RES	125	I1	System Reset
AEN	13	I1	System Bus Address Enable
ALE	14	I1	System Bus Address Latch Enable
IOCS16 #	116	O8	System Bus I/O Chip Select 16
IOCHRDY	118	O8	System Bus I/O Channel Ready
SYSCLK	128	I1	System Clock—Processor Clock Divide by 2
CS4 #	7	O1	Chip Select 4—Normally for External Floppy Disk Controller
CS5 #	8	O1	Chip Select 5—Normally HCS0 # for IDE
CS6 #	9	O1	Chip Select 6—Normally for External Floppy Disk Controller
CS7 #	10	O1	Chip Select 7—Normally for External Floppy Disk Controller
CDAK4 #	102	I1	DMA Acknowledge Forces —CS4 Active
XDDIR	120	I1	X Data Bus Transceiver Direction
XDIRS	5	O1	Modified X Data Bus Transceiver Direction Control Signal—Excludes Real Time Clock and Keyboard Controller Decodes
XDIRX	6	O1	X Data Bus Transceiver Control Signal—Includes All CS Decodes Generated On Chip

Figure 13.13 (Continued)

82341 that are used to implement the serial interfaces are grouped together and marked as COM A and COM B. Tracing the COM A outputs, *request to send port A* ($\overline{\text{RTSA}}$), *data terminal ready* ($\overline{\text{DTRA}}$), and *serial data output port A* ($\overline{\text{SOUTA}}$), shows that they are buffered to RS-232C-compatible voltage levels by an MC1488 line driver and then output at COM A connector J_{53}. Inputs from this RS-232C port are the signals *data set ready port*

Signal Name	Pin Number	Signal Type	Signal Description
COMMON BUS I/O (Continued)			
XDEN #	4	O1	X Data Bus Transceiver Enable
TRI #	126	I4	Three-State Control Input—For All Outputs to Isolate Chip for Board Tests
ICT #	127	I4	In Circuit Test Mode Control

I/O LEGEND

Pin Type	mA	Type	Comment
01	2	TTL	
02	24	TTL	
04	12	TTL-OD	Open Drain, Weak Pull-Up, No VDD Diode
06	4	TTL-TS	Three-State
07	24	TTL-TS	Three-State
08	24	TTL-OD	Open Drain, Fast Active Pull-Up
I1	—	TTL	
I2	—	CMOS	
I4	—	TTL	30 kΩ Pull-Up
I5	—	TTL	Schmitt-Trigger
I02	24	TTL-TS	Three-State
I04	12	TTL-OD	Open Drain, Slow Turn-On
I05	12	TTL-TS	Three-State
I06	24	TTL-TS	Three-State, 30 kΩ Pull-Up

Figure 13.13   (Continued)

A ($\overline{DSRA}$), *clear to send port A* ($\overline{CTSA}$), *ring indicator port A* ($\overline{RIA}$), *serial input port A* (SINA), and *receive line signal detect port A* ($\overline{RLSDA}$). The mnemonic, pin number, and type for each of the signals of the 82341's communication ports are listed in Fig. 13.13.

EXAMPLE 13.24

At what pin of communication connector J_{53} is serial data input? What type of buffer is used to receive this signal?

Solution

Looking at J_{53} on sheet 5 of the circuit diagram in Fig. 13.3, we see that SINA is input at pin 3 of the connector. This signal is buffered by the MC1489 device U_{56}.

The serial data-transfer rates of ports A and B are set by a single on-chip baud-rate generator. The frequency of the clock used for baud-rate generation is set by the 18.432-MHz crystal attached between the $XTAL_1$ and $XTAL_2$ pins of the 82341. This clock signal is scaled within the 82341 to set the receive and transmit data transmission rates.

The parallel printer port signal section is identified as LPT (line printer) on the 82341 in Fig. 13.3. These lines are used to implement a *Centronics parallel printer interface* at connector J_{56}. The meaning of the signals at this parallel printer port are identified in Fig. 13.13. Here we find that data are output in parallel over print data lines PD_0 through PD_7. The printer is signaled that data are available on the PD lines by logic 0 at the $\overline{STB}$ output.

In Fig. 13.3 this signal is found to be output at pin 1 of connector J_{56}. Signals are also input to the microcomputer through the Centronics interface. For example, when the printer is busy and cannot accept additional data, it signals this fact to the microcomputer with the BUSY input. BUSY enters at pin 21 of the connector and is applied to the BUSY input of the 82341 (pin 72).

EXAMPLE 13.25

In the PC/AT microcomputer, what interrupt level is used to service the Centronics printer interface?

Solution

From the circuits on sheet 5 of Fig. 13.3, we find that the IRQP output of the 82341 is sent to the IRQ_7 input of the 82344. Therefore, interrupt level 7 is used to service a printer attached to the parallel printer interface.

The keyboard section of the 82341 provides several functions. First, it produces three control outputs. On sheet 5 of Fig. 13.3, the signals at these outputs are identified as $\overline{RC}$, A20GATE, and TURBREQ. From the pin descriptions in Fig. 13.13, we find that they stand for reset, address bit A_{20} gate, and speed select. Signals $\overline{RC}$ and A20GATE are traced to inputs of the 82346 system controller on sheet 3 of Fig. 13.3. In Fig. 13.7 we find that a 1-to-0 transition at the $\overline{RC}$ input initiates a reset of the 80386DX MPU. Moreover, the system controller uses the A20GATE input to generate the $\overline{BLKA_{20}}$ output. Tracing the $\overline{BLKA_{20}}$ output of the 82346 in Fig. 13.3, we find that it goes two places. First, $\overline{BLKA_{20}}$ is input to the 82344 ISA bus controller. When at the active 0 logic level, it forces LA_{20} and SA_{20} to logic 0. On sheet 1 of Fig. 13.3 we find that $\overline{BLKA_{20}}$ is also input at pin 1 of AND gate U_{14}. Here it is used to gate address bit A_{20} to the 82385DX cache controller and the address-latch circuit.

EXAMPLE 13.26

What is the signal mnemonic for the keyboard port input of the 82341 supplied with jumper J_{51}?

Solution

Jumper J_{51} on sheet 5 of Fig. 13.3 controls the input at pin 92. In Fig. 13.13 the signal name for pin 92 is found to be KCM.

Keyboard-entry information is input to the MPU through a synchronous serial interface made up with signals keyboard clock (KBCLK) and keyboard data (KBDAT). Whenever a keycode has been read, the keyboard controller section of the 82341 issues an interrupt to the 80386DX. In Fig. 13.3 (sheet 5) we see that the interrupt request is output on the IRQ_1 line. The service routine for this interrupt initiates reading of the keycode from the keyboard controller.

The last section of the 82341 we will examine is the *Integrated Drive Electronics* (IDE) hard disk bus interface. IDE is an industry standard interface for connecting hard disk drives to a microcomputer system. In the circuit diagram of Fig. 13.3 (sheet 5), the connector to the IDE interface is labeled J_{54}. Notice that EDI implements a 16-bit parallel data path between the hard disk subsystem and the microcomputer. This path consists of system data bus lines SD_0 through SD_6 and SD_8 through SD_{15}. Data bit 7 is not returned to the MPU; instead, it is supplied to the IDB_7 input at pin 119 of the 82341.

The IDE interface on the 82341 controls the 74ALS245 transceivers used in the data path. For instance, in the circuit diagram, we find that the transceiver for the upper part of the data bus, U_{52}, is enabled by the signal $\overline{IDENH}$. The direction in which data are passed through the transceivers is determined by signal $\overline{IOR}$. In this way we see that hard disk data transfers are actually performed with input/output bus cycles.

A number of control inputs and outputs are also supplied at the IDE interface. Some examples of outputs are address strobe BALE, read/write signals $\overline{IOR}$ and $\overline{IOW}$, and address bits SA_0 through SA_2. These signals are first buffered by the 74ALS244 IC (U_{54}) and then supplied to the IDE interface. IDE interrupt-request signal IDINT is an input from the interface.

EXAMPLE 13.27

At what pin of the IDE connector is the signal BALE output?

Solution

Figure 13.3 (sheet 5) shows that BALE is applied to pin 28 of the IDE connector.

▲ 13.9 82077AA FLOPPY DISK CONTROLLER

The last IC we will discuss is the 82077AA *floppy disk controller*. This device is not part of the 82340 PC/AT chip set, but it is used in the microcomputer of Fig. 13.3. In IBM's original 80286-based PC/AT microcomputer, the interface to the floppy disk drive was provided by an add-on card plugged into an ISA bus slot. Due to the high integration of the 82340 ICs, space is available on the main processor board for providing the floppy disk control function.

The 82077AA is a versatile, highly integrated solution for implementing the floppy disk interface in a PC/AT-compatible microcomputer. For instance, the 82077AA can control up to four drives, supports both 5.25- or 3.5-inch drives, and can transfer data at four rates: 250 kilobits/second (KB/s), 300 KB/s, 500 KB/s, and 1 megabit/second (MB/s). The 82077AA also contains the circuitry needed to implement a *tape-drive controller*. Finally, the buffers for the microprocessor interface and drivers for the disk drive interface are built into the device. For this reason a PC/AT floppy disk interface can be completely implemented without any additional circuitry.

Block Diagram of the 82077AA

Let us begin our study of the 82077AA floppy disk controller with the block diagram of Fig. 13.14. Unlike the 82340 chips we have been examining, the 82077AA is architectured like a traditional peripheral IC. That is, it has a standard microprocessor interface. This interface consists of an 8-bit data bus, control signals such as $\overline{RD}$ and $\overline{WR}$, register-select inputs A_0 through A_2, and a chip-select input $\overline{CS}$. Data, configuration information, and

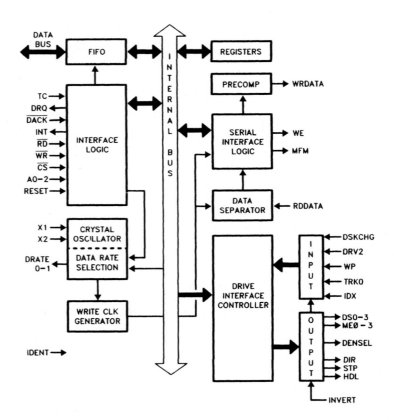

Figure 13.14 Block diagram of the 82077AA. (Reprinted by permission of Intel Corp. Copyright/Intel Corp. 1989)

status information are written into or read from the 82077AA's internal registers through this interface.

Notice in Fig. 13.14 that a first-in, first-out (FIFO) buffer is built into the data bus interface. This FIFO is 16 bytes deep. After reset, the FIFO is disabled. Once the 82077AA is initialized, the FIFO can be either left disabled or enabled under software control. When the FIFO is enabled, all input and output transfers of data, command, and status information go through the FIFO. The FIFO has a programmable threshold level that, when reached, causes an interrupt request to be output. This interrupt request is output on the INT line and can be returned to the MPU to signal that data must be read from or written to the floppy disk controller. In this way we see that incoming or outgoing information is buffered within the floppy disk controller. This buffering results in better bus utilization and higher bus performance.

In Fig. 13.14 we see that the 82077AA has an on-chip *oscillator*. This oscillator is used to synchronize the operation of the circuitry within the 82077AA and to produce the data rate clock. Earlier we pointed out that four data-transmission rates are supported. Notice that the data rate is set by scaling the clock in the *data-rate-selection* block. This scale factor is set under software control.

At the other side of the block diagram in Fig. 13.14, we find the circuitry needed to drive and control the floppy disk interface. This includes the *drive interface controller, input and output buffers, data separator, serial interface logic,* and *precompensation circuit.* The

80386DX PC/AT Microcomputer System Hardware   Chap. 13

drive interface controller section produces all of the control signals needed to interface floppy disk drives to the PC/AT's microcomputer. Notice that all of these inputs and outputs are buffered within the 82077AA. As pointed out earlier, these on-chip buffers eliminate the need for external buffer circuitry in the floppy disk interface.

The data separator is the section of the 82077AA where read data are received. A serial bit stream of data is sent by the floppy disk drive to the RDDATA input. The phase-lock-loop circuitry of the data separator locks onto this serial stream of data and takes a sample of the data during a clock period called the *data window*. The data separator reads the serial bits of data, translates them to parallel form, and loads the parallel byte of data into the FIFO buffer to await transfer to the MPU. Remember that the MPU is not signaled to read data from the FIFO until it is filled to the threshold level.

To assure that reliable serial data transfers take place between the floppy disk drive and microcomputer, the 82077AA compensates for variations in the data read frequency and has a high tolerance to bit jitter. For instance, the frequency at which read data are sent to the controller may change due to drift in the speed of the motor that drives the floppy disk. Bit jitter relates to the shifting of the data bit in the data window. The amount of bit jitter that occurs depends on both the magnetic media being read and the operating characteristics of the drive. To achieve good tolerance to read-frequency deviation and bit jitter, the 82077AA's data separator employs a dual-analog-phase locked-loop design. In fact, this design eliminates the use of the external trimming circuitry that is needed with other floppy disk controller devices.

The write precompensator block, which is identified as PRECOMP in Fig. 13.14, is in the write path from the microcomputer to the floppy disk drive. Notice that write data are output from the precompensator to the drive over the WRDATA signal line. However, the function of the write precompensator actually affects the amount of bit jitter that occurs when reading data from the drive. It turns out that certain bit patterns when read from a floppy disk exhibit more shifting than others. These patterns are known to result in bit jitter. The objective of the write precompensator is to detect these patterns before they are written to the drive and to shift the bits such as to compensate for the expected bit shift. That is, certain bits are automatically made earlier or later relative to surrounding bits. In this way, the bit shift experienced when reading the data back is compensated for and a lower level of bit jitter is achieved. This technique is known as *write precompensation*.

Inputs and Outputs of the 82077AA and Their Use in the PC/AT Microcomputer System

Having introduced the 82077AA floppy disk controller and its block diagram, let us continue by examining its input and output signals and how they are connected in a PC/AT-compatible microcomputer system. The input and outputs signals of the 82077AA are listed in Fig. 13.15. Notice that the signals are divided into three groups: the *host interface signals, disk-control signals*, and *phase-locked-loop signals*.

Earlier we found that the floppy disk controller attaches to an 80386DX microcomputer through its microprocessor interface. In the microcomputer of Fig. 13.3 (sheet 5), we find that the 82077AA's microprocessor interface attaches to the ISA system bus. That is, address inputs A_0 through A_2 are driven by system-address lines SA_0 through SA_2, data bus lines D_0 through D_7 are attached to system data bus lines SD_0 through SD_7, and control inputs $\overline{RD}$ and $\overline{WR}$ are supplied by $\overline{IOR}$ and $\overline{IOW}$, respectively. The functions of these signals are all described in the host interface group in Fig. 13.15. For instance, the address code $A_2A_1A_0$ determines which one of the register within the 82077AA is to be accessed. An

Symbol	Pin #	I/O	Description
HOST INTERFACE			
RESET	32	I	**RESET:** A high level places the 82077AA in a known idle state. All registers are cleared except those set by the Specify command.
$\overline{CS}$	6	I	**CHIP SELECT:** Decodes base address range and qualifies $\overline{RD}$ and $\overline{WR}$ inputs.
A0 A1 A2	7 8 10	I	**ADDRESS:** Selects one of the host interface registers:

A2	A1	A0		Register
0	0	0	R	Status Register A
0	0	1	R	Status Register B
0	1	0	R/W	Digital Output Register
0	1	1	R/W	Tape Drive Register
1	0	0	R	Main Status Register
1	0	0	W	Data Rate Select Register
1	0	1	R/W	Data (FIFO)
1	1	0		Reserved
1	1	1	R	Digital Input Register
1	1	1	W	Configuration Control Register

Symbol	Pin #	I/O	Description
DB0 DB1 DB2 DB3 DB4 DB5 DB6 DB7	11 13 14 15 17 19 20 22	I/O	**DATA BUS:** Data bus with 12 mA drive
$\overline{RD}$	4	I	**READ:** Control signal
$\overline{WR}$	5	I	**WRITE:** Control signal
DRQ	24	O	**DMA REQUEST:** Requests service from a DMA controller. Normally active high, but goes to high impedance in AT and Model 30 modes when the appropriate bit is set in the DOR.
$\overline{DACK}$	3	I	**DMA ACKNOWLEDGE:** Control input that qualifies the $\overline{RD}$, $\overline{WR}$ inputs in DMA cycles. Normally active low, but is disabled in AT and PS/2 Model 30 modes when the appropriate bit is set in the DOR.
TC	25	I	**TERMINAL COUNT:** Control line from a DMA controller that terminates the current disk transfer. TC is accepted only while $\overline{DACK}$ is active. This input is active high in the AT, and PS/2 Model 30 modes and active low in the PS/2™ mode.
INT	23	O	**INTERRUPT:** Signals a data transfer in non-DMA mode and when status is valid. Normally active high, but goes to high impedance in AT, and Model 30 modes when the appropriate bit is set in the DOR.
X1 X2	33 34		**CRYSTAL 1,2:** Connection for a 24 MHz fundamental mode parallel resonant crystal. X1 may be driven with a MOS level clock and X2 would be left unconnected.

Figure 13.15 82077A input/output signals. (Reprinted by permission of Intel Corp. Copyright/Intel Corp. 1989).

Symbol	Pin #	I/O	Description
HOST INTERFACE (Continued)			
IDENT	27	I	**IDENTITY:** Upon Hardware RESET, this input (along with MFM pin) selects between the three interface modes. After RESET, this input selects the type of drive being accessed and alters the level on DENSEL. The MFM pin is also sampled at Hardware RESET, and then becomes an output again. Internal pull-ups on MFM permit a no connect.

IDENT	MFM	INTERFACE
1	1 or NC	AT Mode
1	0	ILLEGAL
0	1 or NC	PS/2 Mode
0	0	Model 30 Mode

AT MODE: Major options are: enables DMA Gate logic, TC is active high, Status Registers A & B not available.
PS/2 MODE: Major options are: No DMA Gate logic, TC is active low, Status Registers A & B are available.
MODEL 30 MODE: Major options are: enable DMA Gate logic, TC is active high, Status Registers A & B available.
After Hardware reset this pin determines the polarity of the DENSEL pin. IDENT at a logic level of "1", DENSEL will be active high for high (500 Kbps/1 Mbps) data rates (typically used for 5.25″ drives). IDENT at a logic level of "0", DENSEL will be active low for high data rates (typically used for 3.5″ drives).

Symbol	Pin #	I/O	Description
DISK CONTROL (All outputs have 40 mA drive capability)			
INVERT	35	I	**INVERT:** Strapping option. Determines the polartity of **all** signals in this section. Should be strapped to ground when using the internal buffers and these signals become active LOW. When strapped to VCC, these signals become active high and external inverting drivers and receivers are required.
ME0 ME1 ME2 ME3	57 61 63 66	O	**ME0–3:** Decoded Motor enables for drives 0–3. The motor enable pins are directly controlled via the Digital Output Register.
DS0 DS1 DS2 DS3	58 62 64 67	O	**DRIVE SELECT 0–3:** Decoded drive selects for drives 0–3. These outputs are decoded from the select bits in the Digital Output Register and gated by ME0–3.
HDSEL	51	O	**HEAD SELECT:** Selects which side of a disk is to be used. An active level selects side 1.
STEP	55	O	**STEP:** Supplies step pulses to the drive.
DIR	56	O	**DIRECTION:** Controls the direction the head moves when a step signal is present. The head moves toward the center if active.
WRDATA	53	O	**WRITE DATA:** FM or MFM serial data to the drive. Precompensation value is selectable through software.
WE	52	O	**WRITE ENABLE:** Drive control signal that enables the head to write onto the disk.

Figure 13.15 (Continued)

Table 1. 82077AA Pin Description (Continued)

Symbol	Pin #	I/O	Description
DISK CONTROL (All outputs have 40 mA drive capability) (Continued)			
DENSEL	49	O	**DENSITY SELECT:** Indicates whether a low (250/300 Kbps) or high (500 Kbps/1 Mbps) data rate has been selected.
DSKCHG	31	I	**DISK CHANGE:** This input is reflected in the Digital Input Register.
DRV2	30	I	**DRIVE2:** This indicates whether a second drive is installed and is reflected in Status Register A.
TRK0	2	I	**TRACK0:** Control line that indicates that the head is on track 0.
WP	1	I	**WRITE PROTECT:** Indicates whether the disk drive is write protected.
INDX	26	I	**INDEX:** Indicates the beginning of the track.
PLL SECTION			
RDDATA	41	I	**READ DATA:** Serial data from the disk. INVERT also affects the polarity of this signal.
HIFIL	38	I/O	**HIGH FILTER:** Analog reference signal for internal data separator compensation. This should be filtered by an external capacitor to LOFIL.
LOFIL	37	I/O	**LOW FILTER:** Low noise ground return for the reference filter capacitor.
MFM	48	I/O	**MFM:** At Hardware RESET, aids in configuring the 82077AA. Internal pull-up allows a no connect if a "1" is required. After reset this pin becomes an output and indicates the current data encoding/decoding mode (Note: If the pin is held at logic level "0" during hardware RESET it must be pulled to "1" after reset to enable the output. The pin can be released on the falling edge of hardware RESET to enable the output). MFM is active high (MFM). MFM may be left tied low after hardware reset, in this case the MFM function will be disabled.
DRATE0 DRATE1	28 29	O	**DATARATE0-1:** Reflects the contents of bits 0,1 of the Data Rate Register. (Drive capability of + 6.0 mA @ 0.4V and − 4.0 mA @ 2.4V)
PLL0	39	I	**PLL0:** This input optimizes the data separator, for either floppy disks or tape drives. A "1" (or V_{CC}) selects the floppy mode, a "0" (or GND) selects tape mode.

Figure 13.15   (Continued)

example is that the address 101 must be output to the device whenever data are read from or written into the data FIFO.

EXAMPLE 13.28

What is the source of the chip-select signal that enables the microprocessor interface of the 82077AA in Fig. 13.3?

Solution

In the circuit diagram, the $\overline{CS}$ input at pin 6 of the 82077AA is traced back to the chip-select output $\overline{CS}_4$ of the 82341 PC/AT peripheral combo IC (U_{50}). This output is labeled $\overline{FDCS}$ for floppy drive chip select.

The microprocessor interface supports DMA transfers of read or write data between the data bus FIFO and memory. The 82077AA's DMA handshake signals are the DMA request (DRQ) output at pin 24 and DMA acknowledge ($\overline{DACK}$) input at pin 3. If the 82077AA in the circuit of Fig. 13.3 needs to request direct access of the ISA bus, it switches the DRQ output to logic 1. This signal is sent to the DRQ_2 input at pin 124 of the 82344 ISA bus controller. When the bus is available, the DMA controller within the 82344 responds that it is ready to initiate the DMA data transfer by making its $\overline{DACK_2}$ output (pin 2) logic 0. This signal is returned through NAND gates in IC U_{501} to the $\overline{DACK}$ input (pin 3) of the floppy disk controller. This completes the DMA request/acknowledge handshake sequence. Now the DMA data transfer operation is performed.

The floppy disk drive attaches to the microcomputer at connector J_{57} on sheet 5 of the circuit diagram in Fig. 13.3. The invert ($\overline{INVERT}$) pin of the 82077AA is an option-select input for the interface. If it is set to logic 1, the external buffer circuitry is needed. On the other hand, setting $\overline{INVERT}$ to 0 means that the internal buffers are in use. This is the reason $\overline{INVERT}$ (pin 27) is wired to ground in the microcomputer of Fig. 13.3.

Let us next look at some of the signals provided at this interface. Earlier we found that read and write data signals are input and output over the RDDATA and WRDATA lines of the 82077AA, respectively. Looking at sheet 5 of Fig. 13.3, we find that the RDDATA is received from the floppy disk drive at pin 30 of connector J_{57} and is input to the data separator of the 82077AA from pin 41. The interface also includes many control inputs and outputs. For example, line ME_0 is the enable signal for the motor connected as drive 0. Logic 1 is output on this line under software control to turn on the motor. An example of a control input is track zero (TRK_0). This input is at its active 1 logic level whenever the drive's read/write head is located in the track 0 position. The microcomputer can identify when the head is positioned at track 0 by polling this input.

EXAMPLE 13.29

What does logic 0 at pin 28 of J_{57} in Fig. 13.3 mean?

Solution

The logic 0 at pin 28 of J_{57} is applied to the WP input of the 82077AA. This means that the diskette in the floppy disk drive is not write protected.

ASSIGNMENTS

Section 13.2

1. Name the three system buses of the original PC/AT.
2. Into what two other buses is the system data bus divided?
3. How much RAM is implemented on the system processor board? What size DRAMs are in use?
4. How much ROM is provided on the system processor board? What size EPROM is used?
5. What functions are performed by the clock-generator block diagram?
6. What range of I/O addresses are dedicated to the master interrupt controller?
7. Which I/O addresses are reserved for DMA controller 2?
8. What I/O address range is assigned to the DMA page register?
9. Which DMA channel is used for cascading DMA controller 1 to DMA controller 2?

10. What functions are performed by timer 0? Timer 1? Timer 2?

11. Which output line is used to enable the speaker? Which input line is used to input the state of the enable signal for the speaker?

12. Over which output line is the parity check circuitry enabled? At which input is the state of the parity check signal read?

13. What are the two sources of the NMI signal?

14. What is assigned to the lowest-priority interrupt request at the master interrupt controller?

15. What is assigned to the highest-priority interrupt request at the slave interrupt controller?

Section 13.3

16. Which devices make up the core of the 80386DX microcomputer in Fig. 13.2?

17. List the numbers and names of the ICs in the 82340 PC/AT chip set.

18. What does ISA stand for?

19. Does the 82341 attach to the local bus of the 80386DX or the ISA bus?

Section 13.4

20. Locate the source of the CLK_2 input of the 80386DX, 80387DX, and 82385DX in the circuit of Fig. 13.3.

21. What affect will logic 0 have at the $\overline{BLKA_{20}}$ input at pin 1 of AND gate U_{14} (see sheet 1 of Fig. 13.3)?

22. To which pins of the 80387DX do the $W/\overline{R}$, $M/\overline{IO}$, and $\overline{ADS}$ outputs of the 80386DX attach?

23. What signals are applied to the $\overline{BE_0}$ through $\overline{BE_3}$ inputs of the 82385DX? What are the sources of these signals?

24. Is the 82385DX in Fig. 13.3 set up for master or slave operation?

25. What signal drives the FLUSH input of the 82385DX?

26. What SRAM device is employed in the cache memory array of the core microcomputer in Fig. 13.3?

27. What is the size of the cache memory array in the core microcomputer of Fig. 13.3?

28. What are the sources of the $\overline{BROYO}$ and $\overline{BROYE}$ inputs of the ready/wait state PAL?

29. What determines the input conditions for which the $\overline{NCA}$ input of the 82385DX is logic 0?

30. Which IC is used to latch CPU address signals CA_{26} through CA_{31}?

31. Which latch IC produces host address signals BA_{10} through BA_{17}?

32. What output of the 82385DX produces $\overline{AOE}$ for the address latches? At what pin of the 82385DX IC is it output?

33. To which input of the 74F374 is BACP output of the 82385DX applied? What is the pin number of this input?

34. What device provides the signal PUE?

35. Which transceiver IC interfaces host data bus lines D_{16} through D_{23} to the CPU data bus?

36. What signal drives the $\overline{\text{G}}$ inputs of the 74F646 transceivers? What is the source of this signal?

Section 13.5

37. What does PQFP stand for?

38. What information can the SD bus multiplexer switch to the SD bus?

39. What are the sources of input data for the SD bus multiplexer?

40. At what pin is the signal $\overline{\text{ROMCS}}$ input to the 82345?

41. Is the signal $\overline{\text{XDREAD}}$ an input or output? What function does it provide?

42. List the names, size, and signals of each of the four data bus interfaces of the 82345.

43. Which data buses provided by the 82345 are part of the ISA bus?

44. Which pins of the 82345 provide the signals SD_0 through SD_{31}?

45. If jumper J_{40} is installed on sheet 4 of Fig. 13.3, what mode of operation is selected?

46. What is the source of the CPST_WRC input of the 82345 on sheet 2 in Fig. 13.3?

47. What does SD_1 stand for?

48. To which pin of the 82346 is the $\overline{\text{LATHI}}$ input of the 82345 attached?

49. What direction of XD bus operation is selected by $\overline{\text{XDREAD}}$ equal 1?

Section 13.6

50. How many pins are on the package of the 82346?

51. If the 80386DX in a PC/AT microcomputer is to operate at 25 MHz, what frequency clock must be applied to the $TCLK_2$ input?

52. Are the internal configuration control registers of the 82346 located in the memory or I/O address space of the PC/AT microcomputer?

53. List the output signals of the numeric coprocessor interface.

54. Which pin of the 82346 is the $D/\overline{C}$ input?

55. In the circuit of Fig. 13.3, which IC is the sources of the $W/\overline{R}$, $D/\overline{C}$, and $M/\overline{IO}$ inputs of the 82346?

56. To which pin of the 82346 must output $\overline{\text{BBE}}_2$ of the 82385DX cache controller be applied?

57. How is the CLK_{2IN} input of the 82346 normally connected?

58. What does $\overline{\text{READYI}}$ stand for? What is the source of this signal in the circuit diagram on sheet 3 of Fig. 13.3?

59. What is the destination of the RESCPU output of the 82346 on sheet 3 of Fig. 13.3?

60. What is the source of the address inputs of the 82346 when the HLDA input is active?

61. Which of the on-board memory system interface signals are applied to inputs of the 82345 data buffer?

62. Which ICs are used to buffer memory address lines MA_0 through MA_{10} to produce buffered memory address lines MA_{0H} through MA_{10H}?

Assignments   **743**

63. Which banks of DRAM SIMs are addressed with the buffered memory address lines?

64. Which banks of DRAM SIMs are supplied by $\overline{RAS_3}$?

65. What size DRAM SIMs can be used in the DRAM array on sheet 6 of Fig. 13.3?

66. Which SIMs make up bank 2 of the DRAM array?

67. If $CASBK_2$, LBE_0, and LBE_1 are at the 1 logic level and the other CAS and LBE signals are at logic 0, which CAS inputs to the DRAM SIM array are active? Over which data bus lines are data carried? Is a byte, word, or double-word data transfer taking place?

68. At what pin of the 82346 is the RES_{387} signal output? Where is this signal sent?

69. Over which data bus lines does the 80386DX MPU access the internal configuration registers of the 82346?

Section 13.7

70. Which circuit block of the 82344 receives the POWERGOOD signal that identifies that the power supply voltage is stable?

71. Make a list of the LSI peripheral devices that are implemented within the 82344 peripheral control block.

72. What are the outputs of the address decoder block of the 82344?

73. At what pins of the 82344 are the $\overline{BE_0}$ through $\overline{BE_3}$ signals located?

74. Is HLDA an input or output of the 82344? In the PC/AT circuit diagrams of Fig. 13.3, where is the HLDA pin of the 82344 connected?

75. If the microcomputer in Fig 13.3 is strapped for 16-bit BIOS, which EPROM sockets are populated? What type EPROM devices are normally used?

76. Over which data lines are information output to the 80386DX MPU when the 82344 is strapped for 16-bit BIOS?

77. If the microcomputer in Fig. 13.3 is configured to provide BIOS in two 27C256 EPROMs, what should be the setting of jumpers J_{30}, J_{31}, and J_{32}?

78. What is the purpose of the ISA bus connectors on the main processor board? How many 16-bit data bus slots are provided?

79. How large an address is implemented at the ISA bus?

80. At what pin of the ISA bus connector is the signal DRQ_0 input?

81. Which signal is output at pin B26 of ISA bus?

82. In the microcomputer of Fig. 13.3, what signal is applied to the VBAT input of the 82344? What is the source of this signal?

83. At what pin of the 82344 is the signal $SDLH/\overline{HL}$ output?

84. What does logic 0 at $\overline{XREAD}$ tell the data buffer?

85. What type of error is signaled to the 82344 when the $\overline{PCK}$ input switches to logic 0? Which interrupt is initiated due to this input?

86. Is $\overline{HIDRIVE}$ an input or output of the 82344? What happens when jumper J_{40} on sheet 4 of the circuits in Fig. 13.3 is installed?

87. Which of the 344/346 interface signals on sheet 4 of Fig. 13.3 is used to request a refresh operation of the 82346?

88. At what pin of the 82341 is the signal SINB output?

89. Which connector implements asynchronous serial port B in the PC/AT microcomputer of Fig. 13.3?

90. Which IC is used to buffer serial output SOUTB of the 82341 in the microcomputer of Fig. 13.3?

91. In the microcomputer of Fig. 13.3, at which pin of the COM B connector is the signal data set ready port B input?

92. At what pin of the 82341 is the select signal ($\overline{\text{SLIN}}$) output to the printer?

93. Is the signal $\overline{\text{ERR}}$ an input or output of the Centronics interface?

94. At what pin of the Centronics interface does the printer report the fact that a paper error has occurred?

95. Trace the destination of the TURBREQ output of the 82341.

96. Is the KKSW pin of the 82341 an input or output of the keyboard section? What does this signal stand for?

97. What level of interrupt is used to service the keyboard in the PC/AT microcomputer of Fig. 13.3?

98. What is the data path of the IDE interface in the microcomputer of Fig. 13.3?

99. Which ICs are the IDE data transceivers?

100. What level interrupt request is used to request service for the hard disk in the microcomputer of Fig. 13.3?

101. At what pin of the IDE connector is the signal $\overline{\text{HCS}}_1$ output?

Section 13.9

102. How many floppy disk drives can be controlled by the 82077AA?

103. What physical size drives can be operated with the 82077AA?

104. List four data-transfer rates supported by the 82077AA.

105. How large is the data bus FIFO of the 82077AA?

106. What is the period during which the 82077AA's data separator reads serial data called?

107. What technique does the 82077AA use to correct for bit jitter?

108. Is the 82077AA device in Fig. 13.3 treated as a memory-mapped or I/O-mapped peripheral?

109. What level interrupt is used to service the floppy disk in the microcomputer of Fig. 13.3?

110. Which DMA channel is used to service the floppy disk controller in the circuit of Fig. 13.3?

111. What frequency crystal is attached to the 82077AA floppy disk controller in Fig. 13.3?

112. How many floppy disk drives are supported with the interface at connector J_{57} in Fig. 13.3?

113. At what pin of connector J_{57} on sheet 5 of Fig. 13.3 is write data output to the floppy disk drive?

114. What does logic 1 at the INDX input of the 82077AA mean? At what pin of connector J_{57} in the microcomputer of Fig. 13.3 is this signal input?

PC/AT Bus Interfacing, Circuit Construction, Testing, and Troubleshooting

▲ 14.1 INTRODUCTION

In the last chapter we learned about the electronics of the PC/AT's main processor board. Here we continue our study of microcomputer electronics with circuits built using the PC/AT's ISA bus signals. This study includes the analysis, design, building, and testing of a variety of bus interface, input/output, and peripheral circuits. The following subjects are covered in this chapter:

1. PC/AT bus-based interfacing
2. The PCμLAB laboratory test unit
3. Experimentation with the on-board circuitry of the PCμLAB
4. Building, testing, and troubleshooting circuits
5. Observing microcomputer bus activity with a digital logic analyzer

▲ 14.2 PC/AT BUS-BASED INTERFACING

In this section, we examine some of the hardware that can be used to experiment with external circuitry in the PC/AT bus-based laboratory environment. That is, we will now work with circuitry that is not already available as part of the PC/AT's main processor board; instead, the circuits will be constructed external to the PC/AT. This includes prebuilt circuits that are readily available on PC/AT add-on boards, such as a serial communication interface, a parallel I/O expansion module, an analog-to-digital (A-to-D) converter, and a digital-to-analog (D-to-A) converter, or custom circuits that are hand-built on special prototyping boards. We call an experimental circuit that is built to test out a function a *prototype*.

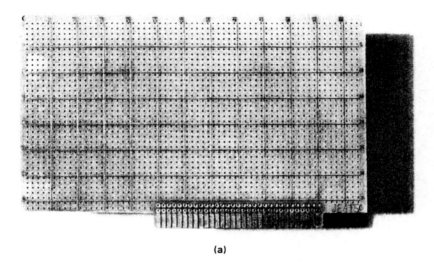

(a)

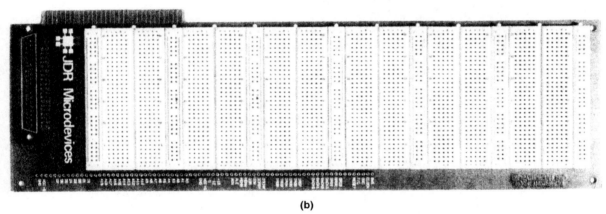

(b)

Figure 14.1 (a) Breadboard card. (b) Solderless breadboard card.

The ISA expansion bus is where additional circuits are added into the microcomputer of the PC/AT. A variety of different methods can be used to implement these circuits. That is, a wide range of hardware is available for building experimental circuits. Figures 14.1(a) and (b) shows just two examples. These cards are known as *breadboards*; that is, a card meant for prototyping circuits. The breadboard card in Fig. 14.1(a) requires the circuit to be constructed on the board by inserting the leads of the devices through the holes; then the leads of the devices are soldered to permanently connect them together. Similar boards are available where the devices are interconnected by wire wrapping instead of with solder. On the other hand, the module shown in Fig. 14.1(b) is what is known as a *solderless breadboard*. Here the components are plugged in and interconnected with jumper wires. Therefore, it is more practical in that the breadboard can be reused many times.

These prototyping modules are intended to be plugged directly into the bus slots of the PC/AT. However, this does not permit easy access to the circuits on the board for testing. One solution to this problem is the *board-extender card* shown in Fig. 14.2. The board extender is plugged into the slot in the PC/AT's main processor board and the card

Figure 14.2 Extender card.

with the experimental circuitry is plugged into the top of the extender card. In this way, the circuitry to be tested becomes more accessible because it is located above the case of the PC/AT.

The PC/AT add-on prototyping cards we just discussed are widely used in industry; however, they require the cover of the PC/AT to be left off and the testing of circuits to take place in close contact to the other circuitry within the PC/AT. In an educational environment, it is beneficial to have the complete experimental environment external to the PC/AT. Moving the breadboard outside permits easier access to the circuits for testing and modification and limits the risk of damage to the PC/AT.

The PCμLAB shown in Fig. 14.3 implements this type of laboratory environment. It is a bench-top laboratory test unit. The illustration in Fig. 14.4 shows that the PCμLAB uses a bus interface module that is permanently left inside the PC/AT. The interface board inside the PC/AT buffers all the bus signals. Cables carry the signals of the ISA expansion bus over to the PCμLAB breadboard unit. This unit has a large solderless breadboarding area for easy construction of circuits and connectors of a single ISA slot for using prebuilt add-on boards. This type of system offers a better solution for an educational microcomputer laboratory and will be used here for discussion.

▲ 14.3 THE PCμLAB LABORATORY TEST UNIT

In the last section we showed how the PCμLAB attaches to the personal computer. Here we examine the features it offers for experimentation in the laboratory. Earlier we indicated that it has a breadboard area for building circuitry and an ISA slot for plugging in a prebuilt board. It also has basic I/O devices such as *switches*, *LEDs*, and a *speaker*, and some *internal I/O interface circuitry*. This built-in I/O circuitry permits exploration of simple

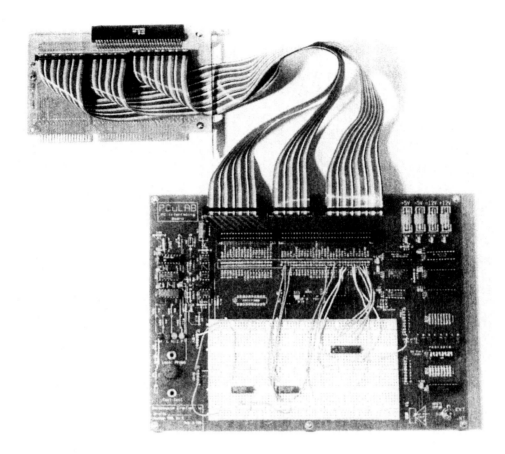

Figure 14.3 PCμLAB. (Reprinted with permission of Microcomputer Directions, Inc., P.O. Box 15127, Fremont, CA 94539)

parallel I/O techniques, such as reading switches as inputs, lighting LEDs as outputs, polling a switch input, and generating tones at the speaker, without having to build any circuitry.

The layout of the top of the PCμLAB is shown in detail in Fig. 14.5. We will begin by identifying the input/output devices. On the right side we find both a block of eight switches, labeled 0 through 7, and a row of eight red LEDs, 0 through 7. The switches can be used to supply inputs, and the LEDs can be used to produce outputs for either the built-in circuits or circuitry constructed on the breadboard area. The INT/EXT switch determines the use of these I/O devices. When it is in the INT (internal) position, they are connected

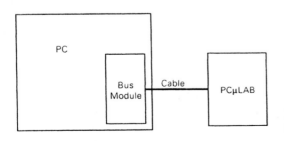

Figure 14.4 PCμLAB system configuration.

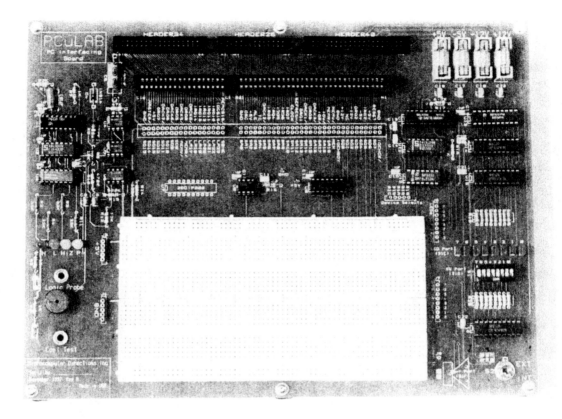

Figure 14.5 PCμLAB top panel. (Reprinted with permission of Microcomputer Directions, Inc., P.O. Box 15127, Fremont, CA 94539)

directly to the on-board circuits. However, moving the switch to the EXT (external) position makes them available for connection to circuits implemented on the breadboard area.

All the signals of the PC/AT's ISA expansion bus are made available at the connectors at the top of the front panel. Here we find that the signals are supplied at the PC/AT slot and a receptacle connector. The slot is the connector into which prebuilt boards are inserted. Figure 14.6 shows the PCμLAB with an add-on card inserted for testing. The receptacle connector is provided to permit connection of the bus signals to circuits built on the breadboard. For ease of use, mnemonics for all signals are labeled next to the connector. The table in Fig. 14.7 identifies the signal name for each of these mnemonics and whether it is an input or output.

Let us next look at the breadboarding area. This area permits the experimenter to build and test custom circuits. Figure 14.8 shows a circuit constructed on the breadboard area of the PCμLAB.

Looking at Fig. 14.5, we see that the breadboard area is implemented with two solderless breadboards. For this reason, it permits installation of two rows of ICs. A drawing of the electrical connection of the wire insertion clips is shown in the PCμLAB circuit layout master of Fig. 14.9. Notice that the column of five vertical clips from a device pin are internally attached together. One is used up when the IC is inserted and the other four are for use in making jumper wire connections to other circuits. The jumpers must be made with wire that is rated 26 AWG (American wire gauge).

Figure 14.6 PCμLAB with add-on card. (Reprinted with permission of Microcomputer Directions, Inc., P.O. Box 15127, Fremont, CA 94539)

At both the top and bottom of the board, there are two horizontal rows of attached wire insertion clips. These four rows of clips are provided for power supply distribution. Two rows are intended to implement the +5-V power supply bus, and the other two are meant to be used as the common ground bus. The power supply for the circuit can be picked up with jumpers from the ISA bus connector or at the separate power supply terminal strip. Notice in Fig. 14.7 that +5 V is available at contacts B_3 and B_{29} of the ISA bus connector.

EXAMPLE 14.1

Which contacts of the ISA bus interface connector can be used as ground points?

Solution

Looking at Fig. 14.7, we see that ground (GND) is provided by contacts B_1, B_{10}, and B_{31} of the I/O interface connector.

Earlier we pointed out that the switches, LEDs, and speaker supply inputs and outputs for the on-board circuitry and can also be connected to circuits built on the breadboard. In both cases, I/O addresses are output over the address bus part of the ISA bus interface, A_0 through A_{15}, and data are input or output over the data bus lines D_0 through D_{15}.

Pin	Name	Type
A1	I/O CH CK	I
A2	D7	I/O
A3	D6	I/O
A4	D5	I/O
A5	D4	I/O
A6	D3	I/O
A7	D2	I/O
A8	D1	I/O
A9	D0	I/O
A10	I/O CH RDY	I
A11	AEN	O
A12	A19	O
A13	A18	O
A14	A17	O
A15	A16	O
A16	A15	O
A17	A14	O
A18	A13	O
A19	A12	O
A20	A11	O
A21	A10	O
A22	A9	O
A23	A8	O
A24	A7	O
A25	A6	O
A26	A5	O
A27	A4	O
A28	A3	O
A29	A2	O
A30	A1	O
A31	A0	O

Pin	Name	Type
B1	GND	
B2	RESET DRV	O
B3	+5 V	
B4	IRQ2	I
B5	−5 V	
B6	DRQ2	I
B7	−12 V	
B8	RESERVED	
B9	+12 V	
B10	GND	
B11	SMEMW	O
B12	SMEMR	O
B13	IOW	O
B14	IOR	O
B15	DACK3	O
B16	DRQ3	I
B17	DACK1	O
B18	DRQ1	I
B19	REFRESH	O
B20	CLOCK	O
B21	IRQ7	I
B22	IRQ6	I
B23	IRQ5	I
B24	IRQ4	I
B25	IRQ3	I
B26	DACK2	O
B27	T/C	O
B28	ALE	O
B29	+5 V	
B30	OSC	O
B31	GND	

Pin	Name	Type
C1	SBHE	O
C2	LA23	O
C3	LA22	O
C4	LA21	O
C5	LA20	O
C6	LA19	O
C7	LA18	O
C8	LA17	O
C9	MEMR	O
C10	MEMW	O
C11	SD08	I/O
C12	SD09	I/O
C13	SD10	I/O
C14	SD11	I/O
C15	SD12	I/O
C16	SD13	I/O
C17	SD14	I/O
C18	SD15	I/O

Pin	Name	Type
D1	MEM CS 16	I
D2	IO CS 16	I
D3	IRQ10	I
D4	IRQ11	I
D5	IRQ12	I
D6	IRQ13	I
D7	IRQ14	I
D8	DACK0	O
D9	DRQ0	I
D10	DACK5	O
D11	DRQ5	I
D12	DACK6	O
D13	DRQ6	I
D14	DACK7	O
D15	DRQ7	I
D16	+5 V	
D17	MASTER	I
D18	GND	

Figure 14.7 ISA bus interface signals.

Figure 14.8 Breadboard circuit. (Reprinted with permission of Microcomputer Directions, Inc., P.O. Box 15127, Fremont, CA 94539)

EXAMPLE 14.2

At what contacts of the ISA bus connector are data bus lines D_0 through D_7 available?

Solution

In Fig. 14.7 we find that the data bus lines are supplied at contact A_2 through A_9 of the connector.

The PCμLAB also has built-in circuit test capability. It has both a continuity tester and a logic probe. Looking at Fig. 14.5, we see that the probes for the continuity tester are inserted into the female connectors identified as *CT*. The continuity tester is useful in debugging circuit connections. That is, it can be used to verify whether or not two points of a circuit are wired together. This is done by attaching one of the probes to the first point in the circuit and then touching the other probe to the second point. If they are connected together, the buzzer sounds. Care must be taken to assure that the power is not applied to the circuit under test while continuity tests are being made; otherwise, the tester circuits may be damaged.

The purpose of the logic probe is not to verify circuit connections; instead, it is used to determine the logic level of signals at various test points in a circuit. The probe that is

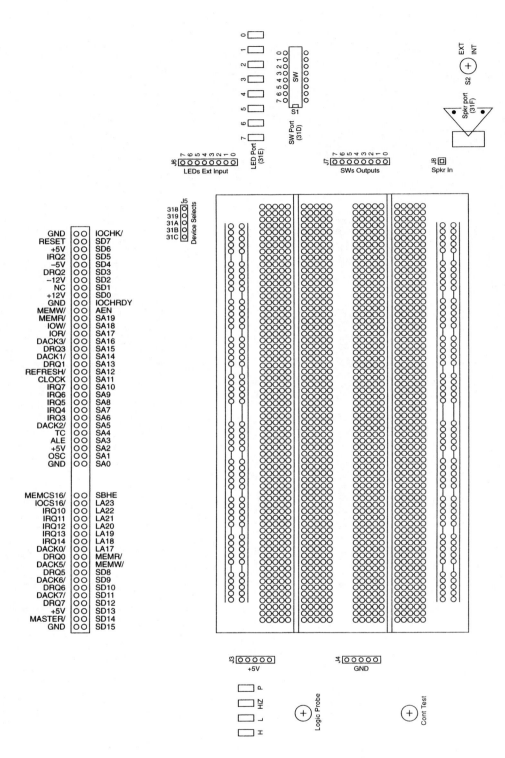

Figure 14.9 Circuit layout master. (Reprinted with permission of Microcomputer Directions, Inc., P.O. Box 15127, Fremont, CA 94539)

754

Logic level	LED	Voltage
1	Red	$V > 2.4$ V
Hi–Z	Amber	0.4 V $< V < 2.4$ V
0	Green	$V < 0.4$ V

Figure 14.10 Logic state voltage levels.

used to input the signal from the circuit is inserted into the LP connector. Then, the probe is touched to the test point in the circuit. Based on the logic level of this signal, the red, green, or amber LED lights. Here red stands for logic 1, green is logic 0, and amber is the high-Z state. The table in Fig. 14.10 shows the voltage levels corresponding to these three states. The second red LED, which is marked P, identifies that the signal at the test point is pulsing. By pulsing, we mean a signal, such as a square wave, that is repeatedly switching back and forth between the 0 and 1 logic levels.

▲ 14.4 EXPERIMENTING WITH THE ON-BOARD CIRCUITRY OF THE PCμLAB

The on-board circuitry of the PCμLAB implements simple parallel input/output interfaces. Having this circuitry built into the experimental board allows us to examine some basic I/O techniques without having to take the time to construct the circuitry. These interface circuits provide the ability to input the settings of the switches, light the LEDs, or sound a tone at the speaker. Earlier we pointed out that this circuitry becomes active when the INT/EXT switch is set to the INT position. In this section, we describe the design and operation of the internal (on-board) circuits. The operation of these circuits is illustrated using several input and output examples.

I/O Address Decoding

Let us begin our study of the on-board circuitry with the address decoder circuit shown in Fig. 14.11(a). Here we find that a 74LS688 parity generator/checker IC (U_{10}), a 74LS138 3-line-to-8-line decoder IC (U_{11}), and a 74LS32 quad-OR gate IC (U_{12}) perform the address decode function. They accept as inputs address bits A_0 through A_9 and the address-enable (AEN) control signal. These inputs are directly picked up from the ISA expansion bus. As shown in Fig. 14.11(b), they correspond to the signals available at contacts A_{22} through A_{30} and A_{11} of the on-board ISA bus connector.

At the other side of the circuit, we find three I/O-select outputs. They are labeled $\overline{IORX31D}$, $\overline{IOWX31E}$, and $\overline{IOWX31F}$ and stand for *I/O read address X31DH, I/O write address X31EH,* and *I/O write address X31FH,* respectively. These signals are used to select between the on-board I/O devices: switches, LEDs, or speaker. For instance, Fig. 14.11(c) shows that output $\overline{IORX31D}$ is used to enable input of the state of the switch settings.

In Fig. 14.11(b) we see that the address bits available at the ISA bus interface are A_0 through A_{23}. However, just the lower 16 address lines A_0 through A_{15} are used in I/O addressing, and many of these address bits are not used in the on-board address-decoder circuit. For this reason, the unused address bits are considered don't-care states. Therefore, the decoded address is

$$A_{15} \ldots A_0 = XXXXXXA_9A_8A_7A_6A_5A_4A_3A_2A_1A_0$$

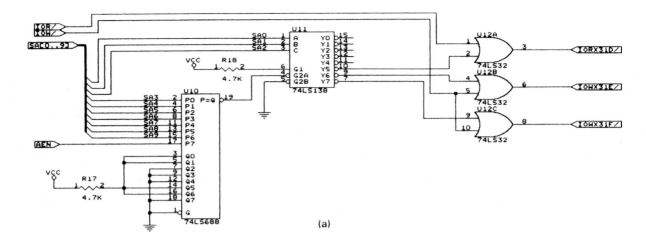

Figure 14.11(a) Address decoder circuit. (Reprinted with permission of Microcomputer Directions, Inc., P.O. Box 15127, Fremont, CA 94539)

Since many address bits are don't-care states, the outputs of the decoder do not correspond to unique addresses. Instead, a large number of I/O addresses decode to produce each chip-select output.

Next we will look at how the higher-order address bits are decoded by the 74LS688 comparator. Looking at Fig. 14.11(a), we see that inputs P_0 through P_7 are supplied by address signals A_3 through A_9 and AEN. This gives

$$P_7 P_6 \ldots P_0 = AEN A_9 A_8 A_7 A_6 A_5 A_4 A_3$$

On the other hand, the Q inputs are set at fixed logic levels and represent

$$Q_7 Q_6 \ldots Q_0 = 01100011_2$$

The circuit within the 74LS688 compares the address information at the P inputs to the fixed code at the Q inputs, and if they are equal, it switches the P = Q output to logic 0. This means that the address on the bus corresponds to an on-board I/O device. This output is applied to the G_{2A} input of the 74LS138 decoder and enables it for operation. In this way, we see that all addresses with

$$A_{15} \ldots A_0 = XXXXXX1100011 A_2 A_1 A_0 \quad \text{along with} \quad AEN = 0$$

map to the PCμLAB's on-board I/O address space.

Once the 74LS138 decoder is enabled, the code on address lines A_0 through A_2 is used to produce the appropriate output. From the circuit diagram in Fig. 14.11(a), we find that the codes that produce the enable signals for the switches, LEDs, and speaker are

$$A_2 A_1 A_0 = 101_2 \quad \text{and} \quad \text{active } \overline{IOR} \quad \text{produces} \quad \overline{IORX31D}$$

$$A_2 A_1 A_0 = 110_2 \quad \text{and} \quad \text{active } \overline{IOW} \quad \text{produces} \quad \overline{IOWX31E}$$

$$A_2 A_1 A_0 = 111_2 \quad \text{and} \quad \text{active } \overline{IOW} \quad \text{produces} \quad \overline{IOWX31F}$$

I/O CH CK	A1	B1	GND
D7	A2	B2	RESET DRV
D6	A3	B3	+5 V
D5	A4	B4	IRQ2
D4	A5	B5	−5 V
D3	A6	B6	DRQ2
D2	A7	B7	−12 V
D1	A8	B8	RESERVED
D0	A9	B9	+12 V
I/O CH RDY	A10	B10	GND
AEN	A11	B11	$\overline{\text{SMEMW}}$
A19	A12	B12	$\overline{\text{SMEMR}}$
A18	A13	B13	$\overline{\text{IOW}}$
A17	A14	B14	$\overline{\text{IOR}}$
A16	A15	B15	$\overline{\text{DACK3}}$
A15	A16	B16	DRQ3
A14	A17	B17	$\overline{\text{DACK1}}$
A13	A18	B18	DRQ1
A12	A19	B19	$\overline{\text{REFRESH}}$
A11	A20	B20	CLOCK
A10	A21	B21	IRQ7
A9	A22	B22	IRQ6
A8	A23	B23	IRQ5
A7	A24	B24	IRQ4
A6	A25	B25	IRQ3
A5	A26	B26	$\overline{\text{DACK2}}$
A4	A27	B27	T/C
A3	A28	B28	ALE
A2	A29	B29	+5 V
A1	A30	B30	OSC
A0	A31	B31	GND

$\overline{\text{SBHE}}$	C1	D1	$\overline{\text{MEM CS16}}$
LA23	C2	D2	$\overline{\text{IO CS16}}$
LA22	C3	D3	IRQ10
LA21	C4	D4	IRQ11
LA20	C5	D5	IRQ12
LA19	C6	D6	IRQ13
LA18	C7	D7	IRQ14
LA17	C8	D8	$\overline{\text{DACK0}}$
$\overline{\text{MEMR}}$	C9	D9	DRQ0
$\overline{\text{MEMW}}$	C10	D10	$\overline{\text{DACK5}}$
SD08	C11	D11	DRQ5
SD09	C12	D12	$\overline{\text{DACK6}}$
SD10	C13	D13	DRQ6
SD11	C14	D14	$\overline{\text{DACK7}}$
SD12	C15	D15	DRQ7
SD13	C16	D16	+5 V
SD14	C17	D17	MASTER
SD15	C18	D18	GND

Figure 14.11(b) Bus signals.

I/O device	Type	Device-Select Signal	Address
Switches	Input	$\overline{\text{IORX31D}}$	031DH
LEDs	Output	$\overline{\text{IOWX31E}}$	031EH
Speaker	Output	$\overline{\text{IOWX31F}}$	031FH

Figure 14.11(c) Output signals.

This results in the device addresses as listed in Fig. 14.11(c). For example, reading of the switches is enabled by any address that is of the form

$$A_{15} \ldots A_0 = XXXXXX1100011101_2 \quad \text{provided} \quad AEN = 0.$$

Some examples of valid addresses are $031D_{16}$, $F31D_{16}$, $FF1D_{16}$, and $0F1D_{16}$. All these addresses make the Y_5 output of the decoder circuit switch to logic 0. Notice that this output is gated with the ISA expansion bus signal $\overline{IOR}$ by OR gate U_{3A}. In this way, the $\overline{IORX31D}$ output can be active only during input (I/O read) bus cycles.

EXAMPLE 14.3

Which output chip select does the I/O address $F71F_{16}$ produce when applied to the input of the circuit in Fig. 14.11(a)? What type of bus cycle must be in progress to produce this chip select output?

Solution

First, the address expressed in binary form is

$$F71F_{16} = 1111011100011111_2$$

Considering the lower 10 bits, we get

$$A_9 \ldots A_0 = 1100011111_2 = 31F_{16}$$

Tracing the circuit, we find that this address-bit combination makes the Y_7 output of U_{11} equal to logic 0. Y_7 is gated with bus signal $\overline{IOW}$ to produce the $\overline{IOWX31F}$ output. Therefore, $\overline{IOWX31F}$ is at its active 0 logic level as long as an output bus cycle is taking place.

Switch Input Circuit

The interface of switches S_0 through S_7 to the data bus is shown in Fig. 14.12. Notice that one contact from each of the eight switches is connected to ground (0 V). The other contact on each switch is supplied to one of the resistors in resistor pack R_{20}, and the other end of each resistor is supplied by V_{cc} (+5 V). The connections between the resistors and the switch contacts are supplied as inputs to the data bus through the 74LS240 inverting buffers of IC U_{16}. For instance, S_0 is supplied from input A_4 of buffer U_{16A} to output Y_4 and onto data bus line D_0. Similarly, the state of switch 7 is passed through the inverter at input A_1 of U_{16B} to data bus line D_7. Because inverting buffers are used in the circuit, logic 0 is applied to the data bus whenever a switch is open and logic 1 is put on the data bus if a switch is closed.

In the circuit diagram, we see that the switch input buffer is enabled by the signal $\overline{IORX31D}$. Earlier we showed that this signal is at its active 0 logic level whenever an input bus cycle is performed to address $031D_{16}$. But, remember that the complete I/O address is not decoded; therefore, many other addresses also decode to enable this buffer.

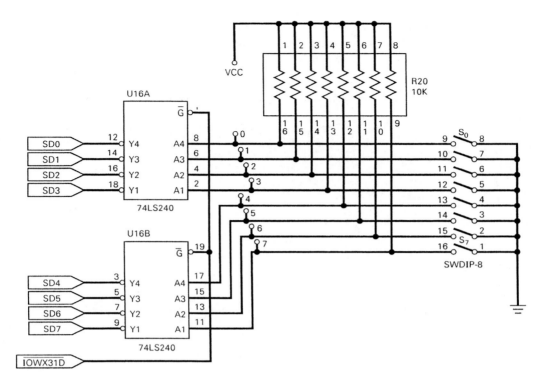

Figure 14.12 Switch input interface circuit. (Reprinted with permission of Microcomputer Directions, Inc., P.O. Box 15127, Fremont, CA 94539)

The state of the switches can be read with an INPUT command. For instance, the DEBUG command

```
I   31D   (↵)
```

causes the settings of all eight switches to be displayed as a hexadecimal byte. In this byte, the most significant bit represents the state of S_7 and the least significant bit that of S_0. Remember that a bit at logic 1 means a closed switch and logic 0 an open switch.

Let us now look at how to read the status of the switches into the accumulator of the MPU. This is done by simply executing an IN instruction. Therefore, after executing the instruction sequence

```
MOV   DX,31DH
IN    AL,DX
```

the switch setting is held in AL.

In practical applications, it is common to want to determine the setting of a single switch. This can be done by additional processing of the byte in AL. For instance, to find the setting of S_7, we can use the instruction

```
AND   AL,80H
```

Execution of this instruction ANDs the contents of AL with the value 80_{16}. Therefore, the result in AL will be 10000000_2 if switch 7 is closed or 00000000_2 if it is open. That is, the zero flag (ZF) will be 1 if SW_7 is open and 0 if it is closed.

EXAMPLE 14.4

Write a program that will poll S_0 waiting for it to be closed. Use a shift instruction to isolate and determine the setting of switch 0. Use a valid address other than $31D_{16}$ to read the setting of the switches.

Solution

The setting of the switches can be input to AL with the instructions

```
           MOV   DX,0FF1DH
    POLL:  IN    AL,DX
```

Here we have used $FF1D_{16}$ as the I/O address for the switch port. Now the setting of switch 0, which is in the bit 0 position, is shifted into the carry flag (CF) with the instruction

```
           SHR   AL,1
```

Finally, the setting of the switch is tested for 0 (open) by testing the carry flag with the instruction

```
           JNC   POLL
```

If CF is 0, the switch is open and the poll loop is repeated. But if the switch is closed, CF is 1; the poll loop is complete; and the instruction following JNC is executed.

LED Output Circuit

Let us next look at the output circuit that drives the LEDs. Figure 14.13 shows the drive circuitry for LEDs 0 through 7. Here we see that the anode side of the individual LEDs are all connected in parallel and supplied by +5 V. On the other hand, the cathode sides of the LEDs are wired through separate resistors of resistor pack R_{19} to the outputs of the 74LS240 LED drive buffer (U_{14}). For example, the cathode of LED 0 connects through the uppermost 330-Ω resistor to the Y_1 output of IC U_{14A}. The inputs to the inverting buffer are supplied by the outputs of the 74LS374 LED port latch (U_{13}).

To light an LED, an output bus cycle must be performed to load logic 1 into the corresponding bit of the LED port latch. As identified earlier, the I/O address accompanying this data and $\overline{IOW}$ must decode to produce logic 0 at $\overline{IOWX31E}$. Notice that this signal is used to clock the data on data bus lines D_0 through D_7 into the 74LS374 latch. The bits at the output of the latch are inverted by the 74LS240 buffer. Logic 0 at any output of the buffer provides a path to ground for the corresponding LED, and thus turns it on.

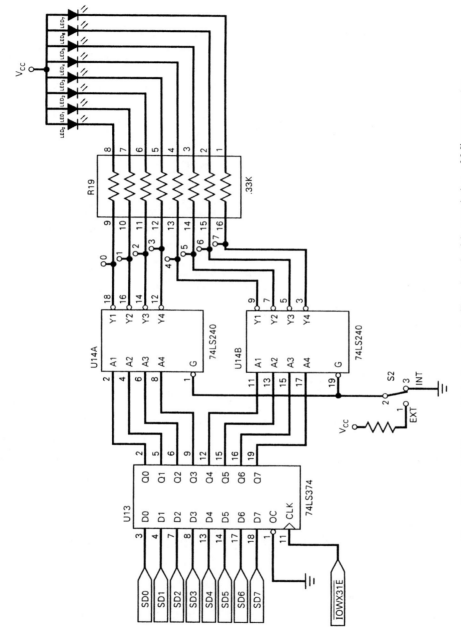

Figure 14.13 LED output interface circuit. (Reprinted with permission of Microcomputer Directions, Inc., P.O. Box 15127, Fremont, CA 94539)

To try out the LEDs on the PCμLAB, we can turn them all on by issuing a single OUT command from DEBUG. This command is

$$O \quad 31E \quad FF \quad (\lrcorner)$$

They can be turned off with the command

$$O \quad 31E \quad 00 \quad (\lrcorner)$$

EXAMPLE 14.5

Write a program that will blink LED 7.

Solution

To turn on LED 7, 80_{16} must be output to the LED port latch, and it is turned off by outputting 00_{16}. Therefore, we begin with the instructions

```
            MOV   DX,31EH
            MOV   AL,80H
ON_OFF:     OUT   DX,AL
```

Next we need to delay for a period of time before turning the LED off. To do this, a count is loaded into CX and a time delay is implemented using a LOOP instruction.

```
            MOV   CX,0FFFFH
HERE:       LOOP  HERE
```

The duration of the time delay can be adjusted by simply changing the value loaded into CX.
After the time delay has elapsed, the value in bit 7 of AL is inverted with the instruction

```
            XOR   AL,80H
```

The new contents of AL equal 00_{16}. This value will cause LED 7 to turn off. Finally, a JMP instruction returns program control to ON_OFF

```
            JMP   ON_OFF
```

and the loop repeats. The complete program is shown in Fig. 14.14.

Speaker-Drive Circuit

The speaker-drive circuit of Fig. 14.15 is an output interface. It is implemented with the 74LS74 data latch device (U_9) and the 75477 speaker driver (U_8). The tone to be sounded at the speaker can be generated under software control.

Let us begin by studying how a tone signal can be generated by the MPU and sent to the speaker. A tone is produced by applying a square wave to the speaker. This signal is output over data bus line D_0 to the D_1 input at pin 2 of U_{9A}. Looking at the circuit

```
            MOV   DX,31EH

            MOV   AL,80H

ON_OFF:     OUT   DX,AL

            MOV   CX,0FFFFH

HERE:       LOOP  HERE

            XOR   AL,80H

            JMP   ON_OFF
```

Figure 14.14 LED blink program. (Reprinted with permission of Microcomputer Directions, Inc., P.O. Box 15127, Fremont, CA 94539)

diagram, we find that the tone is passed from the Q_1 output at pin 5 of U_9 through the EXT/INT switch to the 2A input at pin 7 of U_8. It is then output at pin 6 (2Y) of U_8 and sent through resistor R_{16} to the speaker. The other end of the speaker's coil is connected to $+V_{cc}$.

The square wave can be generated with a program similar to the one we used to blink LED 7. However, the data is output on data bus line D_0 rather than D_7, and it must be accompanied by address $31F_{16}$ instead of $31E_{16}$. This gives the program

```
            MOV   DX,31FH
            MOV   AL,01H
ON_OFF:     OUT   DX,AL
            MOV   CX,0FFFFH
HERE:       LOOP  HERE
            XOR   AL,01H
            JMP   ON_OFF
```

The pitch of the tone can be changed by varying the frequency of the square wave. This is done by adjusting the duration of the time delay; that is, changing the count that is loaded into CX. The lower the count, the higher the pitch.

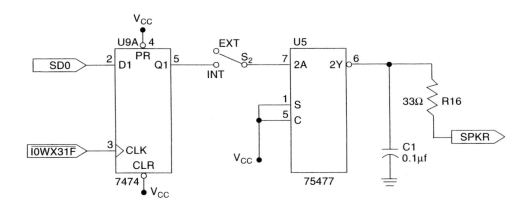

Figure 14.15 Speaker-drive circuit. (Reprinted with permission of Microcomputer Directions, Inc., P.O. Box 15127, Fremont, CA 94539)

Sec. 14.4   Experimenting with the On-Board Circuitry of the PCμLAB   **763**

▲ 14.5 BUILDING, TESTING, AND TROUBLESHOOTING CIRCUITS

In the last section, we examined the on-board circuits of the PCμLAB. Here we turn our attention to circuits that can be built on the breadboarding area of the PCμLAB. First, we look at how circuits are constructed, then how their operation is tested, and finally, troubleshooting techniques that can be used if the circuit does not work.

Building a Circuit

Earlier we showed that the breadboard area is where custom circuits can be built and that it allows for mounting of two rows of ICs. Assuming that a schematic diagram of the circuit to be built is already available, the first step in the process of building the circuit is to make a layout diagram to show how the circuit will be constructed on the breadboard. This drawing can be made on a circuit layout master similar to the one shown in Fig. 14.9.

Figure 14.16(a) is the diagram for a circuit that implements a parallel output port to drive LED_0. This is the circuit we will use to illustrate the method used to breadboard a circuit. Notice that the circuit uses a 74LS138 3-line-to-8-line decoder, 7400 quad 2-input NAND gate, and a 74LS374 octal latch. The 74LS240 inverting buffer, 330-Ω resistor, and LED_0 are supplied by the on-board circuitry of the PCμLAB by using input 0 of receptacle J_6. We begin by marking the IC pin numbers for each of the inputs and outputs into the circuit diagram. For instance, from the pin layout of the 74LS138 in Fig. 14.17, we find that

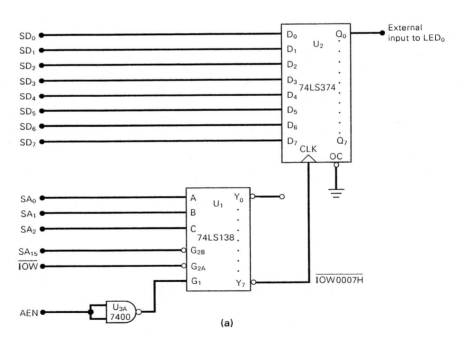

Figure 14.16(a)   LED drive circuit. (Reprinted with permission of Microcomputer Directions, Inc., P.O. Box 15127, Fremont, CA 94539)

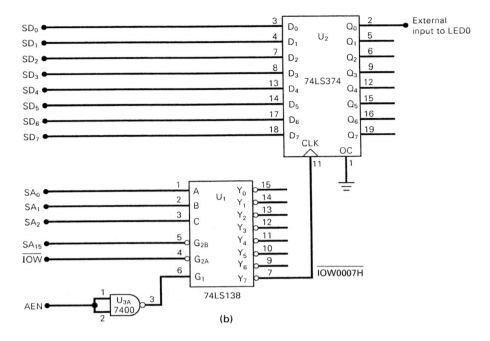

Figure 14.16(b) Schematic with pin numbers marked. (Reprinted with permission of Microcomputer Directions, Inc., P.O. Box 15127, Fremont, CA 94539)

its A, B, and C inputs are at pins 1, 2, and 3, respectively. Moreover, the Y_7 output is identified as pin 7. This is done for each IC to give the circuit shown in Fig. 14.16(b).

Now we are ready to make the layout drawing that shows how the circuits will be laid out on the PCμLAB's breadboard area. To do this, we simply draw the ICs and pin connections onto one of the circuit layout masters. Figure 14.16(c) shows the layout for our test circuit. Looking at this diagram, we see that +5 V is picked up by inserting a jumper between connector J_3 and one of the power bus lines of the breadboard. Ground is supplied in a similar way from connector J_4 to another power bus line. Then +5 V and GND can be jumpered from IC to IC. This completes the power distribution for the circuit.

Let us next look at how the inputs and outputs of the circuit are provided. At the output side, a jumper is used to connect the Q_0 output of the 74LS374 to the 0 input of connector J_6. The data input at pin 3 of the 74LS374 latch is picked up with a jumper to data bus line D_0 at pin A_9 of the ISA bus connector. The output of the 74LS138 decoder (pin 7) that supplies the clock to the latch is jumpered to pin 11 of the 74LS374 IC. Notice that the AEN input is inverted with a NAND gate. The inverter is formed by connecting pins 1 and 2 of the 7400 IC and then applying AEN to pin 1. The inverted output at pin 3 of the 7400 IC is supplied to the G_1 input at pin 6 of the 74LS138 decoder IC.

EXAMPLE 14.6

Use the circuit diagram layout in Fig. 14.16(c) to identify which pins of the ISA bus connector are used to supply the address signals to the A, B, and C inputs of the 74LS138 decoder.

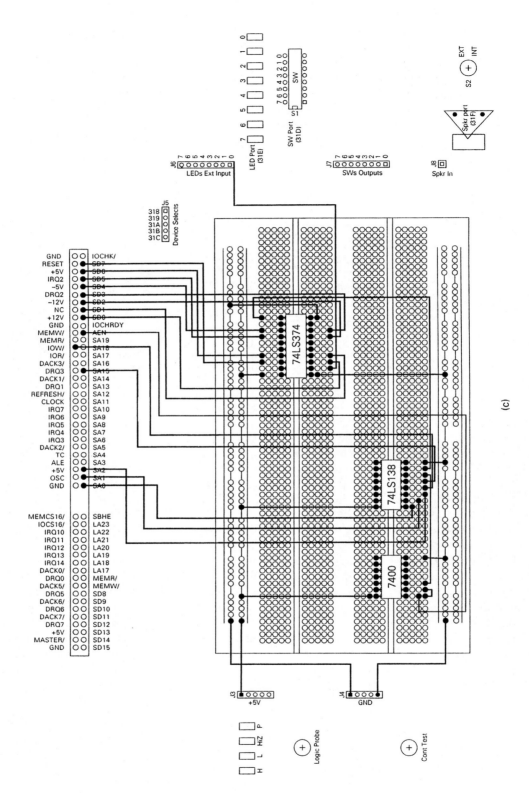

Figure 14.16(c) Completed layout master. (Reprinted with permission of Microcomputer Directions, Inc., P.O. Box 15127, Fremont, CA 94539)

(c)

766

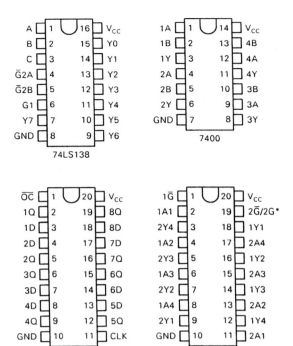

Figure 14.17 Pin layouts for the 74LS138, 7400, 74LS374, and 74LS240.

Solution

In the circuit diagram, we find that the A, B, and C inputs of the decoder are attached to ISA bus address lines A_0, A_1, and A_2, respectively. These signals are picked up at pins A_{31}, A_{30}, and A_{29} of the ISA bus connector.

The layout drawing in Fig. 14.16(c) is our plan for constructing the circuit. This drawing and the schematic diagram marked with pin numbers serve as valuable tools when testing and troubleshooting circuits. Figure 14.8 shows the breadboard of this example circuit.

Testing the Operation of a Circuit

Now that the circuit has been constructed, we are ready to check out its operation. The process of checking out how an electronic circuit works is called *testing*. To test a circuit, we must first know the events that should take place when it is functioning correctly. Usually this means that it will produce certain outputs. These outputs may be a visual event, such as lighting an LED, an audible event, such as sounding a tone, a mechanical event, such as positioning a mechanism, or simply a signal waveshape that can be observed with an instrument. For instance, the function of our example circuit in Fig. 14.16(a) is to light an LED.

To test the operation of our example circuit we can simply turn on the LED or make it blink. However, to do this, the LED output interface must be driven with software. In this way, we see that to test microcomputer interface circuits, they must be driven with software. In fact, they are normally driven by a special piece of software that is specifically written to exercise the interface. This segment of program is sometimes referred to as a *diagnostic program.*

The diagnostic routine does not have to be complex. For instance, to turn on LED_0 in our breadboard circuit, we can simply execute the instructions

```
MOV    DX,0007H
MOV    AL,01H
OUT    DX,AL
```

Note that the I/O address 0007_{16} along with active $\overline{IOW}$ and inactive AEN produce an active low pulse at output Y_7 of the decoder. This pulse is used to clock the data into the octal latch to make the Q_0 output become logic 1. A software routine that will blink LED_0 is as follows:

```
            MOV    DX,0007H
            MOV    AL,01H
ON_OFF:     OUT    DX,AL
            MOV    CX,0FFFFH
HERE:       LOOP   HERE
            XOR    AL,01H
            JMP    ON_OFF
```

The operation of a circuit that produces an electrical output can be tested with instrumentation. For instance, if the circuit we just constructed did not include an LED at the output, we would need to observe the signal at the Q_0 output (pin 2) of the 74LS374 latch. Accessories, such as *IC test clips,* are available to provide easy attachment of instruments to the pins of an IC. Figure 14.18 shows some IC test clips. This type of clip is spring loaded

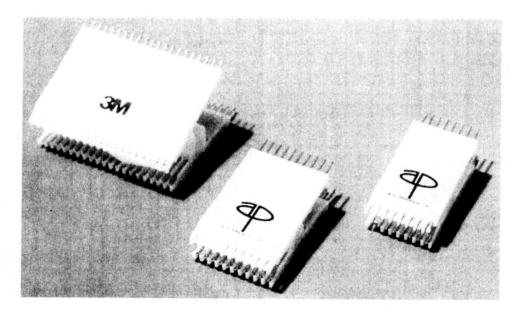

Figure 14.18   IC test clips.

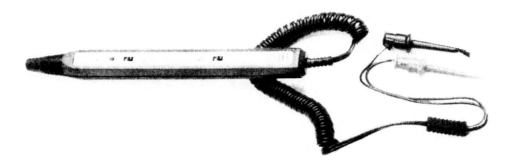

Figure 14.19(a) Logic probe.

and snaps tightly over the top of the IC. Instruments are connected to its pins instead of to those of the IC. For our example of the 74LS374 IC, a 20-pin IC test clip would be attached and then the probe of the instrument clipped onto pin 2 at the top of the test clip.

Various instruments are available to test the electrical signals in a circuit. Figure 14.19(a), (b), and (c) shows three examples, the *logic probe, multimeter,* and *oscilloscope,* respectively. The logic probe is a hand-held instrument that can be used to observe the logic level at a test point in a microcomputer circuit. As we pointed out earlier, a logic probe is built into the PCμLAB. This instrument has the ability to tell whether the signal tested is in the 0, 1, or high-Z logic state or if it is pulsating. The probe is simply touched to the point in the circuit where the signal is to be observed and the logic level is signaled by one of the LEDs.

In microcomputer circuits, the multimeter is useful for measuring static voltage levels. For instance, it can be used to verify that +5 V is applied to each of the ICs. A multimeter can also be used to measure the logic levels at inputs and outputs, but they must be stable voltages, not pulsating signals. In this case, the meter displays the amount of voltage at the test point and from this value we can determine whether the signal is at logic 0, at logic 1, or in the high-Z state.

Most multimeters also have the ability to measure resistance, AC voltage, and AC and DC current. For instance, it could be used to find the amount of current the circuit draws from the V_{cc} supply.

We just mentioned that the logic probe can tell if the signal at a point under test is pulsating. However, in this case, a better instrument for observing the operation of the circuit is an oscilloscope (or *scope,* as it is better known). The scope is the most widely used instrument for observing *periodic signals.* That is, signals, such as a square wave, that have a repeating pattern.

The scope displays the exact waveform of the pulsating signal on its screen. This type of representation gives us much more information. For instance, we can find the value of the high voltage (logic 1), the value of the low voltage (logic 0), how long the signal is at the 0 and 1 logic levels, and the shape of the signal as it transitions back and forth between 0 and 1. Figure 14.20(a) shows the shape of the signal produced at the Q_0 output at pin 2 of the 74LS374 when the diagnostic program that blinks LED_0 is running. The display of the waveform allows us to measure the period (T) of the square wave and calculate its frequency using

$$f = 1/T$$

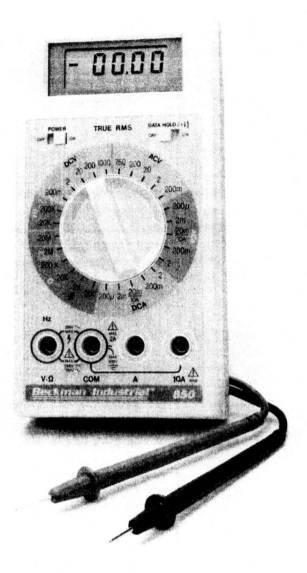

Figure 14.19(b)   Digital multimeter.

Most scopes have the ability to display several signals on the screen at the same time. That is, they have several *signal channels*. The most common scope in use is the *dual-trace scope*, which has the ability to simultaneously display two signals. Figure 14.20(b) shows both the output square-wave and clock input signals of our test circuit. In this example, the scope has been set up to synchronize the display of the square-wave output, which is applied to channel 1, to the clock input at channel 2. For this reason, the waveforms represent their true relationship in time. Notice that a clock pulse is associated with the loading of each

Figure 14.19(c) Oscilloscope.

771

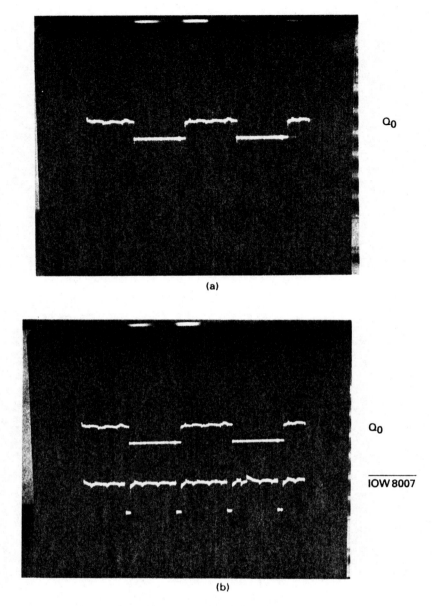

Q_0

Q_0

$\overline{\text{IOW}}\ 8007$

(a)

(b)

Figure 14.20 (a) Q_0 output waveform. (b) Clock input and square-wave output.

logic 0 and logic 1 into the Q_0 output of the latch. This mode of operation is known as using an *external sync*. That is, the sweep of the scope is initiated by an external signal, which in this case is the clock input.

Troubleshooting Microcomputer Interface Circuitry

In the testing of the operation of a circuit, we may find that it does not work. That is, it does not perform the function for which it was designed. In this case, we must identify what is the cause of the malfunction and then correct the problem. The process of finding the cause of a malfunction is called *troubleshooting*, and the process of correcting the problem is known as *repair*. In this section we look at some causes of malfunctions in circuits and then outline methods that can be used to troubleshoot microcomputer interface circuits.

The cause of problems found in malfunctioning circuits depends on the type of circuit being tested. In general, electronic circuits fall into several categories. A first example is a breadboard of a new circuit design. In this case, the circuit may be working correctly but not perform the function for which it was designed. That is, the malfunction may simply be due to the fact that a mistake was made in the design. This may be the most complicated type of failure to find and resolve.

The breadboard of a circuit we use in our laboratory exercises in this book is a second example. Here the circuit is known to operate correctly. For this reason, the most common causes for a circuit not to work are that a wiring mistake was made when the breadboard was built or the software that was written to exercise the hardware has a bug.

A third example is a circuit in an existing electronic system, such as the PC/AT, that has failed. In this case, we know that the system worked correctly in the past but now malfunctions. Therefore, the cause of the problem may not be a mistake in the design, a wiring error, or incorrect software; instead, it is likely to be due to the failure of a component.

The final example is a circuit board, such as the main processor board of the PC/AT, that has just been built on a manufacturing line. Here a wide variety of malfunction causes exist. For instance, a lead of a component may not be correctly soldered; a lead of a device may be short-circuited to a pin on another device with excess solder; the wrong component may have been inserted; or even a component may have been installed in reverse orientation.

Thus we have seen that there are many causes for a circuit to malfunction. In fact, most of the causes we just stated can affect any of the circuits. For instance, a short circuit could occur between the pins of two devices on the main processor board of the PC/AT. This short may have been accidentally created when the system was opened to install an interface board, change the system-configuration DIP switches, or replace another failing subsystem, such as a disk drive. As another example, it is also possible that a bad IC gets installed when building a breadboard of a circuit or even during the manufacturing of a printed circuit board. Finally, an open circuit may occur in a copper trace on the main processor circuit board of the PC/AT, even though the board had been working correctly for a long time. For instance, a crack may have been made by using too much force when inserting an interface board into the expansion bus connector.

Having looked at some of the causes of circuit failures, let us continue by exploring troubleshooting methods. We assume that the circuit is known to have worked previously. This would be the case in troubleshooting a circuit built for one of our laboratory exercises or when repairing an electronic system such as a PC/AT. We will begin with a general procedure that can be used to troubleshoot microcomputer interface circuits.

A general flowchart for testing and troubleshooting a microcomputer interface circuit is shown in Fig. 14.21. Here we will assume that a circuit breadboard and diagnostic software exist. Therefore, the first step is to test the operation of the interface. This is done by exercising the hardware by running the program and observing its operation visually or with instrumentation. If the hardware correctly performs its intended operation, the flowchart's Y path shows that we are done. On the other hand, if it does not work, the N path is taken. That is, we need to troubleshoot the circuit. It is important to remember that circuits may not work due to problems in either software or hardware or both software and hardware.

Figure 14.21 shows that the first step in the troubleshooting part of the process is to identify and describe the symptoms of the failure. It is important to make a clear and concise description of the problem before beginning to examine the software or hardware.

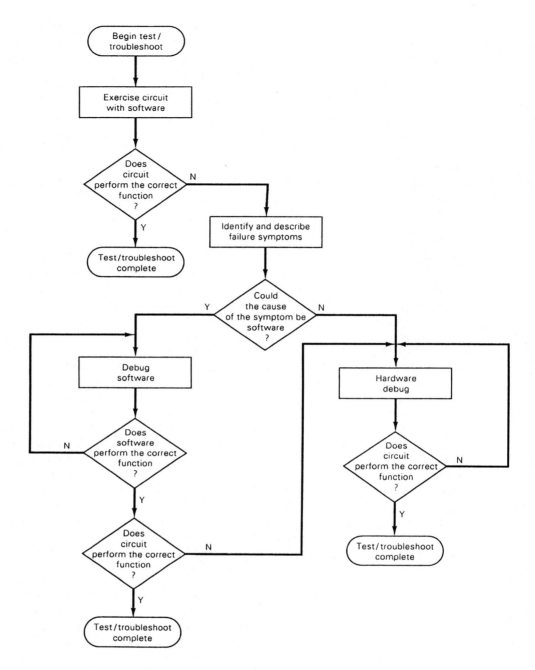

Figure 14.21 General test/troubleshoot flowchart.

For example, in the case of the test circuit in Fig. 14.16(a), running the diagnostic program blinks LED 0. The failure symptom may be that the LED just remains off or it may turn on but not blink.

Now we must decide whether or not software can be ruled out as a cause of the problem. For example, in a laboratory exercise where the program is given, software should not be the cause of the malfunction. Also, in the case of an application program running on a PC/AT, which is known to have no bugs, software may be ruled out. In cases where correct software operation cannot be assumed, the operation of the program should be analyzed before testing the circuitry. That is, as shown in Fig. 14.21, software debugging is the next step. If the program is found to be correct, the N path is followed in the flowchart and then hardware troubleshooting is begun.

When bugs are found in the program, they must be corrected; then the Y path is taken. Here we see that the interface circuit is retested to verify whether the software fixes make it work correctly. If the interface circuit does operate correctly, troubleshooting is complete.

Let us assume that the interface still does not function correctly. Then Fig. 14.21 shows that hardware troubleshooting must begin. After the hardware problems are identified and corrected, the interface circuit is once again tested.

Now that we have covered the general testing and troubleshooting procedure, we continue by looking more closely at the software-debug part of the process. The steps in the software-debug process are identified in the flowchart of Fig. 14.22. Notice that first the programming of any VLSI peripheral ICs in the circuit must be verified to be correct. If they are programmed as part of the program, the programming sequence and command values can be reviewed. In fact, software can be added to read back the contents of the registers (if possible) after programming to verify that they have been updated. In our example circuit of Fig. 14.16(a), there are no peripheral ICs. So this step is not required.

Next, the addresses corresponding to I/O devices or memory locations and data that are to be transferred over the bus must be checked. This will verify that the correct data are transferred and that the data will go to the correct place in the circuit. In the square-wave program we wrote earlier to blink LED_0, the address of the LED latch, which is 0007_{16}, is loaded into DX and the initial data that are to be output to the latch, 01_{16}, are loaded into AL. Both of these values are correct for the circuit under test.

Finally, the flowchart in Fig. 14.22 shows that the last step in the software-debug process is to check the algorithm and its software implementation. This can be done by rechecking the flowchart to confirm that it provides a valid solution to the problem and then comparing the instruction sequence against the flowchart to assure that it implements this algorithm. The instructions of the program can also be traced through to verify correct operation for a known test case.

If software is not the cause of the problem, attention must be turned to the hardware. A general hardware-troubleshooting procedure is outlined in Fig. 14.23. For now we will assume that the circuit undergoing troubleshooting is a breadboard built on the PCμLAB.

The first step identified in the flowchart is to make a thorough visual inspection of the circuit to assure that it is correctly constructed. This includes verifying that the correct IC pin numbers are marked into the schematic diagram; the circuit diagram layout does implement the circuit in the schematic, and that all jumper connections are consistent with those identified in the layout diagram. Second, the mechanical connections of the circuit should be checked to verify that they make good electrical connections—that is, that the pins of the ICs are making good contact with the contacts of the solderless breadboard and that the wire connections provide good continuity between pins of the various ICs. The continuity tester of the PCμLAB can be used to check out these connections.

If the circuit connections are correct, the flowchart shows that the next step is to check out the power supply. That is, we should verify that $+5$ V is applied between the V_{cc} and GND pins of each IC. Here we are normally interested in knowing the exact amount of voltage. For this reason, the voltage measurements are usually taken with a multimeter.

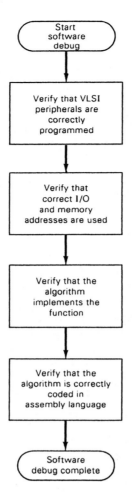

Figure 14.22 Software-debug flowchart.

Since the $+V_{cc}$ supply of TTL ICs is rated at $+5$ V $\pm$ 5%, power supply measurements between $+5.25$ V and $+4.75$ V are satisfactory.

After the circuit connections and power supply have been ruled out as the cause of the malfunction, we are ready to begin checking the operation of the circuit. To be successful at this, we must understand the operation, signal flow through the circuit, and waveshapes expected at select test points. Typically, operation is traced by observing signals starting from the output and working back toward the input in an attempt to identify the point in the circuit up to which correct signals exist. For instance, in our breadboard circuit, we can begin by examining the waveform across the LED with an oscilloscope. This spot is identified as test point 1 (TP1) in the schematic of Fig. 14.24 and corresponds to pin 8 of resistor pack R_{19}. Assuming that a symmetrical square wave is not observed, the probe of the scope can be moved to pin 18 of the 74LS240 IC, test point 2 (TP2). If a square wave is not present there either, the next test point should be the input of the inverter at pin 2 of the 74LS240 (TP3). Assuming that the square-wave signal is again not found, the output at pin 2 of the 74LS374 data latch, TP4, should be checked. This completes tracing of the data path from LED_0 to the data bus.

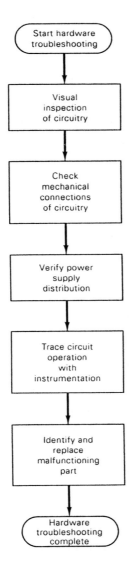

Figure 14.23 Hardware-debug flowchart.

If a square wave is not observed at any test point in the data path from data bus line SD_0 to LED_0, the problem may be in the chip-select decoder circuit. Earlier we found that this circuit produces the clock that loads the data into the 74LS374 latch. Notice that the clock is applied to pin 11 on U_2, which is identified as TP5. This signal is not a symmetrical square wave; instead, it is a repeating pulse that would appear as an asymmetrical square wave on the screen of a scope. Assuming that a pulse is not observed, this would be the reason that data are not being loaded from the data bus into the latch. In this case, the signal path of the latching pulse must be traced back to pin 7 of the 74LS138, TP6, in an attempt to locate the pulse.

Let us assume that no clock pulse is observed at pin 7 of the 74LS138 IC. Then, we should continue by checking for signals $\overline{IOW}$ and $\overline{AEN}$ at inputs $\overline{G}_{2A}$ and G_1, respectively. There should be active low pulses on both $\overline{IOW}$ and $\overline{AEN}$. These points are identified as TP7 and TP8 in the circuit diagram of Fig. 14.24. If that is the case, the only signals that

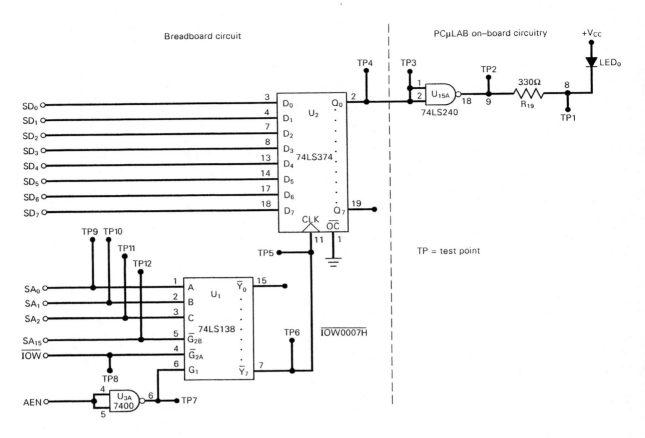

Figure 14.24 Breadboard circuit schematic with test points.

remain to be checked are address lines SA_0, SA_1, SA_2, and SA_{15} connected to A, B, C, and $\overline{G}_{2B}$, respectively. These signals are denoted as TP9, TP10, TP11, and TP12, respectively. To do this the sweep of the scope can be synchronized with $\overline{IOW}$ at $\overline{G}_{2A}$ and the logic levels that exist at inputs A, B, C, and $\overline{G}_{2B}$ observed during this pulse. Assuming that code CBA equals 110 and $\overline{G}_{2A} = 0$, output $\overline{Y}_7$ should be a pulse similar to that at $\overline{G}_{2A}$. Since no pulse was found at pin 7, we have found the source of the problem. The 74LS138 IC is bad and must be replaced.

After the bad IC is replaced, the operation is again observed. If LED_0 blinks, troubleshooting is complete. Otherwise, troubleshooting resumes by verifying that the clock pulse is produced at pin 7 of the 74LS138 and is passed to the clock input of the 74LS374 data latch.

If an oscilloscope is not available, many of the test measurements during the troubleshooting process we just outlined can be made with the logic probe of the PCµLAB. For instance, the signals at the G_1-, $\overline{G}_{2A}$-, and $\overline{G}_{2B}$-enable inputs of the 74LS138 address decoder can be tested. However, the logic probe is not as versatile as the oscilloscope. For example, with the scope we could verify the logic level of the address inputs to the address decoder. This type of synchronous measurement at the time when $\overline{IOW}$ is 0 cannot be made with a logic probe.

Another instrument that is useful in troubleshooting microcomputer interface circuits is a *logic pulser*. The logic pulser can be set to output either a one-shot pulse or a square wave. This instrument can be used to inject a pulse or square wave into the input of a device in the circuit. Figure 14.25 shows a typical logic pulser.

Let us now look briefly at how a logic pulser can be used when troubleshooting the circuit in Fig. 14.24. The pulser would be set to pulse mode of operation and then the pulse injected at test point 2. As long as the connection through the resistor pack is good, the LED_0 should blink. Assuming that this part of the circuit works correctly, the pulser's probe can next be touched to test point 3. Again, the LED should blink. This verifies whether or not the 74LS240 inverter operates correctly. Just as with the oscilloscope, the logic pulser is used to check out the circuit step by step.

Hardware troubleshooting of circuit boards in a manufacturing environment can be quite different. A visual inspection is still performed, but it should look for different things. For instance, the quality of solder joints is checked—that is, whether they have enough solder, are of the correct shape, and that there are no solder shorts between pins of ICs and other components.

After a board has passed visual inspection, it is ready for circuit test. In this case the board is not tested with instruments circuit by circuit as we just described; instead, it is checked out with an automatic test system. The tester is programmed to perform a series of tests on the circuit board. The system provides information on tests that have passed and failed to guide the repair process. In this way, the board is tested and repaired step by step until it is completely functional.

Troubleshooting of an electronic system, such as a PC/AT, that has been working is also different. In this case, we will assume that the symptoms of the problem have been identified; that the software is functional; and that we are investigating a hardware problem. Actually, servicing of a PC/AT is normally a system-level repair. That is, the failing subassembly (main processor board, add-on card, power supply, floppy disk drive, keyboard, or monitor) is identified and replaced.

The hardware troubleshooting procedure outlined in Fig. 14.23 still applies to a system-level repair. Therefore, the first step is a visual inspection; however, the inspection performed is different than that used when examining a breadboard or circuit board that just came off of the manufacturing line and includes a number of mechanical checks. For example, the system should be checked to verify that all cables are securely connected, the setting of the DIP switches and jumpers should be checked to assure that the system configuration is correct, and the surface of the circuit boards can be examined looking for overheated or burned components. An example of a mechanical check is to touch the components to see

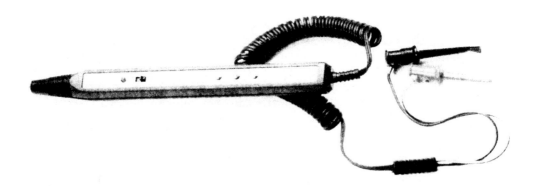

Figure 14.25 Logic pulser.

if any are abnormally hot, but be careful with this because some devices can get very hot. Moreover, to assure that the problem is not due to dirty connector contacts, the cables and add-on boards are removed, their contacts are cleaned, and then they are reseated into the connector.

Next we must begin to check the circuits and subassemblies. If the PC/AT does not come up at all when the power switch is turned on, the first step should be to check the power supply voltages. However, if it boots up and loads the DOS operating system, diagnostic software, such as QAPlus™ by DiagSoft, Inc., can be used to analyze the function of the system. The diagnostic disk is inserted into the floppy disk drive and the diagnostic program is initiated from the keyboard. This program can exercise each of the PC/AT's subassemblies and displays information indicating whether they have passed or failed the diagnostic tests.

Let us assume that the hard disk controller in a PC/AT has failed the diagnostic test. The normal system level repair procedure is to replace the complete controller with another one to quickly get the system back up running. The bad board is usually returned to a circuit-board repair location for IC level troubleshooting and repair. Some diagnostic programs permit IC-level troubleshooting of the dynamic memory subsystem. In the case of a memory failure, the diagnostic program can identify the bad IC. Since the DRAMs in many PC/ATs are socket mounted, the repair may be made by replacing the failing device.

▲ 14.6 OBSERVING MICROCOMPUTER BUS ACTIVITY WITH A DIGITAL LOGIC ANALYZER

Up to this point, we observed signal waveforms in the microcomputer with an oscilloscope. However, this instrument permits viewing of only a limited number of periodic signals at a time. The address, data, and control buses in a microcomputer have many lines. For instance, the ISA data bus alone is 16 bits wide in an 80386DX-based PC/AT. When the microcomputer is running, the data bus could be returning read data or instruction code to the MPU, sending write data to memory, or be in the high-Z state if no bus activity is taking place. Moreover, the data being transferred is rarely the same; therefore, data bus activity is not periodic.

To observe the operation of the ISA data bus signals, we need to see the logic states of all 16 data bits and some of the read/write control signals at the same time. This is not possible with a scope. It is for this type of measurement that an instrument known as a *digital logic analyzer* was developed. Let us now look briefly at what a logic analyzer is and what it is used for in testing microcomputer systems.

The logic analyzer is a modern digital test instrument that is very useful for testing and troubleshooting microcomputer systems. With it, nonperiodic signals, such as those of the address bus, data bus, and control bus can be measured and their waveforms viewed. Figure 14.26 shows a typical logic analyzer. Today, this type of instrument is available with 8, 16, or 32 channels. This means that they are capable of simultaneously sampling and displaying the waveshapes of up to 8, 16, or 32 signals. The probe is a pod that has clips for inputs to each channel. They are attached to the signals that are to be observed. For example, in an 80386SX-based microcomputer, the data lines D_0 through D_{15} and control signals, such as $\overline{RD}$, $\overline{WR}$, DT/$\overline{R}$, IO/$\overline{M}$, $\overline{DEN}$, READY, S_3, and S_4, can be sampled to monitor transfers over the data bus.

The logic analyzer operates differently than an oscilloscope. The oscilloscope immediately displays the voltage of the signal applied at its input. On the other hand, the logic

Figure 14.26   Digital logic analyzer. (Hewlett-Packard Co.)

analyzer samples the voltage at all inputs at a very high rate. This information is stored in memory as a logic 0 or logic 1 and not as a specific voltage level. The waveform of the signal can then be displayed on the screen using the stored data. The user of the instrument has the ability to start the sampling based on the occurrence of a specific event or events indicated by a combination of the logic values of the signals being monitored and continue until the trace buffer memory is full. Moreover, most logic analyzers permit the stored information to be displayed in a variety of ways.

Sample waveforms taken from our test circuit in Fig. 14.16(a) are shown in Fig. 14.27(a). This display shows the address, data transfer, and address decoder output produced when the LED blink diagnostic routine runs. Here we see a timing diagram that clearly illustrates the relationship between the address, decoder output, and the data transfer to the latch. Notice that whenever address decoder output $\overline{IOW0007H}$ switches from logic 0 to logic 1, the byte of data on the data bus, which is 00000001_2, is latched into the 74LS374 device and makes Q_0 switch to logic 1. Figure 4.27(b) shows the signals when the Q_0 output switches from logic 1 back to 0. Both transitions at Q_0 are shown with a single timing diagram in Fig. 4.27(c). Remember that the logic analyzer cannot display signals the way they really would look if observed with an oscilloscope. Since it just saves the logic level (0 or 1) of the signals in memory, waveshapes are shown with sharp transitions between these logic levels. When observed with a scope, the waveshapes may show rise and fall times between the 0 and 1 levels and possibly *ringing, overshoots,* and *undershoots,* around the 0 and 1 logic levels.

A logic analyzer can also be set up to collect just code or data transfers over the data bus. Once this information is stored in memory, it can be *disassembled* into assembly language instructions and displayed on the screen. Figure 14.28 shows an example where a logic analyzer was used to disassemble a series of instructions. This capability permits us to monitor the execution of instructions by the MPU and compare the instruction execution sequence to events observed in the hardware.

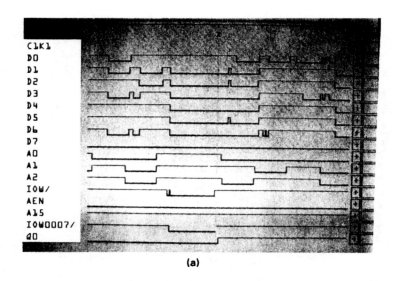

(a)

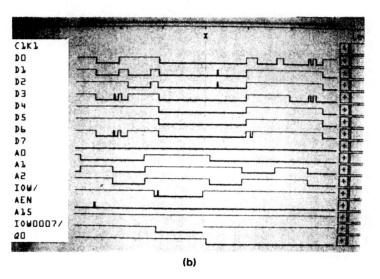

(b)

Figure 14.27 (a) Timing diagram for 0 to 1 transition at Q_0. (b) Timing diagram for 1 to 0 transition at Q_0.

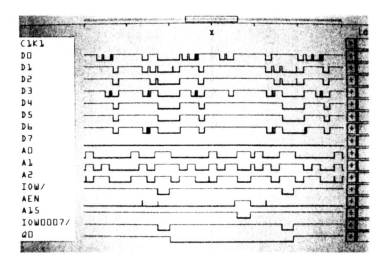

Figure 14.27(c) Timing diagram showing both transitions at Q_0.

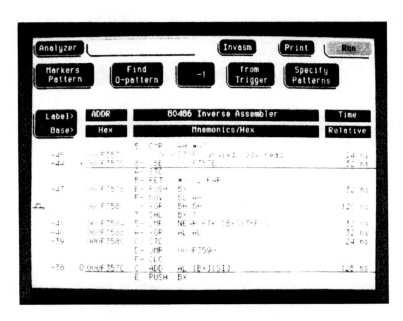

Figure 14.28 Disassembling code with a logic analyzer. (Hewlett-Packard Co.)

ASSIGNMENTS

Section 14.2

1. What is an experimental circuit board that is built to test an electronic function called?

2. What is a breadboard card?

3. What kind of breadboard does not require devices to be soldered in place; instead, they can be plugged in?

4. What is the purpose of an extender card?

5. List the parts of the PCμLAB.

Section 14.3

6. List three types of I/O devices that are built into the PCμLAB.

7. How are the on-board I/O devices set up for use with circuits built on the breadboard area?

8. How many ISA bus expansion slots are provided on the PCμLAB?

9. What size wire jumpers must be used to interconnect circuits on the breadboard area?

10. Which wire-insertion clips are intended for use as the +5 V and GND for power distribution on the solderless breadboard?

11. At which contacts of the ISA bus connector are address lines A_0 through A_{19} available?

12. What is the purpose of the PCμLAB's continuity tester? How does it signal continuity?

13. Identify how the PCμLAB's logic probe signals the 0, 1, and high-Z logic states.

14. How does the logic probe identify that the signal at a test point is switching between the 0 and 1 logic levels?

Section 14.4

15. Which ICs are used in the I/O address decoder circuit of the PCμLAB?

16. Which output of the I/O address decoder circuit is used to enable data output to the LEDs?

17. Which I/O address bits are don't-care states?

18. Does the I/O address $771E_{16}$ activate an output of the I/O address decoder circuit? If so, which output signal is activated?

19. Why are the I/O select outputs of the I/O address decoder circuits produced only during I/O bus cycles?

20. If switches S_0 through S_3 are closed and switches S_4 through S_7 are open, what value will be transferred over the data bus when the switches are read with an IN instruction?

21. Will the command

```
I  FF1D  (↵)
```

read the state of the on-board switches?

22. Write an instruction sequence that will read the state of the switches, and mask off all switch settings but S_0 and S_1. If both switches are found to be closed, a jump is to be initiated to a service routine called SERVE_3.

23. What operation is performed by the instruction sequence that follows?

```
        MOV   DX,31DH
POLL:   IN    AL,DX
        MOV   CL,8
        SHR   AL,CL
        JC    POLL
```

24. Which LED is driven with data from data bus line D_0? From D_7?

25. What will the command O FF1E 0F do?

26. Write a program that will scan the LEDs on the PCμLAB. That is, first light L_0 for a period of time, next turn off L_0 and turn on L_1, and so on until L_7 is lighted. The scan sequence should repeat continuously.

27. Describe the operation performed by the following instruction sequence:

```
        MOV   DX,31EH
        MOV   AL,0H
BIN:    OUT   DX,AL
        MOV   CX,0FFFFH
DELAY:  DEC   CX
        JNZ   DELAY
        INC   AL
        JMP   BIN
```

28. What IC is used to drive the speaker?

29. Which instruction in the tone-generation program given in the section on the speaker-drive circuit needs to be changed to double the frequency of the tone? Write the new instruction.

Section 14.5

30. Write an instruction that reads the switch setting into AL in Fig. 14.29.

31. Mark the pin numbers into the circuit of Fig. 14.29 and then make a layout drawing on a circuit diagram master.

32. What is a program that is used to test the operation of a microcomputer circuit called?

33. What accessory is attached to the top of an IC to make it easier to connect the probe of an instrument?

34. Name three instruments that can be used to test the operation of a microcomputer interface circuit.

35. What information does a logic probe provide about the signal at a test point in a circuit?

36. When used to measure voltage, what information does a multimeter tell about the signal at a test point?

37. What information does an oscilloscope provide about the signal at a test point?

38. What is meant by *periodic signal*?

39. What is the process for determining the cause of a hardware malfunction in a circuit called?

40. If a breadboard of a circuit does not work when tested with a diagnostic program that was just written and never checked on working hardware, should hardware troubleshooting or software debug take place next?

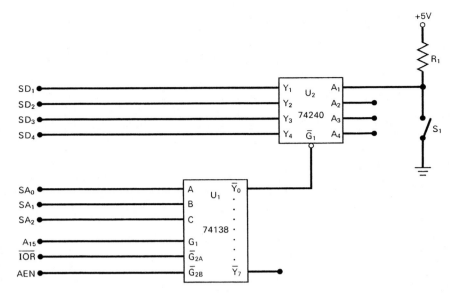

Figure 14.29 Switch input interface circuit.

41. Assume that a diagnostic program is available that has been checked out and verified to correctly test the interface. If a new PC/AT interface module is tested and fails when checked out with this diagnostic program, what is the next step, software debugging or hardware troubleshooting?

42. List three items that should be checked as part of the software-debug process.

43. List three visual inspections that can be made as part of the hardware-troubleshooting process used to determine why a circuit built on the breadboard area of the PCμLAB does not work.

44. During the hardware-troubleshooting process of a breadboard circuit, what should be checked next if the circuit is found to be correctly constructed?

45. Assume that the circuit in Fig. 14.29 is being driven by a software routine that polls the state of switch 0 and that this program is known to operate correctly. What type of signal would you expect to see at test point 1 when the switch is closed and also when open? At test point 2? At test point 3?

46. The results found at test points 1 and 2 of Problem 45 when troubleshooting the circuit are as follows.

Test point	Switch open	Switch closed
1	1	0
2	1	1

What do you think is the problem with the circuit?

Section 14.6

47. List three key groups of signals of the microcomputer system that are, in general, nonperiodic.

48. What instrument is usually used to observe nonperiodic signals in a microcomputer?

49. Compare an oscilloscope and a logic analyzer.

The 80486
Microprocessor Family

▲ 15.1 INTRODUCTION

Up to this point in the book, we have studied the 80386 family of microprocessors. Now we will turn our attention to Intel's second generation of 32-bit MPUs, the *80486 microprocessor family*. In this chapter, we examine the internal, software, and hardware architecture of these newer processors. The focus throughout the chapters is on how the architecture of the 80486 MPUs differ from those of the 80386 family. For this purpose, we have included the following topics in the chapter:

1. The 80486 microprocessor family
2. Internal architecture of the 80486
3. Real-mode software model and instruction set of the 80486SX
4. Protected-mode software architecture of the 80486SX
5. Hardware architecture of the 80486 microprocessor
6. Signal interfaces of the 80486SX MPU
7. Memory and I/O software organization, hardware organization, and interface circuits
8. Nonburst and burst bus cycles
9. Cache memory of the 80486SX
10. High-integration memory/input/output peripheral—R400EX
11. Interrupts, reset, and internal exceptions
12. The 80486DX2 and 80486DX4 microprocessors

Intel's second generation of 32-bit microprocessors, the 80486 family, was introduced in 1989. The first product offered in this family was the *80486DX* MPU. Figure 15.1 shows an 80486DX IC. This device is a full 32-bit microprocessor; that is, its internal registers and external data paths are both 32 bits wide. This device offers a number of advanced software and hardware architecture features as compared to the 80386DX. Two major changes that greatly improve performance are the addition of an on-chip *floating-point math coprocessor* and an on-chip *code and data cache memory*. The 80386 family supports both a math coprocessor and cache memory, but they needed to be implemented external to the device.

The 80486 family maintains real-mode and protected-mode software compatibility with the 80386 architecture. However, important changes have been made in the instruction set. First, and most important, is that the execution speed of most instruction of the instruction set has been improved for the 80486 family. This was done by changing the way in which they are performed by the MPU so that now most of the basic instructions are performed in just one clock cycle. For instance, the move, add, subtract, and logic operations can all be performed in a single clock cycle. With the 80386DX, these same operations took two or more clock cycles to be completed. Finally, a number of new instructions have been added to make the instruction set even more versatile. The result of these architectural changes is an improvement of more than two times in the overall performance for the 80486 family.

Similar to the 80386DX, the 80486DX was followed by an *80486SX* device. However, this time the SX version did not have a 16-bit external architecture. It also is a full 32-bit MPU, although it does not include the on-chip floating-point coprocessor unit. The 80486DX and 80486SX were followed by several new generations of 80486 MPUs which introduced additional architectural features that further enhance the performance of the family. For instance, the *80486DX2* MPU, which is both hardware and software compatible with the 80486DX, has increased performance by a technique known as *clock doubling*. In the 80486DX-4, this internal clock scaling was expanded to permit multiplications of $2\times$, $2.5\times$, or $3\times$.

Let us next briefly compare the levels of performance offered by the 80386 family and 80486 family MPUs. Referring back to the iCOMP™ index chart in Fig. 1.8, we see

Figure 15.1 80486DX IC. (Courtesy of Intel Corp.)

that the 80486SX-20 has an iCOMP™ rating of 78 compared to a rating of 32 for the 80386SX-20. Moreover, the 80486DX-33 is rated at a level of 166, whereas the 80386DX-33 is at 68. In this way, we see that comparable 80486 family members do deliver more than twice the performance. Also, newer members of the 80486 family, such as the 80486DX2-66 (rated at 297 in the iCOMP™ chart), have widened this performance advantage to more than 4×.

EXAMPLE 15.1

What is the iCOMP™ rating of the 80486DX4-100?

Solution

Looking at Fig. 1.8, we find that the iCOMP™ rating of the 80486DX4-100 is 435.

▲ 15.3 INTERNAL ARCHITECTURE OF THE 80486

We already mentioned that the internal architecture of the 80486 family is an improvement over that of the 80386 family. For instance, we said that a floating-point coprocessor and cache memory are now on-chip. These are not the only changes that have been made to improve the performance of the 80486 family. Here we will explore the functional elements within the 80486DX microprocessor's architecture and how they have changed from that of the 80386DX.

A block diagram of the internal architecture of the 80486 family is shown in Fig. 15.2. Similar to the 80386 architecture, we find the execution unit, segmentation unit, paging unit, bus interface unit, prefetch unit, and decode unit. However, the functions of many of these elements have been enhanced for the 80486 family. For example, we already mentioned that the coding of instructions in the control ROM had been changed to permit instructions to be performed in less clock cycles. Some other changes are that the code queue in the prefetch unit has been doubled in size to 32 bytes. This permits more instructions to be held on chip ready for decode and execution. Also an improved algorithm is now used by the translation lookaside buffer in the paging unit. Finally, the bus interface unit has been modified to give the 80486 architecture a much faster and more versatile processor bus.

8086 family microprocessors available before the 80486DX are what are known as *complex instruction set computer* or *CISC* processor. That is, they have a large versatile instruction set that supports many complex addressing modes. In general, the instructions execute in two to many clock cycles. For instance, the 80386's register to register ADD instruction takes two clock cycles, but the same instruction when adding the content of a storage location in memory takes seven clock cycles. Other instructions, for instance, the integer multiply (IMUL) and integer divide (IDIV), take even more clock cycles.

The 80486DX is the first member of this family with a high-performance *reduced-instruction-set computer* (RISC) integer core. A RISC processor is typically characterized as having a small instruction set, limited addressing modes, and single-clock execution for instructions. The MPUs of the 80486 family are best described as *complex reduced-instruction-set computer* (CRISC). This is because a core group of instructions in the 80486's instruction set execute in a single clock cycle. For example, the register-to-register ADD is performed in a single cycle. At the same time, it retains the many complex instructions

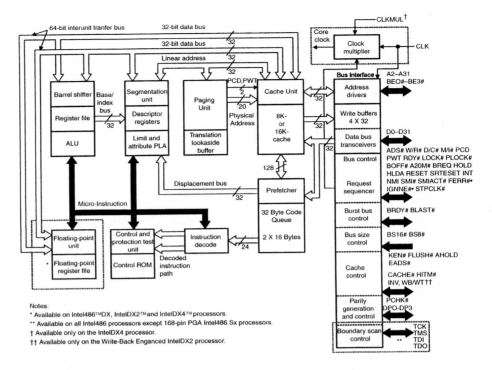

Figure 15.2 Internal architecture of the 80486 MPUs. (Reprinted by permission of Intel Corp. Copyright/Intel Corp. 1994)

and addressing modes that make the instruction set more versatile. However, the number of clock cycles needed to perform many of these complex instructions have also been reduced in the 80486DX. For instance, the register-to-memory ADD is reduced from seven to three clock cycles.

Let us now look more closely at the new elements of the 80486DX's internal architecture. Traditionally, the *floating-point operations* of the microcomputer have been performed by an externally attached *floating-point coprocessor*. With the 80486DX, this function is integrated into the MPU. This *floating-point math unit* supports the processing of the 32-bit, 64-bit, and 80-bit number formats specified in the *IEEE 754 standard* for floating-point numbers. At the same time, it is upward software compatible with the older 8087, 80287, and 80387 numeric coprocessors. The result of this on-chip implementation function is higher-performance floating-point operation. Remember that this unit is not provided in the 80486SX MPU.

Addition of a high-speed *cache memory* to a microcomputer system provides a way of improving overall system performance while permitting the use of low-cost, slow-speed memory devices in the main system memory. During system operation, the cache memory contains recently used instructions, data, or both. The objective is that the MPU accesses code and data in the cache most of the time, instead of from the main memory. Since less time is required to access the information from the cache memory, the result is a higher level of system performance. The internal *cache memory unit* of the 80486DX is 8K bytes in size and caches both code and data.

EXAMPLE 15.2

In Fig. 15.2, identify an architectural element other than the floating-point unit that is not implemented in the PGA-packaged 80486SX MPU.

Solution

Looking at Fig. 15.2, we find that the *Boundary Scan Control* block of the bus interface unit is not available on the 80486SX MPU.

▲ 15.4 REAL-MODE SOFTWARE MODEL AND INSTRUCTION SET OF THE 80486SX

At this point, we will turn our attention to the 80486SX MPU and its real-mode software model. The registers in the software model of the 80486SX are shown in Fig. 15.3. These registers are essentially the same as those shown for the 80386DX in Fig. 2.2. Their organization and functionality are also the same. One exception is control register 0 (CR_0). In the 80386DX, this register has just 1 bit that is active in the real mode. For the 80486SX, 2 other bits, cache disable (CD) and not-write-through (NW), are active. They are used to enable and configure the operation of the on-chip cache memory. Another difference is that three more test registers, TR_3, TR_4, and TR_5, have been added. Of course, the 80486DX has more registers in its real-mode model. The new registers are needed to support the operation of the floating-point coprocessor.

The real-mode instruction set has been enhanced with new instructions for the 80486 family. Figure 15.4(a) shows that the 80486SX's instruction set is simply a superset of that of the 80386DX. A group of new instructions called the *80486 specific instruction set* has been added. The instructions of the 80486 specific instruction set are summarized in Fig. 15.4(b). Since all the earlier instructions are retained in all 80486 family processors and their object code is compatible with the 8086, 8088, 80286, and 80386 processors, upward software compatibility is maintained. Let us now look at the operation of each of the new instructions.

Byte-Swap Instruction: BSWAP

When studying the 80386DX microprocessor, we showed how the bytes of a double word of data were stored in memory. As shown in Fig. 15.5(a), the least significant byte is stored at the lowest-value byte address, which is identified as address *m*. The next more significant bytes are held at address *m* + 1 and *m* + 2. Finally, the most significant byte is saved at the highest-value byte address, *m* + 3. This method of storing information in memory is known as *little endian* organization.

Another method of double-word data organization, which is called *big endian,* is employed by other microprocessor architectures. For instance, Motorola's 68000 family of microprocessors store data in this way. Figure 15.5(b) shows how the bytes of a 32-bit word are arranged in big endian format. Notice that they are stored in the opposite order.

To make it easier for the 80486SX to process data which had been initially created in the big endian format, a special instruction was added to the real-mode instruction set. This *byte-swap* (BSWAP) instruction is provided to convert the organization of the bytes

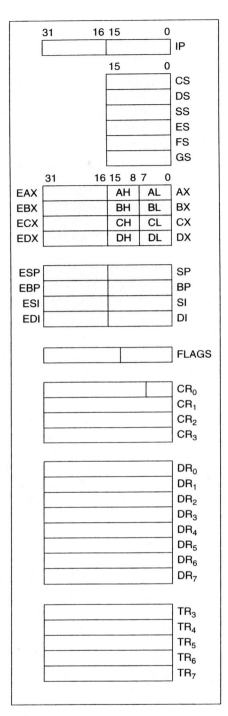

Figure 15.3 Real-mode software model of the 80486SX microprocessor.

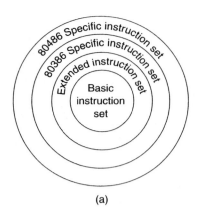

(a)

Mnemonic	Meaning	Format	Operation
BSWAP	Byte swap	BSWAP r32	Reverse the bye order of the 32-bit register.
XADD	Exchange and add	XADD D,S	(D) ←→ (S), (D) ← (S) + (D)
CMPXCHG	Compare and exchange	CMPXCHG D,S	if (ACC) = (D) (ZF) ← 1, (D) ← (S) Else (ZF) ← 0, (ACC) ← (D)

(b)

Figure 15.4 (a) 80486SX instructions set. (b) 80486 specific instruction set.

of a double word of data between the big endian format and little endian format. As shown in Fig. 15.4(b), the instruction has a single 32-bit register as its destination. Therefore, the double word that is to be converted must first be loaded into an internal register of the 80486SX.

An example is the instruction

BSWAP   EAX

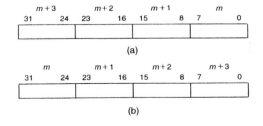

(a)

(b)

Figure 15.5 (a) Little endian memory format. (Reprinted by permission of Intel Corp. Copyright/Intel Corp. 1994) (b) Big endian memory format. (Reprinted by permission of Intel Corp. Copyright/Intel Corp. 1994)

Let us assume that the contents of EAX are in big endian format. Then, when this instruction is executed, the bytes of the double word of data in register EAX are rearranged into the little endian format. For instance, if the big endian contents of the register are

$$(EAX) = 01234567H = 00000001001000110100010101100111_2$$

the new contents of EAX after the byte swap has taken place will be

$$(EAX) = 56743210H = 01110110010101000011001000010000_2$$

Actually, BSWAP will convert the format of data either way. If the bytes of data in EAX are in little endian form when the instruction is executed, it will be changed to big endian form.

EXAMPLE 15.3

Write a program sequence that will read the double-word contents of storage location DS:1000H in memory, rearrange the bytes from big endian to little endian organization, and then return the new value to the original storage location in memory.

Solution

The big endian–format double word of data is read from memory with the instruction

```
MOV   EAX,[1000H]
```

Then, the double word is converted to little endian form by

```
BSWAP   EAX
```

Finally, the double word is returned to memory with the instruction

```
MOV   [1000H],EAX
```

Exchange and Add Instruction: XADD

The second instruction added to the real-mode instruction set of the 80486SX is the *exchange and add* (XADD) instruction. Looking at Fig. 15.4(b), we find that this instruction performs both an add and exchange operation on the contents of the source and destination operands. The source operand must be an internal register, whereas the destination can be either another register or a storage location in memory.

For an example, let us determine the operation of the register-to-register exchange and add instruction

```
XADD   AX,BX
```

We will assume that the contents of registers AX and BX are 1234_{16} and 1111_{16}, respectively. After the exchange and add operation takes place, the sum of these two values ends up in destination register AX:

$$(AX) = 0001001000110100_2 + 0001000100010001_2 = 0010001101000101_2$$
$$= 2345_{16}$$

The original contents of AX are in BX:

$$(BX) = 1234_{16}$$

EXAMPLE 15.4

If the value in EAX is 00000001H and the double word at memory address DS:1000H is 00000002H, what results are produced by executing the instruction

```
XADD   EAX,[1000H]
```

Solution

Execution of the instruction causes the addition

$$(EAX) + (DS:1000H) = 00000001_{16} + 00000002_{16}$$
$$= 00000003_{16}$$

The value that results in the source location is

$$(DS:1000H) = 00000001_{16}$$

and that in the destination is

$$(EAX) = 00000003_{16}$$

Compare and Exchange Instruction: CMPXCHG

The last of the new real-mode instructions is *compare and exchange* (CMPXCHG). This instruction performs a compare operation and an exchange operation that depends on the result of the compare. As shown in Fig. 15.4(b), the compare that takes place is not between the values of the source and destination operand. It is between the content of the accumulator register (AL,AX,EAX) and the corresponding size destination. If the accumulator and destination contain the same value, the zero flag is set to 1 and the content of the source register is loaded into the destination location. Otherwise, ZF is cleared to 0 and the content of the destination is loaded into the accumulator. The destination can be either a register or storage location in memory. The value in the accumulator must be loaded prior to execution of the CMPXCHG instruction.

As an example, let us consider the instruction

```
CMPXCHG   [2000H],BL
```

and assume that register AL contains 11_{16}, register BL contains 22_{16}, and the byte memory location address 2000H contains 12_{16}. When the instruction is executed, the value in AL (11_{16}) is compared to that at address 2000H in memory (12_{16}). Since they are not equal, ZF is made logic 0 and the value 12_{16} is copied from memory into AL. Therefore, after execution of the instruction, the results are

$$(AL) = 12_{16}$$
$$(BL) = 22_{16}$$
$$(2000H) = 12_{16}$$

EXAMPLE 15.5

Assume that the values in registers AL and BL are 12_{16} and 22_{16}, respectively, and that the byte-wide memory location DS:2000H contains 12_{16}, what results are produced by executing the instruction

```
CMPXCHG   [2000H],BL
```

Solution

Since $(AL) = (DS:2000H) = 12_{16}$, the zero flag is set and the value of the source operand is loaded into the destination. This gives

$$(AL) = 12_{16}$$
$$(BL) = 22_{16}$$
$$(2000H) = 22_{16}$$

▲ 15.5 PROTECTED-MODE SOFTWARE ARCHITECTURE OF THE 80486SX

Now that we have examined the differences between the software architectures of the 80386DX and 80486SX in the real mode, let us turn our attention to protected-mode operation. The protected-mode operation of the 80486SX is essentially the same as that of the 80386DX. Enhancements have been made to the register set, system control instruction set, and the page tables. Here we will focus on the differences between the two MPUs.

Software Model

Similar to the real mode, the protected-mode software model of the 80486SX shown in Fig. 15.6 is essentially the same as that given for the 80386DX in Fig. 8.1. That is, the same set of registers exist in both processors, and they serve the same functions relative to

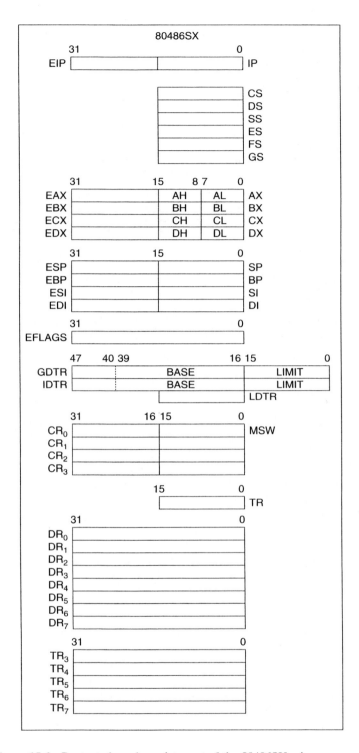

Figure 15.6 Protected-mode register set of the 80486SX microprocessor.

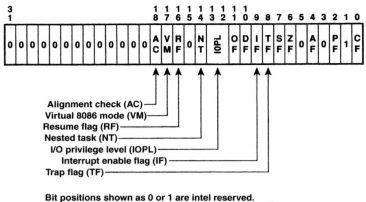

Bit positions shown as 0 or 1 are intel reserved.
Do not use. Always set them to the value previously read.

Figure 15.7 Protected-mode flags register. (Reprinted by permission of Intel Corp. Copyright/Intel Corp. 1992)

protected-mode operation. However, the 80486SX has three additional test registers and new bits defined in both the flags register and control registers. We will continue by examining these new register functions.

Flags Register

Figure 15.7 shows the protected-mode flags (EFLAGS) register of the 80486SX. In this illustration, the system-control flags, bits that affect protected-mode operation, are identified. Comparing these bits to those of the 80386DX in Fig. 8.8, we find that just one new bit has been added. This is the *alignment-check* (AC) flag, which is located in bit position 18. When this bit is set to 1, an alignment check is performed during all memory access operations that are performed at privilege level 3. A double word of data that is not stored at an address that is a multiple of 4 is said to be unaligned. If this double-word storage location is accessed, two memory bus cycles must be performed. The extra bus cycle that is introduced because the data are unaligned reduces overall system performance. The alignment-check feature of the 80486SX can be used to identify when unaligned elements of data are accessed. If an unaligned access takes place, an alignment-check exception, which is exception 17, occurs.

Control Registers

The 80486SX has four control registers just like the 80386DX; however, a number of new bits are now active. The control registers of the 80486SX are shown in Fig. 15.8, and those of the 80386DX are given in Fig. 8.5. Notice that 5 additional bits have been activated in CR_0 of the 80486SX. They are *alignment mask* (AM), *numeric error* (NE), *write protect* (WP), *cache disable* (CD), and *not write-through* (NW). Let us look at a few of these bits in detail.

We just said that the AC system-control flag enabled memory data alignment checks. Actually it takes more than just setting AC to 1 to enable this mode of operation. The AM

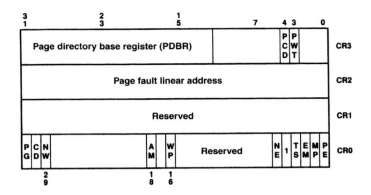

Figure 15.8 Control registers of the 80486SX. (Reprinted by permission of Intel Corp. Copyright/Intel Corp. 1992)

bit, which is bit 18 in CR_0, must also be set to 1. If AM is switched to 0, the alignment check operation is masked out.

Two other bits in CR_0, CD (bit 30) and NW (bit 29), are used to enable and control the operation of the on-chip cache memory. To enable the cache for operation, CD must be cleared to 0. The NW bit enables write through and cache validation cycles to take place when it is set to 0. Therefore, to permit normal cache operation, both of these bits should be cleared to 0.

Some other changes are found in CR_3. Two new bits, *page-level cache disable* (PCD) and *page-level writes transparent* (PWT), have been defined. The state of these bits are output on signal lines PCD and PWT, respectively, during all bus cycles that are not paged. They are used as input signals to the control circuitry for an external cache memory subsystem.

System-Control Instruction Set

The system-control instruction set has been expanded by three instructions for the 80486SX microprocessor. They are *invalidate cache* (INVD), *write-back and invalidate data cache* (WBINVD), and *invalidate translation lookaside buffer entry* (INVLPG). Figure 15.9 shows the format of these instructions and briefly describes their operation.

The first two instructions, INVD and WBINVD support management of the on-chip and external cache memories. When an INVD instruction is executed the on-chip cache is flushed. That is, all of the data that it holds is made invalid. In addition to invalidating the content of the on-chip cache, execution of this instruction also initiates a special bus cycle known as a *flush bus cycle*. External circuitry must detect the occurrence of this cycle and initiate a flush of the data held in the external cache memory subsystem. WBINVD is similar to INVD in that it initiates a flush of the on-chip cache memory; however, it initiates a different special bus cycle, a *write-back bus cycle*. External circuitry must again identify that a write-back cycle has taken place and tell the external cache to write back its content to the main memory.

The INVLPG instruction is used to invalidate a single entry in the 80486SX's internal translation lookaside table register. Notice that the instruction has an operand *m* that identifies which of the 32 table entries is to be marked invalid.

Mnemonic	Meaning	Format	Operation
INVD	Invalidate cache	INVD	Flush internal cache and signal external cache to flush.
WBINVD	Write back and invalidate cache	WBINVD	Flush internal cache, signal external cache to write-back and flush.
INVLPG m	Invalidate TLB entry	INVLPG	Invalidate the signal TLB entry.

Figure 15.9 80486SX specific system-control instructions.

Page Directory and Page Table Entries

The page directory and page tables of the 80486SX are the same size and serve the same function as they did in the 80386DX's protected-mode software architecture. However, a change has been made in the format of the page directory and page table entry. Two additional bits of the 32-bit entry are defined. Let us now look at the function of these two bits.

The format of a page directory/page table entry for the 80386DX is given in Fig. 8.24. Looking at the format of the 80486SX's entry in Fig. 15.10, we find that the 2 new bits are *page cache disable* (PCD) and *page write transparent* (PWT). These 2 bits are used for page-level control of the internal and external caches. To enable caching of a page, the PCD bit in the page table entry must be set to 0. Logic 1 in PWT selects page-level write through operation of the cache for the corresponding page.

When paging is in use, the logic levels of the PCD and PWT bits in the page table entry are output at the PCD and PWT pins of the MPU. This permits control of an external cache memory subsystem.

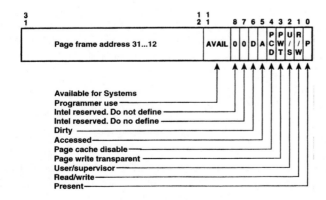

Figure 15.10 80486SX directory and page table entry format. (Reprinted by permission of Intel Corp. Copyright/Intel Corp. 1992)

Earlier in this chapter we pointed out that the 80486 family of microprocessors is Intel Corporation's second generation of 32-bit processors. It brought a higher level of performance and more versatility to the 8086 architecture. Just as for the 80386 family, maintaining compatibility of the 80486 family's hardware architecture to earlier 8086 family MPUs was also less important. This does not mean that the signal interfaces were purposely changed. In fact, many of the 80486's interface signals are the same as those provided on the 80386DX. However, a number of enhancements have been made to the 80486 family that are directed at improving its performance and making the 80486-based microcomputer more versatile. Much of the focus of these enhancements was on the memory interface. For instance, it is now enabled to do dynamic bus sizing down to 8 bits, high-speed burst data transfers over the bus, and write operations are buffered. The addition of these new capabilities has expanded the number of interface signals.

A number of hardware elements that were normally implemented in external circuitry are for the first time added into the MPU with the 80486 family. Examples of these new on-chip hardware functions are parity generation/checking, code/data cache memory, and—in the case of the 80486DX—a floating-point mathematics unit. Addition of these capabilities called for further expansion of the number of interface signals.

The more advanced processes used to manufacture the 80486 family of MPUs permitted the integration of many more transistors into a single IC. The circuitry of the 80486DX is equivalent to approximately 1.2M transistors, four times more than the 80386DX. Actually, the 80486DX was the first IC made by Intel Corporation that contained more than 1 million transistors.

Originally the 80486DX and 80486SX were both manufactured in a 168-lead pin grid array package. The layout of the pins and signals on this package are shown in Figs. 15.11(a) and (b). For example, the pin located at the uppermost left corner, which corresponds to row S and column 1, is address bit A_{27}. Another example is the data bus line D_{20}, which is located at the junction of row A and column 1. Later both devices were made available in a lower-cost 196-lead plastic quad flat package.

EXAMPLE 15.6

What signal is located at pin S17 of the 80486SX's PGA package?

Solution

Looking at Fig. 15.11(a), we find that signal corresponding to this pin is $\overline{\text{ADS}}$.

▲ 15.7 SIGNAL INTERFACES OF THE 80486SX MPU

Let us continue our study of the 80486 family of microprocessors by exploring its signal interfaces. Figure 15.12 is a block diagram showing the signal interfaces of the 80486SX MPU. Many of the 80486SX's interface signals are identical in name, mnemonic, and function to those on the 80386DX. For instance, we find that the 80486SX's 30 address bus lines are

| | | (a) | | | |

Figure 15.11 (a) Pin layout of the 80486SX PGA (Reprinted by permission of Intel Corp. Copyright/Intel Corp. 1992) (b) Signal pin numbering. (Reprinted by permission of Intel Corp. Copyright/Intel Corp. 1992)

Address		Data		Control		N/C	V_{CC}	V_{SS}
A_2	Q14	D_0	P1	A20M#	D15	A3	B7	A7
A_3	R15	D_1	N2	ADS#	S17	A10	B9	A9
A_4	S16	D_2	N1	AHOLD	A17	A12	B11	A11
A_5	Q12	D_3	H2	BE0#	K15	A14	C4	B3
A_6	S15	D_4	M3	BE1#	J16	B10	C5	B4
A_7	Q13	D_5	J2	BE2#	J15	B12	E2	B5
A_8	R13	D_6	L2	BE3#	F17	B13	E16	E1
A_9	Q11	D_7	L3	BLAST#	R16	B14	G2	E17
A_{10}	S13	D_8	F2	BOFF#	D17	B15	G16	G1
A_{11}	R12	D_9	D1	BRDY#	H15	B16	H16	G17
A_{12}	S7	D_{10}	E3	BREQ	Q15	C10	J1	H1
A_{13}	Q10	D_{11}	C1	BS8#	D16	C11	K2	H17
A_{14}	S5	D_{12}	G3	BS16#	C17	C12	K16	K1
A_{15}	R7	D_{13}	D2	CLK	C3	C13	L16	K17
A_{16}	Q9	D_{14}	K3	D/C#	M15	C14	M2	L1
A_{17}	Q3	D_{15}	F3	DP0	N3	G15	M16	L17
A_{18}	R5	D_{16}	J3	DP1	F1	R17	P16	M1
A_{19}	Q4	D_{17}	D3	DP2	H3	S4	R3	M17
A_{20}	Q8	D_{18}	C2	DP3	A5		R6	P17
A_{21}	Q5	D_{19}	B1	EADS#	B17		R8	Q2
A_{22}	Q7	D_{20}	A1	FLUSH#	C15		R9	R4
A_{23}	S3	D_{21}	B2	HLDA	P15		R10	S6
A_{24}	Q6	D_{22}	A2	HOLD	E15		R11	S8
A_{25}	R2	D_{23}	A4	INTR	A16		R14	S9
A_{26}	S2	D_{24}	A6	KEN#	F15			S10
A_{27}	S1	D_{25}	B6	LOCK#	N15			S11
A_{28}	R1	D_{26}	C7	M/IO#	N16			S12
A_{29}	P2	D_{27}	C6	NMI	A15			S14
A_{30}	P3	D_{28}	C8	PCD	J17			
A_{31}	Q1	D_{29}	A8	PCHK#	Q17			
		D_{30}	C9	PWT	L15			
		D_{31}	B8	PLOCK#	Q16			
				RDY#	F16			
				RESET	C16			
				W/R#	N17			

(b)

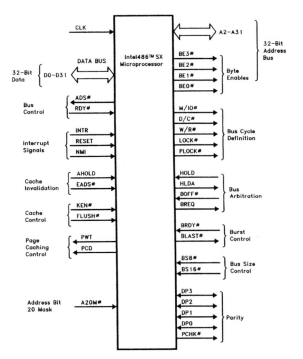

Figure 15.12 Block diagram of the 80486SX. (Reprinted by permission of Intel Corp. Copyright/Intel Corp. 1992)

labeled A_2 through A_{31}, and its 4 byte-enable signals, $\overline{BE}_0$ through $\overline{BE}_3$. However, most of the interfaces have some new signals that are provided to implement enhanced functions. In fact, the interrupt interface is the only interface that is completely unchanged. Here we will focus on the new signals at each of the 80486SX's interfaces.

Memory/IO Interface

Earlier we pointed out that many of the hardware enhancements of the 80486 family are in the memory interface. For this reason, most of the new signal lines of the 80486SX are located at this interface. Let us look at the function of these new signals.

The 80386DX MPU has the ability to configure the data bus as 16 bits instead of 32 bits by activating the $\overline{BS16}$ input. The 80486SX also has this capability; however, another input *bus size 8* ($\overline{BS8}$) also gives the ability to configure the data bus 8 bits wide. If $\overline{BS8}$ is at the active 0 logic level, data are transferred one byte at a time over data bus lines D_0 through D_7.

The bus cycle indication signals $M/\overline{IO}$, $D/\overline{C}$, and $W/\overline{R}$ of the 80486SX are the same as those on the 80386DX, except a change has been made in the coding of the bus cycles. The 80486SX's bus cycle indication codes are shown in Fig. 15.13. Notice that the halt/shutdown bus cycle is identified by the code 001_2, instead of 101_2. The code 101_2 is now reserved. Codes for all other types of bus cycles are unchanged.

An important difference between the memory interface of the 80386DX MPU and that of the 80486SX is that automatic *parity generation and checking* has been added. Parity has been added to each byte of the 80486SX's 32-bit data bus. An even parity bit is automatically generated for each byte of the data written to memory, and on read operations each byte of data is checked for even parity. For this reason, four bidirectional *data parity* (DP_0–DP_3)

M/IO	D/C	W/R	Bus Cycle Initiated
0	0	0	Interrupt acknowledge
0	0	1	Halt special cycle
0	1	0	I/O read
0	1	1	I/O write
1	0	0	Code read
1	0	1	Reserved
1	1	0	Memory read
1	1	1	Memory write

Figure 15.13 80486SX bus cycle indication codes.

lines and a *parity status* ($\overline{\text{PCHK}}$) output have been added into the memory interface. The data parity lines are additional data bus lines that are used to carry parity data to and from memory. On the other hand, the parity status output is used to signal external circuitry whether or not a parity error has occurred on a read operation. Logic 0 at this output identifies a parity error condition.

The 80486SX automatically performs an operation known as *address bit 20 mask*. This is an operation that must be performed in all ISA bus-compatible computers. In an 80386DX-based microcomputer, masking of A_{20} is accomplished with external circuitry. An extra input, *address bit 20 mask* ($\overline{\text{A20M}}$), has been added on the 80486SX to perform this function. Whenever $\overline{\text{A20M}}$ is logic 0, address bit A_{20} is masked out for bus cycles that access internal cache memory or external memory.

Another enhancement in the 80486 family of MPUs is the ability to perform what is known as *burst bus cycles*. A burst bus cycle is a special bus cycle that permits faster reads and writes of data. During a burst bus cycle, the transfer of the first element of data takes place in two clock cycles, and each additional data element is transferred in a single clock cycle. Nonburst bus cycles transfer one data element at a time and require a minimum of two clock cycles for each data transfer. Whenever the 80486SX requires data, it can perform the transfer with normal or burst bus cycles. If the external device can perform burst data transfers, it signals this fact to the MPU. This is done by switching the control signal *burst ready* ($\overline{\text{BRDY}}$), instead of $\overline{\text{RDY}}$, to logic 0.

During all memory bus cycles, the $\overline{\text{BLAST}}$ output signals when the last data transfer takes place. In the case of a normal bus cycle, only one data transfer takes place. This data transfer is marked by $\overline{\text{BLAST}}$ switching to logic 0. For burst cycle, multiple data transfers are performed, but $\overline{\text{BLAST}}$ is active only for the last one.

The 80486SX has a second type of lock signal. This signal, *pseudo-lock* ($\overline{\text{PLOCK}}$), differs from $\overline{\text{LOCK}}$ in that it locks out access to the bus by other devices for more than one bus cycle.

Cache Memory-Control Interface

A new group of interface signals is provided on the 80486 family MPUs to support internal and external cache memory subsystems. These signals include the *cache-enable* ($\overline{\text{KEN}}$) input, *cache-flush* ($\overline{\text{FLUSH}}$) input, the *page cache disable* (PCD) output, the *page write-through* (PWT) output, the *address hold* (AHOLD) input, and the *valid external address* ($\overline{\text{EADS}}$) input. Let us next look briefly at the function of each of these signals.

The external memory subsystem has the ability to tell the 80486SX whether or not a bus cycle is cacheable. It does this by switching the $\overline{\text{KEN}}$ input to logic 0 or 1. Whenever

$\overline{KEN}$ is set to logic 0 during a memory-read bus cycle, the information carried over the bus is copied into the on-chip cache. If $\overline{KEN}$ remains at its inactive 1 logic level, a noncacheable bus cycle takes place.

External circuitry also has the ability to invalidate all of the data in the on-chip cache memory of the 80486SX. This operation is known as a *cache flush*. To flush the on-chip cache, the $\overline{FLUSH}$ input is simply switched to its active 0 logic level for one clock cycle.

PCD and PWT are outputs of the MPU and are used to control external cache. Earlier we pointed out that the programmed logic levels of the page attribute bits in the page entry table, page directory table, or control register 3 are output on these lines when caching is enabled. The logic level of PCD signals the external cache memory subsystem whether or not the page of memory that is being accessed is configured as cacheable. Logic 1 at PWT indicates that write operations to the external cache are performed in a write-through fashion.

The last two signals, AHOLD and $\overline{EADS}$, are used to perform what is known as a *cache-invalidate cycle*. This type of bus cycle is used to maintain consistency between data in the internal cache and external main memory. For instance, if another bus master device modifies data in main memory, a check must be made immediately to see if the contents of this storage location are currently held in the 80486SX's internal cache memory. If they are and the 80486SX would read the value at this address, the wrong value would be accessed from the cache. For this reason, the value in the cache must be invalidated.

The signal AHOLD is used to tristate the address bus during a cache-invalidate cycle. The first step in this cycle is for an external device to apply logic 1 to the AHOLD input of the 80486SX. In response to this input, the address lines are immediately put into the high-Z state. Next, the external device puts the address of the main memory storage location that was modified onto the 80486SX's address bus and then switches $\overline{EADS}$ to logic 1 to signal the MPU that a valid address is on the address bus. In this case, the address lines are inputs to the MPU. If the internal cache subsystem identifies that the contents of this memory location is stored in the cache, the value held in the cache is invalidated. Since the cache entry is no longer valid, consistency is restored between cache memory and main memory.

Bus Arbitration Interface

The DMA interface of the 80386DX is expanded in the 80486SX MPU to make it into what is called the *bus arbitration interface*. Two new signals, *backoff input* ($\overline{BOFF}$) and *bus request output* (BREQ), have been added to the interface. Let us look at the function of each of these signals.

The BREQ output of the 80486SX signals whether or not the MPU is generating a request to use the external bus. When a bus cycle is to be performed, BREQ is switched to logic 0 and remains at this level until the bus cycle is completed. This signal can be used by external circuitry to tell other bus masters that the MPU has a bus access pending.

$\overline{BOFF}$ is similar to the HOLD input of the MPU in that its active logic level tristates the bus interface signals. However, there are two differences between the operation of $\overline{BOFF}$ and HOLD. First, a bus backoff operation is initiated at the completion of the current clock cycle, not at the end of the current bus cycle. Moreover, no hold-acknowledge response is made to external circuitry. When $\overline{BOFF}$ returns to its inactive logic level, the interrupted bus cycle is restarted. This input can be used by external bus masters to quickly take over control of the system bus.

▲ 15.8 MEMORY AND I/O SOFTWARE ORGANIZATION, HARDWARE ORGANIZATION, AND INTERFACE CIRCUITS

The software and hardware organization of the memory and input/output address spaces of the 80486SX microprocessor are identical to those of the 80386DX MPU. Figure 15.14(a) shows the protected-mode software organization of these address spaces. Here we see that it consists of the 4G-byte addresses in the range 00000000_{16} through $FFFFFFFF_{16}$, and the isolated I/O address space is the 64K-byte addresses from 0000_{16} through $FFFF_{16}$. Since the 80486SX has a 32-bit data bus, the memory and I/O address spaces are organized from a hardware point of view 32 bits wide. The organization of the memory-address space as 1M double words is shown in Fig. 15.14(b). Notice that double word zero is located at address 0000000_{16} and that its bytes are selected with byte select signals $\overline{BE}_0$ through $\overline{BE}_3$.

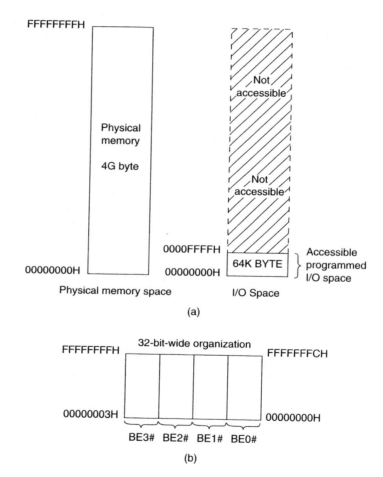

Figure 15.14 (a) Software organization of the memory and I/O address space of the 80486SX. (b) Organized as double words of data.

EXAMPLE 15.7

What is the address of the next-to-highest addressed double-word storage location in Fig. 15.14(b)?

Solution

From Fig. 15.14(b), we see that the highest byte of this word is at address $FFFFFFFB_{16}$. Therefore, the double word is located at address $FFFFFFF8_{16}$.

The data types supported by the 80486SX's instruction set are the same as those supported by the 80386DX. The chart in Fig. 15.15 summarizes these data types. Included are the type, range, precision, and memory organization. For instance, 8-bit, 16-bit, and 32-bit integer numbers are supported. The range of the 16-bit integer is identified as 10^4, and its precision is 16 bits.

EXAMPLE 15.8

What are the range and precision of an 8-bit packed BCD number?

Solution

Looking at Fig. 15.15, we find that the range is numbers 0 through 9, and the precision is 2 digits.

Data Format	Supported by base registers	Range	Precision	Memory organization
Byte	X	0–255	8 bits	7 – 0
Word	X	0–64K	16 bits	15 – 0
Dword	X	0–4G	32 bits	31 – 0
8-Bit integer	X	10^2	8 bits	2's complement, 7 – 0
16-Bit integer	X	10^4	16 bits	2's complement, 15 Sign bit 0
32-Bit integer	X	10^9	32 bits	2's complement, 31 Sign bit 0
8-Bit unpacked BCD	X	0–9	1 digit	One BCD digit per byte, 7 – 0
8-Bit packed BCD	X	0–9	2 digits	Two BCD digits per byte, 7 – 0

Least significant byte

Figure 15.15 80486SX data types.

The 80386DX has the ability to dynamically adjust the physical width of the data bus as either 32 bits or 16 bits. This capability has been expanded further in the 80486SX to enable sizing of the data bus as 32 bits wide, 16 bits wide, or 8 bits wide. The memory interface signals that select the size of the bus are *bus size 16* ($\overline{BS16}$) and *bus size 8* ($\overline{BS8}$). If both these signals are at their active 0 logic level, the bus defaults to 8-bit-wide operation.

EXAMPLE 15.9

How would the data bus of the 80486SX be configured if both $\overline{BS16}$ and $\overline{BS8}$ are wired to +5 V?

Solution

The 80486SX's data bus is permanently set for 32-bit-wide operation.

Figure 15.16 shows a typical application where the memory address space is partitioned in this way. Notice that a single-boot EPROM is used in an 8-bit memory configuration; a memory-mapped peripheral that has a 16-bit data bus in its microprocessor interface is treated as a 16-bit segment of memory; and the main program and data-storage memories are both 32 bits wide.

If the data bus is sized at 16 bits or 8 bits, additional bus cycles may need to be performed to complete a read- or write-data transfer. For instance, if $\overline{BS16}$ is logic 0 when an instruction is executed that initiates an aligned 32-bit read from the memory-mapped peripheral in Fig. 15.16, the data bus is set for 16-bit width, and two consecutive 16-bit

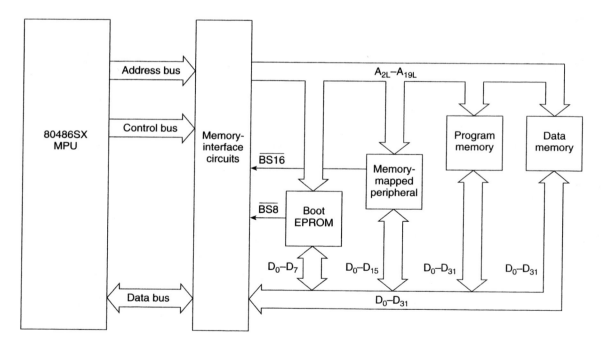

Figure 15.16 8-bit, 16-bit, and 32-bit memory subsystems.

data-read bus cycles will take place. Moreover, when an instruction fetch occurs from the boot ROM, $\overline{BS8}$ is active during this memory operation, and the data transfer is performed with four 8-bit data-read bus cycles.

EXAMPLE 15.10

How many bus cycles are required to read the double word of data from address $F1001_{16}$ in the address space of the memory-mapped peripheral in Fig. 15.15?

Solution

Since the 32-bit word at address $0FFF1_{16}$ is misaligned, three 16-bit read-data transfer bus cycles must take place.

Implementation of a memory and input/output interface that employs dynamic bus sizing, such as that in Fig. 15.16, requires additional address decoding, byte-select logic, and data bus multiplexing circuitry. The block diagram in Fig. 15.17(a) includes the address decode and byte-select logic needed to implement both 8-bit and 16-bit dynamic bus sizing for the memory subsystem in an 80486SX-based microcomputer. All address lines are inputs to the address-decode block. This circuit decodes each address in the range of the memory-mapped peripheral device to activate the $\overline{BS16}$ input of the MPU and tell it that the data bus is configured 16 bits wide. Addresses that correspond to the address range of the EPROM are decoded to produce $\overline{BS8}$ and set the data bus for 8-bit-wide operation.

When the data bus of the 80386SX is operating in the 8-bit or 16-bit mode, the byte-enable signals still identify which byte or bytes of data are being transferred over the bus, but they can no longer be used directly to enable the memory banks. Instead they must be decoded to produce address bits and high and low byte-enable signals. This is the role of the byte-select logic circuit that has been added in the memory interface of Fig. 15.17(a).

Looking at the 16-bit memory interface in Fig. 15.17(a), we see that signals $\overline{BE_0}$ through $\overline{BE_3}$ are decoded by the byte-select logic to produce three new bus-control signals, *byte high enable* ($\overline{BHE}$), *byte low enable* ($\overline{BLE}$), and *address bit one* (A1). The 80486SX is capable of making either 8-bit or 16-bit data transfers through this 16-bit memory interface. Logic 0 at $\overline{BLE}$ tells the memory subsystem that a byte data transfer is taking place over data bus lines D_7 through D_0, and logic 0 at $\overline{BHE}$ indicates that a byte of data is carried over D_{15} through D_8. Whenever a word of data is transferred over the bus, both $\overline{BLE}$ and $\overline{BHE}$ are switched to their active logic level.

The table in Fig. 15.17(b) shows the relationship between the byte-enable signals of the 80486SX and the 8-bit and 16-bit bus-control signals produced at the outputs of the decoder. From this table, we find that output A_1 of the byte-select logic circuit must be at the 0 logic level whenever either $\overline{BE_0}$ or $\overline{BE_1}$ is at logic 0 (L) and if both of them are logic 0 together. This operation is implemented with the NAND gate and inverter connection shown in Fig. 15.17(c).

EXAMPLE 15.11

Use the table of Fig. 15.17(b) to determine the conditions of the byte-enable signals for which the $\overline{BHE}$ output of the byte-select logic circuit must be logic 0. Show a logic circuit that will implement this input-output logic function.

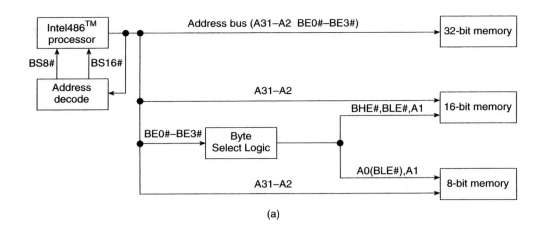

(a)

Intel486™ SX microprocessor				8-, 16-bit Bus signals			Comments
BE3#	BE2#	BE1#	BE0#	A1	BHE#	BHE# (A0)	
H*	H*	H*	H*	x	x	x	x—no active bytes
H	H	H	L	L	H	L	
H	H	L	H	L	L	H	
H	H	L	L	L	L	L	
H	L	H	H	H	H	L	
H*	L*	H*	L*	x	x	x	x—not contiguous bytes
H	L	L	H	L	L	H	
H	L	L	L	L	L	L	
L	H	H	H	H	L	H	
L*	H*	H*	L*	x	x	x	x—not contiguous bytes
L*	H*	L*	H*	x	x	x	x—not contiguous bytes
L*	H*	L*	L*	x	x	x	x—not contiguous bytes
L	L	H	H	H	L	L	
L*	L*	H*	L*	x	x	x	x—not contiguous bytes
L	L	L	H	L	L	H	
L	L	L	L	L	L	L	

BLE# asserted when D0–D7 of 16-bit bus is active.
BHE# asserted when D8–D15 of 16-bit bus is active.
A1 low for all even words; A1 high for all odd words.

Key:
  x = don't care
  H = high voltage level
  L = low voltage level
  * = a nonoccurring pattern of Byte Enables; either none are asserted,
      or the pattern has Byte Enables asserted for noncontiguous bytes

(b)

Figure 15.17 (a) Address decode and byte-select logic for a memory interface with dynamic bus sizing. (Reprinted by permission of Intel Corp. Copyright/Intel Corp. 1994) (b) Relationship between byte-enable signals and 8-bit and 16-bit bus address signals. (Reprinted by permission of Intel Corp. Copyright/Intel Corp. 1994)

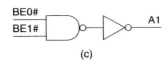

(c)

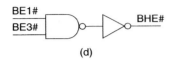

(d)

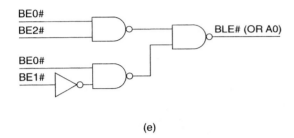

(e)

Figure 15.17 (c) A_1 control signal circuitry. (Reprinted by permission of Intel Corp. Copyright/Intel Corp. 1994) (d) $\overline{\text{BHE}}$ control-signal circuitry. (Reprinted by permission of Intel Corp. Copyright/Intel Corp. 1994) (e) $A_0/\overline{\text{BLE}}$ control-signal circuitry. (Reprinted by permission of Intel Corp. Copyright/Intel Corp. 1994)

Solution

Comparing the $\overline{\text{BHE}}$ output column of the table to the BE inputs, we find that the output is logic 0 if the $\overline{\text{BE}}_1$ input is logic 0 (L), if $\overline{\text{BE}}_3$ is logic 0, or if both $\overline{\text{BE}}_1$ and $\overline{\text{BE}}_3$ are logic 0 together. A logic circuit that implements the $\overline{\text{BHE}}$ function is shown in Fig. 15.17(d).

Let us now look briefly at the 8-bit memory interface in Fig. 15.17(a). Here the byte enables are decoded to make control signals *address bit 0/byte low enable* ($A_0/\overline{\text{BLE}}$) and address bit 1 ($A_1$). The table in Fig. 15.17(b) shows that these outputs of the byte-select logic circuit are produced for the exact same input conditions as for the 16-bit memory interface. A circuit that implements A_1 was developed for the 16-bit bus interface. Figure 15.17(e) shows a circuit that performs the $A_0/\overline{\text{BLE}}$ logic function. The design of the address-decode and byte-swap logic circuits are a good application for a PAL device.

The way in which the 80486SX reads and writes data over the 16-bit data bus when $\overline{\text{BS16}}$ is asserted is different from how these data transfers are performed by the 80386DX. The 80386DX provides data duplication that assures that all data transfers automatically take place across data bus lines D_0 through D_{15}. Data duplication is not implemented on the 80486SX; instead, this function must be performed with external hardware. The table in Fig. 15.18 summarizes how the 80486SX performs data transfers over a 32-bit, 16-bit, and 8-bit data bus. Notice for the code $\overline{\text{BE}}_3\overline{\text{BE}}_2\overline{\text{BE}}_1\overline{\text{BE}}_0 = 1100$, the word data transfer takes place over the low-data bus lines, D_{15} through D_0, whether the data bus is set up either 32 bits wide or 16 bits wide. When the byte-enable code is 0011, the path of the word data transfer in both 32-bit and 16-bit modes is over the upper 16 data bus lines D_{31} through D_{16}. If this data transfer takes place when the data bus is configured 16 bits wide, we expect the data to be carried over D_{15} through D_0, not D_{31} through D_{16}. For this byte-enable code and data transfer, the 80386DX performed data bus duplication and output another copy (a duplicate) of the data on data lines D_{15} through D_0. Since the 80486SX does not do this

BE3#	BE2#	BE1#	BE0#	w/o BS8# /BS16#	w BS8#	w BS16#
1	1	1	0	D7–D0	D7–D0	D7–D0
1	1	0	0	D15–D0	D7–D0	D15–D0
1	0	0	0	D23–D0	D7–D0	D15–D0
0	0	0	0	D31–D0	D7–D0	D15–D0
1	1	0	1	D15–D8	D15–D8	D15–D8
1	0	0	1	D23–D8	D15–D8	D15–D8
0	0	0	1	D31–D8	D15–D8	D15–D8
1	0	1	1	D23–D16	D23–D16	D23–D16
0	0	1	1	D31–D16	D23–D16	D31–D16
0	1	1	1	D31–D24	D31–D24	D31–D24

Figure 15.18 Data-transfer paths for 32-bit, 16-bit, and 8-bit memory interfaces. (Reprinted by permission of Intel Corp. Copyright/Intel Corp. 1994)

data-swap operation, it must be performed with external hardware in the memory interface of the microcomputer. Figure 15.19 shows the addition of *byte-swap logic* to a memory interface with dynamic bus sizing.

When the data bus is set for 16-bit or 8-bit mode, some word and all double-word data transfers require multiple bus cycles. The byte-enable code output during the bus cycles are different. Figure 15.20 shows the value of the byte-enable code for the first bus cycle of the data transfer and that for the next cycle for both 8-bit-wide and 16-bit-wide bus operation. For instance, $\overline{BE_3}\overline{BE_2}\overline{BE_1}\overline{BE_0} = 0000$ stands for an aligned 32-bit data transfer. Assuming the $\overline{BS16}$ is active when this data transfer is initiated, the word transfer that takes place during the second data-transfer bus cycle is accompanied by the byte-enable code 0011. Looking at Fig. 15.18, we see that the data during the first bus cycle is transferred over data bus lines D_{15} through D_0. When the second bus cycle is performed, the data are carried over data bus lines D_{31} through D_{16}. During a read transfer the data from the 16-bit memory system data bus must be switched to D_{31} through D_{16} for input to the MPU.

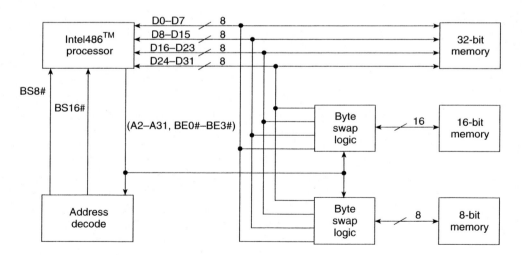

Figure 15.19 Data multiplexing for a memory interface with dynamic bus sizing. (Reprinted by permission of Intel Corp. Copyright/Intel Corp. 1994)

Current				Next with BS8#				Next with BS16#			
BE3#	BE2#	BE1#	BE0#	BE3#	BE2#	BE1#	BE0#	BE3#	BE2#	BE1#	BE0#
1	1	1	0	n	n	n	n	n	n	n	n
1	1	0	0	1	1	0	1	n	n	n	n
1	0	0	0	1	0	0	1	1	0	1	1
0	0	0	0	0	0	0	1	0	0	1	1
1	1	0	1	n	n	n	n	n	n	n	n
1	0	0	1	1	0	1	1	1	0	1	1
0	0	0	1	0	0	1	1	0	0	1	1
1	0	1	1	n	n	n	n	n	n	n	n
0	0	1	1	0	1	1	1	n	n	n	n
0	1	1	1	n	n	n	n	n	n	n	n

"n" means that another bus cycle will not be required to satisfy the request.

Figure 15.20 Byte-enable codes for multiple-cycle data transfers with BS8 and BS16 active. (Reprinted by permission of Intel Corp. Copyright/ Intel Corp. 1994)

On the other hand, during a write cycle the data output on D_{31} through D_{16} must be switched to lines D_{15} through D_0 of the memory system data bus. This switching of data is performed by the data bus byte-swap logic.

EXAMPLE 15.12

Assume that the $\overline{BS8}$ signal is active when an aligned double-word read operation is initiated by the 80486SX. What are the values of the byte-enable signals and over which data bus lines must the data be input to the MPU by the byte-swap logic during each of the bus cycles?

Solution

To perform a 32-bit write operation to memory over an 8-bit memory system data bus takes four read bus cycles. From Figs. 15.20 and 15.18 the byte-enable code and data path for each cycle are found to be as follows:

Bus cycle	Byte-enable code	Data path
1	0000	D_7-D_0
2	0001	$D_{15}-D_8$
3	0011	$D_{23}-D_{16}$
4	0111	$D_{31}-D_{24}$

▲ 15.9 NONBURST AND BURST BUS CYCLES

In our description of the memory interface, we found that the 80486SX's data bus can be dynamically configured for an 8-bit, a 16-bit, or a 32-bit mode of operation. For each of these modes, the 80486SX can perform either of two bus cycles, called a *nonburst bus cycle* or a *burst bus cycle*. Both of these cycles can be made cacheable or noncacheable. Here we look briefly at the bus activity for each of these bus cycles.

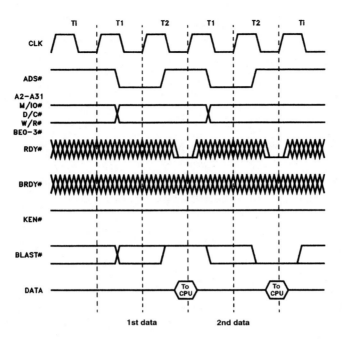

Figure 15.21 Nonburst, noncacheable bus cycle. (Reprinted by permission of Intel Corp. Copyright/Intel Corp. 1992)

Nonburst, Noncacheable Bus Cycle

The timing diagram in Fig. 15.21 shows the sequence of bus activity that takes place as the 80486SX reads or writes data to memory or an I/O device with a *nonburst, noncacheable bus cycle*. Notice that the minimum duration of a bus cycle is two clock cycles. They are identified as T_1 and T_2 in the bus cycle timing diagram.

Looking at Fig. 15.21, we see that early in clock cycle T_1 the address (A_2 through A_{31}), byte enables ($\overline{BE}_0$ through $\overline{BE}_3$), and memory indication signals ($M/\overline{IO}$, $D/\overline{C}$, and $W/\overline{R}$) are made available and latched into external circuitry with the transition of $\overline{ADS}$. Assuming that the data transfer is to take place in a single bus cycle, $\overline{BLAST}$ is switched to the 0 logic level during clock cycle T_2. This tells the external circuitry that the data transfer is to be complete at the end of the current bus cycle. Therefore, at the end of T_2, external circuitry switches $\overline{RDY}$ to logic 0 to tell the MPU that the data transfer is to take place.

A bus cycle can be extended by any number of clock cycles by holding $\overline{RDY}$ at logic 1 during T_2.

Nonburst, Cacheable Bus Cycle

Earlier we pointed out that the $\overline{KEN}$ input determines whether or not a bus cycle is cacheable. Figure 15.22 shows the 80486SX's *nonburst, cacheable bus cycle*. Here we see that the bus cycle starts the same way, but later in T_1 the external circuitry switches the $\overline{KEN}$ input to logic 0. This indicates that the cycle is a cacheable bus cycle. Only memory-read bus cycles are cacheable; therefore, the MPU ignores $\overline{KEN}$ during all write and I/O bus cycles.

Information is stored in the internal cache memory as *lines*, which are 16 bytes wide. Whenever a cacheable read cycle is performed, a complete line of code or data (four double words), instead of a single 32-bit word, is read from memory. Looking at the timing diagram

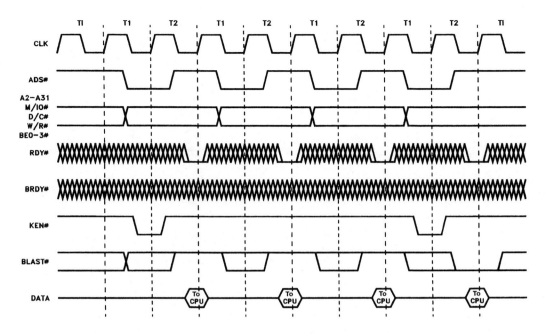

Figure 15.22 Nonburst, cacheable bus cycle. (Reprinted by permission of Intel Corp. Copyright/Intel Corp. 1992)

in Fig. 15.22, we see that four read-data transfers take place. The address is automatically adjusted to point to the appropriate double word after each read operation and then $\overline{RDY}$ is made active. For this reason, the complete bus cycle takes a total of eight clock states. Notice that $\overline{BLAST}$ is not switched to the active 0 logic level until T_2 of the fourth and last double word of data is read.

Burst, Cacheable Bus Cycle

The *burst, cacheable bus cycle* of Fig. 15.23 is similar to the nonburst, cacheable bus cycle we just described in that four read operations are performed. One difference is that during a burst bus cycle external circuitry signals that burst data transfers are to take place by replying with logic 0 on $\overline{BRDY}$ instead of $\overline{RDY}$. A second and very important difference is that only the first data transfer takes two clock cycles. Notice that all four data transfers of the burst bus cycle are completed in just five clock states.

▲ 15.10 CACHE MEMORY OF THE 80486SX

In Chapter 10 we found that addition of a cache memory subsystem to the microcomputer provides a means for improving overall system performance while still permitting the use of low-cost, slow-speed memory devices in main memory. The cache memory is a second small, but very fast, memory section that is added between the MPU and main memory subsystem. The objective is that the MPU accesses code and data in the cache most of the time, instead of from main memory. This results in close to zero-wait-state memory system

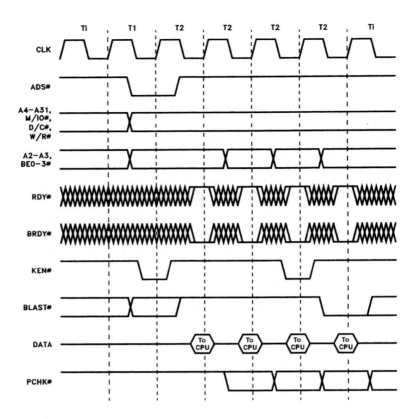

Figure 15.23 Burst, cacheable bus cycle. (Reprinted by permission of Intel Corp. Copyright/Intel Corp. 1992)

operation even though accesses of the main memory require one or more wait states, thus resulting in higher performance for the microcomputer.

As shown in Fig. 15.24(a), the 80486SX's cache differs from the external cache we studied in Chapter 10 in that it is on-chip, that is, internal to the MPU. External cache memories are also widely used in high-performance 80486-based microcomputer systems. Figure 15.24(b) shows the architecture of an 80486SX memory interface that employs an external cache. The internal cache of the 80486SX is called the *first-level cache* and the external cache, a *second-level cache*. Notice in Fig. 15.24(b) that on one side the external cache memory subsystem attaches to the local bus of the MPU, and at the other side it drives the system bus of the microcomputer system. External caches typically range in size from 128KB to 512KB and can be used to cache both data and code. The 80486SX's internal cache also stores both code and data, but is smaller (8KB) in size.

Organization and Operation of the 80486SX's Internal Cache

Let us next examine the organization and operation of the 80486SX's internal cache memory. The 80486SX's internal 8KB cache memory differs from the 82385DX cache controller we studied earlier in that it uses a four-way set associative memory. Therefore, its configuration is similar to that shown for the 82385DX in Fig. 10.53 except that it is arranged into four 2KB banks. Similar to the 82385DX, it employs the LRU replacement

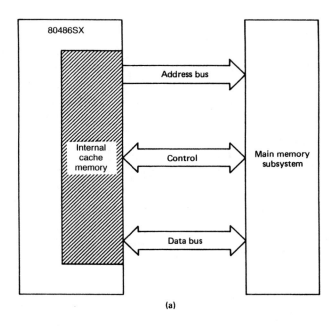

(a)

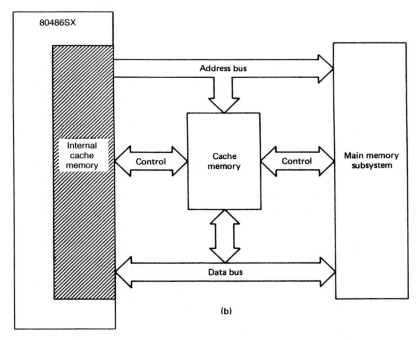

(b)

Figure 15.24 (a) Internal cache of the 80486SX. (b) 80486SX microcomputer with external cache memory.

algorithm to decide which data get replaced. We will begin by determining how data are stored in the cache memory array.

Since the internal cache uses a four-way set associative organization, the data-storage area is partitioned into four separate 2KB areas. Figure 15.25 illustrates how this memory is organized. We will refer to these areas of memory as *SET 0* through *SET 3*. Data are

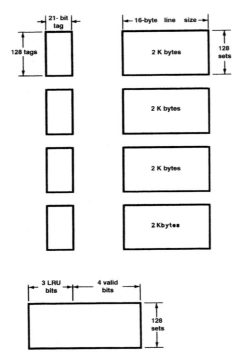

Figure 15.25 Organization of the on-chip cache of the 80486SX. (Reprinted by permission of Intel Corp. Copyright/Intel Corp. 1992)

loaded, stored, validated, and invalidated in 16-byte-wide elements, which are called a *line of data*. The 80486SX's cache does not support filling of partial lines. Therefore, if a single double word of data is to be read from memory and copied into the cache, the MPU must fetch from memory the complete line in which this double word is contained. This is the reason that cacheable read bus cycles initiate four double-word data transfers. That is, they always access code or data a line (16 bytes) at a time so that a complete line of the cache gets filled. In this way, we see that the contents of each 2KB bank in the cache's storage array is further arranged into 128 lines of data.

Associated with each data-storage set is a separate tag directory. The tag directory contains 128 21-bit tag entries, one corresponding to each line of data in the set. Each tag entry includes bits of information about the use of the line and whether or not it currently holds valid information.

When a read operation is initiated by the MPU and the information that is to be accessed is already in the internal cache, the information is obtained without performing external bus cycles. Instead, the information is simply read from the cache memory. However, if a cache miss occurs, a line of code or data must be read from main memory and copied into the cache. The on-chip cache circuitry must determine if there is room in the cache and, if not, which of the current valid lines of information is to be replaced. It does this by checking the information in the tag directory. If the cache is found to contain invalid lines of data or code, one of them can be simply replaced with the new information. On the other hand, if there are no vacant line-storage locations, the least recently used mechanism of the cache automatically checks the use information in the tag to determine which valid line of information will be replaced.

Let us next look at what happens when a data-write operations is performed. Whenever a new value of data is to be written to a storage location in memory, the internal cache circuitry must first be checked to confirm whether or not the contents of this storage location also exist within the cache. If it does, the value must be either invalidated or updated as part of the write operation. Otherwise, a cache data consistency problem will be created.

The 80486SX's on-chip cache is implemented with a write-update method that is known as *write-through*. With this method, all write bus cycles that result in a cache hit automatically update both the corresponding storage location in the internal cache and external memory. That is, write operations to main memory can be viewed as going through the cache. On the other hand, for a cache miss, the data are written only to main memory. Remember that cache write-through operation can be enabled or disabled with the no-write-through (NW) bit of CR_0. However, write-through would not be disabled during normal operation.

Enabling and Disabling Internal Caching

The 80486SX's internal cache is equipped with a variety of methods of controlling the operation of the cache. For instance, the complete cache memory can be turned on or off, the memory-address space can be mapped with cacheable and noncacheable areas, and external circuitry can define any bus cycle as cacheable or noncacheable. Let us briefly review how each of these cache memory controls are implemented on the 80486SX.

Remember that the operation of the cache memory can be enabled or disabled under software control. Logic 1 in the cache fill disable (CD) bit of control register CR_0 can be used to turn off filling of the cache. However, this does not completely disable the cache; it just stops it from being refilled. To completely disable the cache, the no-write-through (NW) bit in CR_0 must also be set to 1 and then the cache must be flushed. The flush operation is needed to remove the stale data that were left in the cache when cache fill was turned off.

Mapping of parts of the memory address space as cacheable or noncacheable can be achieved either through software or hardware. Under software control, each page of the memory address space can be configured as cacheable or noncacheable with the page-level cache-disable (PCD) bit in its page table entry. For instance, to make a page of memory noncacheable, the PCD bit is made 0. The memory address space can be mapped cacheable or noncacheable on a byte-wide basis by external circuitry. The *cache-enable* ($\overline{KEN}$) input can be used to indicate whether or not the data for the current bus cycle should be cached. By decoding addresses in external circuitry and returning logic 1 at $\overline{KEN}$, a part of the address space can be designated as noncacheable.

Flushing the Cache

When the internal cache is flushed, all the line valid bits in the tag directory are invalidated. Therefore, after a *flush* occurs, the cache is empty and will need to be refilled. The cache is flushed whenever the MPU is reset; it can be flushed under software control by executing the invalidate cache (INVD) instruction or with external circuitry by activating the $\overline{FLUSH}$ input.

Cache Line Invalidations

We just described how the complete contents of the cache can be invalidated with a flush operation. It is also possible to invalidate individual lines of information within the cache. This is known as a *cache line invalidation* and is normally done to make the contents of the internal cache consistent with that of external memory.

If an external device changes the contents of a storage location in external memory and the value of this storage location is currently held in the 80486SX's internal cache memory, a cache consistency problem can occur. That is, the corresponding storage location in cache and external memory no longer contain the same value and the value in cache is no longer valid. If the 80486SX were to read this storage location, the incorrect value in the cache would be accessed, and if this data were processed and written back to memory, the results in external memory would then also be wrong.

To protect against inconsistency problems between the contents of the internal cache and external memory, external circuitry needs to initiate a cache line invalidation operation each time an external device modifies the content of a storage location in external memory. This is done by initiating an invalidate bus cycle. The first step in this process is to switch *address hold* (AHOLD) to logic 1. This puts the address bus lines into the high-Z state. Next, the external circuity puts the address of the external memory storage location whose contents were changed onto the address lines and signals the 80486SX that a valid external address is available by switching $\overline{EADS}$ (*external address*) to logic 0. Now the address lines of the MPU act as inputs, instead of outputs. If this address corresponds to an element of data that is currently held in cache, the corresponding cache entry is invalidated. In this way, cache consistency is restored.

▲ 15.11 HIGH-INTEGRATION MEMORY/INPUT/OUTPUT PERIPHERAL—R400EX

High-integration companion ICs are available to enable the implementation of a compact, low-cost, versatile, 80486-based microcomputer solution for embedded applications. Here we will examine the R400EX IC that is manufactured by RadiSys[R] Corporation. The R400EX is compatible with the 80486SX, 80486DX2, and 80486DX4 MPUs and is designed to attach directly to the local bus of these MPUs without the need for any additional interface logic. It contains the memory and input/output peripheral devices needed to implement a PC-compatible embedded microcomputer.

Let us next look briefly at the memory interface and peripheral function implemented within the R400EX. A block diagram of the device is given in Fig. 15.26. Here we find that the device contains many of the peripheral ICs used in a PC/AT-compatible microcomputer. Examples are the 82C54 counter/timer, 82C37 DMA controller, 16C550 UART, and 146818 real-time clock. In addition, other PC-compatible hardware functions, such as a DRAM controller, enhanced IDE drive interface, watchdog timer, nonmaskable interrupt, parity generating and checking logic, and a keyboard/mouse interface, are implemented within the R400EX. These devices and hardware functions are interconnected to implement a flexible PC-compatible microcomputer system.

The R400EX also provides the signals for an ISA bus interface. These signals permit easy implementation of an ISA-compatible system bus for the embedded microcomputer. This bus can be used to connect special-purpose I/O functions and additional memory.

The R400EX is housed in a 208-pin PQFP, as shown in Fig. 15.27.

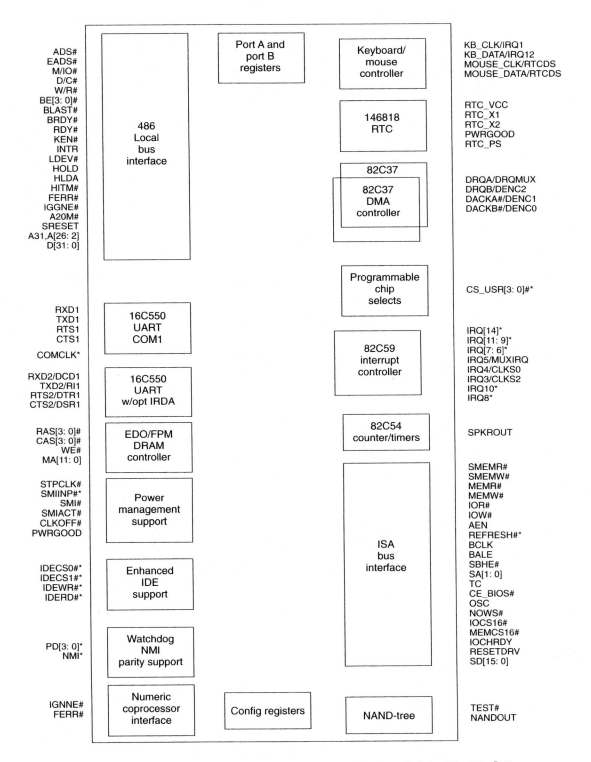

Figure 15.26 Block diagram of the R400EX. (Reprinted by permission of RadiSys[R] Corporation)

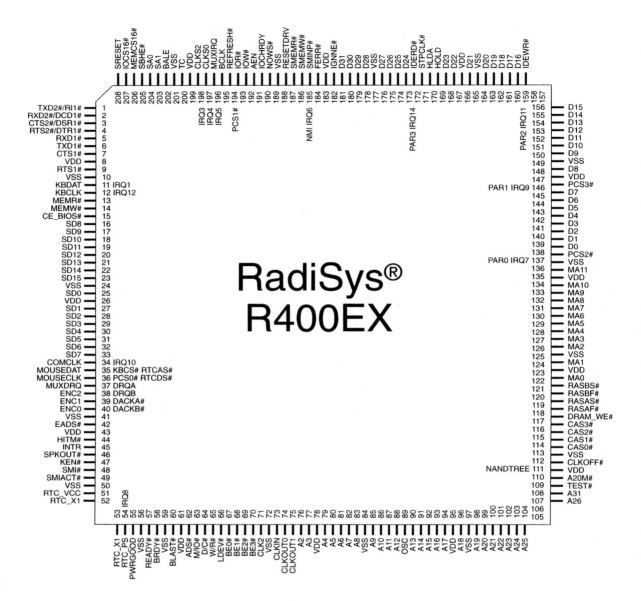

Figure 15.27 Pin layout of the R400EX 80486 MPU companion IC. (Reprinted by permission of RadiSys[R] Corporation)

▲ 15.12 INTERRUPTS, RESET, AND INTERNAL EXCEPTIONS

The interrupt and exception-processing capability of the 80486SX MPU is essentially the same as that of the 80386DX. For example, the hierarchy in which interrupts and exceptions are serviced, priority levels they are assigned, interrupt-vector and interrupt-descriptor tables used to define their starting points, the interrupt–gate descriptor formats used to

822

define protected-mode operation, external hardware interrupt interface, and interrupt-acknowledge bus cycle are identical to those described earlier for the 80386DX. Just as for the 80386DX, the 80486SX microprocessor supports up to 256 interrupts and exceptions. Again, they are grouped into the categories of external hardware interrupts, nonmaskable interrupt, software interrupts, internal interrupts and exceptions, and reset. Here we will briefly review the function of each group.

External Hardware Interrupts and the Nonmaskable Interrupt

Earlier we found that the external hardware interrupts and nonmaskable interrupt (NMI) allowed the MPU to respond quickly to events that occur in external hardware. The 80486SX and 80386DX MPUs service both of these types of interrupt in the same way. Most external devices that are interrupt-driven are serviced with the external hardware-interrupt interface. Some examples of common devices that are interrupt-driven are keyboards, floppy disk drives, hard disk drives, and printers. These devices issue an external hardware-interrupt request by applying logic 1 to the INTR input of the 80486SX. At the completion of execution of the current instruction, the MPU tests the logic level at INTR and, if active, it initiates an interrupt-acknowledge bus cycle to signal that the request for service has been accepted and then passes program control to the selected interrupt-service routine. INTR is a level-sensitive input; therefore, the active request signal must be maintained at logic 1 until acknowledged. However, the request for service must be removed before the service routine is completed; otherwise, it will be detected as still active and acknowledged a second time.

The operation of the 80486SX's nonmaskable interrupt is similar to that just described for external hardware interrupts. Let us just look at the important differences between them. They are that the request for service is applied to the NMI input instead of the INTR input; requests for service at NMI are not masked by the interrupt flag; NMI is an edge-triggered, not level-sensitive, input, and the request for service does not have to be maintained because it is latched within the MPU; NMI requests are serviced immediately, not at the completion of the current instruction; and no interrupt-acknowledge bus cycles are performed.

NMI is normally assigned to the highest-priority external hardware interrupt. Typically this is detection of the power-fail condition. It is used to initiate a service routine that performs an orderly power-down of the microcomputer system. The power-down must be complete before the power supply voltages fall below the system's minimum operating voltage level.

Reset

The reset function of the 80486SX is also similar to that described earlier for the 80386SX MPU. A reset operation is needed to assure that the microcomputer system performs an orderly start-up when power is turned on. To initiate a reset of the 80486SX MPU, the RESET input must be held at its active 1 logic level for 15 clock periods. Waveforms that show the initiation and state of the MPUs signals during reset are given in Fig. 15.28 and the state of the registers after reset is shown in Fig. 15.29.

If the signal AHOLD is at its active logic level for one clock period before the return of RESET to the 0 logic level and maintained for one clock after this edge and also FLUSH is active and A20M is inactive at the falling edge of RESET, the *built-in self-test* (BIST) of the 80486SX is initiated as part of the reset operation. BIST is a diagnostic routine that runs a series of tests on the MPU to verify its correct operation. Just as for

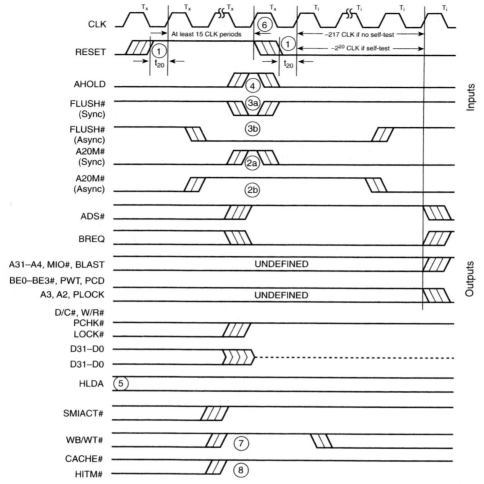

Notes:
1. Reset is an asynchronous input. t_{20} must be met only to guarantee recognition on a specific clock edge.

2a. When A20M# is driven synchronously, it should be driven high (inactive) for the CLK edge prior to the falling edge of RESET to ensure proper operation. A20M# setup and hold times must be met.

2b. When A20M# is driven asynchronously, it should be driven low (active) for two CLKs prior to and two CLKs after the falling edge of RESET to ensure proper operation.

3a. When FLUSH# is driven synchronously, it must be driven low (high) for the CLK edge prior to the falling edge of RESET to invoke the 3-state Output Test Mode. All outputs are guaranteed 3-stated within 10 CLKs of RESET being deasserted. FLUSH# setup and hold times must be met.

3b. When FLUSH# is driven asynchronously, it must be driven low (active) for two CLKs period prior to and two CLKs after the falling edge of RESET to invoke the 3-state Output Test Mode. All outputs are guaranteed 3-stated within 10 CLKs of RESET being deasserted.

4. AHOLD should be driven high (active) for the CLK edge prior to the falling edge of RESET to invoke the Built-In-Self-Test (BIST). AHOLD setup and hold times must be met.

5. Hold is recognized normally during RESET. On power-up HLDA is indeterminate until RESET is recognized by the processor.

6. 15 CLKs RESET pulse width for warm resets. Power-up resets require RESET to be asserted for at least 1 ms after V_{CC} and CLK are stable.

7. WB/WT# should be driven high for at least one CLK before falling edge of RESET and at least one CLK after falling edge of RESET to enable the Enhanced Bus Mode. The Standard Bus Mode will be enabled if WB/WT# is sampled low or left floating at the falling edge of RESET.

8. The system may sample HITM# to detect the presence of the Enhanced Bus Mode. If HITM# is HIGH for one CLK after reset is inactive, the Enhanced Bus Mode is present.

Figure 15.28 80486SX reset signal waveforms. (Reprinted by permission of Intel Corp. Copyright/Intel Corp. 1994)

Register	Initial Value (BIST)	Initial Value (No BIST)
EAX	Zero (pass)	Undefined
ECX	Undefined	Undefined
EDX	0400 + Revision ID	0400 + Revision ID
EBX	Undefined	Undefined
ESP	Undefined	Undefined
EBP	Undefined	Undefined
ESI	Undefined	Undefined
EDI	Undefined	Undefined
EFLAGS	00000002h	00000002h
EIP	0FFF0h	0FFF0h
ES	0000h	0000h
CS	F000h	F000h
SS	0000h	0000h
DS	0000h	0000h
FS	0000h	0000h
GS	0000h	0000h
IDTR	Base = 0, limit = 3FFh	Base = 0, limit = 3FFh
CR0	60000010h	60000010h
DR7	00000000h	00000000h

Figure 15.29 Register values after reset. (Reprinted by permission of Intel Corp. Copyright/Intel Corp. 1994)

the 80386DX, if the diagnostic tests of BIST pass, the EAX register is cleared to zero; if problems are detected, a nonzero value is loaded into EAX.

Software Interrupts and Internal Exceptions

Any of the 256 interrupt vectors of the 80486SX can be assigned to a software interrupt. In our study of the 80386DX microprocessor, we found that software interrupts are initiated by executing the INT *n* instruction. When the software-interrupt instruction is executed, program control is transferred to the beginning of the service routine. Software interrupts cannot be masked out with IF and no external interrupt acknowledge bus cycle are initiated as part of switch in program context. Typically they are used to implement subroutines calls.

The only differences between the exception-processing capability of the 80486SX and 80386DX microprocessors are that one additional internal exception function is defined for the 80486SX and one that was performed by the 80386DX is no longer supported in the 80486SX. Next we look briefly at these two changes in exception processing.

The new exception that is activated for the first time in 80486 family MPUs is called *alignment check exception.* In our description of the software model of the 80486SX earlier in this chapter, we identified two new control bits, the alignment-check (AC) flag in EFLAGS and alignment-mask (AM) control bit in CR_0, that are used to enable address alignment checking for memory-access operations. At that time, we indicated that this function gives the 80486SX the ability to detect an attempt to access an unaligned double word of data.

This exception is detected only for memory accesses initiated while in protected mode and executing user-code privilege level 3. When this option is enabled, any attempt by the program to access an unaligned operand in memory results in an alignment-check exception through exception gate 17.

The 80386DX exception function that is not supported by 80486 family MPUs is exception 9, coprocessor segment overrun.

▲ 15.13 THE 80486DX2 AND 80486DX4 MICROPROCESSORS

The 80486DX2 and 80486DX4 are hardware- and software-compatible upgrades of the 80486DX MPU. They contain a number of architectural enhancements that result in higher performance for the 80486-based microcomputer system. Notice in Fig. 1.8 that the 80486DX2-50 has an iCOMP rating of 231. This value is close to twice the rating of the 80486DX-25 and very close to that of the 80486DX-50.

Two architectural enhancements made in the 80486DX2 are *clock doubling* and *write-back enhanced cache*. 80486DX2 MPUs are driven by what is called a $\frac{1}{2}\times$ *clock*, instead of a $1\times$ clock like the 80486SX and 80486DX. That is, the signal applied to the clock input of a 50-MHz 80486DX2 MPU (80486DX2-50) is actually a 25-MHz signal, or half the clock speed rating of the device. Similarly, the 80486DX2-66 is driven by a 33-MHz Clock signal. On-chip clock multiplying circuitry doubles the input clock frequency to produce a 66-MHz clock that runs the internal circuitry of the MPU. Only the core of the 80486DX2-66 MPU operates at 66 MHz, not the external bus interface. Like the 80486DX-33, the external bus interface of the 80486DX2-66 is operated at 33 MHz. In this way, we see that the 80486DX2 achieves a higher level of performance by running its internal circuitry twice as fast. At the same time, it permits simpler external interface circuit designs by maintaining the clock speed of the external local bus interface at 33 MHz.

The 80486DX2's 8-Kbyte on-chip cache memory is configurable to operate with either the write-through or write-back methods for updating external memory. Actually, the lines of the cache memory array can be individually configured as write-through or write-back. The write-through mode of operation is compatible with that of the 80486SX and 80486DX MPUs, whereas write-back is an improvement that offers higher system-level performance. Unlike write-through operation, write-back updates of external memory are not performed at the same time the information is written into the internal cache. Instead, the updates are accumulated in the on-chip cache memory subsystem and written to memory at a later time. This reduces the bus activity and therefore enhances the microcomputer's performance.

When in write-back mode, the 80486DX2's cache also supports snooping. The ability to snoop is needed to assure that data coherency is maintained between the data in the on-chip cache and the external main memory. A data coherency problem can occur whenever another bus master attempts to read information from or write information into the main memory. For example, if the storage location that is being read in main memory corresponds to an address whose data is cached within the 80486DX2 and modified, but not yet written back to main memory, the information that would be read from main memory is incorrect. In this case, a write-back must be performed to update main memory before the bus master can complete the read cycle. On the other hand, if the bus master writes to a storage location in main memory whose data is also cached, the value held in the cache must be invalidated. In this way, we see that whenever a bus master accesses main memory, a snoop bus cycle must be performed to determine if the content of the storage location to be accessed is cached within the MPU and, if so, initiate a write-back. Otherwise, cache coherency may be lost. The 80486DX2 employs the *MESI* (modify/exclusive/shared/invalid) write-back cache-consistency protocol.

Figure 15.30 is a block diagram that shows the new pin functions defined for the 80486DX2 and 80486DX4 MPUs. Added are four signals, $\overline{\text{CACHE}}$, $\overline{\text{HITM}}$, INV, and $\overline{\text{WBWT}}$, for use in control of the 80486DX2's write-back cache memory. The cacheability

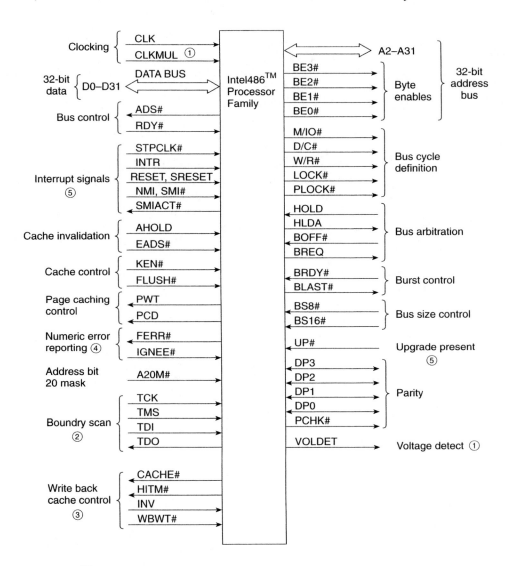

① IntelDX4™ processor only.
② Not on Intel486™ SX processor in PGA package.
③ Pins on Write-Back Enhanced IntelDX2 when in Enhanced Bus Mode/write back cache mode.
④ Not on Intel486™ SX and Intel SX2 processors.
⑤ SMI#, SMIACT#, STPCLK, SRESET, UP#, not on 50-MHz Intel486DX processor.

Figure 15.30 Block diagram of the 80486 MPU, including new pin functions for the 80486DX2 and 80486DX4. (Reprinted by permission of Intel Corp. Copyright/Intel Corp. 1994)

($\overline{\text{CACHE}}$) output switches to its active level for cacheable data reads, instruction code fetches, and data write-backs. Logic 0 at $\overline{\text{CACHE}}$ signals external circuitry that a cacheable read cycle or a burst write-back cycle is taking place.

The hit/miss ($\overline{\text{HITM}}$) output and invalidate request (INV) input are used during snoop bus cycles of the on-chip cache. The $\overline{\text{HITM}}$ output is activated by the cache-coherency protocol of the 80486DX2. If during a snoop bus cycle, the line of information checked for is found to be cached and modified but not yet written back to main memory, this fact is signaled to the external circuitry by logic 1 at $\overline{\text{HITM}}$. If INV is at its active 1 logic level during a snoop cycle for a write to main memory that identifies a line of data that is currently cached on-chip, this line of data is invalidated whether the cache is configured for write-through or write-back operation. However, if a line of data is found in the cache and it has been modified, it is first written back to main memory and then invalidated. INV should be held at logic 0 during snoop cycles for a read of main memory. In this case, the snoop cycle initiates a write-back to main memory but does not invalidate the data in this cache. In this way, cache invalidations are minimized.

The 80486DX2's cache memory is configured for the write-back mode of operation as part of the hardware-reset process. If the write-back/write-through ($\overline{\text{WBWT}}$) input is held at logic 1 for at least two clock periods before and after the falling edge of RESET, write-back configuration is enabled for the cache. The function of the $\overline{\text{FLUSH}}$ input changes when the cache is set up for write-back mode. Now logic 0 at $\overline{\text{FLUSH}}$ initiates a write-back of all lines of data in the cache that are modified before the contents of the cache are invalidated.

With the 80486DX4 came additional architectural enhancements and even higher performance. This device is designed to operate off a 3.3-V dc power supply instead of 5 V dc; the clock-scaling circuitry is enhanced to permit multiplication of the clock by $2\times$, $2.5\times$, or $3\times$; and the size of the cache is increased from 8K bytes to 16K bytes. The result is an

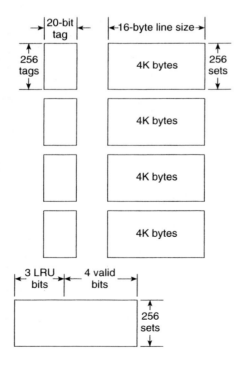

Figure 15.31 Organization of the on-chip cache of the 80486DX4. (Reprinted by permission of Intel Corp. Copyright/Intel Corp. 1994)

increase of the iCOMP™ performance rating for the 80486DX4-100 to 435, almost 50% higher than the 80486DX2-66.

The V_{cc} power supply of the 80486DX4 MPU is rated at +3.3 V dc ± 0.3 V. This does not mean that the 80486DX4 can be used only in 3.3-V system designs. Its inputs are 5-V tolerant and outputs are TTL compatible. Therefore, the 80486DX4 can be interfaced to 5-V logic components even though the internal circuitry is operating at 3.3 V. This is known as a *mixed voltage system*. Mixed voltage system operation is enabled by applying 5 V dc to the V_{cc5} pin.

Earlier we pointed out that the clock-multiplier circuitry in the 80486DX MPU has been enhanced to permit scaling of the clock by 2×, 2.5×, or 3×. The logic level of a new input pin, clock multiplier (CLKMUL), determines the value of the multiplier. The logic level of this input is sampled during hardware reset of the MPU to select the clock multiplier. Logic 1 at this input selects clock-tripled mode of operation. For instance, if the 80486DX4-100 is run by a 33-MHz clock signal, the internal operation of the MPU is 99MHz, and the speed of the local bus is maintained at 33 MHz. This increase in internal speed is the primary cause of higher performance for the 80486DX4-based microcomputer system.

The last architectural change in the 80486DX4 is that the size of the on-chip cache memory is expanded to 16KB. However, unlike the 80486DX2, this cache is write-through only and does not support the write-back mode of operation. Figure 15.31 shows the organization of the cache memory. Expanding the size of the cache results in a higher hit rate. This is another cause for the improved performance obtained with the 80486DX4 MPU.

ASSIGNMENTS

Section 15.2

1. Give two on-chip additions of the 80486DX MPU that result in improvement in performance over the 80386 family.
2. How many clock cycles are required to execute the core instructions (MOV, ADD, SUB, AND, etc.) of the 80486 microprocessor's instruction set?
3. What is the key difference between the 80486DX and 80486SX MPUs?
4. What is the iCOMP™ rating of the 80486SX-33 MPU? The 80486DX-50 MPU?

Section 15.3

5. What is the size of the 80486SX's instruction code queue?
6. What does CISC stand for? RISC? CRISC?
7. List three characteristics of a RISC processor.
8. Is the 80486SX best categorized as that of a CISC, RISC, or CRISC?
9. Identify an input that is available only on the 80486DX4 MPU.

Section 15.4

10. What are the new bits that are active in CR_0 of the real-mode 80486SX?
11. Does the 8086 architecture use little endian or big endian organization for data stored in memory?
12. List the instructions in the 80486 specific instruction set.

13. If the big endian contents of EAX are 0F0F0F0FH, what is the little endian result in the register after executing the instruction SWAP EAX?

14. Write an instruction sequence that will read the big endian double-word elements of a table starting at address BIG_E_TABLE and converts them to little endian format in a table starting at address LIT_E_TABLE. Assume that the number of double-word elements in the table equals COUNT.

15. Write an instruction that will perform an exchange and add operation on the content of the double-word storage location SUM and register EBX. If the original value in EAX is 00000001H and that in EBX is 00000000H, what is the result in SUM if the instruction is executed five times?

16. If the instruction CMPXCHG EBX, ECX is executed when the contents of EBX are $FFFFFFFF_{16}$, of ECX are 00000000_{16}, and of EAX are $0F0F0F0F_{16}$, what results are produced?

17. If the instruction CMPXCHG [DATA], BX is executed when the contents of AX are 1111_{16}, the contents of BX are 2222_{16}, and the word contents of storage location DATA are 1111_{16}, what results are produced?

Section 15.5

18. What is the new flag that is active in the 80486SX's EFLAGS register, and in which bit position is it found?

19. Which bits in the protected-mode CR_0 are used to control the operation of the on-chip cache memory?

20. What instruction should be executed to flush the on-chip cache and initiate a flush bus cycle?

21. What is the difference between the operations performed by the INVD and WBINVD instructions?

22. What is the result of executing the instruction INVLPG 10H?

23. Name the two new active bits in the 80486SX's page table entry.

Section 15.6

24. List three enhancements made in the memory interface of the 80486 MPU.

25. In what two packages is the 80486SX manufactured?

26. What signal is located at pin A16 of the 80486SX's package?

Section 15.7

27. Which bus cycle indication code has changed with the 80486 family?

28. Which lines of the 80486SX's memory/IO interface carry parity bit information? What type of parity is automatically generated?

29. Which output of the 80486SX is used to signal that a parity error has occurred? What logic level signals this error condition?

30. What input signal and logic level does the external circuitry use to tell the 80486SX MPU that it can perform a burst bus cycle?

31. What does $\overline{KEN}$ stand for?

32. Logic 1 at what output of the 80486SX means that a bus cycle is pending?

33. What signal permits an external device to take control of the 80486SX's bus interface at the completion of the current clock cycle?

Section 15.8

34. What are the range and precision of an 8-bit unpacked BCD number?

35. How is the data bus of the 80486SX configured if $\overline{BS16}$ and $\overline{BS8}$ are both at logic 0 during a bus cycle?

36. When using a 16-bit memory subsystem in an 80486SX-based microcomputer system, what additional memory control and address signals must be generated with external logic circuitry?

37. What are the values of the 16-bit bus control and address signals identified for Problem 36 if the byte enable code during a memory write cycle is $\overline{BE}_3\overline{BE}_2\overline{BE}_1\overline{BE}_0 = 1110_2$?

38. For the write-data transfer identified in Problem 37, what size data transfer takes place and over which data bus lines will the data be carried?

39. If the byte-enable code output during a memory write bus cycle to the memory subsystem identified in Problem 36 is $\overline{BE}_3\overline{BE}_2\overline{BE}_1\overline{BE}_0 = 0111_2$, over which data bus lines will the data be output by the MPU? How are these data passed to the memory subsystem?

40. If the byte-enable code output during a memory-read bus cycle for the memory subsystem identified in Problem 36 is $\overline{BE}_3\overline{BE}_2\overline{BE}_1\overline{BE}_0 = 1001_2$, what size data transfer is taking place? Is it an aligned or misaligned data transfer? How many bus cycles are required to perform the data transfer? What is the byte-enable code during each bus cycle? To which data bus lines of the MPU must the data be multiplexed?

Section 15.9

41. How long does it take the 80486SX-25 that is running with no wait states to complete a nonburst, noncacheable bus cycle for an aligned double word of data? With one wait state? For a misaligned double word and with one wait state?

42. What is meant when a read cycle is said to be cacheable? What signal must be supplied to the MPU by the memory subsystem to initiate a cacheable bus cycle? How many bytes of data are transferred during a cacheable bus cycle?

43. If an 80486SX is running at 25 MHz and with no wait states, how long would it take to complete a nonburst, cacheable read bus cycle for an aligned double word of data? Repeat for the MPU using one wait state.

44. What is the maximum number of bytes that can be transferred with a single-burst bus cycle? What signal must be supplied to the MPU by the memory subsystem to initiate a burst bus cycle? How many clock cycles does it take for completion?

45. How long would it take the zero-wait-state bus cycle described in Problem 43 to be completed if it was performed with a burst, cacheable bus cycle?

Section 15.10

46. Is the internal cache of the 80486SX a first-level or second-level cache?

47. What type of cache organization is used for the 80486SX's internal cache?

48. How large is the 80486SX's internal cache? What is the smallest element of data that can be loaded into the cache?

49. What method is used to determine which element of data should be replaced when the 80486SX's cache is full?

50. What write update method is implemented for the internal cache of the 80486SX?

51. How can the 80486SX's on-chip cache be disabled under software control?

52. What happens when the $\overline{\text{FLUSH}}$ input of the 80486SX is switched to logic 0 by external circuitry?

53. How can the internal cache of the 80486SX be flushed under software control?

54. What name is given to the process of invalidating a single line of data in the 80486SX's on-chip cache? What type bus cycle is used to perform this operation?

Section 15.11

55. List three standard peripheral ICs that are implemented in the R400EX.

56. Name three PC-compatible hardware functions implemented in the R400EX.

57. What type of system bus interface is implemented by the R400EX?

Section 15.12

58. How long must the reset input of an 80486SX-25 running at full speed be held at logic 1 to assure that a reset is initiated?

59. What does BIST stand for?

60. What new internal exception is implemented in the 80486SX MPU? What protected mode exception gate is assigned to this exception?

Section 15.13

61. Name two enhancements made to the architecture of the 80486DX2 to improve its performance?

62. What method is employed by the 80486DX2 to assure coherency between data in the on-chip cache and that in external memory?

63. What cache-coherency protocol is employed by the 80486DX2?

64. What output signal and logic level indicate that the line of data checked during a snoop operation is held in the 80486SX's on-chip cache and has been modified but has not yet been written back to main memory?

65. How is the 80486SX's cache put into the write-back mode?

66. Name two enhancements made to the architecture of the 80486DX2 to improve its performance.

67. How many times greater is the performance of the 80486DX4-100 compared to that of the 80386DX-33?

68. What value voltage supply is needed to power the 80486DX4?

The Pentium^R Processor Family

▲ 16.1 INTRODUCTION

In Chapter 15 we studied the software and hardware architecture of the 80486 microprocessor family. We covered their real- and protected-mode software architectures, the 80486 specific and system-control instruction sets, signal interfaces of the 80486SX MPU, memory interface and cache memory, bus cycles, and interrupts/exception processing. Now we will turn our attention to the Pentium^R processor. Again we will focus on the software and hardware architecture differences between this newer processor family and the 80486 family. For this purpose, we have included the following topics in the chapter:

1. The Pentium^R processor family
2. Internal architecture of the Pentium^R processor
3. Software architecture of the Pentium^R processor
4. Hardware architecture of the Pentium^R processor
5. Signal interfaces of the Pentium^R processor
6. Memory subsystem circuitry
7. Bus cycles: nonpipelined, pipelined, and burst
8. Cache memory of the Pentium^R processor
9. Interrupts, reset, and internal exceptions
10. The Pentium^R Pro Processor and Pentium^R Processor with MMX™ Technology

▲ 16.2 THE PENTIUM^R PROCESSOR FAMILY

The Pentium^R processor family, which was introduced in 1993, represents the high-performance end of Intel's 8086 architecture. Just like the 80486 MPUs, the Pentium^R processors are 32-bit MPUs; that is, they have a 32-bit register set and the instructions can process

Figure 16.1 Pentium[R] processor IC. (Reprinted by permission of Intel Corp. Copyright/Intel Corp. 1995)

words of data as large as 32 bits in length. However, the Pentium[R] processor's 32-bit architecture is enhanced with a 64-bit external data bus and a variety of internal data paths that are 64 bits, 128 bits, or 256 bits wide. These large internal and external data paths result in an increased level of performance. A Pentium[R] processor IC is shown in Fig. 16.1.

The Pentium[R] processors' internal, software, and hardware architectures have been enhanced in many other ways. For instance, they employ an advanced *superscaler* pipelined internal architecture. The parallel processing provided by this superscaler pipelining gives the Pentium[R] processors the ability to execute more than one instruction per clock cycle. Improvements to the hardware architecture are that the on-chip cache memory is now divided into separate code and data caches; parity has been added to the address, and, as mentioned earlier, the data bus is expanded to 64 bits. Finally, from a software point of view, the instruction set has been enhanced with new instruction and, more importantly, the performance of the floating-point unit has been greatly increased.

Figure 1.8 shows the performance of the Pentium[R] processor family members relative to those of the 80386 and 80486 families. Notice that the iCOMP index of the entry-level Pentium[R] processor (75 MHz) MPU, which is 610, is about 50% higher than the fastest 80486 MPU, the 80486DX4-100. With the introduction of faster family members, the Pentium[R] processors' performance edge has increased to more than 2×. For instance, the 133-MHz device has an iCOMP rating of 1110.

EXAMPLE 16.1

Which speed Pentium[R] processor is the first device to offer more than a 2× performance increase over the 80486DX4-100?

Solution

In Fig. 1.8, we find that the iCOMP rating of the 80486DX4-100 is 435. Therefore, the Pentium[R] 120 MHz, which has an iCOMP rating of 1000, is the first member of the Pentium[R] processor family that offers more than 2× performance increase over the 80486DX4-100.

A block diagram of the Pentium^R processor's internal architecture is given in Fig. 16.2. Earlier we pointed out a number of important architectural advances introduced with the Pentium^R processor. Three of these are its *superscaler* pipelined architecture, independent code and data caches, and high-performance floating-point unit. Let us next look briefly at each of these architectural features.

Intel uses the term *superscaler* to describe an architecture that has more than one execution unit. In the case of the Pentium^R processor, there are two execution units. These execution units, or *pipelines* as they are also known, process the instructions of the microcomputer program. Each of these execution units has its own ALU, address generation circuity, and data cache interface. They are identified in Fig. 16.2 as the *U pipeline* and the *V pipeline*.

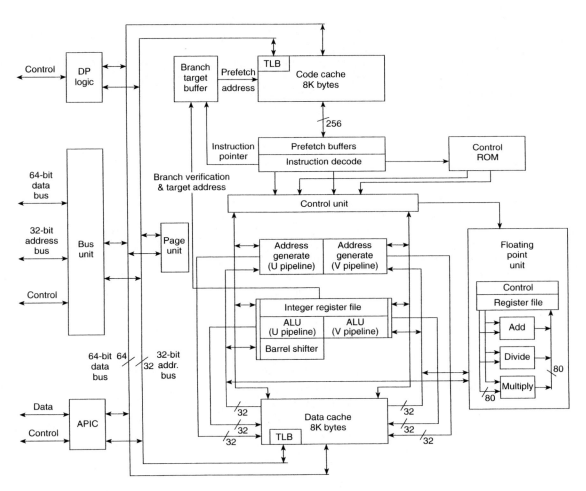

Figure 16.2 Internal architecture of the Pentium^R processors. (Reprinted by permission of Intel Corp. Copyright/Intel Corp. 1994)

In Chapter 15 we found that the execution time of a core group of instructions were reduced to one clock cycle in the 80486 MPU family. With the Pentium[R] processors' dual-pipeline architecture, two instructions can be processed at the same time. Therefore, they have the capability of executing as many as two instructions per clock cycle. For this reason, pipelining makes a significant contribution to the higher level of performance achieved with the Pentium[R] processor family.

The 80486 family of MPUs has an on-chip cache memory that is used to cache both code and data. With the Pentium[R] processor, the on-chip cache memory subsystem has been further enhanced. Its cache memory, like that of the 80486DX4 MPU, has been expanded to 16KB, but it is also partitioned into a separate 8KB code and 8KB data caches. Also, as with the 80486DX4, the write-update method can be configured for either write-through or write-back mode of operation. Notice in Fig. 16.2 that the data cache is accessed independently by the U pipe and V pipe of the ALU. This is known as a *dual-port interface* and permits both ALUs to access data in the cache at the same time. These independent caches result in more frequent use of the cache memory and lead to a higher level of performance for the Pentium[R] processor–based microcomputer system.

All members of the Pentium[R] processor family have a built-in floating-point unit. This floating-point math unit has been further enhanced from that used in the 80486 family. For instance, it employs faster hardwired, instead of microcoded, implementations of the floating-point add, multiply, and divide operations. The result is higher-performance floating-point operation for the Pentium[R] processor. In fact, it can execute floating-point instructions 5 to 10 times as fast as the 80486DX-33 MPU.

▲ 16.4 SOFTWARE ARCHITECTURE OF THE PENTIUM[R] PROCESSOR

The MPUs of the Pentium[R] processor family remain fully software compatible with the 80486 architecture. As do the 80386 and 80486 MPUs, the Pentium[R] processors come up in real mode after reset and can be switched to protected mode by executing a single instruction. Its real- and protected-mode software models and instruction sets are both supersets of those of the 80486 family MPU. They have all the same instructions, functional registers, and register bit definitions. However, as for the 80486 MPU and 80386 MPU before that, a number of new instructions, flags, and control bits have been defined.

Real-mode and Protected-mode Register Sets

The real-mode and protected-mode application register sets of the Pentium[R] processor are essentially the same as those of the 80486SX microprocessor. However, some changes are found in the functionality of both the flags register and the control registers. The EFLAGS register of the Pentium[R] processor is shown in Fig. 16.3. Three additional flag bits, *ID flag* (ID), *virtual interrupt pending* (VIP), and *virtual interrupt flag* (VIF), have been activated. The ID flag can be used to determine if the MPU supports a new instruction called CPUID. If this bit can be set or reset under software control, CPUID is included in the instruction set. The VIP and VIF flag bits are used to implement a virtualized system-interrupt flag for protected-mode multitasking software environments.

The function of the control registers have been expanded in the Pentium[R] processor. Figure 16.4(a) shows that a fifth control register, CR_4, has been added, and the six new control bits are all in this register. The meaning of each of these bits and a brief description

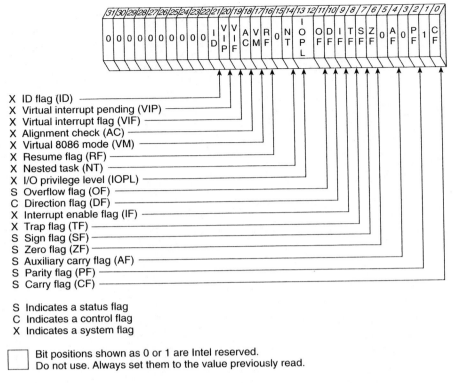

X ID flag (ID)
X Virtual interrupt pending (VIP)
X Virtual interrupt flag (VIF)
X Alignment check (AC)
X Virtual 8086 mode (VM)
X Resume flag (RF)
X Nested task (NT)
X I/O privilege level (IOPL)
S Overflow flag (OF)
C Direction flag (DF)
X Interrupt enable flag (IF)
X Trap flag (TF)
S Sign flag (SF)
S Zero flag (ZF)
S Auxiliary carry flag (AF)
S Parity flag (PF)
S Carry flag (CF)

S Indicates a status flag
C Indicates a control flag
X Indicates a system flag

Bit positions shown as 0 or 1 are Intel reserved.
Do not use. Always set them to the value previously read.

Figure 16.3 EFLAGS registers of the Pentium^R processors. (Reprinted by permission of Intel Corp. Copyright/Intel Corp. 1995)

of its function are given in Fig. 16.4(b). Looking at the functional descriptions for these control bits, we see that they are all used to enable or disable new capabilities of the Pentium^R processor. For instance, the *virtual-8086 mode extensions* (VME) and *protected-mode virtual interrupt* (PVI) bits are used to enable the virtualized system-interrupt capability implemented with the VIP and VIF flag bits in the virtual-8086 mode and protected-mode, respectively.

EXAMPLE 16.2

What is the protected-mode page size when the PSE bit in CR$_4$ is logic 0? Logic 1?

Solution

When PSE is logic 0, the page size is 4KB, and when it is switched to logic 1, the size is increased to 4MB.

Both 80386 and 80486 MPUs also contained debug and test registers in their real- and protected-mode software models. However, these registers are not used in application programming. Another difference in the register models of the Pentium^R processor is that

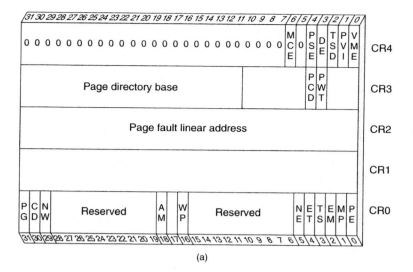

(a)

Bit	Name	Function
0	Virtual-8086 mode Extensions (VME)	Logic 1 enables support for a virtual interrupt flag in virtual-8086 mode.
1	Protected mode Virtual interrupts (PVI)	Logic 1 enables support for a virtual interrupt flag in protected mode.
2	Time-date stamp Disable (TSD)	Logic 1 makes the read from time stamp counter (RDTSC) instruction a privileged instruction.
3	Debugging Extensions (DE)	Logic 1 enables I/O breakpoints.
4	Page size Extensions (PSE)	Logic 1 enables 4M-byte page size.
6	Machine check enable (MCE)	Logic 1 enables the machine-check exceptions.

(b)

Figure 16.4 (a) Control registers of the Pentium^R processors. (Reprinted by permission of Intel Corp. Copyright/Intel Corp. 1995) (b) Function of the CR₄ control bits.

debug registers DR$_4$ and DR$_5$ are now implemented. Also, the test registers no longer exist; instead, their functionality is implemented in a new group of registers called the *model-specific registers*.

For simplicity, we have ignored the registers that are provided for the floating-point math unit in the application software models.

Enhancements to the Instruction Set

Upward software compatibility is maintained in the instruction set of the Pentium^R processor family. That is, it retains all the instructions of the 8086, 8088, 80286, 80386, and 80486 microprocessor families. This means that all software written for a microcomputer constructed with one of these earlier MPUs can be directly run on a Pentium^R processor–based microcomputer.

Mnemonic	Meaning	Format	Operation
CMPXCH8B	Compare and exchange 8 bytes	CMPXCH8B D	if [EDX:EAX] = D 　(ZF)←1, (D)←[EX:EBX] else 　(ZF)←0, [EDX:EAX]←(D)

CPUID	CPU identification	CPUID	if (EAX) = 0H
			　[EAX,EBX,ECX,EDX]←
			　　　Vendor information
			if (EAX) = 1H
			　[EAX,EBX,ECX,EDX]←
			　　　MPU information

RDTSC	Read from time stamp counter	RDTSC	[EDX:EAX]← Time stamp 　　　　counter
RDMSR	Read from model specific register	RDMSR	[EDX:EAX]←MSR(ECX) 　(ECX) = 0H selects MCA 　(ECX) = 1H selects MCT
WRMSR	Write to model specific registers	WRMSR	MSR(ECX)←[EDX:EAX] 　(ECX) = 0H selects MCA 　(ECX) = 1H selects MCT
RSM	Resume from system management mode	RSM	Resume operation from SMM
MOV CR4	Move to/from CR4	MOV CR4,r32 MOV r32, CR4	(CR4)←(r32) (r32)←(CR4)

Figure 16.5 Additions to the instruction set in the Pentium[R] processors.

The instruction set of the Pentium[R] processor is enhanced with three new Pentium[R] processor-specific instructions and four additional system-control instructions. The chart in Fig. 16.5 summarizes the function of each of the new instructions. For instance, the instruction *compare and exchange 8 bytes* (CMPXCHG8B) is an enhancement in the Pentium[R] processor specific instruction set. CMPXCHG8B is a variation of the CMPXCHG instruction, which was first introduced with the 80486 instruction set. It enables a 64-bit compare and exchange operation to be performed. Notice in Fig. 16.5 that when the instruction is executed, the 64-bit value formed from EDX and EAX, which is denoted [EDX:EAX] (where EDX is the more significant double word), is compared to the value of a 64-bit data word (quad-word) in memory. If the two values are equal, ZF is set to 1 and the quad-word value formed from [ECX:EBX] (where ECX is the more significant double word) is written into the quad-word storage location in memory. On the other hand, if the value [EDX:EAX] does not match that in memory, ZF is cleared to 0 and the quad word held in memory is read into [EDX:EAX].

EXAMPLE 16.3

If the contents of EAX, EBX, ECX, and EDX are 11111111_{16}, 22222222_{16}, 33333333_{16}, and $FFFFFFFF_{16}$, respectively, and the content of the quad-word memory storage location pointed to by the address TABLE is $\text{FFFFFFFF}11111111_{16}$, what is the result produced by executing the instruction

$$\text{CMPXCHG8B} \quad [\text{TABLE}]$$

Solution

When the instruction is executed, the value of

$$(\text{EDX:EAX}) = \text{FFFFFFFF}11111111_{16}$$

is compared to the quad-word pointed to by address TABLE. The contents of this memory location are given as

$$(\text{DS:TABLE}) = \text{FFFFFFFF}11111111_{16}$$

Since these two values are equal, the results produced are

$$\text{ZF} \leftarrow 1$$

$$(\text{DS:TABLE}) \leftarrow (\text{ECX:EBX}) = 3333333322222222_{16}$$

Another one of the new Pentium[R] processor specific instructions, *CPU identification* (CPUID), permits software to identify the type and feature set of the PENTIUM[R] processor that is in use in the microcomputer system. The EAX register must be set to either 0 or 1 before executing CPUID. Depending on which value is selected for EAX, different identification information is made available to software. By executing this instruction with EAX equal to 0, the processor identification information provided is as follows:

$$\text{EAX} = 1 \leftarrow \text{Pentium}^R \text{ processor}$$

$$\text{EBX} = \text{vendor identification string} \leftarrow \text{Genu}$$

$$\text{ECX} = \text{vendor identification string} \leftarrow \text{ineI}$$

$$\text{EDX} = \text{vendor identification string} \leftarrow \text{ntel}$$

Notice that the MPU is identified as a Pentium[R] processor and that it is a genuine Intel device. Now, executing the instruction again with EAX equal to 1 provides more information about the MPU. The results produced in EAX, EBX, ECX, and EDX are as follows:

$$\text{EAX}(3:0) \leftarrow \text{stepping ID}$$

$$\text{EAX}(7:4) \leftarrow \text{model}$$

$$\text{EAX}(11:8) \leftarrow \text{family}$$

$$\text{EAX}(31:12) \leftarrow \text{reserved bits}$$

$$\text{EBX} \leftarrow \text{reserved bits}$$

$$\text{ECX} \leftarrow \text{reserved bits}$$

$$EDX(0:0) \leftarrow \text{FPU on-chip}$$

$$EDX(2:2) \leftarrow \text{I/O breakpoints}$$

$$EDX(4:4) \leftarrow \text{time-stamp counter}$$

$$EDX(5:5) \leftarrow \text{Pentium}^R \text{ CPU-style model-specific registers}$$

$$EDX(7:7) \leftarrow \text{machine-check exception}$$

$$EDX(8:8) \leftarrow \text{CMPXCHG8B instruction}$$

$$EDX(31:9) \leftarrow \text{reserved}$$

EXAMPLE 16.4

Figure 16.6 shows the contents of the EAX register after executing the CPUID instruction. What are the family, model, and stepping?

Solution

In Fig. 16.6 we find

$$\text{family} = 0101 = 5 = \text{Pentium}^R \text{ processor}$$

$$\text{model} = 0000 = 0$$

$$\text{stepping} = 0000 = 0$$

The last of the new Pentium[R] processor-specific instructions is *read from time-stamp counter* (RDTSC). The Pentium[R] processor has an on-chip 64-bit counter called the *time-stamp counter*. The value in this counter is incremented during every clock cycle. Executing the RDTSC instruction reads the value in this counter into the register set [EDX:EAX]. EDX holds the upper 32 bits of the 64-bit count, and EAX holds the lower 32 bits. In some software applications, it may be necessary to determine how many clock cycles have elapsed during an event. This can be done by reading the value of the time-stamp counter before and after the event being measured and then determining the number of elapsed clock cycles by forming the difference of the two count readings.

Looking at Fig. 16.5, we find that the four new system-control instructions are *read from model-specific register* (RDMSR), *write to model-specific register* (WRMSR), *resume from system management* (RSM), and a new form of the move instruction (MOV CR4,r32 and MOV r32,CR4) that permits data to be moved directly between control register 4 (CR$_4$) and an internal register. The instructions RDMSR and WRMSR are used to permit software

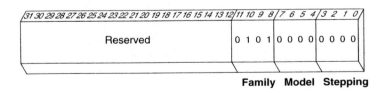

Figure 16.6 EAX after executing the CPUID instruction. (Reprinted by permission of Intel Corp. Copyright/Intel Corp. 1995)

access to the contents of the model-specific registers of the Pentium^R processor. The two model-specific registers that can be accessed with these instructions are the *machine check address register* (MCA) and the *machine check type register* (MCT). To access MCA, the value 0H must be loaded into ECX prior to executing the instruction, or to select MCT, ECX must be loaded with the value 1H. Execution of the instructions causes either a 64-bit read- or 64-bit write-data transfer to take place between the selected model-specific register and the register set [EDX:EAX]. The RSM instruction is used by the Pentium^R processor's system management mode.

System-Management Mode

The Pentium^R processor also has a mode of operation known as *system-management mode* (SMM). This mode is used primarily to perform management of the system's power consumption. SMM is entered by an interrupt request from external hardware, and return to real or protected mode is initiated by executing the RSM instruction.

▲ 16.5 HARDWARE ARCHITECTURE OF THE PENTIUM^R PROCESSOR

Many enhancements have been made to the hardware architecture of the MPUs in the Pentium^R processor family. Some examples of important improvements are that the data bus has been expanded to 64 bits, parity has been provided for the address bus, the on-chip cache memory is partitioned into separate code and data caches, the internal caches support either the write-through or write-back cache-update methods, and pipelined bus cycles have been implemented. These hardware changes play a key role in giving the Pentium^R processor its higher level of performance and simplify the design of Pentium^R processor-based multiprocessor microcomputer systems.

We have seen that continued advances in process technology have enabled Intel Corporation to integrate more and more transistors into their MPUs. The Pentium^R family of microprocessors were originally built on a 0.6-μm manufacturing process but have been moved to an even smaller geometry process. The reduction in transistor size achieved with these more advanced processes has permitted the integration of more than 3 million transistors into the Pentium^R processor.

Because of the process used to manufacture the Pentium^R processor, it must be powered by a 3.3-V power supply, V_{cc}. Its inputs and outputs are also rated at 3.3 V. Output signal lines are TTL-compatible in that they do meet the minimum high-logic level (V_{IHmin}) for a TTL input. For this reason, they may directly drive external interface circuits made with either 3.3-V or 5-V TTL logic devices. On the other hand, the inputs of the Pentium^R processor cannot tolerate more than 3.3 V as the V_{IHmax}. Therefore, input signal lines must be driven from outputs of 3.3-V logic devices or 5-V devices with open-collector outputs that are set up to convert between 5-V and 3.3-V logic levels.

The pin layout of the Pentium^R processor is shown in Fig. 16.7. Here we see that it is housed in a 296-pin staggered-pin grid array (SPGA) package.

EXAMPLE 16.5 ─────────────────────────────

What are the pin locations of the signals D_0 and D_{64}?

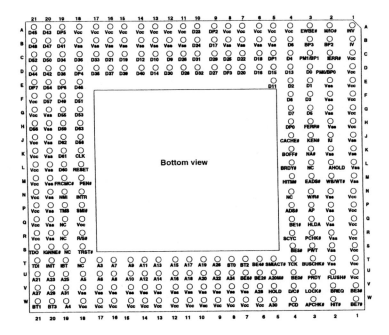

Figure 16.7 Pin layout of the Pentium[R] processor. (Reprinted by permission of Intel Corp. Copyright/Intel Corp. 1995)

Solution

From the pin layout diagram in Fig. 16.7, we find that signal D_0 is at pin D3 and signal D_{63} is at pin H18.

▲ 16.6 SIGNAL INTERFACES OF THE PENTIUM[R] PROCESSOR

A block diagram of the Pentium[R] processor is shown in Fig. 16.8. Although a lot of changes have been made in the hardware architecture of the Pentium[R] processor, many of the interface signals remain the same as those used on 80386 and 80486 family MPUs. For example, similar to the 80386DX and 80486SX, the type of bus cycle is defined by the code output on $M/\overline{IO}$, $D/\overline{C}$, and $W/\overline{R}$, and the interrupt interface consists of INTR, NMI, and RESET inputs. Most of the interface signals first introduced on the 80486 family of MPUs are also provided on Pentium[R] processors. For instance, the signal lines $\overline{A20M}$, $\overline{BOFF}$, $\overline{BRDY}$, $\overline{FLUSH}$, $\overline{KEN}$, PWT, and PCD are all part of the Pentium[R] processor's memory/ IO and cache memory interfaces. Here we will focus on the interface signals first introduced with the Pentium[R] processor family.

Memory/IO Interface

Earlier we pointed out that the memory/IO interface has been improved by making the data bus 64 bits wide and by adding parity on the address bus. The data bus now consists

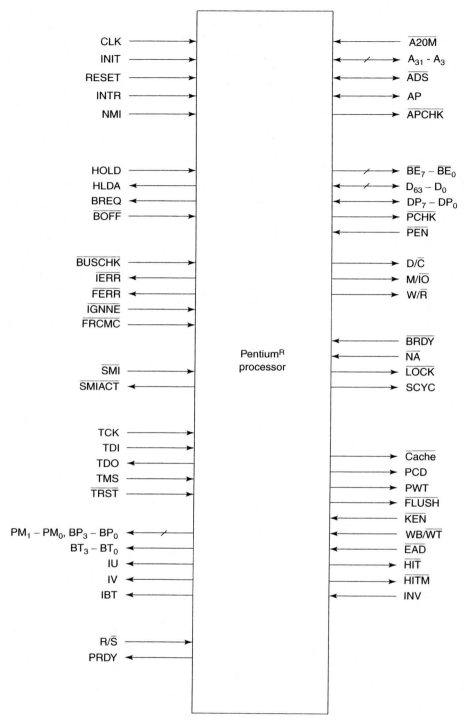

Figure 16.8 Block diagram of the Pentium^R processor. (Reprinted by permission of Intel Corp. Copyright/Intel Corp. 1995)

of bidirectional data lines D_0 through D_{63}. Because of the larger data bus, the number of byte-enable and data parity lines have been increased to 8. In Fig. 16.8, they are labeled $\overline{BE_0}$ through $\overline{BE_7}$ and DP_0 through DP_7, respectively. Extending this bus to 64 bits results in an increased data-transfer rate between the MPU and its memory and I/O subsystem and higher-performance operation.

During all write cycles, the bus interface unit generates a code at DP_0 through DP_7 that produces even parity for each byte of data on D_0 through D_{63}. When a read bus cycle takes place, the data on D_0 through D_{63} and DP_0 through DP_7 are tested for even parity on a byte-wide basis. If a data parity error is detected in any byte of the quad-data word, this fact is signaled to external circuitry with logic 0 at the $\overline{PCHK}$ output. The *parity-enable* ($\overline{PEN}$) input is used to determine whether or not an exception is initiated when a data-read parity error occurs. Logic 0 at this input configures the MPU to initiate an exception automatically whenever a read parity error is detected.

In the Pentium[R] processor, parity generation and checking have been added to the address bus. Whenever an address is output on A_3 through A_{31}, an even parity bit is generated and output at pin *address parity* (AP). In this way, the memory subsystem can perform parity checks on both the data and the address. Adding address parity checking and detection to the memory/IO subsystem results in an increased level of data integrity for the microcomputer system.

The address bus is actually bidirectional. This is because the Pentium[R] processor, like the 80486SX, permits external devices to examine the contents of its internal caches. The operation is enhanced on the Pentium[R] processor with address parity checking. The external system applies what is known as an *inquire address* to the processor on A_5 through A_{31}. This is the address of the cache-storage location to be accessed during the inquire cycle. Logic 0 at the valid external address ($\overline{EADS}$) input, signals the MPU that the address is available. As part of the read operation, a parity-check operation is performed on the inquire address, and if a parity error is detected, it is identified by logic 0 at the address parity check ($\overline{APCHK}$) output.

The Pentium[R] processor's memory/IO interface can also detect whether or not a bus cycle has run to completion correctly. This is known as a *bus error* condition. The *bus check* ($\overline{BUSCHK}$) input is used for this purpose. External circuitry must determine whether or not the current bus cycle is not successfully completed. If the bus cycle is not completed, it switches $\overline{BUSCHK}$ to logic 0. This input is sampled during read and write bus cycles. If an active 0 logic level is detected, the address and type of the failing bus cycle are latched within the MPU. An exception can also be automatically initiated to transfer program control to a service routine for the bus error problem.

Cache Memory-Control Interface

The number of control signals at the cache memory-control interface of the Pentium[R] processor have also been expanded. Comparing the block diagram of Fig. 16.8 to that of the 80486SX in Fig. 15.12, we find that five new signals are provided. They are cacheability ($\overline{CACHE}$), write-back/write-through (WB/$\overline{WT}$), inquire cycle hit/miss indication ($\overline{HIT}$), hit/miss to a modified line ($\overline{HITM}$), and invalidation request (INV). Here we look briefly at how they function in support of the internal cache memory.

The logic level of the $\overline{CACHE}$ output has a different meaning, depending on whether a read or write operation is taking place. This output is switched to logic 0 during bus cycles where the data-read from external memory can be cached. That is, it signals external circuitry that a cacheable data read or cacheable code fetch is taking place. $\overline{CACHE}$ is also made active during write cycles that represent a write-back to external memory of data that was updated in the internal cache.

WB/$\overline{\text{WT}}$ is an input that can be used to define the individual storage locations of the internal cache memory as write-back or write-through. By applying logic 0 or 1 to this input, external circuitry can decide whether the external memory-update method for the addressed storage location is write-back or write-through.

The other three signals, the $\overline{\text{HIT}}$ output, $\overline{\text{HITM}}$ output, and INV input, are involved in the inquire address operation.

Interrupt Interface

Comparing the interrupt interface of the Pentium[R] processor to that of the 80486SX, we find that just one new signal has been added. The function of this input, initialization (INIT), is similar to that of RESET in that it is used to initialize the MPU to a known state. However, in the case of INIT, the content of the internal cache and a number of other registers remain unchanged.

▲ 16.7 MEMORY SUBSYSTEM CIRCUITRY

Now that we have introduced the interface signals, let us turn our attention to the design of a practical memory subsystem for a Pentium[R] processor–based microcomputer system. A block diagram of a typical memory subsystem is shown in Fig. 16.9. This subsystem supports pipelined and burst bus cycles, paging and bank interleaving, and write-back updates from cache. Looking at the block diagram, we see that it consist of the DRAM

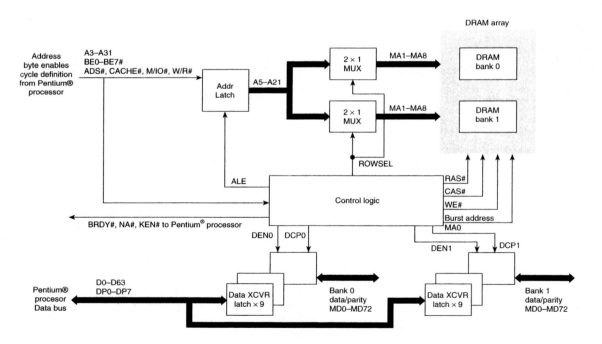

Figure 16.9 Block diagram of a memory subsystem. (Reprinted by permission of Intel Corp. Copyright/Intel Corp. 1995)

data storage array, RAS/CAS address-multiplexing circuitry, data bus transceiver circuitry, and control-logic sections. Next we explore the function and circuitry of each section of this interleaved memory subsystem design.

Interleaved DRAM Memory Array

Let us begin by examining the architecture of an interleaved memory subsystem and the benefit it brings to the microcomputer system. In an interleaved memory design, the data-storage array is partitioned into two separate banks of memory. Typically, one bank is used to store even-addressed words of data and the other, odd-addressed words. The objective is for the MPU to perform consecutive data accesses from storage locations in alternating banks. For instance, if the current memory access is from an even-addressed storage location, the next storage location to be accessed would need to be at an odd address. This second access is called a *bank miss*. That is, the address misses the bank that is currently being accessed. When this is the case, the MPU can access data in the even-addressed bank of DRAMs and at the same time prepare to access a storage location in the odd-addressed bank. The result of this overlapping mode of memory operation is higher performance for the microcomputer system.

As shown in Fig. 16.9, the two banks of memory are labeled bank 0 and bank 1. A more detailed diagram of the memory-array circuit is given in Fig. 16.10. Let us look more closely at the organization of bank 0 in this diagram. Notice that it is formed from four $512K \times 36$-bit DRAM modules. This gives a total storage capacity of 1M, 64-bit data words. The -60 on these devices means that they have a 60-ns access time.

The two modules to the left, which are identified as row A of bank 0, represents $512K \times 72$ bits of storage. To select a storage location within row A, a row and column address are applied as inputs over the nine address lines $B0MA_0$ through $B0MA_8$ synchronous to a strobe at $\overline{RASA_0}$ and $\overline{RASA_1}$ and $\overline{CAS_0}$ through $\overline{CAS_3}$, respectively. The logic level applied at the $\overline{WE_0}$ input signals the memory devices whether a data read or write operation is taking place to the selected storage location. This input or output data transfer is performed over the 64 data bus lines $B0MD_0$ through $B0MD_{63}$ and the eight data parity lines $B0DP_0$ through $B0DP_7$. The organization of the DRAM array in row B of bank 0 is similar to that of row A except that it uses row address strobes $\overline{RASB_0}$ and $\overline{RASB_1}$.

RAS/CAS Address Multiplexer Circuitry

To support interleaved operation, each bank of DRAMs must have its own RAS/CAS address multiplexing circuitry. In Fig. 16.9, we see that address bits A_5 through A_{21} are latched into an address-latch circuit with the ALE signal. Latching of the address permits pipelined bus operation. The latched address is input to the bank 0 and bank 1 multiplexer circuits for conversion into a row and column address for application to the DRAM array.

Typical circuitry that can perform this row and column address multiplexing function is shown in Fig. 16.11. 74FCT573 octal transparent latch devices are employed in this circuit. Notice that address bits A_5 through A_{12} are used to form the column address and A_{14} through A_{21} are used as the row address. The row and column addresses output from the latches are applied to one set of inputs on each of the multiplexers. Use of a transparent latch makes the address information immediately available at the inputs of the multiplexer.

The multiplexer used in the circuit is a 74FCT257 quad 2-line-to-1-line device. The logic level of the ROW SELECT input determines whether the row address or column address is output to the buffers. As shown in the circuit, 74F244 octal buffers are used to form a separate 8-bit Row A and Row B address for each bank. The address bus for row

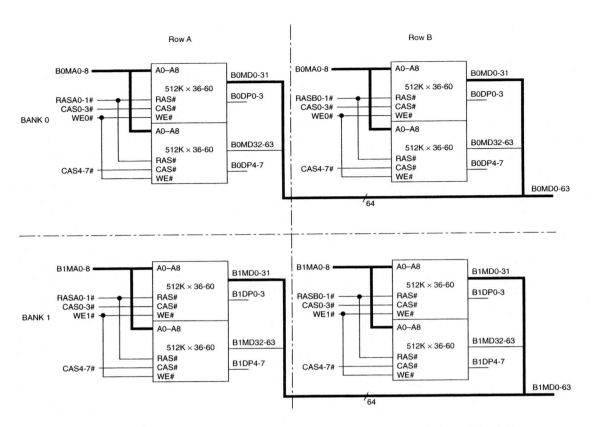

Figure 16.10 Organization of the DRAM array. (Reprinted by permission of Intel Corp. Copyright/Intel Corp. 1995)

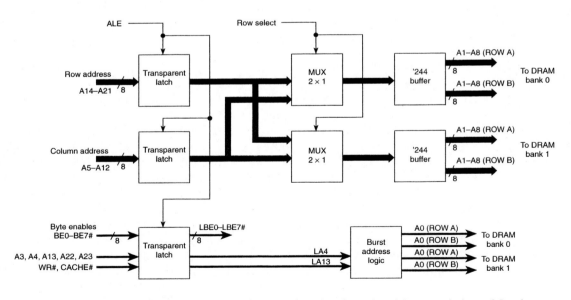

Figure 16.11 RAS/CAS address multiplexer circuits. (Reprinted by permission of Intel Corp. Copyright/Intel Corp. 1995)

A in bank 0 is denoted as A_1–A_8 (Row A). They are applied to the B0MA_{0-8} inputs of the row A DRAMs in Fig. 16.10.

Data Bus Transceiver Circuitry

Figure 16.9 shows that separate data bus transceiver circuits are also required for each of the DRAM banks. A more detailed diagram of the circuitry in the data bus path is given in Fig. 16.12. Notice that the device used is a 74FCT646A bidirectional registered transceiver. At the left side we see that the MPU's 64 data bus lines D_0 through D_{63} and data parity lines DP_0 through DP_7 are applied in parallel to both groups of transceivers. The other side of the upper group of data transceivers are used to supply the data input/outputs of the DRAMs in bank 0. On the other hand, the lower group of transceivers interface bank 1 to the MPU's data bus.

Let us now look at the control signals that determine how the data bus interface works. Control input $W/\overline{R}$ is applied to the DIR input of all transceiver devices in parallel. Its logic level determines the direction in which data are passed through the devices. For instance, when a read cycle is in process, $W/\overline{R}$ is made logic 0. This sets the transceivers to pass data from either bank 0 or bank 1 to the data bus of the MPU. The control logic selects between the data read out of bank 0 and bank 1 by generating the appropriate data enable (DEN) and data clock pulse (DCP) signals. For example, to read data from bank 0, DEN_0 is first set to logic 0 to enable the upper set of transceivers and then on the 0-to-1 transition of DCP_0, the data output from the memory array is latched into the registered transceivers. The quad data word and its parity bits are now available for the MPU to read.

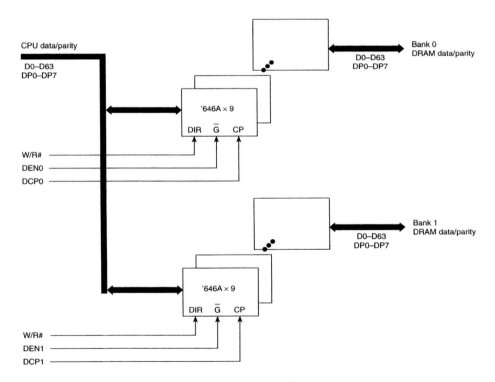

Figure 16.12 Data bus transceiver circuits. (Reprinted by permission of Intel Corp. Copyright/Intel Corp. 1995)

Control Logic Circuit

The control logic section of the memory subsystem is implemented with PAL devices. It accepts the bus cycle definition and other control outputs from the MPU as inputs and generates at its outputs all of the signals necessary to control the operation of the circuits within the memory subsystem. For instance, in Fig. 16.9 we see that it receives signals such as $\overline{ADS}$, $\overline{CACHE}$, M/$\overline{IO}$, and W/$\overline{R}$ as inputs from the MPU. In response to these inputs, it produces outputs to control the RAS/CAS multiplexer circuit, data bus transceiver circuit, DRAM array, and memory bus cycle of the MPU.

We have already identified many of these control signals produced in the control-logic section. For instance, in our discussion of the RAS/CAS multiplexer circuit, we found that the control-logic section provided two signals, ALE and ROWSEL, that are needed to control its operation. Also, notice in Fig. 16.9 that the control logic section is the source of the $\overline{RAS}$, $\overline{CAS}$, and $\overline{WE}$ inputs that control read and write accesses of the DRAM array. The signals produced by the control logic that we have not already examined are those that are returned to the MPU. Figure 16.9 shows that the control logic section supplies $\overline{BRDY}$, $\overline{NA}$, and $\overline{KEN}$ outputs. These signals are returned to the MPU and tell whether a burst, pipelined, or cacheable bus cycle is in progress, respectively.

▲ 16.8 BUS CYCLES: NONPIPELINED, PIPELINED, AND BURST

The bus interface of the Pentium^R processor has been designed to be very versatile and permit a number of different types of data-transfer bus cycles to be performed. Similar to the 80486 family of MPUs, the Pentium^R processor can perform bus cycles with either a single data transfer or burst of data transfers, and these bus cycles can be made either noncacheable or cacheable. Moreover, like the 80386 family, it can perform nonpipelined and pipelined bus cycles. The table in Fig. 16.13 shows the relationship between the bus cycle indication signals and corresponding bus activity. For instance, if M/$\overline{IO}$ D/$\overline{C}$ W/$\overline{R}$ = 010_2 and $\overline{CACHE}$ = 1, a noncacheable 32-bit single-data-transfer I/O read-bus cycle is taking place. Moreover, if M/$\overline{IO}$ D/$\overline{C}$ W/$\overline{R}$ = 100_2 and $\overline{CACHE}$ = 1, a 64-bit code read is in process. They are both examples of single-data-transfer bus cycles. On the other hand, if M/$\overline{IO}$ D/$\overline{C}$ W/$\overline{R}$ = 100_2, $\overline{CACHE}$ = 0, and $\overline{KEN}$ = 0, code is read into the on-chip code cache memory with a 256-bit burst line-fill bus cycle. Let us next look more closely at the different types of read and write bus cycles that can be performed by the Pentium^R processor.

Nonpipelined Read and Write Cycles

The waveforms for the nonpipelined single-data read or write bus cycle are shown in Fig. 16.14(a). Notice that a single-data-transfer read or write takes a minimum of two clock cycles, denoted T1 and T2 in the timing diagram. The read bus cycle starts with an address being output on the address bus accompanied by an address strobe pulse at $\overline{ADS}$. At the same time, W/$\overline{R}$ is switched to logic 0 to identify a read-data transfer. Notice that $\overline{NA}$ and $\overline{CACHE}$ are both left at logic 1 throughout the bus cycle. This means that the bus cycle is nonpipelined and noncacheable. If the Pentium^R processor samples the $\overline{BRDY}$ input late in T2 and finds that it is at its active 0 logic level as shown, the read-data transfer takes place to the MPU and the bus cycle is complete. Otherwise, the bus cycle is extended with additional clock periods until a logic 0 is detected at $\overline{BRDY}$. The bus cycle diagram of Fig. 16.14(b) shows a read and write bus cycle extended with one wait state.

M/IO#	D/C#	W/R#	CACHE#*	KEN#	Cycle description	No. of Transfers
0	0	0	1	x	Interrupt acknowledge (2 locked cycles)	1 transfer each cycle
0	0	1	1	x	Special cycle	1
0	1	0	1	x	I/O read, 32 bits or less, noncacheable	1
0	1	1	1	x	I/O write, 32 bits or less, noncacheable	1
1	0	0	1	x	Code read, 64 bits, noncacheable	1
1	0	0	x	1	Code read, 64 bits, noncacheable	1
1	0	0	0	0	Code read, 256-bit burst line fill	4
1	0	1	x	x	Intel reserved (will not be driven by the Pentium® processor)	n/a
1	1	0	1	x	Memory read, 64 bits or less, noncacheable	1
1	1	0	x	1	Memory read, 64 bits or less, noncacheable	1
1	1	0	0	0	Memory read, 256-bit burst line fill	4
1	1	1	1	x	Memory write, 64 bits or less, noncacheable	1
1	1	1	0	x	256-bit burst writeback	4

* CACHE# will not be asserted by any cycle in which M/IO# is driven low or for any cycle in which PCD is driven high.

Figure 16.13 Types of bus cycles. (Reprinted by permission of Intel Corp. Copyright/Intel Corp. 1995)

EXAMPLE 16.6

What bus cycle indication code is output when a nonpipelined, noncacheable 64-bit data-write bus cycle is in progress?

Solution

From the table in Fig. 16.13, we find that

$$M/\overline{IO}\ D/\overline{C}\ W/\overline{R} = 111_2$$

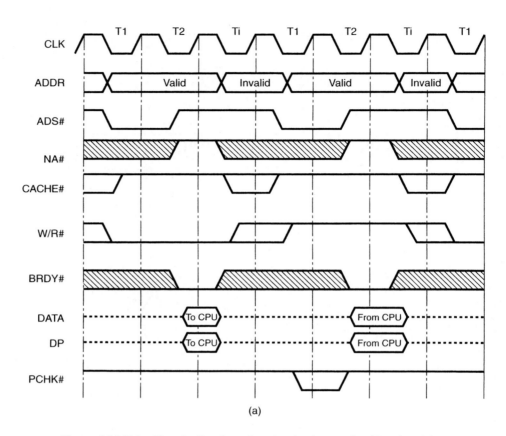

Figure 16.14(a) Nonpipelined read and write bus cycle. (Reprinted by permission of Intel Corp. Copyright/Intel Corp. 1995)

Burst Read and Write Bus Cycles

In Fig. 16.13 we find that the Pentium[R] processor performs only three types of burst bus cycles. They are called a *code-read burst line fill, data-read line fill,* and a *burst write-back.* Each of these cycles represents an update of the cache memory. Notice that a burst bus cycle involves 256 bits of data—that is, transfer of four quad data words.

Figures 16.15(a) and (b) show typical burst read and burst write bus cycles, respectively. Since all cacheable bus cycles are performed as burst cycles, we see that the $\overline{\text{CACHE}}$ output is held at its active 0 logic level throughout the bus cycle. For the burst read bus cycle, logic 0 must be returned to the $\overline{\text{KEN}}$ input of the MPU in clock 2 of the first data transfer. This signals that the memory subsystem will support the current read bus cycle as a burst line fill. $\overline{\text{KEN}}$ is not active during burst write bus cycles. The address and byte enables of the first quad-word of data to be accessed is output by the MPU along with a pulse at $\overline{\text{ADS}}$ at the beginning of the bus cycle. This original address is maintained valid throughout the bus cycle. For this reason, the address must be incremented in external hardware to point to the storage locations for each of the other three quad-word data transfers that follow. Notice that the first 64-bit data transfer of a burst read or write takes place in two clock cycles; however, just one additional cycle is needed for each of the other three data transfers.

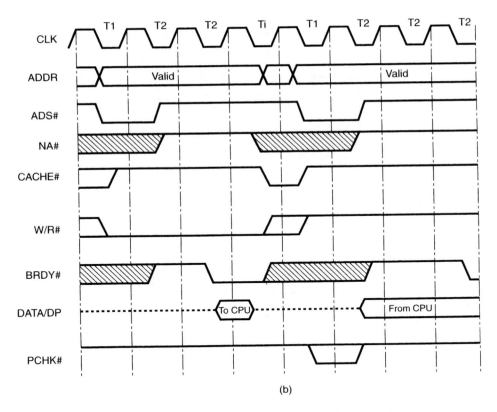

(b)

Figure 16.14(b) Nonpipelined read and write bus cycle with wait states. (Reprinted by permission of Intel Corp. Copyright/Intel Corp. 1995)

EXAMPLE 16.7

What bus cycle indication code is output when a burst write-back bus cycle is performed? What are the values of $\overline{\text{CACHE}}$ and $\overline{\text{KEN}}$ during this bus cycle?

Solution

From the table in Fig. 16.13, we find that

$$\text{M}/\overline{\text{IO}}\ \text{D}/\overline{\text{C}}\ \text{W}/\overline{\text{R}} = 111_2$$

$$\overline{\text{CACHE}} = 0$$

$$\overline{\text{KEN}} = \text{X}$$

Pipelined Read and Write Bus Cycles

The Pentium^R processor is equipped with a *next address* ($\overline{\text{NA}}$) input to enable it to perform pipelined bus cycles. Remember that in a pipelined-mode bus cycle, the address

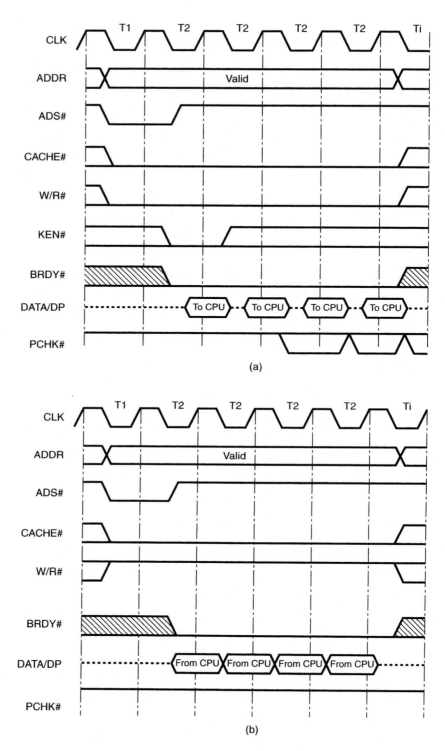

Figure 16.15 (a) Burst read bus cycle. (Reprinted by permission of Intel Corp. Copyright/Intel Corp. 1995) (b) Burst write bus cycle. (Reprinted by permission of Intel Corp. Copyright/Intel Corp. 1995)

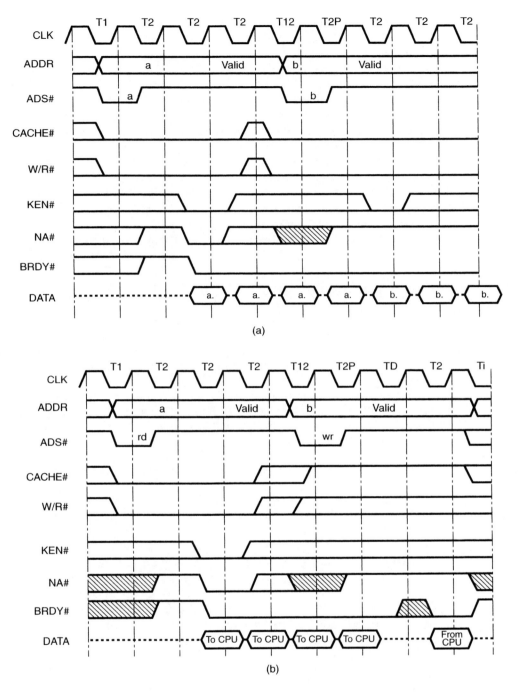

Figure 16.16 (a) Pipelined back-to-back burst read bus cycles. (Reprinted by permission of Intel Corp. Copyright/Intel Corp. 1995) (b) Pipelined back-to-back read and write bus cycles. (Reprinted by permission of Intel Corp. Copyright/Intel Corp. 1995)

for the next bus cycle is output overlapping with the data transfer for the prior bus cycle. Both single data-transfer and burst data-transfer bus cycles can be performed in a pipelined fashion.

Figure 16.16(a) illustrates the bus activity for back-to-back pipelined burst read cycles. First a cacheable burst read is initiated to address a. When $\overline{BRDY}$ switches to logic 0, $\overline{NA}$ also becomes active. Logic 0 at $\overline{NA}$ signals the MPU that the next address, identified as address b, can be output on the address bus. As expected in a pipelined bus cycle, the address for the second cacheable burst read cycle is available to the external memory subsystem prior to the completion of the current burst read bus cycle. Therefore, the valid address is available for a longer period of time, and the memory subsystem can be designed with slower-access-time memory devices.

The waveforms in Fig. 16.16(b) represent pipelined back-to-back read and write cycles. The first cycle is a cacheable burst read and the second is a noncacheable single data-transfer write. Notice that there is one dead clock period between the read and write cycles. It is identified as T_D in the timing diagram. This period of time is needed by the Pentium[R] processor to turn around the bus from input for the read cycle to output for the write cycle.

▲ 16.9 CACHE MEMORY OF THE PENTIUM[R] PROCESSOR

The architecture of the Pentium[R] processor–based microcomputer system supports both an internal and external cache memory subsystem. Here we focus on the architecture, organization, and operation of the internal cache memory of the Pentium[R] processor. Its on-chip cache memory differs from that of the 80486SX in several ways. For instance, there are separate cache memories for storage of data and code; they employ a two-way set associative organization instead of a four-way set associative organization, and two write-update methods, write-through and write-back, are supported for the data cache. The separate caches and write-back capability lead to higher performance for the Pentium[R] processor–based microcomputer.

Organization and Operation of the Internal Cache Memory

Let us begin by examining the organization of the Pentium[R] processor's on-chip data and code cache memories. They both are 8K bytes in size. Since the two-way set associative organization is used, the storage array in each cache memory is organized into two separate 4KB areas called *WAY 0* and *WAY 1*. This organization is shown in Fig. 16.17. Just as with the cache of the 80486SX MPU, updates to the data or code cache of the Pentium[R] processor are always done a line of data at a time; however, its line width is 256 bits (32 bytes), instead of 128 bits (16 bytes). Therefore, WAY 0 and WAY 1 can each hold 256 lines (sets) of data.

WAY 0 and WAY 1 each have a separate tag directory. The tag directory contains one tag for each of the 256 set entries. As shown in Fig. 16.17, a data cache tag entry is formed from a tag address and two MESI (modified-exclusive-shared-invalid) state bits. The MESI bits are used to maintain consistency between the Pentium[R] processor's on-chip data cache and external caches. Notice that the code cache tag supports only 1 bit for storage of MESI state information.

When a data or code read is initiated by the MPU to a storage location whose information is already held in an on-chip cache, a cache hit has occurred, and the information is read from the internal cache memory. On the other hand, if a cache miss results from the read operation, a line-fill cache read takes place from external memory. All line-fill cache reads involve four quad-word (64-bit) data transfers and are performed with burst

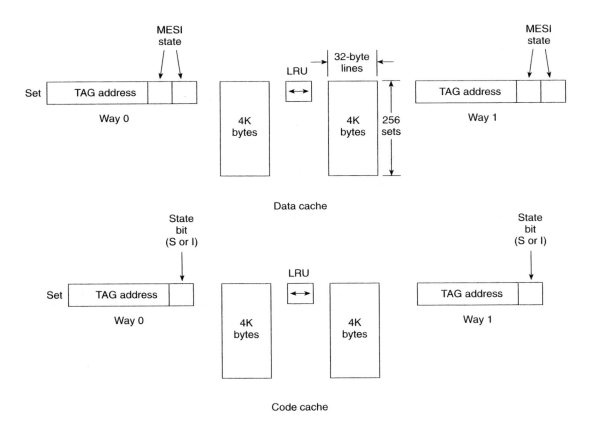

Figure 16.17 Organization of the on-chip cache of the Pentium[R] processor. (Reprinted by permission of Intel Corp. Copyright/Intel Corp. 1995)

bus cycles. Depending on whether the data read is for the code cache or data cache, the data transfer is performed as a code-read or memory-read burst line-fill bus cycle, respectively. If an invalid line exists within the cache, this quad-word of information is loaded into it; otherwise, the LRU algorithm is used to decide which of the valid lines of data will get replaced.

During data-write operations, cache-data consistency must be maintained between the on-chip data cache and external memory. The internal cache circuitry must determine whether or not the data held at the storage location being accessed also exists within the data cache. If a cache hit occurs during a write operation, the corresponding storage location in both the cache and the external memory must be updated.

As identified earlier, the Pentium[R] processor's data cache supports both the write-through and write-back update methods. The 80486SX's on-chip cache provided only the write-through method. With this method, both the line in the internal data cache and its corresponding storage locations in external memory are updated as the write operation is performed. Unlike write-through operations, write-back updates of external memory are not performed at the same time the information is written into the cache. Instead, they are accumulated in the cache memory subsystem and written to memory at a later time. This reduces the bus activity and therefore enhances the microcomputer's performance. The data transfers that take place during a write-back operation to external memory are performed with 256-bit burst write-back bus cycles.

Different areas of the memory address space can be defined as either write-through or write-back and this can be done through software or hardware. For instance, logic 1 at the write-back/write-through (WB/$\overline{\text{WT}}$) input selects write-back operation for the current write update.

The memory address space can also be partitioned into noncacheable and cacheable sections through software or hardware. The way in which this is done on the Pentium[R] processor is identical to how it is performed on the 80486SX MPU. As discussed earlier for the 80486SX MPU, individual pages of memory are made cacheable or noncacheable under software control with the page-level cache-disable (PCD) bit in its page table entry, and the memory address space can be hardware mapped as cacheable or noncacheable on a line-by-line basis with the cache-enable ($\overline{\text{KEN}}$) input.

Enabling, Disabling, and Flushing the On-chip Cache

Just like the cache of the 80486SX, the operation of the Pentium[R] processor's cache memories can be controlled with software and hardware. The table in Fig. 16.18 shows how

CD	NW	Purpose/Description
0	0	**Normal highest performance cache operation.** Read hits access the cache. Read misses may cause replecement. Write hits update the cache. Only writes to shared lines and write misses appear externally. Write hits can change shared lines to exclusive under control of WB/WT#. Invalidation is allowed.
0	1	**Invalid setting.** A general-protection exception with an error code of zero is generated.
1	0	**Cache disabled. Memory consistency maintained. Existing contents locked in cache.** Read hits access the cache. Read misses do not cause replacement. Write hits update cache. Only write hits to shared lines and write misses update memory. Write hits can change shared lines to exclusive under control of WB/WT#. Invalidation is allowed.
1	1	**Cache disabled. Memory consistency not maintained.** Read hits access the cache. Read misses do not cause replacement. Write hits update cache but not memory. Write hits change exclusive lines to be modified. Shared lines remain shared lines after write hit. Write misses access memory. Invalidation is inhibited.

Figure 16.18 On-chip cache operating modes. (Reprinted by permission of Intel Corp. Copyright/Intel Corp. 1995)

the cache disable (CD) and not-write-through (NW) bits of control registers CR_0 affect the operation of the caches. The values of CD and NW can be changed under software control. For instance, the cache memories are enabled for operation by setting both CD and NW to logic 0. Notice that making both CD and NW logic 1 does not totally disable the caches. Instead, as shown in the table for this state, read hits still access valid information held in the cache memory, but misses do not result in an update of the corresponding storage locations in the cache. To completely disable the cache, it must also be flushed after performing the software disable. This can be done with the cache flush ($\overline{\text{FLUSH}}$) input. Switching $\overline{\text{FLUSH}}$ to logic 0 initiates a write-back to external memory of the contents of all modified lines in the data cache, and when this is completed, all the data held in the caches are invalidated.

The contents of the cache memories can also be invalidated under software control. This can be done with the write-back and invalidate cache (WBINVD) instruction. Execution of WBINVD causes a write-back to take place to external memory of any modified lines in the data cache and then the contents of both the code and data caches are invalidated. The table in Fig. 16.18 shows that a software invalidation can be performed only when the NW control bit is logic 0.

Both CD and NW are set to 1 and the internal caches are flushed whenever the MPU is reset by applying logic 1 to the RESET input. Therefore, a hardware reset also completely disables the cache.

▲ 16.10 INTERRUPTS, RESET, AND INTERNAL EXCEPTIONS

Interrupt and exception-processing capabilities have undergone very little change as part of the migration path from the 80386 family of microprocessor to the Pentium^R processor family. In the study of the 80486SX MPU in the last chapter, we found that the only changes made in the exception processing of the 80486SX are that one new exception, alignment-check exception, has been added and that one of the 80386DX's exceptions, coprocessor segment overrun, is no longer supported. The extensions made to the exception handling of the Pentium^R processor are also very small. In fact, just one new exception, *machine-check exception*, has been defined. Moreover, the reset function has been expanded with a second input called *initialization* (INIT). Here we look at these two changes.

Machine-Check Exception

Machine check is a new exception that is first implemented in the Pentium^R processor. This exception is enabled for operation by making the machine-check-enable (MCE) control bit, which is identified as bit position 6 of CR_4 in Fig. 16.4(b), to logic 1 under software control. Two events that can occur in external hardware will cause this type of exception. They are the detection of a parity error in a data read or the unsuccessful completion of a bus cycle. The data-read parity error condition is signaled to the MPU by external circuity with logic 0 at the $\overline{\text{PEN}}$ input and the occurrence of a bus cycle error is identified by logic 0 at the $\overline{\text{BUSCHK}}$ input. If either of these events occur, the address and type information for the bus cycle is latched in the machine check-address (MCA) and machine check-type (MCT) registers, respectively. Figure 16.19 shows the bus cycle–type information saved in MCT. The CHK bit is set to 1 when data are latched into MTC and automatically cleared when the contents are read through software. The exception service routine, which is initiated through exception gate 18, can access this information in an effort to determine

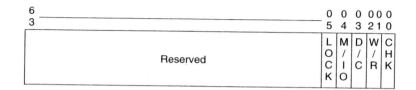

Figure 16.19 Machine check-type register contents. (Reprinted by permission of Intel Corp. Copyright/Intel Corp. 1995)

the cause of the exception. The read from the model-specific register (RDMSR) instruction is used to read the contents of MCA and MCT.

Initialization with RESET and INIT

Two input pins are provided on the Pentium[R] processor for use in initiating a reset operation. They are the *reset* (RESET) line, which is also provided on the earlier 80386DX and 80486SX MPUs, and a new input, *initialization* (INIT), which is first introduced with the Pentium[R] processor family. The active level for both RESET and INIT is logic 1. When active, RESET forces the MPU to terminate program execution and bus activity within 2 clock cycles. On the other hand, INIT performs as an edge-triggered interrupt. When it become active, it is latched within the MPU but not acted upon until the current instruction is completed.

Using these two lines, the MPU can be initialized in several different ways. The table in Fig. 16.20(a) summarizes the different reset modes. Notice that a power-on reset of the MPU is initiated by switching both RESET and INIT to logic 1. This will cause the built-in self-test (BIST) to be run, the code and data caches to be cleared (invalidated), the internal registers to be initialized, and program execution to restart at address $FFFFFFF0_{16}$. The initialized state of the internal registers after a power on reset is shown under the column identified as RESET(BIST) in Fig. 16.20(b).

If RESET is left at logic 0 and INIT is switched to 1, an initialization operation is performed instead of a reset. In this case, BIST is not run, the caches are not flushed, and just some of the registers are initialized. The state of the registers after an initialization are listed in the INIT column of the table in Fig. 16.20(b). Program execution again resumes at address $FFFFFFF0_{16}$.

RESET	INIT	BIST Run?	Effect on Code and Data Caches	Effect on FP Registers	Effect on BTB and TLBs
0	0	No	n/a	n/a	n/a
0	1	No	None	None	Invalidated
1	0	No	Invalidated	Initialized	Invalidated
1	1	Yes	Invalidated	Initialized	Invalidated

Figure 16.20(a) Reset modes. (Reprinted by permission of Intel Corp. Copyright/Intel Corp. 1995)

Storage Element	Reset (No BIST)	Reset (BIST)	INIT
EAX	0	0 if pass	0
EDX	0500+stepping	0500+stepping	0500+stepping
ECX,EBX,ESP,EBP,ESI,EDI	0	0	0
EFLAGS	2	2	2
EIP	0FFF0	0FFF0	0FFF0
CS	selector = F000	selector = F000	selector = F000
	AR = P, R/W, A	AR = P, R/W, A	AR = P, R/W, A
	base = FFFF0000	base = FFFF0000	base = FFFF0000
	limit = FFFF	limit = FFFF	limit = FFFF
DS,ES,FS,GS,SS	selector = 0	selector = 0	selector = 0
	AR = P, R/W, A	AR = P, R/W, A	AR = P, R/W, A
	base = 0	base = 0	base = 0
	limit = FFFF	limit = FFFF	limit = FFFF
(I/G/L)DTR, TSS	selector = 0	selector = 0	selector = 0
	base = 0	base = 0	base = 0
	AR = P, R/W	AR = P, R/W	AR = P, R/W
	limit = FFFF	limit = FFFF	limit = FFFF
CR0	60000010	60000010	Note 1
CR2,3,4	0	0	0
DR3-0	0	0	0
DR6	FFFF0FF0	FFFF0FF0	FFFF0FF0
DR7	00000400	00000400	00000400
Time stamp counter	0	0	Unchanged
Control and event select	0	0	Unchanged
TR12	0	0	Unchanged
All other MSRs	Undefined	Undefined	Unchanged
CW	0040	0040	Unchanged
SW	0	0	Unchanged
TW	5555	5555	Unchanged
FIP,FEA,FCS,FDS,FOP	0	0	Unchanged
FSTACK	0	0	Unchanged
Data and code cache	Invalid	Invalid	Unchanged
Code cache TLB, data cache TLB, BTB, SDC	Invalid	Invalid	Invalid

Note:
CD and NW are unchanged, bit 4 is set to 1, all other bits are cleared.

Figure 16.20(b) Register state after reset or initialization. (Reprinted by permission of Intel Corp. Copyright/Intel Corp. 1995)

EXAMPLE 16.8 _____

What are the differences between a power-on initialization initiated with RESET = INIT = 1 and that produced when RESET is switched to 1 but INIT is left at 0?

Solution

From Figs. 16.20(a) and (b), we find that the only difference is that the BIST is not run.

▲ 16.11 THE PENTIUM^R PRO PROCESSOR
AND PENTIUM^R PROCESSOR WITH MMX™ TECHNOLOGY

▲ 16.11 THE PENTIUM[R] PRO PROCESSOR
AND PENTIUM[R] PROCESSOR WITH MMX™ TECHNOLOGY

Intel Corporation has continued to extend the hardware and software capabilities of the Pentium[R] processor family with new members. For instance, a second generation of devices, the *Pentium[R] Pro processor* was introduced in 1995. This MPU contains advanced features needed by high-performance personal computers, workstations, and servers. Some examples of these capabilities are support for easy, low-cost implementation of multiprocessor systems, and data integrity and reliability functions such as error checking and correction (ECC), fault analysis/recovery, and functional redundancy checking (FRC). The architecture of another family member, the *Pentium[R] processor with MMX™ technology,* has been enhanced to provide higher performance for multimedia and communication applications. The intended use for this device, which was introduced in January 1997, is in desktop and laptop personal computers.

Let us begin by looking more closely at the Pentium[R] Pro processor. As shown in Fig. 16.21, this device is actually two separate die that are housed in a single 387-pin dual-cavity staggered-pin grid-array package (SPGA). One die is the Pentium[R] Pro processor's MPU and the other is a second-level cache. These dies are manufactured using Intel's 0.35-μm BiCMOS process. This process technology uses bipolar transistors to implement high-speed circuitry and CMOS transistors for low power, high density circuitry. The circuitry of the Pentium[R] Pro processor is equivalent to 5.5 million transistors.

The internal architecture of the Pentium[R] Pro processor has been enhanced with what is known as *dynamic execution.* Unlike earlier members of the Pentium[R] processor family, the Pentium[R] Pro processor does not fetch and execute instructions in order. In the dynamic execution architecture, a larger group of instructions are fetched and decoded and made available for execution. They are identified as the *instruction pool* in Fig. 16.22. The traditional instruction-execution phase is replaced by the identified dispatch/execute and retire

Figure 16.21 Pentium[R] Pro processor IC (Reprinted by permission of Intel Corp. Copyright/Intel Corp. 1996)

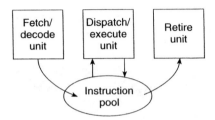

Figure 16.22 Instruction execution of the Pentium[R] Pro processor. (Reprinted by permission of Intel Corp. Copyright/Intel Corp. 1995)

phases in the Pentium[R] Pro processor. This permits instructions to executed out of order but assures that they are put back in their original order when completed. For instance, if an instruction that is being executed cannot be completed because it is waiting for additional data, the Pentium[R] Pro processor looks ahead in the instruction pool and begins working on the execution of other instructions. This is known as *speculative execution. Data-flow analysis* is performed to determine the best order for execution of instructions. That is, instructions are executed based on whether they are ready to be executed, not based on their order in the program. Moreover, due to the larger pool of instructions, more branch operations may be encountered during instruction execution. The *multiple branch prediction* capability of the Pentium[R] Pro processor gives it the ability to predict the flow of the program through several levels of branching and to adjust the instruction execution sequence based on this information. The result is more efficient instruction execution and a higher level of performance.

Some hardware architectural elements that in past family members were implemented with external circuitry are now implemented within the Pentium[R] Pro processor. For instance, the device is available with either a 256KB or 512KB level-2 cache. Unlike an external second-level cache, this internal level-2 cache runs at full speed and results in a higher level of performance. Another hardware element that is now implemented internal to the processor is the advanced programmable interrupt control (APIC).

The performance of Pentium[R] processor family devices is compared with the iCOMP[R] Index 2.0. This new version of the index cannot be compared to the original iCOMP index

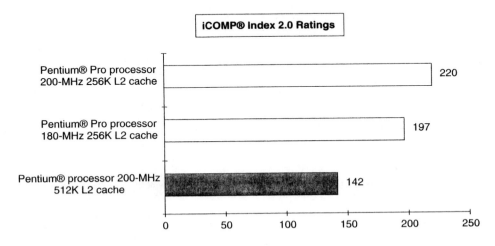

Figure 16.23 Pentium[R] Pro processor iCOMP[R] Index 2.0 ratings. (Reprinted by permission of Intel Corp. Copyright/Intel Corp. 1996)

Figure 16.24 Pentium^R processor with MMX™ technology and iCOMP index 2.0 ratings. (Reprinted by permission of Intel Corp. Copyright/Intel Corp. 1997)

introduced in Chapter 1 because it is formulated based on a different set of benchmarks. The chart in Fig. 16.23 compares the 180-MHz and 200-MHz Pentium^R Pro processors with an internal 256KB level-2 cache to a 200-MHz Pentium^R processor with an external 512KB level-2 cache. Notice that the 200 MHz Pentium^R Pro processor has a rating of 220.

The Pentium^R processor with MMX™ technology is shown in Fig. 16.24. This device is both software and pin-for-pin compatible with the earlier Pentium^R processor family

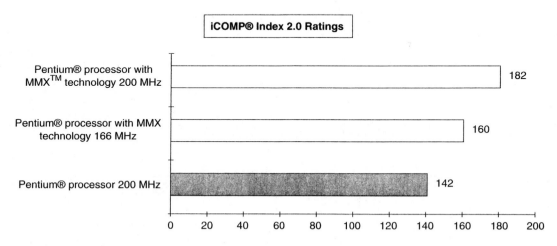

Figure 16.25 Pentium^R processor with MMX™ technology and iCOMP^R index 2.0 ratings. (Reprinted by permission of Intel Corp. Copyright/Intel Corp. 1997)

members. Like the Pentium[R] Pro processor, it is manufactured on Intel's 0.35-μm process technology, but it contains 4.5 million transistors and is made with a single die.

The internal architecture of the Pentium[R] processor with MMX™ technology has been enhanced in a number of ways. First, the instruction set has been expanded with a new group of *MMX™ technology* instructions and data types. They include 57 new instructions and 4 new 64-bit data types. An additional pipeline stage, the fetch stage, has been added into the instruction execution sequence, and its performance is further increased by improvements made in the dynamic branch-prediction capability. Finally, the size of the internal first-level code and data caches has been doubled to 16KB, the caches are four-way set associative instead of two-way set associative, and the write buffers that are used to improve memory-write performance have been expanded from two to four buffers. These enhancements lead to the higher level of performance achieved with the Pentium[R] processor with MMX™ technology. The iCOMP index 2.0 ratings of the 166-MHz and 200-MHz devices are shown in Fig. 16.25.

ASSIGNMENTS

Section 16.2

1. How wide is the Pentium[R] processor's external data bus?
2. What does the term *superscaler* mean?
3. What is the iCOMP rating of the 90 MHZ Pentium[R] processor MPU?

Section 16.3

4. How many pipelines are in a Pentium[R] processor MPU? What are they called?
5. List three improvements made in the Pentium[R] processor's on-chip cache memory over that provided in the 80486SX MPU.
6. How many times faster is the floating-point unit of the Pentium[R] processor compared to that of the 80486DX-33?

Section 16.4

7. Give the mnemonics for the three new flags activated in the Pentium[R] processor's EFLAGS register.
8. What does PSE stand for? How many 32-bit entries are in a page when PSE is set to 1?
9. What capability does the MCE bit of CR_4 enable/disable?
10. List three new real mode instructions supported by the Pentium[R] processor MPU.
11. If the contents of EAX, EBX, ECX, and EDX are 11111111_{16}, 22222222_{16}, 33333333_{16}, and $FFFFFFFF_{16}$, respectively, and the content of the quad-word memory storage location pointed to by the address TABLE is $11111111FFFFFFFF_{16}$, what is the result produced by executing the instruction CMPXCHG8B [TABLE]?
12. Which model specific register is accessed when the RDMSR system control instruction is executed when ECX contains the value 1_{16}?

Section 16.5

13. Approximately how many transistors are used to implement the Pentium[R] processor?
14. What is the nominal value of the Pentium[R] processor power supply?
15. At what pin of the Pentium[R]'s package is data bus line D_{45} located? Address line A_3?

Section 16.6

16. How many byte-enable, data parity, and address parity lines are provided in the memory/IO interface of the Pentium[R] processor?
17. What kinds of parity are supported on the Pentium[R] processor's bus interface? How are parity errors identified?
18. What does logic 0 at the $\overline{\text{BUSCHK}}$ input mean?
19. What logic level is output on $\overline{\text{CACHE}}$ during a cacheable code-fetch bus cycle?
20. What logic level at the WB/$\overline{\text{WT}}$ input means that the write-update method for an external memory storage location is write-back?

Section 16.7

21. How many bytes of data can be stored in the memory array of Fig. 16.10?
22. List the signal names for the address inputs and data input/outputs for bank 1 of the memory array in Fig. 16.10.
23. In the circuit of Fig. 16.11, what signal is used to latch the address information into the 74FCT573 octal latches? What is the source of this signal?
24. How many latches, multiplexers, and buffers are needed to implement the circuit of Fig. 16.11?
25. The 74FCT646A transceiver used in Fig. 16.12 is an octal device. How many ICs are needed to implement the two bank transceiver circuits?
26. What signal is used to control the direction of data transfer through the data bus transceiver circuit of Fig. 16.12? What is the source of this signal?
27. Which data lines of the memory array in Fig. 16.10 are applied to the D_0 through D_{63} and DP_0 through DP_7 lines for bank 0 in the circuit of Fig. 16.12?
28. What kind of devices are used to implement the control logic circuit?

Section 16.8

29. What type of bus cycle is in progress if the bus cycle indication information output by the processor is M/$\overline{\text{IO}}$ D/$\overline{\text{C}}$ W/$\overline{\text{R}}$ $\overline{\text{CACHE}}$ $\overline{\text{KEN}}$ = 0111X?
30. What bus cycle indication information is output when a data-read burst line-fill bus cycle is taking place?
31. What conditions at $\overline{\text{CACHE}}$ and $\overline{\text{KEN}}$ identify that a cacheable burst read bus cycle is taking place?
32. What does logic 0 at $\overline{\text{NA}}$ mean about the current bus cycle?

Section 16.9

33. What type of organization is implemented with the Pentium[R] processor's caches?
34. How much cache memory is provided in the Pentium[R] processor?

35. What cache write-update methods are supported in the Pentium^R processor?

36. How many bits are in the Pentium^R processor's cache line?

37. Are cacheable memory-read and write operations performed with single data-transfer or burst transfer bus cycles? Nonpipelined or pipelined bus cycles?

38. What does logic 0 on WB/$\overline{\text{WT}}$ mean about the current write bus cycle?

39. Describe the operation of the cache memories enabled by making CD = 0 and NW = 0.

40. What must be done in addition to setting CD = NW = 1 to completely disable the cache memory?

41. How does the operation of the INVD instruction differ from that of WBINVD?

Section 16.10

42. What two events in external circuitry can cause a machine-check exception?

43. Tell how the values of the address and bus cycle type information for a machine-check exception can be accessed?

44. What are the key differences between a hardware reset initiated by making RESET and INIT logic 1 and an initialization initiated by making RESET = 0 and INIT = 1?

Section 16.11

45. Which of the new Pentium^R processor family microprocessors is intended for use in workstations and servers?

46. How many transistors are in the Pentium^R Pro processor? The Pentium^R Processor with MMX™ technology?

47. What size level-2 caches is the Pentium^R Pro processor available with?

48. How large are the internal caches of the Pentium^R Processor with MMX™ technology? How is it organized?

49. What is the iCOMP index 2.0 rating of the 200-MHz Pentium^R Processor with MMX™ technology?

Answers to Selected Assignments

▲   CHAPTER 1

Section 1.2

1. Original IBM PC.
3. I/O channel.
5. Industry standard architecture.
7. Mainframe computer, minicomputer, and microcomputer.
9. Very large scale integration.

Section 1.3

11. Microprocessing unit (MPU).
13. Keyboard; mouse and joy stick.
15. Primary storage and secondary storage memory.
17. Read-only memory (ROM) and random access read/write memory (RAM).
19. The DOS program is loaded from the hard disk into RAM and then run. Since RAM is volatile, the disk operating system is lost whenever power is turned off.

Section 1.4

21. 4004, 8008, 8086, 80386DX.
23. Million instructions per second.
25. Drystone program.

27. 30,000, 140,000, 275,000, 1,200,000.

29. Event controller and data controller.

31. 8088, 8086, 80286, 80386DX, 80486DX, and Pentium[R] processor.

33. Upward software compatible means that programs written for the 8088 or 8086 will run directly on the 80286, 80386DX, and 80486DX.

35. Floppy disk controller, communication controller, and local area network controller.

▲ CHAPTER 2

Section 2.2

1. Bus unit, prefetch unit, decode unit, execution unit, segment unit, and page unit.

3. Separate address and data buses.

5. Prefetch unit.

7. Translation lookaside buffer.

Section 2.3

9. Their purpose, function, operating capabilities, and limitations.

11. Machine status word (MSW).

13. 1,048,576 (1M) bytes.

Section 2.4

15. Bytes.

17. $00FF_{16}$; aligned word.

19. *Address*   *Contents*

   0A003H   CD
   0A004H   AB
   Aligned word.

Section 2.5

21. Unsigned integer, signed integer, unpacked BCD, packed BCD, ASCII.

23. (0A000) = 27H
   (0A001) = 01H

25. **(a)** 00000010, 00001001; 00101001
   (b) 00001000, 00001000; 10001000

27. NEXT I

Section 2.6

29. 64K bytes.

31. CS.

33. 256K bytes.

Section 2.7

35. The instruction pointer is the offset address of the next double word of code to be fetched by the 80386DX relative to the current value in CS.

37. IP is incremented by 4 such that it points to the next sequential double word of code.

Section 2.8

39. Accumulator (A) register, base (B) register, count (C) register, and data (D) register.

41. DH and DL.

Section 2.9

43. Stack pointer (SP) and base pointer (BP).

45. SI = source index and DI = destination index.

Section 2.10

47. *Flag* *Type*

Flag	Type
CF	Status
PF	Status
AF	Status
ZF	Status
SF	Status
OF	Status
TF	Control
IF	Control
DF	Control

49. Instructions can be used to test the state of these flags and based on their setting modify the sequence in which instructions of the program are executed.

51. DF.

Section 2.11

53. 16 bits and 20 bits.

55. ES.

57. (a) $? = 0123_{16}$
  (b) $? = 2210_{16}$
  (c) $? = 3570_{16}$
  (d) $? = 2600H$

59. $A000_{16}$

Section 2.12

61. The stack is the area of memory that is used to store information (parameters) temporarily that is to be passed to subroutines and other information such as the contents of IP and CS that are needed to return from a called subroutine to the main part of the program.

63. 128 words.

Section 2.13

65. Separate.

67. Page 0.

▲ CHAPTER 3

Section 3.2

1. Software.

3. 80386DX machine code.

5. Mnemonic that identifies the operation to be performed by the instruction; ADD and MOV.

7. START; ADD EBX TO EAX.

9. Programs written in assembly language or high-level language statements are called source code. The machine-code output of an assembler or compiler is called object code.

11. A real-time application is one in which the tasks required by the application must be completed before any other input to the program occurs that can alter its operation.

Section 3.3

13. **(a)** Describe the problem.
(b) Plan the steps of the solution.
(c) Implement a flowchart and assembly language program for the solution.
(d) Create a source file.
(e) Assemble the source file.
(f) Link the program into a run module.
(g) Execute and/or debug the program.

15. Algorithm; software specification.

19. Assembler.

21. Linker.

23. **(a)** PROG_A.ASM
(b) PROG_A.LST, PROG_A.OBJ, and PROG_A.CRF
(c) PROG_A.EXE and PROG_A.MAP

Section 3.4

25. 80386DX specific instruction set.

Section 3.5

27. An addressing mode means the method by which an operand can be specified in a register or a memory location.

29. Base, index, and displacement.

31.

Instruction	*Destination*	*Source*
(a)	Register	Register
(b)	Register	Immediate
(c)	Register indirect	Register
(d)	Register	Register indirect
(e)	Based	Register
(f)	Indexed	Register
(g)	Based-indexed	Register

▲ CHAPTER 4

Section 4.2

1. $0000001111000010_2 = 03C2H$.

3. **(a)** $00011110_2 = 1EH$. **(b)** $1101001011000011_2 = D2C3H$.
   (c) $00000011000001100011010000010010_2 = 03063412H$.

Section 4.3

5. 24 bytes.

Section 4.4

7. Yes.

9. ```
-R   CX   (↵)
CX   XXXX
:0010   (↵)
```

**11.** `-R  (↵)`

## Section 4.5

**13.**
```
-E CS:0 (↵)
1342:0000 CD. 20. 00. 40. 00. 9A. EE. FE.
1342:0008 1D. F0. F5. 02. A7. 0A. 2E. 03. (↵)
```
After a byte of data is displayed, the space bar is depressed to display the next byte. The values displayed may not be those shown but will be identical to those displayed with the DUMP command.

**15.** `-E  SS:(SP)  0  ......  0  (32 zeros)  (↵)`
   `-`
```

Section 4.6

17. Input command and output command.

19. O   124   5A   (↵)

Section 4.7

21. 4 digit.

Section 4.8

23. ‑E   CS:100   32   0E   34   12   (↵)
  ‑U   CS:100   103   (↵)
  1342:100   320E3412   XOR   CL,[1234]
  ‑W   CS:100   1   50   1   (↵)

Section 4.9

25. ‑A   CS:100   (↵)
  1342:0100   MOV   [DI],DX   (↵)
  1342:0102   (↵)

Section 4.10

27. ‑L   CS:300   1   50   1   (↵)
  ‑U   CS:300   303   (↵)
  ‑R   CX   (↵)
  CX   XXXX
  :000F   (↵)
  ‑E   DS:1234   FF   (↵)
  ‑T   =CS:300   (↵)
  ‑D   DS:1234   1235   (↵)

Section 4.11

29. A syntax error is an error in the rules of coding the program. On the other hand, an execution error is an error in the logic of the planned solution for the problem.

31. Debugging the program.

▲ CHAPTER 5

Section 5.2

1. (a) Value of immediate operand 0110H is moved into AX.
  (b) Contents of AX are copied into DI.
  (c) Contents of AL are copied into BL.
  (d) Contents of AX are copied into memory address DS:0100H.

(e) Contents of AX are copied into the data segment memory location pointed to by (DS)0 + (BX) + (DI).

(f) Contents of AX are copied into the data segment memory location pointed to by (DS)0 + (DI) + 4H.

(g) Contents of AX are copied into the data segment memory location pointed to by (DS)0 + (BX) + (DI) + 4H.

3.
```
MOV   AX,1010H
MOV   ES,AX
```

5. Destination operand CL is specified as a byte and source operand AX is specified as a word. Both must be with the same size.

7.
```
MOVZX   EAX,[DATA_WORD]
```

9. 10,750H + 100H + 10H = 10,860H

11.
```
LDS   AX,[0200H].
```

13.
```
MOV   AX,DATA_SEG   ;Establish the data segment
MOV   DS,AX
MOV   AL,[MEM1]     ;Get the given code at MEM1
MOV   BX,TABL1
XLAT               ;Translate
MOV   [MEM1],AL     ;Save new code at MEM1
MOV   AL,[MEM2]     ;Repeat for the second code at MEM2
MOV   BX,TABL2
XLAT
MOV   [MEM2],AL
```

Section 5.3

15. **(a)** (AX) = 010FH

(b) (SI) = 0111H

(c) (DS:100H) = 11H

(d) (DL) = 20H

(e) (DL) = 0FH

(f) (DS:220H) = 2FH

(g) (DS:210H) = C0H

(h) (AX) = 0400H
   (DX) = 0000H

(i) (AL) = 0F0H
   (AH) = 0FFH

(j) (AL) = 02H
   (AH) = 00H

(k) (AL) = 08H
   (AH) = 00H

17.
```
SBB   AX,[BX]
```

19. (AH) = remainder = 3_{16}, (AL) = quotient = 12_{16}; therefore, (AX) = 0312_{16}.

21. AAS

23. (AX) = 7FFFH, (DX) = 0000H

Section 5.4

25. (a) 0FH is ANDed with the contents of the byte-wide memory address DS:300H.

(b) Contents of DX are ANDed with the contents of the word storage location pointed to by (DS)0 + (SI).

(c) Contents of AX are ORed with the word contents of the memory location pointed to by (DS)0 + (BX) + (DI).

(d) F0H is ORed with the contents of the byte-wide memory location pointed to by (DS)0 + (BX) + (DI) + 10H.

(e) Contents of the word-wide memory location pointed to by (DS)0 + (SI) + (BX) are exclusive-ORed with the contents of AX.

(f) The bits of the byte-wide memory location DS:300H are inverted.

(g) The bits of the word memory location pointed to by (DS)0 + (BX) + (DI) are inverted.

27. AND   EDX,0080H

29. The new contents of AX are the 2's complement of its old contents.

31. MOV   AL,[CONTROL_FLAGS]
AND   AL,81H
MOV   [CONTROL_FLAGS],AL

Section 5.5

33. (a) Contents of DX are shifted left by a number of bit positions equal to the contents of CL. LSBs are filled with 0s, and CF equals the value of the last bit shifted out of the MSB position.

(b) Contents of EDX are shifted left by 7 bit positions. LSBs are filled with 0s and CF equals the value of the last bit shifted out of the MSB position.

(c) Contents of the byte-wide memory location DS:400H are shifted left by a number of bit positions equal to the contents of CL. LSBs are filled with 0s, and CF equals the value of the last bit shifted out of the MSB position.

(d) Contents of the byte-wide memory location pointed to by (DS)0 + (DI) are shifted right by 1 bit position. MSB is filled with 0, and CF equals the value shifted out of the LSB position.

(e) Contents of the double-word-wide memory location pointed to by DS:DI are shifted right by 3 bit positions. MSBs are filled with 0s and CF equals the value of the last bit shifted out at the LSB position.

(f) Contents of the byte-wide memory location pointed to by (DS)0 + (DI) + (BX) are shifted right by a number of bit positions equal to the contents of CL. MSBs are filled with 0s, and CF equals the value of the last bit shifted out of the LSB position.

(g) Contents of the word-wide memory location pointed to by (DS)0 + (BX) + (DI) are shifted right by 1 bit position. MSB is filled with the value of the original MSB and CF equals the value shifted out of the LSB position.

(h) Contents of the word-wide memory location pointed to by (DS)0 + (BX) + (DI) + 10H are shifted right by a number of bit positions equal to the contents of CL. MSBs are filled with the value of the original MSB, and CF equals the value of the last bit shifted out of the LSB position.

35. `SHL    ECX,1`

37. The original contents of AX must have the 4 most significant bits equal to 0.

39. Double-precision shift right.

41. The first instruction reads the byte of control flags into AL. Then all but the flag in the most significant bit location B_7 are masked off. Finally, the flag in B_7 is shifted to the left and into the carry flag. When the shift takes place, B7 is shifted into CF; all other bits in AL move one bit position to the left, and the LSB locations are filled with 0s. Therefore, the contents of AL become 00H.

Section 5.6

43. (a) Contents of DX are rotated left by a number of bit positions equal to the contents of CL. As each bit is rotated out of the MSB position, the LSB position and CF are filled with this value.

   (b) Contents of the byte-wide memory location DS:400H are rotated left by a number of bit positions equal to the contents of CL. As each bit is rotated out of the MSB position, it is loaded into CF, and the prior contents of CF are loaded into the LSB position.

   (c) Contents of the byte-wide memory location pointed to by (DS)0 + (DI) are rotated right by 1 bit position. As the bit is rotated out of the LSB position, the MSB position and CF are filled with this value.

   (d) Contents of the byte-wide memory location pointed to by (DS)0 + (DI) + (BX) are rotated right by a number of bit positions equal to the contents of CL. As each bit is rotated out of the LSB position, the MSB position and CF are filled with this value.

   (e) Contents of the word-wide memory location pointed to by (DS)0 + (BX) + (DI) are rotated right by 1 bit position. As the bit is rotated out of the LSB location, it is loaded into CF, and the prior contents of CF are loaded into the MSB position.

   (f) Contents of the word-wide memory location pointed to by (DS)0 + (BX) + (DI) + 10H are rotated right by a number of bit positions equal to the contents of CL. As each bit is rotated out of the LSB position, it is loaded into CF, and the prior contents of CF are loaded into the MSB position.

45. `RCL    WORD PTR [BX],1`

47.
```
MOV    AX,[ASCII_DATA]
MOV    BX,AX
MOV    CL,08H
ROR    BX,CL
AND    AX,00FFH
AND    BX,00FFH
MOV    [ASCII_CHAR_L],    AX
MOV    [ASCII_CHAR_H],    BX
```

Section 5.7

49. (a) AX = F0F0H, CF = 1.

   (b) AX = F0E0H, CF = 1.

   (c) AX = F0E0H, CF = 1.

51. [DS:100H] = 00F7H and CF = 1.

53. BSR   EAX,DWORD   PTR   [SI]

▲ CHAPTER 6

Section 6.2

1. Executing the first instruction causes the contents of the status register to be copied into AH. The second instruction causes the value of the flags to be saved in memory location (DS)0 + (BX) + (DI).

3. STC; CLC.

5.
```
CLI
MOV    AX,0H
MOV    DS,AX
MOV    BX,0A000H
LAHF
MOV    [BX],AH
CLC
```

Section 6.3

7. (a) The byte of data in AL is compared with the byte of data in memory at address DS:100H by subtraction, and the status flags are set or reset to reflect the result.

(b) The word contents of the data storage memory location pointed to by (DS)0 + (SI) are compared with the contents of AX by subtraction, and the status flags are set or reset to reflect the results.

(c) The immediate data 00001234H are compared with the double word contents of the memory location pointed to by (DS)0 + (DI) by subtraction, and the status flags are set or reset to reflect the results.

9.

	ZF	CF
Initial state	0	0
After MOV BX,1111H	0	0
After MOV AX,0BBBBH	0	0
After CMP BX,AX	0	1

11. ZF = 0

13. If the execution of the preceding instruction has set OF, FF_{16} is written to the memory location pointed to by the value of OVERFLOW.

Section 6.4

15. When an unconditional jump instruction is executed, the jump always takes place. On the other hand, when a conditional jump instruction is executed, the jump takes place only if the specified condition is satisfied.

17. 8-bit; 16-bit; 16-bit; 32-bit.

19. (a) Intrasegment; short-label; the value 10H is placed in IP.

(b) Intrasegment; near-label; the value 1000H is copied into IP.

(c) Intrasegment; Memptr16; the word of data in memory pointed to by (DS)0 + (SI) is copied into IP.

21. ZF, CF, SF, PF, and OF.

23. (CF) = 0 and (ZF) = 0.

25. 0100H

27.
```
     ;N! = 1*2*3*4. . .*(N-1)*N
     ;Also note that 0! = 1! = 1
          MOV   AL,1H       ;Initial value of result
          MOV   CL,0H       ;Start multiplying number
          MOV   DL,N        ;Last number for multiplication
NXT:      CMP   CL,DL       ;Skip if done
          JE    DONE
          INC   CL          ;Next multiplying number
          MUL   CL          ;Result Result * number
          JMP   NXT         ;Repeat
DONE:     MOV   [FACT],AL   ;Save the result
```

29.
```
          MOV   CX,64H      ;Set up the counter
          MOV   AX,0H       ;Set up the data segment
          MOV   DS,AX
          MOV   BX,0A000H   ;Pointer for the given array
          MOV   SI,0B000H   ;Pointer for the +ve array
          MOV   DI,0C000H   ;Pointer for the -ve array
AGAIN:    MOV   AX,[BX]     ;Get the next source element
          CMP   AX,0H       ;Skip if positive
          JGE   POSTV
NEGTV:    MOV   [DI],AX     ;Else place in -ve array
          INC   DI
          INC   DI
          JMP   NXT         ;Skip
POSTV:    MOV   [SI],AX     ;Place in the +ve array
          INC   SI
          INC   SI
NXT:      DEC   CX          ;Repeat for all elements
          JNZ   AGAIN
          HLT
```

31.
```
     ;Assume that all arrays are in the same data segment
          MOV   AX,DATASEG          ;Set up data segment
          MOV   DS,AX
          MOV   ES,AX
          MOV   SI,OFFSET_ARRAYA    ;Set up pointer to array A
          MOV   DI,OFFSET_ARRAYB    ;Set up pointer to array B
          MOV   CX,62H
          MOV   AX,[SI]             ;Initialize A(I-2) and B(1)
          MOV   ARRAYC,AX
          MOV   [DI],AX
          ADD   SI,2
          ADD   DI,2
          MOV   AX,[SI]             ;Initialize A(I-1) and B(2)
          MOV   ARRAYC+1,AX
          MOV   [DI],AX
          ADD   SI,2
          ADD   DI,2
          MOV   AX,[SI]             ;Initialize A(I)
          MOV   ARRAYC+2,AX
```

878

```
            ADD     SI,2
            MOV     AX,[SI]              ;Initialize A(I+1)
            MOV     ARRAYC+3,AX
            ADD     SI,2
            MOV     AX,[SI]              ;Initialize A(I+2)
            MOV     ARRAYC+4,AX
            ADD     SI,2
    NEXT:   CALL    SORT                 ;Sort the 5 element array
            MOV     AX,ARRAYC+2          ;Save the median
            MOV     [DI],AX
            ADD     DI,2
            MOV     AX,ARRAYC+1          ;Shift the old elements
            MOV     ARRAYC,AX
            MOV     AX,ARRAYC+2
            MOV     ARRAYC+1,AX
            MOV     AX,ARRAYC+3
            MOV     ARRAY+2,AX
            MOV     AX,ARRAYC+4
            MOV     ARRAYC+3,AX
            MOV     AX,[SI]              ;Add the new element
            MOV     ARRAY+4,AX
            ADD     SI,2
            LOOP    NEXT                 ;Repeat
            SUB     SI,4                 ;The last two elements of array B
            MOV     AX,[SI]
            MOV     [DI],AX
            ADD     SI,2
            ADD     DI,2
            MOV     AX,[SI]
            MOV     [DI],AX
    DONE:   ---     ---
;SORT subroutine
    SORT:   PUSHF                        ;Save registers and flags
            PUSH    AX
            PUSH    BX
            PUSH    DX
            MOV     SI,OFFSET_ARRAYC
            MOV     BX,OFFSET_ARRAYC+4
    AA:     MOV     DI,SI
            ADD     DI,02H
    BB:     MOV     AX,[SI]
            CMP     AX,[DI]
            JLE     CC
            MOV     DX,[DI]
            MOV     [SI],DX
            MOV     [DI],AX
    CC:     INC     DI
            INC     DI
            CMP     DI,BX
            JBE     BB
            INC     SI
            INC     SI
            CMP     SI,BX
            JB      AA
```

```
                POP     DX            ;Restore registers and flags
                POP     BX
                POP     AX
                POPF
                RET
```

Section 6.5

33. The intersegment call provides the ability to call a subroutine in either the current code segment or a different code segment. On the other hand, the intrasegment call only allows calling of a subroutine in the current code segment.

35. **(a)** Intrasegment; near-proc; a call is made to a subroutine by loading the immediate value 1000H into IP.

   (b) Intrasegment; Memptr16; a call is made to a subroutine by loading the word at address DS:100H into IP.

   (c) Intersegment; Memptr32; a call is made to a subroutine by loading the two words of the double-word pointer located at (DS)0 + (BX) + (SI) into IP and CS, respectively.

   (d) Intersegment; Regptr32; a call is made to a subroutine by loading the two words of the double word in register EDX into IP and CS, respectively.

37. At the end of the subroutine a RET instruction is used to return control to the main (calling) program. It does this by popping IP from the stack in the case of an intrasegment call and both CS and IP for an intersegment call.

39. SS:SP+1 = 10H
   SS:SP   = 00H

41.
```
        ;For the decimal number = D3D2D1D0,
        ;the binary number = 10(10(10(0+D3)+D2)+D1)+D0
                    MOV     BX,0          ;Result = 0
                    MOV     SI,0AH        ;Multiplier = 10
                    MOV     CH,4          ;Number of digits = 4
                    MOV     CL,4          ;Rotate counter = 4
                    MOV     DI,DX
        NXTDIGIT:   MOV     AX,DI         ;Get the decimal number
                    ROL     AX,CL         ;Rotate to extract the digit
                    MOV     DI,AX         ;Save the rotated decimal number
                    AND     AX,0FH        ;Extract the digit
                    ADD     AX,BX         ;Add to the last result
                    DEC     CH
                    JZ      DONE          ;Skip if this is the last digit
                    MUL     SI            ;Multiply by 10
                    MOV     BX,AX         ;and save
                    JMP     NXTDIGIT      ;Repeat for the next digit
        DONE:       MOV     DX,AX         ;Result = (AX)
```

43.
```
        ;Assume that the offset of A[I] is AI1ADDR
        ;and the offset of B[I] is BI1ADDR
                MOV     AX,DATA_SEG   ;Initialize data segment
                MOV     DS,AX
                MOV     CX,62H
                MOV     SI,AI1ADDR    ;Source array pointer
                MOV     DI,BI1ADDR    ;Destination array pointer
                MOV     AX,[SI]
                MOV     [DI],AX       ;B[1] = A[1]
```

```
        MORE:   MOV     AX,[SI]         ;Store A[I] into AX
                ADD     SI,2            ;Increment pointer
                MOV     BX,[SI]         ;Store A[I+1] into BX
                ADD     SI,2
                MOV     CX,[SI]         ;Store A[I+2] into CX
                ADD     SI,2
                CALL    ARITH           ;Call arithmetic subroutine
                MOV     [DI],AX
                SUB     SI,4
                ADD     DI,2
                LOOP    MORE            ;Loop back for next element
                ADD     SI,4
        DONE:   MOV     AX,[SI]         ;B[100] = A[100]
                MOV     [DI],AX
                HLT
        ;Subroutine for arithmetic
        ;(AX) ← [(AX) - 5(BX) + 9(CX)]/4
        ARITH:  PUSHF                   ;Save flags and registers in stack
                PUSH    BX
                PUSH    CX
                PUSH    DX
                PUSH    DI
                MOV     DX,CX           ;(DX) ← (CX)
                MOV     DI,CX
                MOV     CL,3
                SAL     DX,CL
                ADD     DX,DI
                MOV     CL,2            ;(AX) ← 5(BX)
                MOV     DI,BX
                SAL     BX,CL
                ADD     BX,DI
                SUB     AX,BX           ;(AX) ← [(AX) - 5(BX) + 9(CX)]/4
                ADD     AX,DX
                SAR     AX,CL
                POP     DI              ;Restore flags and registers
                POP     DX
                POP     CX
                POP     BX
                POPF
                RET                     ;Return
```

Section 6.6

45. ZF.

47. Jump size $= -126$ to $+129$.

49.
```
                MOV     AL,1H
                MOV     CL,N
                JCXZ    DONE            ;N = 0 case
                LOOPZ   DONE            ;N = 1 case
                INC     CL              ;Restore N
        AGAIN:  MUL     CL
                LOOP    AGAIN
        DONE:   MOV     [FACT],AL
```

Section 6.7

51. DF.

53. (a)
```
CLD
MOV   ES,DS
MOVSB
```
(b)
```
CLD
LODSW
```
(c)
```
STD
CMPSB
```

55.
```
            MOV   SI,OFFSET DATASEG1_ASCII_CHAR        ;ASCII offset
            MOV   DI,OFFSET DATASEG2_EBCDIC_CHAR        ;EBCDIC offset
            MOV   BX,OFFSET DATASEG3_ASCII_TO_EBCDIC    ;Translation
                                                        ;table offset
            CLD                    ;Select autoincrement mode
            MOV   CL,64H           ;Byte count
            MOV   AX,DATASEG1      ;ASCII segment
            MOV   DS,AX
            MOV   AX,DATASEG2      ;EBCDIC segment
            MOV   ES,AX
NEXTBYTE:
            LODSB                  ;Get the ASCII
            MOV   DX,DATASEG3      ;Translation table segment
            MOV   DS,DX
            XLAT                   ;Translate
            STOSB                  ;Save EBCDIC
            MOV   DX,DATASEG1      ;ASCII segment for next ASCII
            MOV   DS,AX            ;element
            LOOP  NEXTBYTE
     DONE:  ---   ---
```

▲ CHAPTER 7

Section 7.2

1. Assembly language statements and pseudo-operation statements.
3. Pseudo-operations give the 80386DX's macroassembler directions about how to assemble the source program.
5. Opcode.
7. A label gives a symbolic name to an assembly language statement that can be referenced by other instructions.
9. Identifies the operation that must be performed.
11. The source operand is immediate data FFH and the destination operand is the CL register.
13. **(a)** A pseudo-operation statement may have more than two operands, whereas an assembly language statement always has two or less operands.
    (b) There is no machine code generated for pseudo-operation statements. There is always machine code generated for an assembly language statement.
15. JMP 11001B; JMP 19H

882

Section 7.3

17. Directive.

19. Define values for constants, variables, and labels.

21. The value assigned by an EQU pseudo-op cannot be changed, whereas a value assigned with the = pseudo-op can be changed later in the program.

23. A block of 128 bytes of memory are allocated to the variable BLOCK_1 and these storage locations are left uninitialized.

25. `SOURCE_BLOCK  DW  16  DUP(?)`

27. This pseudo-op statement defines data segment DATA_SEG so that it is byte aligned in memory and located at an address above all other segments in memory.

29. Module.

31.
```
     PUBLIC  BLOCK
BLOCK  PROC  FAR
       .
       .
       .
     RET
BLOCK  ENDP
```

33. `ORG  1000H`

Section 7.4

35. Line editor; screen editor

37. Deletes lines 15 through 17 of the program.

39. Move, copy, delete, find, and find and replace.

41. Use Save As operations to save the file under the file names BLOCK.ASM and BLOCK.BAK. When the file BLOCK.ASM is edited at a later time, an original copy will be preserved during the edit process in the file BLOCK.BAK.

Section 7. 5

43. Object module: machine language version of the source program.
Source listing: listing that includes memory address, machine code, source statements, and a symbol table.
Cross-reference table: tells the number of the line in the source program at which each symbol is defined and the number of each line in which it is referenced.

45. Looking at Fig. 7.23(b), we see that the error is in the equal to pseudo-op statement that assigns the value 16 to N. The error is that the = sign is left out.

Section 7.6

47. No, the output of the assembler is not executable by the 80386DX in the PC/AT: it must first be processed with the LINK program to form a run module.

49. Object modules.

51. Object modules `[.OBJ]:A:MAIN.OBJ+A:SUB1.OBJ+A:SUB2.OBJ`.

Section 8.2

1. Global descriptor table register, interrupt descriptor table register, task register, and local descriptor table register.

3. Defines the location and size of the global descriptor table.

5. System-segment descriptor.

7. 0FFFH

9. Local descriptor table.

11. CR_0.

13. (MP) = 1, (EM) = 0, and (ET) = 1.

15. Switch the PG bit in CR_0 to 1.

17. 4 K-byte.

19. Selector; selects a task-state segment descriptor.

21. BASE and LIMIT of the TSS descriptor.

23. RPL = 2 bits
TI = 1 bit
INDEX = 13 bits

25. 00130020H

27. Level 2

Section 8.3

29. Selector and offset.

31. 64T bytes, 16,384 segments.

33. Task 3 has access to the global memory address space and the task 3 local address space. But, it cannot access either the task 1 local address space or task 2 local address space.

35. The first instruction loads the AX register with the selector from the data storage location pointed to by SI. The second instruction loads the selector into the code segment-selector register. This causes the descriptor pointed to by the selector in CS to be loaded into the code segment-descriptor cache.

37. 1,048,496 pages; 4096 bytes long.

39. Cache page directory and page table pointers on chip.

Section 8.4

41. 8, BASE = 32-bits, LIMIT = 20-bits, ACCESS RIGHTS BYTE = 8-bits, AVAILABLE = 1 bit, and GRANULARITY = 1 bit.

43. LIMIT = 00110H, BASE = 00200000H.

45. 00200226H

47. R/W = 0 and U/S = 0 or R/W = 1 and U/S = 0.

49. Dirty bit.

Section 8.5

51.
```
LMSW   AX
AND    AX,0FFF7H
SMSW   AX
```

Section 8.6

53. The running of multiple processes in a time-shared manner.

55. Local memory resources are isolated from global memory resources and tasks are isolated from each other.

57. Level 0, level 3.

59. LDT and GDT.

61. Level 0.

63. A task can access data in a data segment at the CPL and at all lower privilege levels. But, it cannot access data in segments that are at a higher privilege level.

65. A task can access code in segments at the CPL or at higher privilege levels. But cannot modify the code at a higher privilege level.

67. The call gate is used to transfer control within a task from code at the CPL to a routine at a higher privilege level.

69. Identifies a task state segment.

71. The state of the prior task is saved in its own task state segment. The linkage to the prior task is saved as the back link selector in the first word of the new task state segment.

Section 8.7

75. Active, level 3.

77. Yes.

▲ CHAPTER 9

Section 9.2

1. CHMOSIII.

3. INTR.

Section 9.3

5. Byte, D_0-D_7, no.

7. I/O data read.

9. 80387DX numeric coprocessor.

Section 9.4

11. F12.

Section 9.5

13. 40 ns.

15. 4; 2; 80 ns.

17. In Fig. 9.11 address n becomes valid in the T_2 state of the prior bus cycle and then the data transfer takes place in the next T_2 state. Also, at the same time that data transfer n occurs, address $n + 1$ is output on the address bus. This shows that during pipelining, the 80386DX starts to address the next storage location that is to be accessed while still reading or writing data for the previously addressed storage location.

19. An extension of the current bus cycle by a period equal to one or more T states because the $\overline{\text{READY}}$ input was tested and found to be logic 1.

Section 9.6

21. 160 ns.

Section 9.7

23.

Types of data transfer	No. of bus cycles
Byte transfer	1
Aligned word transfer	1
Misaligned word transfer	2
Aligned double-word transfer	1
Misaligned double-word transfer	2

25. Higher-addressed byte.

Section 9.8

27. M/$\overline{\text{IO}}$ D/$\overline{\text{C}}$ W/$\overline{\text{R}}$ = 111, all four, $\overline{\text{MWTC}}$.

29. The clock input of the 74F343 is level sensitive and that of the 74F374 is positive edge-triggered.

33. $\overline{\text{DEN}}$ = 0, DT/$\overline{\text{R}}$ = 0.

35. 74F139.

Section 9.9

39. Number of inputs, number of outputs, and number of product terms.

41. GALs are CMOS devices and erasable.

45. 20 inputs; 8 outputs.

Section 9.10

47. Isolated I/O and memory-mapped I/O.

49. Memory-mapped I/O.

51. 64K byte-wide ports, 32K word-wide ports, 16K double-word-wide ports.

Section 9.11

53. Address lines A_2 through A_{15} and $\overline{BE_0}$ through $\overline{BE_3}$ carry the address of the I/O port to be accessed, whereas address lines A_{16} through A_{31} are held at the 0 logic level. Data bus lines D_0 through D_{31} carry the data that are transferred between the MPU and I/O port.

55. Word, D_8 through D_{23}

57. $\overline{IOWR_3}\overline{IOWR_2}\overline{IOWR_1}\overline{IOWR_0} = 0011$

59. 32.

61. The I/O address decoder is used to decode several of the upper I/O address bits to produce the $\overline{I/OCE}$ signals.

The I/O address bus latch is used to latch lower-order address bits, byte-enable signals, and $\overline{I/OCE}$ outputs of the decoder.

The bus-control logic decodes I/O bus commands to produce the input/output and bus-control signals for the I/O interface.

The data bus transceivers control the direction of data transfer over the bus, multiplex data between the 32-bit microprocessor data bus and the 8-bit I/O data bus, and supplies buffering for the data bus lines.

The I/O bank-select decoder controls the enabling and multiplexing of the data bus transceivers.

Section 9.12

63. 80 ns.

65. Five wait states.

Section 9.13

67. Execution of this input instruction causes accumulator AX to be loaded with the contents of the word-wide input port at address $1A_{16}$.

69. Execution of this output instruction causes the value in the lower byte of the accumulator (AL) to be loaded into the byte-wide output port at address $2A_{16}$.

71.
```
MOV   DX,0A000H      ;Input data from port at A000H
IN    AL,DX
MOV   BL,AL          ;Save it in BL
MOV   DX,0B000H      ;Input data from port at B000H
IN    AL,DX
ADD   BL,AL          ;Add the two pieces of data
MOV   [IO_SUM],BL    ;Save result in the memory location
```

73.
```
IN    AL,B0H         ;Read the input port
AND   AL,01H         ;Check the LSB
SHR   AL,1
JC    ACTIVE_INPUT   ;Branch to ACTIVE_INPUT if the LSB = 1
```

75. 15 bytes of data are input from the input port at address $A000_{16}$ and saved in memory locations ES:1001H through ES:100FH.

77. I/O map base; word offset 66H from the beginning of the TSS.

79. 1.

Section 10.2

1. Program-storage memory; data-storage memory.

3. Firmware.

Section 10.3

5. When the power supply for the memory device is turned off, its data contents are not lost.

7. Ultraviolet light.

9. We are assuming that external decode logic has already produced active signals for $\overline{CE}$ and $\overline{OE}$. Next the address is applied to the A inputs of the EPROM and decoded within the device to select the storage location to be accessed. After a delay equal to t_{ACC}, the data at this storage location are available at the D outputs.

11. The access time of the 27C64 is 250 ns and that of the 27C64-1 is 150 ns. That is, the 27C64-1 is a faster device.

13. 1 ms.

Section 10.4

15. Volatile.

17. 32K $\times$ 32 bits (1 MB).

19. Higher density and lower power.

Section 10.5

23. Parity-checker/generator circuit.

25. $\Sigma_{EVEN} = 0$; $\Sigma_{ODD} = 1$.

Section 10.6

29. The storage array in the bulk-erase device is a single block, whereas the memory array in both the boot block and FlashFile is organized as multiple independently erasable blocks.

31. Bulk erase.

33. 28F002 and 28F004

35.

Type	Quantity	Sizes
Boot block	1	16K-byte
Parameter block	2	8K-byte
Main block	4	(1) 96K-byte, (3) 128K-byte

37. Logic 0 at the RY/$\overline{BY}$ output signals that the on-chip write state machine is busy performing an operation. Logic 1 means that it is ready to start another operation.

Section 10.7

39. $\overline{\text{READY}}$.

41. Seven.

Section 10.8

43. $M/\overline{IO} = 1$ and $A_{31} = 1$.

45. 16R8B.

47. $A_{31}\ldots.A_{15}\ldots..A_3A_2 = A_{31} = 0$; $A_{15}..A_2 = 00000000000000_2$ through $11111111111111_2 = 000000H$ through $003FFFH$.

Section 10.9

49. A cache memory is a small, high-speed memory subsystem that is located between the MPU and main memory.

51. 16 KB to 256 KB.

53. 93.2%.

55. Direct-mapped cache and two-way set associative cache.

57. In a direct-mapped cache, the cache memory array is organized as a single bank of consecutively addressed storage locations. On the other hand, in a two-way set associative cache memory, the cache memory array is separated into two equal size banks.

59. A memory update method in which updates to the main memory are performed through the cache and the write operation is performed by the cache controller rather than the MPU.

Section 10.10

61. $\overline{\text{DOE}}$

63. $\overline{CS}_0$ through $\overline{CS}_3$ = cache chip select 0 through 3
CALEN = cache address latch enable
$CT/\overline{R}$ = cache transmit/receive
$\overline{COEA}$ = cache output enable A
$\overline{COEB}$ = cache output enable B
$\overline{CWEA}$ = cache write enable A
$\overline{CWEB}$ = cache write enable B

65. $\overline{BBE}_0$ through $\overline{BBE}_3$.

67. 32 KB; 32.

69. $0001F03_{16} = 0000000000001111100000011_2$.
$TAG = 0000000000001111_2 \rightarrow$ page 15
TAG VALID $= 1 \rightarrow$ tag entry is valid
LINE VALID $= 00000011_2 \rightarrow$ line 0 and line 1

71. BANK A = BANK B = 128 KB.

73. Keeps track of whether BANK A or BANK B contains the least recently used information.

CHAPTER 11

Section 11.2

1. Keyboard interface, display interface, and parallel printer interface.

Section 11.3

3. $A_{15L}A_{14L} \ldots \ldots A_{4L}A_{3L}A_{2L} = 1X \ldots . X111XX_2 = 801C_{16}$ with $X = 0$.

5. Sets all outputs at port 2 (O_{16}–O_{23}) to logic 1.

Section 11.4

7. Port 4.

9.
```
MOV   AX,0A000H     ;Set up the segment start at A0000H
MOV   DS,AX
MOV   DX,8004H      ;Input from port 1
IN    AL,DX
MOV   [0000H],AL    ;Save the input at A0000H
MOV   DX,8008H      ;Input from port 2
IN    AL,DX
MOV   [0001H],AL    ;Save the input at A0001H
```

Section 11.5

11. Handshaking.

13. 74F373 octal latch.

15.
```
PUSH  DX    ;Save all registers to be used
PUSH  AX
PUSH  CX
PUSH  SI
PUSH  BX
  .         ;Program of Example 11.6 starts here
  .
  .
  .         ;Program of Example 11.6 ends here
POP   BX    ;Restore the saved registers
POP   SI
POP   CX
POP   AX
POP   DX
RET         ;Return from the subroutine
```

Section 11.6

17. 24.

19. MODE 0 selects simple I/O operation. This means that the lines of the port can be configured as level-sensitive inputs or latched outputs. Port A and port B can be configured as 8-bit input or output ports, and port C can be configured for operation as two independent 4-bit input or output ports.

890

MODE 1 operation represents what is known as strobed I/O. In this mode, ports A and B are configured as two independent byte-wide I/O ports, each of which has a 4-bit control port associated with it. The control ports are formed from the lower and upper nibbles of port C, respectively. When configured in this way, data applied to an input port must be strobed in with a signal produced in external hardware. An output port is provided with handshake signals that indicate when new data are available at its outputs and when an external device has read these values.

MODE 2 represents strobed bidirectional I/O. The key difference is that now the port works as either input or output and control signals are provided for both functions. Only port A can be configured to work in this way.

21. $D_0 = 1$ Lower 4 lines of port C are inputs.
    $D_1 = 1$ Port B lines are inputs.
    $D_2 = 0$ Mode 0 operation for both port B and the lower 4 lines of port C.
    $D_3 = 1$ Upper 4 lines of port C are inputs.
    $D_4 = 1$ Port A lines are inputs.
    $D_6 D_5 = 00$ Mode 0 operation for both port A and the upper four lines of port C.
    $D_7 = 1$ Mode being set.

23. Control word bits $= D_7 D_6 D_5 D_4 D_3 D_2 D_1 D_0 = 10010010_2 = 92H$.

25.
```
MOV    DX,1000H
MOV    AL,92H
OUT    DX,AL
```

27. To enable $INTR_B$, the INTE B bit must be set to 1. This is done with a bit set/reset operation that sets bit PC_4 to 1. This command is

$$D_7 - D_0 = 0XXX1001$$

29.
```
MOV    AL,03H
MOV    DX,100H
OUT    DX,AL
```

Section 11.7

31. The value at the inputs of port C of I/O device 0 is read into AL.

33.
```
IN     AL,02H    ;READ PORT A
MOV    BL,AL     ;SAVE IN BL
IN     AL,06H    ;READ PORT B
ADD    AL,BL     ;ADD THE TWO NUMBERS
OUT    0AH,AL    ;OUTPUT TO PORT C
```

Section 11.8

35. $\overline{BE_0} = 0$, $A_3 A_2 = 01$, $A_6 A_5 A_4 = 001$, and $A_{14} = 1$; 04014_{16}.

37.
```
MOV    AL,[4002H]    ;Read port A
MOV    BL,[4006H]    ;Read port B
ADD    AL,BL         ;Add the two readings
MOV    [400AH],AL    ;Write to port C
```

Section 11.9

39. Control word $D_7D_6D_5D_4D_3D_2D_1D_0 = 01011010_2 = 5AH$.

41.
```
MOV    DX,100CH    ;Select the I/O location
MOV    AL,5AH      ;Get the control word
MOV    [DX],AL     ;Write it
```

43.
```
MOV    AL,10000000B   ;Latch counter 2
MOV    DX,100CH
MOV    [DX],AL
MOV    DX,1008H
MOV    AL,[DX]        ;Read the least significant byte
```

45. 838 ns; 500 ns.

47. $N = 48_{10} = 30_{16}$.

Section 11.10

49. No.

51. 27.

53.
```
MOV    DX,100DH
OUT    DX,AL
```

55. Assume that the 80386DX microcomputer employs a byte-wide I/O interface and that the DMA controller is located at address $F0_{16}$.

```
MOV    AL,56H
OUT    11CH,AL
```

57. Assume that the 80386DX microcomputer employs a byte-wide I/O interface and that the DMA controller is located at I/O address 5000_{16}.

```
MOV    DX,5020H
IN     AL,DX
```

Section 11.11

59. $M/\overline{IO}A_{31}A_5A_4 = 0000$

61. D_7–D_0

Section 11.12

63. Clock.

65. Simplex: capability to transmit in one direction only.
Half-duplex: capability to transmit in both directions but at different times.
Full-duplex: capability to transmit in both directions at the same time.

Section 11.13

67. Asynchronous character length: 8 bits.
Parity: even.
Number of stop bits: 2.

69. The mode instruction determines the way in which the 8251A's receiver and transmitter are to operate, whereas the command instruction controls the operation. Mode instruction specifies whether the device operates as an asynchronous or synchronous communications controller, how the external baud clock is divided within the 8251A, the length of character, whether parity is used or not and if used then whether it is even or odd, and also, the number of stop bits in asynchronous mode.

The command instruction specifies the enable bits for transmitter and receiver. Command instruction can also be used to reset the error bits of the status register, namely, Parity error flag (PE), Overrun error flag (OE), and Framing error flag (FE).

The 8251 device can be initialized by the command instruction by simply writing a 1 in the bit D of command register. Here, the word *initialization* means returning to the mode-instruction format.

Section 11.14

71. $R_3R_2R_1R_0 = 1011$, $C_3C_2C_1C_0 = 1101$.

Section 11.15

73. 16 8-bit characters, right-entry for display. Strobed input, and decoded display scan for keyboard.

75. Command word 0: to set the mode of operation for keyboard and display.
Command word 1: to set the frequency of operation of 8279.
Command word 2: to read the keycode at the top of FIFO.
Command word 3: to read the contents of display RAM.
Command word 4: to send new data to the display.
Command word 6: to initialize the complete display memory, the FIFO status, and the interrupt-request output line.
Command word 7: to enable or disable the special error mode.

▲ CHAPTER 12

Section 12.2

1. External hardware interrupts, nonmaskable interrupt, internal interrupts and exceptions, software interrupts, and reset.

3. Hardware interrupts, nonmaskable interrupt, software interrupts, internal interrupts and exception, and reset.

5. Higher priority.

Section 12.3

7. Two words; four words.

9. 16-bit segment base address for CS and 16-bit offset for IP.

11. $(IP_{40}) \rightarrow$ (location A0H) and $(CS_{40}) \rightarrow$ (location A2H).

13. **(a)** Active. **(b)** Privilege level 2. **(c)** Interrupt gate. **(d)** B000H:1000H.

Section 12.4

15. 010000_{16}; 512 bytes; 64.

17. 190H through 197H.

19. Assures that the value in DI will only be in the range 0 through 255.

Section 12.5

21.
```
; This is an uninterruptible subroutine
   CLI   ;Disable interrupts at entry point
   .
   .     ;Body of subroutine
   .
   .
   STI   ;Enable interrupts
   RET   ;Return to calling program
```

Section 12.6

23. Interrupt acknowledge.

25. INTR is the interrupt request signal that must be applied to the 80386DX MPU by external interrupt interface circuitry to request service for an interrupt-driven device. When this request is acknowledged by the MPU, it outputs an interrupt-acknowledge bus status code on $M/\overline{IO}$ $D/\overline{C}$ $W/\overline{R}$, and this code is decoded by the bus control logic to produce the $\overline{INTA}$ signal. $\overline{INTA}$ is the signal that is used to tell the external device that its request for service has been granted.

27. The current interrupt request has been acknowledged; put the type number of the highest-priority interrupt on the data bus.

29. D_0 through D_7.

Section 12.7

31. $1 \mu s$.

33. 200 ns.

35. Privilege level is not changed. It remains equal to that of the interrupted code; the current stack remains active ($SS_{OLD}:ESP_{OLD}$); the old flags, old CS, and old EIP are pushed to the stack.

37. EXT = 1 = external interrupt
  IDT = 1 = error is due to interrupt
   TI = 1 = local descriptor table

Section 12.8

39. $D_0 = 0$ ICW_4 not needed
  $D_1 = 1$ Single device
  $D_3 = 0$ Edge-triggered
  and assuming that all other bits are logic 0 gives

$$ICW_1 = 00000010_2 = 02_{16}$$

41. $D_0 = 1$ Use with the 80X86
    $D_1 = 0$ Normal end of interrupt
    $D_3D_2 = 11$ Buffered mode master
    $D_4 = 0$ Disable special fully nested mode
and assuming that the rest of the bits are 0, we get

$$ICW_4 = 00001101_2 = 0D_{16}$$

43.
```
MOV   AL,[0A004H]
```

45.
```
MOV   AL,[0A004H]    ;Read OCW3
MOV   [OCW3],AL      ;Copy in memory
NOT   AL             ;Extract RR bit
AND   AL,2           ;Toggle RR bit
OR    [OCW3],AL      ;New OCW3
MOV   AL,[OCW3]      ;Prepare to output OCW3
MOV   [0A004H],AL    ;Update OCW3
```

Section 12.9

47. 22.
49. 64.

Section 12.10

51. $CS_{80} = A000H$ and $IP_{80} = 0100H$.

Section 12.11

53. Type number 2; $IP_2 \rightarrow 08H$ and $CS_2 \rightarrow 0AH$.
55. Initiate a power-failure service routine.

Section 12.12

57. CLK_2.
59. $UUUU0002_{16}$.
61.
```
RESET:   MOV   AX,0         ;Set up the data segment
         MOV   DS,AX
         MOV   CX,100H      ;Set up the count of bytes
         MOV   DI,0A000H    ;Point to the first byte
NXT:     MOV   [DI],0       ;Write 0 in the next byte
         INC   DI           ;Update pointer, counter
         DEC   CX
         JNZ   NXT          ;Repeat for 100H bytes
         RET                ;Return
```

Section 12.13

63. Divide error.
Debug.
Breakpoint.
Overflow error.
Bounds check.
Invalid opcode.
Coprocessor not available.
Interrupt table limit too small.
Coprocessor segment overrun.
Stack fault.
Segment overrun.
Coprocessor error.

65. Vectors 0 through 31.

67. Any attempt to access an operand that is on the stack at an address that is outside the current address range of the stack segment.

▲ CHAPTER 13

Section 13.2

1. System address bus, system data bus, and system-control bus.

3. 256KB, 128K $\times$ 1-bit DRAMs.

5. System clock, power-on reset, and wait-state generation.

7. 0C0H through 0DFH.

9. Channel 4 on DMA controller 2.

11. OUT_0,; IN_0.

13. R/W memory parity check (PCK) and I/O channel check (IO CH CK).

15. Real-time clock.

Section 13.3

17. 82345 data buffer, 82346 system controller, 82344 ISA controller, and 82341 peripheral combo.

19. ISA bus.

Section 13.4

21. If $\overline{BLKA_{20}}$ is logic 0, address line CA_{20} is forced to the 0 logic level. That is, the logic level at the A_{20} output of the 80386DX is blocked from the CA bus.

23. $\overline{CBE_0}$ through $\overline{CBE_3}$; $\overline{BE_0}$ through $\overline{BE_3}$ outputs of the 80386DX.

25. GND

27. 8K $\times$ 32 bits.

29. The programming of the 16L8 PAL.

31. U_{21}.

33. CK; 11.

35. U_{28}.

Section 13.5

37. Plastic quad flat package.

39. Data at the output of the 16/32 latch and data at the output of the MD latch.

41. Input; the 82346 uses this input to signal the 82345 whether the data transfer over the XD bus is a read (input) or write (output).

43. Peripheral X-data bus and system (slot) data bus.

45. If J_{40} is installed, the $\overline{\text{HIDRIVE}}$ input is active and the bus drivers are set for IOL equal 24 mA.

47. System (slot) data bus bit 1.

49. The XD bus is set for output of data.

Section 13.6

51. 50 MHz.

53. $\overline{\text{ERROR}}_{386}$, $\overline{\text{BUSY}}_{386}$. PEREQ_{386}, RES_{387}, and IRQ_{13}.

55. U_{23}.

57. To the CLK_2 output of the 82346.

59. RESET input at pin C9 of the 80386DX.

61. $\overline{\text{RAMW}}$, $\overline{\text{ROMCS}}$, and LBE_0 through LBE_3.

63. Banks 2 and 3.

65. $256\text{K} \times 9$, $1\text{M} \times 9$, or $4\text{M} \times 9$.

67. $\overline{\text{CAS}}_8$ and $\overline{\text{CAS}}_9$; MD_0 through MD_{15}; word.

69. Peripheral data bus lines XD_0 through XD_7.

Section 13.7

71. (2) 82C59A, (2) 82C37A, and (1) 82C54.

73. Pins 69, 70, 71, and 72.

75. U_{37} and U_{38}; 27C256.

77. $J_{30} = \text{OUT}$, $J_{31} = \text{IN}$, and $J_{32} = \text{don't care}$.

79. 24 bits.

81. $\overline{\text{DACK}}_2$.

83. Pin 102.

85. Parity error for an on-board memory array access; NMI.

87. OUT_1.

Section 13.8

89. J_{55}.

91. Pin 2.

93. Input.

95. TURBREQ is one input of AND gate U_{14} on sheet 7. Here it is ANDed with the signal turbo switch (TSWITCH) to produce the signal TURBO. The TURBO output is used to light the TURBO LED and is supplied to the TURBO input at pin 115 of the 82346 system controller.

97. Level 1.

99. U_{52} and U_{53}.

101. Pin 38.

Section 13.9

103. 5.25-inch and 3.5-inch.

105. 16 bytes.

107. Write precompensation.

109. Level 6.

111. 24 MHz.

113. Pin 22.

▲ CHAPTER 14

Section 14.2

1. Prototype circuit.

3. Solderless breadboard.

5. Bus interface module, ISA expansion bus cables, breadboard unit.

Section 14.3

7. The INT/EXT switch must be set to the EXT position.

9. 26 AWG.

11. A_{31} through A_{12}.

13. Logic 0 lights the green LED; logic 1 lights the red LED; and the high-Z level lights the amber LED.

Section 14.4

15. 74LS688, 74LS138, and 74LS32.

17. A_{10} through A_{15}.

19. The select outputs of the 74LS138 are gated with either $\overline{IOR}$ or $\overline{IOW}$ in 74LS32 OR gates. These signals are active only during an I/O cycle.

21. Yes.

23. The setting of switch 7 is polled waiting for it to close.

25. Lights LEDs 0 through 3.

27. The LEDs are lit in a binary counting pattern.

29. Change MOV CX,0FFFFH to MOV CX,7FFFH.

Section 14.5

33. IC test clip.

35. Whether the test point is at the 0, 1, or high-Z logic state, or if it is pulsating.

37. Amount of voltage, duration of the signal, and the signal waveshape.

39. Troubleshooting.

41. Hardware troubleshooting.

43. **1.** Check to verify that correct pin numbers are marked into the schematic diagram.
2. Verify that the circuit layout diagram correctly implements the schematic.
3. Check that the ICs and jumpers are correctly installed to implement the circuit.

45.

Test Point	Switch Open	Switch Closed
1	1	0
2	1	0
3	Pulse	Pulse

Section 14.6

47. Address bus, data bus, and control-bus signals.

49.

Oscilloscope	Logic analyzer
Requires periodic signal to display.	Can display periodic or nonperiodic signals.
Small number of channels.	Large number of channels.
Displays actual voltage values.	Displays logic values.
Generally does not store the signals for display.	Stores signals for display.
Simple trigger condition using a single signal.	Trigger signal can be a combination of a number of signals.

▲ CHAPTER 15

Section 15.2

1. On-chip floating-point math coprocessor and code and data cache memory.

3. The 80486SX does not have an on-chip floating-point math coprocessor.

Section 15.3

5. 32 bytes.

7. Small instruction set, limited addressing modes, and single-clock execution for instructions.

9. CLKMUL.

Section 15.4

11. Little endian.

13. F0F0F0F0H.

15. XADD [SUM], EBX

EAX	SUM	
00000001H	00000000H	
00000001H	00000001H	1st execution
00000001H	00000002H	2nd execution
00000002H	00000003H	3rd execution
00000003H	00000005H	4th execution

17. Since the contents of the destination operand, memory location DATA, and register AX are the same, ZF is set to 1 and the value of the source operand, 2222_{16}, is copied into destination DATA.

Section 15.5

19. Cache disable (CD) and not write-through (NW).

21. WBINVD initiates a write-back bus cycle to update external memory and then a flush of the internal cache.

23. Page-cache disable (PCD) and page-write transparent (PWT).

Section 15.6

25. 168-lead PGA; 196-lead PQFP.

Section 15.7

27. Halt/shutdown bus cycle; 001_2.

29. $\overline{\text{PCHK}}$; logic 0.

31. Cache enable.

33. $\overline{\text{BOFF}}$.

Section 15.8

35. 8 bits wide.

37. $\overline{\text{BHE}} = 1$, $\overline{\text{BLE}} = 0$, and $A_1 = 0$.

39. D_{31}–D_{24}; The byte-swap logic transfers the byte to data line D_7–D_0 for transfer to the memory subsystem.

Section 15.9

41. 8 ns; 12 ns; 24 ns.

43. 32 ns; 48 ns.

45. 20 ns.

Section 15.10

47. Four-way set associative.
49. 8K bytes; line of data = 128 bits (16 bytes).
51. Set both the CD and NW bits in CR0 to logic 1.
53. By executing the INVD instruction.

Section 15.11

55. 82C54, 82C37, and 16C550.
57. ISA bus.

Section 15.12

59. Built-in self-test.

Section 15.13

61. Clock doubling and write-back cache.
63. Modify/exclusive/shared/invalid (MESI) protocol.
65. The $\overline{\text{WBWT}}$ input must be held at logic 1 for at least two clock periods before and after a hardware reset.
67. $435/68 = 6.4$

▲ CHAPTER 16

Section 16.2

1. 64 bits.
3. 735.

Section 16.3

5. Separate code and data caches; write-through or write-back update methods; dual port ALU interface.

Section 16.4

7. ID, VIP, and VIF.
9. Machine-check exception.
11. $ZF \leftarrow 0$; $(EDX:EAX) \leftarrow (TABLE) = 11111111FFFFFFFF_{16}$.

Section 16.5

13. 3 million transistors.
15. D_{45} is at pin A21; A_3 is at pin T17.

Section 16.6

17. Data parity and address parity; $\overline{\text{APCHK}} = 0$ address parity error and $\overline{\text{PCHK}} = 0$ data parity error.

19. $\overline{\text{CACHE}} = 0$.

Section 16.7

21. 8MB.

23. ALE; In Fig. 16.9 we find that ALE is generated by the control-logic section.

25. Eighteen 74FCT646A ICs.

27. $\text{B0MD}_{0\text{-}63} \rightarrow \text{D}_0\text{-}\text{D}_{63}$ and $\text{B0DP}_{0\text{-}7} \rightarrow \text{DP}_0\text{-}\text{DP}_7$.

Section 16.8

29. I/O write.

31. $\overline{\text{CACHE}} = \overline{\text{KEN}} = 0$

Section 16.9

33. Two-way set associative.

35. Write-back and write-through.

37. Burst bus cycles; either nonpipelined or pipelined.

39. Read hits access the cache, read misses may cause replacement, write hits update the cache, writes to shared lines and write misses appear externally, write hits can change shared lines to exclusive under control of WB/$\overline{\text{WT}}$, and invalidation is allowed.

41. INVD does not initiate a write-back to external memory before invalidating the contents of the caches.

Section 16.10

43. These values can be read from the MCA and MCT model-specific registers of the MPU by using the RDMSR instruction. To read the address, ECX must be loaded with 0H prior to executing the instruction and to read the type information ECX must be loaded with 1H before the instruction is executed.

Section 16.11

45. Pentium[R] Pro processor.

47. 256KB and 512KB.

49. 182.

Bibliography

Bradley, David J., *Assembly Language Programming for the IBM Personal Computer*. Upper Saddle River, N.J.: Prentice Hall, 1984.

Ciarcia, Steven. "The Intel 8086," *Byte*, November 1979.

Coffron, James W. *Programming the 8086/8088*, Berkeley, Calif.: Sybex Inc. 1983.

Intel Corporation. *AP-478 An Example Memory Subsystem for the Pentium*^{R} *Processor*. Santa Clara, Calif.: Intel Corporation, 1993.

———. *Components Data Catalog*. Santa Clara, Calif.: Intel Corporation, 1980.

———. *80386 Microprocessor Hardware Reference Manual*. Santa Clara, Calif.: Intel Corporation, 1987.

———. *80386 Programmer's Reference Manual*. Santa Clara, Calif.: Intel Corporation, 1987.

———. *80386 System Software Writer's Guide*. Santa Clara, Calif.: Intel Corporation, 1987.

———. *iAPX86,88 User's Manual*. Santa Clara, Calif.: Intel Corporation, July 1981.

———. *Introduction to the 80386*. Santa Clara, Calif: Intel Corporation, September 1985.

———. *i486*™ *Microprocessor Family Programmer's Reference Manual*. Santa Clara, Calif.: Intel Corporation, 1992.

———. *i486*™ *Microprocessor Hardware Reference Manual*. Santa Clara, Calif.: Intel Corporation, 1990.

———. *Intel486*™*SX Microprocessor Data Book*. Santa Clara, Calif.: Intel Corporation, 1992.

———. *MCS-86*™ *User's Manual*. Santa Clara, Calif: Intel Corporation, February 1979.

———. *Memory*. Santa Clara, Calif.: Intel Corporation, 1989.

———. *Microprocessor*. Santa Clara, Calif.: Intel Corporation, 1989.

———. *Microprocessor and Peripheral Handbook*, vols. 1 and 2. Santa Clara, Calif.: Intel Corporation, 1989.

903

———. *Pentium*[R] *Processors and Related Products.* Santa Clara, Calif.: Intel Corporation, 1995.

———. *Pentium*[R] *Processor Family Developer's Manual*, vols. 1, 2, and 3. Santa Clara, Calif.: Intel Corporation, 1995.

———. *Peripheral.* Santa Clara, Calif.: Intel Corporation, 1989.

———. *Peripheral Design Handbook.* Santa Clara, Calif.: Intel Corporation, April 1978.

Morse, Stephen P. *The 8086 Primer.* Rochelle Park, N.J.: Hayden Book Company, 1978.

National Semiconductor Corporation. *Series 32000 Databook.* Santa Clara, Calif.: National Semiconductor Corporation, 1986.

Rector, Russell, and George Alexy, *The 8086 Book.* Berkeley, Calif.: Osborne/McGraw-Hill, 1980.

Scanlon, Leo J. *IBM PC Assembly Language.* Bowie, Md.: Robert J Brady, 1983.

Schneider, Al. *Fundamentals of IBM PC Assembly Language.* Blue Ridge Summit, Pa.: Tab Books, 1984.

Singh, Avtar , and Walter A. Triebel. *IBM PC/8088 Assembly Language Programming.* Upper Saddle River, N.J.: Prentice Hall, 1985.

———. *The 8088 Microprocessor: Programming, Interfacing, Software, Hardware, and Applications.* Upper Saddle River, N.J.: Prentice Hall, 1989.

———. *The 8086 and 80286 Microprocessors: Architecture, Software, and Interface Techniques.* Upper Saddle River, N.J.: Prentice Hall, 1990.

Strauss, Ed, *Inside the 80286.* New York: Brady Books, 1986.

Texas Instruments Incorporated. *Programmable Logic Data Book.* Dallas, Tex.: Texas Instruments Incorporated. 1990.

Triebel, Walter A. *Integrated Digital Electronics*, 2d ed. Upper Saddle River, N.J.: Prentice Hall, 1985.

———. *The 80386DX Microprocessor: Hardware, Software, and Interfacing.* Upper Saddle River, N.J.: Prentice Hall, 1992.

Triebel, Walter A., and Alfred E. Chu. *Handbook of Semiconductor and Bubble Memories.* Upper Saddle River, N.J.: Prentice Hall, 1982.

Triebel, Walter A., and Avtar Singh. *The 8088 and 8086 Microprocessor: Programming, Interfacing, Software, Hardware, and Applications*, 2d ed. Upper Saddle River, N.J.: Prentice Hall, 1997.

Willen, David C., and Jeffrey I. Krantz. *8088 Assembler Language Programming: The IBM PC.* Indianapolis, Ind.: Howard W. Sams, 1983.

Index